AF601874

Springer Monographs in Mathematics

This series publishes advanced monographs giving well-written presentations of the "state-of-the-art" in fields of mathematical research that have acquired the maturity needed for such a treatment. They are sufficiently self-contained to be accessible to more than just the intimate specialists of the subject, and sufficiently comprehensive to remain valuable references for many years. Besides the current state of knowledge in its field, an SMM volume should ideally describe its relevance to and interaction with neighbouring fields of mathematics, and give pointers to future directions of research.

More information about this series at http://www.springer.com/series/3733

Władysław Narkiewicz

The Story of Algebraic Numbers in the First Half of the 20th Century

From Hilbert to Tate

Springer

Władysław Narkiewicz
University of Wrocław
Wrocław, Poland

ISSN 1439-7382 ISSN 2196-9922 (electronic)
Springer Monographs in Mathematics
ISBN 978-3-030-03753-6 ISBN 978-3-030-03754-3 (eBook)
https://doi.org/10.1007/978-3-030-03754-3

Library of Congress Control Number: 2018960727

Mathematics Subject Classification (2010): 11Rxx, 11-03, 01A60

This Springer imprint is published by the registered company Springer Nature Switzerland AG
The registered company address is: Gewerbestrasse 11, 6330 Cham, Switzerland

To the memory of my wife

Preface

The aim of this book is to give a survey of results in the theory of algebraic numbers achieved in the first half of the twentieth century and may be viewed as a companion to my previous book *Rational Number Theory in the 20th Century* in which the part of number theory dealing with rational numbers has been treated. It is an attempt to fulfil the wish of H. S. Vandiver expressed in 1960 in his paper [4185], and perhaps it might be helpful in preventing rediscoveries.

Chapter 1 gives a concise presentation of the beginnings of the theory of algebraic numbers. One finds here first a description of the work on special cases of algebraic integers done by Gauss, Dirichlet and Eisenstein, followed by Kummer's work on cyclotomic fields. Then the creation of the general theory by Kronecker and Dirichlet is treated, and the chapter concludes with a short description of the related work of other mathematicians, including Hermite, Minkowski, Frobenius and Stickelberger.

In Chap. 2 one finds a presentation of the work of Hilbert, who in his report on algebraic numbers summarized the state of their theory at the end of the nineteenth century, as well of Hensel, who created p-adic and $\mathfrak{p}$-adic numbers, which turned out to be an indispensable tool in future research. In the last part of the chapter the first steps towards creation of the class-field theory, characterizing Abelian extensions of algebraic number fields, are described.

Chapter 3 covers the first twenty years of the twentieth century. In the first section we present its central subject, the use of analytic methods in the theory of algebraic numbers. This has been initiated by Landau, who established the Prime Ideal Theorem giving asymptotics for the number of prime ideal with bounded norms. The next big achievement was Hecke's proof of the continuation of the Dedekind zeta-function to a meromorphic function on the plane, and the study of several generalizations of Dirichlet L-functions to number fields, developed by him and Landau. The second section presents the results dealing with the algebraic structure, and the last section is devoted to other results achieved in the beginning of the twentieth century.

The central themes of Chap. 4 are the creation of the modern ideal theory by Emmy Noether, and the establishment of fundamental theorems of class-field theory by Takagi and Artin. Also various other questions were considered at that time, for example the first results in the additive theory of algebraic numbers obtained by Rademacher.

In Chap. 5 we present first the progress in the study of the structure of number fields, the central subject being the existence of normal and normal integral bases, and then consider some additive questions, mainly on sums of squares. The next section concentrates on the simplification of the class-field theory by Hasse and Chevalley, and the following sections concern i.a. the class-number and class-group of quadratic fields, the question of the existence of the Euclidean algorithm, the distribution of algebraic integers on the complex plane and infinite extensions of number fields. Chapter 6 covers The Forties, the main results being obtained by Brauer and Siegel.

In all chapters one will find also some selected information about the subsequent developments of the arising problems.

I am very grateful to my friends Kálman Győry and Andrzej Schinzel for reading the draft of the book and providing several comments and suggestions. I thank also the referees of the book for several important hints.

I am very grateful to the Springer staff for helpful cooperation in preparing the publication. My specials thanks go to Ms Elena Griniari and Ms Angela Schulze-Thomin.

Wrocław, Poland Władysław Narkiewicz

Contents

Chapter 1
The Birth of Algebraic Number Theory

1.1 The Beginning

1.1.1 Euler

1. Algebraic numbers used in early mathematical research were essentially defined by various expressions involving radicals. It seems that the first serious application of them to arithmetical questions appeared in a paper of Euler[1] [1131] of 1765, where continued fractions of quadratic surds are used to find solutions of the Pell equation

$$x^2 - dy^2 = 1 .$$

Later [1133] Euler applied numbers of the form $a + \sqrt{d}$ with $a, d \in \mathbf{Z}$ to deal with equations of the form

$$x^4 + kx^2y^2 + y^4 = z^2 .$$

A similar approach can also be found in [1134, 1135].

In Sect. 169 of the second part of his book [1136] Euler considered divisibility properties of numbers $a + bi$ with integral a, b, and in later sections he did the same for the numbers $a + b\sqrt{-c}$ with integral $c > 0, a, b$. It seems that he assumed that these numbers have arithmetical properties similar to those of the usual integers, as he wrote in Sect. 191:

" *Denn*[2] *wenn z.B.* $x^2 + cy^2$ *ein Cubus seyn soll, so kann man sicher schliessen, dass auch die beyden irrationalen Factoren davon, nämlich* $x + y\sqrt{-c}$ *und*

---

[1]Leonhard Euler (1765–1823), professor in St. Petersburg and Berlin. See [8, 1174, 4051, 4190].

[2]"*Because if e.g.,* $x^2 + cy^2$ *is a cube, then one can safely infer that its two irrational factors, namely* $x + y\sqrt{-c}$ *and* $x - y\sqrt{-c}$*, must be cubes, because they are co-prime as the numbers* x *and* y *do not have a common divisor.*"

W. Narkiewicz, *The Story of Algebraic Numbers in the First Half of the 20th Century*, Springer Monographs in Mathematics,
https://doi.org/10.1007/978-3-030-03754-3_1

$x - y\sqrt{-c}$ *Cubos seyn müssen, weil dieselben unter sich untheilbar sind, indem die Zahlen x und y keinen gemeinschaftlichen Theiler haben.*"

This implication is correct only in the case when the law of unique factorization holds in the set $\{x + y\sqrt{-c} : x, y \in \mathbf{Z}\}$.

In 1875 Pépin[3] [3236] provided correct formulations and proofs for Euler's use of integers of quadratic fields in [1136].

1.1.2 Gauss

1. A part of the theory of quadratic forms with integral coefficients developed by Lagrange[4] in [2399] and by Gauss[5] in his book "Disquisitiones Arithmeticae" [1394] can be translated into the language of quadratic number fields. This has been pointed out in 1847 by Kummer[6] [2350], who wrote:

"*Die*[7] *ganze Theorie der Formen vom zweiten Grade, mit zwei Variabeln, kann nämlich als Theorie der complexen Zahlen von der Form* $x + y\sqrt{D}$ *aufgefasst werden*".

This has been made explicit much later by Dedekind[8] in [848] (see Sect. 1.3.3).

Gauss dealt in [1394] with quadratic forms $f(x, y) = ax^2 + 2bxy + cy^2 \in \mathbf{Z}[x, y]$ having middle coefficient even and defined their *determinant* $\Delta(f)$ by putting $\Delta(f) = b^2 - ac$. He called a form f *primitive* if $(a, b, c) = 1$ and *properly primitive* if $(a, 2b, c) = 1$. Forms with $(a, b, c) = 1$ and even a and c were called *improperly primitive*. It was later shown by Kronecker[9] (Section VIII in [2299]) that Gauss's approach works also for forms $f(x, y) = ax^2 + bxy + cy^2$ without restricting the middle coefficient to be even. Since then it became customary to consider the *discriminant* $d(f) = b^2 - 4ac$, which for forms with even middle coefficient equals $4\Delta(f)$.

Gaussian theory of quadratic forms over $\mathbf{Z}$ has been later generalized to the case of other base rings. See the papers of Speiser[10] [3868], R. König[11] [2207], Lubelski[12] [2662], Kaplansky[13] [2119], Butts[14] and Estes[15] [532], Butts and Dulin [1023], Shyr [3767], Pfeuffer [3277], Earnest

[3]Jean François Théophile Pépin (1826–1904), Jesuit, teacher of mathematics.

[4]Joseph Louis Lagrange (1736–1813), worked in Turin, Berlin and Paris. See [2082].

[5]Carl Friedrich Gauss (1777–1855), professor in Göttingen. See [1028, 2124].

[6]Ernst Eduard Kummer (1810–1893), professor in Breslau and Berlin. See [1782, 2410].

[7]"*The whole theory of forms of second degree in two variables can be considered as the theory of complex numbers of the form* $x + y\sqrt{D}$."

[8]Richard Dedekind (1831–1916), professor in Zürich and Braunschweig. See [836, 1718, 2429, 3107].

[9]Leopold Kronecker (1823–1891), professor in Berlin. See [1258, 2181, 4317].

[10]Andreas Speiser (1885–1970), professor in Zürich. See [1199].

[11]Robert König (1885–1979), professor in Tübingen, Münster, Jena and Munich.

[12]Salomon Lubelski (1902–?), worked in Warsaw, died in a Nazi concentration camp, date unknown.

[13]Irving Kaplansky (1917–2006), professor in Chicago and Berkeley. See [212].

[14]Hubert S. Butts (1923–1999), professor at the Louisiana State University.

[15]Dennis Ray Estes (1941–1999), professor at the University of Southern California.

and Estes [1036, 1037], Towber [4073], M. Kneser[16] [2185] and Wood [4444]. For a survey of modern work see Earnest [1035].

The class-number $H(D)$ of properly primitive quadratic forms with a given determinant D is defined as the number of equivalence classes of such forms f under the action of the group $SL(2, \mathbf{Z})$ (for $D < 0$ one considers only positive-definite forms): two forms $f(x, y), g(x, y)$ are equivalent if for some $a, b, c, d \in \mathbf{Z}$ with $ad - bc = 1$ one has

$$g(x, y) = f(ax + by, cx + dy) .$$

The finiteness of the class-number was established by Lagrange (who considered, as did later Kronecker, forms $ax^2 + bxy + cy^2$ without restricting the parity of b), as well as by Gauss (Sects. 174 and 185 of [1394]). If $H'(d)$ denotes the number of equivalence classes of quadratic forms $ax^2 + bxy + cy^2$ with discriminant $d = b^2 - 4ac$, satisfying $(a, b, c) = 1$, then $H(D) = H'(4D)$.

These results on class-numbers of quadratic forms have interpretations in algebraic number theory. Recall that the class-group $H(K)$ of an algebraic field K is defined by $H(K) = I(K)/P(K)$, and the narrow class-group $H^*(K)$ equals $H^*(K) = I(K)/P^*(K)$, where $I(K)$ is the group of fractional ideals of K, $P(K)$ is the group of principal fractional ideals of K, and $P^*(K)$ is the group of principal fractional ideals having a totally positive generator. The class-numbers of number fields are defined by

$$h(K) = \#H(K), \quad h^*(K) = \#H^*(K).$$

It has been shown later by Dedekind [848] that if K is an imaginary quadratic number field, then its class-number $h(K)$ equals $H'(d(K))$, $d(K)$ denoting the discriminant of K. In the case when K is a real quadratic field the class-number $H'(d(K))$ equals $h^*(K)$, the narrow class-number of K. If K has a unit of negative norm then the class-numbers $h(K)$ and $h^*(K)$ are equal, otherwise one has $h^*(K) = 2h(K)$. The same applies to the class-numbers of orders (which are subrings of $\mathbf{Z}_K$ containing $\mathbf{Z}$) in K. If d is the discriminant of a quadratic field K, then one writes usually $h(d), h^*(d)$ instead of $h(K), h^*(K)$.

2. In Sect. 234 of [1394] defined Gauss the composition of forms, which led in Sect. 249 to composition of form classes[17] and showed that the set of classes has the properties used much later to define the notion of a group.

In part IX of Sect. 306 we find a remark indicating the possibility of presenting the class-group as a product of cyclic groups, but no proof was given. Gauss remarked that he will consider this question on another opportunity but this never happened.

---

[16]Martin Kneser (1928–2004), professor in München and Göttingen, son of Hellmuth Kneser. See [3943].

[17]The original definition of the composition is rather complicated. A simple way of defining it was provided in 1912 by Speiser [3869]

The first proof of the decomposition of the class-group into cyclic groups was given by Schering[18] [3600] in 1869, and next year this result has been extended to arbitrary finite Abelian groups by Kronecker [2288]. See also the paper of Frobenius[19] and Stickelberger[20] [1263]. Simpler proofs were later provided by Remak[21] [3438] in 1921, Mathewson [2773] in 1929, Korselt[22] [2226] and Franz[23] [1232] in 1931, Rado[24] [3369] in 1951, L. Fuchs [1293] in 1953 and Schenkman[25] [3598] in 1960.

A discussion of the ideas which led Gauss to define the composition of binary quadratic forms has been presented by Weil[26] [4358] in 1986.

For modern approach to Gaussian composition see Edwards [1049], Fenster and Schwermer [1179] and Bhargava [336]. The last author provided in [337, 338] generalizations to the cubic and quartic case.

3. In Sect. 261 and 286 Gauss established the *theorem on the genera* which implies for quadratic fields that the genus group $\mathfrak{G}(K) = H^*(K)/H^*(K)^2$ is isomorphic to $C_2^{\omega(d(K))-1}$, $\omega(n)$ denoting the number of distinct prime factors of n.

Other proofs of this result have been given (in the language of quadratic forms) by Dirichlet[27] [963] in 1839, Arndt[28] [109] in 1858, Kronecker [2287] in 1864, Mertens[29] [2824] in 1905, and Reiner[30] [3426] in 1945. An elementary proof using the language of quadratic fields has been given in 1992 by Nemenzo and Wada [3083]. The modern approach to this theorem can be found in the paper [3767] by Shyr. See also Gogia and Luthar[31] [1452] and Lemmermeyer [2530].

4. Sect. 303 of [1394] brings a short table of negative determinants $-D$ with small $H(-D)$. In particular it lists five determinants with $H(-D) = 1$, namely for $D = 1, 2, 3, 4$ and 7, eight with $H(-D) = 3$ ($D = 11, 19, 23, 27, 31, 43, 67$ and 163), four with $H(-D) = 5$, six with $H(-D) = 7$, 15 with $H(-D) = 2$, 44 with $H(-D) = 4$, 17 with $H(-D) = 8$ and four with $H(-D) = 16$. Gauss expressed the belief that his lists are complete. He conjectured also that there are only finitely many determinants with a given class-number.

[18] Ernst Schering (1833–1897), professor in Göttingen. See [2168].

[19] Ferdinand Georg Frobenius (1849–1917), professor in Zürich and Berlin. See [1721].

[20] Ludwig Stickelberger (1850–1936), professor in Freiburg in Breisgau. See [1748].

[21] Robert Remak (1888–1942), dozent in Berlin.

[22] Alwin Reinhold Korselt (1864–1947), teacher in Plauen.

[23] Wolfgang Franz (1905–1996), professor in Frankfurt. See [517].

[24] Richard Rado (1906–1989), professor in Sheffield, Kings College London and Reading. See [2548, 3490].

[25] Eugene Schenkman (1922–1977), professor at the Purdue University.

[26] André Weil (1906–1998), professor in Sao Paulo, Chicago and at IAS in Princeton. See the special issue of Notices AMS, vol 46/4, 1999.

[27] Peter Gustav Lejeune-Dirichlet (1805–1859), professor in Breslau, Berlin and Göttingen. See [2879].

[28] Peter Friedrich Arndt (1817–1866), professor in Berlin.

[29] Franz Mertens (1840–1927), professor in Kraków, Graz and Vienna. See [930].

[30] Irving Reiner (1924–1986), professor at the University of Illinois in Urbana-Champaign. See [2035].

[31] Indar Singh Luthar (1932–2006), professor at the Panjab University.

The last conjecture has been established in the case when the class-number is prime to 6 by Joubert[32] [2080] in 1860, who applied the theory of elliptic functions. He showed in particular that Gauss's lists of determinants D with $H(D) = 5$ and $H(D) = 7$ are complete. Translated into the language of quadratic fields the result of Joubert asserts that there are only finitely many imaginary quadratic fields with even discriminant and a given class-number prime to 6. An elementary and effective proof of Joubert's result was provided by Shanks[33] [3744] in 1969.

A particular case of Joubert's result has been considered in 1903 by Landau[34] [2416] who showed in an elementary way that one has $H(-D) = 1$ if and only if $D \in \{1, 2, 3, 4, 7\}$, and Lerch[35] [2562] provided an even simpler proof. This implies that if K is an imaginary quadratic field with class-number 1 and $4 \mid d(K)$, then either $K = \mathbf{Q}(i)$ or $K = \mathbf{Q}(\sqrt{2})$.

To interpret Gauss's table in terms of class-numbers of quadratic fields one has to have in mind the following equality, relating the class-numbers of Gaussian forms and Kronecker forms:

For negative square-free $D \equiv 1$ mod 4 one has

$$H(D) = H'(4D) = \begin{cases} H'(D) & \text{if } D \equiv 1 \bmod 8\ , \\ 3H'(D) & \text{if } D \equiv 5 \bmod 8\ , \end{cases} \tag{1.1}$$

This formula follows from the case $p = 2$ of the equality

$$H'(p^2D) = \frac{H'(D)}{\varepsilon(D)} \left(p - \left(\frac{D}{p} \right) \right) , \tag{1.2}$$

where p is prime, $D < 0$ is square-free and

$$\varepsilon(D) = \begin{cases} 1 & \text{if } D < -4\ , \\ 2 & \text{if } D = -4\ , \\ 3 & \text{if } D = -3\ . \end{cases}$$

This result in a more general form (but with H' replaced by H) relating $H(a^2D)$ to $H(D)$ for arbitrary a, occurs explicitly for the first time in the paper [964] by Dirichlet. Gauss gave in §253–256 of [1394] a rather complicated description of this relation. They both treated also the case of D positive, giving a formula in which σ depends on solutions of certain Pell's equations.

For other proofs see Lipschitz[36] [2608], Dedekind [842, 844], Kronecker [2299] (who established (1.2) in the form presented by as above), Weber[37] [4319, 4320], Mertens [2818] and Lerch (p. 368 in [2564]). An elementary proof of (1.2) and its analogue for $h(p^2D)$ with positive D has

---

[32] P. Charles Joubert (1825–1907), jesuit, teacher at Ste. Genévieve in Paris.

[33] Daniel Shanks (1917–1966), professor at the University of Maryland. See [4403].

[34] Edmund Landau (1877–1938), professor in Göttingen. See [1626, 2188].

[35] Matiaš Lerch (1860–1922), professor in Prague, Fribourg and Brno. See [801, 3336].

[36] Rudolf Lipschitz (1832–1903), professor in Breslau and Bonn. See [2225].

[37] Heinrich Weber (1842–1913), professor in Heidelberg, Zürich, Königsberg, Charlottenburg, Marburg, Göttingen and Strassburg. See [4249].

been given in 1935 by Pall[38] [3213], who applied his results on the number of representations of integers by quadratic forms established in [3211, 3212]. See also Sect. 6.5.

Using (1.1) one sees that Gauss's list implies $h(-d) = 1$ for $d = 4, 7, 8, 11, 12, 19, 43, 67$ and 163. The question whether this list contains all negative discriminants with class-number one was usually called the "Gauss class-number problem". For later development see Sects. 5.3.1 and 6.2.1.

5. In Sect. 302 of [1394] Gauss conjectured that one has

$$\Phi(x) := \sum_{D \le x} H(-D) \approx \frac{4\pi}{21\zeta(3)} x^{3/2} - \frac{2}{\pi^2} x \,. \tag{1.3}$$

Because of the relation between the class-number of quadratic forms and the number of representations of a positive integer as a sum of three squares discovered by Gauss ([1394], §291) the evaluation of the error term in this formula is connected with the error term in the problem of lattice points in three-dimensional spheres.

The first result dealing with the conjecture (1.3), which can be also interpreted in the language of quadratic fields, was obtained in 1865 by Lipschitz [2609] who established

$$\Phi(x) = \left(\frac{4\pi}{21\zeta(3)} + o(1)\right) x^{3/2} \,.$$

This showed that the first term in Gauss's conjecture is correct. Later Mertens [2814] provided another proof, providing the error term $O(x)$. His paper, as well as a paper of Hermite[39] [1825], gives also asymptotics for the sum of the class-numbers of all, not necessarily primitive, quadratic forms of negative determinants $-d$ with $d \le x$. The main term of this sum equals $2\pi x^{3/2}/9$.

In 1912 Landau [2421] used a method of Pfeiffer [3275] to confirm Gauss' conjecture by establishing the formula

$$\Phi(x) = \frac{4\pi}{21\zeta(3)} x^{3/2} - \frac{2}{\pi^2} x + R(x) \,, \tag{1.4}$$

with

$$R(x) = O\left(x^{5/6} \log x\right) \,.$$

Five years later I.M. Vinogradov[40] improved the error term in this formula first to $O(x^{5/6} \log^{2/3} x)$ [4224], and then to $O(x^{3/4} \log^2 x)$ [4225]. For further development see Sect. 6.2.1.

6. At the end of Sect. 304 one finds the analogue of the conjecture (1.3) in the case of positive determinants. If for $D > 0$ the pair $(X, Y) = (t(D), u(D))$ forms the minimal positive solution of the Pellian equation

$$X^2 - DY^2 = 1 \,,$$

and $\eta_D = t(D) + u(D)\sqrt{D}$, then Gauss asserted that the mean value of the product

$$H(D) \log(\eta_D)\sqrt{D}$$

in the interval $[1, x]$ is asymptotically equal to $c_1\sqrt{x} - c_2$ with some constants c_1, c_2. He noted that c_1 seems to be close to $2^{1/3}$ but later changed his mind and in the Appendix to [1394] asserted c_1 to be $2\pi^2/7\zeta(3)$. This modification would imply

---

[38] Gordon Pall (1907–1987), professor at McGill University.

[39] Charles Hermite (1822–1901), professor in Paris. See [3128, 3283].

[40] Ivan Matveevič Vinogradov (1891–1983), professor in Moscow. See [601].

$$\sum_{D\leq x} H(D)\log(\eta_D) \approx \frac{4\pi^2}{21\zeta(3)} x^{3/2} . \tag{1.5}$$

This conjecture has been established in 1944 by Siegel[41] [3786]. See Sect. 6.2.1.

Around 1837 Gauss stated in [1396] certain formulas for the class-number of quadratic forms which were established by Dirichlet in 1838 (see Sect. 1.1.3). They are equivalent to the formulas for the class-number of quadratic fields proved later by Dedekind (see (1.33) and (1.34).

7. In Sect. 306 Gauss defined a determinant to be regular if its *principal genus* formed by the set of all squares of classes consists of powers of a single class, i.e. forms a cyclic group. In the case of an irregular determinant D he defined its irregularity exponent $i(D)$ as the ratio $a(d)/b(d)$, where $a(d)$ is the number of classes in the principal genus and $b(d)$ is the cardinality of the largest cyclic subgroup of the principal genus. He noted examples of negative determinants with $i(D) = 2, 3$ and asked whether there exists a negative $D < 10\,000$ with $i(D) > 3$. He asked also whether there exists a positive non-square determinant D with odd $i(D)$.

This first question has been positively answered in 1882 by Pépin [3239], who showed $i(-6075) = 9$. Later Perott [3244, 3245] showed that for every prime p there are infinitely many irregular positive determinants divisible by p. The second question has been answered positively in 1936 by Pall [3214] who found that $D = 62\,501$ has the required property. Later Shanks [3744] provided the smaller example $D = 32\,009$.

Gauss conjectured also that if d is a determinant of the form $-(216k + 27)$ and $-(1000k + a)$ with $a = 75$ and 675, then d is irregular of exponent 3, with exception of $d = -27$ and $d = -75$, and this has been confirmed in 1890 by Mathews[42] [2766].

A modern presentation of Gauss's theory of quadratic forms has been given in 1990 by Ribenboim [3455].

8. The theory of the cyclotomic equation $x^p = 1$ with prime p, developed in Sect. 339–366, a forerunner of the Galois theory in a special case, can be regarded as the first step towards the theory of cyclotomic fields. After giving in Sect. 341 a rather complicated proof of the irreducibility of the p-th cyclotomic polynomial

$$\Phi_p(X) = \frac{X^p - 1}{X - 1} = X^{p-1} + X^{p-2} + \cdots + X + 1 ,$$

Gauss studied the *periods* (f, r) defined in the following way: denote by g a fixed primitive root mod p and for every factorization $p - 1 = ef$ and positive r put

$$(f, r) = \sum_{j=0}^{f-1} \zeta_p^{rg^{je}} , \tag{1.6}$$

[41] Carl Ludwig Siegel (1896–1981), professor in Frankfurt and Göttingen. See [917, 1871, 3661].

[42] George Ballard Mathews (1861–1922), professor in Bangor and lecturer in Cambridge. See [325].

where $\zeta_p = \exp(2\pi i/p)$ is the p-th primitive root of unity. Note that the period (f, r) generates the unique subfield of degree e of the cyclotomic field $\mathbf{Q}(\zeta_p)$.

Section 355 of [1394] contains a result which translated into the language of quadratic fields states that if f is even, then the field generated by the period (f, r) is real, and in the next section it is shown that the period $((p-1)/2, 1)$ is a root of the polynomial

$$x^2 + x + \eta \frac{p+\eta}{4} ,$$

where

$$\eta = \begin{cases} -1 & \text{if } p \equiv 1 \bmod 4, \\ 1 & \text{if } p \equiv 3 \bmod 4 . \end{cases}$$

In Sect. 357 it is shown that for every odd prime p one can write

$$4\Phi_p(X) = Y^2(X) + \eta p Z^2$$

where $Y(X), Z(X)$ are polynomials having rational integral coefficients. These results imply that the field $\mathbf{Q}(\zeta_p)$ contains the quadratic subfield $\mathbf{Q}(\sqrt{-\eta p})$. Section 358 contains the minimal polynomial for the periods $((p-1)/3, r)$ (see also Eisenstein[43] [1068]). A way of obtaining the minimal polynomial of the period $((p-1)/4, r)$ has been indicated by Gauss in the first part of [1395].

In Sect. 359–366 Gauss used the periods to prove that the roots of $X^p - 1$ can be expressed by radicals of order smaller than p, and as a corollary he showed that if $n = 2^a p_1 \cdots p_r$, where $a \geq 0$ and the p_i's are distinct primes of the form $2^N + 1$ (*Fermat primes*), then a regular n-gon can be constructed with the use of compass and ruler. He asserted in Sect. 365 the necessity of this condition but did not prove it.

The first proof of the necessity was provided in 1837 by Wantzel[44] [4297], who established a general criterion for constructibility by compass and ruler. His proof had a lacuna, pointed out by Loewy[45] ([2619], footnote on p. 108) in 1918 (see also Lützen [2671]). It seems that the first correct proof was provided by J. Petersen[46] ([3265], Chap. 7) in 1878. See also Pierpont[47] [3290], Loewy [2620] and Bauer[48] [230].

In 1897 a description of the construction of a regular n-gon with $n = 65\,537$ has been published by Hermes[49] [1812] (a large box containing the manuscript can be seen in the library of the mathematical department of the Göttingen University).

An elementary way of finding the minimal polynomial of the number $\cos(2\pi/n)$, generating the maximal real subfield of the p-th cyclotomic field with prime p, has been proposed in 1894

[43]Gotthold Eisenstein (1823–1852), dozent in Berlin. See [351, 4357].

[44]Pierre Wantzel (1814–1848), worked at L'École Polytechnique in Paris. See [552, 3554].

[45]Alfred Loewy (1873-1935), professor in Freiburg in Breisgau. See [1229].

[46]Julius Petersen (1839-1910), professor in Copenhagen. See [2672].

[47]James P. Pierpont (1866–1938), professor at Yale. See [3187].

[48]Mihály Bauer (1874–1945), professor in Budapest.

[49]Johann Gustav Hermes (1846–1912), teacher in Königsberg, Linden and Osnabrück.

by Dickson[50] [931]. For the case of prime powers this has been made much later by Aranés and Arenas [102].

In 1937 Lévy[51] [2585] conjectured that for prime p the polynomial $(x^p - 1)/(x - 1)$ is not a product of two polynomials with real nonnegative coefficients. This has been established in the same year by Krasner[52] and Ranulac [2268] and by Raĭkov[53] [3372].

9. The first systematic treatment of a ring of algebraic integers was made in 1832 by Gauss in the second part of [1395], where *complex integers* $a + bi$ (with $a, b \in \mathbf{Z}$) were introduced. Gauss wrote in Sect. 30 that he thought about this subject already in 1805, and soon became convinced in the necessity of extending arithmetics to the set of complex integers.

He defined and described prime numbers in the set $\mathbf{Z}[i]$ of complex integers, pointed out the existence of four units $\pm 1, \pm i$, and proved the unique factorization theorem, not using the Euclidean algorithm, but applying the unique factorization in $\mathbf{Z}$ instead. He defined congruences, described residue classes with respect to primes, gave a way of performing this task in the case of an arbitrary modulus, and in Sect. 46 the Euclidean algorithm has been shown to hold in $\mathbf{Z}[i]$.

The number of steps in the Euclidean algorithm in $\mathbf{Z}[i]$ has been studied in 1848 by Dupré [1030], who earlier did this in the case of rational integers [1029]. This question has been considered for arbitrary Euclidean imaginary quadratic fields by A. Knopfmacher and J. Knopfmacher[54] [2187] in 1991. For further results on Euclidean fields see Sect. 1.4.

Later polynomial congruences were considered and the existence of primitive roots for primes was established with an application to the proof of the quadratic reciprocity law in $\mathbf{Q}(i)$.

Other proofs of the quadratic reciprocity in $\mathbf{Q}(i)$ were given by Busche[55] [526] in 1890, Bonaventura [391] in 1892 (who in [390] proved it in the field $\mathbf{Q}(\sqrt{2})$) and Hilbert[56] in 1894 [1833]. A generalization to other quadratic fields with class-number one was provided by Dörrie[57] [992] in 1898 (cf. Dintzl[58] [951]), and in 1900 K.S. Hilbert [1847] considered this question in $\mathbf{Q}(\zeta_8)$ and $\mathbf{Q}(\zeta_{16})$. See also Rückle[59] [3528]. For all quadratic fields the quadratic reciprocity law has been established in 1919 by Hecke[60] [1737] and Vel'min[61] [4204, 4205], and the case of arbitrary fields has been settled by Hecke in his book [1742] (see Sect. 4.2.1). A proof for imaginary quadratic fields has been given in 1995 by Bayad [256] who applied elliptic curves and elliptic functions. Some corrections to his arguments were provided later by Hayashi [1723].

---

[50] Leonard Eugene Dickson (1874–1945), professor in Chicago. See [52, 1178].

[51] Paul Lévy (1886–1971), professor in Paris. See [4039].

[52] Marc Krasner (1912–1985), professor in Clermont-Ferrand and at the Université Paris VI.

[53] Dmitriĭ Abramovič Raĭkov (1905–1980), professor in Moscow.

[54] John Knopfmacher (1937–1999), father of Arnold Knopfmacher, professor at the University of the Witwatersrand. See [3710].

[55] Edmund Busche (1861–1916), teacher in Bergedorf and Hamburg. See [3463].

[56] David Hilbert (1862–1943), professor in Königsberg and Göttingen. See [3420, 3425, 4379].

[57] Heinrich Dörrie (1873–1955), teacher in Biedenkopf and Wiesbaden.

[58] Erwin Dintzl (1878–1972).

[59] Gottfried Martin Rückle (1879–1929).

[60] Erich Hecke (1887–1947), professor in Göttingen and Hamburg. See [3266, 3662].

[61] Vladimir Petrovič Vel'min (1885–1974), professor in Warsaw, Rostov and Kiev. See [2094].

In §67 of [1395] Gauss formulated the biquadratic reciprocity law in $\mathbf{Z}[i]$. To do this Gauss first defined the biquadratic character of a with respect to a prime p in $\mathbf{Z}[i]$ not dividing a as the power of i congruent to $a^{(N(\pi)-1)/4}$ mod π. He did not introduced any notation for this notion, but in 1844 Eisenstein in [1069] denoted it by

$$\left[\frac{a}{\pi}\right],$$

switching in [1070] to $[a;\pi]$.

Now one uses usually the form

$$\left(\frac{a}{\pi}\right)_4 \equiv a^{(N(\pi)-1)/4} \bmod \pi\ , \tag{1.7}$$

and calls it the *biquadratic power residue symbol.*

The reciprocity law is stated in §67 in the following form. Let a, b be complex primes congruent to 1 mod $(1+i)^3$ (such primes were called by Gauss *primary*). If at least one of the numbers a, b is congruent to 1 mod 4, then the biquadratic character of a with respect to b equals the biquadratic character of b with respect to a, otherwise these characters differ by the factor -1.

Its proof was promised to appear in the third part of [1395]. Unfortunately Gauss never fulfilled his promise, and proofs were given later by Eisenstein [1069, 1070, 1072, 1073] and Jacobi[62] [2008] (in some particular cases an elementary approach to this problem has been used by Dirichlet [954] in 1828).

Eisenstein [1070] presented Gauss's reciprocity law in a form similar to Legendre's formulation of the quadratic reciprocity law. Using (1.7) it takes the form

$$\left(\frac{a}{b}\right)_4 = (-1)^{(\Re a-1)(\Re b-1)/4}, \tag{1.8}$$

with a, b being primary primes. If $\pi = A + Bi$ is a primary prime, then

$$\left(\frac{1+i}{\pi}\right) = i^{(A-B-B^2-1)/4}\ .$$

It can be checked that the exponent $(\Re a - 1)(\Re b - 1)/4$ in (1.8) can be replaced by $(N(a) - 1)(N(b) - 1)/16$.

Proofs of the Gaussian biquadratic reciprocity law were given also by Pépin [3238] in 1880, and by Busche in 1886 [525]. See also Dintzl [949] and Busche [528]. For modern approach see the papers by Kubota [2338], Burde [516], Shiratani [3763, 3764], Watabe [4305] and K.S. Williams [4415]. See also Chap. 9 in the book [1965] by Ireland and Rosen.

This law in certain other fields was later considered by Lietzmann[63] [2596, 2599] and Bohniček[64] [378].

---

[62] Carl Gustav Jacob Jacobi (1804–1851), professor in Königsberg and Berlin. See [2210, 2211].

[63] Walther Lietzmann (1880–1959), teacher in Barmen, Jena and Göttingen.

[64] Stjepan Bohniček (1872–1956), professor in Zagreb.

In 1867 Mathieu[65] [2774] showed that the consequences of the Gaussian biquadratic reciprocity law to rational integers can be also obtained with the use of quadratic extensions of finite fields instead of complex integers.

The sixth chapter of Lemmermeyer's book [2527] is devoted to the biquadratic reciprocity.

Jacobi believed that Gauss was led to his study of complex integers by considering the problem of rectification of the lemniscate arcs. He wrote in [2009]:

"...*ich*[66] *glaube nicht, dass zu einem so verborgenen Gedanken die Arithmetik allein geführt hat, sondern dass er aus dem Studium der elliptischen Transcendenten geschöpft worden ist, und zwar der besonderen Gattung derselben, welche die Rectification von Bogen der Lemniscata giebt.*"

Indeed, in Sect. 335 of [1394] Gauss mentioned that his methods are also applicable to problems concerning transcendental functions related to the integral

$$\int_0^{\alpha} \frac{dx}{\sqrt{1-x^4}} ,$$

giving the length of lemniscate arcs. This was achieved in 1828 by Abel[67], who used complex integers in Sect. 40 of his paper [16] on elliptic functions to show that the division of the lemniscate into p parts for prime p by compass and ruler can be achieved if $p = 2$ or p is a Fermat prime. For primes $p = 2, 3$ and 5 this has been shown already in 1718 by Fagnano.[68]

A modern proof of Abel's result and its converse was given in 1981 by Rosen [3514]. For a generalization see the paper of Cox and Shurman [788].

In 1843 Liouville[69] [2604] deduced from Abel's theorem on Abelian equations [17] that the equation of the division of the lemniscate into n parts is solvable by radicals. A further study of this equation was made by Eisenstein [1075] in 1850.

The problem of division of the lemniscate was later treated by Kiepert[70] [2146], Kohl [2200], Schwering[71] [3713–3717], Mathews [2770, 2772] and Mitra [2891]. In 2014 Cox and Hyde [787] presented a modern approach, using class-field theory and complex multiplication to determine the Galois group of the field generated by the n-division point of the lemniscate (it equals the multiplicative group of residue classes mod n in the ring $\mathbf{Z}[i]$). They gave also an elementary proof of this result.

Gauss realized that higher reciprocity laws are connected with generalizations of the ring $\mathbf{Z}[i]$, writing in a footnote in Sect. 30 of [1395]:

---

[65] Émile Mathieu (1835–1890), professor in Besançon and Nancy. See [1018].

[66] "*...I do not believe that only arithmetics led to such a hidden idea, but that it was drawn from a study of elliptical transcendentals, from a special kind of them giving the rectification of the arc of the lemniscate.*"

[67] Niels Henrik Abel (1802–1829), docent in Christiania (Oslo). See [2486, 3188, 3942].

[68] Giulio Carlo Fagnano dei Toschi (1682–1766). See [160].

[69] Joseph Liouville (1809–1882), professor in Paris. [2670].

[70] Ludwig Kiepert (1841–1934), professor in Hannover.

[71] Karl Schwering (1846–1925), teacher in Köln.

"*Theoria*[72] *residuorum cubicorum simili modo superstruenda est considerationi numerorum formae* $a + bh$, *ubi* h *est radix imaginario aequationis* $h^3 - 1 = 0$, *puta* $h = -\frac{1}{2} + \sqrt{\frac{3}{4}}i$, *et perinde theoria residuorum potestatum aliorum quantitatum imaginarium postulabit*".

The influence of Gauss' work on the creation of algebraic number theory was discussed by O.Neumann [3101–3103].

1.1.3 Dirichlet

1. Numbers of the form $a + b\sqrt{5}$ with $a, b \in \mathbf{Z}$ were used in 1828 by Dirichlet [955] in his proof of the non-existence of solutions for a family of equations of the form

$$x^5 + y^5 = az^5$$

with fixed a. He considered in particular the case $a = 1$ and succeeded in showing the non-existence of solutions satisfying $5 \nmid xyz$ (the first case of Fermat's Last Theorem (FLT) for the exponent 5), a result weaker than that of Legendre[73] [2510] who proved FLT for the exponent 5 in 1823, but whose paper appeared only four years later. In 1832 Dirichlet [957] used the numbers $a + b\sqrt{-7}$ to establish Fermat's theorem for the exponent 14, which was superseded in 1840 by Lamé[74] [2407] who proved FLT for the exponent 7 (cf. the comments of Cauchy[75] [608], and a simplification due to Lebesgue[76] [2495] who succeeded to apply elementary methods to prove Fermat's theorem in this case).

In 1832 Dirichlet [956] proved the quadratic reciprocity law in $\mathbf{Z}[i]$ and showed in [968] that if a, b are co-prime complex integers, then there are infinitely many primes of the form $ax + b$ with $x \in \mathbf{Z}[i]$, an analogue of his famous result dealing with this question in the case of rational integers [959, 960].

In a footnote [956, p. 370] Dirichlet wrote:

"*On*[77] *peut, au lieu des expressions de la forme* $t + u\sqrt{-1}$, *considerér celles de la forme* $t + u\sqrt{a}$, *a étant sans diviseurs carré. Les expressions de se genre, considérées*

[72] "*The theory of cubic residues should be similarly based on numbers of* $a + bh$, *where* h *is a certain complex root of the equation* $h^3 - 1 = 0$, *e.g.,* $h = -\frac{1}{2} + \sqrt{\frac{3}{4}}i$, *and in exactly the same way the theory of residues for higher exponents needs the introduction of other imaginary quantities.*"

[73] Adrien-Marie Legendre (1752–1833), professor in Paris.

[74] Gabriel Lamé (1795–1870), professor in St. Petersburg and at École Polytechnique in Paris. See [4242].

[75] Augustin Louis Cauchy (1789–1851), professor in Paris. See [280, 4123].

[76] Victor Amédée Lebesgue (1791–1875), professor in Bordeaux.

[77] "*One could consider instead of expressions of the form* $t + u\sqrt{-1}$ *also expressions of the form* $t + u\sqrt{a}$, *a being without square factors. Expressions of this kind considered from the same point of view satisfy theorems analogous to those which are subject of this memoire and possibly can be proved in a similar way.*"

dans la même point de vue, donnent lieu à des théorèmes analogues à celui qui fait l'objet de ce mémoire et susceptibles d'une démonstration toute semblable."

This was the first occurrence of integers of an arbitrary quadratic field. It seems, however, that Dirichlet did not suspect that the arithmetical properties of these new numbers in many cases differ essentially from those of the Gaussian complex integers.

In 1874 and 1899 Mertens [2814, 2820] gave new proofs of Dirichlet's theorem on primes in progressions in the ring of integers of $Q(i)$. The analogous result for primes in $\mathbf{Q}(\sqrt{-3})$ has been established by E.Fanta [1161] in 1901. See Sect. 3.1.2 for the case of arbitrary number fields.

2. In 1838 Dirichlet [962] applied analytical methods to establish a formula for the class-number of properly primitive Gaussian quadratic forms of negative odd prime determinants. For prime $p \equiv 3 \bmod 4$ he obtained

$$H(-p) = \frac{2\sqrt{p}}{\pi}\left(1 - \frac{1}{2}\left(\frac{2}{p}\right)\right)\sum_{n=1}^{\infty}\left(\frac{n}{p}\right)\frac{1}{n},$$

and for prime $p \equiv 1 \bmod 4$ he showed

$$H(-p) = \frac{2\sqrt{p}}{\pi}\sum_{2\nmid n>0}\left(\frac{n}{p}\right)\frac{1}{n}.$$

Later [963, 964] he obtained similar formulas also for composite discriminants, both negative and positive, establishing on the way in [964] formulas for the value of L-functions at $s = 1$ for quadratic characters.

These results were later extended by Kronecker (Sect. VIII of [2299]) to the case of properly primitive quadratic forms $ax^2 + bxy + cy^2$ of given discriminant, without assuming $2 \mid b$.

It has been noted later by Dedekind (§186 of [848]) that this formula gives also the number of ideal classes in imaginary quadratic fields and a similar approach permits to do the same also for real quadratic fields (see Sect. 1.3.3).

Later certain simplifications of Dirichlet's proof were provided by Hermite [1821] in 1862 and Pépin [3235] in 1873.

3. In 1840 Dirichlet [964] asserted that any primitive quadratic binary form f represents infinitely many primes, and there are such primes in any progression containing infinitely many values of the form f. This implies the existence of prime ideals in any fixed ideal class in a quadratic field and shows that one can find them in appropriate residue classes. The proof of the first assertion[78] was given by Mertens [2814] in 1874 for positive-definite forms and H. Weber [4310] in 1882 in the general case, and the second result was established in 1888 by A.Meyer[79] [2844] (see also Mertens [2819, 2821]). This result is usually called the *Dirichlet-Weber* theorem. Its quantitative form was proved in 1896 by de la Vallée-Poussin[80] [4121, 4122].

[78] Already in 1863 Schering found a proof but it was published only in 1909 [3601].

[79] Arnold Meyer (1844–1896), professor in Zürich. See [2451].

[80] Charles de la Vallée-Poussin (1866–1962), professor in Louvain. See [519].

The proof of the quantitative form of the Dirichlet–Weber theorem has been simplified in 1915 by Landau [2426]. See also the dissertation of Bernays[81] [305].

Elementary proofs were given by Briggs [450] in 1954 and by Ehlich[82] [1056] in 1959.

In a later paper ([969], announced in [967]) Dirichlet studied arithmetics in $\mathbf{Z}[i]$ more closely. He introduced the name *norm* for the product $N(a+bi) = (a+bi)(a-bi)$ of two conjugated integers of $\mathbf{Z}[i]$, related it to the number of residues mod $a+bi$, proved the unique factorization property using the Euclidean algorithm, determined the form of complex primes and introduced the analogue of Euler's function $\varphi(n)$. He presented also a theory of quadratic forms with coefficients in $\mathbf{Z}[i]$, culminating in a formula for the class-number of quadratic forms with given discriminant. In the proof of this formula the infinite series

$$\sum_{\alpha\in\mathbf{Z}[i]} \frac{1}{N(\alpha)^x} = \prod_{\pi} \left(1 - \frac{1}{N(\pi)^x}\right)^{-1}$$

(α and π running over non-associated integers, resp. primes of $\mathbf{Z}[i]$) appeared, the first example of the Dedekind zeta-function for a non-rational field.

A modification of Dirichlet's formula for this class-number has been given in 1880 by Bachmann[83] [171].

Further progress in the theory of quadratic forms over $\mathbf{Z}[i]$ has been later obtained by Bianchi[84] [345–347] and Mathews [2767, 2768, 2771]. A modern presentation of reduction of quadratic forms over $\mathbf{Z}[i]$ was given by A.L. Schmidt (Chap. 6 in [3636]) in 1975. An algorithm for checking their equivalence has been given by Wolfskill [4438–4440].

At the end of his paper Dirichlet gave a formula for the class-number of biquadratic field containing $\mathbf{Q}(i)$, showing that if $K = \mathbf{Q}(\sqrt{m}, i) = \mathbf{Q}\left(\sqrt{m}, \sqrt{-m}\right)$ with $m > 0$, then

$$h(K) = ch(m)h(-m), \tag{1.9}$$

$h(d)$ denoting the class-number of $\mathbf{Q}(\sqrt{d})$ and $c \in \{1, 1/2\}$ (one has $c = 1/2$ if and only if the fundamental units of $\mathbf{Q}(\sqrt{m})$ and K coincide).

He wrote about (1.9): "*il*[85] *parait difficile de les établir par des considérations purement arithmétiques*", but in 1894 Hilbert [1833] succeeded in finding such proof. His paper contains a thorough study of quadratic extensions of $\mathbf{Q}(i)$.

The equality (1.9) has been generalized to the case $K = \mathbf{Q}(\sqrt{a}, \sqrt{b})$ by Bachmann [168] in 1867 (see also Amberg [61]), and to $K = \mathbf{Q}(\sqrt{m_1}, \dots, \sqrt{m_n})$ by Herglotz[86] [1810] in 1922. In the biquadratic case algebraic proofs have been given in case $a = -1$ by Hilbert [1833] in 1894 and

[81] Paul Isaac Bernays (1888–1977), professor in Göttingen, worked later at ETH in Zürich.

[82] Hartmut Ehlich (1931–2001), professor in Bochum.

[83] Paul Bachmann (1837–1920), professor in Breslau and Münster. See [1797].

[84] Luigi Bianchi (1856–1928), professor in Pisa. See [1287].

[85] "*It seems difficult to establish this by purely arithmetic considerations.*"

[86] Gustav Herglotz (1881–1953), professor in Leipzig. See [4064].

S.Kuroda[87] [2383] in 1943. Kubota [2335, 2336] in 1953-1956 applied class-field theory to give proofs in the general case. An elementary proof has been provided in 1972 by Halter-Koch [1599]. See also Lubelski [2660, 2661] and Reichardt[88] [3415]). For a generalization to the compositions of pure extensions of odd prime degree see Sect. 6.5.

4. In a letter to Liouville in 1840 Dirichlet [965] announced the first result[89] dealing with units in rings generated by an arbitrary algebraic integer. He stated it in the following form:

Let $f(X)$ be a monic irreducible polynomial of degree N with integral coefficients, and let $\alpha, \beta, \dots, \xi$ be its zeros, at least one of them being real. Then there exist infinitely many polynomials

$$g(X) = x_0 + x_1 X + \cdots + x_{N-1} X^{N-1} \tag{1.10}$$

with integral coefficients such that one has

$$g(\alpha) g(\beta) \cdots g(\xi) = 1 \; . \tag{1.11}$$

Two years later Dirichlet [970] stated that in the case $N \geq 3$ this result holds also for polynomials without real zeros.

Stated in modern terms this asserts that if α is an algebraic integer having at least one real conjugate, then the order $\mathbf{Z}[\alpha]$ has infinitely many units. This result was a forerunner of Dirichlet's theorem on the structure of units in algebraic number fields, presented in 1846 [971] with a sketch of its proof[90] (in an earlier paper he stated this theorem for cubic α [966]). He formulated it in terms of solutions of (1.11), but translated into modern language his theorem asserts that the multiplicative group U_K of units of the order $\mathbf{Z}[\alpha]$, where α is an algebraic integer, is a product of a finite group E_K, consisting of roots of unity contained in $\mathbf{Z}[\alpha]$ and r copies of the infinite cyclic group, where $r = r_1 + r_2 - 1$ with $r_1, 2r_2$ being the number of real resp. non-real embeddings of K into the complex field.

An exposition of Dirichlet's results on units has been presented in 1864 by Bachmann in his habilitation thesis at Breslau University [167].

Dirichlet's proof works also for units in arbitrary rings $\mathbf{Z}_K$, which in general are not of the form $\mathbf{Z}[\alpha]$. It has been simplified in 1883 by Kronecker [2297] and Molk[91] [2900].

[87] Sigekatu Kuroda (1905–1972), professor in Nagoya and at the University of Maryland. See [2556].

[88] Hans Reichardt (1908–1991), professor in Berlin. See [2197].

[89] It seems that the first mention of units in a non-quadratic field occurs in the letter of Jacobi to Legendre of 27 May 1832 ([2], 275–277 and [3287], 80–83) in which he announces a result equivalent to the existence of infinitely many units in biquadratic fields.

[90] The complete proof appeared in his lectures, see [972], §166 in the second edition.

[91] Jules Molk (1857–1914), professor in Besançon and Nancy. See [4236].

A simpler proof of Dirichlet's unit theorem, working also for the unit group of orders, has been given in 1928 by van der Waerden[92] [4133]. An analogue of Dirichlet's theorem for the field of all algebraic numbers, established in 2014 by Fili and Miner [1185]. For a generalization to S-units see Sect. 5.2.2.

1.2 First Steps

1.2.1 *Eisenstein*

1. The formulation of the cubic reciprocity law appeared for the first time in a paper of Jacobi [2007] in 1827, who presented its proof[93] in his lectures at the Königsberg University in 1836 [2008]. For the first published proof one had to wait until 1844, when Eisenstein [1065, 1068], who at that time was still a student,[94] followed Gauss' advice and used the ring $\mathbf{Z}(\varrho) = \{a + b\varrho : a, b \in \mathbf{Z}\}$ with $\varrho = \zeta_3$ to prove the cubic reciprocity law in the following form:

For $a \in Z[\varrho]$ and a prime $\pi \in \mathbf{Z}[\varrho]$ with $\pi \nmid 3a$ put

$$\left(\frac{a}{\pi}\right)_3 = \zeta_3^k \equiv a^{(N(\pi)-1)/3} \pmod{\pi}$$

If π_1, π_2 are primes in $\mathbf{Z}[\varrho]$ of distinct norms, satisfying $\pi_1 \equiv \pi_2 \equiv 2 \bmod 3$, then

$$\left(\frac{\pi_1}{\pi_2}\right)_3 = \left(\frac{\pi_2}{\pi_1}\right)_3 .$$

Moreover one has

$$\left(\frac{1-\varrho}{\pi}\right)_3 = \varrho^{-m} , \tag{1.12}$$

where

$$m = \begin{cases} (1+p)/2 & \text{if } \pi = p \in \mathbf{Z} , \\ (1+a)/3 & \text{if } \pi = a + b\varrho \equiv 2 \bmod 3, b \neq 0 . \end{cases}$$

Eisenstein did not develop in detail arithmetical properties of $\mathbf{Z}[\varrho]$, writing in the introduction to his paper:

---

[92] Bartel Leendert van der Waerden (1903–1996), professor in Leipzig, Amsterdam and Zürich. See [1240, 3721]

[93] Jacobi published his proof in [2010] in 1846.

[94] It is remarkable that in 1844 Eisenstein published 22 papers and three lists of problems in the Journal für reine und angewandte Mathematik.

" *Die*[95] *Elementarsätze der Theorie der ganzen complexen Zahlen von der Form* $a + b\varrho$, *wo* ϱ *eine imaginäre Cubicwurzel der Einheit bezeichnet, finden sich zwar noch nirgends aufgezeichnet; indessen glauben wir, weger der grossen Analogie, welche zwischen diesen complexen Zahlen und den gewöhnlich sogenannten complexen Zahlen von der Form* $a + b\sqrt{-1}$ *herrscht, diese Sätze, in soweit sie sich auf die Theilbarkeit der Zahlen durcheinander, Zerlegbarkeit in einfache Faktoren, Theorie der complexen Primzahlen, u.s.w. beziehen, hier als bekannt voraussetzen zu dürfen*".

Other proofs of the cubic reciprocity law were later given by Dantscher[96] [813] and Pépin [3237] in 1877 (cf. [3238]), Gegenbauer[97] [1398] in 1880, Koschmieder[98] [2227] in 1921, Zhuravskiĭ [4470] in 1927, Habicht [1579] in 1960, Joly [2063] in 1972, Hayashi [1722] in 1974, and Friesen, Spearman and K.S. Williams [1257] in 1986. A proof using formal groups has been given in 2013 by Demchenko and Gurevich [885]. See also Chap. 9 in the book [1965] by Ireland and Rosen. A simple proof of (1.12) was provided by K.S. Williams [4412, 4416]. Cf. Dintzl [950].

In 1890–1892 Gmeiner[99] [1441–1443] proved the reciprocity law for sixth powers, but it has pointed out later by Hasse[100] ([1664], p. 76) that this law is a consequence of the quadratic and cubic reciprocity laws.

A new approach to Eisenstein's proofs of the biquadratic and cubic reciprocity laws based on θ-functions has been given in 1961 by Kubota [2337].

2. In the same year 1844 Eisenstein [1066, 1067] gave a formula for the number of classes of quadratic forms over $\mathbf{Z}[\zeta_3]$, showing in [1067] that the number of non-associated integers in $\mathbf{Z}[\zeta_3]$ of norm M equals $\sum_{d|M}\chi(d)$, with $\chi(n) = \left(\frac{-3}{n}\right)$. This paper contains also an occurrence of a particular case of Dedekind's zeta-function $\zeta_K(s)$ presented in the form

$$S(s) = \sum_{\alpha} \frac{1}{N(\alpha)^s} \quad (s > 1)\ ,$$

the sum being extended over non-associated integers of the field $\mathbf{Q}(\zeta_3)$. Eisenstein obtained the product formula for it and proved the equality

$$S(s) = \zeta(s)L(s, \chi)\ ,$$

where $L(s, \chi)$ is the Dirichlet L-function associated with χ.

---

[95] "*Although elementary theorems of the theory of integral complex numbers of the form* $a + b\varrho$, *where* ϱ *denotes an imaginary cubic root of unity are to be found nowhere, we believe that in view of the great analogy between these complex numbers and the usually called complex integers of the form* $a + b\sqrt{-1}$, *we may assume the knowledge of these theorems as far as they deal with divisibility, factorization into simpler factors, the theory complex primes, etc.*"

[96] Victor von Dantscher (1847–1921), professor in Graz.

[97] Leopold Gegenbauer (1807–1894), professor in Czernowitz, Innsbruck and Vienna. See [2189].

[98] Lothar Koschmieder (1890–1974), professor in Brno, Graz, Aleppo, Tucumán, Baghdad and Tübingen.

[99] Josef Anton Gmeiner (1862–1927), professor in Prague and Innsbruck.

[100] Helmut Hasse (1898–1979), professor in Halle, Marburg, Göttingen, Berlin and Hamburg. See [1238].

In the long paper [1071] Eisenstein applied arithmetics in the ring $\mathbf{Z}[\zeta_3]$ to study representations of integers by cubic forms, using implicitly integers of cyclic cubic fields of prime conductor, i.e. contained in cyclotomic fields $\mathbf{Q}(\zeta_p)$ with prime p congruent to unity mod 3. He believed at that time that to build a theory of algebraic numbers one should first have a theory of forms in several variables. In a letter to Stern[101] (first letter in [1940]) he wrote:

"*Auch*[102] *Jacobi ist ganz meiner Ansicht, dass die Theorie der allgemeinen complexen Zahlen erst durch eine vollständige Theorie der höheren Formen ihre Vollendung erhalten kann*".

His results were later reformulated in the language of cyclic cubic fields in two particular cases by Nowlan [3135] in 1926, and in the general case by Latimer[103] [2475, 2477] in 1929–1930.

Eisenstein's work on cubic irrationalities was limited to cyclic fields. General cubic fields occur implicitly in a paper of Arndt [108], published in 1857 and dealing with binary cubic forms of positive discriminant. His main result implies the finiteness of the class-number of totally real cubic fields.

In a later paper Eisenstein [1074] utilized arithmetics in $\mathbf{Z}[i]$ and in the order $\mathbf{Z} \oplus 3\omega\mathbf{Z} \oplus 3\omega^2\mathbf{Z}$ (with $\omega = \zeta_7 + \zeta_7^{-1}$) in the maximal real subfield of $\mathbf{Q}(\zeta_7)$ to determine the numbers x in representations of primes $p \equiv 3 \bmod 8$ by the form $x^2 + 2y^2$, and of primes $p \equiv 2, 4 \bmod 7$ by the form $x^2 + 7y^2$. In this paper one finds also a special case of the octic reciprocity law.

The proof of the octic reciprocity law in the general case has been obtained in 1889 by Goldscheider [1466].[104] Another proof has been later given by Bohniček [379]. For a discussion and further references see Chap. 9 of Lemmermeyer's book [2527].

In 1850 considered Eisenstein [1077, 1078] the reciprocity law for l-th powers with odd primes l. His result deals with the cyclotomic field $K = \mathbf{Q}(\zeta_l)$ and concerns the l-th power residue symbol defined by

$$\left(\frac{a}{b}\right)_l = \zeta_l^k \equiv a^{(N(b)-1)/l} \pmod{b}, \tag{1.13}$$

where $a, b \in \mathbf{Z}_K$ and $0 \leq k < l$.

In the case when a is a rational integer not divisible by l, and $b \equiv r \bmod (1-\zeta_l)^2$ with $r \in \mathbf{Z}$ (Eisenstein called such integers *primary*[105]), Eisenstein established the equality

$$\left(\frac{a}{b}\right)_l = \left(\frac{b}{a}\right)_l .$$

---

[101] Moritz Abraham Stern (1807–1894), professor in Göttingen. See [3529].

[102] *"Also Jacobi supports my view that the theory of general complex numbers can be completed only by the complete theory of higher forms"*

[103] Clairborne Green Latimer (1893–1960), professor at Tulane, Kentucky University and Emory.

[104] Lemmermeyer wrote on p. 311 of his book [2527]: "*Given that his paper appeared in a rather obscure journal, it is surprising that it has been noticed at all*".

[105] Later such integers were called semi-primary (see [1836], §115.)

This was used later by Furtwängler[106] [1329, 1333] in the proof of his reciprocity law. Later [1343] he gave another proof of Eisenstein's law. It is reproduced with all details in the third volume of Landau's book [2444]. See also Chap. 11 in Lemmermeyer's book [2527] and Chap. 14 in the book [1965] by Ireland and Rosen.

In 1908 A.E. Western[107] [4369] extended Eisenstein's reciprocity law to odd prime powers l^k (for the case $k = 2$ see also Furtwängler [1354]), and in 1927 Hasse [1658] generalized it to exponents divisible by 8.

In 1909 Furtwängler ([1338], §12) gave a new proof of Eisenstein's reciprocity law, and three years later used it to deduce a criterion for the truth of Fermat's Last Theorem [1342].

In 1927 Fueter[108] [1308, 1309] extended Eisenstein's law to the fields $\mathbf{Q}(\sqrt{d}, \zeta_l)$ (with $d < 0$).

1.2.2 Kummer

1. Factorizations of primes congruent to unity mod n into factors in $\mathbf{Z}[\zeta_n]$ were first studied by Jacobi in 1839 [2009]. He showed that in the case $n = 8$ every such prime is a product of four factors and asserted that the same holds for $n = 5$ and 12 (see also [2010]). Kummer became interested in this problem and tried to show that if p is a prime, then every prime q, congruent to unity mod p, is a product of $p - 1$ elements of $\mathbf{Z}[\zeta_p]$,

$$q = \pi_1 \cdots \pi_{p-1}. \tag{1.14}$$

He presented a proof of this assertion in a paper submitted on 20 April 1844 to the Monatsberichte of the Berlin Academy. It is reproduced in Appendix I to the paper [1044] of Edwards. Unfortunately, this proof contains an error just at its beginning (see Edwards [1043, 1044] and Bölling [383]), and the submission was retracted.

In his first published paper on algebraic numbers [2347], in which he dealt with integers of the cyclotomic field $\mathbf{Q}(\zeta_p)$ with prime p, Kummer checked numerically his assertion for primes $p \leq 19$ and noted that it fails for $p = 23$. This implied that the unique factorization does not hold in $\mathbf{Z}[\zeta_{23}]$. The proof for $p = 5, 7$ is contained also in an unpublished manuscript of Kummer (see Bölling [384] where this manuscript is reproduced), and other proofs for $p = 5$ were given in 1850 by Hermite [1814], who utilized his theory of minima of quadratic forms and in 1882 by Schwering [3712]. Hermite applied also his method to show that every prime congruent to 6 mod 7 is a product of three factors in the cubic field of discriminant 49.

Reuschle[109] [3447] published in 1875 a table of factorization of primes $p < 1000$ in cyclotomic fields of small degrees (for its description see Kronecker [2290]).

In 1892 H.W.L. Tanner[110] [4004] presented an algorithm for computing prime factors of primes $p \equiv 1 \bmod 5$ in $\mathbf{Q}(\zeta_5)$,

---

[106] Philipp Furtwängler (1869–1940), professor in Bonn, Aachen and Vienna. See [1885, 1915].

[107] Alfred Edward Western (1873–1961), worked as a solicitor. See [2863].

[108] Rudolf Fueter (1880–1950), professor in Zürich. See [3874].

[109] Carl Gustav Reuschle (1812–1875), teacher in Tübingen and Stuttgart. See [3106].

[110] Henry William Lloyd Tanner (1851–1915), professor in Cardiff.

Kummer's paper [2347] contains also the assertion that every unit of $\mathbf{Z}[\zeta_p]$ is a product of a power of ζ_p and a real unit.

In 1845 Kummer published another paper on algebraic numbers [2348] in which he considered **Z**-linear combinations of Gaussian periods (1.6), (i.e. integers of a subfield of the p-th cyclotomic field with odd prime p). He showed that if $p-1=ef$ and R is the set of all **Z**-linear combinations of the periods (f,r) $(r=1,2,\dots,e)$ (hence R coincides with the ring of integers of the unique subfield K of degree e of $\mathbf{Q}(\zeta_p)$), then every prime q which is an e-th power residue mod p divides an element of R, not all of whose coefficients are divisible by p.

2. In March 1847 Lamé [2408, 2409] claimed to have a proof of Fermat's Last Theorem, based of properties of numbers in $\mathbf{Z}[\zeta_p]$, and Kummer in a letter to Liouville [2349] pointed out that Lamé tacitly assumed unique factorization in that domain which fails in certain cases. The discussion in the Paris Academy concerning Lamé's work was presented by Edwards [1043, 1045] (see also Ribenboim [3454]).

In the same year Kummer [2350, 2351] introduced "ideal complex numbers" in the ring $\mathbf{Q}[\zeta_p]$ with prime p. If q is a prime congruent to 1 mod p, then with every solution ξ of the congruence $\xi^p \equiv 1 \bmod p$ Kummer associated an ideal prime factor $\Phi(\xi)$ and defined an element $f(\zeta_p)=\sum_{j=0}^{p-1} a_j\zeta_p^j$ to be divisible by $\Phi(\xi)$ if $p \mid f(\xi)$. For other primes q the definition is more complicated (see Chap. 4 in the book [1045] of Edwards for details). Kummer defined multiplication of ideal numbers, showed that numbers with the same ideal prime factors differ by a unit, deduced the unique factorization into prime ideal numbers and defined the partition of ideal numbers into classes. After introducing multiplication of these classes he established the validity of the factorization (1.14) for all primes $q \equiv 1 \pmod p$ into $p-1$ prime ideal factors. He proved moreover that every ideal number can be considered as a complex number whose certain power belongs to $\mathbf{Z}[\zeta_p]$, showed that the number h_p of classes is finite, and determined for $p \leq 47$ the minimal integer n such that for every ideal number α its nth power becomes a complex number in $\mathbf{Z}[\zeta_p]$ (in modern language n is the exponent of the class-group of the field $\mathbf{Q}(\zeta_p)$). He defined the exponent of a prime ideal number π as the order mod p of the unique rational prime divisible by π and showed on p. 357 of his paper is that every class contains a product of prime ideal numbers of exponent one (this means that every class of ideals contains products of prime ideals of degree one).

In §12 of [2351] one finds an assertion which can be regarded as the first result dealing with Galois structure of the class-group of $\mathbf{Q}(\zeta_p)$: let r be a fixed primitive root mod p and s the generator of the Galois group satisfying $s(\zeta_p)=\zeta_p^r$. Let r_i be the smallest positive residue mod p of r^{-i}, and put

$$q_i=\frac{1}{p}(rr_i-r_{i-1}) \quad (i=0,1,\dots,p-2)\,.$$

Then for each prime ideal $\mathfrak{p}$ of degree one the product

$$\prod_{i=0}^{p-2} s^i(\mathfrak{p})^{q_i}$$

is principal.

This result has been generalized in 1952 by MacKenzie [2682] to arbitrary cyclotomic extensions of the rationals, and Yokoyama [4455] provided in 1964 a generalization to normal extensions $K/\mathbf{Q}$ containing a cyclotomic field.

Note that in his first papers on the field $\mathbf{Q}(\zeta_p)$ Kummer wrote the integers in the form $\sum_{j=0}^{p-1} x_j \zeta_p^j$, and only in 1851 [2358] switched to the form $\sum_{j=0}^{p-2} x_j \zeta_p^j$.

It is possible that Kummer was led to his idea of ideal numbers by the previous work of Jacobi [2009] (see Bölling [385] and Lemmermeyer [2531]). See also O.Neumann [3104] for Kummer's motivation.

3. A formula for the number h_p of classes of ideal numbers in the field $\mathbf{Q}(\zeta_p)$ has been given by Kummer in the second part of [2352]. He presented the details later in [2354]. The main tool was the function

$$F(x) = \sum_a \frac{1}{N(a)^x} \quad (x > 1)\ ,$$

where a runs over all non-associated (i.e. not differing by a unit factor) ideal numbers of $\mathbf{Z}[\zeta_p]$, and $N(a)$ denotes the norm of a. It coincides with the Dedekind zeta-function of the p-th cyclotomic field. He introduced also the class zeta-functions

$$F_A(x) = \sum_{a \in A} \frac{1}{N(a)^x}$$

for classes A of ideal numbers, proved the equality

$$F(x) = \zeta(x) \prod_\chi L(x, \chi), \tag{1.15}$$

(where χ runs over non-principal characters mod p, and $L(x, \chi)$ is the corresponding Dirichlet L-function) and showed that the limit

$$c = \lim_{x \to 1+} (x-1) F_A(x)$$

does not depend on A, thus

$$\lim_{x \to 1+} (x-1) F(x) = h_p c\ .$$

He used Dirichlet's unit theorem to compute the value of c for the principal class A_0, being equal to the set of all pairwise non-associated integers of the field.

On the other hand the equality (1.15) leads to

$$\lim_{x\to 1+} (x-1)F(x) = \prod_{\chi\neq\chi_0} L(1,\chi)\,,$$

and using the explicit value of c Kummer arrived at the formula

$$h_p = \frac{p^{(p-2)/2}}{R(p)2^{(p-3)/2}\pi^{(p-1)/2}} \prod_{\chi\neq\chi_0} L(1,\chi)\,, \tag{1.16}$$

where $R(p)$ denotes the regulator of the field $\mathbf{Q}(\zeta_p)$, which is a determinant formed by logarithms of fundamental units and their conjugates. To present the product of values of Dirichlet's L-functions at 1 in a more explicit form utilized Kummer the formulas for $L(1,\chi)$ established earlier by Dirichlet [960].

An important role in the proof played real algebraic units of the form

$$\left(\frac{(1-\zeta_p^{ga})(1-\zeta_p^{-ga})}{(1-\zeta_p^{a})(1-\zeta_p^{-a})}\right)^{1/2} = \left|\frac{1-\zeta_p^{ga}}{1-\zeta_p^{a}}\right|,$$

where g is a fixed primitive root mod p and $a = 1, 2, \ldots, p-1$. These units were called later *cyclotomic units*[111] (see e.g. Hilbert's book [1836]). Kummer used this name (*Kreistheilungs-Einheit* in German) in the case $a = 1$ ([2357], p. 162).

Kummer described also formulas for the class-number of subfields of $\mathbf{Q}(\zeta_p)$, and in the case of the maximal real subfield his formula implies the equality between h_p^+ and the index of the group of cyclotomic units in the group of all units.

In 1952 Hasse ([1700], §10) defined cyclotomic units for real Abelian fields, and his definition has been modified next year by Leopoldt [2550] (see also R.Greenberg (Sect. 5 of [1511] in the case of cyclic fields of prime degree, and Gillard [1430])). For arbitrary Abelian fields this has been done by Sinnott [3804] under the name *circular units*. Another definition in the case of real fields has been given later by Thaine [4048]. In 1990 Lettl [2571] proved that the two definitions of circular units give the same set of units. See Kučera [2340] for an overview of various kinds of cyclotomic or circular units.

Bases for the group of cyclotomic units were constructed by Gold and J.M. Kim [1457], Kučera [2339, 2341] and M. Conrad [764].

Kummer presented the class-number h_p as the product of two factors

$$h_p = h_p^- h_p^+, \tag{1.17}$$

calling them the *first factor*, resp. the *second factor* of h_p (earlier these factors were denoted differently by different authors. The notation h_p^-, h_p^+ is of much later vintage). He showed that the second factor equals the class-number of the maximal real subfield of $\mathbf{Q}(\zeta_p)$, and the first can be written as

[111] Nowadays the name "cyclotomic unit" is used for the elements of the group generated by $\pm\zeta_p$ and $1-\zeta_p^a$ $(a = 1, 2, \ldots, p-1)$.

$$h_p^- = \frac{P}{(2p)^{m-1}}, \tag{1.18}$$

where $m = (p-1)/2$, and

$$P = \prod_{j=1}^{m} F\left(\zeta_{p-1}^{2j-1}\right),$$

where

$$F(X) = \sum_{j=0}^{p-2} g_j X^j,$$

and g_j is the smallest positive residue mod p of g^j, g being a fixed primitive root mod p.

He showed that h_p^- is a positive integer and proved that if p divides h_p, then it divides h_p^-.

It has been asserted by Kummer (on p. 114 of [2354]) with a fallacious proof that if $K \subset \mathbf{Q}(\zeta_p)$, then $h(K) \mid h_p$. The correctness of this assertion (also for subfields of $\mathbf{Q}(\zeta_q)$ with prime power q) for the narrow class-number h^* has been established in 1908 by Furtwängler [1337], who proved also that if $K \subset L$ are subfields of $\mathbf{Q}(\zeta_q)$ with a prime power q, then $h^*(K)$ divides $h^*(L)$. He showed also by the example $K = \mathbf{Q}(\sqrt{-5})$, $L = K(i)$ that this may fail if q is not a prime power.

A simple proof of Kummer's assertion based on class-field theory has been provided in 1932 by Herbrand[112] [1803], who noted that his proof works also for the class-number $h(K)$. It has been shown by Latimer [2479] in 1933 that Furtwängler's result holds for a large class of cyclic extensions of $\mathbf{Q}$. See also Värmon [4192].

At the end of [2354] Kummer announced a similar formula for the class-number for a family of subfields of $\mathbf{Q}(\zeta_p)$ and pointed out that as a special case one obtains Dirichlet's formulas for the class-number of quadratic forms.

4. In a letter to Kronecker, written in December 1849 [2353] mentioned Kummer "*den*[113] *noch zu beweisenden Satz*" asserting that the second class-number h_p^+ is not divisible by p. He tried unsuccessfully to prove it, but four years later wrote to Kronecker [2362]: "*...eines*[114] *meiner Hauptresultate, auf welches ich seit einem Vierteljahr gebaut hatte, daß der zweite Faktor der Klassenzahl $\frac{D}{\Delta}$ niemals durch λ theilbar ist, falsch ist oder wenigstens unbewiesen*".

This conjecture is usually called the *Vandiver conjecture* or *Kummer-Vandiver conjecture*. Vandiver[115] wrote in 1946: "*About twenty five years ago I conjectured that this number was never*

[112] Jacques Herbrand (1908–1931). See [943].

[113] "*the theorem yet to be proved*".

[114] "*one of my main results, on which I built since a quarter of the year, asserting that the second factor of the class-number $\frac{D}{\Delta}$ is never divisible by λ, is false, or at least unproved*".

[115] Harry Schultz Vandiver (1882–1973), professor at the University of Texas. See [772].

divisible by l" ([4181], p. 576), adding that the same assertion was also conjectured by Furtwängler in 1928. See Sect. 4.4.2.

5. As the first important application of his theory presented Kummer [2352, 2356] the proof of Fermat's Last Theorem[116] for all prime exponents p satisfying $p \nmid h_p$. In the first version he had to assume additionally that every unit congruent mod p to a rational integer is a p-th power of another unit, but then [2355] he showed that this is a consequence of the assumption on p.

Two proofs of the last result have been later given by Hilbert ([1836], Th. 156; see also Satz 969 in Landau's book [2444]). In 1934 Vandiver [4172] observed that Furtwängler's results on class-fields established in [1336] lead to a similar assertion:

If $p \nmid h_p$, $\zeta_p \in K$, and η is a unit of K congruent to a p-th power mod $(1-\zeta_p)^p$, then η is itself a p-th power.

He provided also a generalization.

6. Primes p satisfying $p \nmid h_p$ are called *regular*. In [2355] Kummer showed that a prime p is regular if and only if p does not divide the numerators of the Bernoulli numbers[117] $B_2, B_4, \dots, B_{p-3}$, defined by the equality

$$\frac{z}{e^z - 1} = 1 + \sum_{n=1}^{\infty} \frac{B_n}{n!} z^n .$$

(A list of the first 498 Bernoulli numbers has been prepared by Plouffe [3313].)

Because of the equality

$$\zeta(1-n) = -\frac{B_n}{n}$$

for positive integers n Kummer's criterion can be expressed using values of Riemann's zeta-function.

Kummer succeeded [2375] in finding the first nine irregular primes:

$$37, 59, 67, 101, 103, 131, 137, 149, 157 ,$$

and conjectured that in a large initial interval there are approximately twice as many regular primes as irregular.

Later Hensel[118] ([1782], p. 32) recalled a remark of Kummer that the ratio $1:2$ should replaced by $1:1$, and Siegel [3792] gave arguments for this ratio to be rather $1:(\sqrt{e}-1)$.

A simpler proof of Kummer's characterization of regular primes in terms of Bernoulli numbers was later provided by Kronecker [2279]. His proof is reproduced in Hilbert's report [1836] (Lemma

[116] For expositions of Kummer's work on Fermat's theorem see Bachmann [173] and Edwards [1043–1045]

[117] Kummer actually used another numeration of Bernoulli numbers, omitting the vanishing terms with odd indices > 1

[118] Kurt Hensel (1861–1941), professor in Berlin and Marburg, edited the *Journal für reine und angewandte Mathematik* from 1901 on. See [1693].

28). Later Mirimanoff[119] [2883] found another condition equivalent to $p \mid h_p$, utilizing cyclotomic units. Another proof of Kummer's result has been given in 1919 by Vandiver [4143], who expressed the residue of h_p^- mod p^n in terms of Bernoulli numbers, generalizing Kummer's result for $n = 1$ (simpler proofs of Vandiver's congruence were later found by Hasse [1706] and Slavutskiĭ [3825]. See also Inkeri[120] [1964]).

It is not known whether there are infinitely many regular primes. On the other hand there are infinitely many irregular primes, as shown in 1915 by K.L. Jensen [2051] (for his proof see also Vandiver [4184]). Another proof gave in 1954 Carlitz[121] [577]. The question whether for every $N > 1$ every residue class mod N contains infinitely many irregular primes still awaits an answer (see the papers of Montgomery [2918], Metsänkylä [2838, 2840] and Yokoi [4454]).

In 1937 Vandiver [4175] determined all irregular primes in the interval $(307, 617)$. The first computer search for irregular primes has been done in 1954 by D.H. Lehmer[122], E. Lehmer[123] and Vandiver [2517]. They used the SWAC[124] computer to find all irregular primes below 2000. The same computer has been later used to extend the range to $p < 2521$ (Vandiver [4183] with the help of Selfridge[125] in 1954) and to $p < 4002$ (Selfridge, Nicol and Vandiver [3727] in 1955). After twenty years this range has been extended to 30 000 (W. Johnson [2058]), in 1978 to 125 000 (Wagstaff [4262]), in 1987 to 150 000 (J.W. Tanner and Wagstaff [4005]), in 1993 to $4 \cdot 10^6$ (Buhler, Crandall[126], Ernvall, Metsänkylä [505]) and in 2001 to $12 \cdot 10^6$ (Buhler, Crandall, Ernvall, Metsänkylä, Shokrollahi [506]). Ten years later Buhler and Harvey [508]) extended the search to $163 \cdot 10^6$, and in 2017 Hart, Harvey and Ong [1635] reached $2^{31} > 2 \cdot 10^9$.

It has been shown in 2015 by Luca, Pizarro-Madariaga and Pomerance [2668] that there are $\geq (1 + o(1)) \log\log x / \log\log\log x$ irregular primes $\leq x$.

In 1973 R. Greenberg [1510] generalized Kummer's criterion by establishing a similar necessary and sufficient condition for the divisibility of the class-number of $K(\zeta_p)$ by p in terms of the values of Dedekind zeta-function $\zeta_K(s)$ at negative rational integers for a large class of fields K.

7. In 1851 Kummer [2358] presented a broad exposition of his theory [2358], covering the results obtained in [2351, 2354, 2355]. One finds there the values of h_p^- for primes $p < 100$ (later [2375] he extended this list to $p \leq 163$). Note that the value of h_{19}^- given in [2358] is erroneous and has been corrected by Kummer in [2360] (it equals $7^2 \cdot 79241$, and not $7^2 \cdot 29 \cdot 3851$).

On p. 473 he observed that h_p^- grows extraordinarily quick and stated the following assertion:

For prime p the number h_p^- is asymptotically equal to

$$L(p) = 2^{-(p-3)/2} \pi^{-(p-1)/2} p^{(p+3)/4} . \tag{1.19}$$

For later work on this conjecture see Sect. 6.2.2.

---

[119] Dmitry Mirimanoff (1861–1945), professor in Geneva. See [4182].

[120] Kuusta Adolf Inkeri (1908–1997), professor in Turku.

[121] Leonard Carlitz (1907–1999), professor at the Duke University. See [437].

[122] Derrick Henry Lehmer (1907–1999), son of Derrick Norman Lehmer, husband of Emma Lehmer, professor at Berkeley. See [454].

[123] Emma Lehmer (1906–2001), wife of Derrick Henry Lehmer. See [455].

[124] Standard Western Automatic Computer, constructed in 1950, with 9472-bit memory. See Corry [773].

[125] John Lewis Selfridge (1927–2010), professor at the University of Illinois and the Northern Illinois University.

[126] Richard Crandall (1947–2012), professor at Reed College in Portland.

In 1964 Schrutka v. Rechtenstamm [3691] computed h_p^- for all primes $p \leq 257$. Six years later Newman [3108] transformed (1.18) to a form which permitted to compute h_p^- for all primes $p < 200$. The range $200 < p < 521$ was covered by D.H. Lehmer and Masley [2518] in 1978, and in 1992 Fung, Granville and H.C. Williams [1325] reached 3000. Two algorithms for computing h_p^- were analysed by Jha [2052] in 1995, one of them with running time of the order $O(p^2 \log p)$ under Generalized Riemann Hypothesis[127] (GRH). In 1999 Shokrollahi [3765] went up to $p < 5\,000$. His computation covered the ratio $h(K)/h(K^+)$ for all complex Abelian fields with conductor below 10 000.

To the study of h_p^- returned Kummer in 1870 [2374], when he considered the question in which cases every totally positive unit of the p-th cyclotomic field is a square of a unit. He showed that this happens if and only if h_p^- is odd and showed moreover that if h_p^+ is even, then h_p^- is also even, but not conversely, as shown by the examples $p = 29$ and 113. For further results on the first factor see Sect. 3.3.2.

The paper [2374] contains also the proof of divisibility of h_p^+ by 3 for $p = 219$ and 257, and of $2 \mid h_{163}^+$., as well as the assertion that h_{937}^+ is even.

It has been shown by H. Weber [4312] that in the field $\mathbf{Q}(\zeta_{2^n})^+$ every totally positive unit is a square. In 1952 Hasse (Satz 6 and 7 in [1700]) gave a simpler proof and established the same assertion for real Abelian fields of degree 2^n whose discriminant is a prime power. In 1967 Armitage and Fröhlich[128] [107] showed this in the case when K is a real cyclic extension of prime degree p, the order of 2 mod p is even, and $2 \nmid h(K)$. This has been extended by Adachi [20] in 1969, and M.-N. Gras [1496] and G. Gras and M.-N. Gras [1493]. In 1975 G. Gras [1488] showed that if K is Abelian of odd degree, then the class-number $h(K)$ is even if and only if if there exists in K a totally positive unit, congruent to a square mod 4, which is not a square itself. Later larger classes of real Abelian fields have been treated by Hughes and Mollin[129] [1918] in 1983 and M.-H. Kim and Lim [2149].

It has been proved by Garbanati [1384] in 1976 that this property fails for $\mathbf{Q}(\zeta_n)$ in the case when n is not a prime power (cf. [1383]).

The computation of h_p^+ presents considerable difficulties.[130] Kummer showed only that for $p = 5$ and 7 one has $h_p^+ = 1$ [2358], and in 1885 Wolfskehl[131] [4437] proved $h_p^+ = 1$ also for $p = 11, 13$. Later Minkowski[132] [2872] used a geometrical approach to obtain $h_p^+ = 1$ for $p = 13, 17$ and 19 in a simple way. For later results on h_p^+ see Sect. 3.3.2.

8. In 1857 Kummer extended his result on Fermat's theorem to the case when h_p is not divisible by p^2, covering the three irregular primes below 100, namely $p = 37, 59$ and 67 ([2366] (for $p = 37$ a simpler proof was provided by Mirimanoff [2884] in 1893). In [2356] he formulated his result in a stronger form, asserting the non-existence of solutions of the Fermat equation

---

[127] By *Generalized Riemann Hypothesis* we understand the assertion that Dedekind zeta-functions do not have zeros in the half-plane $\Re s > 1/2$.

[128] Albrecht Fröhlich (1916-2001), professor at the King's College, London. See [362, 4036].

[129] Richard Anthony Mollin (1947–2014), professor in Calgary.

[130] Up to now (2017) this has been done only for $p < 151$ (J.C. Miller [2862]).

[131] Paul Wolfskehl (1856–1906), lectured in Darmstadt. See [191].

[132] Hermann Minkowski (1864–1909) professor in Bonn, Königsberg, Zürich and Göttingen. See [1844].

$$x^p + y^p = z^p$$

in non-zero integers of $\mathbf{Q}(\zeta_p)$ for regular primes p, but assumed in his proof that the integers x, y, z do not have a common ideal factor, reducing thus its generality. Actually his proof works only if either x, y, z are rational or the class-number of $\mathbf{Q}(\zeta_p)$ equals 1. Nevertheless Kummer's assertion is true, and the first correct proof was given by Hilbert in §172 of [1836].

Proofs of Kummer's results on Fermat's equation were later simplified by Hilbert [1836] and Mirimanoff [2886]. See also F. Bernstein[133] [313, 314] and Furtwängler [1339].

In 1940 Vandiver [4179] showed that if Fermat's theorem fails for a prime p in the first case, then $p \mid E_{p-3}$, the sequence E_n of *Euler numbers* (Chap. 9 in the second part of Euler's book [1130]) being defined by

$$\frac{2}{e^t + e^{-t}} = \sum_{n=0}^{\infty} E_n \frac{t^n}{n!}. \tag{1.20}$$

It follows from a congruence found by E. Lehmer [2522] that for odd primes p the condition $p \mid E_{p-3}$ is equivalent to the divisibility of $\sum_{j \le [p/4]} 1/j^2$ by p. There are three primes (149, 241 and 2 946 901) having this property below 10^7 (Ernvall and Metsänkylä [1120], Meštrović [2836]).

A similar result relating the first case of Fermat's theorem for the exponent $2p$ with Euler numbers has been proved in 1950 by Gut[134] [1561].

There is a series of papers by Vandiver on applications of cyclotomic fields to Fermat's theorem [4144, 4146, 4147, 4149, 4151, 4152, 4154, 4155, 4157–4161, 4169, 4175, 4176, 4186–4188], which included also various simplifications and modifications of Kummer's arguments. He presented summaries of his results in [4153, 4156, 4162, 4163, 4165, 4168].

In 1995 A. Wiles and Taylor [4038, 4400] established Fermat's theorem in the case of rational integers. For later work in other number fields see Sect. 3.4.

9. Kummer's paper [2366] contains the following unproved assertion about a congruence in $\mathbf{Z}(\zeta_p)$:

Let $f, g \in \mathbf{Z}[X]$ have their degrees $\le p-2$, put $a = f(\zeta_p), b = g(\zeta_p)$ and assume $(ab, p) = 1$ and $a \equiv b \bmod p^{n+1}$. Moreover let $F(x) = \log f(e^x), G(x) = \log g(e^x)$ and for any integer $m > 0$ denote by $D_m F, D_m G$ the value of the m-th derivative of F resp. G at $x = 0$. Then for k not divisible by $p-1$ one has

$$D_{kp^n} F \equiv D_{kp^n} G \pmod{p^{n+1}}.$$

He used it in the case when $p | h_p$ but $p^2 \nmid h_p$ to distinguish between principal and non-principal ideals $\mathfrak{p}$ with principal $\mathfrak{p}^p$.

A special case of Kummer's assertion, which is sufficient to the attempted application has been established in 1922 by Vandiver [4148, 4167], and in the general case it has been proved in 1952 by Dénes [891] (cf. [893]).

In the paper [2369] (formula (7) on p. 119) Kummer showed that if $f \in \mathbf{Z}[X]$ is of degree $p-2$ and $(f(\zeta_p), p) = 1$, then the value of the $(p-1)$-th derivative of $\log f(e^x)$ at $x = 0$ is congruent mod p to

---

[133] Felix Bernstein (1878–1956), professor in Göttingen, New York and Syracuse. See [1246].

[134] Max Gut (1898–1988), professor in Zürich.

$$\frac{f(1)^{p-1} - N(f(\zeta_p))}{p}.$$

This result has been of importance in the later development of reciprocity laws (see §131 of Hilbert's report [1836], and §21 of Hasse [1664]).

A simpler proof of this result has been given in 1919 by Vandiver [4142].

10. In the next years Kummer worked on the reciprocity laws for prime exponents and presented his results in a series of papers [2359, 2363, 2367–2370] published between 1850 and 1861. In [2357] he stated a form of the reciprocity law for the exponent p in the p-th cyclotomic field in the case when p is a regular prime and checked it numerically for ideal divisors of small primes in the cases $p = 5, 7$ and 23.

He called an integer α of the field *primary* if it is congruent mod $(1 - \zeta_p)^2$ to a rational integer and $|\alpha|^2$ is congruent mod p to a rational integer, and showed that every element α can be made primary by multiplying it by a suitable unit u: $\hat{\alpha} = u\alpha$. For such elements α and arbitrary ideal prime number $\mathfrak{p}$ he defined

$$\left(\frac{\alpha}{\mathfrak{p}}\right) \equiv \hat{\alpha}^{(N\mathfrak{p}-1)/p} \equiv \zeta_p^{\kappa} \pmod{p}$$

with some $\kappa = \kappa(\alpha) \in [0, p-2]$.

If $\mathfrak{q}$ is an ideal prime number and $\beta = \mathfrak{q}^h$ is a number of the field, then he defined the power residue symbol mod $\mathfrak{p}$ by

$$\left(\frac{\mathfrak{q}}{\mathfrak{p}}\right) = \zeta_p^{\mu},$$

where $h\mu \equiv \kappa\hat{\beta} \pmod{p}$. He stated his reciprocity law in the form

$$\left(\frac{\mathfrak{q}}{\mathfrak{p}}\right) = \left(\frac{\mathfrak{p}}{\mathfrak{q}}\right).$$

This generalized the cubic reciprocity law in the form established by Jacobi [2006] and extended the work of Eisenstein who conjectured in [1077] a form of the reciprocity law for p-th powers, and established a particular case of it in [1078]. Proofs of the supplementary reciprocity laws were given by Kummer in [2359, 2368], and after establishing his law in certain special cases in [2363] he gave the first complete proof in 1858–1859, using the theory of ideal numbers in a family of fields, called now *Kummer extensions*, which, using modern notation, are of the form $\mathbf{Q}(\zeta_p, \sqrt[p]{a})$, where $a \in \mathbf{Q}(\zeta_p)$ and the polynomial $X^p - a$ is irreducible over $\mathbf{Q}(\zeta_p)$. Further proofs are contained in [2370, 2376]).

The paper [2359] contains explicit formulas for the p-th power character of cyclotomic units in $\mathbf{Q}(\zeta_p)$.

See also Vandiver [4150, 4166, 4170].

A simplified exposition of Kummer's reciprocity law and the theory of Kummer extensions was given in 1897 by Hilbert [1836] (see Sect. 1.1.2). A proof of the reciprocity law for $p = 5$ has been given by D. Grant [1482] in 1996, and for $p = 7$ by Clement Fernández, Echarri Hernández and Gómez Ayala [720] in 2011, using arithmetics on algebraic curves (see also D. Grant [1483].

The book of Lemmermeyer [2527] presents a detailed early history of reciprocity laws and has an excellent bibliography.

11. In 1853 Kummer [2360] considered primes which are regular in the sense of Sect. 306 of Gauss' [1394] (see Sect. 1.1.2 above), and presented the results of his computations for $p < 100$ which led to the regularity of every prime $p < 100$ except $p = 29$ (with class-group C_2^3) and $p = 41$ (with class-group C_{11}^2), but did not give the proof in all cases. For $p = 31$ he gave a proof later in [2373], and the first proof for $p = 71$ has been provided in 1998 by Schoof [3688]. Kummer's paper contains also a necessary condition for regularity. Cf. also Kummer [2358].

A modern proof of a more general form of Kummer's condition has been given in 1966 by Iwasawa[135] (Section 5 in [1990]).

12. Ideal numbers in cyclotomic fields $\mathbf{Q}(\zeta_n)$ with composite n were introduced by Kummer in 1856 [2364]. His paper starts with the observation that the first $\varphi(n) - 1$ powers of ζ_n form an integral basis, without using that name (a simple proof of this result has been given in 1939 by M. Bauer [252]). Then periods in $\mathbf{Q}(\zeta_n)$ were introduced in the following way:

Let $q \nmid n$ be a prime. For $k = 1, 2, \dots, n$ put $d = (k, n)$ and let m be the order of $q \bmod n/d$. The period π_k corresponding to k is defined by

$$\pi_k = \sum_{j=0}^{m-1} \zeta_n^{kq^j} .$$

The question of the existence of vanishing periods has been studied and resolved in 1863 by Lazarus Fuchs[136] [1289] (some inaccuracies in the argument were corrected in Sect. 9 of [1290]).

Another proof has been given in 1981 by Evans [1138]. Minimal polynomials for the periods were studied by several authors. For details in the case $((p-1)/4, r)$ see Chap. 16 of Bachmann's book [169]. For more recent papers see Dickson [939], Myerson [3007], Gurak [1546–1549], Evans [1139], D.H. Lehmer and E. Lehmer [2515, 2516] and S. Gupta and Zagier [1541].

The class-number formulas in this case were established by Kummer a few years later in [2371, 2372]. In [2372] he established also an analogue of (1.15) for arbitrary cyclotomic fields, replacing the factors $L(\chi, s)$ by $L(\chi^*, s)$, χ^* being the primitive character corresponding to χ.

As in the case of prime n presented Kummer the class-number h_n as a product of two factors, $h_n = h_n^{(1)} h_n^{(2)}$, the second factor was said to be the class-number of the maximal real subfield of $\mathbf{Q}(\zeta_n)$:

---

[135] Kenkichi Iwasawa (1917–1998), professor in Tokyo, at MIT and in Princeton. [722].

[136] Lazarus Fuchs (1833–1902), professor in Berlin, Greifswald, Göttingen and Heidelberg. See [1613].

Wenn[137] *n eine beliebige ganze Zahl ist, und ω eine primitive n-te Wurzel der Einheit, so ist der zweite Faktor für sich selbst die Klassenzahl der, aus den Grössen*

$$\omega+\omega^{-1}, \omega^2+\omega^{-2}, \omega^3+\omega^{-3}, \ldots$$

gebildeten complexen Zahlen, ... ([2371] p. 1051).

In reality Kummer's $h_n^{(2)}$ is the narrow class-number of that field. This forces $h_n^{(1)}$ to be non-integral in certain cases, as noted by Kummer in [2371], where the first factors for $n \leq 100$ were computed. Later [2372] he showed that the denominator of this number is a power of 2, and in the same year Kronecker [2286] showed that it equals either 1 or 2.

If one uses the usual class-number instead of the narrow class-number, and defines h_n^+ and h_n^- accordingly, then h_n^- is always an integer. This follows from a more general assertion, stating that if L/K is a quadratic extension of a totally real field K, and L is totally complex (i.e. L is a CM-field), then $h(K)$ divides $h(L)$ (see e.g. theorem 4.10 in Washington's book [4304]).

The cyclotomic units in $\mathbf{Q}(\zeta_n)$ were defined as units lying in the group generated by $\pm\zeta_n$ and $1-\zeta_n^k$ for $k=1,2,,\ldots n-1$. In the case when n is a prime power, then the index of the group of cyclotomic units in the group of all units is equal to h_n^+.

13. Kummer did not generalize his theory of ideal numbers to more general fields but was aware of the possibility of doing this. We see from his letter to Kronecker in March 1853 [2361] that at first he was not sure about that, writing "*...von*[138] *den idealen Primfactoren der complexen Zahlen, welche aus den Wurzeln einer beliebigen irreduziblen Gleichung des n^{ten} Grades gebildet sind, habe ich selbst keine recht klare Vorstellung*". However a few months later [2277] after presenting some assertions about the class-group of $\mathbf{Q}(\zeta_p)$, he stated: "*Diese*[139] *Sätze, welche auch Gauss für die quadratischen Formen gegeben hat, gelten ganz allgemein für alle Systeme nicht äquivalenter idealer Zahlen, welche man bildet kann, auch wenn die complexen Zahlen nicht aus den Wurzeln der Gleichung $\alpha^\lambda=1$, sondern aus der Wurzeln irgend einer algebraischen Gleichung gebildet werden.*"

The class-number formula for subfields K of $\mathbf{Q}(\zeta_n)$ in the case when the extension $\mathbf{Q}(\zeta_n)/K$ is cyclic has been established in 1866 by Lazarus Fuchs [1290]. The assertion that he found the class-number formula for all subfields of cyclotomic fields which sometimes is given in the literature (e.g. on p. 1 of [1700] and in [1304]) is certainly incorrect.

[137] *If n is an arbitrary integer, and ω is a primitive n-th root of unity, then the second factor is by itself the class-number of the complex numbers constructed from the numbers $\omega+\omega^{-1}, \omega^2+\omega^{-2}, \omega^3+\omega^{-3}, \ldots$".*

[138] *...about the ideal prime factors of complex numbers, built of roots of an arbitrary irreducible equation of nth degree I do not have any clear idea.*

[139] *These theorems, which were given by Gauss for quadratic forms hold also for all systems of non-equivalent ideal numbers, which one can construct from complex numbers built from roots of arbitrary algebraic equations, not only from roots of the equation $\alpha^\lambda=1$.*

A way of generalizing Kummer's class-number formulas to all Abelian fields has been proposed in 1917 by Hecke [1735]. The first such generalization has been obtained by Beeger[140] [266, 267] in 1919–1920. Another formula was provided in 1929 by Gut [1551], who earlier [1550] gave a new proof of Dirichlet's formulas for quadratic fields, utilizing the functional equation of $\zeta_K(s)$. In the special case of cyclic cubic fields a formula was given in 1930 by Latimer [2476]. Later Latimer [2478] gave a simpler form of Gut's formula in the cyclic case.

For arbitrary normal extensions of the rationals the ideal numbers have been introduced in 1865 by Selling[141] [3728], and the biquadratic case $\mathbf{Q}(\sqrt{d}, \sqrt{D})$ has been treated in 1867 by Bachmann [168].

14. Not every argument in Kummer's papers was correct (see Weil's comments in his introduction to Kummer's "Collected Papers" [4356] and Edwards [1044, 1045]), but the lacuna were removed either by Kummer himself in 1857 [2365] or in a later work by Vandiver [4145, 4148, 4153] in the twenties. Modern descriptions of Kummer's ideal numbers were given by Edwards [1045, 1046], Lemmermeyer [2531] and Soublin [3859].

1.3 Establishing the Theory

1.3.1 Kronecker

1. Kronecker made his doctorate in 1845 [2276] with a study of units in cyclotomic fields, proving in particular Dirichlet's unit theorem for the field $\mathbf{Q}(\zeta_p)$ with prime p. He returned to this subject in 1857 [2282] when he showed that every unit in the cyclotomic field $\mathbf{Q}(\zeta_n)$ is a product of a real unit and a root of unity of order n if n is a prime power, and of order $2n$ or $4n$, according to the parity of n, otherwise. This extended Kummer's result [2347] dealing with the case of prime n.

In 1854 he proved [2278] the irreducibility of the n-th cyclotomic polynomial

$$\Phi_n(X) = \prod_{1 \leq k < n, (k,n)=1} (X - \zeta_n^k) .$$

This was earlier asserted by Cauchy [611]. In 1850 Serret[142] [3741] claimed incorrectly that this is a direct consequence of the irreducibility of $\Phi_{p^k}(X)$ for prime p and $k \geq 1$, proved by him. Soon simpler proofs of Kronecker's result were provided by Dedekind [837] in 1857 and Arndt [110] in 1858 (for its modification see Lebesgue [2496]).

[140]Nicolaas George Wijnand Henri Beeger (1884-1965), teacher.

[141]Eduard Selling (1834–1920), professor in Würzburg.

[142]Joseph Alfred Serret (1819–1885), professor in Paris. See [2074].

Later other proofs were given by Mertens [2823, 2827] in 1905 and 1908, M. Bauer [227] in 1916, Grandjot[143] [1480] in 1924, Späth [3864] in 1927, Schur[144] [3701] and Landau [2445] in 1929, Levi[145] [2580] in 1934, Toepken [4067] in 1937 and Skolem[146] [3817] in 1949 (cf. Weintraub [4363]). A simple proof for prime power n has been given in 1934 by Plemelj[147] [3312].

In 1928 Weisner [4364] described quadratic fields $\mathbf{Q}(\sqrt{d})$ (with $d \neq 1$ square-free) in which $\Phi_m(X)$ is reducible. For prime m this is hidden in Sect. 357 of Gauss [1394], and for odd $m = pq$ with prime p, q this can be deduced from Dirichlet's remarks in the last lines of his paper [961]. Other special cases were treated by Cauchy [609] and Genocchi[148] [1409]. A proof has been also given later by Nagell[149] [3034]). Factorizations of cyclotomic polynomials in quadratic fields were considered by Petersson[150] [3267, 3268] and Pumplün [3345]. It turned out that the coefficients of the factors are related to Eisenstein series [3268].

There is a large literature dealing with arithmetical properties of cyclotomic polynomials. Early papers were listed in the report [941], and for later papers up to 1974 see Apostol[151] [98]. A survey has been presented by Vaughan [4199] in 1984. From later results let us quote a few. If $a_m(n)$ is the m-th coefficient of the n-th cyclotomic polynomial, then put

$$A(n) = \max_m |a_m(n)|, \quad L(m) = \log \max_n |a_m(n)| \ .$$

In 1974 Erdős[152] and Vaughan [1117] established

$$L(m) \gg (m/\log m)^{1/2} \ ,$$

and in 1985 Montgomery and Vaughan [2919] proved

$$m^{1/2} \log^{-1/4} m \ll L(m) \ll m^{1/2} \log^{-1/4} m \ .$$

In 1990 Maier [2708] got the existence of a sequence n_j of density 1 with $\lim_{j\to\infty} A(n_j) = \infty$, and in 1993 found for any N a sequence m_j of positive density with $A(m_j) > m_j^N$ for sufficiently large j. Three years later [2709] he showed that for any function $f(n)$ tending to infinity one has $A(n) < n^{f(n)}$ for almost all n (see also Maier [2710], Konyagin, Maier and Wirsing [2212]). On the other hand Bachman [166] and N. Kaplan [2108] showed that for infinitely many prime triples p, q, r one has $A(pqr) = 1$, and in [2109] N. Kaplan presented a way to construct sequences of integers with $A(n) = 1$.

In 1988 Kaminski [2096] gave a new characterization of cyclotomic polynomials by showing that if $F(X) \in \mathbf{Z}[X]$ is monic, distinct from X, and for infinitely many n the number $F(\zeta_n)$ is a unit, then F is a cyclotomic polynomial.

Extensive computations of $A(n)$ were performed by Arnold and Monagan [113] in 2011.

2. In 1857 Kronecker showed [2281] that if $f(X) \neq X$ is an irreducible monic polynomial with rational integral coefficients whose all zeros lie in in the unit disk,

---

[143] Karl Grandjot (1900–1979), professor at the University of Chile. See [1563].

[144] Issai Schur (1875–1941), professor in Bonn and Berlin. See [2079].

[145] Friedrich Wilhelm Levi (1888–1966), professor in Leipzig, Calcutta, Bombay, and the Freie Universität Berlin. See [1295].

[146] Thoralf Skolem (1887–1963), professor in Oslo and Bergen. [3033].

[147] Josip Plemelj (1873-1967), professor at the TU Wien, in Czernowitz and Ljubljana.

[148] Angelo Genocchi (1817–1889), professor in Torino. See [3232].

[149] Trygve Nagell (1895-1988), professor in Uppsala. See [597].

[150] Hans Petersson (1902–1984), professor in Hamburg and Münster. See [4434].

[151] Tom Mike Apostol (1923–2016), professor at Caltech.

[152] Paul Erdős (1913–1996), professor in Budapest. See [164, 165, 1594].

then these zeros must be roots of unity. Kronecker actually assumed that the zeros lie on the unit circle, but his proof works without changes for the disk.

Later other proofs were given by Spencer [3875] in 1977 and Greiter [1512] in 1978. Generalizations to several variables have been given in 1981 by Boyd [411, 412] and Smyth [3837].

Kronecker proved also that if all roots of f are real and non-zero, then either f has a root of the form $2\cos(\pi r)$ with rational r, or it has a root outside the interval $[-2, 2]$.

For further results and generalizations see Sect. 4.5.2.

3. The first mention of the Kronecker–Weber theorem appeared in Kronecker's paper [2277] where he asserted that if a polynomial with rational coefficients has a cyclic[153] Galois group, then its roots can be expressed as linear combinations of roots of unity with rational coefficients. He repeated this later [2291] in the modern form: every Galois extension of the rationals having an Abelian Galois group is a subfield of a cyclotomic field. The first proof (not quite exact) was published by H. Weber in 1886 [4312] and the first complete proof was provided by Hilbert ten years later [1834, 1836].

In the first years of the new century other proofs of the Kronecker–Weber theorem were given by Mertens [2822, 2825], H. Weber [4328, 4330] (his proof in [4329] is incomplete) and Steinbacher [3911]. Later Delon[154] [874, 876] gave a simple proof in the case of cyclic fields of prime.

The Kronecker–Weber theorem implies that every Abelian extension of $\mathbf{Q}$ is a subfield of a composite of fields with at most one ramified prime. An analogue of this assertion for relative Abelian extensions has been established in 1931 by Neiss [3081] in an elementary way.

In 1951 Šafarevič[155] [3546] gave a proof of the analogue of the Kronecker–Weber theorem in the local case and then deduced the global case. His proof was exposed in the books by Narkiewicz [3065] and Washington [4304]. Other proofs were given by Čebotarev[156] [618], M.J. Greenberg [1509], Riese [3475], Speiser [3871] and Lemmermeyer [2528]. Modern versions of the proofs given by Kronecker and H. Weber were presented by O. Neumann [3105].

4. A list of eight recursion formulas for the class-number of positive-definite quadratic forms (hence also for the class-number of quadratic orders) has been presented by Kronecker in 1857 and 1860 [2283, 2284]. Here is the simplest of them: for positive m one has

$$3H(-m) + 6\sum_{j=1}^{\sqrt{m}} H(-m + j^2)$$

$$= \tau(m) + 3\sum_{d|m} \eta(d)d + 3\left(\tau_{4,1}(m) - \tau_{4,3}(m)\right) + 2\left(\tau_{3,1}(m) - \tau_{3,2}(m)\right) ,$$

---

[153] Note that at that time Kronecker used the word "Abelian" to mean "cyclic". Cf. p. 237 of the paper by Petri and Schappacher [3274].

[154] Boris Nikolajevič Delone (Delaunay) (1890–1980), professor in Leningrad and Moscow.

[155] Igor Rostislavovič Šafarevič (1923–2017), professor in Moscow.

[156] Nikolaĭ Grigorievič Čebotarev (1894–1947), professor in Kazan. In publications in German language he spelled his name Tschebotareff and Tschebotaröw. See [881].

where $\tau(m)$ denotes the number of divisors of m, $\tau_{k,l}$ is the number of divisors of d congruent to l mod k and

$$\eta(d) = \operatorname{sgn}\left(d - \sqrt{m}\right) .$$

He pointed out that these formulas can be obtained from the theory of elliptic functions but gave no proofs. They were later provided by Hermite [1822, 1823] and Joubert [2080] (see also Liouville [2606, 2607], Petr[157] [3271] and Smith's report [3832]). Six of these formulas were proved by Kronecker [2298] in 1883 (see also [2289]). Several similar formulas were later found (see, e.g. Gierster [1425–1427], G. Humbert [1922], Hurwitz[158] [1927, 1928] and Petr [3272]). They were surveyed by Cresse in Chap. VI of the third volume of Dickson's history [935].

An elementary proof of Kronecker's relations has been given in 1913 by Uspensky [4115] (see also Cresse [792]). In 1925-1926 Uspensky [4116, 4117] published a series of seven papers in which he presented an elementary way of obtaining class-number relations providing proofs for relations obtained earlier by various authors with the use of elliptic functions.

A geometric proof for one of Kronecker's formulas has been found recently by Popa and Zagier [3335].

In 1875 Kronecker [2289] expressed certain series related to the class-numbers of quadratic forms by theta-functions, and in 1884 new proofs were provided by Hermite [1824].

5. In 1880 Kronecker [2293] established the following result about polynomial congruences which later induced Frobenius to prove his density theorem (see Sect. 1.3.4).

If $\Phi(X)$ is a polynomial with rational integral coefficients having m irreducible factors, and ν_p denotes for prime p the number of solutions of the congruence

$$\Phi(x) \equiv 0 \pmod{p} ,$$

then the limit

$$\lim_{\varepsilon \to 0} \left(\sum_p \frac{\nu_p}{p^{1+\varepsilon}} - m \log(1/\varepsilon) \right)$$

exists.

Note that if $\Phi(X)$ is irreducible, then $\nu(p)$ equals the number of prime ideals of degree 1 dividing p in the field generated by a root of Φ, except for finitely many primes p.

6. In 1882 Kronecker [2294] (cf. [2296]) published a long paper in which he presented a theory of algebraic number fields and function fields.

In the first three sections he defined fields and finite field extensions using his own terminology ("Rationalitäts-Bereich" [rationality region] = field, and "Gattung" [genus] = field extension) and in §4 a method of factorization of polynomials in any

---

[157] Karel Petr (1868–1950), professor in Prague. See [2233].

[158] Adolf Hurwitz (1859–1919), brother of Julius Hurwitz, professor in Königsberg and Zürich. See [1845, 3203, 4457].

finite number of variables into irreducibles has been given (it has been pointed out by Mignotte and Ştefănescu [2852] that the idea of this algorithm is much older, appearing already in the papers by N. Bernoulli[159] [309] and Schubert[160][3692]).

In the case of one variable it was later shown by Runge[161] [3534] that Kronecker's method is of practical use, and Mandl[162] [2732] provided some simplifications. Later Hancock[163] [1615] and Mandl [2733] showed that this method is also applicable in the case of several variables.

Kronecker's factorization algorithm has been later simplified by Hausmann [1720]. Another method based on the consideration of coefficients of the polynomial instead of its values, has been proposed in 1953 by Kelly [2135]. Kempfert [2137] showed in 1969 how to factorize $f \in \mathbf{Z}[X]$ using factorizations of $f \bmod p$ for prime p. There is a huge literature dealing with factorizations of polynomials over finite fields, starting with the paper of Ward[164] [4299] published in 1935. A survey has been given in Chap. 4 of the book [2595] by Lidl and Niederreiter.

In 1982 A.K. Lenstra, H.W.Jr. Lenstra and Lovász [2539] found an algorithm for factorizing $f \in \mathbf{Z}[X]$ in polynomial time, generalized later to multivariate polynomials by A.K. Lenstra [2536]. These results were also generalized to polynomials with coefficients in algebraic number fields by A.K. Lenstra [2537, 2538].

In the next sections Kronecker studied integral elements in case when the considered rationality region is formed either by rational numbers or by rational functions in N variables with rational coefficients, introduced the discriminant of an extension and in §9 proved the divisibility property of discriminants of consecutive extensions. The first part of the treatise contains a presentation of Galois theory and some of its applications.

The second part of [2294] brings the study of the divisibility and factorization properties based on the notion of divisors, presented in a rather complicated form.

A readable exposition of Kronecker's ideas has been given by H. Weber in the second volume of his book "Lehrbuch der Algebra" [4321], whose first edition[165] appeared in 1896 (an earlier exposition has been provided by Molk [2901] in 1885).

H. Weber based his exposition on the notion of *functionals*, which are rational functions in finitely many variables with coefficients in the given field. A functional with rational coefficients is called a rational functional and one can write it uniquely in the form $\Phi(\bar{x}) = af(\bar{x})/g(\bar{x})$, where f, g are polynomials with integral coefficients without common factor, and $a > 0$ is rational. The number a is called the *absolute value of* Φ, and if $a = 1$, then Φ is called an integral rational functional. A functional Ψ with coefficients in a field $K \neq \mathbf{Q}$ is called integral if it is a root of a polynomial

---

[159] Nicolas I Bernoulli (1687–1759), professor in Padua and Basel.

[160] Friedrich Theodor von Schubert (1758–1825), great grandfather of Sofija Kovalevskaja, worked in St. Petersburg.

[161] Carl Runge (1856–1927), professor in Hannover and Göttingen. See [785].

[162] Max Mandl (1859–1910), teacher in Prostějov and Lublana.

[163] Harris Hancock (1867–1944), professor in Chicago and Cincinnati. See [2922].

[164] Morgan Ward (1901–1963), professor at Caltech. See [2514].

[165] See the report on Weber's book by Pierpont [3291].

$$T^n + \sum_{j=0}^{n-1} A_j T^j$$

whose coefficients are integral rational functionals. Integral functionals dividing 1 are called unit functionals, and two integral functionals are called associated if their ratio is a unit functional. The first main theorem (§141) asserts the unique factorization in the set of integral functionals, and in §151 it is shown that Dedekind's ideals in the ring of integers of K are in one-to-one correspondence with classes of associated integral functionals of K. At the end of §143 it is observed that starting with the field of rational functions of one complex variable in place of $\mathbf{Q}$ one obtains in this way a starting point of the theory of algebraic functions in one complex variable.

It is interesting to note how H. Weber tried to convince the reader about legality of using rational functions. In §136 he wrote: "*Die*[166] *Variablen, die in der Theorie der algebraischen Zahlen verwendet werden, haben nicht die Bedeutung von Zeichen für veränderliche Zahlenreihen, ..., sondern sind sie lediglich Rechensymbole ohne eine selbständige Bedeutung*."

In 1903 J. Kőnig[167] published a book [2206] containing a clear and detailed exposition of Kronecker's theory (see the comments on it by Gray [1505]). See also the thesis of Hancock [1616].

Dedekind's comments on Kronecker's paper were published in 1982 by Edwards, O. Neumann and Purkert [1050]. A modern presentation of Kronecker's approach was given by Edwards [1048]. See also del Corso [864].

For Kronecker's results which led to the notion of a *class-field* see Sect. 2.3.1.

1.3.2 Geometrical Approach: Hermite and Minkowski

1. Around 1850 Hermite considered in a series of papers [1813, 1815, 1816] the equivalence classes of binary forms $f(X, Y) = \sum_{i=0}^{n} a_i X^i Y^{n-i}$ of fixed degree n with coefficients in $\mathbf{Z}$ under the action of $SL_2(\mathbf{Z})$. He showed first that if the polynomial $g(X) = f(X, 1)$ has all roots real, then there are only finitely many equivalence classes of such forms having a given determinant, defined by

$$a_0 \min_{x_1,\dots,x_n} \frac{\Psi(x_1 \dots, x_n)^{n/4}}{\sqrt{x_1 x_2 \cdots x_n}} \quad ,$$

where

[166]"*The variables used in the theory of algebraic numbers do not have the meaning of variable sequences of numbers, ..., but are only symbols for calculation, without having any independent meaning.*"

[167]Julius [Gyula] Kőnig (1849–1913), professor in Budapest.

$$\Psi(x_1 \ldots, x_n) = \frac{1}{2} \sum_{i=1}^{n} \sum_{j=1}^{n} x_i x_j (\alpha_i - \alpha_j)^2 ,$$

with $\alpha_1, \alpha_2, \ldots, \alpha_n$ being the roots of g and $x_1, x_2, \ldots, x_n$ ranging over all positive reals. The same assertion has been obtained also in the case when $g(X)$ has non-real roots, but in that case the definition of the determinant had to be modified. This implies the corresponding assertion for the induced equivalence classes of polynomials in $\mathbf{Z}[X]$.

Later Hermite [1817] formulated the same assertion with the determinant replaced by the discriminant,[168] and in 1857 proved this in the case of coefficients in $\mathbf{Z}[i]$, and showed also that in these results one does not have to fix the degree, hence there are only finitely many equivalence classes of polynomials with rational integral coefficients having the same discriminant [1819]. This implies in particular that the minimal discriminant of an algebraic field of degree n tends to infinity with n, and there are only finitely many fields with the same discriminant.

Another proof of the last result, based on Minkowski's theorem on linear forms (*Linearformensatz,* §36-37 of Minkowski's book [2874]) has been given in 1918 by Schur [3698], and in 1931 Nagell [3022] provided a generalization, showing that there are only finitely many finitely generated rings of algebraic integers having the same discriminant.

Much more precise results were later obtained for polynomials and binary forms using stronger equivalence. In 1972 Birch and Merriman [361] showed that there are only finitely many equivalence classes of binary forms defined over $R = \mathbf{Z}_K$ with given degree ≥ 3 and discriminant being invertible in the ring R_S under the action of the group $GL_2(R_S)$, S being a finite set of prime ideals, and the same happens for binary forms over R of given degree and discriminant under the action of $GL_2(R)$. They considered also equivalence of algebraic numbers, calling α, β equivalent if their difference is a rational integer, and showed that there are only finitely many equivalence classes of algebraic integers having a fixed discriminant (for numbers of degree ≤ 4 this has been proved earlier by Nagell [3036, 3040]). Their proofs were ineffective. In an effective form this result has been obtained in 1973 by Győry [1565]. For quantitative versions, some generalization and various applications see Győry [1566–1568]. An effective version of the result of Birch and Merriman [361] on equivalence classes of binary forms has been established in 1991 by Evertse and Győry [1147]. See also Evertse [1142], Berczés, Evertse and Győry [293], Győry [1575], and the recent book [1150] by Evertse and Győry.

2. In another paper of 1854 Hermite [1818] considered approximation of complex numbers by elements of the field $\mathbf{Q}(i)$ (see also Sect. 39 in the book [2874] by Minkowski). He showed that for every complex z there are infinitely many co-prime $p, q \in \mathbf{Z}[i]$ with

$$\left| z - \frac{p}{q} \right| \leq \frac{1}{\sqrt{2}|q|^2}. \tag{1.21}$$

For the field $\mathbf{Q}\left(\sqrt{-3}\right)$ a similar result has been proved in 1887 by A. Hurwitz in [1929] with the use of continued fractions in complex quadratic fields introduced by him (see also Minkowski [2878] and Chap. 6 in [2880]).

[168] After defining the discriminant Hermite wrote on p. 335: "*...laquelle les géométres anglais ont donné le nom de discriminant.*" (*"which the English geometers have given the name discriminant"*).

J. Hurwitz[169] [1942] showed in 1902 that elements of quadratic extensions of the field $K = \mathbf{Q}(i)$ have a periodic continued fraction with denominators in $\mathbf{Z}[i]$. This has been also proved a few years later by Mathews [2771]. Another proof has been provided by Arwin [142] in 1926, whose argument worked also in the case of fields $\mathbf{Q}\left(\sqrt{-d}\right)$ for $d = 2, 3, 7, 11$. Later Arwin [143] considered this question in some other classes of fields. See also Hofreiter [1884].

A modern approach to continued fractions with denominators in $\mathbf{Z}[i]$ has been presented in 1975 by A.L. Schmidt [3636]. For further development see A.L. Schmidt [3638] and Nakada [3044, 3045],

For the field $K = \mathbf{Q}\left(\sqrt{-D}\right)$ (with square-free $D > 0$) let $\gamma(D)$ denote the *Hurwitz constant* (after A. Hurwitz [1930] who determined the corresponding constant for the field of rationals) the smallest number c such that for every complex z and infinitely many $p, q \in \mathbf{Z}_K$ one has

$$\left| z - \frac{p}{q} \right| \le \frac{c}{|q|^2},$$

thus (1.21) gives $\gamma(1) \le 1/\sqrt{2}$.

The sequence of approximations of a complex number by elements of $\mathbf{Q}(i)$ introduced by Hermite has been studied by L.R. Ford[170] [1215, 1216], who also made a deep study of continued fractions with elements in $\mathbf{Z}[i]$. In 1925 [1217] he proved $\gamma(1) = 1/\sqrt{3}$, and another proof of this result has been given later by Perron[171] [3252, 3253], who showed also $\gamma(3) = 1/\sqrt[4]{13}$ [3254], and $\gamma(2) = 1/\sqrt{2}$ [3256]. In the last paper he obtained also for $D > 3$ the bounds

$$\gamma(D) \le \begin{cases} \sqrt{2D}/\pi & \text{if } D \equiv 3 \bmod 4\,, \\ 2\sqrt{2D}/\pi & \text{otherwise,} \end{cases} .$$

These bounds were improved in 1937 by Hofreiter[172] [1883], who also proved $\gamma(7) = 1/\sqrt[4]{8}$. Later research determined a few new values of $\gamma(D)$: $\gamma(11) = 2/\sqrt{5}$ (Descombes and Poitou[173] [898] in 1950), $\gamma(19) = 1$ (Poitou [3326] in 1953), $\gamma(5) = \gamma(6) = 1$ (Vulakh [4252] in 1995), and $\gamma(15) = 1/\sqrt{(2)}$ (Vulakh [4253] in 1999).

In 2010–2012 Vulakh [4257] determined the values of $\gamma(D)$ for $D = 30, 33, 34, 57$ and 62 [4256, 4257].

The analogue of the Markov spectrum in imaginary quadratic fields has been also considered. See Poitou [3326] for $D = 3$ and A.L. Schmidt [3635, 3637, 3639] for $D = 2, 3, 7$ and 11 and Vulakh for $D = 5, 6$ and 15 [4254, 4255].

In §39 of his book [2874] considered Minkowski simultaneous approximations of complex numbers by elements of $\mathbf{Q}(i)$.

This has been later generalized to the case of imaginary quadratic fields with class-number 1 by Hofreiter [1882]. In this paper he computed also the absolute smallest discriminants of quadratic extensions of the fields considered. Other imaginary quadratic fields were by him treated in [1886].

[169] Julius Hurwitz (1857–1919), brother of Adolf Hurwitz, Privatdozent in Basel. See [3203].

[170] Lester Randolph Ford (1886–1967), professor at the Illinois Institute of Technology.

[171] Oskar Perron (1880–1975), professor in Tübingen, Heidelberg and Munich. See [1230, 1758].

[172] Nikolaus Hofreiter (1904–1990), professor in Vienna. See [1872].

[173] Georges Poitou (1926–1989), professor in Lille and at the Université Paris-Sud.

3. In 1861 Hermite [1820] showed that every separable extension L/K of degree five has a generator whose minimal polynomial is of the form $X^5 + aX^3 + bX + c$, and a similar result for extensions of degree 6 (assuming $\mathrm{char}(K) \neq 2$) has been obtained in 1867 by Joubert [2081].

For modern proofs of these results see Coray [769] and Kraft [2237]. It has been shown by Reichstein [3418] in 2014 that Joubert's result fails in the case of fields of characteristic 2.

Analogues for extensions with larger degrees were discussed by Reichstein [3417]. See also Reichstein and Youssin [3419].

4. Denote by $M(r_1, r_2)$ the minimal value of $|d(K)|$ for fields K of signature (r_1, r_2), and put

$$M(n) = \min\{M(r_1, r_2) : r_1 + 2r_2 = n\} .$$

We mentioned already in Sect. 1.3.2 that in 1857 Hermite [1819] established

$$\lim_{n\to\infty} M(n) = \infty ,$$

and Kronecker ([2294], p. 21) asserted in 1882 the truth of $|d(K)| > 1$ for all algebraic number fields K. This assertion, which implies the non-existence of unramified extensions of $\mathbf{Q}$, has been established in 1891 by Minkowski [2872], who in [2873] proved the bound

$$M(r_1, r_2) \geq \left(\frac{\pi}{4}\right)^{2r_2} \left(\frac{n^n}{n!}\right)^2 , \tag{1.22}$$

implying in view of Stirling's formula the inequality

$$M(n) \geq \left(\frac{11}{12}\right)^2 \left(\frac{\pi e^2}{4}\right)^n \frac{1}{2\pi n}$$

(in [2872] a weaker bound has been given).

5. Minkowski obtained his bounds for $M(r_1, r_2)$ using a new geometrical approach to the theory of algebraic numbers in which rings of integers and ideals in an algebraic number field of degree n were interpreted as lattices in the n-dimensional space. He utilized it to prove the finiteness of class-groups of algebraic number fields, asserted by Kronecker on p. 64 in [2294]. He showed also in [2872] that every ideal class in a field K of degree n contains an ideal of norm not exceeding

$$\frac{n!}{n^n} \left(\frac{4}{\pi}\right)^{r_2} \sqrt{|d(K)|} .$$

For large n this bound has been improved by Zimmert [4474] in 1981, and better evaluations for $n \leq 10$ were given by de la Maza [863] in 2002.

An exposition of Minkowski's approach has been given in his book [2874].

Another proof of the inequality $|d(K)| > 1$ was later provided by H. Weber and Wellstein[174] [4331] in 1913, Schur (for totally real fields) [3698] in 1918, Landau [2440] in 1922, Müntz[175] [2998] in 1923, Mordell[176] [2929] in 1931 and Calloway [555] in 1955. For totally real fields a proof has been given in 1922 by Siegel [3772] with the use of an identity for which another proof has been later given by Mordell [2927].

The computation of exact values of $M(r_1, r_2)$ is easy for small n: one has $M(2, 0) = 5$, $M(0, 1) = 3$, $M(1, 1) = 23$ and $M(3, 0) = 49$. The case of quartic fields has been settled in 1929 by J. Mayer [2792] ($M(4, 0) = 725$, $M(2, 1) = 275$, $M(0, 2) = 117$), and for quintic fields by Hunter [1925] in 1957 ($M(1, 2) = 1609$, $M(3, 1) = 4511$, $M(5, 0) = 14\,641$). The seventies and eighties brought progress for the next degrees: sextic fields were treated by Kaur [2125] ($M(6, 0) = 300\,125$), Pohst [3316] ($M(4, 1) = 92\,779$, $M(2, 2) = 28\,037$) and J.J. Liang and Zassenhaus[177] [2592] ($M(0, 3) = 9747$), and septic fields were considered by Pohst [3315] ($M(7, 0) = 20\,134\,193$), and Diaz y Diaz [923, 924, 926] ($M(5, 1) = 2\,306\,559$, $M(3, 2) = 612\,233$. $M(1, 3) = 184\,607$.

In case of degree 8 only the values of $M(8, 0) = 282\,300\,416$ (Pohst, Martinet, Diaz [3319]) and $M(0, 4) = 1\,257\,728$ (Diaz [925]) are known, and $M(9, 0) = 9\,685\,993\,193$ has been determined by K. Takeuchi [3984].

Put

$$D = \liminf_{n\to\infty} M(n)^{1/n}, \quad D_0 = \liminf_{n\to\infty} M(n, 0)^{1/n}, \quad D_1 = \liminf_{2|n\to\infty} M(0, n/2)^{1/n}.$$

Minkowski's bound implies for large n the inequality $D > \pi e^2/4$, and $D_0 > e^2$. In 1914 Blichfeldt[178] [374] improved this to $D \geq \pi e$ and obtained in 1939 [375]

$$D_0 \geq 2\pi e^{3/2} = 28.159\ldots,$$

and on the other hand in 1938 Scholz[179] [3676] established for infinitely many values of n the upper bound

$$M(n)^{1/n} \ll \left(\frac{\log\log n}{\log n}\right)^2.$$

In 1950 Rogers[180] [3488] established

$$D_0 \geq 16e^3\pi^2 = 32.561\ldots,$$

and in 1960 Mulholland [2994] obtained $D \geq 15.775$.

The conjecture $D = \infty$ has been shown to be incorrect in 1964 by Golod and Šafarevič [1472] who established $D \leq 4404.5$ as a consequence of their proof of the existence of infinite class-field towers (see Sect. 4.3.1). This has been improved in 1978 to $D \leq 92.369$ by Martinet [2752], who also showed $D_0 \leq 1059$.

In 2001 Hajir and Maire [1590] obtained $D_1 \leq 83, 9$, and next year they got $D_0 \leq 954.3$ and $D \leq D_1 \leq 82.2$ [1591].

---

[174] Josef Wellstein (1869–1919), professor in Giessen and Strassburg.

[175] Chaim Herman Müntz (1884–1956). See [3194].

[176] Louis Joel Mordell (1888–1972), professor in Manchester and Cambridge. See [596, 825].

[177] Hans Zassenhaus (1912–1991), professor in Hamburg, at the McGill University, the University of Notre Dame and the Ohio State University. See [3318].

[178] Hans Frederik Blichfeldt (1873–1945), professor at Stanford University. See [940].

[179] Arnold Scholz (1904–1942), professor in Kiel. See [4024].

[180] Claude Ambrose Rogers (1920–2005), professor in Birmingham and at the University College, London.

An important step in bounding $M(r_1, r_2)$ from below has been made in 1975 by Odlyzko [3137], who applied an identity of Stark (Lemma 3 in [3896]) permitting the evaluation of the discriminant of K in terms of zeros of the Dedekind zeta-function $\zeta_K(s)$, and established for large n the lower bound

$$M(r_1, r_2) \gg a^{r_1} b^{r_2} ,$$

with $a = 55, b = 21$ [3138] and $a = 60.1, b = 22.2$ [3139]. He showed also that the assumption of Generalized Riemann Hypothesis (GRH) permits to take $a = 136, b = 34.5$ [3138], and $a = 188.3$, $b = 41.6$ [3139] (see also [3137]). In 1976 Poitou [3327] obtained

$$D_0 \geq 60.8, \; D_1 \geq 22.3 ,$$

and under GRH

$$D_0 \geq 215.3, \; D_1 \geq 44.7 .$$

Minimal discriminants of primitive (i.e. not containing proper subfields $\neq \mathbf{Q}$) sextic fields with given Galois group and signature were determined by D. Ford [1210], D. Ford and Pohst [1211, 1212] and D. Ford, Pohst, Daberkow and Haddad [1213]. For non-primitive sextic fields this has been done by Bergé[181], Martinet and Olivier [297] and Olivier [3149–3151]. One finds there also lists of fields having small discriminants.

The same task for imprimitive octic fields has been realized by H. Cohen, Diaz y Diaz and Olivier [734] in 1999 for fields with quartic subfields, and by Fieker and Klüners [1184] in the remaining case. For imprimitive nonic fields this has been done in 1993 by Fujita [1318], and a list of such fields with small discriminants was provided in 1995 by Diaz y Diaz and Olivier [927]. The minimal discriminant of solvable nonics has been found by J.W. Jones [2066] in 2013. For certain classes of fields of degree 10 see the paper of Driver and J.W. Jones [999].

Fields K with small degree n having small $|d(K)|^{1/n}$ were studied by Martinet [2753, 2755], J.W. Jones and Roberts [2069] and J.W. Jones and Wallington [2071].

Surveys were presented by Poitou [3328] in 1976, Martinet [2754] in 1980 and Odlyzko [3137] in 1990. The last paper lists also several open questions related to discriminant evaluations.

6. Another geometrical method was used by Klein[182] in his lectures [2167] in the case of quadratic fields (he published a summary in 1893 [2166]). In this approach the theory of ideals is based on consideration of lattices associated with suitable quadratic forms.

Much later Klein's method was generalized to arbitrary fields of finite degree by Furtwängler [1350], who earlier [1328] treated the case of cubic fields.

A geometric approach was also utilized by Bianchi [347, 348] to study the group $SL_2(\mathbf{Z}_K)$ for imaginary quadratic fields K. He showed that this group is finitely generated, and this has been extended to all algebraic number fields by A. Hurwitz [1933]. In this paper A. Hurwitz treated two non-zero elements α, β of K as equivalent, if $\alpha = A/B, \beta = C/D$ with $A, B, C, D \in \mathbf{Z}_K$ and the ideals (A, B) and (C, D) lie in the same ideal class, which happens if and only if the pairs A, B and C, D lie in the same orbit under the action of $SL_2(\mathbf{Z}_K)$.

An analogue of the last result for classes mod $\mathfrak{f}$ has been established later by Fueter [1313].

7. An important result about units has been obtained by Minkowski [2876] in 1900. After proving the simple but useful lemma stating that if an $n \times n$ matrix

---

[181] Anne-Marie Bergé (1939–2008), professor in Bordeaux. See [2756].

[182] Felix Klein (1849–1925), professor in Erlangen, München, Leipzig and Göttingen. See [784].

$[a_{ij}]$ with real entries satisfies $a_{ij} < 0$ for $i \neq j$, $a_{ii} > 0$ and $\sum_{j=1}^{n} a_{ij} > 0$, then its discriminant does not vanish, he used it to give an algorithm leading to a maximal system of independent units. As a corollary he showed that in a normal extension $K/\mathbf{Q}$ one can always find a unit such that its $r(K)$ conjugates are multiplicatively independent, hence generate a subgroup of $U(K)$ of finite index. In the case of $K = \mathbf{Q}(\zeta_p)$ with prime p this has been established already by Kummer ([2358], pp. 389–391). Such unit is now called *Minkowski unit*, and if the index equals 1, then one speaks about a *strong Minkowski unit*. See 5.1.3 for later results on Minkowski units.

A simpler proof of the lemma of Minkowski, due to Schur, appears in a paper of Rohrbach[183] [3493], who provided also an algebraic proof of an extension of this lemma established in 1929 by Tambs-Lyche[184] [3988]. A very simple proof of the lemma has been later found by Artin[185] [124]. See also Furtwängler [1358].

8. In his book [2874] established Minkowski the following theorem in case $n = 2$: *If*

$$L_i(X_1, \ldots, X_n) = \sum_{j=1}^{n} a_{ij} X_j \quad (i = 1, 2, \ldots, n)$$

are real linear forms with determinant $D = \det[a_{ij}] \neq 0$, *then for every real* $y_1, \ldots, y_n$ *there exist rational integers* $x_1, \ldots, x_n$ *satisfying*

$$\prod_{i=1}^{n} |L_i(x_1, \ldots, x_n) - y_i| \leq \frac{D}{2^n} \cdot \tag{1.23}$$

It has been conjectured that this theorem holds for all $n \geq 2$, but at this moment it has been established only for $n \leq 8$ (Remak [3439, 3440] for $n = 3$, Dyson [1033] for $n = 4$, Skubenko[186] [3819] and Bambah and Woods [184] for $n = 5$, McMullen [2809] for $n = 6$, and Hans-Gill, Madhu and Ranjeet [1621, 1622] for $n = 7, 8$.

The truth of this conjecture would have the following consequence for algebraic number fields:

If K *is a totally real number field, then for every* $x \in K$ *there exists* $y \in \mathbf{Z}_K$ *with*

$$|N(x - y)| \leq \frac{|d(K)|^{1/2}}{2^n}. \tag{1.24}$$

This inequality is of importance in the study of Euclidean fields.

In 1899 Minkowski [2875] gave an algorithm which becomes periodic if and only if a given number is algebraic of a given degree. Later [2877] he applied it in the theory of Diophantine approximations.

---

[183] Hans Rohrbach (1903–1993), professor in Mainz.

[184] Ralph Tambs-Lyche (1890–1991), professor in Oslo.

[185] Emil Artin (1898--1962), professor in Hamburg, Princeton and at Notre Dame University and Indiana University. See [685, 1026].

[186] Boris Faddeevič Skubenko (1929–1993). See [81].

1.3.3 Dedekind

1. A translation of Kummer's theory of ideal numbers into the language of ideals, which became later commonly accepted has been made by Dedekind. In the supplements[187] [839, 844, 848] to the consecutive editions of Dirichlet's lectures [972] starting with the second as well in his later papers [841–843, 845, 850, 853, 855] he presented the principal notions of the theory of algebraic number fields[188] in the general case.

In §163 of [839] Dedekind introduced ideals as subsets of the set of algebraic integers in a field K closed under addition, subtraction and multiplication by integers of K. Although in the announcement of [839] in [838] he defined the multiplication of ideals and asserted the uniqueness of representations of ideals as products of prime ideals, the text of [839] did not mention this. One finds there only the definition of divisibility of ideals ($I \mid J$ defined by $J \subset I$) and a proof of the unique representation of ideals as the least common multiple of powers of prime ideals. The norm $N(I)$ of an ideal I has been defined as the number of residue class mod I, and a proof of the equality $N(IJ) = N(I)N(J)$ has been given.

In §167 appeared the Dedekind zeta-function

$$\zeta_K(s) = \sum_I \frac{1}{N(I)^s}$$

for an arbitrary number field. Dedekind computed the limit

$$\lim_{s\to 1}(s-1)\zeta_K(s)$$

in two different ways, and this led him to a formula for the class-number. Note that in [839] the definition of the class-number was different from the currently used, as in the case when the field K did not have units of negative norms a principal ideal belonged to the principal class only if it had a positive norm. Later, in §184 of [848], the now standard definition appeared (see Sect. 1.3.3 below).

2. In 1877 Dedekind [842] introduced *orders* in rings $\mathbf{Z}_K$ of algebraic integers of a field K. He defined them in the following way: he considered modules, i.e. subsets M of $\mathbf{Z}_K$ closed upon addition and subtraction and containing a basis of deg K elements linearly independent over the rationals, defined the set $\{a \in K : aM \subset M\}$ to be the order of M, and observed that every such order is a subring of $\mathbf{Z}_K$ containing the rational integers, and conversely, every such subring is the order of a suitable module (later, in §172 of [844], he used this property for the definition of an order, without relating them to modules). The *conductor* $\mathfrak{f}(\mathfrak{o})$ of an order $\mathfrak{o}$ has been defined as the ideal of $\mathbf{Z}_K$ contained in $\mathfrak{o}$ and having the minimal norm (hence $\mathfrak{f}(\mathfrak{o})$ is the largest ideal of $\mathbf{Z}_K$ contained in $\mathfrak{o}$). Dedekind considered only ideals I of $\mathfrak{o}$ co-prime with

[187] X Supplement in the second edition and XI Supplement in the next two.

[188] In a paper published in 1882 jointly with H. Weber [857] Dedekind extended this theory to fields of algebraic functions.

the conductor, i.e. satisfying $I + \mathfrak{f} = \mathfrak{o}$ (now they are usually called *regular ideals*), called the number of residue classes $\mathfrak{o}$ mod I the norm $N_{\mathfrak{o}}(I)$ of I, and showed that the map $\varphi : I \mapsto I\mathbf{Z}_K$ is multiplicative and gives a norm-preserving one-to-one correspondence between ideals of $\mathfrak{o}$ prime to $\mathfrak{f}$ and ideals of $\mathbf{Z}_K$ prime to $\mathfrak{f}$ (now one would say that φ is a norm-preserving isomorphism of the semi-groups of ideals of $\mathfrak{o}$ and $\mathbf{Z}_K$ co-prime to the conductor). Then Dedekind called two ideals I, J of $\mathfrak{o}$ equivalent if there exists $a \in K$ of positive norm with $J = aI$, showed that the equivalence classes form a finite Abelian group, which we shall denote by $H(\mathfrak{o})$, proved that the map $H(\mathfrak{o}) \longrightarrow H(K)$ is surjective and established a formula for the ratio $|H(\mathfrak{o})|/h(K)$. He presented two proofs for this formula, one elementary and another analytic. This generalized previous results of Gauss [1394] and Dirichlet [963] who considered this question for class-numbers of binary quadratic forms of given discriminant, which, as proved by Dedekind in §165 of [839], are equal to the number of ideal classes in corresponding orders in quadratic fields.

3. In the paper [843] published in 1878 Dedekind applied the theory of congruences to the factorization of rational primes in extensions. He showed there that if $F(X) \in \mathbf{Z}[X]$ is irreducible, θ is one of its zeros, $K = \mathbf{Q}[\theta]$, p is a prime not dividing the *index* $i(\theta)$ of θ (defined as the index of $\mathbf{Z}[\theta]$ in the ring $\mathbf{Z}_K$ of integers of K), and

$$F(X) \equiv F_1^{e_1}(X) \cdots F_g^{e_g}(X) \pmod{p}, \tag{1.25}$$

where the polynomials F_j are irreducible mod p, then

$$p\mathbf{Z}_K = \prod_{j=1}^{g} \mathfrak{p}_j^{e_j}, \tag{1.26}$$

where $\mathfrak{p}_j$ is a prime ideal of degree $\deg F_j$.

A simpler proof of this theorem has been later given by Ore[189] [3166]. Another proof has been given by Engstrom [1100] in 1930.

A similar result, valid for all primes, has been obtained later by Hensel [1768]. Let $\omega_1, \ldots, \omega_n$ be an integral basis of K, and put

$$F(X; x_1, \ldots, x_n) = N_{K/\mathbf{Q}}(X - \sum_{j=1}^{n} x_j\omega_j) \, .$$

The factorization (1.26) holds if and only if

$$F(X; x_1, \ldots, x_n) \equiv \prod_{j=1}^{g} P_j(X; x_1, \ldots, x_n)^{e_j} \pmod{p} \, ,$$

[189] Öystein Ore (1899–1968), professor at Yale. See [11].

where $P_1, \ldots, P_n$ are distinct polynomials irreducible mod p.

The paper [843] contains also a study of the *field index* $i(K)$, defined as the greatest common divisor of all indices of algebraic integers generating the field K. This study has been continued by Hensel [1769] (see Sect. 2.2).

4. In 1879 the third edition of Dirichlet's lectures [972] appeared containing Dedekind's supplement [844]. In it one finds the multiplication of ideals and its properties, culminating in §173 with the proof of the unique factorization of ideals into prime ideals. Dedekind pointed out that the main difficulty in the proof consisted in the establishing of the assertion that to every ideal I one can find an ideal J such that their product IJ is principal.

Much simpler proofs of this assertion were later found by A. Hurwitz [1931] in 1894 and Dedekind [850] in 1895.

In the manuscript of [844] there is a proof[190] of unique factorization of ideals into primary ideals[191], defined as ideals having only one prime ideal divisor.

He considered also orders $\mathfrak{o} \subset \mathbf{Z}_K$ and stated[192] that every ideal of $\mathfrak{o}$ can be written in a unique way as the product of ideals having only one prime ideal divisor.

The formula $[M : K] = [M : L][L : K]$ for the field extensions $K \subset L \subset M$ appeared for the first time in an unpublished paper of Dedekind (see [3596], Sect. 6), written in 1855. He mentioned this result in his report on Bachmann's book [169] [840], and published a proof in 1879 in §164 of [844]. Another proof was given by A. Kneser[193] [2180] in 1887. In this paper studied also Kneser the question of the degree of the field generated by $\sum_{j=1}^{r} a_j\theta_j$, where θ_j are given algebraic numbers.

Related questions were later considered by Landsberg [2450], Loewy [2621] and M. Bauer [246, 247].

5. In 1882 Dedekind [845] gave a proof of the discriminant theorem, which he formulated already in 1871 [838]:

A prime p divides the discriminant $d(K)$ if and only if it is ramified, i.e. in the factorization of the ideal $p\mathbf{Z}_K$ there occurs a prime ideal with exponent > 1.

In §13 of [845] defined Dedekind the *different* $D_{K/\mathbf{Q}}$ of a number field, showed that it is generated by the set $\{F'_\alpha(a) : \alpha \in \mathbf{Z}_K\}$ ($F_\alpha(X)$ denoting the minimal polynomial of α), proved the equality

$$N_{K/\mathbf{Q}}(D_{K/Q}) = d(K)\mathbf{Z}_K ,$$

and established the *different theorem*, stating that if $\mathfrak{p}$ is a prime ideal in K dividing the prime p, and $p\mathbf{Z}_K = \mathfrak{p}^e I$ (with $\mathfrak{p} \nmid I$), then $D_{K/Q}$ is divisible by $\mathfrak{p}^{e-1}$. Moreover, if $p \nmid e$ (the tame case), then $\mathfrak{p}^e \nmid D_{K/Q}$. The discriminant theorem is an immediate

[190] This proof is included in §172 of [844] in Dedekind's collected papers [856].

[191] Dedekind called them "einartige Ideale".

[192] In [856] the editors inserted into the text of [844] a proof of this assertion taken from Dedekind's manuscripts.

[193] Adolf Kneser (1862–1930), professor in Dorpat, Berlin and Breslau, father of Hellmuth Kneser. See [2228].

consequence of the different theorem. At the end of [845] Dedekind promised to extend later his results to the case of relative extensions, but he made only the first step towards this goal in [851], the final generalization being done by Hilbert [1831].

Other proofs of the different or the discriminant theorem were given by Hensel [1768], Mertens [2817], Hilbert (Th. 31 in [1836]), Landsberg[194] [2449] in 1897, M. Bauer [239], Hecke [1742] in 1923 (Hecke's proof was given only for Galois extensions, but it can be modified to cover also the general case, as shown by Narkiewicz and Schinzel [3071] in 1969), Cebotarev [628] in 1935 and Artin [129, 130] in 1959 and 1967.

The different theorem implies that if $p\mathbf{Z}_K = \prod_{i=1}^{g} \mathfrak{p}_i^{e_i}$ and $p \nmid \prod_{i=1}^{g} e_i$, i.e. p is at most tamely ramified in $K/\mathbf{Q}$, then

$$p^a \parallel d(K) \quad \text{with} \quad a = \sum_{i=1}^{g} f_i(e_i - 1), \tag{1.27}$$

f_i being the degree of $\mathfrak{p}_i$.

See Sect. 4.1.2 for the determination of the exponent a in the general case.

In a footnote in §7 of [845] stated Dedekind the following necessary and sufficient condition[195] for an ideal $\mathfrak{f}$ of $\mathbf{Z}_K$ to be the conductor of an order:

If $\mathfrak{p}$ is a prime ideal dividing $\mathfrak{f}$ and $\mathfrak{f} = \mathfrak{p}I$, then every rational integer contained in I lies in $\mathfrak{f}$.

A proof of this condition has been given on p. 445 of Weber's paper [4323]. In the same paper Weber showed that the class-number of an order in $\mathbf{Q}(\sqrt{d})$ having discriminant df^2 coincides with the class-number of primitive binary quadratic form of discriminant df^2.

This condition has been later rediscovered by Furtwängler [1347]. Another proof has been given in 2014 by Lettl and Prabpayak [2572].

A generalization to arbitrary Dedekind domains has been given in 1927 by Grell[196] [1516]. See also Reinhart [3428].

In 1929 W. Weber [4333] simplified Dedekind's theory of orders in the case of quadratic fields, stressing the role of invertible ideals, and in the next year he applied this to the study of representations of integers by quadratic forms [4334]. Later he studied invertible ideals in a more general situation [4335].

6. In 1894 the fourth edition of Dirichlet's lectures [972] appeared, containing an extended XIth supplement [848] of Dedekind.

In §159 described Dedekind arithmetics of integers in the Gaussian field $\mathbf{Q}(i)$ pointing out the existence of the Euclidean algorithm. He noted also that the same happens for integers in the fields $\mathbf{Q}(\sqrt{D})$ for $D = -11, -7, -3, -2, 2, 3, 5, 13$.

---

[194]Georg Landsberg (1865–1912), professor in Heidelberg, Breslau and Kiel.

[195]Already in [842] he mentioned the existence of such condition, pointing out that its determination is not particularly difficult.

[196]Heinrich Grell (1903–1974), professor in Berlin. See [5].

In 1906 Birkhoff[197] [364] used a geometrical method to prove this assertion in the imaginary case.

Dedekind observed also that there is no unique factorization for integers for $d = -5$ and stated that although there is no Euclidean algorithm in the case $d = -19$ the corresponding integers obey the unique factorization law.

A simple proof of the last assertion has been given in 1975 by K.S. Williams [4413].

In §160 number fields are defined and in the next sections the main notions of their theory are introduced and studied, in particular the notion of an isomorphism occurs, called "permutation" by Dedekind. In §173-174 algebraic integers are defined and their divisibility properties are explained. In §175 Dedekind established the existence of an integral basis of the ring $\mathfrak{o}$ of integers of the field K and defined the discriminant of K. In §177 he defined ideals of $\mathfrak{o}$, and called an ideal I to be divisible by J, when $I \subset J$. In §179 prime ideals were defined, and a proof of the uniqueness of factorization of ideals into prime ideals was presented. The next section introduced congruences with respect to ideals and in §181 ideal classes were defined, and it was shown that their number is finite. In §182 relations between the theory of ideals in a field of degree n and the theory of splitting forms of degree n in n variables were presented. We find here a.o. the assertion that if $\omega_1, \dots, \omega_n$ is an integral basis of a field K, and $\alpha_1, \dots, \alpha_n$ is a basis of an ideal I, then one has

$$\det[\alpha_i^{(j)}]^2 = N(I)^2 d(K) .$$

An analogous result for relative extension has been proved in 1948 by L. Fuchs [1292].

Particular attention gave Dedekind to the case $n = 2$. In this case he associated with an ideal $I = a\mathbf{Z} + b\mathbf{Z}$ the form

$$f_I(X, Y) = \frac{N(aX + bY)}{N(I)} ,$$

and showed that if the ideals I, J lie in the same equivalence class, then the forms f_I, f_J are equivalent under the action of the group $SL_2(\mathbf{Z})$. In the case of negative discriminant d one considers only positive-definite forms, and for positive d one takes in account the narrow equivalence of ideals, two ideals I, J being equivalent if and only if one has $aI = bJ$ with some totally positive integers a, b of the field. More generally, if d is a form discriminant, but not a fundamental discriminant, then there is a one-to-one correspondence between narrow classes of forms of discriminant d and ideal classes in the order of discriminant d in the field $\mathbf{Q}\left(\sqrt{d}\right)$. This implies that several results of Gauss's book [1394] can be translated into statements about quadratic number fields.

An exposition has been presented by H.H. Mitchell [2890] in 1926.

[197] George David Birkhoff (1884–1944), professor at Princeton and Harvard. See [4394].

Units were considered in §183, where Dirichlet's unit theorem has been proved for the first time in full generality, not only for rings of the form $\mathbf{Z}[a]$. Dedekind defined the *regulator* of the field K by

$$R(K) = \left|\det\left[\log(F_i(\varepsilon_j))\right]_{i,j}^{(r)}\right| ,$$

(where $r = r_1 + r_2 - 1$, ε_i are fundamental units and $F_1, \dots, F_r$ are embeddings of K in $\mathbf{C}$ chosen so that $\overline{F_i} \neq F_j$ for $i \neq j$) and showed that $R(K)$ does not depend on the choice of the set of fundamental units.

With the use of the unit theorem established Dedekind in §184 the *Ideal Theorem for ideal classes* in the ring of integers of a field K in the following form:

$$\lim_{x\to\infty} \frac{\#\{I \in A : N(I) \le x\}}{x} = \kappa(K), \tag{1.28}$$

with

$$\kappa(K) = \frac{2^{r_1}(2\pi)^{r_2} R(K)}{w(K)\sqrt{|d(K)|}}, \tag{1.29}$$

where n is the field degree, A is a fixed ideal class, $w(K)$ is the number of roots of unity contained in K, $d(K)$ is the discriminant and $R(K)$ is the regulator.

As a consequence he obtained the *Ideal Theorem*:

$$A_K(x) := \#\{I : N(I) \le x\} = (1 + o(1))h(K)\kappa(K)x. \tag{1.30}$$

In the case $K = \mathbf{Q}(i)$ the evaluation of $A_K(x)$ forms the circle problem due to the equality

$$A_{Q(i)}(x) = \#\{a, b \in \mathbf{Z} : a^2 + b^2 \le x\} .$$

The first asymptotics in this case was obtained already by Gauss in a paper published posthumously [1396].

In 1946 Wintner [4429] expressed the coefficient of the main term of (1.30) as a product over rational primes p in the following way:

$$h(K)\kappa(K) = \prod_p \left(1 - \frac{1}{p}\right) \prod_{\mathfrak{p}|p} \left(1 - \frac{1}{N(\mathfrak{p})}\right)^{-1} .$$

In Theorem IV in this section the zeta-function

$$\zeta_K(s) = \sum_I \frac{1}{N(I)^s} = \prod_{\mathfrak{p}} \frac{1}{1 - N(\mathfrak{p})^{-s}} \quad \text{for } s > 1 \tag{1.31}$$

has been considered (denoted by Dedekind by $\Omega(s)$), and the equality

$$\lim_{s\to 1}(s-1)\zeta_K(s) = h(K)\kappa(K) \tag{1.32}$$

has been deduced from (1.30).

Dedekind did not consider his zeta-function at complex variables. It seems that it was Landau [2414] who initiated the study of the behaviour of $\zeta_K(s)$ in the complex plane. See Sect. 3.1.1.

7. The equality (1.32) is used in §185 to prove Kummer's formula for the class-number of cyclotomic fields. On the way Dedekind noted that since the prime p generates in the field $\mathbf{Q}(\zeta_p)$ an ideal which is the $(p-1)$-th power of a prime ideal, the degree of that field equals $p-1$, and so the polynomial $(X^p-1)/(X-1)$ is irreducible over $\mathbf{Q}$.

This idea has been later used by Perron [3247] to produce a series of irreducibility criteria, generalizing those of Eisenstein-Schönemann[198] [1075, 3682], Königsberger[199] [2209] and Netto[200] [3086]. See also M. Bauer [222, 225], Ore [3167, 3169, 3170] and Mac Lane[201] [2686].

8. In §186 one finds the proof of the formula for the class-number of quadratic fields, which essentially is a translation of Dirichlet's formula for the class-number of quadratic forms into the language of quadratic fields:

If K is a quadratic field of discriminant d, then for negative d one has

$$h(K) = \frac{w\sqrt{|d|}}{2\pi} L(1, \chi_d) , \tag{1.33}$$

with

$$w = \begin{cases} 4 & \text{if } d = -4 , \\ 6 & \text{if } d = -3 , \\ 2 & \text{if } d < -4 , \end{cases}$$

whereas for positive d

$$h(K) = \frac{\sqrt{d}}{2\log\varepsilon} L(1, \chi_d) , \tag{1.34}$$

where $\varepsilon > 1$ is the fundamental unit of the field $\mathbf{Q}\left(\sqrt{d}\right)$ and

$$L(s, \chi_d) = \sum_{n=1}^{\infty} \frac{\chi_d(n)}{n^s}$$

is Dirichlet's L-function associated with the character

$$\chi_d(n) = \left(\frac{d}{n}\right) ,$$

[198] Theodor Schönemann (1812–1868), teacher in Brandenburg.

[199] Leo Königsberger (1837–1921), professor in Greifswald, Heidelberg, Dresden and Vienna. See [394].

[200] Eugen Netto (1848–1918), professor in Strassburg, Berlin and Giessen. See [1693].

[201] Saunders Mac Lane (1909–2005), professor at Harvard and in Chicago. See [2690].

Kronecker's extension of the Legendre symbol.

A similar formula has been also established for $h^*(K)$ the narrow class-number of K.

For later results on this topic see Sect. 3.3.1..

9. In a letter to Frobenius of 8 June 1882 Dedekind[202] proved the *conductor-discriminant formula*[203] for Abelian extension of the rationals:

Let K be an Abelian field, and let m be its *conductor*, i.e. the minimal integer with $K \subset K_m = \mathbf{Q}(\zeta_m)$. Since the Galois group of $K_m/\mathbf{Q}$ is isomorphic to the group $G(m)$ of residue classes mod m prime to m, the field K corresponds to a subgroup H of $G(m)$. Denote by $X(K)$ the group of characters mod m trivializing on H, and for $\chi \in X(K)$ let f_χ be its conductor (Dedekind used the word "exponent"). The conductor-discriminant formula is formed by the following two equalities:

$$d(K) = (-1)^{r_2(K)} \prod_{\chi \in H} f_\chi, \quad LCM\{f_\chi : \chi \in X(K)\} = m. \tag{1.35}$$

This formula has been later generalized to relative Abelian extensions. See Satz 16 and footnote 44 in Hasse's report [1655].

10. An extension L/K is called a *pure extension* if it is generated by a root of a polynomial $X^n - a \in K[X]$, assumed to be irreducible over K. Reducible polynomials of this form were described by Capelli[204] [561–563] and Wendt[205] [4368]. The case $K = \mathbf{Q}$ has been dealt earlier by Vahlen[206] [4119], and the case of prime n goes back to Abel [15]. A very simple proof of Abel's result has been given by Mertens [2816] in 1891.

A new proof of Capelli's theorem has been given by Nagell [3025] in 1939 (note however that Theorems 2 and 3 in this paper are incorrect). For generalizations see Sect. 2.1 in the book [3621] by Schinzel.

Mathews [2769] described in 1892 prime ideals in $\mathbf{Q}(\sqrt[3]{2})$, and formulas for the discriminant in pure extensions of $\mathbf{Q}$ having prime degree were provided in 1897 by Landsberg [2448]:

Let p be a prime. If $K = \mathbf{Q}(\omega)$ with $\omega = \sqrt[p]{a}$, then

$$d(K) = \varepsilon^{p-1}(-1)^{(p-1)(p-2)/2} p^A \prod_{q|a} q^{p-1},$$

where q runs over primes, ε denotes the sign of a, and

---

[202] See also [846].

[203] Lemmermeyer notes in his book [2527] (p. 125) that a part of its fame "*is due to its name in German, where it is often called the Führerdiskriminantenproduktformel*".

[204] Alfredo Capelli (1855–1910), professor in Palermo and Naples. See [4069].

[205] Ernst Adolf Wendt (1872–1946).

[206] Karl Theodor Vahlen (1869–1945), professor in Greifswald and Berlin. See [3799].

$$A = \begin{cases} p & \text{if } a^{p-1} \not\equiv 1 \bmod p^2 \text{ or } p|a , \\ p-2 & \text{otherwise} . \end{cases}$$

In the case $p \nmid a$ Landsberg showed that an integral basis of K is formed by

$$1, \omega, \omega^2, \dots, \omega^{p-2}, \Omega ,$$

where

$$\Omega = \begin{cases} (\omega - a)^{p-1}/p & \text{if } a^{p-1} \equiv 1 \bmod p^2 , \\ \omega^{p-1} & \text{otherwise} . \end{cases}$$

Proofs of the formula for the discriminant were also given later by Westlund[207] [4374] in 1910, Berwick[208] in his book [327] in 1927, Wegner[209] [4340] in 1932 and Hasse [1686] in 1937. Westlund constructed also integral bases in all cases. Wegner's method was applied later by Tietze[210] [4063] to construct integral bases in $\mathbf{Q}(\sqrt[p]{a})$ in the case $a^{p-1} \not\equiv 1 \bmod p^2$. In 1931 Wegner [4338, 4339] showed that if K is a field of prime degree p, and all splitting primes are congruent to 1 mod p, then $K = \mathbf{Q}(a^{1/p})$ with some rational a. He showed later [4341] that there are infinitely many primes $q \equiv 1 \bmod p$ which do not split in K. The number of fields $\mathbf{Q}(a^{1/p})$ with given discriminant has been given by D.C. Mayer [2791] in 1993.

Discriminants and integral bases for arbitrary pure extensions of the rationals were given in 1927 by Berwick [327]. For the case $(a, n) = 1$ see also Okutsu [3148]. Factorization of prime ideals in pure extensions of arbitrary number fields was treated by Mann[211] and Vélez [2736] and Vélez [4201–4203].

In 1900 a long paper of Dedekind [853] appeared, presenting the theory of pure cubic extensions $K = \mathbf{Q}(\sqrt[3]{m})$. Dedekind proved that if $m = ab^2$ with square-free ab, then

$$d(K) = \begin{cases} -27(ab)^2 & \text{if } 9 \mid a^2 - b^2 , \\ -3(ab)^2 & \text{otherwise} , \end{cases}$$

and the factorization of primes p has the following form:00

$$3\mathbf{Z}_K = \begin{cases} \mathfrak{p}^3 & \text{if } 3 \mid ab \text{ or } 9 \mid a^2 - b^2, \\ \mathfrak{p}_1^2 \mathfrak{p}_2 & \text{if } 3 \nmid ab \text{ and } 9 \nmid a^2 - b^2 . \end{cases}$$

If $p \neq 3$ divides m, then $p = \mathfrak{p}^3$, and if $p \nmid 3m$, then

[207] Jacob Westlund (1868–1947, professor at Purdue University.

[208] William Edward Hodgson Berwick (1888–1944), professor in Bangor. See [819].

[209] Udo Wegner (1902–1989), professor in Darmstadt, Heidelberg, Saarbrücken and Stuttgart.

[210] Heinrich Franz Friedrich Tietze (1880–1964), professor in Erlangen and Munich. See [3260].

[211] Henry Berthold Mann (1905–2000), professor at the Ohio State University, Univ. of Wisconsin and Univ. of Arizona. See [3152].

$$p\mathbf{Z}_K = \begin{cases} \mathfrak{p}_1\mathfrak{p}_2 & \text{if } p \equiv 2 \bmod 3, \\ \mathfrak{p}_1\mathfrak{p}_2\mathfrak{p}_3 & \text{if } p \equiv 1 \bmod 3,\ \left(\frac{ab^2}{p}\right)_3 = 1, \\ \mathfrak{p} & \text{if } p \equiv 1 \bmod 3,\ \left(\frac{ab^2}{p}\right)_3 \neq 1. \end{cases}$$

He considered also the zeta-function $\zeta_K(s)$ (calling it "*Dirichletsche*[212] *Idealfunction*") and used (1.28) and (1.29) to show

$$\lim_{s\to 1}(s-1)\zeta_K(s) = \frac{2\pi}{3}\frac{\log\varepsilon}{\eta ab}h(K),$$

with $\varepsilon > 1$ being the fundamental unit of K, and

$$\eta = \begin{cases} 3 & \text{if } 9 \nmid a^2 - b^2, \\ 1 & \text{otherwise}. \end{cases}$$

He presented also the ratio $\zeta_K(s)/\zeta(s)$ as a Dirichlet L-series with a cubic character.

An analogue of the last result for arbitrary cubic fields has been given in 1930 by Jaeger [2021]. Integral bases for pure cubic extensions of $\mathbf{Q}(\zeta_3)$ were constructed in 1970 by Wada [4258]. Pure cubic extensions of other quadratic fields were treated by K. Nagata [3009] in 1985. A recent exposition of results obtained in [853] has been presented by Lemmermeyer [2532]. A modern description of the theory of pure cubic fields has been given by Barrucand and Cohn [196].

11. In the paper [855], published posthumously, gave Dedekind the following simple condition for an algebraic number field K to have class-number 1:

For each pair $\alpha, \beta \neq 0$ in $\mathbf{Z}_K$ there exist $\mu, \nu \in \mathbf{Z}_K$ such that $(\mu\mathbf{Z}_K, \nu\mathbf{Z}_K) = 1$ and

$$|N(\alpha\mu + \beta\nu)| < |N(\beta)|.$$

This condition was also found by Rabinowitsch[213] [3359] in 1913. A similar condition has been used in 1927 by Hensel [1798] to show that if a domain R has the unique factorization property, then the same holds in the ring $R[X]$.

Dedekind's result was later put in a more general form by Hasse [1660]. It has been rediscovered later by Kutsuna [2388] and Queen [3350, 3351].

The history of Dedekind's creation of his theory of algebraic numbers has been presented by Edwards [1046, 1047].

---

[212] "*Dirichlet's ideal function*"

[213] Georgij Rabinowitsch = Georg Yuri Rainich (1886–1968), professor at the University of Michigan.

1.3.4 Frobenius and Stickelberger

1. In 1896 Frobenius [1259] made a thorough study of the relations between prime ideals in a Galois extension and the Galois group. He showed[214] that if p is unramified in an extension $K/\mathbf{Q}$ and factorizes into prime ideals of degrees $f_1, \dots, f_r$, then the Galois group of K, considered as a permutation group, contains a permutation consisting of cycles of orders $f_1, \dots, f_r$. He proved also the following result, known as the *Frobenius density theorem.*

Let K be a normal extension of the rationals of degree n with Galois group G treated as a subgroup of S_n, and let $f_1, f_2, \dots, f_k$ be given positive integers summing up to n. If δ denotes the density of those rational primes which are products of k prime ideals in K of degrees $f_1, f_2, \dots, f_k$, then the product δn equals the number of elements of G which are products of k cyclic permutations of $f_1, f_2, \dots, f_k$ elements.

Another proof of the Frobenius density theorem has been given in 1926 by A. Hurwitz [1938]. It has been pointed out by Gassmann ([1393], §2) in his comments to [1938] that this theorem holds also for relative extensions with only minor changes in the proof.

With every prime ideal $\mathfrak{p}$ in K dividing the prime p Frobenius associated an element $g_{\mathfrak{p}}$ of the group G satisfying

$$g_{\mathfrak{p}}(a) \equiv a^p \pmod{\mathfrak{p}}$$

(the *Frobenius automorphisms*). If $\mathfrak{p}'$ is another prime ideal dividing p, then the elements $g_{\mathfrak{p}}$ and $g_{\mathfrak{p}'}$ are conjugated and this associates with every unramified prime p a conjugacy class $\mathfrak{C}_p$ in the Galois group.

Later the Frobenius automorphism was defined also for relative Galois extensions L/K by the condition

$$g_{\mathfrak{P}}(a) \equiv a^{N(\mathfrak{p})} \pmod{\mathfrak{P}}$$

where $\mathfrak{P}$ is a prime ideal of $\mathbf{Z}_L$ dividing $\mathfrak{p}\mathbf{Z}_L$. This automorphism played later an important role (see, e.g. Sect. 4.3.2, 4.3.2).

At the end of [1259] one finds the following conjecture:

If $\mathfrak{C}$ is a conjugacy class in G, then there are infinitely many rational primes p satisfying $\mathfrak{C}_p = \mathfrak{C}$. The density of the set of those primes is proportional to the number of elements of $\mathfrak{C}$.

It has been shown by Artin (Theorem 4 in [118]) in 1924 that this conjecture in the quantitative form

$$\#\{p \le x : \mathfrak{C}_p = A\} = \frac{\#A}{n} \operatorname{li} x + O\left(x \exp(-c\sqrt{\log x})\right), \tag{1.36}$$

where n is the degree and the constant $c > 0$ depends on the extension is a consequence of his conjectured reciprocity law. The analogous assertion holds also for relative Galois extensions (cf. Deuring [909]). (see Sect. 4.3.2)

[214] Actually the proof of this result was found by Dedekind, who communicated it to Frobenius in June 1882. See [847].

The conjecture of Frobenius has been proved in 1923 by Čebotarev [617, 620] (see Sect. (4.3.2)).

2. In 1897 Stickelberger [3934] noted that the discriminant $d(K)$ of an algebraic number field K is congruent to 0 or 1 mod 4. He showed also that if $p \nmid d(K)$ is a prime with

$$p\mathbf{Z}_K = \prod_{j=1}^{g} \mathfrak{p}_j \,,$$

then

$$\left(\frac{d(K)}{p}\right) = (-1)^{\sum_j (f_j-1)} = (-1)^{n-g} \,,$$

where f_j denotes the degree of the prime ideal $\mathfrak{p}_j$. An equivalent result for discriminants of polynomials has been earlier established by Pellet [3233, 3234] who showed that if D is the discriminant of a polynomial $f \in \mathbf{Z}[X]$, and p is an odd prime not dividing D, then

$$\left(\frac{D}{p}\right) = (-1)^{n-r} \,,$$

where r denotes the number of irreducible factors of $f(X)$ mod p.

Another proof of these results was given by Voronoĭ[215] [4245] in 1904, Hensel [1777] in 1905, Lasker[216] [2474] in 1916, Skolem [3818] in 1952 and Cvetkov [805] in 1983. In 1953 Carlitz [572] rediscovered the case $p = 2$.

In 1955 Dalen [812] proved the analogue for local fields (see also Swan [3965] and Dalawat [811]). For generalizations see Barrucand and Laubie [197] and Movahhedi and Zahidi [2993].

Mirimanoff and Hensel [2888] showed in 1905 how this theorem implies the quadratic reciprocity law. A very simple proof of the congruence $d(K) \equiv 0, 1$ mod 4 has been given in 1929 by Schur [3702].

In an earlier paper [3933], devoted to an extension of results of Eisenstein and Kummer on reciprocity laws, Stickelberger made a study of the set of residue classes mod $\mathfrak{p}$ in an algebraic number field. This paper contains also a result, which turned out later to be of importance:

Let $K = \mathbf{Q}(\zeta_n)$ and for a prime to n denote by g_a the element of $G_n = Gal(K/\mathbf{Q})$ with $g_a : \zeta_n \mapsto \zeta_n^a$. Moreover put

$$\Theta_K = \frac{1}{n} \sum_{(a,n)=1} a \cdot g_a^{-1} \in \mathbf{Q}[G_n] \,,$$

and define the Stickelberger ideal by

$$St_K = \Theta_K \mathbf{Z}[G_n] \cap \mathbf{Z}[G_n] \,.$$

---

[215] Georgiĭ Fedoseevič Voronoĭ (1868–1908), professor in Warsaw. See [3925].

[216] Emanuel Lasker (1868–1941), world chess champion in 1894–1921.

Then St_K annihilates the class-group of K, i.e. for $A \in St_K$ the ideal I^A is principal for every ideal I of K . In the case of prime n this theorem has been proved in 1847 by Kummer (Sect. 12 of [2351]).

A new proof has been given in 1975 by Fröhlich [1277].

For a cyclotomic field K with Galois group G put

$$\mathbf{Z}[G]^{-1} = (1-j)\mathbf{Z}_K, \quad St_K^{-1} = St_K \cap \mathbf{Z}[G]^{-1},$$

where j denotes the complex conjugation. It has been shown in 1962 by Iwasawa [1989] that if $K = \mathbf{Q}(\zeta_{p^k})$ with a prime p, then the index of St_K^{-1} in $\mathbf{Z}[G]^{-1}$ equals $h_{p^k}^-$ (another proof has been given in 1981 by Skula [3820]). Later Sinnott [3803] showed that in the case when n is not a prime power this index equals $2^s h_n^-$, with $s = 2^{\omega(n)-2} - 1$.

In 1980 Sinnott [3804] defined a generalization of the Stickelberger ideal to Abelian extensions of the rationals.

C.G.Schmidt [3641] and Iimura [1957] obtained a generalization of Iwasawa's index-formula for arbitrary Abelian fields. For other generalizations of Iwasawa's formula see A. Endo [1092, 1093, 1096] and Ichimura [1949],

1.4 Other Results

1. In 1840 Cauchy [610] established congruences mod p for the difference between the number of quadratic residues and non-residues mod p in the interval $(0, p/2)$, which imply the congruence

$$h(-p) \equiv c(p)B_{(p+1)/2} \pmod{p},$$

with

$$c(p) = \begin{cases} 2 & \text{if } p \equiv 7 \bmod 8, \\ -6 & \text{if } p \equiv 3 \bmod 8. \end{cases}$$

See also Pepin [3235], A. Hurwitz [1934], Friedmann[217] and Tamarkin[218] [1256].

In 1895 A. Hurwitz [1934] obtained several congruences relating the residue of $h(d)$ mod p for prime divisors p of d to coefficients of power series of suitable trigonometric functions. In the particular case $d = -p = 4k - 1$ he established

$$h(-p) \equiv \frac{(-1)^k}{2}\alpha_k \pmod{p},$$

where α_k is defined by

[217] Aleksandr Aleksandrovič Friedmann (1888–1925), professor in Perm. See [1183].

[218] Jacob David [Yacov Davidovič] Tamarkin (1888–1945), professor in Perm, St. Petersburg and at the Brown University. See [1851].

$$\tan x = \sum_{k=1}^{\infty} \frac{\alpha_k}{(2k-1)!} x^{2k-1} .$$

A similar formula holds in the case $d = -4p$ with $p \equiv 1 \bmod 4$, the function $\tan x$ being replaced by $1/\cos x$. At the end of [1934] one finds also the determination of the residue of $h(d) \bmod 2^r$ for $r = 2, 3$ (see also Glaisher[219] [1440], Lerch [2565]) and Plancherel[220] [3306]).

See also Gut and Stünzi [1562] and Slavutskiĭ [3824]. An analogous congruence in the case of real quadratic fields has been established in 1948 by Kiselev [2157] (see (6.19)).

2. A theory of algebraic numbers using essentially local considerations was constructed by Zolotarev[221] [4478, 4479]. An exposition of it has been presented in 1893 by Sochocki[222] [3842] (for some simplifications see Rychlik[223] [3538, 3539, 3541]). Zolotarev's method was applied later by Markoff[224] [2741, 2742] and Voronoĭ [4243] to study cubic fields (Markoff restricted his attention to pure cubics). A comparison of the methods of Dedekind and Zolotarev has been presented by Ivanov[225] in his M.A. thesis [1977] in 1891.

An exposition of Zolotarev's approach was given by Čebotarev in the second volume of [627] (see also [630]). In 1930 Engstrom [1100, 1101] discussed the relation between the results on theorem on factorization of rational primes in Zolotarev's and Dedekind's approaches (cf. Engstrom [1101]).

In 1924 Grave[226] [1503, 1504] utilized Zolotarev's method to substantiate his approach to algebraic numbers, based on the generalization of the greatest common divisor.

Voronoĭ's construction of integral bases in cubic fields has been given in the book [3853] by Sommer[227], appearing in 1907. A simpler proof has been given by Bergström[228] [301]. See also Delone and Faddeev[229] [882]. A modern proof has been given by Alaca and K.S. Williams [42] in 2004. A generalization of Voronoĭ's construction for fields of larger degree has been given in 1949 by Epelbaum [1109]. A modern description of Zolotarev's theory has been given by Piazza [3281, 3282].

3. At the end of his letter to Jacobi [1815], written in 1850, Hermite pointed out that his method of reduction of n-ary forms should permit the description of all units of $\mathbf{Z}[\sqrt[3]{m}]$, and this has been done in 1869 by Zolotarev [4477], who determined the fundamental unit in this case. For arbitrary cubic fields this has been done in 1894 by Voronoĭ [4243] (it has been presented in 1940 by Delone and Faddeev in their book

[219] James Whitbread Lee Glaisher (1848–1928), worked in Trinity College, Cambridge.

[220] Michel Plancherel (1885–1967), professor in Fribourg and Zürich.

[221] Egor Ivanovič Zolotarev (spelled also Zolotareff) (1847–1878), professor in St. Petersburg. See [2390, 3206].

[222] Julian Sochocki (1842–1927), professor in St. Petersburg.

[223] Karel Rychlik (1885–1968), professor in Prague. See [1943].

[224] Andrei Andreyevič Markov (1856–1922), professor in St. Petersburg. See [1523, 3917].

[225] Ivan Ivanovič Ivanov (1862–1939), professor in St. Petersburg. See [2389].

[226] Dmitriĭ Aleksandrovič Grave (1863–1939), professor in Kharkov and Kiev. See [880, 977].

[227] Julius Sommer (1871–1943), professor at the Technische Hochschule in Danzig.

[228] Harald Bergström (1908–2001), professor in Göteborg.

[229] Dmitriĭ Konstantinovič Faddeev (1907–1989), professor in Leningrad. See [57, 3553].

[882]). In 1896 Voronoĭ [4244] gave another method of finding the fundamental units in cubic fields using a generalization of continued fractions. He gave also a way to compute the class-number and an algorithm to check whether two given ideals are equivalent.

A geometrical interpretation of Voronoĭ's algorithm has been presented in 1923 by Delone (Delaunay, N.) [877]. Simplifications of this algorithm in the case of pure cubic fields have been presented in the eighties by H.C. Williams, Cormack and Seah [4405], H.C. Williams [4401, 4402], and by H.C. Williams, Dueck and Schmid [4406]. In 1985 Buchmann [487, 488] generalized Voronoĭ's algorithm to all fields with unit rank ≤ 2 and applied in the case of orders in a totally complex quartic field [489].

An extension of Zolotarev's approach to arbitrary cubic orders has been given in 1931 by Uspensky [4118].

For a generalization of Voronoĭ's algorithm to function fields see Scheidler and Stein [3597].

A way of finding units in cubic fields using ternary quadratic forms has been given in 1880 by Charve [657].

4. In 1880 Poincaré[230] [3321] based ideal theory in both imaginary and real quadratic fields on the theory of two-dimensional lattices developed earlier by Bravais[231] [436]. Later [3322, 3323] he presented a way to determine all ideals and principal ideals of a given norm in an arbitrary number field.

An elementary description of arithmetics on the field $\mathbf{Q}(\zeta_5)$ has been presented in 1896–1900 by Gmeiner [1444–1446].

5. Certain arithmetical functions defined on integers of the field $\mathbf{Q}(i)$ were considered by Mertens [2814] in 1874. For $\alpha \in \mathbf{Z}[i]$ let $\Phi(\alpha)$ be the generalization of Euler's function to algebraic number fields, giving the number of reduced residue classes mod α, and $\tau(\alpha)$ the number of divisors of α. Mertens proved

$$\sum_{N(\alpha)\leq x} \Phi(\alpha) = \frac{3}{\pi\mathfrak{L}}x^2 + O(x^{3/2}),$$

where

$$\mathfrak{L} = \sum_{n=0}^{\infty} \frac{(-1)^n}{(2n+1)^2} = \sum_{n=1}^{\infty} \frac{\chi_4(n)}{n^2},$$

with χ_4 being the non-principal character mod 4, and

$$\sum_{N(\alpha)\leq x} \tau(\alpha) = \pi^2 x \log x + cx + O(x^{3/4})$$

with a constant c. He obtained also a similar evaluation for the number of square-free divisors. See also Busche [527].

---

[230] Henri Poincaré (1854–1912), professor in Paris. See [3284] and the volume 38 (1921) of Acta Mathematica.

[231] Auguste Bravais (1811–1863), professor in Lyon and Paris.

Several other arithmetical functions in $\mathbf{Z}[i]$ were considered in 1885–1890 by Gegenbauer [1399–1402]. Similar results for functions defined in $\mathbf{Z}[\zeta_3]$ were proved in 1904 by Axer[232] [154].

In 1910 Schleser [3627] provided some extensions of Axer's results.

Let $\Pi(x)$ be the number of primes π of $\mathbf{Z}[i]$ satisfying $N(\pi) \leq x$. It has been shown in 1892 by Poincaré [3324] that with some constants $0 < a < b$ one has for large x the inequalities

$$a\frac{x}{\log x} \leq \Pi(x) \leq b\frac{x}{\log x}. \tag{1.37}$$

He conjectured that a similar result holds for all fields, and this has been later confirmed by Landau [2414]. See Sect. 3.1.1.

6. In 1877 Brill[233] [451] made the simple but useful observation about the sign of the discriminant of a polynomial f. He showed that it is positive if and only if the number of complex roots of f is divisible by 4. This implies the corresponding assertion for the discriminants of number fields.

In 1883 Netto [3084] showed that the discriminants of Abelian extensions of the rationals having odd degree are squares, and one year later he showed [3085] that the discriminant of a cyclic extension of prime degree p is a $(p-1)$-th power.

In 1895 Furtwängler [1327] attached to every ideal I of $\mathbf{Z}_K$ an n-ary form F_I in the following way: if I is generated by $a_1, a_2, \dots, a_n$, and c is the greatest common divisor of coefficients of

$$G_I(X_1, \dots, X_n) = N\left(\sum_{j=1}^{n} a_j X_j\right),$$

then $F_I = G_I/c$. He used the fact that multiplication of ideals corresponds to the composition of the associated forms to show that a suitable power of an ideal is a principal ideal.

7. In 1893 H. Weber gave the first axiomatic definition of a field in [4318], where he presented his approach to Galois theory. Earlier authors considered only fields whose elements were numbers, and Weber's paper was the first in which infinite and finite fields were treated in a uniform way.

8. A field K is called *norm-Euclidean* (earlier such fields were called simply *Euclidean*) if its ring of integers is an Euclidean domain with $\Phi(x) = |N_{K/\mathbf{Q}}(x)|$. In such fields the analogue of the Euclidean algorithm for rational integers holds, and they have class-number one.

A domain R is called an *Euclidean domain* if there exists a mapping Φ of R into the set of non-positive integers such that $\Phi(x) = 0$ holds only for $x = 0$, and for $a, b \in \mathbf{Z}_K$, $b \neq 0$ there exist $q, r \in \mathbf{Z}_K$ with $\Phi(a - bq) < \Phi(r)$. It seems that this definition appeared for the first time in

[232] Alexander Axer (1880–1948), teacher in Zürich.

[233] Alexander v.Brill (1842–1935), professor in Darmstadt, München and Tübingen. See [1187, 2622].

1949 in the paper [2989] of Motzkin, earlier the condition $\Phi(ab) = \Phi(a)\Phi(b)$ has been additional assumed. Later this definition has been extended to the case when Φ attains values in a well-ordered set (see M. Nagata[234] [3010, 3011] and Samuel [3570]).

We noted already that Gauss [1395] showed that the field $\mathbf{Q}(i)$ is norm-Euclidean, Dirichlet [969] did this for $\mathbf{Q}(\zeta_3)$, Eisenstein [1075] for $\mathbf{Q}(\zeta_8)$, and the proof for $\mathbf{Q}(\zeta_p)$ for $p = 5, 7$ is contained in a manuscript of Kummer written in 1844 (see Bölling [384]). The cases $p = 3, 4$ were rediscovered in 1847 by Wantzel [4298], who also asserted that the same holds for all cyclotomic fields $\mathbf{Q}(\zeta_n)$, but it has been noted by Cauchy [611] that his proof fails for $n = 7$ (actually Wantzel's argument is incorrect already in the case $n = 4$, as noted by H.W.Jr. Lenstra [2544]). In [611] Cauchy gave a sketchy argument[235] for the norm-Euclidicity of $\mathbf{Q}(\zeta_n)$ for $n = 5, 7, 8, 9, 12$ and 15 (Bölling wrote ([384] p. 280) that for $p = 5$ Cauchy's argument can be saved).

The first correct proof for $n = 5$ seems to be that given by Uspensky[236] [4113, 4114] in 1906. Elementary proofs were given by Branchini [423] in 1923 and Chella[237] [667] in 1924. The last paper contains also new proofs for $n = 7, 8, 9$, as well as for $n = 16$.

In the seventies Lakein [2402] and Masley [2759] showed that $\mathbf{Q}(\zeta_n)$ is norm-Euclidean for $n = 8, 12$, and H.W.Jr. Lenstra [2541] did this for $n = 11, 20$ and 24. A proof for $n = 16$ has been given in 1977 by Ojala [3146].

Now one knows that this holds also for $n = 13$ (McKenzie [2807]), and the only cyclotomic fields with class-number one for which the Euclidicity is left undecided are $\mathbf{Q}(\zeta_{17})$ and $\mathbf{Q}(\zeta_{19})$.

It has been shown in 1965 by Godwin[238] [1448] that the maximal real subfield of $\mathbf{Q}(\zeta_m)$ with $m = 11$ is norm-Euclidean, and in 2000 Cerri [635] obtained this assertion for $m = 16$ and 32.

In 1895 A. Hurwitz [1932] proved that for every algebraic number field K there exists a constant $m = m(K)$ such that for every $a \in K$ there exist $g \leq m$ and $b \in \mathbf{Z}_K$ with

$$|N(ga - b)| < 1 ,$$

which can be regarded as an analogue of the Euclidean algorithm. He used this result to give a simple proof of the finiteness of the class-number. Later he established the bound $m(K) \leq \sqrt{|d(K)|}$ [1937].

For later work on Euclidean fields see 5.4.3 and 6.4.

9. In 1899 A. Hurwitz [1935] defined an analogue of the Bernoulli numbers for the field $K = \mathbf{Q}(i)$ by putting

$$E_n = \frac{4(4n)!}{(2\omega)^{4n}} \zeta_K(4n) ,$$

where $\zeta_K(s)$ is the Dedekind zeta-function of K and

---

[234] Masayoshi Nagata (1927–2008), professor in Kyoto. See [2898].

[235] Lenstra wrote in [2544]: "*His proof is probably not correct, but its sketchiness makes this difficult to confirm.*"

[236] James Victor Uspensky [Jakov Viktorovič Uspenskiĭ] (1883–1947), professor in St. Petersburg and at Stanford University.

[237] Tito Chella (1881–1923), professor in Pisa.

[238] Herbert James Godwin (1916–2009), professor at the Royal Holloway College, London.

$$\omega = 2\int_0^1 \frac{dt}{\sqrt{1-t^4}}\,.$$

He showed that the sequence

$$\frac{2^{4n}E_n}{4n(4n-2)!}$$

is the sequence of coefficients of a Laurent series of an elliptic function.

For generalizations see Dintzl [952, 953], Matter [2777], Naryškina[239] [3075, 3076] and Katz [2122].

1.5 Remarks

1. Most results in the nineteenth century dealing with the theory of algebraic numbers were published in Germany and France. H.J.S. Smith[240] [3833] pointed out in 1876 in his speech as retiring president of the London Mathematical Society that number theory has been neglected in England, and described the progress made on the continent in the study of algebraic numbers.[241] It seems that the first English papers concerning with that subject wrote Cayley[242] [614–616], who studied the periods (his work has been later pursued by H.W.L. Tanner [4002, 4003], Carey [567] and Burnside[243] [523] [see Upadhyaya [4102] for some corrections]).

In the Western Hemisphere this subject did not attract many followers. The first paper on this topic, dealing with a simplification of a formula of Cayley in [614], written by Scott [3718] appeared in *American Journal of Mathematics* in 1886, and next year a survey was presented by Hathaway[244] in a rather original way [1717]. The first paper dealing with the theory of algebraic numbers published in *Annals of Mathematics* appeared in 1914 and contained the assertion that every finite Abelian group is the Galois group of a normal extension of the rationals (G.A. Miller[245] [2859]). Some years earlier Pierpont [3292, 3293] published his introductory lectures on Galois theory.

2. We mentioned already Dirichlet's lectures [972] whose second edition, published in 1871, contained Dedekind's supplement [839] dealing with the theory of

[239] Ekaterina Alekseevna Naryškina (1895–1940).

[240] Henry John Stephen Smith (1826–1883), professor in Oxford.

[241] In his survey on number theory, published in six parts in the Report of the British Association for the years 1859–1865 Smith [3833] presented also the first steps of the theory of algebraic numbers [3832]

[242] Arthur Cayley (1821–1895), professor in Cambridge. See [3127].

[243] William Burnside (1852–1927), professor in Greenwich. See [1219].

[244] Arthur Stafford Hathaway (1855–1934), professor at the Cornell University and in Terre Haute.

[245] George Abram Miller (1863–1951), professor at the Cornell University, at Stanford and the University of Illinois at Urbana-Champaign. See [422].

algebraic numbers. One year later Bachmann's book [169] appeared with an exposition of Gaussian theory of circle division. One finds in it proofs of the cubic and biquadratic reciprocity laws, as well as a chapter presenting Kummer's ideal numbers in $\mathbf{Q}(\zeta_p)$ for prime p. The theory of quadratic fields has been presented in 1892 by H. Weber in his book on elliptic functions [4316].

A presentation of the state of art in algebraic number theory has been made in 1897 by Hilbert in his report [1836] which we shall discuss in the next chapter. In the algebra textbook by Netto [3087], appearing in 1900, algebraic numbers and their fields serve as tools in the study of polynomial equations, and in the second volume of Weber's treatise [4321], published in 1896, one finds a broad exposition of Kronecker's theory of algebraic numbers.

Chapter 2
The Turn of the Century

2.1 David Hilbert

2.1.1 First Results

1. The first paper of Hilbert related to algebraic number theory appeared in 1892 [1830]. Its main result, now known as *Hilbert's irreducibility theorem*, asserts that if $F(X_1, \ldots, X_n)$ is an irreducible polynomial over a finite extension K of the rationals, then for any fixed set S of $s \leq n-1$ variables and infinitely many choices $X_i = a_i \in K$ for $i \in S$ the resulting polynomial remains irreducible over K. In the case $S = \{2, 3, \ldots, n\}$ one can also obtain that the Galois group of the resulting polynomial is isomorphic with the Galois group of F over the field $K(X_2, \ldots, X_n)$.

Fields in which Hilbert's irreducibility theorem holds are called *Hilbert fields*.

It has been pointed out by Schinzel in [3621] that Hilbert's proof is valid only for normal extensions of the rationals. He quoted Franz [1231] for the first correct proof in the general case. Franz showed also that if Hilbert's theorem holds in a field K, then it holds also in every finite separable extension of K, and if K is infinite then it holds also in every purely transcendental extension of K (an exposition has been given by S.Lang[1] [2462]). The separability assumption has been later removed by Inaba [1962].

Hilbert's proof has been simplified in 1929 by Dörge [991]. A different proof in the case $n = 2$ has been given in 1939 by Eichler[2] [1057]. In 1955 Gilmore and A. Robinson[3] [1436] used model theory to give a new proof (cf. Roquette [3500]).

An effective version of Hilbert's has been established in 1979 by Sprindžuk [3876–3878].

It has been shown by Schinzel [3614] that the elements a_i in Hilbert's theorem can be arbitrarily chosen from suitable arithmetical progressions. A quantitative version of this result has been established by S.D. Cohen [740] in 1981 (see Castillo and Dietmann [607] for an improvement of the error term).

See also Sect. 4.5.4.

---

[1] Serge Lang (1927–2005), professor at Columbia University and at Yale. See [2075].

[2] Martin Eichler (1912–1992), professor in Münster, Marburg and Basel. See [2186].

[3] Abraham Robinson (1918–1974), professor in Toronto at the Hebrew University, UCLA, and Yale. See [2681].

W. Narkiewicz, *The Story of Algebraic Numbers in the First Half of the 20th Century*, Springer Monographs in Mathematics,
https://doi.org/10.1007/978-3-030-03754-3_2

Hilbert used his theorem to show that for every $n \geq 2$ there exist infinitely many polynomials of degree n whose Galois group is the symmetric group S_n, and proved also the same assertion for the alternating group A_n. He showed moreover that for every n there exist infinitely many fields of degree n having no proper subfield $\neq \mathbf{Q}$.

2. In 1894 Hilbert [1831] provided a new proof of the unique factorization property for ideals in number fields. In the same year he studied [1832] normal extensions of **Q**, introduced the decomposition, inertia group and ramification groups, and developed their properties. If G denotes the Galois group of a normal extension $K/\mathbf{Q}$, and $\mathfrak{p}$ is a prime ideal of K, then he defined

$$G_{-1} = \{g \in G : \ g(\mathfrak{p}) = \mathfrak{p}\} \quad \text{(Decomposition group)} ,$$

$$G_0 = \{g \in G : \ g(x) \equiv x \pmod{\mathfrak{p}}\} \quad \text{(Inertia group)} ,$$

and for $n = 1, 2, \dots$

$$G_n = \{g \in G : \ g(x) \equiv x \pmod{\mathfrak{p}^{1+n}}\} \quad \text{(Ramification groups)} ,$$

with x ranging over the integers of K. The fields corresponding to the these groups according to Galois theory were called by Hilbert the decomposition, inertia and ramification fields, respectively.

He showed that G_0 is a normal subgroup of G_{-1} with a cyclic factor group of cardinality equal to the order of $\mathfrak{p}$, and G_1 is a normal subgroup of G_0 with cyclic factor group of order dividing $N(\mathfrak{p}) - 1$ (note that Hilbert did not use explicitly factor groups, describing the arising situation in a rather cumbersome way).

As an application Hilbert proved that the exact power of a prime ideal dividing the different (called by him "Grundideal", i.e. the fundamental ideal) and the discriminant of K can be expressed by a formula involving the cardinalities of these groups.

The inertia group has been also defined by Dedekind [849] who showed that its knowledge for a prime ideal $\mathfrak{p}$ in a normal field K can be used to determine the factorization of prime ideals lying below $\mathfrak{p}$ in subfields of K. He noted that his approach works also for relative extensions.

A modern approach to this problem has been presented in 1931 by Herbrand [1802].

In another paper of 1894 [1833] studied Hilbert quadratic extensions $K/\mathbf{Q}(i)$. He determined the integral basis and discriminant of K, described its prime ideals and extended Gaussian genus theory to $K/\mathbf{Q}(i)$. This led to a new proof of the quadratic reciprocity law in $\mathbf{Q}(i)$ and Dirichlet's formula (1.9).

3. Hilbert's paper [1834] brings the first complete proof of the Kronecker–Weber theorem. After noting that it is sufficient to show that every cyclic field of prime power degree is a subfield of a cyclotomic field Hilbert constructs three families of cyclic fields: the first consists of the fields $\mathbf{Q}(\zeta_{p^m})$ for prime $p \neq 2$, the second is formed by maximal real subfields of $\mathbf{Q}(\zeta_{2^m})$, and the third contains cyclic subfields of degree l^m of the field $\mathbf{Q}(\zeta_p)$, where $p \equiv 1 \bmod l^m$ is a prime. He applies an inductional argument to show that every cyclic field of prime power degree is contained in a composition of $\mathbf{Q}(i)$ and fields from these three families.

2.1.2 *Zahlbericht*

1. In 1897 appeared Hilbert's report [1836] on algebraic numbers which heavily influenced the development of the subject in years to come, although much later it underwent criticism (see the introduction by Lemmermeyer and Schappacher to the English translation of [1836] and Sect. 5 in the article [2533] by Lemmermeyer).

In the first two parts Hilbert introduced the fundamental notions and theorems of the theory of ideals in algebraic number fields. At the end of the second part, in Sec 54 one finds Theorem 90, quoted often in the literature by its number:

If the extension L/K of an algebraic number field K is cyclic and g is the generator of its Galois group, then every element $a \in L$ with $N_{L/K}(a) = 1$ can be written in the form

$$a = \frac{b}{g(b)},$$

with some integer $b \in K$.

In modern language Hilbert's Theorem 90 expresses the triviality of the first cohomology group $H^1(G, L^*)$ for cyclic group G acting on the multiplicative group of a field L. In 1919 Speiser [3872] proved that for every finite Galois extension L/K with Galois group G and for $m = 1, 2, \dots$ the equality[4] $H^1(G, GL_m(L)) = 1$ holds, which for $m = 1$ reduces to $H^1(G, L^*) = 1$. The last result is often attributed to Noether[5] [3126]. She published a proof of it, but pointed out that it is actually due to Speiser. See Lorenz [2630] for a discussion.

Speiser's proof was simplified by Schur [3699]. For generalizations to division rings see N.Jacobson[6] [2016] (corollary on p. 47) and Lam and Leroy [2406]. An analogue in the theory of derivations in fields was given by N. Jacobson [2015] in 1937.

2. The next result, Theorem 91, deals with the structure of units in a relative cyclic extension L/K of odd prime degree p. Let σ be a fixed generator of the Galois group, $H = \{u \in U(L) : u^l \in K\}$, let r be the rank of units in K, and for $A(X) = \sum_{j=0}^{N} c_j X^j \in \mathbf{Z}[X]$ and $a \in K^*$ define[7]

$$a^{A(\sigma)} = a^{c_0} \sigma(a)^{c_1} \dots \sigma^N(a)^{c_N}.$$

Theorem 91 asserts the existence of units $\varepsilon_1, \varepsilon_2, \dots, \varepsilon_{r+1}$ of L (with r being the rank of units) such that if $F_1(X), \dots, F_{r+1}(X) \in \mathbf{Z}[X]$, then the unit

---

[4]Speiser formulated to in a more elementary but equivalent way.

[5]Emmy Noether (1882–1935), daughter of Max Noether, worked in Göttingen and Bryn Mawr. See [929, 4137].

[6]Nathan Jacobson (1910–1999), professor at the University of North Carolina, John Hopkins University and Yale. See [289].

[7]This way of writing has been first used in Kronecker's thesis [2276].

$$\eta = \prod_{j=1}^{r+1} \varepsilon_j^{F_j(\sigma)}$$

can be written in the form

$$\eta = uv^{1-\sigma}$$

with $u \in H$ and $v \in U(L)$ only if for $j = 1, 2, \ldots, r+1$ one has $1 - \zeta_p \mid F_j(\zeta_p)$.

Such sets $\{\varepsilon_1, \ldots, \varepsilon_r\}$ are called a *fundamental set of relative units* of L/K.

A similar assertion holds also for a quadratic extension L/K, provided that one has $r_1(L) = 2r_1(K)$ (see the last sentence in §55 of [1836]).

This result has been utilized in 1942 by Niven[8] [3116] in his work on quadratic diophantine equations in quadratic fields.

It has been shown in 1939 by Lednev [2497] that the assertion of Theorem 91 holds for cyclic extensions of arbitrary degree.

Later the notion "*relative unit for* L/K" acquired different meanings. Hasse [1696] in 1950 used it for units u of L with $N_{L/K}(u) = 1$, Leopoldt [2551] called a unit u of L a relative unit if for all proper subfields K of L one had $N_{L/K} = \pm 1$, and some authors (see, e.g. Brunotte and Halter-Koch [484] and Odai [3136]) assumed that $N_{L/K}(u)$ is a root of unity.

In Theorem 92 Hilbert showed that in every cyclic extension L/K exists a unit of norm 1, for which the number b in Theorem 90 is not a unit.

3. Theorem 94 is also quoted often by its number:

If K/k is a cyclic unramified extension of an odd prime degree p, then there exists a non-principal ideal I in k which becomes principal in K. The ideal I^p is principal in k; hence the class-number of k is divisible by p.

This shows that at least one class of $H(k)$ trivializes in K.

This result led later Hilbert to the conjecture that in the maximal unramified Abelian extension of K all ideal classes of K trivialize (*Principal Ideal Theorem, Hauptidealsatz*) (see Sect. 4.3.2).

4. The third part of Hilbert's report has been devoted to the theory of quadratic fields. We find here the translation of Gauss's theory of genera of quadratic forms into the language of quadratic fields, based on the norm residue symbol defined for rational integers n, m (with non-square m) and prime p by

$$\left(\frac{n, m}{p}\right) = 1$$

if for $k = 1, 2, \ldots$ one has

$$n \equiv N_{K/\mathbf{Q}}(x_k) \pmod{p^k}$$

with some integer x_k of the field $K = \mathbf{Q}(\sqrt{m})$, and

[8] Ivan Morton Niven (1915–1999), professor at the University of Illinois, Purdue University and the University of Oregon.

$$\left(\frac{n, m}{p}\right) = -1$$

otherwise. An ideal class X is said to be in the principal genus if for all prime divisors ℓ of the discriminant of $\mathbf{Q}(\sqrt{m})$ one has

$$\left(\frac{N(I), m}{\ell}\right) = 1$$

for unramified ideals $I \in X$. Two classes X, Y lie in the same genus if XY^{-1} belongs to the principal genus, and the group $G(K)$ of genera is the factor group $H(K)/G_0(K)$, $G_0(K)$ being the principal genus, but, as we already noted, Hilbert for some reasons avoided the use of factor groups.

Theorem 100 presents a translation to the language of quadratic fields of Gauss's theorem on the genera. In the case of imaginary fields Gauss's theorem implies that $G(K)$ has $2^{\omega(d(K))-1}$ elements ($\omega(n)$ being the number of distinct prime factors of n), whereas for real fields $K = \mathbf{Q}(\sqrt{m})$ with m square-free this equality holds if and only if for every prime divisor p of the discriminant of K one has

$$\left(\frac{-1, m}{p}\right) = 1.$$

Otherwise $G(K)$ has $2^{\omega(d(K))-2}$ elements.

This implies that the 2-ranks of $H(K)$ and $H^*(K)$ coincide if and only if all prime factors of $d(K)$ are congruent to unity mod 4.

For other proofs of the last assertion see Kaplan [2111] and Nemenzo and Wada [3083].

A part of the proof of Theorem 100 is formed by Theorem 102 showing that a rational integer m lies in $N_{K/\mathbf{Q}}(K)$ if and only if for every prime p there exists $a_p \in \mathbf{Z}_K$ with

$$m \equiv N_{K/\mathbf{Q}}(a_p) \pmod{p} .$$

This is one of the first examples of the local–global principle of Hasse.

Theorem 103 implies that the principal genus coincides with the set of squares of ideal classes, Theorem 113 asserts that every genus of a quadratic field contains infinitely many prime ideals, and Theorem 114 gives the class-number formulas (1.33) and (1.34). In §83 Hilbert noted that analogous results hold also for narrow ideal classes, even with simpler proofs.

It has been shown by Fueter [1296] in 1903 that if $d(K) = \prod_{i=1}^{r} d_r$ is the factorization of the discriminant of K into prime power discriminants of quadratic fields, then the Galois group of the field

$$L = \mathbf{Q}(\sqrt{d_1}, \dots, \sqrt{d_r}) \tag{2.1}$$

is isomorphic with the narrow genus group of K, and a prime ideal of K splits in L if and only if it belongs to the principal narrow genus of K.

5. The fourth part of [1836] dealt with cyclotomic fields. It starts with the principal properties of $\mathbf{Q}(\zeta_m)$ (integral basis, discriminant, prime factorization), the assertion that $1, \zeta_n, \ldots, \zeta_n^{\varphi(n)-1}$ forms an integral basis of the field $\mathbf{Q}(\zeta_n)$ (Theorem 124) being deduced from the easier prime power case, established in Theorem 121.

A direct proof has been found in 1984 by Lüneburg [2669]).

It followed a proof of the Kronecker–Weber theorem (Theorem 130) based on [1834].

6. In §105 Hilbert showed (Theorem 132) that every Abelian extension of the rationals with co-prime degree and discriminant has an integral basis consisting of conjugates of an element. He called such basis a *normal basis* (NIB). The first step in the proof was done in Lemma 20 which showed that each subfield of an Abelian field with a NIB also has such a basis. Since cyclotomic fields have obvious normal integral bases it remained to apply the Kronecker–Weber theorem. As Hilbert's theory of ramification groups implies that tamely ramified Abelian extensions of $\mathbf{Q}$ coincide with subfields of fields $\mathbf{Q}(\zeta_d)$ with square-free d this theorem implies that every tamely ramified Abelian field has a NIB.

Hilbert's Theorem 132 has been earlier stated by Dedekind in a letter to Frobenius of 8 July 1896 [852]. Dedekind explained in it his research dealing with the notion of the group determinant[9] $\det\left[X_{gh^{-1}}\right]$ and wrote:

"*Auf*[10] *den Begriff der allgemeinen Gruppen-Determinante bin ich zuerst bei dem Studium der Discriminante eines beliebigen Normalkörper Ω geführt, indem ich solche (sehr nützliche) Basen von Ω betrachtete, die aus den Conjugirten einer einzigen Zahl ω bestehen (bisweilen besitzt auch das System $\mathfrak{o}$ aller ganzen Zahlen in Ω eine solche Basis, z.B. wenn ω eine m^{te} Einheitswurzel, und m durch kein Quadrat theilbar ist, und dasselbe gilt dann auch von allen Divisoren von Ω, z.B. allen quadratischen Körpern von ungerader Grundzahl)*".

Dedekind never published a proof.

Normal integral bases were then applied in §106–112 to the study of Abelian fields of degree l and discriminant p^{l-1} (both l and p being primes). In §111 introduced Hilbert particular normal integral bases in these fields (the *Lagrange normal bases*) formed by the periods of $\mathbf{Q}(\zeta_l)$. Then a proof of the Eisenstein reciprocity law for lth power residues has been presented (§113–§115) as well as the class-number formula for $\mathbf{Q}(\zeta_m)$ (§116–118).

For later results about normal integral bases see Sects. 3.2.3 and 5.1.2.

7. In the last part exposed Hilbert two approaches to the theory of Kummer extensions of $\mathbf{Q}(\zeta_l)$ with prime l, one based on the work of Kummer, and the other avoiding the computations occurring in the first version.

---

[9] For the history of this notion see K. Conrad [763].

[10] "*I was led for the first time to the notion of a general group determinant during my study of discriminants of a general normal field Ω, when I considered such (very useful) bases on Ω which consist of conjugates to a single number ω (sometimes the system of all integers of Ω has such a basis, e.g. when ω is an mth root if unity and m is not divisible by a square, and the same applies also to all subfields of Ω, e.g. to all quadratic fields with an odd fundamental number*".

In §129–133 Hilbert presented his theory of norm residues in Kummer extensions. Let l be a prime, $k = \mathbf{Q}(\zeta_l)$, let $\mu \in k$ be not an lth power in k, and put $K = k(\mu^{1/l})$. For a prime ideal $\mathfrak{p}$ of k call a non-zero $a \in \mathbf{Z}_k$ a *norm residue* mod $\mathfrak{p}$, if for every $m \geq 1$ there is some $b_m \in \mathbf{Z}_K$ with

$$a \equiv N_{K/k}(b_m) \pmod{\mathfrak{p}^m} .$$

It is shown in Theorem 150 that if $\mathfrak{p}$ does not ramify in K/k, and $a \notin \mathfrak{p}$, then a is a norm residue mod $\mathfrak{p}$, and if $\mathfrak{p}$ ramifies and does not divide l, then for every $m \geq 1$ exactly $1/l$ of residue classes mod $\mathfrak{p}^m$ consist of norm residues mod $\mathfrak{p}$. If $\mathfrak{p}$ ramifies and equals the unique prime ideal divisor $\mathfrak{l}$ of l, then this holds for every $m > l$.

In the case $l \notin \mathfrak{p}$ the norm residue symbol is defined in §131 in the following way: Let $a, b \in \mathbf{Z}_k$, write $a\mathbf{Z}_k = \mathfrak{p}^\alpha$, $b\mathbf{Z}_k = \mathfrak{p}^\beta$, write

$$\frac{a^\beta}{b^\alpha} = \frac{A}{B}$$

with $(AB\mathbf{Z}_k, \mathfrak{p}) = 1$, and put

$$\left(\frac{a, b}{\mathfrak{p}}\right) = \left(\frac{A}{\mathfrak{p}}\right)_l \left(\frac{B}{\mathfrak{p}}\right)_l^{-1} ,$$

where $\left(\frac{x}{\mathfrak{p}}\right)_l$ is the lth power character in k (see (1.13)). In the case $l \in \mathfrak{p}$ Hilbert gives a rather more complicated definition. In §133 it is shown (Theorem 151) that $a \in \mathbf{Z}_k$ is a norm residue mod $\mathfrak{p}$ if and only if

$$\left(\frac{a, \mu}{\mathfrak{p}}\right) = 1.$$

This theory has been applied by Hilbert to the proof of Kummer's reciprocity law in cyclotomic fields $K_l = \mathbf{Q}(\zeta_l)$ for regular primes l (Theorem 161). To formulate it one needs the notion of a *primary integer*, going back to Kummer [2368]. Let $\mathfrak{L}$ be the unique prime ideal of K_l dividing l. An integer α of K_l, prime to $\mathfrak{L}$, is called primary if it is congruent to a rational integer mod $\mathfrak{L}^2$, and its absolute value is congruent mod $\mathfrak{L}^{l-1}$ to a rational integer (note that in Kummer's papers the second condition is weaker, and he assumes only the congruence mod $\mathfrak{L}^2$). Theorem 157 shows that every $\alpha \in \mathbf{Z}_K$ prime to $\mathfrak{L}$ becomes primary after multiplication by a suitable unit. If h_l is the class-number of K_l, and h' is defined by $h'h_l \equiv 1 \bmod l$, then the ideal $\mathfrak{p}^{hh'}$ is principal, and one of its generators is a primary integer, say $\pi_\mathfrak{p}$. The lth power residue symbol for prime ideals in K_l is defined by

$$\left(\frac{\mathfrak{p}}{\mathfrak{q}}\right)_l = \left(\frac{\pi_\mathfrak{p}}{\pi_\mathfrak{q}}\right)_l . \tag{2.2}$$

Theorem 161 gives the reciprocity law in the following form:
If l is a regular prime, and $\mathfrak{p} \neq \mathfrak{q}$ are prime ideals $\neq L$, then one has

$$\left(\frac{\mathfrak{p}}{\mathfrak{q}}\right)_l = \left(\frac{\mathfrak{q}}{\mathfrak{p}}\right)_l . \tag{2.3}$$

For regular primes l one finds in Theorem 163 the equality

$$\prod_{\mathfrak{p}} \left(\frac{\alpha, \beta}{\mathfrak{p}}\right) = 1, \tag{2.4}$$

the product taken over all prime ideals of K.

In §172 presented Hilbert a proof of Kummer's theorem of Fermat's Last Theorem in $\mathbf{Q}(\zeta_p)$ for regular prime exponents p, and in §173 he showed the impossibility of $X^4 + Y^4 = Z^2$ in $\mathbf{Q}(i)$. For later results on FLT in number fields see Sect. 3.4.

8. In §9 of [1836] Hilbert pointed out that the theory of primitive roots for powers of prime ideals in algebraic number fields was missing. This lacuna has been filled in 1899 by the following theorem of Wiman[11] [4425]:

Let $\mathfrak{p}$ be a prime ideal and let f be its degree, i.e. $N(\mathfrak{p}) = p^f$. The prime ideal power $\mathfrak{p}^n$ has a primitive root only in the following cases:

(i) $f \geq 2$ and $n = 1$,
(ii) $f = 1, p \nmid 2d(K), n = 1, 2, 3, \ldots$
(iii) $f = 1, p \mid d(K), p \neq 2, n = 1, 2,$
(iv) $f = 1, p = 2 \mid d(K), n = 1, 2, 3,$
(v) $f = 1, p = 2 \nmid d(K), n = 1, 2.$

If I is not a prime ideal, then it has a primitive root only if

$$I = \mathfrak{p}^n \prod_{j=1}^{k} \mathfrak{q}_j ,$$

where the prime ideals $\mathfrak{p}, \mathfrak{q}_1, \ldots, \mathfrak{q}_k$ are distinct, $2 \in \prod_{j=1}^{k} \mathfrak{q}_j$, the ideal $\mathfrak{p}^n$ has a primitive root, and for $i \neq j$ one has $(2^{f_i} - 1, 2^{f_j} - 1) = 1$, with f_i, f_j being the degrees of $\mathfrak{q}_i$ respectively of $\mathfrak{q}_j$.

This result was later rediscovered several times (see Ranum [3384], Myller-Lebedeff[12] [3008] [for quadratic fields], Westlund [4375] and Albis-Gonzaléz [54]). The structure of the multiplicative groups mod $\mathfrak{p}^n$ was determined by Wolff [4436] in 1905 and rediscovered by Takenouchi [3981, 3982] in 1913 and Nakagoshi [3047] in 1979. It can be also deduced from Hensel's results on units in $\mathfrak{p}$-adic fields [1788, 1789].

The question of the minimal norm $\nu_{\mathfrak{p}}$ of a primitive root mod $\mathfrak{p}$ has been considered in 1983 by Hinz, who first showed it to be $O(N(\mathfrak{p}^{1/2+\varepsilon}))$ for every $\varepsilon > 0$ [1856], applying a generalization of

[11] Anders Wiman (1865–1959), professor in Uppsala. See [3029].

[12] Vera Myller-Lebedeff (1880–1970), professor in Iaşi.

the Pólya–Vinogradov inequality to algebraic number fields, and then used Burgess' method [518] of evaluating character sums to get $\nu_\mathfrak{p} = O(N(\mathfrak{p}^{1/4+\varepsilon}))$ [1857]. Later [1858] he showed

$$\sum_{N\mathfrak{p}\leq x} \nu_\mathfrak{p} = O\left(x \log^2 x (\log\log x)^a\right) ,$$

with $a = 2(r_1(K) + r_2(K) + 1)$ (see also [1860]). The similar problem for $\mathfrak{p}^2$ was considered by him in [1861, 1862].

Under Generalized Riemann Hypothesis these bounds can be essentially improved. Generalizing a theorem of Ankeny[13] and Chowla[14] [93, 95], who considered the rational case, T.Z. Wang and Gong [4292] showed in 2010 (improving upon Y. Wang and C. Bauer [4296]) that GRH implies the bound

$$\nu_\mathfrak{p} = O_K\left((m \log m)^4 \log^2 N(\mathfrak{p})\right) ,$$

m denoting the number of prime factors of $N(\mathfrak{p} - 1)$.

2.1.3 *After the Zahlbericht*

1. The main results of [1833] about quadratic extensions of $\mathbf{Q}(i)$ were extended in 1899 by Hilbert [1838, 1839] to quadratic extensions of arbitrary totally imaginary algebraic number field having odd class-number. He established there the quadratic reciprocity law for the fields considered, and presented the theory of the norm residue symbol for such extensions with a direct definition of the norm residue symbol. If $\beta \in \mathbf{Z}_k$ is not a square in k and $K = k(\sqrt{\beta})$, then for $\alpha \in \mathbf{Z}_K$ he put

$$\left(\frac{\alpha, \beta}{\mathfrak{p}}\right) = \begin{cases} 1 & \text{if } \beta \text{ is a norm residue mod } \mathfrak{p}, \\ -1 & \text{otherwise.} \end{cases} \tag{2.5}$$

If β is a square in k, then one puts

$$\left(\frac{\alpha, \beta}{\mathfrak{p}}\right) = 1 .$$

After developing the main properties of this symbol and constructing the theory of genera in the considered fields Hilbert established (Theorem 60 of [1839]) the equality (2.4) and showed (Theorem 65) that if $\beta \in \mathbf{Z}_K$ is not a square, and $K = k(\sqrt{\beta})$, then $\alpha \in \mathbf{Z}_K$ lies in $N_{L/K}(L^*)$ if and only if it is a norm residue for every prime ideal $\mathfrak{p}$. This in turn is equivalent to the solvability of the equation

$$\alpha x^2 + \beta y^2 = 1$$

in k, as well as to the solvability of the congruence

[13] Nesmith Corbett Ankeny (1927–1993), professor at M.I.T.

[14] Sarvadaman Chowla (1907–1995), professor in Delhi, Benares, Waltair, Lahore, at the University of Kansas, University of Colorado and Pennsylvania State University. See [159].

$$\alpha x^2 + \beta y^2 \equiv 1 \pmod{\mathfrak{p}^n}$$

for every $n \geq 1$ (the equivalence of the two last conditions was established by Hilbert two years earlier in [1835], where he determined all polynomials $f \in \mathbf{Z}[X]$ with discriminant ± 1).

In the case when k has real conjugates $k_1 = k, k_2, \dots, k_r$ Hilbert stated in [1838] without proof the quadratic reciprocity law in the form

$$\prod_{\mathfrak{p}} \left(\frac{\alpha, \beta}{\mathfrak{p}} \right) = (-1)^t, \tag{2.6}$$

where t denotes the number of fields k_i in which the corresponding conjugates of α, β are both negative.

A simplification of Hilbert's proof of the quadratic reciprocity law in this case has been provided in 1915 by Mayr [2793].

It has been shown later (see Hasse [1664]) that the equality (2.6) can be written in the form (2.4), taking in account also infinite primes, responsible for the term $(-1)^t$ in (2.6).

The results of [1838] were extended by Hilbert to the case of quadratic extensions of arbitrary base fields k with odd class-number in [1837, 1843]. In particular the proof of (2.6) has been given in the case $h(k) = 1$, with a remark that only small changes are necessary to prove this result for fields with odd $h(k)$.

At the end of [1843] Hilbert stated several assertions concerning arbitrary Abelian extensions. We shall discuss the fate of these assertions in Sect. 2.3. Hilbert pointed out that the truth of these assertions would lead to the quadratic reciprocity law for arbitrary algebraic number fields The first proof of this law has been given in 1923 by Hecke in his book [1742] (see Sect. 4.2.1).

2. In 1899 Hilbert published a book on the foundation of geometry [1841]. In it one finds the assertion that totally positive numbers in an algebraic number field K are sums of four squares of elements of K. Hilbert wrote "*Der*[15] *Beweis dieses Satzes bietet erhebliche Schwierigkeiten dar; er beruht wesentlich auf der Theorie der relativquadratischen Zahlkörper...*", but gave no proof, indicating only that it is based on a condition for solving the equation $\alpha X^2 + \beta Y^2 + \gamma Z^2 = 0$ in algebraic number fields.

The first steps towards a proof were made by Meissner who showed in 1903–1905 [2810–2812] that in the field $\mathbf{Q}(i)$ every irrational element is a sum of two squares, and in other quadratic fields every totally positive element is a sum of 5 squares. He obtained also bounds for the number of square summands needed to represent totally positive elements in pure extensions of $\mathbf{Q}$ having an odd degree.

For further development see Sect. 4.2.2.

[15] *"The proof of this theorem presents considerable difficulties; it depends essentially on the theory of relative quadratic extensions".*

3. During the International Congress of Mathematicians held in Paris in August 1900 Hilbert gave a talk [1842] in which he presented 23 mathematical problems. Some of them were taken from the theory of algebraic numbers.

The last part of the eighth problem asked for a generalization of results concerning the distribution of primes to the case of prime ideals in a fixed algebraic number field. For the first steps in this problem see Sect. 3.1.1.

The ninth problem dealt with the reciprocity law for lth powers (with prime l) in an arbitrary algebraic number field, and in the 11th problem Hilbert considered the theory of quadratic forms in any finite number of variables in algebraic number fields (see Sects. 4.3.3 and 5.4.1).

In the 12th problem he proposed to generalize the Kronecker–Weber theorem on Abelian extensions of the rationals to the case when the base field is an arbitrary algebraic number field (see Sect. 2.3.1).

For a survey of the influence of Hilbert's problems see the books [460] and [58]. The story of the ninth problem has been presented by Tate [4009], and the development around the twelfth problem was the subject of a paper of Langlands [2469].

Hilbert's research concerning class-fields is presented below in Sect. 2.3.

2.2 Kurt Hensel

2.2.1 Field Index and Monogenic Fields

1. The first result of Hensel contained in his thesis [1763] answered a question of Kronecker concerning prime divisors of the index of an algebraic number field.

If α is an integer of an algebraic field K, then the ring $\mathbf{Z}[\alpha]$ generated by α is a subring of $\mathbf{Z}_K$ having a finite index $i(\alpha) = [\mathbf{Z}_K : \mathbf{Z}[\alpha]]$. If $d(\alpha)$ denotes the discriminant of the minimal polynomial of α and $d(K)$ is the discriminant of K, then

$$d(K) = d(\alpha)i^2(\alpha) ,$$

a formula established by Dedekind in §175 of [848].

The greatest common divisor $i(K)$ of indices of elements of the ring of integers of a field K is called the *field index* (in the older literature it was called the *common non-essential discriminant divisor*, "ausserwesentlicher Discriminantenteiler" in German).

The first study of the index $i(\alpha)$ was made by Dedekind [843] in 1878. He showed there a simple way to determine the factorization of a rational prime p in K in the case when for some $\alpha \in \mathbf{Z}_K$, generating K one has $p \nmid i(\alpha)$ (see (1.26)), and gave also the following condition for the divisibility of $i(\alpha)$ by p:

If $F(X) \in \mathbf{Z}[X]$ *is the minimal polynomial for* α,

$$F(X) \equiv \prod_{j=1}^{g} P_j^{e_j}(X) \pmod{p}\,.$$

with $P_1, \dots, P_g$ being polynomials irreducible and distinct mod p, then put

$$M(X) = \frac{1}{p}\left(F(X) - \prod_{j=1}^{g} P_j^{e_j}(X)\right)\,.$$

One has $p \mid i(\alpha)$ if and only if $M(X)$ is divisible by some $P_i(X)$ with $e_i \geq 2$.

In 1977 Uchida [4097] reformulated this condition in the following way:

One has $p \mid i(\alpha)$ if and only if the minimal polynomial of α is contained in the square of a maximal ideal of $\mathbf{Z}[X]$.

For generalizations of Dedekind's result see Albu [55], Charkani and Deajim [656], del Corso [865], Eršov [1123], Khanduja and Kumar [2142, 2143].

Dedekind proved also ([843], §4) the equivalence of the two following conditions for prime p:

(i) *There exists $\alpha \in \mathbf{Z}_K$ with $p \nmid i(\alpha)$.*

(ii) *If $f_1, f_2, \dots, f_g$ are the degrees of prime ideals dividing $p\mathbf{Z}_K$, then for each $j = 1, 2, \dots, g$ there exists a polynomial $F_j(X) \in \mathbf{Z}[X]$ of degree f_j, irreducible mod p, such that the polynomials $F_1, \dots, F_g$ are pairwise incongruent mod p.*

He used this result to show ([843], §5) that in the field generated by a root of the polynomial $X^3 - X^2 - 2X - 8$ the indices $i(\alpha)$ have a non-trivial common divisor (they all are even).

2. In 1882 Kronecker [2294] considered the indices of subfields of the 13th cyclotomic field. He proposed later to Hensel the study of this question in his thesis [1763], presented to the faculty in 1884. Hensel showed there that if p is a prime unramified in $K/\mathbf{Q}$, $F(X)$ is the minimal polynomial for an element generating K,

$$F(X) \equiv \prod_{j=1}^{s} F_j(X) \pmod{p}$$

holds with polynomials $F_j(X)$ irreducible mod p, and λ_r denotes the number of F_j's having degree r, then p divides $i(K)$ if and only if for at least one r the inequality

$$\lambda_r > \frac{1}{r}\sum_{d|r} \mu(d) p^{r/d} \tag{2.7}$$

holds.

The right-hand side of (2.7) equals the number of irreducible polynomials over $\mathbf{F}_p$ having degree f.

3. To the problem of the field index returned Hensel [1769, 1771] in 1894. He showed that the set of indices of integers of a field of degree d coincides with the set of absolute values of integers represented by a form of degree $d(d-1)/2$ in $d-1$ variables. It is called the *index form*.

In [1769] Hensel used (2.7) to show that if K is a cyclic cubic field of prime conductor p (i.e. K is a subfield of the cyclotomic field $\mathbf{Q}(\zeta_p)$), then $i(K) = 1$, except when p can be written in the form $p = x^2 + 32y^2$, in which case one has $i(K) = 2$.

The prime divisors p of $i(K)$ must be smaller than the degree d of K. The first explicit proof of this occurs in a paper of Żyliński[16] [4481] in 1913, but already in 1907 M. Bauer considered the inequality $p < d$ as evident in the paper [226] in which he showed that if p is a prime with $p < d$, then there is a field K of degree d with $p \mid i(K)$. He proved there also that every splitting prime $p < n$ divides $i(K)$. A proof of $p < d$ for primes $p \mid i(K)$ has been also given by Nagell (Theorem 2 in [3037]).

Jointly with Hensel's criterion this implies that for cubic fields K one has $i(K) \neq 1$ if and only if the prime 2 splits in K. Other conditions for $i(K) \neq 1$ in the case of cubic fields were given in 1914 by Levi [2579], who established a correspondence between cubic rings and cubic forms.

For the case of cubic fields see also Tornheim [4070], Nagell [3037] and Llorente and Nart [2614]. The paper of Nagell contains also a characterization of prime divisors of $i(K)$ for quartic K.

In 1933 Carlitz [570] presented a classification of Abelian fields with odd conductor and in certain cases determined the prime divisors of the field index.

In 1936 Bungers [515] showed in a simple way that every prime divides $i(K)$ for a suitable field K and established the existence of infinite many quartic fields with $3 \mid i(K)$.

It has been shown in 1937 by M. Hall Jr.[17] [1596] that there exist pure cubic fields K with arbitrary large $\min_{\alpha \in \mathbf{Z}_K} i(a)$.

The first study of prime power decomposition of $i(K)$ has been done by Engstrom in 1930 [1099]. He showed that the exponent a in $p^a \parallel i(K)$ for fields of degree $d \leq 7$ is determined by the factorization of p in K, and this fails for fields of larger degrees, confirming a conjecture of Ore [3182]. In the case of splitting primes Engstrom gave an explicit formula for a valid for all fields. In 1982 Śliwa [3829] proved that if p is unramified, then a depends only on the type of factorization of p. Further progress was made by Nart [3074], who reduced the problem to a local question. His approach was utilized later by del Corso and Dvornicich [866–869] who showed i.a. that this method permits to determine the exponent a for tamely ramified Galois fields.

For quartic fields Engstrom [1099] proved that the field index is of the form $2^a 3^b$ with $a \leq 2$, $b \leq 1$. Later Gaál, Pethő and Pohst [1373] showed that every such value can be realized by infinitely many biquadratic fields $\mathbf{Q}(\sqrt{a}, \sqrt{b})$. They gave also algorithms for the minimal index in such fields. Later Pethő and Pohst [3269] showed that for any integer N the index of a normal field with Galois group C_2^k is divisible by N, provided k is sufficiently large.

4. An algebraic number field K is called *monogenic* if its ring of integers has a power integral basis, i.e. a basis of the form $1, \alpha, \alpha^2, \ldots, \alpha^{d-1}$, d denoting the degree of K. This happens if and only if the index form of K represents 1 or -1.

---

[16] Eustachy Żyliński (1889–1951), professor in Lwów and Gliwice.

[17] Marshall Hall Jr. (1910–1990), professor at the Ohio State University and CalTech.

If K is monogenic, then $i(K) = 1$, but the converse does not hold. The field $K = \mathbf{Q}\left(\sqrt[3]{175}\right)$ is a simple example of the case when $i(K) = 1$, but K is not monogenic, as its index form equals $5X^3 - 7Y^3$ and never assumes the values ± 1.

It is immediate that all quadratic fields are monogenic, and the monogenity of cyclotomic fields has been established by Kummer (for prime n in [2358], and for all n in [2364]). For other classes of fields monogenity seems to be rather an exception.

We list now the main results dealing with the problem of characterization of monogenic fields, obtained mostly in the last part of the century.

It has been proved by Győry in 1976 [1567] that if $f_K(x_1, \dots, x_{d-1})$ is the index form for a field K of degree d, then the equation

$$f_K(x_1, \dots, x_{d-1}) = a$$

can have only finitely many integral solutions and gave effective bounds for them. This implies that up to translations by rational integers there are at most finitely many $\alpha \in \mathbf{Z}_K$ with $\mathbf{Z}[\alpha] = \mathbf{Z}_K$, and a full set of representatives can be effectively determined. For improvements of these bounds see Győry and Papp [1576], Győry [1571], as well as two recent books [1149, 1150] by Evertse and Győry.

There are infinitely many monogenic cyclic cubic fields (Dummit, Kisilevsky [1027]; criteria were given by M.-N. Gras [1494, 1495]) as well as non-cyclic cubic fields (Spearman and K.S. Williams [3867]). The same is true for biquadratic fields (Nakahara [3049]; a criterion was given by M.-N. Gras and Tanoé [1501]), pure quartic fields (Funakura [1323]), quartic fields whose Galois closure has the dihedral Galois group (Gaál [1362], Huard, Spearman, K.S. Williams [1914]; a criterion was given by Kable [2086]; cf. also Gaál, Nyul [1370]) and quartic fields whose Galois closure has the alternating group A_4 as Galois group (Gaál [1362], Spearman [3865]). A criterion in the case of cyclic quartic fields was given by M.-N. Gras [1498]. She showed also that there are only two such imaginary monogenic fields, and Nakahara [3048] produced infinitely many cyclic quartic fields with $i(K) = 1$ but not monogenic.

In 2012 Kedlaya [2133] showed that there exist infinitely many monogenic fields of any signature, and in 2016 Bhargava, Shankar and X. Wang [342] proved that the number of monogenic fields K of degree n with Galois group S_n and $|d(K)| \leq x$ exceeds $c(n)x^{1/2+1/n}$ with some $c(n) > 0$.

In 1925 Hancock [1618] showed that the maximal real subfields of $\mathbf{Q}(\zeta_n)$ with odd n are monogenic. In the general case this has been made in 1976 by J.J. Liang [2591]. Another proof has been given by Yamagata and Yamagishi [4449].

M.-N. Gras [1500] showed in 1986 that if $p \geq 5$ is a prime, and K is a monogenic cyclic field of degree p, then $q = 2p + 1$ is a prime, and K is the maximal real subfield of $\mathbf{Q}(\zeta_q)$. A similar result for cyclic extensions of degree $2p$ ($p \geq 5$, prime) was obtained two years later by Cougnard [775]. More generally, if $d \geq 4$, then there are only finitely many monogenic Abelian extensions of degree d prime to 6 (M.-N. Gras [1499]). The last paper contains also a determination of all monogenic Abelian fields of prime conductor.

Monogenity of relative extensions has been studied by Payan [3231], G. Gras [1490], Cougnard [776–778], Ph.Cassou-Noguès, M.J. Taylor [602–605], Fleckinger [1200, 1201], Cougnard, Fleckinger [782], Motoda, Nakahara, Shah [2985], Kable [2086] and Gaál and Nyul [1370].

In 1988 Bremner [441] conjectured that in the fields $K = \mathbf{Q}(\zeta_p)$ with prime p every power base is generated by an element conjugated to some element equivalent either to ζ_p or to $\sum_{j=1}^{(p-1)/2} \zeta_p^j$, two numbers being treated as equivalent if their difference is a rational integer. He established his conjecture for $p = 7$. In the case $p = 5$ this was showed earlier by Nagell [3038]. A criterion for checking this conjecture for regular primes was given by Robertson [3478] who applied it successfully for $p = 11, 13, 19, 23$. For numerical results for $p \leq 100$ see Miller-Sims and Robertson [2864]. In 2001 Robertson [3479] considered the analogous question for the fields $\mathbf{Q}(\zeta_{2^m})$, with $m \geq 2$, and showed that the generator of a power basis in these fields is equivalent to ζ_m.

In 2010 Ranieri [3382] proved that if K is an imaginary Abelian field, with conductor prime to 6, having a power integral basis $1, \alpha, \ldots$, then either α is equivalent to a root of unity (two numbers α, β being called equivalent if one has $\pm\beta = \alpha + c$ with $c \in \mathbf{Z}$), or the sum $\alpha + \overline{\alpha}$ is an odd rational integer, and Robertson [3480] showed that the same holds also for $K = \mathbf{Q}(\zeta_n)$, where $n = 3d, 4d$ and $(d, 6) = 1$. This was earlier known for cyclotomic fields $\mathbf{Q}(\zeta_q)$ with prime power $q = p^m$ satisfying $(h_p^+, p(p-1)/2) = 1$ (Gaál, Robertson [1377]) , and the assumption about the class-number was removed by Ranieri [3381] in 2008.

The determination of all power integral bases in cubic fields was reduced in 1989 by Gaál and Schulte [1378] to a Thue equation. Using this they computed all such bases for cubic fields with discriminants between -300 and 3137. A similar approach works also for quartic fields (Gaál, Pethő, Pohst [1371, 1372, 1374–1376], Koppenhöfer [2214]). For extensions of larger degree the approach via Thue equations does not work, but then the problem can be reduced to unit equations. This has been applied for quintic fields by Gaál and Győry [1369]), for certain classes of sextic fields by Gaál [1363], Járási [2036] and Bilu, Gaál and Győry [355]) and for nonic fields by Gaál [1366]. The case of relative cubic extensions was treated by Gaál [1367] in 2001.

The book of Gaál [1368] as well as Chap. 7 of the book [1150] by Evertse and Győry is devoted to power integral bases. For surveys see Gaál [1364, 1365].

Let $g(K)$ denote the minimal number of generators of the ring $\mathbf{Z}_K$. Monogenic fields are characterized by the equality $g(K) = 1$. A method of determining $g(K)$ was provided by Pleasants[18] [3310] if $g(K) > 2$, and by Győry ([1568, 1570], cf. p. 120 of [1572]) in the general case. Another proof has been given in 2015 by Kravchenko, Mazur and Petrenko [2274] who also provided a generalization to finite-dimensional algebras over number fields.

An algorithm determining a minimal set of generators, based on results of Kravchenko, Mazur and Petrenko [2273], can be found in Chap. 11 of [1150].

2.2.2 Discriminants

1. In [1766, 1773] considered Hensel integral bases and discriminants of fields which are composites of two fields and determined the prime power divisors of their discriminants for primes which are not wildly ramified (see also Wahlin[19] [4263]). He observed also that if K is the composite of the fields K_1 and K_2 having co-prime discriminants, then $\deg K = \deg K_1 \deg K_2$, and if $a_1, \ldots, a_m$ is an integral basis of K_1 and $b_1, \ldots, b_n$ is an integral basis of K_2, then the set $\{a_i b_j\}$ is an integral basis of K. It follows from a result of Hensel on determinants [1767] that in this case one has $d(K) = d(K_1)^n d(K_2)^m$.

A simpler proof of Hensel's results was provided in 1921 by M. Bauer [236]. In 1927 Ore [3181] generalized this to the case when the prime p is wildly ramified in at most one of the fields K_1, K_2, assuming that the degree of the composite field equals the product of the degrees of K_1 and K_2.

In 1894 Hensel [1768] applied Kronecker's method to obtain a simpler proof of Dedekind's discriminant theorem. He showed also that if $K/\mathbf{Q}$ is an extension of degree n, a rational prime p has the factorization

[18] Peter Arthur Barry Pleasants (1939–2008), lecturer in Cardiff, the University of New England, Macquarie University and the University of the South Pacific.

[19] Gustav Eric Wahlin (1880–1948), professor at the University of Missouri.

$$p\mathbf{Z}_K = \prod_{j=1}^{g} P_j^{e_j}$$

with $p \nmid e_1 \dots e_g$ (the tame case), and the norm of the product $P_1 P_2 \dots P_g$ equals p^r, then the maximal power of p dividing the discriminant $d(K)$ equals p^{n-r}. He obtained this by considering the linear form $w(x_1, \dots, x_n) = \sum_{j=1}^{n} x_j \omega_j$ with $\omega_1, \dots, \omega_n$ being an integral basis of K, and looking for the smallest degree of a polynomial

$$F(t) = t^m + u_{m-1}(x_1, \dots, x_n)t^{m-1} + \dots + u_0(x_1, \dots, x_n)$$

(with u_j being forms over $\mathbf{Z}$) satisfying

$$F(w(x_1, \dots, x_n)) \equiv 0 \pmod{P}$$

for a prime ideal P. His main tool was the assertion that if $w_1 = w, w_2, \dots, w_n$ are forms conjugated to w, and $V(x_1, \dots, x_n)$ denotes their Vandermonde determinant, then

$$V^2(x_1, \dots, x_n) = d(K)\Delta(x_1, \dots, x_n) ,$$

where Δ is a form over $\mathbf{Z}$ whose coefficients do not have any common integral divisor >1. A weaker form of this result was obtained earlier by Kronecker ([2294], §25). Later Hensel [1770] presented a simpler proof.

In [1769] Hensel proved that if $\omega_1, \dots, \omega_n$ is an integral basis of K,

$$F(X; t_1, \dots, t_n) = N\left(X - \sum_{j=1}^{n} t_j \omega_j\right)$$

and

$$p\mathbf{Z}_K = \prod_{j=1}^{r} \mathfrak{p}_j^{e_j} ,$$

then there exist unique polynomials $P_1, \dots, P_r \in Z[X; t_1, \dots, t_n]$ which are monic and irreducible mod p such that

$$F \equiv \prod_{j=1}^{n} P_j^{e_j} \pmod{p} .$$

In 1919 M. Bauer [232] established that the same congruence holds with suitable polynomials also with respect to arbitrary powers of the prime p.

2.2.3 p-Adic Numbers

1. In the first paper of Hensel published in Crelles Journal [1764] Kronecker's theory of forms was used to study the set of residue classes with respect to a prime ideal $\mathfrak{p}$ in an algebraic number field, creating essentially a part of the theory of finite fields. In his next paper [1765] he proved a result which translated in modern terminology asserts that every finite field has a normal basis over the prime field $\mathbf{F}_p$ (this assertion has been stated without proof already in 1850 by Eisenstein on p. 622 of [1076]. See also Schönemann [3683]).

In 1899 Hensel [1772] announced a new approach to the theory of algebraic numbers and published the details three years later [1774]. Motivated by power series expansions in the theory of algebraic functions of a complex variable he associated with every prime ideal $\mathfrak{p}$ and each non-zero number α of a fixed algebraic number field K a sequence of congruences. Let π be a fixed element of $\mathbf{Z}_K$, divisible by $\mathfrak{p}$ but not by $\mathfrak{p}^2$, and let Ω be a fixed complete set of residues mod $\mathfrak{p}$ containing 0. If $\alpha \in \mathfrak{p}^r \setminus \mathfrak{p}^{r+1}$, then Hensel wrote

$$\alpha \equiv c_r \pi^r \pmod{\mathfrak{p}^{r+1}},$$

with non-zero $c_r \in \Omega$, and continued

$$\alpha \equiv c_r \pi^r + c_{r+1}\pi^{r+1} \pmod{\mathfrak{p}^{r+2}},$$

$$\dots \quad \dots \quad \dots$$

$$\alpha \equiv c_r \pi^r + c_{r+1}\pi^{r+1} + \cdots + c_m \pi^m \pmod{\mathfrak{p}^{m+1}},$$

$$\dots \quad \dots \quad \dots$$

He used to present this sequence in the form of a power series in π,

$$\alpha = \sum_{j=r}^{\infty} c_j \pi^j, \tag{2.8}$$

disregarding the question of its convergence[20]. This approach permitted him to obtain new proofs of his results about the field discriminant and different obtained in [1770] and [1771]. He showed in particular, confirming the belief of Dedekind expressed at the end of [845], that if $\mathfrak{p}$ is a prime ideal in K lying over the prime p, its ramification index e equals $p^s q$ with $p \nmid q$, and the different of $K/\mathbf{Q}$ is exactly divisible by $\mathfrak{p}^r$, then $r < e(s+1)$.

---

[20]In a later paper Hensel [1779] tried, rather unsuccessfully, to reconcile the usual convergence of series with $\mathfrak{p}$-adical convergence.

Other proofs of the last result were given later by M. Bauer [231, 235, 239] and Ore [3175] (see also M. Bauer [242] and Ore [3179]). An exact formula for r was given in 1926 by Ore [3177], who also gave a necessary condition for the existence of a field K with given $r \in [1, e(s+1)]$. A generalization to relative extensions has been provided by Ore in [3176].

Note that in [1774] only elements of K were developed in such $\mathfrak{p}$-adic series (2.8), and there is no mention of series corresponding to elements (algebraic or not) lying outside K. These general series appear for the first time in the paper [1775] published in 1904, containing the definition of the p-adic field $\mathbf{Q}_p$ (denoted by Hensel by $K(p)$) formed by formal series $\sum_{n=r}^{\infty} c_n p^n$ with $r \in \mathbf{Z}$, its integers and units and an exposition of the involved arithmetics. Hensel showed i.a. that a p-adic number is rational if and only if the coefficients of its series form a periodic sequence from some point on. In the final part of that paper the theory of polynomials and rational functions in one variable with p-adic coefficients is developed, culminating in the theorem on the unique factorization of polynomials[21]. One finds there also the proof of the first version of *Hensel's lemma* in the following form:

Let $F(X)$ be a polynomial with integral p-adic coefficients, and let p^δ be the highest power of p dividing the discriminant of F. The polynomial F is reducible in the p-adic field if and only if there exist polynomials f, g with

$$F(X) \equiv f(X)g(X) \pmod{p^{1+\delta}} .$$

In the book [1780] (p. 71) gave Hensel another version of that lemma:

If $F(X), f_0(X), g_0(X) \in \mathbf{Z}_p[X]$, $p^\varrho \parallel \mathfrak{R}(f_0, g_0)$ (where $\mathfrak{R}$ denotes the resultant), $n > 2\varrho$ and

$$F(X) \equiv f(X)g(X) \pmod{p^n} ,$$

then

$$F(X) = f(X)g(X) ,$$

with $f, g \in \mathbf{Z}_p[X]$, $\deg f = \deg f_0$, $\deg g = \deg g_0$ and f_0, g_0 are congruent to f and g, respectively, mod $p^{n+1-\varrho}$.

A simplification of the proof was presented in 1910 by Dickson [932].

Note that the following elementary version of Hensel's lemma has been established already in 1846 by Schönemann [3682]:

If $F(X), f(X), g(X) \in \mathbf{Z}[X]$ are monic and

$$F(X) \equiv f(X)g(X) \pmod{p} ,$$

then for every $n \geq 1$ there are polynomials $f_n(X), g_n(X) \in \mathbf{Z}[X]$ with

$$F(X) \equiv f_n(X)g_n(X) \pmod{p^n}$$

and

$$f_n(X) \equiv f(X) \pmod{p}, \quad g_n(X) \equiv g(X) \pmod{p} .$$

[21] Hensel's proof gives actually the unique factorization property for polynomials in one variable over an arbitrary field.

It has been pointed out by G. Frei (p. 185 of [1242]) that a version of Hensel's lemma has been known already to Gauss.

2. One of the first applications of Hensel's theory has been made in 1906 by Dumas[22] [1024], who studied polynomials with p-adic coefficients and applied p-adic numbers and Newton polygons to the proof of certain irreducibility criterias for polynomials, for example the following extension of the Eisenstein–Schönemann criterion:

If $f(X) = \sum_{j=0}^{n} a_j X^j \in \mathbf{Z}[X]$ *and* p *is a prime satisfying* $p \nmid a_0$,

$$\frac{v_p(a_i)}{i} > \frac{v_p(a_n)}{n} \quad (i = 1, 2, \ldots, n-1),$$

and if $p \mid a_n$, *then* $(v_p(a_n), n) = 1$, *then* f *is irreducible over* $\mathbf{Q}$.

For a recent generalization see Bonciocat [392].

3. The results of [1775] were extended by Hensel to number fields in [1776], where he associated with a field K and a suitable element π the field of formal series of the form

$$\sum_{j=n}^{\infty} \alpha_j \pi^j$$

with $n \in \mathbf{Z}$ and α_j taken from a set of residues mod π of integers of K. He showed that the resulting field L is a finite extension of the p-adic field, as defined in [1775], and proved that every non-constant polynomial over L has a root in a suitable finite extension.

As an application Hensel gave in [1777] simple proofs of the determination of the sign and quadratic character mod p of field discriminants, given earlier by Pellet [3233] and Stickelberger [3934].

A survey of his theory presented Hensel [1778] on a meeting of DMV in Meran. One finds there the first mention of analytical methods in $\mathbf{Q}_p$, in particular the properties of the p-adic[23] exponential function $\exp_p(x)$. These properties were developed by Hensel in Chap. 7 of his book [1783], and its analogue in $\mathfrak{p}$-adic fields has been considered later in [1785].

Hensel defined the p-adic exponential function $\exp_p(x)$ in $\mathbf{Q}_p$ as the unique function $E(x) \neq 0, 1$ satisfying $E(x+y) = E(x)E(y)$ and having a convergent power series

$$\exp_p(x) = 1 + x + \sum_{n=2}^{\infty} a_n x^n$$

[22]Gustave Dumas (1872–1955), professor in Zürich and Lausanne. See [896].

[23]The name "p-adic numbers" seems to appear for the first time in Hensel's paper [1784], in which he considered quadratic extensions of $\mathbf{Q}_p$.

in a neighbourhood of $x = 0$. He showed that one has $a_n = 1/n!$ and noted that for odd p this series converges for x divisible by p, whereas in the case $p = 2$ this condition has to be replaced by $2^2 \mid x$.

He defined also in [1783] the p-adic logarithm by

$$\log_p(1+x) = \sum_{n=1}^{\infty} \frac{(-1)^{n+1}}{n} x^n ,$$

convergent for x divisible by p for odd p, and by 2^2 for $p = 2$, and established the equality

$$\exp_p(\log_p(1+x)) = 1 + x$$

for these x.

Since the series of the exponential function $\exp_p(x)$ in $\mathbf{Q}_p$ does not converge at $x = 1$, there is no analogue of the number e in this case. In 1925 Wahlin [4269] defined a $\mathfrak{p}$-adic number e as the root of the equation

$$x^p = \sum_{j=0}^{\infty} \frac{p^j}{j!} .$$

Some properties of the p-adic logarithm were studied by Disse [974, 975]. It has been pointed out by Hasse in the review of [974] in the Jahrbuch that the description of roots of unity in $\mathfrak{p}$-adic fields given there is not correct.

In 1932 Mahler[24] [2694] proved that for a p-adic number α the numbers α and $\exp_p(\alpha)$ cannot be both algebraic over $\mathbf{Q}$.

In 1908 Hensel published his first book[25] [1780] on p-adic and $\mathfrak{p}$-adic numbers. He showed there a.o. that every element of the rational p-adic field can be written in the form $\varepsilon w p^n$, where w is a root of unity of order dividing $p-1$ and ε is a principal unit, i.e. $\varepsilon \equiv 1 \bmod p$ (the analogous result for $\mathfrak{p}$-adic fields he proved later in [1785]). In the introduction[26] he promised to study in the second volume the size of p-adic numbers, having in mind future applications to the theory of algebraic numbers, but this promise was never fulfilled. It is unclear what kind of size he had in mind.

In his book presented Hensel a way of using Newton's approximation method to the solution of p-adic equations (a simplification of this approach was obtained in 1923 by Rella[27] [3435]). Hensel showed also that every extension can be obtained by making first an unramified extension and then a fully ramified extension, the latter generated by a root of Eisensteinian equation.

Hensel's second book [1783], published in 1913, contained an elementary presentation of the theory of p-adic numbers.

---

[24] Kurt Mahler (1903–1988), professor in Manchester and Canberra. See [405, 598, 725, 4128].

[25] It was planned to have two volumes, but only the first volume appeared.

[26] [1780], p. VII.

[27] Tonio Rella (1888–1945), professor in Wien and Graz.

In a series of later papers Hensel modified and extended his theory. So in [1781] he constructed $\mathfrak{p}$-adic fields, defined as extensions of p-adic fields generated by a root of an irreducible polynomial. He defined $\mathfrak{p}$-adic integers and units, showed that every non-zero element is of the form $\eta\pi^a$, where π is a fixed prime element and η is a unit, and defined the ramification index. The main result of this paper states that there are only finitely many extensions of a given degree of the field $\mathbf{Q}_p$. He wrote in a footnote on p. 186 that the results of Steinitz[28] influenced his work on the subject of his paper.

In [1784] presented Hensel details of his theory in the case of quadratic fields. The same approach to arbitrary fields was used in 1915 by Wahlin [4265].

In [1785–1788] Hensel generalized and made more precise his result about multiplicative presentation of elements of a p-adic field. He called 1-units ("*Eins-Einheiten*") the units congruent to 1 mod $\mathfrak{p}$ (now they are called *principal units*), and showed that every non-zero element of a $\mathfrak{p}$-adic field K containing $\mathbf{Q}_p$, can be written in the form

$$\pi^a w^b \prod_{j=1}^{m} \eta_j^{A_j}, \tag{2.9}$$

where π is a fixed prime element of K, m is the degree of $[K : \mathbf{Q}_p]$, w is the generator of the group of roots of unity in K, $\eta_1, \dots, \eta_m$ are fixed principal units, and $A_1, A_2, \dots, A_m$ are p-adic integers.

The definition of exponentiation η^A for principal units η and $A \in \mathbf{Z}_p$ has been given by Hensel in [1785, 1786] in a rather cumbersome way using the exponential function, but in [1787] he observed that for $A = \sum_{j=0}^{\infty} a_j p^j$ one can put simply

$$\eta^A = \lim_{n\to\infty} \eta^{A_n}$$

with A_n being the nth partial sum of the series for A.

The equality (2.9) implies that the group $U_1(K)$ of principal units is a free $\mathbf{Z}_p$-module with $[K : \mathbf{Q}_p]$ generators.

In the case when $K/\mathbf{Q}_p$ is a normal extension certain simplifications to Hensel's theory were introduced by Rella [3433], who also studied the action of the inertia group on the group U_k formed by elements $\eta \equiv 1 \bmod \mathfrak{p}^k$. For the structure of the group of principal units see also the papers of Wahlin [4266, 4271]. For further research on this topic see Sect. 5.4.6.

In 1918 Hensel [1791] considered extensions of $\mathfrak{p}$-adic fields as factor rings of polynomial rings. He showed i.a. that an element of such extension is integral if and only if it has an integral norm and proved also that every finite Galois extension of a $\mathfrak{p}$-adic field has a solvable Galois group.

In 1916 Hensel [1788, 1789] determined the structure of the group $G(\mathfrak{p}^n)$ of reduced residue classes mod $\mathfrak{p}^n$ for an arbitrary prime ideal $\mathfrak{p}$ and $n = 1, 2, \dots$, and

[28] Ernst Steinitz (1871–1928), professor in Breslau and Kiel. See [3507].

used this result in [1790] to determine the solvability of binomial equations $X^n = a$ in $\mathfrak{p}$-adic fields. The case of prime n has been earlier studied by Wahlin [4264].

The factorization of primes in Kummerian fields $\mathbf{Q}\left(\zeta_p, a^{1/p}\right)$ (with $a \in \mathbf{Q}(\zeta_p)$) was determined already by Kummer ([1836, 2369], §128), but Hensel showed in [1792, 1793] how to prove it with the use of $\mathfrak{p}$-adic numbers. His method was extended by Rella [3434] to the case of extensions $K(a^{1/p})/K$ in the case, when K does not contain the pth roots of unity.

4. In 1912 Fraenkel[29] [1225] presented a rather complicated axiomatical approach[30] to p-adic numbers, and a few years later Hensel's construction of these numbers was put in a more general setting by Kürschak[31] [2387] who introduced the notion of a *valuation* in an arbitrary field K, and defined as a nonnegative real-valued function $v(x)$ on K, obeying the following three conditions:

(a) $v(x) = 0$ holds if and only if $x = 0$.

(b) $v(xy) = v(x)v(y)$,

(c) $v(x \pm y) \leq v(x) + v(y)$.

If v satisfies the condition

(d) $v(x \pm y) \leq \max\{v(x), v(y)\}$,

then it is called a *non-Archimedean valuation*.

Kürschak showed in particular that if the extension L/K is finite, then every valuation of K can be extended to L, and another proof was given in 1924 by Rychlik [3542].

Kürschak's paper is now regarded as the beginning of valuation theory.

5. In 1916 Rychlik [3537] showed how one can introduce p-adic numbers in the same way as G. Cantor[32] [560] did for the reals by completing the field of rational numbers with respect to the p-adic valuation. In that paper one finds also a generalization of valuations, the condition (b) in Kürschak's definition being replaced by

$$v(xy) \leq v(x)v(y).$$

This notion, called now *pseudo-valuation*, has been rediscovered several years later by Mahler [2700–2703], who described all such functions v in the case of algebraic number fields and their rings of integers.

In 1923 Rychlik [3540] gave a construction of a continuous function without derivative in $\mathbf{Q}_p$, and in [3542] presented a version of Hensel's lemma.

6. The question whether the algebraic closure of $\mathbf{Q}_p$ is complete, posed by Kürschak ([2387], p. 217), got a negative answer when Ostrowski[33] [3197, 3198] proved that a separable extension L/K of a complete field K is complete if and only if its degree is finite. In his next paper [3199] he described all valuations of

---

[29] Adolf Fraenkel (1801–1965), professor in Marburg, Kiel and Jerusalem.

[30] Roquette wrote in [3502] that "Fraenkel's paper is completely forgotten today".

[31] József Kürschak (1864–1933), professor at the Technical University in Budapest

[32] Georg Cantor (1845–1918), professor in Halle. See [3346].

[33] Alexander Ostrowski (1893–1986), professor in Basel. See [1061, 2045].

the rational field and all archimedean valuations in arbitrary fields, observing that in any field a non-archimedean valuation induces a prime divisor. The first complete proof of the description of non-archimedean valuations of algebraic number fields has been given in 1932 by Artin [125].

In 1924 Rychlik [3542] showed that if K is an algebraically closed valued field, then its completion is also algebraically closed. Rella [3436] considered valuations of arbitrary domains and applied them to study irreducibility of polynomials.

In 1934 Ostrowski [3201, 3202] developed the arithmetic properties of arbitrary valued fields, including a general theory of ramification groups for arbitrary extensions. In 1936 Mac Lane [2685] studied extensions of valuations in separable extensions of fields.

Valuations with values in an arbitrary group were introduced in 1932 by Krull[34] [2316].

The first book describing the valuation theory has been published in 1950 by Schilling[35] [3611]. Later appeared the books by Endler[36] [1089] in 1972, Ribenboim [3456] in 1999 and Engler and Prestel [1098] in 2005. Its history has been presented in the book [3502] by Roquette.

7. Hilbert's theory of ramification groups in the case of $\mathfrak{p}$-adic fields has been presented in 1922 by H. Bauer [238, 244]. He used it to give a new proof of Hensel's result [1791] showing that all Galois extensions of $\mathfrak{p}$-adic fields are solvable. In [241] M. Bauer showed that if $K/\mathbf{Q}$ is a finite extension, $\mathfrak{p}$ is a prime ideal in K lying over the prime p, and e is the ramification index of $\mathfrak{p}$, then the completion $K_\mathfrak{p}$ coincides with $K\mathbf{Q}_p$, and one has $[K_\mathfrak{p} : \mathbf{Q}_p] = ef$, where f is the degree of $\mathfrak{p}$. In 1924 he presented [243] a construction of the algebraic closure of $\mathbf{Q}_p$, and another construction has been proposed by Strassmann[37] [3939] in 1926.

Hilbert's theory of ramification groups has been extended to arbitrary fields with a valuation by Krull [2314, 2316, 2328, 2329] and Deuring[38] [906]. For ramification groups of infinite extensions see Sect. 5.4.5.

2.3 The Beginnings of Class-Field Theory

2.3.1 Kronecker's Jugendtraum

1. The idea of the class-field occurs for the first time in Kronecker's paper [2294] published in 1882. In Sect. 19 of it, devoted to applications of his general theory to the case of algebraic number fields, he recalled a result stated in his earlier paper [2283] showing that the theory of complex multiplication associates with every imaginary

---

[34]Wolfgang Krull (1899–1971), professor in Freiburg, Erlangen and Bonn. See [3078, 3663].

[35]Otto Schilling (1911–1973), professor at the Purdue University.

[36]Otto Endler (1929–1988), professor at the IMPA in Rio de Janeiro.

[37]Reinhold Strassmann (1893–1944), worked in an insurance company in Marburg. Killed in Auschwitz.

[38]Max Deuring (1907–1984), professor in Jena, Marburg, Hamburg and Göttingen. See [1060, 3501].

quadratic field $K = \mathbf{Q}(\sqrt{-m})$ an extension L/K having the propriety that its degree equals the class-number of K and moreover every ideal of K becomes principal in L. On basis of numerical examples he conjectured in 1862 [2285] that the discriminant of the extension L/K is a unit; i.e. L/K is unramified (see also [2295] where he studied cubic Abelian extensions of the field $K = \mathbf{Q}(\sqrt{-31})$ and found that the extension of K generated by a root of $X^3 - 10X + \sqrt{-31}(X^2 - 1)$ has these properties). In his book on elliptic functions and algebraic numbers [4316], published in 1891, H. Weber called the field L the "Classenkörper" (class-field) of determinant $-m$. This book, whose second addition appeared in 1898 as the third volume of the treatise [4321], contains a broad presentation of the theory of complex multiplication.

2. Applying the Kronecker–Weber theorem and Galois theory one can derive a description of all Abelian extensions of $\mathbf{Q}$. An analogous result in the case of an imaginary quadratic base field was first mentioned in Kronecker's paper [2291] and has been made more precise in the letter of Kronecker to Dedekind dated 15 March 1880 [2292]. He stated there that all Abelian extensions of an imaginary quadratic field can be generated by "*transformation equations of elliptic functions with singular moduli*" (this assertion is called usually "*Kronecker's Jugendtraum*[39]"). He expressed also his belief that he can prove this assertion, provided no new difficulties will occur. Kronecker's assertion was not very precise and has been understood in various ways. Hilbert in the statement of his 12th problem in [1842] interpreted it as the assertion that every Abelian extension of an imaginary quadratic field K is generated by values of the modular function $j(z)$ at elements of K. The function $j(z)$ occurring here classifies elliptic curves over the complex field and is defined for z in the upper half-plane by

$$j(z) = \frac{1728 g_2^3(z)}{\Delta(z)} = e^{-2\pi i z} + 744 + 196884 e^{2\pi i z} + \cdots ,$$

where

$$\Delta(z) = g_2^3(z) - 27 g_3^2(z) ,$$

where

$$g_2(z) = 60 \sum_{m=-\infty}^{\infty} \sum_{n \neq 0} \frac{1}{(mz+n)^4} ,$$

and

$$g_3(z) = 140 \sum_{m=-\infty}^{\infty} \sum_{n \neq 0} \frac{1}{(mz+n)^6} .$$

A simple example of Kronecker's assertion has been presented in 1903 by Mirimanoff [2885], who considered the extensions $\mathbf{Q}(\zeta_3, \sqrt[3]{m})/\mathbf{Q}(\zeta_3)$ with $m = -3k^2$.

---

[39] *Kronecker's Youth Dream.*

In 1885 H. Weber [4311] studied some aspects of the theory of elliptic functions and in Chap. 4 showed that to every class of properly primitive quadratic forms of discriminant $d < 0$ one can associate a value $j(\omega)$ of the modular function at some $\omega \in K$, which is an algebraic integer of degree $H(d)$. He called these values the "*class invariants*" ("Classeninvariante"). He studied these numbers in Chap. 2 of [4314] presenting some methods for their determination, and in [4315] considered the special case when there is only one form in each genus. These results have direct implications to the study of imaginary quadratic fields and their orders, presented by H. Weber in his book [4316].

An early survey of problems around "Kronecker's Jugendtraum" was prepared in 1900 by H. Weber [4326].

Hilbert's reformulation of Kronecker's assertion turned out to be incorrect, even if one would add suitable roots of unity to the generators. This has been noticed in 1914 by Fueter [1303], who earlier [1297–1299, 1301] presented an incorrect argument (based on his 1903 thesis [1296] written under Hilbert's supervision) in favour of the extended assertion. He presented the counterexample $\mathbf{Q}\left(\sqrt[4]{1+2i}\right)/\mathbf{Q}(i)$, noted that his previous proof is correct in the case of Abelian extensions of odd degree and formulated without great details the corresponding result for even degree, involving values of certain other complex functions. Another proof in the case of odd degree was provided by Takagi[40] [3975]. In 1926 Fueter published his book [1306], presenting in it his approach (see also Fueter [1307]).

Abelian extensions of $\mathbf{Q}(i)$ were described in the thesis of Takagi [3973] as fields generated by roots of certain polynomials related to the division of the lemniscate (as noted by Iwasawa in a footnote on p. 343 in [1996] the first lemma in §9 is false, but its use in the proof of the main result can be avoided). Another proof was provided in 1909 by Mertens [2828].

In 1916 Takenouchi [3983] showed that Abelian extensions of $\mathbf{Q}(\zeta_3)$ are subfields of the fields of division of Jacobi's elliptic function $sn(z,k)$, defined as the inverse function of

$$u(z,k) = \int_0^z \frac{dw}{\left((1-w^2)(1-k^2w^2)\right)^{1/2}},$$

defined for $0 < k^2 < 1$ (see Gudermann[41] [1537] and Chap. 22 of Whittaker[42] and Watson[43] [4395]).

These fields were also treated by Bindschedler [356] in 1923. The case of an arbitrary imaginary quadratic base field was finally settled by Takagi [3976] in 1920 as a consequence of the class-field theory (see Sect. 4.3.1).

Important simplification were introduced by Hasse [1657, 1669] in 1927. The final result can be described in the following way:

If K is an imaginary quadratic field, and $\mathbf{Z}_K$ denotes its ring of integers, then there exist elliptic curves $E = \mathbf{C}/\Lambda$ (where $\Lambda = \mathbf{Z} \oplus \mathbf{Z}\tau$ with $\tau \in \mathbf{Z}_K$ and $\Im\tau > 0$ is a lattice in the complex plane) having $\mathbf{Z}_K$ for its endomorphism ring. For every ideal class X in $\mathbf{Z}_K$ there exists such a curve E with Λ being a fractional ideal belonging to X. The value $j(\tau)$ depends only on the class X. If $X_1, X_2, \dots, X_h$ are all ideal classes of K, and for $i = 1, 2, \dots, h$ the numbers τ_i are chosen so that $\mathbf{Z} \oplus \mathbf{Z}\tau_i \in X_i$, then the numbers $j(\tau_i)$ are called the singular moduli. They are algebraic integers conjugated over K and generating the maximal unramified Abelian extension[44] *of K.*

---

[40] Teiji Takagi (1875–1960), professor in Tokyo. See [1891, 1996, 2897].

[41] Christoph Gudermann (1798–1852), professor in Münster, teacher of Weierstrass.

[42] Edmund Taylor Whittaker (1873–1956), professor in Dublin and Edinburgh. See [2795].

[43] George Neville Watson (1886–1965), professor in Birmingham. See [3383].

[44] Another proof of the last assertion has been given later by Eichler [1058].

If $K = \mathbf{Q}(\sqrt{-d})$ with square-free $d > 3$, then every Abelian extension of K is generated by some values of $j(z)$ and

$$\tau(z) = \wp(z)g_2(z)g_3(z)/\Delta(z) \tag{2.10}$$

(with $\wp(z)$ being the Weierstrass $\wp$-function attached to one of the elliptic curves corresponding to K) for $z \in K$. For $d = 1, 3$ the function $\tau(z)$ is to be replaced by $\wp(z)^2 g_2^2(z)/\Delta(z)$ in the case $d = 1$ and by $\wp(z)^3 g_3(z)/\Delta(z)$ in the case $d = 3$.

Some generalizations of Hasse's result were provided by Franz [1234] and Söhngen [3849]. In 1936 Sugawara [3962, 3963] showed that certain Abelian extensions of an imaginary quadratic field can be generated by one value of the function τ given by (2.10) (cf. Sugawara [3960]).

Note that Kronecker, Fueter and Takagi did not use the function $j(z)$ in their research, formulating their results in the language of Jacobi's elliptic functions. The use of the function $j(z)$ in the theory of elliptic functions was first made by Klein [2165], and its first application to complex multiplication seems to appear in 1885 in the paper [3285] of Pick[45].

Much later a purely algebraic approach to these results was invented by Deuring [913, 914], who utilized the theory of elliptic function fields developed by him in [912] and Hasse [1679, 1683].

For a thorough analysis of various versions of Kronecker's assertion see the paper of Schappacher [3593]. The history of research on complex multiplication has been presented in the book of Vlăduţ [4235].

For modern expositions of the classical part of the theory of complex multiplication see Deuring [915], Borel[46], S. Chowla, Herz[47], Iwasawa, Serre [396] and Schertz [3605] as well as Chaps. 8–14 in S. Lang [2458] and Chap. 2 in Silverman [3802].

2.3.2 Heinrich Weber

1. In 1886 published H. Weber [4312] the first part of his work on Abelian fields. He recalled first the main properties of Abelian groups and their characters, developed earlier by Schering [3600], Kronecker [2288], Frobenius and Stickelberger [1263] and H. Weber [4310]. He applied them to show that every Abelian field is a composite of cyclic[48] fields of prime power degree and used Gaussian periods to describe subfields of cyclotomic fields $\mathbf{Q}(\zeta_n)$ which are not subfields of $\mathbf{Q}(\zeta_m)$ with $m < n$ (this was done earlier by Kronecker [2280] in the case of cyclic cyclotomic fields). In the next section considered Weber the field $\mathbf{Q}(\zeta_{2^n})$, constructed a system of fundamental units, gave a formula for its class-number and showed that it is odd for all n. The last section contains a proof of the Kronecker–Weber theorem, which we mentioned already in the preceding chapter.

In the second part of his treatise [4313] Weber showed how to construct all Abelian fields with a given Galois group, extending the work of Kronecker [2288] who did this for cyclic groups

2. We noted already in Sect. 1.3.3 that Dedekind established in §184 of [848] the Ideal Theorem (formula (1.30)) and deduced the formula (1.32). Although he

---

[45] Georg Pick (1859–1942), professor in Prague.

[46] Armand Borel (1923–2003), professor in Zürich and at IAS in Princeton. See [114].

[47] Carl Samuel Herz (1930–1995), professor at the Cornell University and McGill University. See [1].

[48] Cyclic groups and fields were called *regular* by Weber.

considered the zeta-function $\zeta_K(s)$ only at the real line his result actually gives its residue at $s = 1$.

In the same section the function

$$L(s,\chi) = \sum_I \frac{\chi(I)}{N(I)^s} = \prod_{\mathfrak{p}} \frac{1}{1-\chi(\mathfrak{p})N(\mathfrak{p})^{-s}}$$

for characters χ of the class-group has been considered.

The first evaluation of the error term in the Ideal Theorem has been obtained in 1897 by H. Weber [4324] who applied a formula for the number of lattice points in an n-dimensional sufficiently regular body, established[49] by him in [4322] to establish for the numbers $A_K(x)$, $A_K(x; X)$ of ideals with norm $\leq x$ lying in $\mathbf{Z}_K$, respectively in the ideal class $X \in H(K)$, the equalities

$$A_K(x) = h(K)\kappa(K)x + O\left(x^{1-1/n}\right) \tag{2.11}$$

and

$$A_K(x; X) := \#\{I \in X : N(I) \leq x\} = \kappa(K)x + O\left(x^{1-1/n}\right), \tag{2.12}$$

n denoting the degree of K and $\kappa(K)$ given by (1.29).

The last equality implies the uniform distribution of ideals in classes $X \in H(K)$.

It has been pointed out in 1945 by Wintner[50] [4428] that the form of the main term in (2.11) and in other similar formulas can be obtained from the Tauberian theorem of Ikehara [1959], which states that if the Dirichlet series

$$f(s) = \sum_{n=1}^{\infty} \frac{a_n}{n^s}$$

has nonnegative coefficients, converges for $\Re s > 1$ and satisfies

$$f(s) = \frac{c}{s-1} + g(s)\,,$$

where $g(s)$ is a function continuous at the line $\Re s = 1$, then for $x \to \infty$ one has

$$\sum_{n \leq x} a_n = (c + o(1))x\,.$$

In 1958 Rieger [3469] found an elementary proof of (2.12) with weaker but more explicit error term. An explicit bound for the error term in (2.12) has been provided by M.R. Murty and Van Order [3004].

3. In 1897–1898 H. Weber published three papers [4323–4325] providing the basis for subsequent research concerning arbitrary finite Abelian extensions culminating in the establishment of the class-field theory by Takagi (see Sect. 4.3.1).

---

[49] He pointed out that a similar result under somewhat different assumptions appears on p. 62 in the book [2874] of Minkowski.

[50] Aurel Friedrich Wintner (1903–1958), professor in Baltimore. See [1636].

The first paper begins with some elementary results on Abelian groups and contains the proof of an assertion, which is equivalent to the statement that if Z is the infinite cyclic group, then any subgroup of Z^n is isomorphic to Z^m with $m \leq n$. This is followed by a study of orders consisting of all integers of a given field which are congruent to a rational integer modulo a fixed ideal $\mathfrak{f}$. Particular attention was devoted to quadratic fields and the relation with Gaussian theory of quadratic forms.

In the second paper Weber considered groups $H \subset G$ of fractional ideals in a normal extension K of the rationals, assuming that they satisfy the following conditions:

(a) The group G contains all but finitely many prime ideals of $\mathbf{Z}_K$.

(b) The subgroup H contains only principal ideals and is of finite index h in G.

(c) For every ideal $\mathfrak{m}$ the number $T_\mathfrak{m}$ of integral ideals $I \in H$ satisfying $N(I) \leq t$ and divisible by $\mathfrak{m}$ satisfies

$$T_\mathfrak{m}(t) = \frac{ct}{N(\mathfrak{m})} + O\left(t^{1-1/n}\right),$$

where $n = [K : \mathbf{Q}]$, and $c > 0$ does not depend on $\mathfrak{m}$.

(d) There exists an extension K'/K of degree $\leq h$ such that all prime ideals $\mathfrak{p} \in H$ of first degree are products of distinct prime ideals in K', whereas there remaining prime ideals $\mathfrak{p} \in G$ do not have this property.

Weber proved that these assumptions imply that the extension K'/K has degree h, and every residue class of G/H contains infinitely many prime ideals. He showed also that the conditions (a)–(c) are satisfied in the case, when G consists of all ideals prime to a fixed rational integer k, and $H \subset G$ is closed under congruences mod k; i.e. if $a\mathbf{Z}_K \in H$ and $a \equiv b \bmod k$, then $b\mathbf{Z}_K \in H$.

The field K' appearing in d) was later called the *class-field* of K.

Later Fueter [1297, 1298] generalized[51] Weber's approach, introducing the *ray class-groups* mod $\mathfrak{f}$ for ideals $\mathfrak{f}$ of $\mathbf{Z}_K$ in the following way: if $\mathfrak{f}$ is a given ideal, $G(\mathfrak{f})$ is the group of all ideals prime to $\mathfrak{f}$, and $G_0(\mathfrak{f})$ is its subgroup consisting of principal ideals having a generator $\alpha \equiv 1 \bmod \mathfrak{f}$, then the factor group $G(\mathfrak{f})/G_0(\mathfrak{f})$ he called *Strahlklassengruppe* mod $\mathfrak{f}$ (*ray class-group*). These groups, with a modification consisting in assuming the generator α to be totally positive, played a fundamental role in the establishing of the class-field theory by Takagi. Their detailed theory has been presented in Hasse's report [1655, 1656] (see Sect. 4.3).

The assumptions (a)–(d) permitted H. Weber to deduce the analogue of Dirichlet's theorem about progressions in the case of algebraic numbers:

If $a, b \in \mathbf{Z}_K$, and the ideals generated by a and b are relatively prime, then there exist infinitely many elements $\pi \equiv b \bmod a$ which generate principal prime ideals.

In his proof he followed Dirichlet's idea, using the L-functions associated with characters of the factor group G/H:

$$L(s, \chi) = \sum \frac{\chi(I)}{N(I)^s}, \quad (s > 1)$$

[51] Note that in these papers as well as in his thesis ([1296], p. 5) Fueter used the name "Strahl" (i.e. "ray") for any multiplicative group consisting of numbers.

the sum taken over all ideals $I \in \mathbf{Z}_K \cap G$. The analogue of Dirichlet's theorem for the field $\mathbf{Q}(i)$ was presented in detail by H. Weber later in [4327], and for $\mathbf{Q}\left(\sqrt{-3}\right)$ this has been done by Bresslau [442].

In the third paper [4325] H. Weber showed how the classical theory of divisibility of elliptic functions, which was exposed earlier by him in [4316], permits to prove the existence of the field K', appearing in d), in the case of imaginary quadratic fields K.

In 1989 G. Frei [1239] described the work of H. Weber related to class-field theory.

2.3.3 *Hilbert's Class-Field*

1. In Hilbert's work the word *class-field* is used for the first time at the end of §58 of [1836], where after the proof of Theorem 94 (see Sect. 2.1.2) he wrote: "*Wegen*[52] *der engen Beziehung, die nach Satz 94 der Körper K zu gewissen Idealklassen des Körpers k aufweist, wird K ein Klassenkörper des Körpers k genannt*".

He defined class-fields of arbitrary fields in 1898 [1837], stating the following assertions for an arbitrary finite extension of $k/\mathbf{Q}$ and its narrow class-group $H^*(K)$ (we changed a little the formulation to abide to the terminology now used):

There exists a unique Abelian unramified extension K/k with the following properties:

I. *The Galois group $Gal(K/k)$ is isomorphic to the narrow class-group $H^*(k)$ of k.*

II. *The field K contains all Abelian unramified extensions of k.*

III. *The form of factorization of a prime ideal $\mathfrak{p}$ of k in K depends only on the class in $H^*(k)$ containing $\mathfrak{p}$.*

Hilbert called the field K the *class-field* of k. Today one uses the name *Hilbert class-field* of k.

Before stating the next assertion Hilbert defined an integer $a \in K$ to be *ambiguous*, if it is totally positive and the principal ideal generated by it is invariant under the action of $Gal(K/k)$.

IV. *The principal ideal generated by an ambiguous element $a \in K$ is of the form $I\mathbf{Z}_K$, where I is an ideal of $\mathbf{Z}_k$, and the lifting to K of any ideal I of $\mathbf{Z}_k$ has an ambiguous generator.*

V. *Every ambiguous element of K can be written uniquely (up to unit factors) as a product of ambiguous elements which are not products of two non-unit ambiguous elements.*

VI. *The last two conditions characterize the class-field K among all Abelian extensions of k.*

He established in Satz 10 and 11 of [1837] these assertions for quadratic fields k, satisfying $h(k) = h^*(k) = 2$ or 4, and repeated them at the end of [1843]:

---

[52] "*Because of the close relation, which according to theorem 94, exists between the field K and certain ideal classes on k, the field K will be called the class-field of the field k*".

"*Wir*[53] *stellen...folgende Theoreme auf, die im Vorstehenden für gewisse besondere Fälle bewiesen worden sind, deren vollständiger Beweis jedoch, wie ich überzeugt bin, auf Grund der von mir angegebenen Methoden gelingen muss*".

2. The first steps towards the proof of Hilbert's assertions were made in 1903 by Furtwängler [1330]. He showed that if p is an odd prime and the p-Sylow subgroup $H_p(k)$ of the class-group of k is cyclic, then there exists an unramified Abelian extension K/k with Galois group isomorphic to $H_p(k)$ having the following properties: $p \nmid h(K)$, K contains all unramified Abelian p-extensions of k, and if $h(k) = qp^a$ with $p \nmid q$, then for every ideal I of k the ideal $I^q\mathbf{Z}_K$ is principal. This established **I** in the case when $H(k)$ is a cyclic p-group. At the end of the paper it has been noted that in the case of a totally imaginary field k the proof works also for $p = 2$.

In his next paper [1331] Furtwängler obtained **I** in the case when $\zeta_p \in k$ and $H(k) = C_p^m$, and in [1332] the existence of the class-field satisfying **I** has been established for all fields with the proviso that if $p = 2$ and k is not totally imaginary, then the class-group $H(k)$ should be replaced by the narrow class-group $H^*(K)$.

In [1334, 1335] Furtwängler showed that the class-field of k satisfies the condition **II** and in [1336] gave a uniform proof of **I**. In the last paper it is also shown that the existence of class-fields for imaginary quadratic fields is a consequence of the theory of complex multiplication of elliptic functions.

In certain special cases the existence of the class-field has been also established by F. Bernstein [310–312].

The assertion **III** dealing with factorizations of prime ideals in the class-field of k was established by Furtwängler in 1911 [1340].

The assertion **IV** which implies that all ideals of k are becoming principal in the class-field (Principal Ideal Theorem, "*Hauptidealsatz*") was established by Furtwängler [1330] in 1903 in the case when the class-group of k is a cyclic p-group and k contains pth roots of unity. His paper [1336] contained the proof in the case of arbitrary cyclic class-groups, and later [1345] he proved this also in the case when the class-group was a product of two cyclic groups and the class-number was cube-free. See Sect. 4.3.2 for the general case.

A purely group-theoretical proof of the result in [1336] has been later given by Taussky[54] [4021].

In 1910–1913 Hecke showed [1729–1731] that one can utilize modular functions in two variables to construct class-fields of real quadratic fields. This approach has been later continued by Sugawara [3961].

In 1919 Furtwängler [1348, 1349] considered ring class-fields defined in the following way: let O be an order in $\mathbf{Z}_K$, and let f be its conductor. The ring class-group H is defined as the $I_f(O)/PI_f(O)$, where $I_f(O)$ is the group of fractional ideals of O prime to f, and $PI_f(O)$ is its subgroup consisting of principal ideals. The ring class-field is the extension L/K satisfying the condition (d) of Weber's

[53] "*We present the following theorems, which in certain special cases were established in the preceding, but whose complete proofs, I am sure, could be obtained on the basis of methods showed by me.*"

[54] Olga Taussky-Todd (1906–1995), wife of John Todd, worked at the National Bureau of Standards and CalTech. See [1620, 1873, 2162].

paper [4324], mentioned in Sect. 2.3.2. Furtwängler determined the ring class-fields L of imaginary quadratic fields K in the case when the degree L/K is a prime.

In 1931 Hasse [1674, 1675] presented a simpler approach to ring class-fields in the case of imaginary quadratic fields.

For further development see Sect. 4.3.

Chapter 3
First Years of the Century

3.1 Analytic Methods

3.1.1 Edmund Landau

1. The first results concerning the eighth problem of Hilbert on the distribution of prime ideals were obtained by Landau [2414] in 1903. He presented first the main results of the theory of Dirichlet series (these series were earlier considered by J.L.W.V. Jensen [2050] in 1884, Pringsheim[1] [3340] in 1890 and E. Cahen[2] [547] in 1894) and in particular applied partial summation to obtain the following sufficient condition for a continuation of a function defined by a Dirichlet series to a larger region:

If the series

$$F(s) = \sum_{n=1}^{\infty} \frac{a_n}{n^s}$$

converges for $\Re s > 1$ and one has

$$\sum_{n \le x} a_n = cx + O(x^{\alpha})$$

with non-zero c and $\alpha < 1$, then $F(s)$ can be continued to the half-plane $\Re s > \alpha$ with a pole at $s = 1$ as its unique singularity.

---

[1]Alfred Pringsheim (1850–1941), professor in München, father-in-law of the writer Thomas Mann. See [3257].

[2]Eugéne Cahen (1865–1941), teacher in Paris.

W. Narkiewicz, *The Story of Algebraic Numbers in the First Half of the 20th Century*, Springer Monographs in Mathematics,
https://doi.org/10.1007/978-3-030-03754-3_3

It has been shown earlier by Phragmén[3] [3280] that under these assumptions the function $F(s) - a/(s-1)$ can be continued to the disc of radius $(1-\alpha)/2$ around $s = 1$.

Using Weber's result (2.11) Landau obtained the continuation of $\zeta_K(s)$ to a regular function in the half-plane $\Re s > 1 - 1/n$ (with $n = [K : \mathbf{Q}]$), its only singularity being the simple pole at $s = 1$. He established also the equality

$$\sum_{N(I)\leq x} \frac{1}{N(I)} = h(K)\kappa(K)\log x + c + O(x^{-1/n})$$

with $\kappa(K)$ given by (1.29) and

$$c = \lim_{s\to 1}\left(\zeta_K(s) - \frac{h(K)\kappa(K)}{s-1}\right).$$

He proved also the non-vanishing of $\zeta_K(s)$ on the line $\Re s = 1$ using Hadamard's ([1581], p. 202) theorem asserting that if a function $\Phi(s)$ is regular for $\Re s > c > 0$, $\log \Phi(s)$ has a Dirichlet series with nonnegative coefficients convergent for $\Re s > c$, $\Phi(s)$ can be extended to a region Ω on the left of the line $\Re s = c$, and the only singularity of $\Phi(s)$ in Ω is a simple pole at $s = c$, then $\Phi(c+it)$ does not vanish for $t \neq 0$. He provided also another proof of $\zeta_K(1+it) \neq 0$, generalizing the second Vallée-Poussin's [4120] proof in the case of Riemann zeta-function.

Landau pointed out that if K is either a quadratic field or a cyclotomic field $\mathbf{Q}(\zeta_q)$ with a prime power q, then the continuation of $\zeta_K(s)$ to a meromorphic function on the plane follows from earlier results of Kinkelin[4] [2153] and A. Hurwitz [1926] about Dirichlet L-functions. He pointed out that this can be also achieved for fields in which the factorization of rational primes is known.

2. In the same paper Landau showed that $\zeta_K(s)$ does not have any zeros $s = \sigma + it$ in the region

$$\sigma > 1 - \frac{c}{\log^9(3+t)},$$

proved the bounds

$$\zeta_K(1+it) = O(\log^7|t|), \quad \frac{1}{\zeta_K(1+it)} = O(\log^7|t|) \tag{3.1}$$

for $|t| > 1$, showed that the equality

$$\zeta_K(s) = \prod_{\mathfrak{p}} \frac{1}{1 - N(\mathfrak{p})^{-s}}$$

holds on the line $\Re s = 1$ ($s \neq 1$), and established the equality

---

[3] Lars Edward Phragmén (1863–1937), professor in Stockholm. See [568].

[4] Hermann Kinkelin (1832–1913), professor in Basel.

$$\sum_{N(I)\leq x}\frac{1}{N(I)^{1+it}}=\zeta_K(1+it)+\frac{h(K)\kappa(K)}{itx^{it}}+O(x^{-1/n}).$$

In the case of Riemann's zeta-function this equality has been proved by Kinkelin [2153] in 1862.

In 1921 Weyl[5] [4377] showed the error term in (3.1) to be $O(\log t)$ for $t\geq c>0$, and for Abelian K this was improved to $O(\log t/\log\log t)$ by Walfisz[6] [4275] in 1927.

In 1968 Mitsui[7] [2895] and Sokolovskiĭ [3850] established the bound

$$\zeta_K(1+it)=O(\log^{2/3}t)\quad(t>0),\tag{3.2}$$

as a consequence of their proofs of the non-vanishing of $\zeta_K(s)$ in the region

$$\Omega=\{s=\sigma+it:\sigma\geq 1-C\log^{-2/3}t(\log\log t)^{-1/3},\quad t\geq c>0\},\tag{3.3}$$

with c, C depending on the field. In the case of $K=\mathbf{Q}$ this has been established in 1958 by I.M. Vinogradov [4231] and Korobov[8] [2223]. Effective values of the constants C, c in (3.3) have been obtained in 1978 by Bartz [201], who got $C=a/n^{11}|d(K)|^2$, $c=4$ with $n=[K:\mathbf{Q}]$ and an absolute constant a.

The bound (3.2) in the case of Riemann's zeta-function is due to Richert[9] [3459].

Explicit constants in (3.2) and in the bounds for $\zeta_K(s)$ at points close to the line $\Re s=1$ were later provided by Staś[10] [3902, 3903]. See also Kadiri [2092].

3. In the next section of [2414] one finds the first results dealing with the distribution of prime ideals. Let $\pi_K(x)$ be the number of prime ideals $\mathfrak{p}$ of an algebraic number field K satisfying $N(\mathfrak{p})\leq x$. If K is either a quadratic field or a cyclotomic field, then the equality (Prime Ideal Theorem)

$$\pi_K(x)=(1+o(1))\frac{x}{\log x}\tag{3.4}$$

follows from the quantitative form of Dirichlet's about primes in progressions, established in 1896 by Hadamard[11] [1581] and Vallée-Poussin [4120], but at that time nothing has been known about the growth of $\pi_K(x)$ in other fields.

In 1874 Mertens [2814, 2815] applied the properties of the Möbius function $\mu(n)$ to obtain several properties of the counting function $\pi(x)$ of rational primes and obtained analogous results for prime numbers of the field $\mathbf{Q}(i)$ ([2814]). Landau generalized his results to arbitrary number fields. He followed Mertens introducing the analogue $\mu_K(I)$ of the Moebius function by

---

[5]Hermann Weyl (1885–1955), professor at ETH in Zürich, Göttingen and Princeton. See [686, 3068, 3109].

[6]Arnold Walfisz (1892–1962), professor in Tbilisi. See [2269].

[7]Takayoshi Mitsui (1929–1997), professor at the Gakushuin University in Toshima.

[8]Nikolaĭ Mihailovič Korobov (1917–2004), professor in Moscow. See [6].

[9]Hans-Egon Richert (1924–1993), professor in Marburg and Ulm.

[10]Włodzimierz Staś (1925–2011), professor in Poznań.

[11]Jacques Salomon Hadamard (1865–1963), professor in Bordeaux and Paris. See [2586, 2794].

$$\mu_K(I) = \begin{cases} 1 & \text{if } I = \mathbf{Z}_K, \\ (-1)^r & \text{if } I \text{ is a product of } r \text{ distinct prime ideals,} \\ 0 & \text{if } I \text{ is not square-free,} \end{cases} \tag{3.5}$$

and using elementary methods established the equalities

$$\sum_{N(\mathfrak{p})\le x} \frac{\log N(\mathfrak{p})}{N(\mathfrak{p})} = \log x + O(1) ,$$

and

$$\sum_{N(\mathfrak{p})\le x} \frac{1}{N(\mathfrak{p})} = \log\log x + C(K) + O\left(\frac{1}{\log x}\right)$$

with a certain constant $C(K)$.

Defining

$$\vartheta_K(x) = \sum_{N(\mathfrak{p})\le x} \log(N\mathfrak{p}) \tag{3.6}$$

and

$$\psi_K(x) = \sum_{N(\mathfrak{p}^n)\le x} \log(N\mathfrak{p}) \tag{3.7}$$

he established the bounds

$$x \ll \vartheta_K(x) \ll x, \quad x \ll \psi_K(x) \ll x ,$$

generalizing Čebyšev's[12] [632, 633] results for rational primes.

This implied

$$a\frac{x}{\log x} \le \pi_K(x) \le b\frac{x}{\log x} \quad (x > x_0) ,$$

with certain positive a, b, depending on K, confirming the conjecture (1.37) of Poincaré [3324]. This was the first step towards the Prime Ideal Theorem. He showed also that one has

$$\liminf_{x\to\infty} \frac{\pi_K(x)\log x}{x} \le 1, \quad \limsup_{x\to\infty} \frac{\pi_K(x)\log x}{x} \ge 1 ,$$

hence if the limit

$$\lim_{x\to\infty} \frac{\pi_K(x)\log x}{x}$$

exists, then it equals 1.

Moreover Landau proved

[12] Pafnutiĭ Lvovič Čebyšev (1821–1894), professor in St. Petersburg. See [3341].

$$\sum_{N\mathfrak{p}\le x} \frac{1}{N\mathfrak{p}} = \log\log x + C(K) + O\left(\frac{1}{\log x}\right),$$

with

$$C(K) = \log\alpha + \gamma - \sum_{k=2}^{\infty} \frac{1}{k} \sum_{N\mathfrak{p}\le x} \frac{1}{N(\mathfrak{p}^k)},$$

where $\gamma = 0.5772\ldots$ denotes the Euler constant and $\alpha = h(K)\tau(K)$.

4. In the final part of [2414] considered Landau arithmetical functions defined on ideals of a given field K, generalizing previous results of Mertens [2814], who considered the case $K = \mathbf{Q}(i)$. He dealt in particular with the number $\tau_K(I)$ of ideals dividing I and with $\varphi_K(x)$, the Euler function of K, giving the number of residue classes mod I, prime to I. He used Weber's method outlined in [4322] to establish the asymptotic formulas

$$\sum_{N(I)\le x} \tau_K(I) = h^2\kappa^2 x \log x + Cx + O\left(x^{1-1/2n}\right), \tag{3.8}$$

and

$$\sum_{N(I)\le x} \varphi_K(I) = \frac{h\kappa}{2\zeta_K(2)} x^2 + O\left(x^{2-1/n}\right), \tag{3.9}$$

with $n = [K:\mathbf{Q}]$, $C = 2\alpha h\kappa - h^2\kappa^2$,

$$\alpha = \lim_{s\to 1}\left(\zeta_K(s) - \frac{h\kappa}{s-1}\right),$$

where $\kappa = \kappa(K)$ is defined in (1.29) and $h = h(K)$.

He proved also similar asymptotical formulas for the sums

$$\sum_{N(I)\le x} 1/\varphi_K(I), \quad \sum_{N(I)\le x} \Phi_K(I),$$

where $\Phi_K(I)$ is the number of square-free divisors of I.

The equality (3.8) has been generalized in 1957 by Rieger [3466] who showed that if $\mathfrak{f}$ is an ideal of a field K of degree n and X is an ideal class mod $\mathfrak{f}$, then

$$\sum_{I\in X, N(I)\le x} \tau_K(I) = c_1 x \log x + c_2 x + O\left(x^{1-1/(n+1)}\right),$$

where

$$c_1 = \frac{h^2\kappa^2}{h_{\mathfrak{f}}} \prod_{\mathfrak{p}|\mathfrak{f}} \left(1 - \frac{1}{N(\mathfrak{p})}\right)^2,$$

with $h_{\mathfrak{f}}$ being the number of ideal classes mod $\mathfrak{f}$, and c_2 depends only on K and $\mathfrak{f}$.

5. In his next paper [2415] published in the same year gave Landau a new proof of the Prime Number Theorem not appealing to Hadamard's theory of entire functions, basing his arguments on bounds for the ratio $\zeta'(s)/\zeta(s)$ in the region $\{\sigma+it : \sigma \geq 1-c/\log^9 t, t \geq 10\}$ with $c = 10^{-6}$.

He applied a similar method to the study of $\zeta_K(s)$ and using his results in [2414] proved the Prime Ideal Theorem, the analogue of the Prime Number Theorem for algebraic number fields:

One has

$$\vartheta_K(x) = x + O\left(x \exp(-\log^{1/12} x)\right) ,$$

and

$$\pi_K(x) = \mathrm{li}(x) + O\left(x \exp(-\log^{1/13} x)\right), \tag{3.10}$$

the implied constants depending on K.

In a short paper published in 1904 Landau [2417] considered the Dirichlet series $\sum_{n=1}^{\infty} c_n/n^s$ of $\zeta_K(s)/\zeta(s)$ and showed that it converges at $s=1$ to $h(K)\kappa(K)$.

In the case when K is a cubic field K of negative discriminant the coefficients c_n were used by Carlitz [569] in 1931 to construct a modular function.

In 1906 considered Landau [2418] relations between the following five assertions established by him in [2415]:

(i) $\vartheta_K(x) = (1+o(1))x$,

(ii) $\pi_K(x) = (1+o(1))\frac{x}{\log x}$,

(iii) $\sum_I \frac{\mu_K(I)}{N(I)} = 0$,

(iv) $\sum_{N(I)\leq x} \mu_K(I) = o(x)$,

(v) $\sum_I \frac{\mu_K(I)\log N(I)}{N(I)} = -1$.

6. It follows from the Dirichlet–Weber theorem on prime values of binary quadratic forms that if K is an imaginary quadratic field, then for ideal classes $X \in H(K)$ one has

$$\sum_{\mathfrak{p}\in X} \frac{1}{N(\mathfrak{p})^s} = \frac{1}{h} \log \frac{1}{s-1} + g(s). \tag{3.11}$$

In 1904 F. Bernstein (Theorem 3 in [312]) established (3.11) in the case when K is a totally imaginary field containing ζ_h, where $h = h(K)$, and conjectured that this equality holds for all algebraic number fields. This has been proved in 1907 by Furtwängler ([1336], Satz 11) as a consequence of the existence of Hilbert's class-field, established by him. As a corollary the existence of infinitely many prime ideals in every class of ideals has been established.

7. In 1907 Landau [2419] deduced uniform distribution of prime ideals in narrow ideal classes from the assumption of the divergence of the series

$$\sum_{\mathfrak{p}\in E} \frac{1}{N(\mathfrak{p})}, \tag{3.12}$$

with E denoting the principal class. He observed that this assumption is satisfied if either K is quadratic, or the class-number $h(K)$ is odd, and only at the end of his paper noted that for every number field it is a consequence of Furtwängler's result, quoted above.

His main result stated that if $\pi_X(x)$ denotes the number of prime ideals having norm $\leq x$ and lying in class X, then one has

$$\pi_X(x) = \frac{x}{h(K)} + O\left(x \exp(-c \log^{\alpha} x)\right) \tag{3.13}$$

with some positive α, depending on K. This immediately implies the same assertion also for the usual ideal classes.

In his proof Landau utilized the L-functions

$$L(s,\chi) = \sum_{I} \frac{\chi(I)}{N(I)^s} \quad (\Re s > 1)$$

associated with characters χ of the class-group and showed that for non-principal χ their Dirichlet series converges in the half-plane $\Re s > 1 - 1/[K:\mathbf{Q}]$, and in the case of principal χ it can be prolonged to that half-plane with a simple pole at $s = 1$. The divergence of the series in (3.12) was applied to show that the product

$$\prod_{\chi} L(s,\chi)$$

has a simple pole at $s = 1$, which he used to show $L(1,\chi) \neq 0$ for non-principal characters.

Note that Landau defined narrow ideal classes in a way differing from that used later, as he considered two fractional ideals to be equivalent if their ratio is a principal ideal having a generator of positive norm.

In the proof of (3.13) the bound (3.1) has been improved to

$$|\zeta_K(1+it)| = O(\log t)$$

for $t > \exp(2[K:\mathbf{Q}])$, and it was shown that the same bound applies for $s = \sigma + it$ with $\sigma \in (1 - 1/\log t, 2)$.

At the end of [2419] Landau showed that a result analogous to (3.13) holds also for other partitioning of ideal into classes, satisfying the conditions stated by H. Weber [4324], which we described in Sect. 2.3.2. In particular (3.13) holds also for narrow classes in the sense now used.

It was observed by Landau in 1908 [2420] that his improvement of the error term in the Prime Number Theorem for progressions obtained in that paper leads *mutatis mutandi* to the equality (3.13) with any $\alpha < 1/8$, and the results about quadratic forms obtained by Bernays in his thesis [305] imply that for quadratic fields this holds even with $\alpha = 1/8$. In 1914 Landau [2424] showed that for imaginary quadratic and pure cubic fields one can take $\alpha = 1/2$.

8. In 1910 Bohr and Landau [382] proved that for $|t| > 16$ in the region

$$\left\{\sigma + it : |1 - \sigma| < \frac{1}{\log\log\log t}\right\}$$

$\zeta_K(s)$ attains every complex value with at most one exception.

This was later strengthened by Walfisz [4276] who showed that this holds for certain regions of arbitrary small measure, lying close to the line $\Re s = 1$.

9. In a series of papers [2423, 2427, 2430, 2441], starting in 1912, Landau applied Perron's[13] formula (Perron [3249])

$$\sum_{n \le x} a_n = \frac{1}{2\pi i} \int_{a-i\infty}^{a+i\infty} f(s) \frac{x^s}{s} ds, \tag{3.14}$$

for Dirichlet series

$$f(s) = \sum_{n=1}^{\infty} \frac{a_n}{n^s},$$

convergent for $\Re s > c$, with positive $a > c$, in the case when $f(s)$ can be prolonged to a meromorphic function with a functional equation of the Riemannian type, having only finitely many poles in each strip $A \le \Re s \le B$. He obtained bounds for the difference

$$\sum_{n \le x} a_n - A(x),$$

where $A(x)$ denotes the sum of residues of $f(s)x^s/s$ in some strip, depending on the form of the functional equation. In [2423] he showed in particular that the error term in the Ideal Theorem for quadratic fields is $O\left(x^{1/3}\right)$ (which in the case of the field $Q(i)$ was earlier established by Sierpiński[14] [3800]), and for cyclotomic fields $\mathbf{Q}(\zeta_m)$ equals $O\left(x^{1-c_m}\right)$ with $c_m = 1/(1 + \varphi(m))$.

A modern approach to the evaluation of error terms in asymptotical formulas for the sum of coefficients of a Dirichlet series satisfying functional equations with multiple Γ-factors has been presented by Chandrasekharan and Narasimhan [647–649] in the sixties. See also Hafner [1585], Redmond [3409], Lau [2483], Ramachandra[15] [3375], Friedlander and Iwaniec [1254], J. Liu and Ye [2612] and de Roton [897].

10. The coefficients of the expansion of $\zeta_K^k(s)$ in a Dirichlet series,

---

[13] Oskar Perron (1880–1975), professor in Tübingen, Heidelberg and München. See [1230, 1758].

[14] Wacław Sierpiński (1882–1969), professor in Lwów and Warsaw. See [3617].

[15] Kanakanahalli Ramachandra (1933–2011), professor at the Tata Institute and the National Institute of Advanced Studies in Bangalore. See [3001].

$$\sum_{n=1}^{\infty} \frac{\tau_K^{(k)}(I)}{N(I)^s} = \zeta_K^k(s) \quad (\Re s > 1) \ ,$$

give the number of writing I as a product of k factors.

The problem of evaluating the sum of coefficients in this series in the case $\zeta_K(s) = \zeta(s)$ has been considered already in 1881 by Piltz [3294], hence is called usually the *Piltz problem*.

In [2422] Landau generalized (3.8), establishing for $k \geq 2$ the formula

$$\sum_{N(I) \leq x} \tau_K^{(k)}(I) = x \sum_{j=0}^{k-1} c_j \log^j x + c + R(x), \tag{3.15}$$

where

$$R(x) = O\left(x^{1-1/kn} \log^{k-2} x\right) \ ,$$

with $n = [K : \mathbf{Q}]$ and real $c_{k-1} > 0, c_{k-2}, \ldots, c_0, c$, depending on K and k.

For later results on the Piltz problem see Sect. 4.2.4.

3.1.2 Erich Hecke and the New L-Functions

1. A breakthrough in the theory of Dedekind zeta-function $\zeta_K(s)$ occurred in 1917 when Hecke [1732] showed that it can be extended to a meromorphic function in the complex plane, its only singularity being a simple pole at $s = 1$. He proved also the functional equation

$$F(s) = F(1 - s) \tag{3.16}$$

for the function

$$F(s) = A^s \Gamma(s/2)^{r_1} \Gamma(s)^{r_2} \zeta_K(s),$$

where $r_1, 2r_2$ are the numbers of real and complex embedding of K, respectively, and

$$A = \sqrt{|d|} 2^{-r_2} \pi^{-n/2} \ ,$$

with d being the discriminant of K and n denoting its degree. Hecke gave details of his argument only in the case of cubic fields with $r_1 = 1$ and then sketched the necessary changes needed in the general case. His main tool was a new kind of theta-series attached to fractional ideals I, defined in the case of a not totally real cubic field K by

$$\theta(x, y; I) = \sum_{a \in I} \exp\left(-\frac{\pi(xa_1^2 + 2y|a_2|^2)}{(|d(K)|N(I)^2)^{1/3}}\right) \ ,$$

(a_j denoting the jth conjugate of a with real a_1), and satisfied the functional equation

$$\theta(x, y; I) = \frac{1}{\sqrt{xy^2}}\theta\left(\frac{1}{x}, \frac{1}{y}; \frac{1}{D_K I}\right),$$

D_K being the different of K.

Hecke utilized this theta-series in a way analogous to that used by Riemann[16] [3474] in his proof of the functional equation for $\zeta(s)$.

He observed that in the case of Abelian fields the functional equation for $\zeta_K(s)$ can be deduced from known functional equations of Dirichlet's L-functions, writing that in this case "*Die*[17] *wirkliche Aufstellung der Funktionalgleichung für die* $\zeta_k(s)$ *ist meines Wissens bisher nicht geleistet, bietet aber keine prinzipielle Schwierigkeiten*". He also pointed out that the same can be done for class-fields of quadratic fields using the results of Vallée-Poussin [4121, 4122] simplified by Landau [2426].

Other proofs of (3.16) were later given by Siegel [3771, 3773] in 1922 (in [3771] only the case of totally real fields has been treated, and the proof given in [3773] avoided the use of Dirichlet's unit theorem), Müntz [2999] in 1924, Mordell [2929] (based on Poisson's sum formula) in 1931 and Tate [4007] in 1950 (see Sect. 6.1.1). The paper of Müntz applied the method used by him earlier [2997] in the case of Riemann's zeta-function and contains also the proof of the functional equation for zeta-functions of ideal classes, defined by

$$\zeta_K(X, s) = \sum_{I \in X} \frac{1}{N(I)^s},$$

X being an ideal class in K.

A modern version of Hecke's proof was presented in Chap. 7 of Neukirch's[18] book [3099].

As an application of his result Hecke [1733] gave a simple proof of Furtwängler's theorem [1336, 1340] about the relative discriminant and prime ideal factorization in the Hilbert class-field of an imaginary quadratic field.

3. In 1917 Hecke [1734] considered L-functions for characters of the group of ideal classes mod $\mathfrak{f}$, where $\mathfrak{f}$ is an ideal in K. Let us recall that two ideals I, J lie in the same class mod $\mathfrak{f}$, if $(IJ, \mathfrak{f}) = 1$, and the fractional ideal IJ^{-1} is principal. The so-defined ideal classes form a finite group $H_{\mathfrak{f}}(K)$, and its cardinality $h_{\mathfrak{f}}(K)$ is related to $h(K)$ by the equality

$$h_{\mathfrak{f}}(K) = \frac{h(K)\varphi_K(\mathfrak{f})}{\alpha_K(\mathfrak{f})},$$

where $\varphi_K(\mathfrak{f})$ denotes the number of residue classes mod $\mathfrak{f}$ prime to $\mathfrak{f}$ and $\alpha_K(\mathfrak{f})$ is the number of residue classes mod $\mathfrak{f}$ containing units.

---

[16]Bernhard Riemann (1826–1866), professor in Göttingen. See [2487].

[17]"*The actual writing of the functional equation for* $\zeta_k(s)$ *is, as far I know, not yet done but presents no principal difficulties*".

[18]Jürgen Neukirch (1937–1997), professor in Regensburg.

If χ is a character of $H_{\mathfrak{f}}(K)$, then the corresponding L-function is defined for $\Re s > 1$ by

$$L(s,\chi) = \sum_{(I,\mathfrak{f})=1} \frac{\chi(I)}{N(I)^s}. \tag{3.17}$$

Using again theta-functions Hecke showed that for non-principal χ the function $L(s,\chi)$ can be continued to an entire function satisfying a functional equation. As a consequence he obtained the existence of infinitely many prime ideals in every class of $H_{\mathfrak{f}}(K)$. It was of importance that he did not need class-field theory to achieve this result, as the earlier proof given in the case $\mathfrak{f} = \mathbf{Z}_K$ by Furtwängler [1336] used that theory. As a particular case the analogue of Dirichlet's theorem on primes in progression was established for all algebraic number fields.

Hecke noted in [1734] that similar results can be proved also for the group $H^*_{\mathfrak{f}}(K)$ of narrow ideal classes mod $\mathfrak{f}$ (two ideals I, J prime to $\mathfrak{f}$ belonging to the same class if IJ^{-1} is principal and has a totally positive generator), but wrote "*Ich*[19] *habe die Rechnung nicht durchgeführt*".

3. In 1917 Landau [2428] used Hecke's functional equation (3.16) to improve the error term $R(x)$ in the Ideal Theorem (formula (2.11)). He proved

$$R(x) = O\left(x^{1-2/(n+1)}\right) \tag{3.18}$$

and

$$R(x) = \Omega\left(x^{1/2-1/2n-\varepsilon}\right) \tag{3.19}$$

for every $\varepsilon > 0$ with $n = [K : \mathbf{Q}]$. For imaginary quadratic fields this was obtained by him five years earlier [2421], and according to Landau [2432] the real quadratic case has been treated in the dissertation of Hammerstein [1614].

Later Landau [2436] made more precise the dependence of the implied constant on the discriminant d by proving

$$R(x) = O\left(x^{1-2/(n+1)}|d|^{n+1}\log^n|d|\right)$$

with the implied constant depending only on the degree of the field.

It has been noted much later by Berndt [306] that Landau's method permits to reduce by 1 the exponent of the logarithm in the last formula. For imaginary quadratic fields Ayoub [158] obtained for every $\varepsilon > 0$ the bound

$$R(x) = O_\varepsilon\left((|d|x)^{1/3+\varepsilon}\right) + O\left(|d|^{1/2+\varepsilon}x^\varepsilon\right),$$

which is better for small x.

Another proof of (3.18) has been given by Chandrasekharan and Narasimhan [648] in 1962. They removed also ε from the exponent in (3.19). In 1971 Berndt [308] showed that if K is neither quadratic nor a totally real cubic field, then

---

[19] "*I did not perform the calculation*".

$$R(x) = \Omega\left((x\log x)^{1/2-1/2n}\right),$$

and for quadratic fields he sketched in [306] the proof of

$$R(x) = \Omega_{\pm}\left((x\log\log x\log\log\log x)^{1/4}\right).$$

See also Joris [2076–2078].

It has been shown by Müller [2995] in 1988 that the error term in the Ideal Theorem for cubic fields is $O(x^a)$ with any $a > 43/96 = 0.4479\ldots$, and the first essential improvement of Landau's upper bound in (3.18) in the general case has been achieved in 1993 by Nowak [3134] who showed that for fields of degree $n \geq 7$ it is $O(x^a \log^b x)$ with

$$a = 1 - \frac{2}{n} + \frac{3}{2n^2}, \quad b = 2/n,$$

and for $3 \leq n \leq 6$ one has

$$a = 1 - \frac{2}{n} + \frac{8}{5n^2 + 2n}, \quad b = \frac{10}{5n+2}.$$

For $n > 9$ this has been improved in 2010 by Lao [2471], who showed that one can take any $a > 1 - 3/(n+6)$. The best result for $4 \leq n \leq 10$ has been obtained in 2015 by Bordellès [395].

4. The error term in the Prime Ideal Theorem (3.10) has been replaced in 1918 by Landau ([2432], Satz LXXXV) by

$$O\left(x\exp\left(-\frac{c}{\sqrt{n}}\sqrt{\log x}\right)\right),$$

where $n = [K:\mathbf{Q}]$ and c does not depend on K. In the same paper it is shown that the error term is not $O(\sqrt{x}/\log x)$ (Satz CXX), generalizing the result of E. Schmidt[20] [3642], who proved this for the error term of the Prime Number Theorem.

In the case of quadratic fields this error term has been improved in 1936 by Čudakov[21] [795] to

$$O\left(x\exp\left(-c\log^{\mu} x\right)\right)$$

for any $\mu < 1/2+1/42 = 0.5238$, as a result of his extension of the zero-free region for Dirichlet's L-functions. He wrote: "*We can extend both theorems to algebraic fields of any order ...*", but this promise has not been fulfilled.

It has been shown in 1959 by Staś [3901] that the size of the error term in the Prime Ideal Theorem is equivalent to the size of the zero-free region of $\zeta_K(s)$ (see also Staś and Wiertelak [3904]). This generalized a result of Turán[22] [4080–4082] concerning the Prime Number Theorem.

The best known evaluation of the error term in the Prime Ideal Theorem

$$\pi_K(x) = \mathrm{li}(x) + O\left(x\exp(-c\log^{3/5} x(\log\log x)^{-1/5})\right). \tag{3.20}$$

has been obtained in 1968 by Mitsui [2895] and Sokolovskiĭ [3850], as a consequence of their proofs of the non-vanishing of $\zeta_K(s)$ in the region (3.3).

The best known explicit evaluation of the difference

---

[20] Erhard Schmidt (1876–1959), professor in Zürich, Erlangen, Breslau and Berlin. See [948, 3494].

[21] In [795] his name is written "Tchudakoff".

[22] Paul Turán (1910–1976), professor at the Budapest University. See [1114, 1593].

$$\sum_{N(\mathfrak{p}^n)\leq x} \log(N\mathfrak{p}) - x$$

under GRH has been given recently by Grenié and Molteni [1519].

5. The functions defined by the Dirichlet series

$$\sum_I \frac{\chi(I)}{N(I)^s}$$

associated with characters χ of the group $H_\mathfrak{f}^*(K)$ were studied first by Landau [2432] in 1918 (therefore they are usually called *Hecke–Landau zeta-functions*). He denoted them by $\zeta(s, \chi)$ and applied theta-functions to establish their functional equations.

A new proof of the functional equation for the Hecke–Landau zeta-functions has been given in 1960 by Tatuzawa [4017].

In the same paper Landau obtained also an analogue of the Pólya–Vinogradov inequality (Pólya [3333], I.M. Vinogradov [4227]) for non-principal characters mod $\mathfrak{f}$:

$$\sum_{N(I)\leq x} \chi(I) = O\left(x^{1-2/(n+1)}\right), \tag{3.21}$$

with $n = [K : \mathbf{Q}]$ (Satz 95).

Landau [2436] made more precise the error term in (3.21) by showing that it is

$$\leq c(n)\Delta^{1/(n+1)} \log^n \Delta x^{1-2/(n+1)},$$

where $\Delta = |d(K)|N(\mathfrak{f})$ and $c(n)$ depends only on the degree n of the field.

Another generalization of the Pólya–Vinogradov inequality to algebraic number fields occurs in a paper of Hinz [1856], who considered sets A of integers α of K satisfying

$$0 < \alpha^{(i)} \leq x_i \quad (1 \leq i \leq r_1),$$

$$|a^{(i)}|^2 \leq x_i \quad (r_1 < i \leq r_1 + r_2),$$

$\alpha^{(i)}$ being the conjugates of α, from each pair of complex conjugated taken only one. He proved the bound

$$\sum_{\alpha\in A} \chi(\alpha) = O\left(N(I)^{1/(2(r+2))} X^{1-1/(r+2)+\varepsilon}\right),$$

for all $\varepsilon > 0$ and every non-principal character χ mod I, X being the product of x_i's and $r = r_1 + r_2 - 1$. In the case of totally real fields he showed that the factor $X^{1-1/(r+2)+\varepsilon}$ can be replaced by X^ε. This improved the bound $O\left(N(I)^{1/2}X^\varepsilon + N(I)\right)$, obtained in 1979 by K.-C. Lee [2501]. A stronger result in this case has been obtained in 1993 by Söhne [3848] who showed that the error term is $O(|d^{n/2}(K)|N(I)^{1/2} \log^n(|d(K)|X)$, with $n = [K : \mathbf{Q}]$. See also Rausch [3392].

Landau used his results to prove the Ideal Theorem for narrow ideal classes, showing that for any class $X \in H_\mathfrak{f}^*(K)$ one has

$$\#\{I \in X : N(I) \leq x\} = \frac{\varphi_K(\mathfrak{f})}{N(\mathfrak{f})} \frac{h(K)}{h_{\mathfrak{f}}^*(K)} x + O\left(x^{1-2/(n+1)}\right) \tag{3.22}$$

with $n = [K : \mathbf{Q}]$. A more explicit form of the error term has been given by Landau in [2436].

The bounds in (3.22) were later improved by Nowak [3134].

If K is a field of discriminant d, χ is a non-real character of the group of residues mod $\mathfrak{f}$, prime to $\mathfrak{f}$ and

$$L(s, \chi) = \sum_{n=1}^{\infty} \frac{\chi(I)}{N(I)^s}$$

is the corresponding L-function, then it has been shown in 1919 by Landau [2437] that one has

$$\frac{1}{|L(1, \chi)|} \leq c(n) \log^n k (\log\log n)^{3n/k} ,$$

where $k = |d| N(\mathfrak{f})$. This generalized an earlier result of Gronwall[23] [1524] who considered the rational case.

In 1919 Landau [2438] established the existence of a function $\lambda(n)$ with the property that every Hecke–Landau zeta-function $\zeta(s, \chi)$ corresponding to a field of degree n has a zero ϱ in the half-plane $\Re s \geq 1/2$ satisfying $|\Im\varrho| \leq \lambda(n)$.

The analogue of Turán's [4080] result relating the error term in the Prime Number Theorem to zero-free regions for Riemann's zeta has been obtained in the case of Hecke–Landau zeta-functions by Staś and Wiertelak [3905] in the seventies. In a later paper [3906] they proved a similar assertion for the size of the difference

$$\psi(x, A) - \psi(x, B) ,$$

where $A \neq B$ are classes of $H_{\mathfrak{f}}^*(K)$ and

$$\psi(x, X) = \sum_{\mathfrak{p}^m \in X, N(\mathfrak{p}^m) \leq x} \log N(\mathfrak{p}) .$$

The lack of zeros (with at most one exception) of the Hecke–Landau ζ-functions in regions similar to the region given by (3.3) has been established by Hinz [1853, 1854], who improved earlier results of Mitsui [2892] and Fogels [1204].

The involved constants were made effective in the eighties by Bartz [202–205]. See also Lagarias, Montgomery and Odlyzko [2397], Ahn and Kwon [29].

Bounds for the exceptional real zeros were obtained by Fogels [1205, 1206] in 1962 and by Hinz and Lodemann [1867] in 1994.

In 1986 Bartz and Staś [207] obtained for non-principal characters $\chi \in H_{\mathfrak{f}}^*(K)$ in a field K of degree n the bound

$$|\zeta_K(\sigma + it)| \leq a(K, \mathfrak{f}) t^{b(1-\sigma)^{3/2}} \log^{2/3} t$$

for $1 - 1/(n+1) \leq \sigma \leq 1$, $t \geq 1.1$, with explicit constants a, b depending on K and $\mathfrak{f}$. This generalized the result of Richert [3459], who proved this for Riemann's zeta-function. This has been later improved by Bartz [204].

[23]Thomas Haakon Gronwall (1877–1932), worked in Princeton and New York. See [1850].

In 1989 Bartz and Fryska [206] obtained effective bounds for the density of zeros of $\zeta(s, \chi)$.

In the paper [2432] Landau proved also the Prime Ideal Theorem for narrow ideal classes:

If X is a class in $H_{\mathfrak{f}}^*(K)$, then

$$\#\{\mathfrak{p} \in X : N(\mathfrak{p}) \leq x\} = \frac{\mathrm{li}(x)}{h_{\mathfrak{f}}^*(K)} + O\left(x \exp\left(-\frac{c\sqrt{\log x}}{\sqrt{n}}\right)\right),$$

with $n = [K : \mathbf{Q}]$.

This gave an asymptotic for the counting function of the set of principal prime ideals having a totally positive generator π such that

$$\pi \equiv \alpha \pmod{\mathfrak{f}}$$

(where $\alpha \in K$ is an integer prime to $\mathfrak{f}$), a generalization of Dirichlet's Prime Number Theorem to algebraic number fields established in a qualitative form by Hecke [1738] in 1920.

Other proofs of the Prime Ideal Theorem for narrow ideal classes were given by Rieger [3470] and Ahern [24].

In 1950 Holzer [1890] used this theorem to obtain a proof of Legendre's ([2509], p. 49) criterion for the solvability of the equation $ax^2 + by^2 + cz^2 = 0$, giving optimal upper bounds for the size of the solution in case when it exists. Elementary proofs of these bounds were later provided by M. Kneser [2184], Mordell [2938], K.S. Williams [4408] and Cochrane and P. Mitchell [729].

6. In 1918 Hecke [1736] introduced a new kind of characters called by him "*Charaktere*[24] *nach den Einheiten*", which in the second part of his paper [1738] were renamed "*Grössencharaktere*".[25] They were defined in the following way:

Let $\varepsilon_1, \varepsilon_2, \ldots, \varepsilon_r$ be a system of fundamental units of a field K having degree n, and for $\alpha \in K$ let $\{a^{(1)}, \ldots, a^{(r+1)}\}$ be the set of its conjugates, from each pair of complex conjugated numbers containing only one.

Put

$$A = \begin{bmatrix} 1 & |\log \varepsilon_1^{(1)}| & \ldots & |\log \varepsilon_r^{(1)}| \\ \ldots & \ldots & \ldots & \ldots \\ \ldots & \ldots & \ldots & \ldots \\ 1 & |\log \varepsilon_1^{(r+1)}| & \ldots & |\log \varepsilon_r^{(r+1)}| \end{bmatrix},$$

and let $e_1/n, e_2/n, \ldots, e_{r+1}/n$ be the first row of the matrix A^{-1}. For given rational integers $m_1, m_2, \ldots, m_r$ and non-zero $\alpha \in K$ put now

$$\lambda(\alpha) = \exp\left(2\pi i \sum_{i=1}^{r+1} m_i \sum_{j=1}^{r+1} e_j^{(i)} \log |\alpha^{(j)}|\right).$$

---

[24] *Characters according to units.*

[25] *Characters of magnitude.*

The function $\lambda(\alpha)$ is a character of the multiplicative group of K, satisfying $\lambda(\varepsilon\alpha) = \lambda(\alpha)$ for every unit ε, and thus depends only on the ideal generated by α. Using the fact that if h is the class-number of K, then for every ideal I its power I^h is principal, Hecke extended λ to a character of the group of fractional ideals of K. In the case when

$$m_j = \begin{cases} 1 & \text{if } j = i, \\ 0 & \text{otherwise,} \end{cases}$$

the obtained character is denoted by λ_i.

Finally let $\mathfrak{f}$ be an ideal of $\mathbf{Z}_K$, and let χ be a character of the class-group mod $\mathfrak{f}$. With every product $\chi(I)\lambda(I)$ Hecke associated the L-function

$$L(s, \chi\lambda) = \sum_I \frac{\chi(I)\lambda(I)}{N(I)^s} \ .$$

With the use of theta-series he showed that these L-functions are meromorphic in the plane, entire and non-vanishing at $s = 1$ for non-principal $\chi\lambda$, and satisfy a functional equation. He used this to establish the following property of the characters λ_i:

If $|z_i| = 1$ *for* $i = 1, 2, \ldots, r$, *then for every* $\varepsilon > 0$ *there exist in every ideal class infinitely many prime ideals* $\mathfrak{p}$ *of first degree, satisfying*

$$|\lambda_i(\mathfrak{p}) - z_i| < \varepsilon \ .$$

In the case of real quadratic fields with a negative fundamental unit introduced Hecke some modified new characters, for which the corresponding L-function also satisfied a functional equation. He applied this to show that if $f(X, Y)$ is an indefinite quadratic form with a non-square discriminant, then there are infinitely many primes $p = f(x, y)$ with the point (x, y) lying in a given angle with vertex at the origin.

7. Two years later presented Hecke [1738] another approach to the characters $\chi\lambda$, associating with a field K the set of ideal numbers defined in the following way. Let $X_1, \ldots, X_m$ be a set of generators of the class-group of K, and let c_j be the order of X_j. For each j choose an ideal $I_j \in X_i$, let β_j be a generator of the principal ideal $I_j^{c_j}$, and put $\alpha_j = \sqrt[c_j]{\beta_j}$. The elements

$$\rho \prod_{j=1}^{m} \alpha_j^{n_j} ,$$

where $\rho \in K$ and $n_1, \ldots, n_m$ are rational integers, form the set of *Hecke ideal numbers*. Using these numbers Hecke defined a family of characters encompassing the characters introduced in [1736], associated with them a corresponding L-function in the same way as before, and showed that they satisfy a functional equation. He called the new characters "*Grössencharakters*" and showed that if χ is a non-principal character, then

$$\sum_{N\mathfrak{p}\leq x} \chi(\mathfrak{p}) = O\left(x\exp(-c\sqrt{\log x}\right) ,$$

with $c > 0$ depending on χ. He established also a kind of uniform distribution mod 1 for values of $(\lambda_1(\mathfrak{p}), \dots, \lambda_r(\mathfrak{p}))$ for the sequence of prime ideals $\mathfrak{p}$ lying in a fixed class mod $\mathfrak{f}$. As a corollary he obtained in particular an asymptotic evaluation of the number $N_f(x;\alpha)$ of rational primes $p \leq x$ of the form $p = f(x, y)$, where f is a quadratic form with a non-square discriminant, and (x, y) lies in a fixed angle α with vertex at the origin (a simpler proof in the case of positive-definite forms was presented in 1969 by Knapowski[26] [2179]). He showed in particular that there is an infinite sequence of irrational Gaussian primes $\pi_n = x_n + iy_n$ satisfying

$$y_n^2 = o(N(\pi_n)). \tag{3.23}$$

In the case of real quadratic fields Hecke's result about uniform distribution has been made more precise by Rademacher[27] [3364] in 1935. He used this [3365, 3367] to obtain asymptotics for the number of totally positive prime numbers π in a real quadratic field, satisfying the conditions $0 < \pi \leq x, 0 < \pi' \leq y$ (with π' being the conjugate of π) and lying in a residue class modulo a fixed ideal I. More generally he got asymptotics for sums of the form

$$\sum_{0<\alpha\leq x, 0<\alpha'\leq y} f(\alpha, \alpha') ,$$

where $f(X, Y)$ is a function satisfying $f(\varepsilon\alpha, \varepsilon'\alpha') = f(\alpha, \alpha')$ for every totally positive unit ε, and $f(\alpha, \alpha') = O\left(N(\alpha)^c\right)$ for some real c. In the case when α runs over elements in a fixed residue class mod I the error term depended on I, and in 1957 Schulz-Arenstorff [3694] obtained error terms depending only on the field.

Analogous results for arbitrary algebraic number fields were obtained in 1952 by Kubilius[28] [2332], who earlier [2331] considered the case of the field $\mathbf{Q}(i)$, making the relation (3.23) more precise. He formulated his results in the language of ideal numbers, following Hecke, and the same approach prevailed in the later research which led to essential improvements in the error terms of the asymptotic formulas for primes in regions. The case of imaginary quadratic fields has been considered by Bulota[29] [512], Kovalčik [2234] and Maknis[30] [2715–2717]. This problem in arbitrary fields has been dealt with by Mitsui [2892] and Urbjalis[31] [4111]. For totally real fields see Urbjalis [4110, 4112].

It has been shown by Ankeny [89] that under GRH one can replace in Hecke's result (3.23) $N(\pi_n)$ by $\log N(\pi_n)$. Now it is known that on the right-hand side of (3.23) one can have $O((N(\pi_n^{0.238}))$ (Harman and P. Lewis [1632]. Previously M.D. Coleman [758] obtained this with the exponent 0.3262).

In 1990 M.D. Coleman [757] showed that L-functions with Hecke characters do not vanish in regions comparable to (3.3).

---

[26] Stanisław Knapowski (1931–1967), professor in Poznań and Miami. See [4083].

[27] Hans Rademacher (1892–1969), professor in Breslau and the University of Pennsylvania. See [79].

[28] Jonas Kubilius (1921–2011), professor in Vilnius. See [2737].

[29] Kęstutis Antanas Bulota (1929–1990), professor in Vilnius.

[30] Mindaugas Maknys [Maknis] (1944–1992), professor in Vilnius.

[31] Jonušas Adolfas Urbelis [Urbjalis] (1927–2015), professor in Kaunas.

Hecke's ideal numbers were later used in the theory of arithmetical functions (Grotz [1529, 1530], Rausch [3391], Helmut Weber [4332]) and in the theory of modular forms (Herrmann [1827] and Styer [3944]).

Hecke stated on p. 122 of his book [1742] that if $\hat{K}$ is the smallest field containing all Hecke ideal numbers of a field K, then $[\hat{K} : K] = h(K)$. It seems that the first proof of this equality has been published in 1995 by Albu and Nicolae [56], but it can be also obtained as a simple consequence of the main result of Schinzel [3620].

A modern definition of Hecke characters has been given in 1950 by Tate [4007] (see Sect. 6.1.1).

8. The influence of real zeros of Dedekind zeta-functions on the class-number of imaginary quadratic fields has been observed first by Hecke. He showed that if the zeta-function of $\mathbf{Q}\left(\sqrt{-d}\right)$ does not vanish in the interval $[1 - a/\log d, 1]$, then one has

$$h(-d) \geq c(a)\frac{\sqrt{d}}{\log d} \tag{3.24}$$

with $c(a) > 0$. His proof appeared in Landau's paper [2435] in 1918. Landau noted that a slightly weaker result (having an extra factor $\sqrt{\log\log d}$ in the denominator) follows from Dirichlet's class-number formula and an earlier result of Gronwall [1524], who in 1913 obtained lower bounds for $|L(1, \chi_d)|$, where $d > 0$ and $\chi_d(n) = \left(\frac{-d}{n}\right)$.

A simpler proof of (3.24) was given by Landau [2443] in 1927, and another proof has been given in 1935 by Mahler [2699].

In 1919 Landau showed [2438] that if for all imaginary quadratic fields K the function $\zeta_K(s)$ does not vanish in a neighbourhood of 1, independent on K, then for large d one has

$$\frac{\sqrt{d}}{\log\log d \log\log\log d} \ll h(-d) \ll \sqrt{d}\log\log d \log\log\log d,$$

and in 1928 Littlewood[32] [2610] showed that GRH implies the bounds

$$\frac{\sqrt{d}}{\log\log d} \ll h(-d) \ll \sqrt{d}\log\log d. \tag{3.25}$$

It has been showed later that the lower bound in (3.25) holds unconditionally for infinitely many d (S. Chowla [699] in 1934), and the same holds also for the upper bound (Linnik[33] [2600] and Walfisz [4278]. See also S. Chowla [702, 703], Bateman,[34] S. Chowla and Erdős [213] and Barban [187]). In 2002 Conrey and Soundararajan [766] showed that for a positive proportion of discriminants the corresponding function $L(s, \chi_d)$ has no zeros in (0, 1); thus the assumption of Hecke's theorem on (3.24) is satisfied for these d's.

---

[32] John Edensor Littlewood (1885–1977), professor in Cambridge. See [520].

[33] Yurij Vladimirovič Linnik (1915–1972), professor in Leningrad. See [1947, 2730].

[34] Paul Trevier Bateman (1919–2012), professor at the University of Illinois in Urbana-Champaign. See [12].

9. In 1918 Landau [2433] proved that for all algebraic number fields K of degree n one has

$$h(K) = O\left(\sqrt{|d(K)|}\log^{n-1}(|d(K)|)\right), \tag{3.26}$$

and showed that if a field of degree n contains infinitely many units, then there exists a unit $u \in K$, not a root of unity, such that its conjugates $u^{(i)}$ satisfy

$$|u^{(i)}| \le A_n\sqrt{|\, d(K)\,|}\log(|d(K)|)\ .$$

In another paper in the same year [2434] he discovered that the discriminants of imaginary quadratic fields with a fixed class-number are rare, by showing that if $0 < d_1 < d_2 < \cdots$ is an infinite sequence of discriminants with fixed $h(-d_i)$, then for every $M > 0$ for sufficiently large i one has

$$d_{i+1} > d_i^M, \tag{3.27}$$

and the same holds in the case when the weaker condition

$$h(-d_i) < d_i^{1/2-\delta}$$

is satisfied with some fixed $\delta > 0$. In [2435] he obtained the same assertion under the still weaker assumption

$$h(-d_i) < \frac{\sqrt{d_i}}{\log d_i}\ .$$

For later results dealing with the size of the class-number of quadratic fields see Sect. 4.4.1.

10. If A is an ideal class in K, then the zeta-function of A is defined for $\Re s > 1$ by

$$\zeta_K(s;\, A) = \sum_{I \in A} \frac{1}{N(I)^s}. \tag{3.28}$$

The limit

$$\mathcal{K}_A = \lim_{s\to 1} \zeta_K(s;\, A) - \frac{\kappa(K)}{(s-1)} \tag{3.29}$$

is of importance in studying the L-functions of characters of the class-group in view of the equality

$$L(1, \chi) = \sum_{A} \chi(A)\mathcal{K}_A.$$

In the case of imaginary quadratic fields the value of this limit can be obtained from *Kronecker's limit formula* [2299], which for a positive-definite quadratic form $Q(X, Y)$ expresses the limit

$$\lim_{s\to 1}\left(\sum_{m,n\in Z\setminus\{0\}}\frac{1}{Q(m,n)^s}-\frac{2\pi}{\sqrt{|d|}}\frac{1}{s-1}\right)$$

by Dedekind's η-function

$$\eta(z)=\exp(\pi iz/12)\prod_{n=1}^{\infty}(1-\exp(2\pi inz)) \tag{3.30}$$

in the following way:

If I is an ideal in the class A^{-1} and $(1, u+iv)$ is its basis with $v>0$, then

$$\mathcal{K}_A=\frac{1}{w\sqrt{|d(K)|}}\left(4\pi(\gamma-\log(2v)-2\log(|\eta(u+iv)|)\right)\ ,$$

w denoting the number of roots of unity in K and γ being the Euler constant.

For modern approach see the book of Siegel [3791].

Kronecker's limit formula has been used in 1910 by Fueter [1300] to obtain a class-number formula for Abelian extensions of imaginary quadratic fields (for modern approach see Chap. 21 in S. Lang's book [2458]).

A formula for (3.29) in the case of real quadratic fields has been given in 1917 by Hecke [1735].

It has been later simplified by Herglotz [1811], Siegel [3791] (see also Meyer [2845]), and its simplest form has been presented by Zagier [4459]:

$$\mathcal{K}_A=\frac{2\log\varepsilon}{\sqrt{d}}\left(2C-\frac{1}{2}\log d\right)+\frac{1}{\sqrt{d}}\int_{-\log\varepsilon}^{\log\varepsilon}\left(\log\left(\frac{e^v+e^{-v}}{w-w'}\right)-\log\left|\eta\left(\frac{w+iw'e^{-v}}{1+ie^{-v}}\right)\right|^4\right)dv,$$

where d is the discriminant of K, $\varepsilon>1$ its fundamental unit, and $1, w$ form a basis of an ideal lying in the class A^{-1}.

Zagier's paper [4459] expresses also (3.29) by the dilogarithm function. A generalization to L-functions for ideal classes in arbitrary fields has been obtained in 1974 by Goldstein [1470].

Hecke's paper [1735] presents also a way of obtaining a formula for (3.29) for arbitrary fields and contains a promise to present a proof in another paper. Hecke never returned to this subject.

3.2 Structure

3.2.1 Steinitz

1. In 1910 Steinitz published his fundamental paper [3914] on the theory of fields. He based his approach on the abstract definition of a field given earlier by H. Weber

[4318] and introduced the notions of prime fields and characteristics, as well as of various types of field extensions: simple, finite, algebraic and transcendental. He showed that if K is a field and the polynomial $f \in K[X]$ is irreducible, then the factor field $K[X]/f(X)K[X]$ provides an extension of K in which f has a zero. He defined algebraically closed fields and showed that every field can be extended to an algebraically closed field, which is unique up to isomorphisms, noting that his proof in the general case is based on the axiom of choice. Finally he showed that every extension L/K can be realized in two steps: first one makes, if necessary, a purely transcendental extension which is then followed by an algebraic extension. These results gave a solid basement for algebraic number theory.

Induced by Steinitz's treatment of the theory of fields Fraenkel [1226–1228] attempted to construct a similar theory for rings and their extensions. It seems that he was the first to give a definition of an abstract ring.

A similar attempt for commutative rings has been made later by Krull [2301, 2304, 2305] based on the theory of ideals developed by E. Noether [3121] (see Sect. 4.1.1).

Commutative domains R in which every non-zero proper ideal can be uniquely represented as a product of prime ideals were considered by Sono [3856, 3857] in 1919. He gave a necessary and sufficient condition for this property to happen: there is no infinite ascending sequence of distinct ideals, there is no infinite descending sequence of distinct ideals in the residue rings R/I, and if M is a maximal ideal in R, then there is no ideal between M^2 and M.

The first two conditions coincide with the first two conditions given later by E. Noether [3123] (see Sect. 4.1.1), and the first and the last conditions imply that R is integrally closed in its field of fractions (see S. Mori and Dodo [2949]).

Such rings are now called *Dedekind domains*. All rings of integers of algebraic number fields having finite degree are Dedekind domains.

2. In 1910 Hecke [1729] quoted Speiser's dissertation [3868] for the assertion that with suitable ideals $I_1, I_2, \dots, I_n$ $(n = [L : K])$ of $\mathbf{Z}_K$ one has the isomorphism of $\mathbf{Z}_K$-modules $\mathbf{Z}_L$ and $\bigoplus_{j=1}^n I_j$.

Actually Speiser considered only the case $n = 2$, and the first proof was given in 1912 by Steinitz [3915, 3916], as a consequence of his main result which shows that every torsion-free finitely generated $\mathbf{Z}_K$-module is isomorphic to a direct sum of ideals of $\mathbf{Z}_K$, and a finitely generated torsion $\mathbf{Z}_K$-module is a direct sum of modules of the form $\mathbf{Z}_K/I$, where I is a non-zero ideal. His results imply also that torsion-free $\mathbf{Z}_K$-modules

$$M = \bigoplus_{j=1}^{m} A_j, \quad N = \bigoplus_{j=1}^{n} B_j$$

with $A_i, B_j \neq \mathbf{Z}_K$ are isomorphic if and only if $m = n$ and the products $\prod_{j=1}^m A_j$ and $\prod_{j=1}^n B_j$ lie in the same ideal class. Moreover every finitely generated torsion-free $\mathbf{Z}_K$-module can be written in the form

$$M = \mathbf{Z}_K^m \oplus I$$

with an ideal I, and its isomorphism class is determined by m and the ideal class of I. He showed also (in §4 of [3915]) that if L/K is an extension of degree n, then the $\mathbf{Z}_K$-module $\mathbf{Z}_L$ has a set of generators of at most $n+1$ elements. It has been known earlier (Wahlin [4263]) that if $[K : \mathbf{Q}] = n$ and $[L : K] = m$, then $\mathbf{Z}_L$ can be generated as a $\mathbf{Z}_K$-module by mn elements. In the case $m = n = 2$ this has been shown in 1907 by Sommer in his book ([3853], p. 299).

In his papers Steinitz used the language of matrices, generalizing his results obtained earlier for $\mathbf{Z}$-modules [3913].

In 1932 Krull [2315] presented Steinitz's results in a more general form using modern language. A simplified version of Steinitz's approach was given by Franz [1233] in 1934, who two years later applied it to topological questions [1235].

Later Steinitz's results were carried over to arbitrary Dedekind domains by Chevalley[35] [680] in 1936 and Kaplansky [2118] in 1952. A simple proof gave Asano [144] in 1950. See also Artin [127] and O'Meara [3153].

3. The class in $H(K)$ of the product $s(L/K) = \prod_{i=1}^n I_j$ of ideals occurring in the decomposition

$$Z_L = \bigoplus_{j=1}^n I_j$$

is called the *Steinitz class* of the extension L/K.

One says that a finite extension L/K of degree n has a *relative integral basis* if the ring $\mathbf{Z}_L$ is a free $\mathbf{Z}_K$-module; i.e. there exist elements $\omega_1, \omega_2, \dots, \omega_n$ such that every element $a \in \mathbf{Z}_K$ can be written in an unique way in the form

$$a = \sum_{j=1}^n x_j \omega_j \text{ with } x_j \in \mathbf{Z}_K. \tag{3.31}$$

Similarly one says that an ideal I of $\mathbf{Z}_L$ has a relative basis, if there exist $\omega_1, \dots, \omega_n$ in I such that for all $a \in I$ the equality (3.31) holds. The existence of a relative integral basis for the extension L/K is equivalent to the triviality of the Steinitz class $s(L/K)$. For the problem of existence of such bases see Sect. 3.2.3.

The range $R(K, G)$ of $s(L/K)$ for L running over all Galois extensions of K with Galois group G has been first considered in the case $G = C_2$ by Fröhlich [1267], who established the equality $R(K, C_2) = H(K)$. Later McCulloh [2797] considered the case when $G = C_N$ and K contains ζ_N and showed that in this case $R(K, G)$ is a subgroup of $H(K)$, which he described. He proved also that every class of $R(K, C_N)$ can be realized by a tamely ramified extension. In 1971 Long [2623] showed that if N is a prime, then also in the case when K does not contain ζ_N the set $R(K, C_N)$ forms a group, and described it. He described also the set $R_t(K, C_N)$ of Steinitz classes realized by tamely ramified extensions, and in [2624, 2625] did this in the case when N is a prime power. A description of $R(K, G)$ for Abelian G follows from the results of McCulloh [2799], who also

[35] Claude Chevalley (1909–1984), professor at the Columbia University and in Paris. See [944, 2004].

showed that for Abelian G the set $R_t(K, G)$ is always a group. It has been conjectured (see, e.g. Conjecture 3 in [543]) that the last assertions hold for all groups G, and this has been established in several cases by Bruche and Sodaïgui [472], Byott, Greither and Sodaïgui [543], Byott and Sodaïgui [544–546], Carter and Sodaïgui [590], Cobbe [726–728], Farhat and Sodaïgui [1162] and Sodaïgui [3845–3847]. A description of $R_t(K, G)$ for certain classes of non-Abelian groups has been given by Bruche [471], Carter [585–588], Sodaïgui [3843, 3844] and Soverchia [3863].

3.2.2 Galois Groups

1. We mentioned already in Sect. 2.1.1 that Hilbert [1830] showed in 1892 that there are infinitely many extensions of the rationals with Galois group isomorphic to S_n and A_n. This implies that every finite group can be realized as the Galois group of an extension of some algebraic number field.

In 1899 Runge [3535] proved that for a given a_1 there can be only finitely many polynomials $X^n + a_1X^{n-1} + \cdots + a_n \in \mathbf{Z}[X]$ with Galois group distinct from S_n, except when the coefficients $a_2, \ldots, a_n$ are related by certain equalities.

In his book H. Weber ([4321], 2nd ed., vol. 1, §160) gave a simple sufficient condition for a polynomial of a prime degree p to have S_p for its Galois group. This was extended in 1922 by Furtwängler [1351] to composite degrees.

Various proofs of the existence of polynomials over $\mathbf{Q}$ with symmetric Galois groups were given by M. Bauer [224, 225, 240] in 1906, 1907 and 1923, Mertens [2829] in 1911, Schur [3700] in 1920 and Čebotarev [621] in 1926.

In his proof in [224] M. Bauer applied *Puiseux numbers* associated with a prime ideal $\mathfrak{p}$ in K and an irreducible polynomial $f(X) = X^n + \sum_{j=0}^{n-1} a_jX^{n-j} \in \mathbf{Z}_K$. He defined these numbers in the following way: let P be the polygon in the plane with vertices (j, r_j) $(j = 0, 1, \ldots, n-1)$ where r_j is given by $\mathfrak{p}^{r_j} \parallel a_j$. The Puiseux numbers are defined as the slopes of the line segments of P.

Relations between Puiseux numbers and prime ideal factorization of $\mathfrak{p}$ in $L = K(\alpha)$, where α is a root of f, were established by M. Bauer in an earlier paper [223]. He showed there that the set T of all Puiseux numbers associated with $\mathfrak{p}$ and f coincides with set of ratios f_j/e_j, where e_j, f_j are defined by

$$\mathfrak{p}\mathbf{Z}_L = \prod_{j=1}^{r} \mathfrak{P}_j^{e_j}, \quad \alpha\mathbf{Z}_L = I_\alpha \prod_{j=1}^{r} \mathfrak{P}_j^{f_j},$$

where $(I_\alpha, \mathfrak{p}\mathbf{Z}_L) = 1$ (see also M. Bauer [245] and Ore [3174]).

In 1930 Schur [3703] proved that the Galois group of the nth Laguerre polynomial

$$L_n(X) = \frac{e^X}{n!}\frac{d^n(X^ne^{-X})}{dX^n}$$

equals A_n if $4 \mid n$ and equals S_n otherwise. He showed also that the polynomial $\sum_{j=0}^{n} X^j/j!$ has symmetric Galois group.

Other constructions of S_n-extensions of $\mathbf{Q}$ were later provided by Perron [3251] in 1923, Dörge[36] [990] in 1927, Schur [3704] in 1930 and 1931 (cf. Schulz [3693]), Mac Lane [2684] and Vassiliou [4197] in 1936 and Preuss and F.K. Schmidt[37] [3339] in 1951. In 1929 Furtwängler [1355] presented a construction of S_n-extensions for any algebraic number field.

In 1933 van der Waerden [4136] proved that almost all polynomials of degree $n \geq 3$ in $\mathbf{Z}[X]$ have S_n for their Galois group, and in 1936 [4139] he showed that if $F_n(H)$ denotes the number of monic polynomials in $\mathbf{Z}[X]$ of degree n having height[38] $\leq H$ and whose Galois group is a proper subgroup of S_n, then

$$F_n(H) \ll_n H^{n-c_n/\log\log H}$$

with $c_n = 1/6(n-2)$. He conjectured the bound $F_n(H) = O_n(H^{n-1})$.

His result has been improved in 1973 by Gallagher [1380] to

$$F_n(H) \ll_n H^{n-1/2} \log^{1-c_n} x$$

with $c_n = 1/\sqrt{2\pi n}$. For the next improvement in the general case one had to wait until 2010 when Zywina [4482] removed the logarithm from Gallagher's result for sufficiently large n.

A close approximation to van der Waerden's conjecture for cubic polynomials has been obtained in 1979 by Lefton [2507], who proved

$$F_3(H) \ll_\varepsilon H^{2+\varepsilon}$$

for every $\varepsilon > 0$. In 2006 Dietmann [945] obtained a similar result for quartic polynomials establishing

$$F_4(H) \ll_\varepsilon H^{3+\varepsilon}\,,$$

and in 2013 [946] he obtained for $n \geq 3$ the bound

$$F_n(H) \ll_{n,\varepsilon} H^{n-2+\sqrt{2}+\varepsilon}$$

for every $\varepsilon > 0$.

For similar investigations see S.D. Cohen [738–740] and S. Davis, Duke and Sun [833].

In 1935 Wegner [4342, 4343] and in 1940 Vassiliou [4198] constructed families of trinomials with symmetric Galois group. For later results on Galois groups of trinomials see Osada [3195, 3196], S.D. Cohen [741], Komatsu [2205], Movahhedi [2991], Movahhedi and Salinier [2992], S.D. Cohen, Movahhedi, Salinier [743, 744], Bensebaa, Movahhedi, Salinier [291, 292] and Bishnoi and Khanduja [365].

Nakagawa [3046] showed in 1988 that all polynomials of degree n with square-free discriminant have S_n for their Galois group.

2. Let G be a subgroup of the symmetric group S_n, and let K be a field. In her talk at a conference in Vienna in 1913 E. Noether [3118] considered the following assertion:

If G acts on the field $L = K(X_1, \dots, X_n)$ as a group of permutations of the variables, and $M = L^G$ is its field of invariants, then the extension M/K is purely transcendental.

Noether pointed out that the truth of this assertion would imply the possibility of describing all Galois extensions of K with Galois group G using finitely many parameters. She presented a proof in 1917 [3119]. Her result has been applied in the same

[36] Karl Dörge (1899–1975), professor in Köln.

[37] Friedrich Karl Schmidt (1901–1977), professor in Jena, Münster and Heidelberg. See [2379].

[38] The height of a polynomial $P(X) = \sum_{j=0}^{n} a_j X^j \in \mathbf{Z}[X]$ equals $\max_i |a_i|$.

year in the thesis of Seidelmann [3724] who gave such a description for cubic and quartic extensions with a given Galois group different from S_n. For example, cyclic cubic extensions are generated by roots of the polynomials $X^3 - (a^2+3b^2)(3X-2a)$ (with $a, b \in K$), provided they are irreducible over K. For quartic fields this question has been also studied later by Garver[39] [1388, 1390], who in [1389] gave a simple proof of Seidelmann's result in the case of cyclic cubic extensions. A similar description for the quaternion group H_8 was given by Mertens [2830, 2832] (see also Bucht [491] and Gröbner[40] [1522]). Cyclic extensions were considered by Mertens in [2831].

A construction of cyclic extensions of degree 8 of arbitrary fields has been described in 1933 by Albert[41] [46], who in [47] described cyclic extensions of prime degree, and in [48] did this for cyclic extensions of prime power degree.

A parametrization of quintic cyclic fields has been given in 1935 by Hull [1919] (see also Hashimoto and Hoshi [1637]), and the case of dihedral groups of orders $2n \leq 6$ and the Frobenius group F_{20} has been settled by Hashimoto and Hoshi [1638] in 2005.

2. The question whether Noether's condition holds for a given field K and all finite groups is usually called the *Noether problem.* It has been pointed out by H.W.Jr. Lenstra in [2540] that Noether did not conjecture that her problem has always a positive answer, and noted that this question has been formulated in a special case already in Chap. 17 of the second edition of Burnside's book [521], published in 1911.

In the case of the complex field Fischer[42] [1188, 1189] showed in 1915 that if G is an Abelian subgroup of $GL_n(\mathbf{C})$, then the fields L^G and L are isomorphic.

It has been shown by Masuda [2764, 2765] that Noether's condition holds for the cyclic group C_n for $n \leq 7$ and $n = 11$ provided $\mathrm{char}(K) \nmid n$.

In 1969 Swan [3966] showed that if the answer is positive for the field $\mathbf{Q}$ and cyclic group C_p with prime p, then there is a principal ideal of norm p in the cyclotomic field $\mathbf{Q}(\zeta_{p-1})$, and used this to give the first counterexamples in the case $K = \mathbf{Q}$ with $p = 47$, 113 and 233 (cf. Martinet [2749] and S. Endô and Miyata [1097]). In the case of Abelian groups a necessary and sufficient condition for a positive answer was found by H.W. Jr. Lenstra [2540] (cf. Kervaire [2141]).

The converse to Swan's theorem has been established in 1970 by Voskresenskiĭ [4247], who in an earlier paper [4246] gave another proof for Swan's counterexample, and in 1973 presented a survey of Noether's problem [4248].

Quite recently Plans [3309] showed that Swan's condition holds if and only if the class-number of $\mathbf{Q}(\zeta_{p-1})$ equals unity, confirming a suggestion of H.W. Jr. Lenstra [2545].

It has been shown by Bogomolov [376] that for every prime power p^k with $k \geq 6$ there is a group of p^k elements for which Noether's condition fails for the complex field (an example with $k = 9$ has been earlier constructed by Saltman [3567]). Later Moravec [2923] showed that the same holds also for a group of 3^5 elements. On the other hand for all groups of 2^5 elements Noether's condition over $\mathbf{C}$ is satisfied (Chu, Hu, Kang, Prokhorov [711]).

---

[39] Raymond Joseph Garver (1901–1935), professor at the University of California at Los Angeles. See [4396].

[40] Wolfgang Gröbner (1899–1980), professor in Innsbruck.

[41] Abraham Adrian Albert (1905–1972), professor at the University of Chicago. See [2017, 2120].

[42] Ernst Sigismund Fischer (1899–1983), professor in Erlangen and Köln.

Positive answers to Noether's question in various particular cases were given by Breuer [443–447] in 1925–1932, Furtwängler [1352] in 1925, Hasse [1658] in 1927 and Gröbner [1521, 1522] in 1932–1934.

The case of finite groups generated by reflections in finite-dimensional vector spaces has been resolved in the fifties by Shephard and J.A. Todd[43] [3752] and Chevalley [684] (cf. Flatto [1198] and Teranishi [4046]).

Later other classes of groups were considered by Matsuda [2775], Haeuslein [1582], Miyata [2899] and Takiff [3987] in 1971, Hajja [1592] in 1983. In the beginning of the twenty-first century this subject has been treated by Ahmad, Hajja and Kang [25], Chu and Kang [712], Chu, Hu and Kang [710], Kang [2102–2104], Plans [3307, 3308], Michailov [2850], Kang, Michailov and Zhou [2105] and Kang and Zhou [2106].

For further development of the problem of the existence of extensions with given Galois group see Sect. 4.5.1.

3.2.3 Discriminants and Integral Bases

1. In 1901 Reid[44] [3422] published tables giving the discriminant, class-number, integral basis and some units for cubic fields generated by roots of X^3+aX+b with $|a|, |b| \leq 10$.

An elementary proof of the formula for the discriminants of cyclotomic polynomials has been presented by Rados [3370] in 1906.

Another elementary proof appeared in a paper of E. Lehmer [2521] in 1930.

2. It has been observed early that not every finite extension has a relative integral basis. For quadratic extensions it has been noted in 1909 by Speiser [3868] that if the different $D_{L/K}$ is principal, then L/K has a relative integral basis. Three years later Hecke showed in his thesis [1730] that if K is totally real, and L/K is a totally imaginary quadratic extension, then an ideal I of $\mathbf{Z}_L$ has a relative integral basis if and only if $N_{L/K}(I)$ lies in the ideal class X of K defined in the following way: if $L = K(\sqrt{\Delta})$, then the ideal $\sqrt{\Delta}D_{L/K}^{-1}$ is a lifting of an ideal A of K, and X is the ideal class containing A.

A necessary and sufficient condition for the existence of a relative integral basis in a given finite extension L/K was given by Artin [127]. He showed that if L/K is generated by $a \in \mathbf{Z}_L$, then there exists a fractional ideal I in K whose square equals $d(L/K)/d_{L/K}(a)$, and an integral basis exists if and only if I is principal. Here $d_{L/K}(a)$ denotes the discriminant of the minimal polynomial of a over K. A proof has been also given in 1974 by Fujisaki [1316].

In 1958 Mann [2735] showed that a quadratic extension L/K has a relative integral basis if and only if $L = K(\sqrt{\alpha})$ with $d(L/K) = \alpha\mathbf{Z}_K$ and deduced that if every quadratic extension of K has an integral basis, then $h(K) = 1$. Another proof has been given in 1960 by Fröhlich [1267]. For simple examples see MacKenzie and Scheuneman [2683], Cvetkov [806] and Example 5.3.2 in book [2867] by Milnor and Husemoller.

---

[43] John Arthur Todd (1908–1994), Reader in Cambridge. See [150].

[44] Legh Wilber Reid (1867–1961), professor at Haverford College.

In 1960 Fröhlich [1267] defined the idelic[45] discriminant $\partial(L/K)$ and used it in [1268] to show that the ideal class containing the relative discriminant $d(L/K)$ equals the square of the Steinitz class $s(L/K)$, and in [1269] obtained a characterization of extensions having relative integral bases (see also Fröhlich [1270]).

Simpler conditions for particular classes of extensions were provided by McCulloh [2796] in 1963 (for Kummer extensions of prime degree) and [2798] in 1971 (for extensions with Frobenius Galois group[46]), Bird and C.J. Parry [363] in 1976 (L—biquadratic, K—quadratic subfield of L), Thome [4054] in 1988 (for composites of Kummer extensions of prime degree), Hymo and C.J. Parry [1944, 1945] in 1990–1992 (for L quartic cyclic or pure and K being its quadratic subfield).

There are cases, when a relative integral basis exists automatically. This happens, e.g. when $h(K) = 1$, or if K and L are Galois extensions of the rationals with Galois group $C_{p^s}^n$ respectively $C_{p^s}^m$ with $m > n$ and p is an odd prime (X.K. Zhang [4469]). The same holds also for $p = 2$ under certain additional assumptions.

3. Theorem 132 in Hilbert's report [1836] implies that an at most tamely ramified Abelian extension $K/\mathbf{Q}$ has a normal integral basis (NIB). In 1916 Speiser observed that if an extension $K/\mathbf{Q}$ has a normal integral basis, then it is at most tamely ramified. Speiser's result is hidden in the first lines of §6 of [3870]. The assertion that an at most tamely ramified Abelian extension $K/\mathbf{Q}$ has a NIB is now called the *Hilbert–Speiser theorem*.

Let G be a finite group. A field K is called a *Hilbert–Speiser field* (HS-field) for the group G if every tame Galois extension L/K with Galois group G has a normal integral basis. It has been shown in 1969 by Martinet [2748] that the rational field is a HS-field for the dihedral group D_{2p} with odd prime p, and later [2750] gave the first example of a tame extension of $\mathbf{Q}$ without NIB. It is an extension with the quaternion Galois group H_8. In 1972 Fröhlich [1275] showed that there are infinitely many such extensions with group H_8 [1274], and also with group H_{4p^m} for odd primes $p \equiv 3 \bmod 4$.

In 1974 Fröhlich, Keating and S. Wilson [1283] proved that $\mathbf{Q}$ is an HS-field for D_{2^n} (an explicit construction of a NIB in the case $n = 2$ has been provided by Cougnard [779], and in 1976 Fröhlich [1278] showed this also for groups $\Omega(p, q)$, generated by elements σ, ω satisfying

$$\sigma^p = \omega^q = 1, \quad \omega\sigma\omega^{-1} = \sigma^r ,$$

where p is an odd prime, q divides $p - 1$, and r is the order of q mod p. The same assertion for the alternating group A_4 has been established in 1974 by Reiner and Ullom [3427]. Another proof in the case $G = A_4$ has been provided in 2006 by Cougnard [781] who also gave an algorithm for the construction of NIB in this case.

In 2002 Cougnard and Queyrut [783] gave a criterion for the existence of a NIB in quaternionic fields of degree 12, utilizing only the ramification properties.

It was shown by Greither, Replogle, Rubin and Srivastav [1514] that the field of rationals is the only field which is a HS-field of type C_p for every prime p.

Imaginary quadratic HS-fields for cyclic groups C_p with prime p were determined by Carter [589] and Ichimura and Sumida-Takahashi [1954, 1955]. Such fields exist only for $p \leq 7$. It has been later shown that the only non-quadratic CM-field which is an HS-field for C_p with $p \geq 5$ is $\mathbf{Q}(\zeta_{12})$ (Ichimura [1953]). Some other classes of fields were treated by Ichimura [1950, 1951], Greither and Johnston [1513] and Yoshimura [4456].

For more results on normal integral bases see Sect. 5.1.2.

---

[45]For ideles see Sect. 5.2.2.

[46]If a group G has a non-trivial subgroup H whose intersections with its conjugates are trivial, then G is called a Frobenius group.

3.2.4 Units

1. Fundamental units of real quadratic fields can be obtained from fundamental solutions of the equations $x^2 - dy^2 = \pm 4$ (Dedekind, §186 in [848], H. Weber, pp. 399–400 in the first edition of the first volume of his treatise [4321]). Tables of these solutions for small values of d were prepared by Legendre [2509] in 1798, Degen[47] [858] in 1817 and Bickmore [349] in 1893. They were extended in 1912 by Whitford[48] [4392, 4393] up to $d = 997$ and for $d \equiv 5 \bmod 8$ up to 1997. A list of errors in these tables has been prepared by Cunningham [799] in 1916.

For an extension of Cunningham's list see D.H. Lehmer [2511].

The first upper bound for the regulator $R(d) = \log \varepsilon$ of a real quadratic field of discriminant d, $\varepsilon > 1$ being the fundamental unit, has been obtained in 1903 by Lerch ([2563], p. 391) who showed

$$R(d) = O\left(\frac{\sqrt{d}\log d}{h(d)}\right) \ll \sqrt{d}\log d\,.$$

It seems that this result has been overlooked for some time as in 1913 Remak [3437] modified Dirichlet's proof (see pp. 371–375 in the fourth edition of [972]) of the existence of solutions of the Pell equation without the use of continued fractions, and obtained the bound

$$R(d) \le 16d^{3/2}\log(1 + 2\sqrt{d}) \ll d^{3/2}\log d$$

as a special case of the bound for the minimal solution of Pell's equation $X^2 - dY^2 = \pm 1$. This has been soon improved by Perron [3250] who used continued fractions and obtained

$$R(d) \le (1 + o(1))d\log d\,.$$

In 1916 Schmitz [3659] proved

$$R(d) \le (8 + o(1))d\,,$$

and two years later Schur [3697] deduced from Pólya's inequality

$$\left|\sum_{n\le x}\chi(n)\right| \le \sqrt{k}\log k \tag{3.32}$$

for non-principal characters $\chi \bmod k$ (the *Pólya–Vinogradov inequality*, Pólya [3333], I.M. Vinogradov [4226]) and the class-number formula the explicit bound

[47] Carl Ferdinand Degen (1766–1825), professor in Copenhagen.

[48] Edward Everett Whitford (1865–1946), instructor at the College of the City of New York.

$$R(d) \le \frac{1}{2}(\log d + 2\log\log d + 2)\sqrt{d}, \tag{3.33}$$

(see also Landau [2433]).

In 1942 Hua [1908] established the inequality

$$\sum_{n=1}^{\infty} \left(\frac{d}{n}\right) < \frac{1}{2}\log d + 1\,,$$

and used it to remove the term $2\log\log d$ in (3.33). Later, in 1964, Y. Wang [4293] proved

$$R(d) \le (c + o(1))\sqrt{d}\log d$$

for any $c > 1/4$.

All these results were obtained as special cases of bounds for solutions of Pell's equation.

It has been conjectured by Hooley [1893] that for almost all d one has $R(d) > d^c$ for every $c < 1/2$.

In 1971 Yamamoto [4451] showed that for infinitely many d one has $R(d) \ge c\log^3 d$ (see also C. Reiter [3429] for another proof). In 2002 Golubeva [1474] established $R(d) \gg \log^c d$ with $c = 2$ on a set of d having positive density, and this has been extended to all $c < 3$ by Fouvry and Jouve [1222] in 2013 (cf. Fouvry [1221]). In 2016 J. Park [3219] proved $R(d) \gg \log^2 d$ for almost all d.

The first bound for the regulator $R(K)$ of a field of arbitrary degree n has been obtained in 1918 by Landau [2433]:

$$R(K) = O\left(\sqrt{D}\log^{n-1} D\right)\,,$$

where $D = |d(K)|$.

An elementary but weaker bound has been given in 1931 by Remak [3441] (see also Remak [3443, 3444]).

2. In 1902 Minkowski [2878] presented an algorithm leading to a unit in cubic fields with negative discriminant.

In 1936 Bullig [509] showed that this algorithm leads to the fundamental unit, and later [510, 511] gave a geometrical algorithm leading to fundamental units in totally real cubic fields.

In 1918 D.N. Lehmer[49] [2519] observed that one can use a generalization of continued fractions proposed by Jacobi in the paper [2011] (published posthumously in 1868) to study cubic irrationalities. In 1873 Bachmann [170] showed that this algorithm is not always periodic, and in 1907 Perron [3248] presented a generalization of the algorithm which could be applied to study algebraic numbers of arbitrary degree.

Later Lehmer (see footnote 49) [2519, 2520] and his students J.B. Coleman [755, 756] and Daus [816–818] considered the question in which cases Jacobi's algorithm in the cubic case is periodic.

[49] Derrick Norman Lehmer (1867–1938), father of Derrick Henry Lehmer, professor at Berkeley. See [3347].

In 1969 L. Bernstein and Hasse [321] applied Perron's generalization of Jacobi's algorithm to construct units in certain families of pure fields of arbitrary degree, using earlier results of L. Bernstein [315–317] showing periodicity of the algorithm in these cases. For later research on this subject see the book by L. Bernstein [318] as well as L. Bernstein [319, 320], Stender [3923], Perron [3258, 3259] and Thomas [4053]. For the metrical theory of Jacobi's algorithm see the book by Schweiger [3711].

Tables for cyclic cubic fields giving fundamental units for 45 fields were prepared in 1957 by Cohn and Gorn [753]. For real fields with $d(K) \leq 20\,000$ this has been done in 1957 by Godwin and Samet [1451], who also computed the class-numbers.

In 1911 Châtelet[50] [659] presented a way to associate matrices with algebraic numbers and applied reduction procedures of matrices for construction of units in algebraic number fields, giving details in the case of quadratic and cubic fields. Later [660] he interpreted division of ideals in terms of associated matrices.

An algorithm for fundamental units in cubic fields has been presented in 1913 by Berwick [324].

A generalization of that algorithm to arbitrary fields has been given in 1978 by Rudman and Steiner [3530].

A method of calculating units in pure cubic fields presented Pocklington[51] [3314] in 1928.

In 1960 Godwin [1447] considered totally real cubic fields, defined for its integer α_1 with conjugates a_2, a_3

$$S(\alpha_1) = \sum_{i<j} (\alpha_i - a_j)^2 ,$$

and conjectured that if $\varepsilon \neq \pm 1$ is a unit with minimal $S(\varepsilon)$, η is a unit distinct from $\pm\varepsilon^n$ ($n \in \mathbf{Z}$) with minimal $S(\eta)$, and $S(\varepsilon) > 9$, then ε, η are fundamental units. This has been shown to be true, with two explicit exceptions, for cyclic fields in 1980 by M.-N. Gras [1497], and in 1987 Ennola [1104] did this in the general case, apart of finitely many possible exceptions.

A similar method, with $S(\alpha)$ replaced by the trace of α, has been used in 1982 by Cusick [802] to produce a strong Minkowski unit (i.e. a unit which with one of its conjugates forms a set of fundamental units) in the cyclic case. Two years later Godwin [1450] showed that this approach leads to fundamental units for every totally real cubic field. An improvement has been made by Cusick [803], and this led to a table of fundamental units for these fields with discriminant ≤ 6885 (Cusick and Schoenfeld[52] [804]). Another replacement of $S(\alpha)$ used Matveev [2781] in 1993.

A table of fundamental units and the class-number for cubic fields with $-20\,000 < d(K) < 0$ has been described in 1973 by Angell [82], who three years later did the same for totally real cubic fields with $d(K) < 100\,000$ [83] (cf. Llorente and Oneto [2615]). The limit $d(K) < 500\,000$ has been reached in 1985 by Ennola and Turunen [1108].

For cubic fields with $-10^6 < d(K) < 0$ a table has been described by Fung and H.C. Williams [1326].

In 1907 Mertens [2826] found all roots of unity in $\mathbf{Q}(\zeta_p, \sqrt{-d})$ with prime p and $d > 0$.

---

[50] Albert Châtelet (1883–1960), professor in Lille, Caen and at the Sorbonne.

[51] Henry Cabourn Pocklington (1870–1952), fellow of the Royal Society and St. John's College at Cambridge. He worked as a teacher of physics in Leeds. See [3520].

[52] Lowell Schoenfeld (1920–2002), professor at the Pennsylvania State University and in Buffalo.

3.2.5 Splitting Primes

1. Denote by $P_0(L/K)$ the set of prime ideals of K splitting in the extension L/K, and by $P(L/K)$ the set of all unramified prime ideals of K having in L at least one prime ideal factor of degree 1 (note that for normal extensions these two sets coincide).

In 1903 M. Bauer [219] observed that if $f, g \in \mathbf{Z}[X]$ are irreducible, then their roots generate the same normal extension if and only if f and g split mod p for the same prime numbers p, with the exception of finitely many primes. The sufficiency of this condition in the case of polynomials of prime degree has been earlier asserted by Kronecker [2293]. In [220] M. Bauer deduced from his result a characterization of cyclotomic polynomials. Another proof of this characterization has been given in [221], as a consequence of the assertion that if $K/\mathbf{Q}$ and $L/\mathbf{Q}$ are both normal, then a rational prime splits in the composite field KL if and only if it splits in K and in L, up to finitely many exceptions. In 1916 he proved [228] that if $L/\mathbf{Q}$ is normal, then for every extension $K/\mathbf{Q}$ the condition $P(L/\mathbf{Q}) \subset P(K/\mathbf{Q})$ is equivalent to $K \subset L$ (another proof has been given in 1935 by Deuring [910]).

The question, whether one can omit the normality of $K/\mathbf{Q}$ in the last theorem (replacing "$K \subset L$" by "K is isomorphic to a subfield of L"), got a negative answer in 1926 by Gassmann[53] [1393] who constructed two non-isomorphic fields K, L of the same degree 180 in which all primes factorize in the same way with finitely many exceptions. He gave also a necessary and sufficient condition for the symmetric difference $P(K) \triangle P(L)$ to be finite (such fields are now called *Kronecker equivalent*), noting that this can hold only if the smallest normal fields containing K and L coincide. A characterization of Kronecker extensions using prime ideal factorizations has been given in 1995 by Lochter [2618].

Much later Schinzel [3615] characterized pairs L/K, M/K of extensions with finite difference $P(M/K) \setminus P(L/K)$. He called an extension L/K *Bauerian* if for every extension M/K with finite $P(M/K) \setminus P(L/K)$ the field M contains a field isomorphic to L over K and gave a description of such extensions. This implied that all extensions $K/\mathbf{Q}$ of degree ≤ 4 are Bauerian. A large class of non-normal Bauerian extensions of $\mathbf{Q}$ was found by D.J. Lewis,[54] Schinzel and Zassenhaus [2588]. See also Schinzel [3618] and Bilhan [352].

Jehne [2042] studied Kronecker equivalent extensions and showed i.e. that there exist infinite families of Kronecker equivalent fields, solving a problem of Nakatsuchi [3055]. In [2043] he studied Kronecker classes of atomic extensions (i.e. extensions L/K having no intermediate fields). The study of classes of Kronecker equivalent extensions was pursued by N. Klingen [2171, 2172] who published in 1998 a book on this subject [2173].

In 1930 Hasse [1662] (see also th. VI on p. 144 of [1664]) used Čebotarev's theorem (see (4.3.2)) to prove a generalization of Bauer's theorem and showed its use in the determination of class-fields.

Obviously two fields having the same Dedekind zeta-function are *Kronecker equivalent* over $\mathbf{Q}$. Such extensions are called *arithmetically equivalent*. This notion has been introduced in 1977 by R. Perlis [3240], who showed that for such fields the degrees, discriminants, signatures and normal closures coincide and the unit groups are isomorphic. Later he showed [3241] that for primes not dividing the degree the Sylow p-groups of the class-groups of such fields coincide. In 1985 he proved also [3242] that their trace forms $Tr(X^2)$ are rationally equivalent. He showed also in a paper with Stuart [3941] that two fields K, L are arithmetically equivalent if and only if for almost all primes p the number of prime ideals lying over p in K and L coincides.

---

[53] Fritz Gassmann (1899–1990), professor at the ETH in Zürich.

[54] Donald John Lewis (1926–2015), professor at Ann Arbor.

Arithmetically equivalent fields may differ in several aspects. In 1994 de Smit and R. Perlis [901] presented such fields with distinct class-numbers (see also de Smit [899, 900]), and in 2000 Coykendall [789] gave an example of such fields K, L with $N_{K/\mathbf{Q}}(K) \neq N_{L/\mathbf{Q}}(L)$.

A field K is called *solitary* if it is arithmetically equivalent only to its conjugate fields. Bauer's theorem implies that normal fields are solitary, and it has been shown by R. Perlis [3240] that any field of degree ≤ 6 is solitary, whereas there exist non-solitary septic fields.

3.2.6 Reciprocity

1. The first generalization of the reciprocity law given by Hilbert's Theorem 161 (2.3) has been obtained in 1902 by Furtwängler [1329, 1333], who closely followed Hilbert's method, replacing the cyclotomic field $\mathbf{Q}(\zeta_l)$ with odd prime l by any field K containing ζ_l and satisfying $l \nmid h(K)$. He defined an integer α of K to be primary if it is congruent to an lth power mod $(1-\zeta_l)^l \mathbf{Z}_K$. A prime ideal $\mathfrak{p}$ of K was called primary if for all units ε of K one had

$$\left(\frac{\varepsilon}{\mathfrak{p}}\right)_l = 1 .$$

With such $\mathfrak{p}$ he associated an element $\pi_{\mathfrak{p}}$ in the same way, as Hilbert did, and using the definition (2.2) he established the equality

$$\left(\frac{\mathfrak{p}}{\mathfrak{q}}\right)_l = \left(\frac{\mathfrak{q}}{\mathfrak{p}}\right)_l^n ,$$

with some n, prime to l, for prime ideals $\mathfrak{q}$ and primary prime ideals $\mathfrak{p}$, prime to l. In certain cases he was able to show that one can take $n = 1$ in the last equality.

It has been showed in 1905 by Lietzmann [2597, 2598] that an analogue of Eisenstein's reciprocity law for arbitrary exponents would be a consequence of such law for prime powers.

In 1909 Furtwängler [1338] removed the assumption $l \nmid h(K)$ and showed that if K contains the lth root of unity (with $l \neq 2$), and α, β are co-prime integers of K satisfying $(\alpha\beta, 1-\zeta_l) = 1$, then one has

$$\left(\frac{\alpha}{\beta}\right)_l = \left(\frac{\beta}{\alpha}\right)_l ,$$

provided at least one of the numbers α, β is primary. In [1343] he filled a gap in his proof in [1338] (for a simplified proof see Furtwängler [1354]) and in [1344] obtained a similar result for $l = 2$, completing Hilbert's work on this subject, which was restricted to totally complex fields K. In 1926 he sketched [1353] the proof of Eisenstein's reciprocity law for the exponent l^2 in the case of fields containing ζ_{l^2} and promised to publish the proof of its analogue for arbitrary prime powers. This never happened, possibly because in the meantime Artin proved his general reciprocity law, but in 1927 Hasse [1658] presented an exposition of Furtwängler's approach, and used it to prove the analogue of Eisenstein's reciprocity law for all exponents divisible by 8.

2. The paper of A.E. Western [4370] contains some remarks on mth power reciprocity in $\mathbf{Q}(\zeta_m)$ for $m = 2^3, 2^4$ and 3^2, as well as a criterion for primes p such that 2 is an 2^3th power residue mod p.

According to Hasse [1704] this criterion has been formulated already in 1856 by Reuschle [3446]. A similar criterion for the exponent 2^4 has been conjectured by Cunningham[55] [798] in 1895 and proved by Aigner[56] [31] in 1939. A simpler proof provided Beeger [268] in 1948, and later Whiteman[57] [4390, 4391] used Jacobi sums in $\mathbf{Q}(\zeta_{2^n})$ to deduce these criteria for $n = 3, 4$. In 1958 Hasse [1704] gave a formula for

$$\left(\frac{2}{\alpha}\right)_{2^n}$$

for any odd $\alpha \in \mathbf{Z}[\zeta_{2^n}]$, and Shiratani [3762] obtained a similar formula, replacing 2 by an odd prime.

In 1922 Takagi [3977] showed that Furtwängler's reciprocity law can be obtained as a consequence of his construction of the class-field theory (see Sect. 4.3). In 1923 he presented [3978] an explicit formula for the lth power reciprocity law in the lth cyclotomic field with prime l.

3.3 Class-Number

3.3.1 *Quadratic Fields*

1. In the first part of the century the work on class-numbers has been done mostly in the language of quadratic forms, but the obtained results can be also interpreted in the language of quadratic fields.

The use of the formulas (1.33) and (1.34) for the computation of the class-number is rather awkward. In 1903, 1905 and 1906 three long papers of Lerch [2563–2565] appeared in which he transformed these formulas into a more user-friendly form. In particular he showed in [2563] that if $d > 0$, then for any $u > 0$ one has

$$h(d)R(d) = 2\sqrt{d/\pi}\sum_{m=1}^{\infty}\frac{1}{m}\left(\frac{d}{m}\right)\int_{m\sqrt{\pi u/d}}^{\infty}e^{-x^2}dx + \sum_{m=1}^{\infty}\left(\frac{d}{m}\right)\int_{\pi m^2/du}^{\infty}\frac{dx}{xe^x},$$

where $R(d)$ is the regulator of $\mathbf{Q}(\sqrt{d})$. Lerch transformed this formula into a form which permitted to calculate the class-number for large d in the case when a good

[55] Allan Joseph Champneys Cunningham (1842–1928), Fellow of King's College, London. See [4371].

[56] Alexander Aigner (1909–1988), professor in Graz.

[57] Albert Leon Whiteman (1915–1995), professor at the University of Southern California. See [1473].

approximation to $R(d)$ is known. He showed on this on the examples $d = 9817$ and 3436. In the case of negative d he established for $t > 0$ the equality

$$h(d) = \frac{w\sqrt{|d|}}{2\pi} \sum_{n=1}^{\infty} \frac{1}{n} \left(\frac{d}{n}\right) \exp(-2\pi nt/|d|) + \frac{1}{\pi} \sum_{n=1}^{\infty} \left(\frac{d}{n}\right) \tan^{-1}(t/n) .$$

A simple proof of these equalities based on a variant of Poisson's summation formula has been given in 1928 by Mordell [2928].

The papers [2564, 2565] contain also several congruences for class-numbers. One finds there i.e. the determination of $h(-pqr) \bmod 8$ for odd primes p, q, r with either $p \equiv q \equiv r \equiv 3 \bmod 4$ or $p \equiv q \equiv -r \equiv 1 \bmod 4$.

In 1905 A. Hurwitz [1936] found a rapidly convergent series giving the class-number for Gaussian quadratic forms.

In 1911 Lerch [2566] simplified the proof of (1.33), and a proof using theta-function has been found in 1918 by Mordell [2924].

An elementary proof in the case when the discriminant is not congruent to 1 mod 8 has been found in 1928 by B.A. Venkov[58] [4207, 4208] (a German version appeared in 1931 [4209]). See also R.W. Davis [831, 832]. An elementary proof covering also the cases excluded by Venkov has been found in 1978 by Orde [3164] (for an exposition of Orde's proof see Chap. 5 of [3066]).

A simple modern proof of Dirichlet's formulas has been given by Stark [3900] in 1993.

Several other related papers published in the early years of the twentieth century are listed in Chap. 6 of the third volume of Dickson's book [935].

2. We noted already in Chap. 1 that in 1903 Landau [2416] used the language of quadratic forms to prove that the only imaginary quadratic fields with even discriminant and class-number 1 are the fields $\mathbf{Q}(-d)$ with $d = 1, 2$. A simpler proof has been given in the same year by Lerch [2562], who at the end of his paper stated that if $-d$ is an odd negative discriminant with $h(-d) = 1$, then both d and $(d+1)/4$ must be primes, and the polynomial $X^2 + X + d$ attains prime values at integers from the interval $[0, m]$, where

$$m = \left[\frac{\sqrt{d/3} - 1}{2}\right] .$$

This was an early forerunner of the result of Rabinowitsch [3358, 3359], who showed in 1912 that an imaginary quadratic number field with discriminant $-d \neq -3, -4, -8$ has class-number one if and only if the polynomial

$$F_d(X) = X^2 - X + \frac{1-d}{4}$$

attains prime values at integers of the interval $[1, (1-d)/4 - 1]$ (the necessity of this condition occurs also in a paper of Frobenius [1261]).

[58] Boris Alekseevič Venkov [Wenkov] (1900–1962), father of B.B. Venkov, professor in Leningrad.

Already in 1772 Euler [1132] noted that the polynomial $F_{41}(X+1) = X^2+X+41$ produces primes for the first 40 consecutive nonnegative integers and called this behaviour "*plus remarcable*".

Later several other proofs were given for the result of Rabinowitsch: Connell [761] in 1962, Szekeres[59] [3971] in 1974, Ayoub and S. Chowla [161] in 1981. A similar condition for $h(-p) = 2$ has been given in 1974 by Hendy [1759].

The *Ono number* p_d of an imaginary quadratic field $\mathbf{Q}\left(\sqrt{-d}\right)$ with square-free d has been introduced by Ono (in his unpublished lectures) as the maximal number of prime divisors of values of the polynomial $f_d(X)$ given by (3.34) in the interval $[0, |d|/4-1]$. The theorem of Rabinowitsch implies that $h(-d) = 1$ holds if and only if $p_d = 1$, and it has been shown in 1986 by R. Sasaki [3579] that $h(-d) = 2$ is equivalent to $p_d = 2$. R. Sasaki proved also the inequality $h(-d) \geq p_d$ and observed that $h(-21) = 4 > 3 = p_{21}$. Later J. Cohen and Sonn [737] conjectured that for prime $q \equiv 3 \bmod 4$ the equality $p_q = 3$ implies $h(-p) = 3$, and H. Gu, D. Gu and Y. Liu [1536] deduced this conjecture from the Generalized Riemann Hypothesis (see also Guo and Qin [1539]).

The authors of [737] deduced from GRH the relation $p_d \to \infty$ and quoted Rudnick for the observation that

$$\lim_{d\to\infty} \frac{p_d}{h(-d)} = 0$$

(this has been also established by Sairaiji and Shimizu [3556]); therefore the equality $p_d = h(-d)$ can hold only for finitely many d. Louboutin [2650] found in 2009 144 such d's and showed that it can be at most one more whose existence would contradict GRH.

It follows from (3.35) that GRH implies

$$p_d \geq c\frac{\log d}{\log\log d},$$

with some $c > 0$, and Sairaiji and Shimizu [3555] showed that the one can take

$$c = \frac{\log\log(163)}{\log(163)} = 0.3196\ldots .$$

In the nineties Mollin [2905, 2909] gave a Rabinowitsch-type criterion for imaginary quadratic fields with class-group of exponent ≤ 2.

In 1976 Möller [2902] considered for square-free $d > 0$ the polynomials

$$f_d(X) = \begin{cases} X^2+d & \text{if } d \equiv 1 \bmod 4, \\ X^2+X+(1+d)/4 & \text{if } d \equiv 3 \bmod 4, \end{cases} \tag{3.34}$$

and showed that if we put

$$m(d) = \max\{\Omega(f_d(x)) : 0 \leq x \leq A(d)\},$$

where

$$A(d) = \begin{cases} d & \text{if } d \equiv 1 \bmod 4, \\ (d-7)/4 & \text{if } d \equiv 3 \bmod 4, \end{cases}$$

then $m(d) \leq D(H(K)) \leq h(K)$, where $D(G)$ denotes the Davenport constant[60] of the group G. He showed also that GRH implies

[59] George Szekeres (1911–2005), professor in Adelaide and the University of New South Wales. See [1429].

[60] The Davenport constant of a finite Abelian group G is the smallest integer m with the property that every sequence of m elements of G has a subsequence with vanishing sum.

$$m(d) \gg \frac{\log d}{\log\log d}. \tag{3.35}$$

On the other hand for infinitely many d one has

$$m(d) \le c\frac{\log d}{\log\log d}$$

with some constant c.

In 1980 Kutsuna [2388] proved that if $d = 1+4m > 0$ is square-free and $-X^2+X+m$ attains prime values in $[1, \sqrt{m}-1]$, then $h(d) = 1$. Seven years later Mollin [2903, 2904] showed that if $K = \mathbf{Q}\left(\sqrt{d}\right)$ with $d = m^2+r \equiv 1 \bmod 4$ where $r = \pm1, \pm4$, then $h(K) = 1$ holds if and only if the polynomial $-x^2+x+(1-d)/4$ assumes prime values in the interval $[2, \sqrt{d-1}/2-1]$. See also Mollin and H.C. Williams [2913], Mollin [2907].

In 1998 Mollin and Goddard [2912] gave a Rabinowitsch-type criterion for real quadratic fields with class-group of exponent 2.

A polynomial $f(X) = X^2+X-m$ (with $m > 0$) has been called a *Rabinowitsch polynomial* by Byeon and Stark [541, 542] if there is a sequence of consecutive $[\sqrt{m}]$ nonnegative integers x with $|f(x)|$ being either 1 or a prime. They showed that there are only finitely many such polynomials, and the class-number of $\mathbf{Q}\left(\sqrt{4m+1}\right)$ equals 1. In 2011 Byeon and J. Lee [540] determined all such polynomials.

For other results on the connection between polynomial values and the class-number in complex or real quadratic fields see R. Sasaki [3580], Halter-Koch [1603], Mollin [2908, 2910, 2911], Srinivasan [3883, 3884], Granville and Mollin [1486] and book [2906] by Mollin.

Gauss's conjectures on $H(d)$ imply that the only imaginary quadratic fields $\mathbf{Q}\left(\sqrt{-d}\right)$ with $h(-d) = 1$ occur for $d = 3, 4, 7, 8, 11, 19, 43, 67$ and 163. In 1911 Dickson [933] listed all negative discriminants of quadratic forms below 1 500 000 and confirmed this conjecture in that range. For further development of this question see Sect. 4.4.1.

3.3.2 *Cyclotomic Fields*

1. Recall that h_n, h_n^-, h_n^+ denote the class-number and the first and second factors of the class-numbers of $\mathbf{Q}(\zeta_n)$.

Already in 1886 H. Weber [4312] showed that for $n = 2, 3, \dots$ the class-number h_{2^n} of $\mathbf{Q}(\zeta_{2^n})$ is odd. This has been generalized in 1911 by Furtwängler [1341] who showed that for a prime p one has $p \nmid h_{p^n}$ if and only if p is a regular prime. He obtained this as a corollary of the assertion that if $p \nmid h_{p^n}$, then the every unit of $\mathbf{Q}(\zeta_{p^n})$ is a norm of a unit of $\mathbf{Q}(\zeta_{p^{n+1}})$.

A simpler proof of Furtwängler's result gave Moriya[61] [2954] in 1930, and in the case $p = 2$ another proof has been given by Hasse [1703]. In 1956 Iwasawa [1983] generalized Furtwängler's theorem by showing that if p is a regular prime and K is a subfield of $\mathbf{Q}(\zeta_{p^n})$, then $p \nmid h(K)$.

In 1976 Washington [4301] showed $2 \nmid h_{5^n}$ for all n, and $2 \nmid h_{7^n}$ has been established by Ichimura [1952] in 2009. K. Horie [1895] proved in 2002 that if $g \ge c(p) = 1.5p^2 \log p$ is a prime primitive root mod p^2, then $g \nmid h_{p^n}$ holds for $n = 1, 2, \dots$. In 2005 he showed [1896] that

---

[61] Mikao Moriya (1906–1982), professor at Nakayama University.

for $p \leq 7$ this holds with $c(p) = 1$, and later this has been extended by K. Horie and M. Horie [1898–1900] to $p \leq 23$. In [1898] it has been also shown that $2 \nmid h_{p^n}$ holds for $p \leq 13$ and all n.

2. In 1903 Westlund [4373] showed

$$h^-_{p^{n-1}} \mid h^-_{p^n} ,$$

applying a method used in case $p = 2$ by H. Weber (in the 2nd edition of [4321], vol. 2). In the case of the second factor he showed that the denominator of the ratio $h^+_{p^n} / h^+_{p^{n-1}}$ must be a power of p.

It has been shown in 1972 by Metsänkylä [2839] that if $p^{n-1} \notin \{3, 4, 5, 8, 9, 16\}$, then

$$h^-_{p^{n-1}} < h^-_{p^n} .$$

A conjecture stated by D. Davis in 1978 [830] is equivalent to the assertion that if q and $p = 2q + 1$ are primes, then h^-_p is odd. He provided a criterion for checking this, and a similar criterion has been given by Stevenhagen [3929] who used it to show that if $2 \mid h^-_p$, then h^-_p is divisible by 2^{95}. In the case when 2 is inert in the maximal real subfield of $\mathbf{Q}(\zeta_p)$ the conjecture of Davis, and even the stronger assertion $2 \nmid h_p$, has been established by Estes [1125] in 1989. For similar results with the number 2 replaced by other primes below 10 000 see Jakubec [2024, 2025, 2029], Jakubec and Trojovský [2031] and Trojovský [4075].

3. Let p be an odd prime, and for $1 \leq a, b \leq (p-1)/2$ put $c(a, b) = ab' \bmod p$, where $1 \leq b' \leq p - 1$ denotes a solution of $bb' \equiv 1 \pmod{p}$. The determinant

$$D_p = \det [c(a, b)] \tag{3.36}$$

has been considered in 1913 by Maillet [2711], who conjectured that it never vanishes, and in 1914 Malo [2729] asked whether one always have $D_p = (-p)^{(p-3)/2}$.

Malo's conjecture turned out to be incorrect when in 1955 Carlitz and Olson [582] established[62]

$$D_p = (-p)^{(p-3)/2} h^-_p ,$$

which permits an elementary calculation of h^-_p. They also noted that replacing $c(a, b)$ by

$$\left[\frac{ab'}{p}\right] - \left[\frac{(a-1)b'}{p}\right]$$

one obtains a matrix D'_p having determinant h^-_p. In 1961 Carlitz [581] used (3.36) to obtain the bound

$$h^-_p \leq \begin{cases} (m-1)! & \text{if } p = 4m + 1, \\ \sqrt{m}(m-1)! & \text{if } p = 4m + 3. \end{cases}$$

Later Metsänkylä [2841] showed that over $\mathbf{Z}_p$ the matrix D'_p is equivalent to a matrix with zeros over the diagonal, and the number of its non-unit diagonal entries is equal to the index of irregularity of p, defined as the number of $i \in \{1, (p-1)/2\}$ with the Bernoulli number B_{2i} divisible by p.

---

[62]This result has been also found by S. Chowla and Weil, but they did not publish it.

Inkeri [1964] found in 1955 another determinant equal to h_p^-, and his formula was generalized to $h_{p^n}^-$ by Metsänkylä [2837], and to $h_K^- = h(K)/h(K^+)$, the first factor of the class-number of complex Abelian K by Hirabayashi [1868].

In 1970 Newman[63] [3108] gave still another determinant expressing h_p^-. Later Skula [3820] found a similar formula for $h_{p^n}^-$, and in 2008 Hirabayashi [1869] generalized this to h_K^-.

Various generalizations and other determinants related to class-numbers were considered by Kühnová [2345], Tateyama [4012], K. Wang [4285], Okada [3147], Fujisaki [1317], A. Endô [1094, 1095], Dohmae [984], Sands and Schwarz[64] [3572], Tsumura [4077], Girstmair [1438], Kanemitsu and Kuzumaki [2101], Kučera [2342] and Jung and Ahn [2083]. A survey was presented by Kanemitsu and Kuzumaki [2100].

4. A conjecture, often attributed to H. Weber, states that for all n one has

$$h_{2^n}^+ = 1. \tag{3.37}$$

H. Weber himself established $h_{2^4}^+ = 1$ and conjectured $h_{2^6}^+ > 1$ (see p. 868 in the 2nd edition of the second volume of Weber's book [4321]).

In 1960 Cohn [748] pointed out that Weber's method can be used to obtain $h_{2^5}^+ = 1$. He showed also that either (3.37) holds for $n = 6$ or $h_{2^6}^+$ is a prime in the interval [1601, 83 921]. In 1969 H. Bauer [216] established (3.37) for $n = 6$ and for $n = 7$ this has been achieved in 1982 by[65] van der Linden [4125] who also deduced it in the case $n = 8$ from GRH. An unconditional proof in this case has been provided in 2014 by J.C. Miller [2860] who showed also that in the case $n = 9$ it follows from GRH.

A support for (3.37) has been provided by Fukuda and Komatsu. They showed first [1319] that for every prime p there is an explicit number $m(p)$ such that if $p \nmid h_{2^{m(p)}}^+$, then $p \nmid h_{2^n}^+$ for all n, and then applied this to show that any prime divisor of $h_{2^n}^+$ must exceed $1.8 \cdot 10^8$ [1320], and even 10^9 [1321].

It has been also conjectured that $h_{3^n}^+ = 1$, and this has been supported by the results of Morisawa who showed that the possible prime divisors of $h_{3^n}^+$ exceed 10^4 [2950] and even $4 \cdot 10^5$ [2951]. Later Fukuda, Komatsu and Morisawa extended this bound to 10^9 [1322]. This conjecture is known to be true for $n = 3$ (H. Bauer [216]) and $n = 4$ (Masley [2762]). One has also $h^+(p^2) = 1$ for $p = 5, 7$ (H. Bauer [216]).

It has been shown in 1965 by Ankeny, S. Chowla and Hasse [97] that $h_p^+ > 1$ holds for all primes p of the form $4n^2 + 1$ with n having an odd prime divisor. In 1977 S.D. Lang [2464] proved this for primes $p = a^2q^2 + 4$, where q is a prime and a is odd. Now one knows that for primes $p \leq 151$ one has $h_p^+ = 1$ (J.C. Miller [2862]), and under GRH this holds for $p \leq 241$ with exception of $p = 163, 191, 229$.

[63] Morris Newman (1924–2007), professor of the University of California at Santa Barbara. See [1459].

[64] Wolfgang Schwarz(1934–2013), professor in Freiburg and Frankfurt/M.

[65] In case $n = 7$ the published proof is incomplete, but can be repaired. See the review of [4125] by Diaz y Diaz in Math. Reviews 84e:12005.

3.4 Other Questions

1. In 1909 Thue[66] [4058] improved the old result of Liouville [2605] on approximations of algebraic numbers by rationals, showing that if α is algebraic of degree $n \geq 2$, then for every $c > 1 + n/2$ the inequality

$$\left|\alpha - \frac{p}{q}\right| \leq \frac{1}{q^c} \tag{3.38}$$

can have only finitely many solutions.

As a corollary showed Thue that if $F(X, Y) \in \mathbf{Z}[X, Y]$ is an irreducible form of degree ≥ 3, then the equation $F(x, y) = 0$ can have only finitely many solutions $x, y \in \mathbf{Z}$. This implies that if K is a cubic field, and $\theta \in \mathbf{Z}_K$ is not rational, then in the set $\{x + y\theta : x, y \in \mathbf{Z}\}$ there are only finitely many elements of a given norm.

An essential improvement has been later provided by Siegel [3768] who showed that (3.38) holds with $c > 2\sqrt{n}$ (the *Thue–Siegel theorem*). He obtained also a corresponding result for approximation of algebraic numbers by algebraic numbers of smaller degree. He showed (Satz 1) that if $H(\alpha)$ denotes the height of the minimal polynomial for α over $\mathbf{Z}$ and ξ is a fixed algebraic number of degree $d \geq 2$ over a field K, then for every $\varepsilon > 0$ the inequality

$$|\xi - \alpha| \leq \frac{1}{H(\alpha)^{t+\varepsilon}}, \tag{3.39}$$

with

$$t = \min_{s<d}\left(\frac{d}{1+s} + s\right)$$

can hold only for finitely many numbers α generating K. This implies the inequality

$$|\xi - \alpha| > \frac{c}{H(\alpha)^{2\sqrt{d}}}, \tag{3.40}$$

with some $c = c(\xi, K) > 0$.

The second result (Satz 2) asserts that if α runs over all algebraic numbers of a fixed degree $m < n$, then

$$|\xi - \alpha| > \frac{c_1}{H(\alpha)^{2m\sqrt{n}}} \tag{3.41}$$

holds with a certain positive $c_1 = c_1(\xi, m)$.

In the same year Siegel [3770] showed that some stronger inequalities than (3.40) and (3.41) hold with possibly infinitely many but extremely rare exceptions.

In 1929 A. Brauer[67] [424] showed that the right-hand side of (3.41) may be replaced by

$$c(\alpha, n/m)/H(\xi)^{n/m},$$

which in certain cases improves Siegel's result.

It has been shown by Dyson [1032] in 1947 and Gelfond [1406] in 1948 that the exponent $2\sqrt{d}$ in (3.40) can be replaced by $\sqrt{d}$, and in 1955 Roth [3525] stated that it can be replaced by any number $\lambda > 2$, the constant c depending on ξ, K and λ, and proved this assertion for $K = \mathbf{Q}$. A

[66] Axel Thue (1863–1922), professor in Oslo. See [366, 483].

[67] Alfred Theodor Brauer (1894–1985), brother of Richard Brauer, professor at the University of North Carolina. See [1917, 3496].

proof in the general case has been presented in the second volume of the book [2577] by LeVeque. This result can be also obtained as a corollary to the subspace theorem (see below).

The results in [3768] had important applications in the theory of Diophantine equations. One of them (Satz 4) implies the following assertion:

If m is a given integer and n is sufficiently large, then for every algebraic number ξ of degree n there are only finitely many numbers of given norm in the set $\{\sum_{j=0}^{m} x_j\xi^j \,:\, x_j \in \mathbf{Z}\}$.

The following generalization has been conjectured later: if K is an algebraic number field of degree n, $m < n$ and $\alpha_1 = 1, \alpha_2, \dots, \alpha_m \in K$ are $\mathbf{Q}$-linearly independent, then there are only finitely many elements of the $\mathbf{Z}$-module $M = \bigoplus_{j=1}^{m} \alpha_j \mathbf{Z}$ with a given norm, except when M is degenerated; i.e. the $\mathbf{Q}$-linear space generated by M contains a subspace V, and there exists $\beta \in K$ such that βV is a field, not equal to $\mathbf{Q}$ or to an imaginary quadratic field.

For $m = 3$ this conjecture has been established in 1967 by W.M. Schmidt [3648] as an application of his results on simultaneous approximation to algebraic numbers [3647]. In the case of imaginary Abelian fields with degrees not divisible by 3 and 4 it has been proved for $m = 4$ by Győry [1564]. The general case has been settled in 1971 by W.M. Schmidt [3650, 3651] who used the following *subspace theorem* established by him in [3649, 3650] (see also [3655], p. 275).

If $L_1, \dots, L_n$ are linear forms in n variables with algebraic coefficients, then for every $\delta > 0$ there is a finite set $\{T_1, \dots, T_m\}$ of proper subspaces of $\mathbf{Q}^n$ such that if $\bar{x} \in \mathbf{Z}^n$ satisfies

$$\left| \prod_{j=1}^{n} L_j(\bar{x}) \right| < \frac{1}{|\bar{x}|^{\delta}} ,$$

then $\bar{x} \in \bigcup_{i=1}^{m} T_i$.

A survey of various applications of this theorem has been given in 2011 by Bugeaud [498].

Siegel showed also ([3768] Satz 7) that for an algebraic number ξ of degree n and given m there exist only finitely many polynomials $f \in \mathbf{Z}[X]$ with $\deg f = m$ satisfying

$$0 < |f(\xi)| \le H(f)^{-c}, \tag{3.42}$$

where

$$c = m^2 \left(\frac{n}{s+1} + s \right) - 1 + \varepsilon ,$$

with arbitrary $1 \le s < m$ and $\varepsilon > 0$.

As a corollary (Zusatz 6 to Satz 7) he proved that if N_α is the maximal norm of a prime ideal dividing $f(\alpha)$, where f is a fixed polynomial in $\mathbf{Z}_K[X]$ having at least two distinct zeros, and $\alpha \in \mathbf{Z}_K$, then N_α can be bounded only for finitely many α. In particular such polynomials can represent only finitely many units (see Sect. 4.1.3).

2. The impossibility of Fermat's equation

$$X^n + Y^n = Z^n \quad (XYZ \neq 0) \tag{3.43}$$

for $n = 5$ in $\mathbf{Q}(\sqrt{5})$ follows in view of $\mathbf{Q}\left(\sqrt{5}\right) \subset \mathbf{Q}(\zeta_5)$ from results of Kummer [2352, 2356]. Proofs using only elementary properties of quadratic fields have been found by Plemelj [3311] in 1912 and Tschakaloff[68] [4076] in 1926.

It has been proved in 1912 by Bohniček [380, 381] that (3.43) with $n = 2^k$ has no solution in the field $\mathbf{Q}(\zeta_{2^{k+1}})$ for $k \ge 3$.

[68] Lubomir Tschakaloff (1886–1963), professor in Sofia.

The case $n = 3$ has been considered in 1913 by Fueter [4321], who showed that if $d < 0$, $d \equiv 2 \bmod 3$ and the class-number of $\mathbf{Q}\left(\sqrt{d}\right)$ is not divisible by 3, then (3.43) has no non-trivial solutions in (in [1311] he gave another proof), and noted that in certain quadratic fields (imaginary and real) there are infinitely many solutions. Two years later Burnside [522] established in a completely elementary way the following necessary and sufficient condition for solvability in $\mathbf{Q}\left(\sqrt{d}\right)$: the equation $dX^2 + 12Y^3 + 3 = 0$ has a rational solution (see also Duarte [1001] and Mirimanoff [2887]).

In 1935 Holzer[69] [1889] showed that the cubic Fermat equation has no solutions prime to 3 in $\mathbf{Q}(\sqrt{m})$ with $m \equiv 5 \bmod 8$. A further study of that equation in quadratic fields has been done in the fifties by Aigner [32–36]. In 2013 M. Jones and Rouse [2072] applied a form of the Birch–Swinnerton-Dyer conjecture to describe quadratic fields $\mathbf{Q}\left(\sqrt{d}\right)$ in which FLT holds in the cubic case. Their result is similar to the description of congruent numbers done by Tunnell [4079] and utilizes the number of representation of d and $d/3$ by certain ternary quadratic forms.

In 1934 Aigner [30] showed that for $n = 4$ the equation (3.43) has no solutions in quadratic fields except for $\mathbf{Q}\left(\sqrt{-7}\right)$, where

$$\left(1+\sqrt{-7}\right)^4 - \left(1+\sqrt{-7}\right)^4 = 2^4 .$$

is essentially the only solution. Another proof has been given by Faddeev [1157].

In 1957 [37] Aigner showed that in cases $n = 6, 9$ there are no solutions in quadratic fields.

Holzer [1889] generalized Wieferich's[70] [4397] criterion ($2^{p-1} \equiv 1 \bmod p^2$) for a family of quadratic fields. He used also class-field theory to give a proof of Kummer's result on (3.43) in the case of regular prime exponents.

In 1935 Morishima[71] [2953] showed that if Fermat's equation with prime exponent p in $\mathbf{Q}(\zeta_p)$ has solutions prime to p (the so-called *first case*), then for all $m \le 31$ one has

$$m^{p-1} \equiv 1 \pmod{p^2} .$$

(This was known earlier in the case of rational solutions (Morishima [2952]).

In 1939 Rosser[72] [3522] showed that for odd primes $p < 8\,332\,403$ Fermat's equation in $\mathbf{Q}(\zeta_p)$ does not have solutions in the first case. That range has been enlarged in 2001 to $p < 7.568 \cdot 10^{17}$ by Kolyvagin [2202, 2203], who extended Morishima's criterion up to $m \le 89$ (see also [2201, 2204]). Another extension provided Anglés [84] in 2001.

In 1950 Dénes [890] showed that if $p^5 \nmid h_p$, then the first case of Fermat's theorem holds for the exponent p in the field $\mathbf{Q}(\zeta_p)$.

Another criterion for the first case of Fermat's equation in cyclotomic fields has been given in 1989 by Terjanian [4047].

Let K be a number field. A prime p is called K-regular, if it does not divide the class-number of $K(\zeta_p)$. In 1984 Hao and C.J. Parry [1623] showed that if K is a real quadratic field and p is a K-regular prime splitting in K, then the second case of Fermat's theorem for p holds in K.

Several authors studied algebraic points of small degree on the Fermat curve $F_p : X^p + Y^p = Z^p$ for prime p. In the case $p = 3, 5, 7, 11$ Gross and Rohrlich [1525] showed in 1978 that the only

[69]Ludwig Holzer (1891–1968), professor in Vienna, Graz and Rostock.

[70]Arthur Wieferich (1884–1954), teacher of mathematics.

[71]Taro Morishima (1903–1989), professor in Tokyo.

[72]John Barkley Rosser (1907–1989), professor at the University of Wisconsin in Madison.

points of degree $\leq (p-1)/2$ are $(0,1,1)$, $(1,0,1)$, $(-1,1,0)$, $(\zeta_6, \zeta_6^{-1}, 1)$ and $(\zeta_6^{-1}, \zeta_6, 1)$. In 1994 Debarre and Klassen [835] showed that for $p \geq 7$ the set of points of degree $\leq p-2$ is finite (such points are called *low-degree points*), and found all points with degree ≤ 6 on F_5. They mentioned also that all known low-degree points lie on the line $Y = 1 - X$ (they considered the Fermat curve in the form $X^n + Y^n = 1$), and later Klassen and Tzermias [2164] formulated the conjecture that every low-degree point lies on the line $X + Y = Z$. Their paper contains also a list of points of degree ≤ 6 on F_5, and a list of points of degree ≤ 5 on F_7 has been provided in 1998 by Tzermias [4085]. The search for low-degree points has been continued in the twenty-first century by Sall [3565] and Tzermias [4086, 4087].

In 2004 Jarvis and Meekin [2039] established FLT for the field $\mathbf{Q}\left(\sqrt{2}\right)$, and recently Freitas and Siksek [1244] extended this result to $\mathbf{Q}\left(\sqrt{d}\right)$ for $3 \leq d \leq 23$ with exception of $d = 5, 17$. In [1243] they applied the theory of modular forms to show that if $d \equiv 3, 6, 10$ or $11 \bmod 16$, then FLT holds for prime exponents $p > n(K)$ with $n(K)$ effectively computable. They showed also that this holds for all totally real fields in which a generalization of the Eichler–Shimura conjecture (asserting that for any Hilbert newform F of level N and parallel weight 2, having rational eigenvalues, there is an elliptic curve over K having the same L-function as F) holds.

An analogue for quadratic fields of Wendt's [4367] criterion for the first case of FLT has been proved in 2015 by Kraus [2270]. He showed also [2271] that if K is a field with $p \nmid h(K(\zeta_p))$ and p is inert in K, then the second case of FLT holds in K for the exponent p. In contrast to the results quoted above the proof does not use modularity.

3. In 1919 Pólya[73] [3334] studied the set $Int(\mathbf{Z}_K)$ of polynomials over an algebraic number field K, mapping $\mathbf{Z}_K$ into $\mathbf{Z}_K$, and called a sequence of polynomials $p_n \in Int(\mathbf{Z}_K)$ a *regular basis*, if for $n = 0, 1, \ldots$ one has $\deg p_n = n$, and the p_n's span $Int(\mathbf{Z}_K)$ as a $\mathbf{Z}_K$-module. He proved that if $h(K) = 1$, then a regular basis exists, showed that such basis exists in a quadratic field K if and only if all prime ideals dividing $d(K)$ are principal and presented a characterization of fields with a regular basis. Fields for which a regular basis exists are now called *Pólya fields*. It has been shown by Ostrowski [3200] that K is a Pólya field if and only if for every prime power q the product of all prime ideals of norm q is principal. He deduced that cyclotomic fields of prime power order are Pólya fields.

Much later, in 1982, Zantema [4462, 4463] observed that all cyclotomic fields and their maximal real subfields are Pólya fields. He showed also that if $K/\mathbf{Q}$ is Galois with group G, then K is a Pólya field if and only if the cardinality of the first cohomology group $H^1(G, U)$ of units of K equals the product of ramification indices of primes. An algorithm leading to a regular basis in the case it exists was given by Gerboud [1410].

Further results about Pólya fields were obtained by Heidaryan and Rajaei [1749], Leriche [2567–2569].

The theory of integral-valued polynomials in arbitrary domains was presented in the book [550] by P.-J. Cahen and Chabert, who recently presented a survey [551]. See also [3069].

4. In the first years of the twentieth century many papers appeared dealing with elementary algebraic proprieties of number fields.

[73] George Pólya (1887–1985), professor at ETH in Zürich and at Stanford University. See [59, 4033].

It has been observed in 1903 by G. Fontené[74] [1208] that adjoining to the field $\mathbf{Q}\left(\sqrt{-5}\right)$ the numbers $\left(x + y\sqrt{-5}\right)/\sqrt{2}$ with $x - y$ even leads to a set of numbers having the unique factorization property. A simple proof has been provided by E. Cahen [548] who observed that a similar approach can be realized for all quadratic fields. Actually this has been earlier shown by Klein in his lectures (see pp. 179–180 in [2167]) using a geometrical approach to the theory of quadratic fields and forms.

In 1908 Bennett [290] considered factorizations in $\mathbf{Q}\left(\sqrt{-6}\right)$, noting in particular that every prime congruent to 1 or 7 mod 24 has unique factorizations there.

Bickmore and O. Western [350] prepared in 1911 a table giving a prime factor in the field of eighth roots of unity for every prime $p < 25\,000$ congruent to unity mod 8.

Factorizations of primes in cyclic cubic fields with discriminant of the form p^2 with a prime p conductor were described by Westlund [4376].

Elementary observations about the way of factorizing of prime ideals in composites of two Galois extensions were given by M. Bauer [229, 233, 234] (see also [253, 254]).

Factorization of prime ideals in extensions of the form $K(\zeta_p)/K$ with prime p was treated by Rella [3432] in 1919.

5. All solutions of the congruence $x^{\varphi_K(P)} \equiv 1 \bmod P^n$ for a prime ideal P were determined in 1903 by Westlund [4372] in 1903.

In 1909 H. Weber [4329] made a study of cyclic extensions of prime power degree. One finds there on p.39 that if K/k and L/k are normal extensions of degrees m respectively n, and $[KL : k] = N, [K \cap L : k] = M$, then $MN = mn$. It has been observed in 1911 by Frobenius [1260] that it suffices to assume that one of the extensions $K/k, L/k$ is normal. He also noted that Weber's result is equivalent to a statement about reduction of polynomials given as Theorem 13 on p. 269 in Jordan's[75] treatise on permutations [2073].

In 1911 Châtelet [658] presented a description of Abelian cubic fields utilizing integers of the field $\mathbf{Q}(\zeta_3)$. Later [662, 663] he announced similar descriptions of arbitrary Abelian fields.

In the same year Carver[76] [593] described a way of presenting ideals in quadratic number fields.

In 1913 R. König [2208] gave a detailed description of the relations between quadratic fields and binary quadratic forms.

In 1917 Furtwängler [1346] presented a criterion for checking whether a given number is algebraic of a given degree.

---

[74]Georges Fontené (1848–1923), teacher in Belfort, Douai, Rouen and at Collége Rollin in Paris. See [449].

[75]Marie Ennemond Camille Jordan (1838–1922), professor in Paris. See [1852, 2494].

[76]Walter Buckingham Carver (1879–1961), professor at the Cornell University.

Another criterion for positive numbers, based on an analogue of the Euclidean algorithm given by Brun[77] [480, 481], has been given in 1920 by Pipping[78] [3297, 3298]. Later he provided a simplification [3299, 3300]. See also Pisot [3304].

Simple properties of sums of two quadratic residues with respect to an odd prime ideal in a quadratic field were considered in 1919 by Matsumoto [2776]. His paper was the first paper on algebraic number theory published in the Tôhoku Mathematical Journal.

In 1918 H.H. Mitchell [2889] made the following observation: let l be a prime which is not a Fermat prime, let $2e \mid l-1$ with $m=(l-1)/2e$ odd, and let k be a subfield of $\mathbf{Q}(\zeta_l)$ of degree $2e$. If q is a prime with $q^m \equiv 1 \bmod l$ and in k one has $q^{h_l^-} = \mathfrak{A}\overline{\mathfrak{A}}$, (the bar denoting complex conjugation), then the ideal $\mathfrak{A}$ is principal.

This has been later generalized by H.S. Grant [1484] to the case when l is a prime power.

3.5 Books

1. In the first years of the twentieth century several books presented introductions to the theory of algebraic numbers: in 1900 Hilbert [1840] gave a short survey in the German mathematical encyclopedia, and in 1905 Bachmann [172] based his presentation of Dedekind's ideal theory, using at certain places also Kronecker's approach. In the book of Sommer [3853], published in 1907, one finds a theory of quadratic and cubic fields as well as of quadratic extensions of quadratic fields.

In 1908 the third volume of Weber's book [4321] appeared, being actually a changed and extended second edition of his book [4316], published in 1891. Its first part is analytical and brings a theory of θ-functions, elliptic functions and modular functions. The second part is devoted to quadratic fields, culminating in the proof of Dirichlet's class-number formulas and genera theorems. In the third part the theory of complex multiplication is presented, and in the fourth one finds the definition and some simple properties of the class-field. A sample theorem (p 612):

"If there exists a class-field L for the field K and $[L:K] \le h(K)$, then every ideal class in K contains infinitely many prime ideals".

In 1917 the book of Fueter [1305] appeared, in which he gave an introduction to the theory of cyclotomic fields $\mathbf{Q}(\zeta_p)$ with prime p.

In 1918 Landau published his book [2431], in which he presented the use of analytical methods in the theory of algebraic numbers. The book culminates with proofs of the Prime Ideal Theorem and the Ideal Theorem with error term $O(x^c)$ with $c = 1-2/(n+1)$, n denoting the degree of the field in question.

---

[77] Viggo Brun (1885–1978), professor in Oslo. See [3720].

[78] Nils Johan Pipping (1890–1982), professor in Turku.

In 1913 Châtelet [661] published an introduction to number theory, devoting three of its chapters to the theory of algebraic numbers.

It seems that the first textbook on algebraic numbers in the Western Hemisphere was written by Reid [3423] in 1910.

Chapter 4
The Twenties

4.1 Structure

4.1.1 *Ideal Theory*

1. The similarity between the theories of algebraic functions and algebraic numbers has been noted already in the paper of Dedekind and H. Weber [857] published in 1882, in which the authors presented a new theory of algebraic functions based on the notion of the ideal (see Haffner [1583]). This new approach has been omitted in a broad presentation of this theory in the report written in 1892 by Brill and M. Noether[1] [453], and therefore E. Noether prepared in 1919 a completion of that report, taking into account the ideal-theoretic methods [3120]. She based her approach on the theory of fields, presented earlier by Steinitz [3914], and pointed out several common points in the theories of algebraic numbers and algebraic functions.

In 1921 E. Noether published her first paper [3121] on the general theory of commutative rings, considering decomposition of ideals in commutative rings in which every ascending chain $I_1 \subset I_2 \subset \cdots \subset I_n \subset \dots$ of distinct ideals is finite.[2] She defined the divisibility $I \mid J$ by $J \subset I$ and defined the least common multiple (LCM) of the ideals $I_1, I_2, \dots, I_m$ as their intersection:

$$LCM(I_1, I_2, \dots, I_m) = \bigcap_{j=1}^{m} I_j. \tag{4.1}$$

The greatest common divisor (GCD) of ideals has been defined by

[1]Max Noether (1844–1921), professor in Heidelberg and Erlangen, father of Emmy Noether. See [452, 2675].

[2]Rings having this property are now called *Noetherian rings*.

W. Narkiewicz, *The Story of Algebraic Numbers in the First Half of the 20th Century*, Springer Monographs in Mathematics,
https://doi.org/10.1007/978-3-030-03754-3_4

$$GCD(I_1, I_2, \ldots, I_m) = \sum_{j=1}^{m} I_j. \tag{4.2}$$

An ideal I has been called irreducible if it is not of the form $I = LCM(J_1, J_2)$ with $J_1, J_2 \neq I$, and Noether showed (Satz II) that every ideal I can be written in the form (4.1), where I_j are irreducible ideals, none of which contained in the LCM of the other ideals I_j.

She noted that in general this representation is not unique by showing an example in the ring of polynomials in two variables (footnote on p. 33 in [3121]), but proved that the number of terms is uniquely defined by I (Satz IV). She called an ideal I primary if it satisfies the following condition: if $ab \in I$ and $a \notin I$, then some power of a lies in I and showed (Satz V) that a primary ideal I is divisible by a unique prime ideal $\mathfrak{P}$ and a certain power of $\mathfrak{P}$ is divisible by I. Then she established that all irreducible ideals are primary and the prime ideals corresponding to ideals I_j in (4.1) are uniquely determined by I. In §8 it is shown that if a Noetherian ring R has a unit element, then the least common multiple of ideals $I_1, I_2, \ldots, I_m$ satisfying $GCD(I_i, I_j) = R$ for $i \neq j$ equals their product. This applies in particular to orders in algebraic number fields. A simpler proof of this theorem has been found by Krull [2302], who later presented an axiomatic approach to the theory of ideals [2303]. See also Sono [3858].

This generalized the results obtained with the use of elimination theory by Lasker [2473], Macaulay[3] [2674] and Schmeidler[4] [3633, 3634] in the case of rings of polynomials.

2. In 1925 Krull [2307] studied commutative rings in which the condition $I \subset J$ for ideals implies the existence of an ideal J' satisfying $I = JJ'$. He called them *regular multiplication rings* (*reguläres Multiplikationsring*). Now they are called rather *multiplication rings*. Noetherian domains having this property coincide with Dedekind domains.

Characterizations of multiplication rings have been given later by Prüfer [3343], Mott [2986], Gilmer and Mott [1435] and D.D. Anderson [73, 74]. For further results on these rings see the papers by Akizuki [38], S. Mori [2941–2943], Krull [2318] and Griffin [1520].

3. In 1926 E. Noether [3123] continued her study of ideal theory in Noetherian domains, providing simpler proofs for the main results of [3121]. She established also a description of Dedekind domains by showing that they are characterized by the following three conditions:

(a) R is Noetherian.
(b) Every descending chain $I_1 \supset I_2 \supset \cdots \supset I \neq 0$ of distinct ideals of R containing a fixed non-zero ideal I is finite.
(c) R is integrally closed in its field of fractions.

---

[3] Francis Sowerby Macaulay (1862–1937), teacher at St Paul's School in London. See [182].

[4] Werner Schmeidler (1890–1969), professor at the Technische Hochschule Breslau and Technische Universität Berlin.

It has been shown later by Krull [2309] that the condition (*b*) may be replaced by

(*d*) Every non-zero prime ideal is maximal.

In this form the characterization of Dedekind domains appeared in the second volume of the book [4134] of van der Waerden published in 1931.

Later several other conditions characterizing Dedekind domains R have been established. We mention here only a selection of them:

(*i*) *Fractional ideals form a group under multiplication* (Krull, p. 13 in [2317]).
(*ii*) *Every non-zero proper ideal is a product of prime ideals* (Matusita [2779]).
(*iii*) *The domain is Noetherian, and for its maximal ideals M there are no ideals $I \neq M, M^2$ with $M^2 \subset I \subset M$* (I.S. Cohen [736]).
(*iv*) *The domain is Noetherian, and the lattice of its ideals is distributive, i.e. for all ideals I_1, I_2, I_3 one has $I_1 \cap (I_2 + I_3) = (I_1 \cap I_2) + (I_1 \cap I_3)$* (I.S. Cohen [736]).
(*v*) *For proper ideals I the factor ring R/I is a unique factorization ring* (C.U. Jensen [2046]).
(*vi*) *Every non-zero ideal is an intersection of finitely many powers of prime ideals* (Butts and Gilmer [533]. Cf. D.D. Anderson and E.W. Johnson [76]).
(*vii*) *The domain is Noetherian, and if $I \subset J$ are ideals with $IJ^n = J^{n+1}$ for some $n \geq 1$, then $I = J$* (Hays [1724]).

A simple proof of *(ii)* has been given by I.S. Cohen[5] [736] in 1950. For other characterizations of Dedekind domains see Kubo [2334], Matusita [2778], I.S. Cohen [736], Asano [145], L. Fuchs [1294], Ishikawa [1975], Butts [531], Butts and Wade [536], Gilmer [1433], Koyama, Nishi and Yanagihara [2236], Quadri and Irfan [3348], Huckaba and Papick [1916], Lequain [2561], Man [2731] and D.F. Anderson, H. Kim and J. Park [78].

In 1940 S. Mori [2947] characterized commutative rings in which every ideal is a product of prime ideals (Z.P.I.-rings). These rings were later studied by Gilmer [1432], Levitz [2583, 2584] and E.W. Johnson [2056]. Mori gave also a description of commutative rings in which every principal ideal is a product of prime ideals [2944–2946, 2948]. Such rings are now called π-rings (see D.D. Anderson and Matijevic [77] and D.D. Anderson [75]).

4. From the characterization of domains with unique factorization of ideals established by E. Noether in [3123] it follows that if O is a proper order in an algebraic number field K (i.e. $O \neq \mathbf{Z}_K$), then there exists a primary ideal I of O which is not a power of the prime ideal $\mathfrak{p}$ containing it.

In [3122] E. Noether called a primary ideal proper if it is not a prime ideal and generalized to orders the result of Dedekind on prime divisors of the field discriminant by showing that a rational prime p divides the discriminant of O if and only if in the factorization of the ideal pO into primary ideals at least one of the components is proper. An analogous result holds also for relative extensions.

A theory of orders in Dedekind domains has been developed in 1927 by Grell [1516] who applied his earlier results about relations between ideals in a ring and in its subring [1515]. He showed i.a. that a prime ideal $\mathfrak{p}$ in an order $\mathfrak{o}$ divides the conductor of $\mathfrak{o}$ if and only if $\mathfrak{p}$ contains a primary ideal which is not a power of $\mathfrak{p}$.

5. To show that the integral closure S of a Dedekind domain R in a finite extension of its quotient field is Dedekind one needs to show that S satisfies the conditions *(a)* and *(b)* or *(d)* above. For separable extensions this has been done by E. Noether [3123] in 1926. For the inseparable case see Sect. 5.1.1.

[5] Irvin Sol Cohen (1917–1955), professor at the Massachusetts Institute of Technology.

6. We noted already in Sect. 1.3.2 that in 1921 Furtwängler [1350] generalized to arbitrary number fields a geometrical approach used by Klein [2166, 2167] in the quadratic case.

Another approach to the theory of algebraic numbers has been presented in 1925 by Prüfer[6] [3342]: let K be an algebraic number field. With every infinite system of congruences

$$x \equiv a_j \pmod{b_j} \quad (j = 1, 2, \dots)$$

(with $a_j, b_j \in \mathbf{Z}_K$), whose each finite subsystem has a solution in $\mathbf{Z}_K$, Prüfer associated an ideal number and introduced a ring structure in the set of numbers so obtained. The resulting ring contained $\mathbf{Z}_K$, and although it had divisors of zero it turned out to be possible to use it in developing the theory of K and to prove that the ring of ideal numbers has the unique factorization property. He applied the same method also for fields of algebraic functions in one variable over an algebraic number field. Later von Neumann[7] [4241] presented another way to construct Prüfer ideal numbers, similar to Cantor's way of defining real numbers. He extended Prüfer's theory to the field of all algebraic numbers, using this time the approach used in [3342].

4.1.2 Integral Bases, Discriminants, Factorizations

1. In 1921 M. Bauer [236] gave a simple proof, not using ideals, of the fact that if the discriminants of K and L are co-prime, then $[KL : \mathbf{Q}] = [K : \mathbf{Q}][L : \mathbf{Q}]$.

In 1922 Speiser [3873] presented a way of determining the degree of prime ideal factors of unramified primes in the splitting field of a polynomial $F \in \mathbf{Z}[X]$, based on the observation that the degree of a prime ideal divisor of a prime p equals the minimal integer f such that $F(X)$ is a product of linear factors in the finite field of p^f elements.

A description of factorizations of primes in a cubic field, based on the theory of $\mathfrak{p}$-adic numbers, presented Wahlin [4267] in 1922.

In 1925 Värmon [4191] determined the discriminants of fields with Galois groups C_{p^n} and C_p^n (for prime p) and gave formulas for the number of such fields with given discriminant.

2. A method for finding all cubic fields of given discriminant has been presented in 1925 by Berwick [326]. He showed also that the number of cubic fields with a given discriminant can be arbitrarily large.

Another proof gave later Hasse [1663]. In 1930 Nagell [3020] proved that the same holds for fields of any fixed degree ≥ 3. The particular case of cubic cyclic fields has been treated by Cohn [745] in 1954.

[6] Heinz Prüfer (1896–1934), professor in Münster. See [273].

[7] John von Neumann (1903–1957), professor at the Institute in Princeton. See [4101]

Let $N(x)$, $P(x)$ denote the number of non-isomorphic cubic fields K of negative, respectively positive discriminants with $|d(K)| \leq x$. In 1969 Davenport[8] and Heilbronn [826] considered the conjecture

$$N(x) = (a + o(1))x, \quad P(x) = (b + o(1))x \tag{4.3}$$

with non-zero a, b and established for large x the inequalities

$$c_1 x \leq N(x) \leq c_2 x, \quad c_3 x \leq P(x) \leq c_4 x$$

with positive c_1, c_3. Two years later they established (4.3) with

$$a = \frac{1}{4\zeta(3)}, \quad b = \frac{1}{12\zeta(3)}$$

[827].[9]

An extension of this result to relative cubic extensions has been given in 1988 by Datskovsky and Wright [815]. They noted also on p. 125 that it should be possible to obtain a second term of the form $cx^{5/6}$ in these formulas. An explicit form of this conjecture has been proposed in 2001 by Roberts [3477], and it has been established in 2013 by Taniguchi and Thorne [3990] and Bhargava, Shankar and Tsimerman [341]. Earlier, in 2010, Belabas, Bhargava and Pomerance [276] proved the error terms in (4.3) to be $O(x^c)$ for any $c > 7/8$.

Denote by $N_d(x)$ the number of non-isomorphic extensions $K/\mathbf{Q}$ of degree d and $|d(K)| \leq x$. Moreover let $N_d(G, x)$ denote the number of extensions $K/\mathbf{Q}$ of degree d with $|d(K)| \leq x$ having G for the Galois group of the normal closure.

In the case of Abelian fields it has been shown by Mäki [2714] in 1985 that if G is an Abelian group of d elements, then

$$N(G, x) = \begin{cases} c(G)x^{\alpha(G)} P_G(\log x) + O_\varepsilon\left(x^{\beta(G)+\varepsilon}\right) & \text{if } d \not\equiv 2 \pmod 4, \\ c(G)x^{2/N} + O_\varepsilon\left(x^{\beta(G)+\varepsilon}\right) & \text{if } d \equiv 2 \pmod 4, \end{cases}$$

where $c(G) > 0$, $\alpha(G) = p/((p-1)d)$, p being the smallest prime divisor of d, $\beta(G) < \alpha(G)$, $P(X)$ is a polynomial of degree $(p^r - 1)/(p-1) - 1$, and r is the rank of the p-Sylow subgroup of G.

The analogue for relative Abelian extensions has been established in 1989 by Wright [4447].

Earlier results dealt with the groups C_p^m and C_{p^m} (see Baily [175, 176], Sarbasov [3576], Urazbaev [4103–4108] and X.K. Zhang [4467, 4468]).

It is conjectured that one has

$$N_d(x) = (c(d) + o(1))x \tag{4.4}$$

with $c(d) > 0$, but this has been established only for $d \leq 5$.

Upper and lower bounds for $N_4(G, x)$ with $G = S_4$ and $G = D_4$ were obtained in 1980 by Baily [175], and in 2002 H. Cohen, Diaz y Diaz and Olivier [735] established

$$N_4(D_4, x) = (c_1 + o(1))x .$$

This has been followed by

$$N_4(S_4, x) = (c_2 + o(1))x$$

with

$$c_2 = \frac{5}{24} \prod_p \left(1 + \frac{1}{p^2} - \frac{1}{p^3} - \frac{1}{p^4}\right),$$

[8] Harold Davenport (1907–1969), professor in Bangor, at the University College London and in Cambridge. See [2939, 2940, 3489, 3491].

[9] It follows from the remarks made by Cohn in [745] that Davenport and Heilbronn obtained these inequalities at least 15 years earlier.

obtained by Bhargava [339] in 2005 utilizing his parametrization of quartic rings [338]. He obtained also a similar result for quartic fields of fixed signature. This implied the proof of (4.4) for quartic fields. Later Belabas, Bhargava and Pomerance [276] showed that the error term in the last equality is $O(x^a)$ for any $a > 23/24$.

In 2005 Kable and Yukie [2088] used their construction of quintic rings presented in [2087] to establish the bound

$$N_5(x) = O\left(x^{1+\varepsilon}\right)$$

for every $\varepsilon > 0$, and five years later Bhargava [340] proved (4.4) for quintic fields, showing

$$N_5(x) = \frac{13}{120}\prod_p\left(1+\frac{1}{p^2}-\frac{1}{p^4}-\frac{1}{p^5}\right)x + o(x) \ ,$$

and getting asymptotics for quintics of a given signature. In 2014 Shankar and Tsimerman [3743] showed the error term to be $O(x^a)$ with $a = 199/200$. See also Larson and Rolen [2472] for an improvement in the case of $N_5(D_5, x)$.

An asymptotic formula for $N_6(S_3, x)$ has been given by Bhargava and Wood [343] and Belabas and Fouvry [277]. A power-saving bound for the error term has been given by Taniguchi and Thorne [3991].

In 2002 Malle [2725, 2726] conjectured that the number of normal extensions L/K of degree n with $N_{K/Q}(d(L/K) \leq x$ and Galois group which is permutation isomorphic to a fixed subgroup G of S_n equals

$$(c(K, G) + o(1))x^{a(K,G)} \log^{b(K,G)-1} x \ ,$$

with explicitly given exponents. For $n = 3, 4, 5$ and $G = S_n$ this conjecture has been confirmed by the results quoted above, and in 2004 Klüners and Malle [2178] established it for nilpotent groups. However in 2005 Klüners has shown that Malle's conjecture fails for the wreath product of C_2 and C_3. For lower bounds in the case of fields with dihedral group see Klüners [2176].

In the general case one knows only upper bounds for $N_d(x)$. In 1995 W.M. Schmidt [3656] proved

$$N_d(x) = O\left(x^{(d+2)/4}\right) \ ,$$

and in 2006 Ellenberg and Venkatesh [1081] improved this to

$$N_d(x) \ll c(d)x^{\alpha(d)}$$

with $\alpha(d) = \exp(c\sqrt{\log d})$, and obtained a similar bound for relative extensions of degree d.

3. In 1927 Berwick gave in his book [327] a method of constructing integral bases based on Newton polygons. In its last chapter he applied his method to the case of arbitrary pure extensions of the rationals. His results imply the form of factorization of the discriminants of these extensions.

Another method of finding integral bases and bases of ideals in arbitrary fields has been given by N.R. Wilson [4423, 4424] in 1927. If $K = \mathbf{Q}(\theta)$ with $\theta \in \mathbf{Z}_K$ is of degree n, then for $m \leq n - 1$ an integer

$$\alpha = \sum_{j=0}^{m} a_j\theta^j$$

with rational a_j's is called reduced if $a_j \in (-1/2, 1/2]$, and $1/a_m$ is a positive rational integer. If the denominators of a_j's are all powers of the same prime p, and

$1/a_m = p^t$ with the maximal possible t, then α is called a maximal reduced integer. After proving that the set of maximal reduced integers is finite, Wilson showed how one can use them to construct an integral basis of K and illustrated his method on the example of cubic fields.

4. It has been observed in 1926 by Nowlan [3135] that in cubic fields of discriminants 49 and 81 a rational prime p is a product of three prime ideals if and only if it is a cubic residue modulo the discriminant. Latimer [2475] showed that in cyclic cubic fields this condition is sufficient in the case $p \nmid 6d(K)$, and if $d(K)$ is either 81 or the square of a prime, then it is also necessary. Later he used this result to establish a class-number formula for cyclic cubic fields [2476].

Connections between factorization of primes in extensions K/k of prime degree and solvability of the Galois group of their normal closure have been studied by F.K. Schmidt [3643] in 1929.

In 1932 Herbrand (the second theorem[10] 4 in [1802] in the Galois case, Lemma 2 in [1804] in the general case) did this for arbitrary finite extensions.

5. In a series of papers [3165, 3172, 3173, 3175, 3177, 3179, 3182] Ore applied Newton polygons to study integral bases, discriminants and prime factorizations in algebraic number fields.

In [3165] he presented first the theory of these polygons, which he then applied to extend Dedekind's formula (1.26) to the case of primes dividing the index in the case[11] when certain auxiliary polynomials were without multiple roots (he noted on p. 255 that the same method is applicable also in the case of relative extensions). Later [3172] he has been able to show that this method can be applied to every extension, provided one selects an appropriate generator.

In the general case Ore established the following result (Theorem 24 in [3165]): *If $f(X) \in \mathbf{Z}[X]$ is monic and irreducible and for a prime p one has*

$$f(X) \equiv \prod_{j=1}^{r} f_j^{e_j}(X) \pmod{p},$$

where $f_1, \dots, f_r$ are distinct polynomials irreducible mod p, then in the field K generated by a root of f one has

$$p\mathbf{Z}_K = \prod_{j=1}^{r} I_j,$$

where

$$I_j = p\mathbf{Z}_K + f_j(X)^{e_j}\mathbf{Z}_K$$

[10] There are two Theorems 4 in Herbrand's paper.

[11] In the introduction Ore stated that his method works in all cases, but this had not been substantiated by the content of the paper.

are distinct proper ideals of $\mathbf{Z}_K$.

In §6 of [3165] showed Ore that for given prime p, integer $n > 1$ and positive integers $e_1, \dots, e_r, f_1, \dots, f_r$ satisfying

$$\sum_{j=1}^{r} e_j f_j = n$$

there exists a field K of degree n such that

$$p\mathbf{Z}_K = \prod_{j=1}^{r} \mathfrak{p}_j^{e_j}$$

holds with f_j being the degree of the prime ideal $\mathfrak{p}_j$ (in the case $r = n, e_1 = \cdots = e_n = f_1 = \cdots = f_n = 1$ this has been proved earlier by M. Bauer [237]). A way of finding a polynomial whose root generates a field having this property has been presented by Ore [3168] in 1924. In 1926 he extended his theorem to the case of several primes [3178], and Hasse [1652, 1653] generalized this result to relative extensions.

Further development of the theory of Newton polygons and their application to number fields presented Ore in 1928 in [3182].

Ore's method of describing prime ideal factors of a prime in extensions has been generalized to arbitrary fields with valuations by Mac Lane [2685] in 1936.

In [3173] applied Ore the same methods to determine the discriminant of a field and its factorization into prime powers and in [3175] did the same for the different (see also Ore [3181]).

In 1926 Ore [3179] applied Schönemann's [3682] theory of polynomials mod p^N to show (Satz 3) that if f is the minimal polynomial for an integer a generating the field K, p is a prime with $p^\delta \parallel d(K)$, and for some $\alpha \geq 1 + \delta$ one has

$$f(X) = \prod_{j=1}^{t} f_j(t) \pmod{p^\alpha}$$

with f_j mod p^α irreducible, then

$$p\mathbf{Z}_K = \prod_{j=1}^{s} \mathfrak{p}_j^{e_j}$$

with $N(\mathfrak{p}_j^{e_j}) = p^{n_j}$, where $n_j = \deg f_j$ for $j = 1, 2, \dots, t$.

Ore's theorem resembles the result of Dedekind (1.26), but covers also the case of primes dividing the index of a. On the other hand it does not determine the exponents e_j, but this lacuna has been filled out in the second part of Ore's paper [3180], in which also a generalization to relative extensions has been provided (a simpler proof has been later given by M. Bauer and Čebotarev [255]). This result has been used to present a new approach to the ramification theory, the special case of Galois fields being treated later in [3184, 3185], where Hilbert's ramification groups were analysed in great detail.

Ore's results on factorization of primes [3180] have been later utilized by Gut [1554] for the determination of the degrees of prime ideal factors of unramified prime ideals in relative extensions.

In Satz 1 of Chap. 3 in [3179] the result of Dedekind [845] concerning the conductor $f(O)$ of an order $O = \mathbf{Z}(a)$ in an algebraic number field K has been made more precise. Dedekind proved the equality

$$F'(a) = f(O)D_{K/\mathbf{Q}} ,$$

where $F(X)$ is the minimal polynomial of a, and $D_{K/\mathbf{Q}}$ is the different of $K/\mathbf{Q}$. Ore showed that $f(O)$ can be written in the form $f(O) = \prod_p f_p(a)$ with explicitly given local factors $f_p(a)$.

In 1936 M. Bauer [250] gave a simpler proof of this result and in [249] made more precise a result of Ore (Theorem 6 in [3179]) dealing with divisibility of the index of algebraic integers by prime powers.

On p. 339 of [3179] one finds a new proof of Dedekind's characterization of the field index $i(K)$, and in Theorems 7 and 8 the factorization of the different is described, and this leads to a generalization of Dedekind's result (1.27).

6. The maximal power p^δ of a prime p dividing the discriminant of a field of a given degree n has been determined in Satz 9 of Ore's paper [3178]:

If $n = \sum_{j=1}^r c_j p^{a_j}$ *with* $0 \le a_1 < a_2 < \cdots < a_r$ *and* $0 \le c_j < p$, *then*

$$\delta = \delta(p, n) = \sum_{j=1}^r c_j(a_j + 1)p^{a_j} - r .$$

An analogous result holds also for relative extensions (see Satz 14 in [3180]).

Earlier Hilbert (Theorem 80 of [1836]) proved that for Galois fields the exponent δ is bounded by a value dependent only on n. It can be deduced from results of Hensel [1774] (see Schur [3705]) that the same holds for all fields with the explicit bound

$$\delta(p, n) \le (a + 1)n - 1 , \tag{4.5}$$

where $p^a \parallel n$.

In [3183] Ore introduced a generalization of the decomposition group of prime ideals, obtained by associating to an arbitrary ideal $\mathfrak{A}$ of $\mathbf{Z}_K$ the subgroup of the Galois group of $K/\mathbf{Q}$ leaving $\mathfrak{A}$ invariant.

For later developments see Sect. 5.1.2.

4.1.3 Units

1. Unit equations occur for the first time in 1913 in a paper of Jacobsthal[12] [2018], who gave sufficient conditions for the system

$$\alpha_i X + \beta_i Y = \gamma_i \quad (i = 1, 2, \ldots, r)$$

(with given algebraic integers $\alpha_i, \beta_i, \gamma_i$) to have solutions which are units not necessarily lying in the field generated by the coefficients.

It follows from a result of Siegel (Zusatz 6 to Satz 7 in [3768]) that if $f(X) \in \mathbf{Z}_K[X]$ has at least two distinct zeros, then the set $\{f(\alpha) : \alpha \in \mathbf{Z}_K\}$ can contain only finitely many units. Applying this to the polynomial $f(X) = X(aX - c)$ one obtains that for fixed non-zero $a, b, c \in \mathbf{Z}_K$ the equation

$$au + bv = c \tag{4.6}$$

can have only finitely many unit solutions[13] u, v.

Unit solutions of (4.6) in case $a = b = c = 1$ are called *exceptional units.*

Siegel's result implies that in a given field there can be only finitely many exceptional units. In 1928 Nagell determined (Hilfsatz IV in [3018]) all such units $0 < u < 1$ in cubic fields of negative discriminant and later found all exceptional units in fields with unit rank one: in quadratic fields and cubic fields of negative discriminant in [3027], and in totally complex quartic fields in [3028, 3035, 3039] (see also [3041]). In [3042] he determined all exceptional units in $\mathbf{Q}(\zeta_7)^+$ and $\mathbf{Q}(\zeta_9)^+$ and showed that there are at least $3\varphi(n)$ exceptional units in $\mathbf{Q}(\zeta_n)$ for $n \neq 2^k$. In [3041] he showed that for $n \geq 5$ there exists a field of degree n having at least $6n - 9$ exceptional units.

It seems that the above consequence of Siegel's result has not been well known, as in 1961 S. Chowla [705] applied LeVeque's[14] [2577] generalization of Roth's[15] theorem to prove that the equation $u - v = 1$ in units of a given field has only finitely many solutions, stating that this solves a problem of J. Robinson.[16] The same result[17] has also been established in 1964 by Nagell [3035].

Exceptional units in cubic fields were studied by Ennola[18] [1105], and Niklasch and Smart [3112] considered them in quartic fields. In 1996 D. Grant [1481] showed that there are at least $(p-1)^2/2$ exceptional units in $\mathbf{Q}(\zeta_p)$.

In 2000 Wildanger [4399] found a quick algorithm for solving (4.6) which he used to determine all exceptional units in several fields having degrees as large as 22. He showed in particular that the field $\mathbf{Q}(\zeta_{22})$ contains 131 274 exceptional units.

---

[12] Ernst Jacobsthal (1882–1965), professor in Berlin and Trondheim. See [3726].

[13] This argument occurs on p. 207 of [3768] in the case of the equation $u - v = c$.

[14] William Judson LeVeque (1923–2007), professor at the University of Michigan. See [2790].

[15] Klaus Friedrich Roth, (1925–2015), professor at the University College and Imperial College in London.

[16] Julia Robinson (1919–1985), professor at Berkeley. See [3421].

[17] Actually for the equivalent equation $u + v = -1$.

[18] Veikko Ennola (1932–2013), professor in Turku.

Exceptional units are of importance in arithmetic of dynamical systems. See, e.g. Canci [556], Halter-Koch and Narkiewicz [1609, 1610], Morton and Silverman [2977], as well Sect. 10.3 of the book [1149] by Evertse and Győry.

Equation (4.6) has also only finitely many solution in S-units for S finite. In the case of $K = \mathbf{Q}$ this has been shown in 1933 by Mahler (Folgerung 2 in [2695]), who deduced it from the finiteness of the number of points with S-integral coefficients on algebraic curves of genus ≥ 1 established by him in case of the rational number field. He conjectured that the same holds also in other finite extensions of the rationals.

Mahler's conjecture has been proved by D.J. Lewis and Mahler [2587], as well as by S. Lang [2454] in 1960, and by LeVeque [2578] in 1961.

Lang's result is more general and implies that if K is any field of zero characteristic, and the group $\Gamma \subset K^*$ is finitely generated, then for $a, b \in K^*$ the equation $ax + by = 1$ has only finitely many solutions $x, y \in \Gamma$.

In 1974 Győry [1566] used Baker's method of bounding from below linear combinations of logarithms of algebraic numbers (see, e.g. Baker's book [181]) to obtain explicit upper bounds for the houses $\overline{|u|}, \overline{|v|}$ of solutions of the homogeneous version of Eq. (4.6), depending on the degree, class-number and discriminant of the field. In 1979 Győry [1569] generalized his result to the case of S-unit solutions. In the case $K = \mathbf{Q}$ such result has been also established by Kotov and Trelina [2231]. See also Lemma 6.2 in the book of Sprindžuk[19] [3879].

This result has been improved in 1996 by Bugeaud and Győry [501], and further improvements were made by Bugeaud [496] in 1998 and by Győry and K. Yu [1577] in 2006. A generalization to arbitrary finitely generated domains over $\mathbf{Z}$ has been made by Evertse and Győry [1148] in 2013.

An upper bound for the number of S-unit solutions of (4.6) has been given in 1979 by Győry [1569]. In 1984 Evertse [1140] gave the bound $3 \cdot 7^{n+2s}$ with $n = \deg K$ for S-unit solutions with $s = |S|$ (he wrote that the bound $c \cdot 20^s$ for the number of S-unit solutions of $x + y = 1$ has been obtained independently by Silverman in the preprint [3801]). Note that the obtained bound does not depend on the coefficients of the equation.

Next year Evertse and Győry [1145] obtained a similar bound for the number of solutions of equations considered by S. Lang in [2454] (see also Győry [1574]).

In 2007 Konyagin and Soundararajan [2213] showed that already in $\mathbf{Q}$ Eq. (4.6) can have at least $\exp(s^\beta)$ S-unit solutions with $s = |S|$.

In [1141] Evertse showed that there are only finitely many solutions in S-units of the equation

$$u_1 + u_2 + \cdots + u_n = 0, \tag{4.7}$$

such that no proper subsum of the left-hand side vanishes (non-degenerate solutions). This generalized a previous result in the case of $K = \mathbf{Q}$ proved independently by Schlickewei [3628] and E. Dubois and Rhin [1012]. A similar result has been obtained in 1982 by van der Poorten and Schlickewei [4129], who stated also in that the more general equation

$$a_1u_1 + \cdots + a_nu_n = a_0 \tag{4.8}$$

where a_i lie in a finitely generated extension $K/\mathbf{Q}$ and the u_i's lie in a finitely generated subgroup Γ of K^* has only finitely many solutions with no vanishing subsums of the left-hand side. Proofs were given later by Evertse and Győry [1146] (with an non-effective bound $O(C(n, \Gamma))$ for the number of such solutions with the implied constant not depending on the coefficients a_i) and van der Poorten and Schlickewei [4130]. For $n \geq 3$ this bound can be very large already for $K = \mathbf{Q}$ as shown by Evertse, Győry, C.L. Stewart and Tijdeman [1152]. In the same paper they proved that

[19] Vladimir Genad'evič Sprindžuk (1936–1987), professor in Minsk.

for most triples $a, b, c \in K$ Eq. (4.6) has at most two solutions in S-units. In the case of $K = \mathbf{Q}$ a stronger assertion has been established in 1990 by Brindza[20] and Győry [456].

In 1990 Schlickewei [3629, 3630] made the bound $C(n, \Gamma)$ effective, first for $K = \mathbf{Q}$, and then in the case when K is an algebraic number field and Γ is the group of S-units. He obtained this as a consequence of his generalization of the quantitative subspace theorem to number fields [3631]. The improvement of the subspace theorem by Evertse and Schlickewei [1153] made possible to obtain the bound $\exp\left((6n)^{3n}(1+r)\right)$ in the case of Eq. (4.8), where a_i are non-zero elements of an arbitrary algebraically closed field K of characteristic 0, and $u_i \in \Gamma^n$, where $\Gamma \subset K^*$ is a group having r generators (Evertse, Schlickewei and W.M. Schmidt [1154]). A further improvement of the subspace theorem has been obtained in 2013 by Evertse and Ferretti [1144]).

Surveys on unit equations and their applications were given by Evertse, Győry, C.L. Stewart and Tijdeman [1151] in 1988 and by Győry [1573] in 1992. A survey on the number of exceptional units was given by Niklasch [3111] in 1997.

A book presenting the theory of unit equations and its applications, written by Evertse and Győry [1149], and having an important list of references appeared in 2015.

2. One of the results in Siegel's paper [3769] (Satz 8) shows that for every totally real field K there is a rational integer $m = m(K)$ such that for every $\alpha \in \mathbf{Z}_K$ the number $m\alpha$ is a sum of units of K.

In 1964 B. Jacobson [2013] showed that all integers in $\mathbf{Q}(\sqrt{d})$ for $d = 2, 5$ are sums of distinct units and conjectured that this cannot happen in other quadratic fields. His conjecture has been later established by Śliwa [3827]. There are infinitely many quartic fields in which every integer is a sum of distinct units (Belcher [278]), as well infinitely many cubic fields with positive discriminant (Belcher [279]). For quartic fields see Hajdu and Ziegler [1588] and Dombek, Masáková and Ziegler [986].

The question whether there exists a field in which every integer is a sum of a bounded number of units has been answered in the negative (Hajdu [1587], Jarden and Narkiewicz [2038]). Earlier Ashrafi and Vámos [147] showed that this cannot happen in quadratic and cubic fields. Dombek, Hajdu and Pethő [985] proved a generalization in which units are replaced by integers of bounded norm.

Asymptotic formulas for the number of non-associated integers α of a field with $|N(\alpha)| \leq x$ which are sums of m units have been given in 2009 by C. Fuchs, Tichy and Ziegler [1288]. The case of real quadratic fields has been earlier treated by Filipin, Tichy and Ziegler [1186]. A generalization to S-integers and S-units has been obtained by C. Frei, Tichy and Ziegler [1237].

Tichy and Ziegler [4062] described complex cubic fields in which every integer is a sum of units. The same question has been considered by Ziegler [4471], Pethő and Ziegler [3270] and Ziegler [4472] for quartic fields.

C. Frei [1236] showed that every algebraic number field has a finite extension in which every integer is a sum of units.

It has been shown in 2014 by Dombek, Hajdu and Pethő [985] that if K is not a CM-field, then every integer of K can be written as a linear combination of units of K with coefficients belonging to the set $\{1, 1/2, \dots, 1/m\}$ for some m depending only on K.

A survey on problems dealing with sums of units has been given by Barroero, C. Frei and Tichy [194].

3. A method of finding units in pure cubic fields has been proposed in 1923 by Wolfe [4435]. In 1926 Pierce [3289] used recursive formulas for the approximation of the smallest root of a cubic equation to obtain fundamental units for some classes of pure cubic fields.

---

[20] Béla Brindza (1958–2003), professor in Debrecen.

The relation of the group of units of a field K to its subgroup generated by units of proper subfields of K has been studied in 1929 by Pollaczek[21] [3331].

In 1927 Stein [3910] applied continued fractions in the field $\mathbf{Q}(i)$, introduced by J. Hurwitz [1941, 1942], to the determination of the fundamental unit in quadratic extensions of $\mathbf{Q}(i)$.

4.2 Analytical Methods

4.2.1 *Quadratic Reciprocity Law*

1. Hecke's [1737] proof of quadratic reciprocity in quadratic number fields was based on a generalization of Gaussian sums to number fields. If K is quadratic with discriminant d, $\omega \in K$ and $\omega \mathbf{Z}_K = I/J$ $((I, J) = 1)$, then the quadratic Gaussian sum $G(\omega)$ is defined by

$$G(\omega) = \sum_{x \bmod J} \exp\left(2\pi i T\left(\frac{x^2\omega}{\sqrt{d}}\right)\right),$$

$T(a)$ being the trace of a. He established a reciprocity theorem for these sums in the case of real quadratic fields, relating $G(\omega)$ to $G(-1/4\omega)$ (another proof, valid also for imaginary quadratic fields, has been given in 1922 by Mordell [2926]), and mentioned that an analogous result holds in all totally real fields. A proof for all fields K, not necessarily totally real, has been given in Hecke's book ([1742], §54, pp. 56–58).

2. The quadratic reciprocity law for arbitrary number fields has been established by Hecke in his book [1742] in 1923. The main tools in his proof were suitably defined θ-functions. In the case of a totally real field K of degree n these functions were defined in the following way:

Let $I = \sum_{j=1}^{n} \alpha_j \mathbf{Z}$ be an ideal of K, denote by $\alpha^{(i)}$ $(i = 1, 2, \ldots, n)$ the conjugates of $\alpha \in K$, put $u = [u_1, u_2, \ldots, u_n]$, $t = [t_1, t_2, \ldots, t_n]$ with real u_i and positive t_i, and

$$z_i = \sum_{j=1}^{n} \alpha_j^{(i)} u_j \quad (i = 1, 2, \ldots, n).$$

The θ-function associated with I defined Hecke by

$$\theta(t, \bar{z}; I) = \sum_{\mu \in I} \exp\left(-\pi \sum_{j=1}^{n} t_j (\mu^{(j)} + z_j)^2\right),$$

[21] Leo Felix Pollaczek (1892–1981) worked in a telephone company in Germany and later in France.

and in the case $z_1 = \cdots = z_n = 0$ established the relation

$$\theta(t, \bar{0}, I) = \frac{1}{N(I)\sqrt{|d(K)|}\sqrt{t_1 t_2 \cdots t_n}} \theta(1/t, 0, 1/ID), \tag{4.9}$$

$D = D_{K/\mathbf{Q}}$ denoting the different of K. In the general case the θ-function looks more complicated.

3. In the case of arbitrary number fields Hecke defined the Gaussian sum by

$$C(\omega) = \sum_{x \bmod I} \exp\left(2\pi i T(x^2\omega)\right) ,$$

where the ideal I is defined by

$$\omega \mathbf{Z}_K = \frac{J}{ID} \quad (I, J) = 1 .$$

In [1739] Hecke determined the sign of quadratic Gaussian sums in number fields and later provided another proof of his result based on the theory of modular forms [1744].

Gaussian sums in real quadratic fields were later utilized by Kloosterman[22] [2174] in his study of theta-series and modular forms. Another proof of Hecke's reciprocity theorem for Gaussian sums has been given in 1936 by Kunert [2378].

Much later, in 1951, Hasse [1697] defined Gaussian sums with arbitrary characters in the following way:

If I is an ideal in $\mathbf{Z}_K$ and χ is a character of the multiplicative group of the factor ring $\mathbf{Z}_K/I$, then for $a \in (ID_{K/\mathbf{Q}})^{-1}$ the Gaussian sum $\tau_a(\chi)$ is defined by

$$\tau_a(\chi) = \sum_{x \bmod I} \chi(x) \exp\left(2\pi i T_{K/\mathbf{Q}}(ax)\right) ,$$

$T_{K/Q}(a)$ being the trace of a.

Next year Hasse [1701] defined Gaussian sums associated with characters of Galois groups of algebraic number fields, studied their properties and stated several conjectures. Some of them were later proved by him in [1702].

In 1953 Lamprecht[23] [2412] defined Gaussian sums in arbitrary finite commutative rings. An explicit formula for quadratic characters in this case has been given in 2002 by Szechtman [3968].

A further generalization was provided by Lakkis [2403–2405].

4. For the Gaussian sums Hecke established in [1742] a reciprocity formula using the relation (4.9) and its generalization to fields which are not totally real, and this quickly led to the quadratic reciprocity law which we present here in its simplest case:

In the case when integers α, β have odd norms, generate co-prime ideals, and at least one of them is a quadratic residue mod 4, then this reciprocity law has the form

[22]Hendrik Douwe Kloosterman (1900–1968), professor in Leiden. See [3881].

[23]Erich Lamprecht (1926–2003), professor in Saarbrücken.

$$\left(\frac{\alpha}{\beta}\right)\left(\frac{\beta}{\alpha}\right) = (-1)^g$$

where

$$g = \sum_{j=1}^{r_1} \frac{\operatorname{sgn} \alpha_j - 1}{2} \cdot \frac{\operatorname{sgn} \beta_j - 1}{2},$$

α_j, β_j being the conjugates of α, β in real embeddings of the field.

This has been generalized by Hasse [1647] in 1924, who showed that if $\alpha \equiv \beta \equiv 1 \bmod 2$, then

$$\left(\frac{\alpha}{\beta}\right)\left(\frac{\beta}{\alpha}\right) = (-1)^{g+t},$$

where t denotes the trace of $(\alpha - 1)(\beta - 1)/4$.

A simpler proof of the last result has been given in 1960 by Siegel [3790]. In this paper established Siegel the reciprocity law for multiple Gauss sums, stated with an incomplete proof in 1903 by Krazer[24] [2275]. A particular case has been established earlier by Braun [435]. See also Shiratani [3760] and H. Reiter [3430].

In 1964 Weil [4353] presented a proof of Hecke's quadratic reciprocity law based on the theory of unitary representations of locally compact Abelian groups, later simplified by Cartier [591]. A proof using Fourier transform has been given in 1982 by Auslander, Tolimieri and Winograd [151]. See also the book by M.C. Berg [296]. For evaluations of Gaussian sums occurring in Hecke's proof see Braun [435] and Appendix 4 in the book [2867] by Milnor and Husemoller.

For proofs of the quadratic reciprocity law in imaginary quadratic fields see Fueter [1310] and Herglotz [1809].

A discussion of some of these proofs has been given in the book of Lemmermeyer [2527]. See also Kloosterman [2174], Skolem [3809].

4.2.2 Sums of Powers

1. Hilbert's assertion that every totally positive number of an algebraic number field K is a sum of four squares of elements of K has been established for quadratic fields by Landau in 1919 [2439]. He showed also that every totally positive number of K can be written as a sum of squares of elements of K (another proof was given in 1927 by Artin [119]). For cubic fields Hilbert's statement has been established in 1921 by Mordell [2925].

The proof of the four squares theorem for all K has been found in 1921 by Siegel (Satz 1 in [3769]). He used Hilbert's theory of quadratic extensions [1837–1839, 1843], Furtwängler's reciprocity theorem [1333], as well as Hecke's number field analogue of Dirichlet's Prime Number Theorem [1734]. His paper contains also the

[24] Adolf Krazer (1858–1926), professor at the Technische Hochschule in Karlsruhe.

necessary and sufficient conditions for representability of a totally positive integer as a sum of two (Satz 6) or three (Satz 7) squares.

2. Denote by J_m the set of integers of $K_m = \mathbf{Q}(\sqrt{m})$ which are sums of squares of integers of K_m, and let $r_{m,s}(a)$ be the number of representations of $a \in J_m$ as a sum of s squares of integers of K_m.

In 1906 G. Humbert[25] [1920, 1921] applied elliptic functions to the study of $r_{m,2}(a)$ and $r_{m,3}(a)$ for $m = 2, 3, 5$. He gave a formula for $r_{5,3}(a)$, and in 1924 Kirmse [2156] obtained formulas for $r_{m,4}(a)$ for $m = 2, 3, 5, 13$. When $m = 5$ and a is odd (i.e. $(a, 2) = 1$) he showed

$$r_{5,4}(a) = 8 \sum_{I|a} N(I) ,$$

which is an analogue of Jacobi's formula in the case of rational integers. The case $m = 5$ has been also treated in 1928 by Götzky [1479] using Hilbert's modular forms in two variables. His result implies that in $\mathbf{Q}(\sqrt{5})$ every totally positive integer is a sum of four squares.

An elementary proof of the result of Götzky has been found by Cohn[26] and Pall [754] in 1962, and in 1941 Maass[27] [2673] showed that already three squares suffice.

In 1922 Siegel [3774] generalized the circle method of G.H. Hardy[28] and Ramanujan[29] [1628] and established an asymptotic formula for the number $r_s(\alpha)$ of representations of totally positive integers α of a real quadratic field $K = \mathbf{Q}(\sqrt{m})$ as sums of $s \geq 5$ squares of integers of K, provided that in case of $m \equiv 2, 3 \bmod 4$ one has $\alpha = a + b\sqrt{m}$ with $a, b \in \mathbf{Z}$ and even b. His formula implies the existence of a number $r = r(K)$ such that for every totally positive integer α of K the number $r^2\alpha$ is a sum of 5 squares of integers.

In 1923 he applied [3776] again the circle method to prove that for every field K there is a positive rational integer r such that for every totally positive integer α of K the number $r^2\alpha$ is a sum of four squares lying in K. Earlier (Satz 9 in [3769]) he obtained this for totally real fields with "four" replaced by "finitely many". The same result for fields which are not totally real has been proved independently by Kirmse [2156] using the arithmetics of quaternions. Siegel noted that the same method leads to a similar result for representations by an arbitrary quaternary quadratic form. See Sect. 5.4.1 for later progress.

2. One finds in [3769] the first result dealing with sums of higher powers in number fields. In Satz 2 Siegel showed that for each m every totally positive number of K is a sum of $c(m, K)$ mth powers of totally positive elements of K.

---

[25] Marie Georges Humbert (1859–1921), professor at École Polytechnique in Paris. See [2493].

[26] Harvey Cohn (1923–2014), professor at the University of Arizona, CUNY and Stanford.

[27] Hans Maass (1911–1992), professor in Heidelberg. See [524].

[28] Godfrey Harold Hardy (1877–1947), professor in Oxford and Cambridge. See [4066].

[29] Srinivasa Aiyangar Ramanujan (1887–1920), Fellow of Trinity College, Cambridge. See [80, 1625].

In 1922 Kamke[30] [2099] showed that under certain assumptions about the polynomial $f(X) \in K[X]$ an analogous result holds for sums of values of f at totally positive numbers. Earlier [2097, 2098] he obtained this for the rational field (the *Waring–Kamke problem.*

In 1953 Ayoub [155] considered the Waring–Kamke problem and gave an asymptotical for the number of representations for $s \geq n(2^k + n) + 1$, where $k = \deg f$ and $n = \deg K$. Later Körner [2219, 2220] obtained asymptotics in the case when

$$s \geq \max\{2^k + 1, 2k^2(\log(nk^2 \log k) + 2.5n)\} .$$

His result was obtained using a generalization of I.M. Vinogradov's mean value theorem ([4228], see also Hua[31] [1910], Karatsuba[32] [2121], Stečkin[33] [3909], Wooley [4446], and K. Ford and Wooley [1214]) to algebraic number fields.

An elementary approach to the Waring–Kamke problem has been presented in 1956 by Rieger [3465].

4.2.3 Sums of Primes

1. In 1923 Rademacher [3360] obtained an analogue of Brun's [479] approximation to Goldbach's conjecture by showing that if K is a totally real field, $\alpha \in \mathbf{Z}_K$ is even, i.e. is divisible by every prime ideal of the first degree dividing 2 and $|N(\alpha)|$ is sufficiently large, then α can be written as a sum of two integers of K having each at most seven prime ideal factors.

Next year Rademacher considered quadratic fields assuming that the L-functions of Hecke's "Grössencharakters" do not vanish in the half-plane $\Re s > 3/4$. In the case of a real quadratic field he obtained in [3361] an upper bound for the number $N_r(\alpha)$ of representations of a totally positive integer α as a sum of $r \geq 3$ primes (i.e. integers generating prime ideals) and established an asymptotical formula of the form

$$N_r(\alpha) = (c(r, \alpha) + o(1))\, N(\alpha)^{r-1}$$

with explicitly given $c(r, \alpha) > 0$ in the case when α and r have the same parity. For imaginary quadratic fields [3362] he obtained a similar assertion for the sum

$$\sum_{\pi_1+\cdots+\pi_r=\alpha} \prod_{j=1}^{r} \log(N(\pi_j)) \cdot \exp\left(-\frac{2}{|\alpha|}\sum_{j=1}^{r}|\pi_j|\right) . \tag{4.10}$$

[30] Erich Kamke (1890–1961), professor in Tübingen. See [4284].

[31] Loo-Keng Hua (1910–1985), professor in Beijing. See [4295].

[32] Anatoliĭ Alekseevič Karatsuba (1937–2008), professor in Moscow. See [653].

[33] Sergeĭ Borisovič Stečkin (1920–1995), professor in Sverdlovsk and Moscow. See [331].

These results imply in particular that if Hecke's L-functions do not vanish for $\Re s > 3/4$, then every totally positive and odd (i.e. prime to 2) integer of large norm in a quadratic field is a sum of three primes.

In his third paper on this subject [3363] considered Rademacher totally real fields of arbitrary degree, and making the same assumption about zeros of Hecke L-functions obtained an asymptotic formula for a sum analogous to (4.10) in which additionally one could prescribe the residue classes of π_i modulo a fixed ideal.

The same assertion for $r = 2$ and remaining totally positive integers has been proved by Whiteman [4389] for real quadratic fields in 1940.

In 1955 Tatuzawa [4015] used Rademacher's approach to show without any unproved assumptions that every $\alpha \in \mathbf{Z}_K$ which lies in the additive group generated by primes of K can be written as a sum of a bounded number of primes, the bound depending on K. In 1960 Mitsui [2893] and Körner [2215] established unconditionally Rademacher's asymptotical formula in the case of real quadratic fields (see also Körner [2216, 2217, 2221]). Mitsui showed also that almost all even integers in an algebraic number field are sums of 2 primes.

In 1964 A.I. Vinogradov[34] [4220] improved essentially the result of Rademacher's paper [3360], showing the possibility of $\alpha = \beta+\gamma$ with $\Omega_K(\beta) \leq 2, \Omega_K(\gamma) \leq 3$ ($\Omega_K(a)$ denoting the number of prime ideal factors of a) for even α of sufficiently large norm. Much later Hinz [1855] obtained this as a consequence of his generalization of the Bombieri–Vinogradov theorem to algebraic number fields. In [1864] he showed this with $\Omega_K(\beta) = 1$ and $\Omega_K(\gamma) \leq 3$, and for totally real fields he proved this assertion with $\Omega_K(\beta) = 1$ and $\Omega_K(\gamma) \leq 2$ [1865], which is an analogue of J.R. Chen's[35] result ([672], see also Ross [3521] and Chap. 11 in the book [1595] by Halberstam[36] and Richert) for rational integers.

4.2.4 *Piltz Problem*

1. In 1924 Landau [2441] showed that the error term $R(x)$ in formula (3.15) in the Piltz problem for algebraic number fields is $\Omega\left(x^{1/2-1/2kn}\right)$. In the case $k = 1$ this has been done two years earlier by Walfisz in his dissertation [4272].

Landau's result has been improved in 1927 by Walfisz and Szegő[37] [3969, 3970] to

$$\Omega\left((x\log x)^{1/2-1/2kn}(\log\log x)^{k-1}\right) .$$

An expansion of the error term in (3.15) using a generalization of Bessel functions has been given by Walfisz in [4273, 4274] (for some simplifications see Landau [2442]).

Improvements of these Ω-bounds were later provided by Berndt [307, 308] in 1971 and Redmond [3408] in 1979. In 1983 Hafner [1586] showed

---

[34] Askold Ivanovič Vinogradov (1929–2005), professor in the Steklov Institute.

[35] Chen Jing Run (1933–1996), professor in Beijing.

[36] Heini Halberstam (1926–2014), professor in Dublin, Nottingham, and at the University of Illinois at Urbana-Champaign.

[37] Gabor Szegő (1895–1985), professor in Berlin, Königsberg, at the Washington University in St. Louis and at Stanford. See [149].

$$R(x) = \Omega\left((x\log x)^{1/2-1/2kn}(\log\log x)^{c(k,n)}\exp(-B(\log\log\log x)^{1/2})\right),$$

with $c(k,n) = (1/2+o(1))k\log k$, and B depending on K and k. In 2005 Girstmair, Kühleithner, Müller and Nowak [1439] enlarged the exponent $c(k,n)$ and replaced the exponential factor by a negative power of $\log\log\log x$.

The first improvement of Landau's upper bound (3.15) has been made in 1962 by Chandrasekharan and Narasimhan [648] who proved

$$R(x) = O\left(x^{1-1/kn}\right).$$

For some variants of the Piltz problem in algebraic number fields see Grotz [1530] and Rausch [3388, 3391].

In 1925/1926 Suetuna[38] [3945, 3947, 3949] determined the maximal order of $\tau_K^{(k)}$ and certain similar functions. He showed also that if $A_K(m)$ denotes the number of ideals of norm m in the field K and $\Psi_K(x)$ is the number of integers $m \leq x$ with $A_K(m) \neq 0$, then for $r = 1, 2, \ldots$ one has

$$\Psi_K(x) = \sum_{j=1}^{r}\frac{c_j x}{\log^{j-1/n} x} + O\left(\frac{x}{\log^{r+1-1/n} x}\right),$$

where n denotes the degree of K, and the constants $c_1 \neq 0, c_2, \ldots$ depend on K (his proof depended on an assumption satisfied only in certain special cases but later, in [3951], he was able to remove it with the use of Artin's reciprocity). He obtained also [3946] an asymptotic expansion for the sum $\sum_{m\leq x} A_K^k(m)$ for Abelian K. In [3948] he showed that if a Dirichlet series with nonnegative coefficients is the product of two Dirichlet L-functions, then it is either a square of $\zeta(s)$ or it equals $\zeta_K(s)$ for a quadratic field K. He showed also an analogue for products of three L-functions.

In 1952 Ankeny [88] obtained a generalization of the last result for the case of more than three factors. His proof had to be corrected, as one of his lemmas was inexact. This has been noticed by Iwasaki [1979] who provided a simpler proof. A generalization to Artin L-functions has been presented in 1990 by Funakura [1324].

In 1928 Suetuna [3951] gave bounds for the ratio $f_K(m) = \tau_n(m)/A_K(m)$. In [3953] and in [1716] (the second paper written with Hasse) he studied various divisor functions in algebraic number fields.

4.2.5 *Values of Zeta-Functions*

1. In 1922–1924 Hecke [1740, 1741, 1743] utilized his new characters to a problem in the theory of Diophantine approximations and to construction of new classes of modular forms in two variables (see also Behnke[39] [270, 271]).

[38] Zyoiti Suetuna (1898–1970), professor in Kyushu University and in Tokyo University.

[39] Heinrich Behnke (1898–1979), professor in Münster. See [1502].

In the last paper of that series he asserted (Satz 3 in [1743]) that if X is an ideal class in a real quadratic number field K and $\zeta_K(s; X)$ is its the zeta-function defined by (3.28), then for $m = 1, 2, \dots$ one has

$$\zeta_K(2m, X) = r_K(X, m)\pi^{2m}\sqrt{d(K)} \tag{4.11}$$

with rational $r_K(X, m)$. This generalized a similar result for the values of $\zeta_K(s)$ at even positive integers, which has been earlier established by Siegel at the end of [3774]. For the proof of (4.11) Hecke referred to a forthcoming dissertation, which has never been written.

Usually one attributes to Hecke's paper the conjecture that (4.11) holds for all totally real fields, with the term π^{2m} replaced by π^{mn} with n being the degree of the field. A sketch of the proof of this conjecture has been given in 1937 by Siegel ([3782], p. 546), and another proof has been provided in 1961 by H. Klingen [2170] with the use of Hilbert's modular forms whose Fourier series have constant term $\zeta_K(2m)$. Effective determinations of the rational number $r_K(X, m)$ were given by H. Lang [2452] in 1968 for quadratic fields and by Siegel [3794] in 1969 for arbitrary totally real fields. Their denominators in the quadratic case were studied by H. Lang [2453] in 1972, and explicit formulas for $\zeta_K(-n)$ (for odd n) in this case were given by H. Cohen [731].

An exposition of Siegel's formulas and a presentation of their relations with the theory of modular forms have been presented in 1976 by Zagier [4460].

In 1964 C. Meyer [2846] obtained an analogue of (4.11) for the value at $s = 2$ of the ring class zeta-functions of a real quadratic field K associated with rings O_f, where f is a positive rational integer and the ring O_f consists of elements α of K such that for some rational r the ratio $(\alpha - r)/f$ is f-integral, and the ring class-group is the factor group of fractional ideals of O_f prime to f by the group of principal ideals of O_f prime to f, and its zeta-function is defined by

$$\zeta(s, X) = N(I)^s \sum_{\gamma} \frac{1}{|N(\gamma)|^s},$$

where I is a fixed ideal in X^{-1}, and γ runs over all non-associated elements of I.

The case of arbitrary positive even arguments has been treated in 1969 by Barner [190] who obtained explicit formulas for them.

In 1970 Siegel [3795] proved that the values of the zeta-functions for classes mod $\mathfrak{f}$ are rational at negative rational integers. In the case of $\mathbf{Q}$ this has been done already in 1882 by A. Hurwitz [1926].

The results of H. Klingen [2170] and Siegel [3795] were used later by Coates and Sinnott [723] for the construction of p-adic L-functions associated with characters of the Galois group of an Abelian extension of a real quadratic field. This paper contains also congruences for the values of $\zeta_{\mathfrak{f}}(X, s)$, the zeta-function of the ideal class X mod $\mathfrak{f}$, at negative integers. In [724] the authors used this for obtaining some integrality properties of these values.

Later formulas for values at 1 and/or negative integers of various L-function and zeta-functions for algebraic number fields were obtained by several authors. The case of Artin L-functions has been treated by Stark [3898], who in 1975 presented a conjectural formula for their value at $s = 1$, expressing it as a product of certain known parameters of the involved extension, an algebraic number $\theta(\chi)$ and a determinant $R(\chi)$, an analogue of the regulator. He established it in the case of rational characters. A stronger form of this conjecture has been stated by Stark [3899] in 1976. Stark conjectures were discussed in the book [4011] of Tate.

Values at negative rational integers of L-functions for characters of the group $H_{\mathfrak{f}}^*(K)$ for totally real K were determined by Shintani [3759] in 1976, making explicit the result of Siegel [3795] (see also Zagier [4461], Hida [1828] and Kramer [2239]).

4.3 Class-Field Theory

4.3.1 Takagi

1. The main theorems of class-field theory giving a description of all Abelian extensions of an algebraic number field K were established by Takagi [3975, 3976]. He showed that every such extension corresponds to a subgroup of the group $H_{\mathfrak{f}}^*(K)$ of narrow ideal classes mod $\mathfrak{f}$, with a suitably chosen ideal $\mathfrak{f}$, the *conductor* of $\mathfrak{H}$.

His definition of a class-field was only formally different from that used by H. Weber: Let L/K be a finite Galois extension of K. If $\mathfrak{f}$ is an ideal of $\mathbf{Z}_K$ and $\mathfrak{H}$ is the subgroup of the narrow class-group $H_{\mathfrak{f}}^*(K)$ of K formed by classes containing norms $N_{L/K}(I)$ of ideals I of $\mathbf{Z}_L$ prime to $\mathfrak{f}\mathbf{Z}_L$, then Takagi called L the class-field associated with $\mathfrak{H}$ if one has

$$i(\mathfrak{H}) := [H_{\mathfrak{f}}^*(K) : \mathfrak{H}] = [L : K] .$$

Using analytical tools, in particular the L-functions associated with characters of the factor group $H_{\mathfrak{f}}^*(K)/\mathfrak{H}$, Takagi established the inequality

$$i(\mathfrak{H}) \leq [L : K] .$$

Then he showed that if L, L' are class-fields associated with groups $\mathfrak{H}, \mathfrak{H}'$, then $\mathfrak{H} \subset \mathfrak{H}'$ holds if and only if $L' \subset L$. This implies that with every group $\mathfrak{H} \subset H_{\mathfrak{f}}^*(K)$ there can be associated at most one class-field.

The proof of the existence of a class-field L associated with a given subgroup $\mathfrak{H}$ of $H_{\mathfrak{f}}^*(K)$ has been first reduced to the case when this index is a prime power, then to the case when it is a prime and finally to the case when the index equals a prime p with ζ_p lying in K. In the last case Takagi gave an explicit construction based on the theory of cyclic extensions. The proof showed that L/K is Abelian with Galois group isomorphic to $H_{\mathfrak{f}}^*(K)/\mathfrak{H}$, and the set of prime ideal divisors of its discriminant equals the set of prime ideal divisors of $\mathfrak{f}$.

In the next step Takagi showed if L/K is Abelian, then L is the class-field to a certain group $\mathfrak{H}$. For cyclic extensions of prime degree this is achieved by an explicit construction using the fact that the discriminant of such extension is of the form $\mathfrak{f}^{p-1}$ (Satz 79 in [1836]). One of the auxiliary results in the construction is the following *principal genus theorem*:

Let L/K be a cyclic extension of prime power degree l^m, and let $\mathfrak{d}$ be its discriminant. Then there exists an ideal $\mathfrak{f}$ in K whose prime ideal divisors divide $\mathfrak{d}$ and an ideal $\mathfrak{F}$ in L dividing $\mathfrak{f}$ such that the set of classes in $H_{\mathfrak{F}}^*(L)$ whose norms lie in the unit class of $H_{\mathfrak{f}}^*(K)$ forms a group $PG(L/K)$, called the principal genus of L, and the factor group $Gen(L/K) = H_{\mathfrak{f}}^*(L)/PG(L/K)$ equals

$$\{A^{1-\sigma} : A \in H_{\mathfrak{F}}^*(L)\} ,$$

where σ is the generator of $Gal(L/K)$.

Later Herbrand [1807] generalized this result to arbitrary cyclic extensions, and E. Noether [3126] used the theory of linear algebras to obtain a generalization to arbitrary Galois extensions.

In the final step Takagi determined the factorization of prime ideals in the class-field corresponding to $\mathfrak{H}$. He showed that if $\mathfrak{p}$ is an unramified prime ideal in K and f is the minimal exponent with $\mathfrak{p}^f \in \mathfrak{H}$, then the prime ideal factors of $\mathfrak{p}$ have order f. In the case of ramified $\mathfrak{p}$ Takagi's definition has been rather cumbersome, and it has been simplified by Hasse [1655] to the following form: let $H_\mathfrak{p}$ be the smallest group of conductor prime to $\mathfrak{p}$ containing $\mathfrak{H}$, and let $e_\mathfrak{p} = [H_\mathfrak{p} : \mathfrak{H}]$. Then

$$\mathfrak{p}\mathbf{Z}_L = (\mathfrak{P}_1 \cdots \mathfrak{P}_g)^{e_\mathfrak{p}} .$$

It follows from Takagi's theory that a field with class-number 1 cannot have unramified Abelian extensions. This assertion fails for non-Abelian extensions as shown by the example with $K = \mathbf{Q}(\sqrt{2869})$ and $L = K(\alpha)$, where α is a root of $X^5 - X - 1$ (Fujisaki [1315]). In this case $h(K) = 1$ and the extension L/K is unramified. It has been shown in 1962 by Fröhlich [1273] that all symmetric groups S_n can be realized as Galois groups of unramified extensions of algebraic number fields. In 1970 Uchida [4088, 4089] and Yamamoto[40] [4450] proved the existence of infinitely many quadratic fields having an unramified extension with a given alternating Galois group (see also Elstrodt, Grunewald[41] and Mennicke [1088] and Kedlaya [2133]).

The theorem of Golod and Šafarevič [1472] implies that if the p-Sylow subgroup of the class-group of a field K is sufficiently large, then K has an infinite unramified extension.

In 2000 Maire [2713] showed a method to obtain fields with class-number one having infinite unramified extensions. Such is, for example, the biquadratic field $\mathbf{Q}(\sqrt{a}, \sqrt{b})$ with $a = 17\,601\,097$, $b = 17\,380\,678\,572\,159\,893$. Later Brink [457] found simpler examples, e.g. $a = 36\,497$, $b = 290\,357$.

In 1922 Takagi [3977] proved that Furtwängler's reciprocity law can be obtained as a consequence of his construction of the class-field theory, and in the next year he presented [3978] an explicit formula for lth power reciprocity in the lth cyclotomic field (see Hasse [1664]). Such formula for the l^nth power reciprocity has been later given by Rothgiesser [3526].

2. An exposition of Takagi's result gave Hasse [1655] in 1926, presenting next year all details of the proof in [1656]. Some simplifications were later provided by Hasse and Scholz [1715].

In Sect. 8 of [1655] Hasse showed that Takagi's results lead to a simple proof of the analogue of Dirichlet's Prime Number Theorem established by Hecke [1734] and extended by Landau [2432].

Three important open problems were pointed out by Hasse in Sect. 11 of [1655]. The first two were stated already by Hilbert (Hilbert's 12th problem and the Principal Ideal Theorem), and the third, formulated by Furtwängler,[42] and usually called the *class-field tower problem* (*Klassenkörperturmproblem* in German), asked whether every tower of fields

[40]Yoshihiko Yamamoto (1941–2004), professor in Osaka.

[41]Fritz Grunewald (1949–2010), professor in Bonn and Düsseldorf. See [3723].

[42]It seems that it has been formulated in print on p. 46 of [1655].

$$K_0 \subset K_1 \subset \dots \subset K_n \subset \dots ,$$

in which K_n is the absolute class-field of K_{n-1}, is finite. Hasse pointed out that the truth of the last assertion would imply the possibility of embedding an arbitrary algebraic number field of finite degree into a field with class-number one.

In 1929 Scholz [3664] showed that there exist fields K with arbitrary long class-field towers. He obtained this by choosing K to be the composite of fields $K^l_{p_1}, K^l_{p_2}, \dots, K^l_{p_{n+1}}$, where l is a fixed prime, $p_1, p_2, \dots, p_{n+1}$ are suitable chosen primes congruent to 1 mod l, and K^l_p denotes the unique subfield of $\mathbf{Q}(\zeta_p)$ having degree l.

In 1962 Fröhlich [1272] determined pure fields $\mathbf{Q}(\sqrt[p]{m})$ (with an odd prime p) having non-trivial class-fields of lengths 2.

The first layers of the 2-class-field towers of quadratic fields were studied in 1964 by Koch [2192] and later by Kisilevsky [2161], Lemmermeyer [2525], Benjamin, Lemmermeyer and Snyder [286–288], Bush [529] and Steurer [3926].

It has been shown in 1964 by Golod and Šafarevič [1472] that if G is a finite p-group, and

$$d(G) = \dim_{\mathbf{F}_p} H^1(G, \mathbf{F}_p), \quad r(G) = \dim_{\mathbf{F}_p} H^2(G, \mathbf{F}_p) ,$$

then

$$r(G) > \left(\frac{d(G)-1}{2}\right)^2 , \tag{4.12}$$

and this implied that if ρ denotes the unit rank of a field K, γ_p is the minimal number of generators of the p-Sylow subgroup of $H(K)$, and

$$\gamma_p \ge 3 + 2\sqrt{\rho + 2}, \tag{4.13}$$

then the p-class-field tower of K is infinite. This holds in particular for the quadratic field $\mathbf{Q}(\sqrt{4\,849\,845})$, establishing the existence of infinite class-field towers.

Soon Vinberg [4216] and, independently, Gaschütz (see Roquette [3499] and Chap. 7 in the book [1532] of Gruenberg) replaced the assumption (4.12) by

$$r(G) > \left(\frac{d(G)}{2}\right)^2 , \tag{4.14}$$

and this implied that if for a quadratic field K one has $\omega(d(K)) \ge 3 + 2\sqrt{\varrho + 3} + \varrho$, then K has an infinite 2-class-field tower. This happens, e.g. for $K = \mathbf{Q}\left(\sqrt{-30\,030}\right)$. Other proofs have been given by Koch [2193] (see also [2194]) and Schoof [3687] (see also Lubotzky [2665] and J.S. Wilson [4422]).

On the other hand Kostrikin[43] [2229] found for every n a p-group group G with $d(G) = 2^n$ and $r(G) \le \frac{1}{3}(d^2(G) - 1) + d$, Koch [2195] replaced here the coefficient 1/3 by 11/35 (with $d(G) = 6^n$), and in 1981 Wisliceny [4430] showed that (4.14) is asymptotically best possible.

Several sufficient conditions for the infinitude of the class-field tower were given in 1972 by Furuta [1360]. They were applied in the case of cyclic fields of prime degree by T. Takeuchi [3985, 3986], and later Shparlinski [3766] used them for cyclotomic fields.

In 1974 Koch and B.B. Venkov[44] [4210] considered quadratic fields with finite p-class-field tower and described the Galois groups of their maximal unramified p-extensions. Their results

[43] Alekseĭ Ivanovič Kostrikin (1929–2000), professor in Moscow.

[44] Boris Borisovič Venkov (1934–2011), son of B.A. Venkov, professor in St. Petersburg and Aachen.

imply that if the p-rank of $H(K)$ is at least 3 for an odd prime p, then K has an infinite p-class-field tower.

In 1980 Schmithals [3657] showed the existence of infinitely many complex quadratic fields K with $\omega(d(K)) = 3$ and infinite 2-class-field tower, and in 1986 Schoof [3687] obtained the same assertion for real quadratic fields. He showed also that there are infinitely many quadratic K (real and imaginary) with $\omega(d(K)) = 2$ and infinite 2-class-field tower and constructed such examples with prime discriminant (e.g. $d(K) = 39\,345\,017$ and $-3\,321\,607$).

The conjecture that imaginary quadratic fields K with rank $H_2(K) = 4$ have an infinite 2-class-field tower is still open, although it has been confirmed for several families of fields (see Hajir [1589], Maire [2712], Benjamin [285], Sueyoshi [3958], Gerth [1420, 1421] and Mouhib [2990]).

In 1965 Brumer [475] obtained a lower bound for the minimal number $r(K)$ of generators of the class-group of K in the case, when K/Q is Galois:

$$r(K) \geq \frac{\omega(d(K))}{\omega(n)} - 2n, \tag{4.15}$$

where $n = [K : \mathbf{Q}]$. This implies that if $K_1, K_2, \ldots$ is a sequence of Galois fields of fixed degree, satisfying

$$\lim_{n\to\infty} \omega(d(K_n)) = \infty\,,$$

then

$$\lim_{n\to\infty} h(K_n) = \infty\,,$$

a result proved in 1963 by Brumer and Rosen [478].

Brumer's proof utilized cohomology, but in 1969 Roquette and Zassenhaus [3511] applied elementary methods for the proof of the inequality

$$\gamma_p \geq t_p(K) - c(n)\,,$$

giving a lower bound for the value γ_p in (4.13), with explicit $c(n)$ and t_p being the number of rational primes q such that the ramification indices of all prime ideals of K dividing p are divisible by q. This implies (4.15), also in the case of non-Galois extensions $K/\mathbf{Q}$. This bound has been improved in 1970 by Connell and Sussman [762], and in 1980 Schmithals [3658] gave a generalization.

3. A formula for the prime ideal factorization of the conductor of the ideal group $\mathfrak{H}$ associated with Abelian K/k follows from the result of Sugawara [3959], who proved it in the case of a cyclic extension of a prime power degree. He used analytical tools, and later an arithmetical proof has been provided by Iyanaga [1998]. See also Vassiliou [4196].

4.3.2 Artin

1. The thesis of Artin [116, 117] dealt with arithmetical and analytical theory of function fields of positive characteristics. Artin presented in it the theory of quadratic extensions of the field $\mathbf{F}_p(X)$ of rational functions over the field $\mathbf{F}_p$ of p elements.[45] He introduced the corresponding zeta-function and formulated the analogue of Riemann's conjecture, which he was able to establish in several special cases.

[45] Artin considered also the case when the base field is an arbitrary finite field, but did not publish his results, and his notes on this subject appeared much later, in 2000 [131].

For function fields of genus one this conjecture has been established by Hasse [1679, 1683–1685] in 1936.

In the general case this has been done by Weil [4351]. Other proofs were later given by Roquette [3497, 3498].

Later Stepanov [3918–3922] invented an elementary approach to establish the conjecture in several important cases, and it has been shown by Bombieri [387, 388] and W.M. Schmidt [3652, 3653] that this method can be modified to yield the proof of the complete conjecture. An exposition has been given in the book [3654] by W.M. Schmidt.

The history of the conjecture was presented by Roquette [3503, 3504, 3506, 3508].

The algebraic theory of arbitrary separable finite extensions of $\mathbf{F}_p(X)$ has been studied by Sengenhorst[46] [3731], and Rauter [3393] carried over Hilbert's ramification theory to these extensions. In [3394] he made a study of cyclotomic extensions. The analytic theory for these fields has been developed by F.K. Schmidt [3645] in 1931, who also established in [3646] the class-field theory in this case.

2. In his first paper concerning algebraic number fields [115] considered Artin the question, whether the ratio $\zeta_L(s)/\zeta_K(s)$, where K is a subfield of L, is entire. This was already known in the case of pure cubic extensions of the rationals (Dedekind [853]), as well as for Abelian extensions L/K, due to the factorization

$$\zeta_L(s) = \zeta_K(s) \prod_{\chi} L(s, \chi) ,$$

(where χ runs over all non-principal primitive characters of the class-group corresponding to the extension L/K), which is a consequence of the results of Takagi [3976]. Artin established a positive answer for a class of metabelian[47] extensions, in particular for those having a square-free degree. A lemma in [115] treated the decomposition and inertia groups as permutation groups. This has been later generalized by van der Waerden [4138] who also pointed out that this result is contained implicitly already in Dedekind's paper [849] (see also M. Bauer [251]).

3. In 1924 Artin [118] introduced a new kind of L-functions and developed several their properties. Let G be the Galois group of a Galois extension L/K. Following Frobenius [1259] he associated with each unramified prime ideal $\mathfrak{p}$ of K a conjugacy class $F_{L/K}(\mathfrak{p})$ of Frobenius automorphisms associated with $\mathfrak{p}$. If now $A_\mathfrak{p}$ denotes an element of $F_{L/K}(\mathfrak{p})$, and $T : G \longrightarrow GL_n(\mathbf{C})$ is a finite-dimensional complex representation of G with character χ, then Artin's L-function associated with T is defined for $\Re s > 1$ by

$$L(s, \chi; L/K) = \exp\left(\sum_{\mathfrak{p} \nmid d(L/K)} \sum_{m=1}^{\infty} \frac{\chi(A_\mathfrak{p}^m)}{mN(\mathfrak{p})^{ms}} \right) . \tag{4.16}$$

This can be also written in the multiplicative form

[46] Paul Sengenhorst (1894–1968). See [272].

[47] Let us recall that a group is metabelian if its commutator group is Abelian.

$$L(s,\chi;L/K)=\prod_{\mathfrak{p}\nmid d(L/K)}\det\left[E-N(\mathfrak{p})^{-s}T(A_{\mathfrak{p}})\right],$$

E being the unit matrix.

It follows from (4.16) that an equality $\chi=\chi_1+\chi_2$ implies

$$L(s,\chi;L/K)=L(s,\chi_1;L/K)L(s,\chi_2;L/K),$$

and this leads to the product formula

$$\zeta_K(s)=\prod_{\chi}L(s,\chi;L/K)^{n(\chi)}, \tag{4.17}$$

where χ runs over characters of all irreducible representations of G and $n(\chi)$ is the dimension of the corresponding representation.

Artin showed also that every character can be written as a linear combination with rational coefficients of characters induced by cyclic subgroups, and this implied that every L-function can be written as a product of powers with rational exponents of L-functions corresponding to Abelian characters. In the case of Abelian extensions L/K he conjectured but was unable to show that his new functions coincide with Hecke's L-functions corresponding to characters of the class-group mod $\mathfrak{f}$, for which L is the class-field in the sense of Takagi. He noted that this would follow from the following assertion ([118], Satz 2), which has been later called *Artin's reciprocity law*:

Let L/K be an Abelian extension with Galois group G, and let H be an ideal class-group in K such that L is the class-field corresponding to H. If $\mathfrak{p}$ is a prime ideal in K unramified in L/K, then $F_{L/K}(\mathfrak{p})$ depends only of the class $X_{\mathfrak{p}}$ of $\mathfrak{p}$ in H, and the mapping

$$X_{\mathfrak{p}}\mapsto F_{L/K}(\mathfrak{p})$$

induces an isomorphism of H and G.

Artin gave also in [118] the proof of this assertion for extensions which are composites of extensions of prime degrees, using the reciprocity law for power residues in the form given by Takagi [3977]. Assuming the truth of his reciprocity law Artin deduced the possibility of prolonging his L-functions to the complex plane and proved that they satisfy a functional equation without writing it explicitly. He conjectured that these functions are entire with exception of the case when χ is the principal character. This conjecture would imply the truth of his conjecture on divisibility of Dedekind zeta-functions.

The assertion that Artin's L-function is entire for extensions with Abelian and dihedral Galois groups is a consequence of Artin's reciprocity law, and Artin established it for monomial representations.

In 1939 Aramata [101] showed for a class of linear Galois groups that the corresponding Artin L-functions are meromorphic.

In 1974 Deligne and Serre [873] showed that for certain holomorphic forms of weight one on $GL(2)$ the corresponding L-function is equal to the Artin function $L(s, \varrho)$ of a two-dimensional Galois representation ϱ. This has been later generalized by Rogawski and Tunnell [3487].

Looking for a converse result Langlands [2470] formulated in 1980 a stronger form of Artin's conjecture (*strong Artin conjecture*) on L-functions, asserting that if ϱ is an irreducible n-dimensional representation of $Gal(L/K)$, then $L(s, \varrho)$ is equal to the L-function of a cusp form on $GL_n(K)$. He succeeded to show this in the case of two-dimensional representations of A_4, and later Tunnell [4078] did the same for the group S_4.

For certain representations of A_5 this conjecture has been established by Buhler [504] in 1978 and Ramakrishnan [3376] in 1989. See also Kiming [2150, 2151] and Kiming and X.D. Wang [2152].

An earlier result of Weil [4354] showed that in the two-dimensional case the strong Artin conjecture for $L(s, \varrho)$ follows from the Artin conjecture for all functions of the form $L(s, \varrho \otimes \chi)$, χ running over all Dirichlet characters, and this has been used in 2003 by Booker [393] who proved that in this case for every ϱ the strong Artin conjecture for $L(s, \varrho)$ is a consequence of Artin conjecture.

It follows from results of Weil [4354], Jacquet, Piatetski-Shapiro,[48] Shalika[49] [2019, 2020] and Langlands [2470] that the truth of Artin conjecture for all Galois representations in dimension 2 or 3 implies the strong Artin conjecture for that dimension.

The strong Artin conjecture for two-dimensional representations of A_5 has been treated by Buzzard, Dickinson, Shepherd-Barron and R. Taylor [537] in 2001, Buzzard and Stein [538] in 2002 and R. Taylor [4037] in 2003. It is a consequence of the proof of Serre's modularity conjecture for odd two-dimensional representations, whose final steps were achieved in 2009 by Khare, Wintenberger [2144, 2145] and Kisin [2163].

For certain four-dimensional representation the strong Artin conjecture has been established in 2002–2004 by Martin [2746, 2747] and Ramakrishnan [3377]. For other cases see Calegari [553] and S. Sasaki [3581].

An exposition of the theory of Artin L-functions has been given by Martinet [2751].

4. In [118] as well as in [122] Artin studied multiplicative relations between zeta-functions of subfields of a Galois extension $K/\mathbf{Q}$, showing that in the case of the rational field primitive L-functions are multiplicatively independent ([118], Satz 5).

Multiplicative relations between zeta-functions of a field and his subfields induced by (4.17) were used in 1933 by Nehrkorn [3079] to get relations between the class-groups of an Abelian field and its cyclic subfields. This generalized Dirichlet's formula (1.9) as well as its extension to fields with Galois group C_p^2 with an odd prime p made by Pollaczek[50] [3331] in 1929, and a further extension to Abelian extensions of prime power degree obtained by Scholz [3667] in 1930. In 1933 Scholz [3670] studied also the case of sextic fields with group S_3. Relations between Artin L-functions were studied in a particular case by Suetuna [3954–3956].

In 1950 S. Kuroda [2384] considered extensions K/k with group C_p^n and established the equality

$$\frac{h(K)}{h(k)} = p^a[E(K) : E'(K)] \prod_j \frac{h(K_j)}{h(k)} \tag{4.18}$$

with explicit a, $E(K)$ being the group of units of K, $E'(K)$ the group generated by units of proper subfields of K, and K_j running over subfields of K, cyclic over k (in the case $p = 2$ he assumed that

[48] Ilya Piatetski-Shapiro (1929–2009), professor in Moscow, Tel Aviv and at Yale. See [730].

[49] Joseph Shalika (1941–2010), professor at John Hopkins University.

[50] Felix Pollaczek (1892–1981) worked in Berlin and Paris as engineer.

infinite primes are unramified in K/k). This formula appears already in Nehrkorn's paper [3079] in a less explicit form.

In 1951 R. Brauer[51] [431] used his theorems about characters [428, 429] to describe multiplicative relations between zeta-functions. In particular he showed

$$\zeta_K(s) = \prod_H \zeta_{L^H}^{c_G(H)}(s), \tag{4.19}$$

where L/K is a Galois extension with group G, H runs over all cyclic subgroups of G, L^H is the subfield of L whose elements are fixed by H, and

$$c_G(H) = \frac{1}{[G:H]} \sum_{H^*} \mu([H^*:H]) ,$$

H^* running over all cyclic subgroups of G containing H. A simpler proof of Brauer's result has been given in 1979 by Walter [4282], who also generalized (4.18) to arbitrary Galois extensions. The biquadratic case has been analysed by Lemmermeyer [2524] who also noted that for $p = 2$ Walter's formula is inexact in the case $\zeta_8 \in K$. For this case see also R.I. Berger [299].

In 1997 Boltje [386] showed that the results of S. Kuroda and Walter can be obtained using cohomological Mackey functors introduced by Dress [995, 996] (see also Thévenaz [4049] and Bley and Boltje [373]).

Similar formulas for dihedral extensions of degree $2p$ with prime p were given by Halter-Koch [1600] and in the general case by Castela [606].

For other classes of fields see C.J. Parry [3226] (bicubic fields) and Walter (Frobenius extensions and composites of radical extensions of the same degree) [4281, 4283]).

In 2011 Browkin, Brzeziński and Xu [463] described groups G for which the zeta-function of K does not appear in (4.19) (see also [464] for a similar question in case of groups having pq elements with prime p, q).

5. In a later paper [122] Artin simplified the definition of his L-functions and provided an explicit form of their functional equation. He used here the description of the group-theoretical structure of the relative discriminants obtained by him in [123], in particular the notion of the conductor of a non-Abelian character introduced there. The norms of these conductors occur as factors in the functional equation of Artin L-functions which has the form

$$\Lambda(\chi, s) = W(\chi)\Lambda(\overline{\chi}, 1-s) ,$$

where $\Lambda(\chi, s)$ equals the product of $L(\chi, s)$, certain Γ-factors and some constants related to the extension L/K and χ. The coefficient $W(\chi)$ (called the *Artin root number*) is of absolute value 1.

In 1956 Dwork[52] [1031] obtained a factorization of the Artin root number into local factors,

$$W(\chi) = \prod_{\mathfrak{p}} W_{\mathfrak{p}}(\chi_{\mathfrak{p}})$$

[51] Richard Dagobert Brauer (1901–1977), brother of Alfred Brauer, professor in Toronto, Ann Arbor and at Harvard. See [1168, 1508, 3495].

[52] Bernard Morris Dwork (1923–1998), professor at the Johns Hopkins University and Princeton University. See [2123].

where $\chi_{\mathfrak{p}}$ is the restriction of χ to the decomposition group of $\mathfrak{p}$. These local factors were given up to the sign, which was later determined by Langlands [2468] and Deligne [872]. See also Tate [4010].

It has been shown in 1973 by Fröhlich and Queyrut [1284] that if χ is a character over **R**, then $W(\chi) = 1$.

Artin showed also in [122] that every character of a finite group can be written as a linear combination with rational coefficients of characters induced by cyclic subgroups, and this implied that a certain power of an L-function is meromorphic.

6. The way to the proof of Artin's reciprocity law was paved by a result of Čebotarev [620] appearing in 1926 (actually this paper is the German translation of [617] published in Russian three years earlier). He established the conjecture of Frobenius about densities (see Sect. 1.3.4) by proving the following result:

If $K/\mathbf{Q}$ is a Galois extension with group G and A is a conjugacy class in G, then the set

$$\mathcal{P}(A) = \{p : F_{K/\mathbf{Q}}(p\mathbf{Z}) = A\}$$

is infinite, and one has

$$\sum_{p\in\mathcal{P}(A)} \frac{1}{p^s} = \left(\frac{\#A}{[K:\mathbf{Q}]} + o(1)\right) \log\left(\frac{1}{s-1}\right) .$$

A version of the proof not utilizing infinite series was given by Čebotarev in [623], who earlier [622] provided a similar modification of the proof of Frobenius density theorem. Although Čebotarev considered only extensions of the rationals, after suitable modifications his proof works also for Galois extensions of arbitrary algebraic number fields.

In the same year Schreier[53] [3690] simplified the proof, and further simplifications were provided by Scholz [3668], who reformulated Čebotarev's theorem as an assertion about prime divisors of irreducible polynomials. An exposition of Čebotarev's result (already for relative extensions) was provided in 1930 by Hasse ([1664], pp. 133–138).

An effective bound in Čebotarev's theorem was provided by Lagarias and Odlyzko [2398]. They showed that if GRH holds for $\zeta_L(s)$, then the error term it is bounded by

$$c_1\left(\frac{\#A}{[L:K]}\sqrt{x}\log(|d(L)|x^{n(L)}) + \log(|d(L)|x^{n(L)})\right) ,$$

where c_1 is an absolute constant and $n(L) = [L : \mathbf{Q}]$. Unconditionally they established for $x \geq \exp(10n(L)\log^2(|d(L)|))$ the bound

$$\frac{\#A}{[L:K]}\operatorname{li}\left(x^{\beta}\right) + c_2 x \exp\left(-c_3\sqrt{\frac{\log x}{n(L)}}\right),$$

where β is the hypothetical zero of $\zeta_L(s)$, satisfying

[53] Otto Schreier (1901–1929), professor in Rostock. See [2813].

$$1 - (4\log(d(L))^{-1} \leq \Re\beta \leq 1, \quad | \Im\beta | \leq (4\log(d(L))^{-1} ,$$

and c_2, c_3 are absolute constants (a bound for β has been given in 1974 by Stark [3896]). See also Bartz [205].

An upper bound of the form

$$N(\mathfrak{p}) = O\left(d(L)^M\right)$$

for the prime ideal $\mathfrak{p} \in \mathcal{P}(A)$ of smallest norm has been given in 1979 by Lagarias, Montgomery and Odlyzko [2397]. This generalized Linnik's theorem on the minimal prime in an arithmetical progression [2601, 2602].

It has been shown in 1991 by H.W. Lenstra Jr. and Stevenhagen [2547] that in certain cases one can establish Čebotarev's theorem by algebraic methods.

The analogue of Čebotarev's theorem has been also established in the case of function fields. An elementary proof has been given by Jarden [2037] in 1982, and effective proofs were obtained by Ishibashi [1969] and V.K. Murty and Scherk [3006].

A uniform proof valid for all global fields has been given in 1963 by Serre [3734] and in 1989 by Bilhan [353]. See also M. Fried [1253] and Halter-Koch [1604].

In 1927 Artin [120] was able to provide a proof of his reciprocity theorem in full generality, using certain ideas from Čebotarev's paper [620]. He wrote on the first page of [120]: "*Einen*[54] *der Grundgedanken des Beweises, die Verwendung von Kreiskörpererweiterungen, verdanke ich der wichtigen Arbeit von Herrn Tschebotareff*". As an application he presented a very simple proof of the reciprocity law for mth powers in the following form:

Let $m > 1$ be a rational integer, and assume that k is a field containing all mth roots of unity. For $\alpha \in \mathbf{Z}_k$ and a prime ideal $\mathfrak{p}$ prime to α and the relative discriminant Δ of $k(\sqrt[m]{\alpha})/k$ put

$$\left(\frac{\alpha}{\mathfrak{p}}\right) = \zeta_m^\lambda ,$$

where λ is defined by

$$\alpha^{(N(\mathfrak{p})-1)/m} \equiv \zeta_m^\lambda \pmod{\mathfrak{p}} ,$$

and define the symbol

$$\left(\frac{\alpha}{I}\right)$$

by multiplicativity to all ideals I of k prime to $\alpha\Delta$. Then the value of the last symbol depends only on the class of I in the group of ideal classes for which K is the class-field.

For the history of Artin's reciprocity law see G. Frei [1241].

A detailed exposition of Artin's reciprocity law and its applications to various reciprocity laws gave Hasse [1664] in 1930.

[54] *I owe to the important paper of Mr. Tschebotareff one of the fundamental ideas of the proof, the use of extensions of cyclotomic fields.*

7. The first step towards Artin's conjecture on divisibility of Dedekind zeta-functions has been made in 1930 by Suetuna [3952], who showed that if the extension L/K is Galois, and $\zeta_L(s) \neq 0$ in the half-plane $\Re s > \sigma$ for some $1/2 \leq \sigma < 1$, then $\zeta_K(s) \neq 0$ holds in that half-plane.

In 1931 Aramata [99] used a hint of Suetuna and established the conjecture in the case when L/K is Galois and no two cyclic subgroups of its Galois group have a non-trivial intersection. Two years later he was able to establish the conjecture for all Galois extensions [100]. See (6.1.1) for further development.

8. In [121] Artin showed that using his reciprocity law one is able to reduce the proof of the Principal Ideal Theorem, conjectured by Hilbert, to the proof of a purely group-theoretical assertion. This assertion has been established at the same time by Furtwängler [1356].

This proof has been later simplified by Magnus[55] [2693], and other proofs were given by Iyanaga [2001] (see Witt[56] [4433] for some remarks) and Schumann [3695, 3696]. Later new proofs were given by Bergström [303] and Borevič[57] [397]. A cohomological interpretation of earlier proofs presented Kawada [2130] in 1968.

There are situations in which ideals become principal already in proper subfields of Hilbert's class-field. In 1932 Furtwängler [1357] showed that if $H(K) = C_2^n$, then there is a set of generators $X_1, \dots, X_n$ of $H(K)$ such that ideals from each X_i principalize in a suitable unramified quadratic extension of K. In the case when the class-group equals C_p^n with odd prime p Taussky [4020] had given sufficient conditions for this to happen. In 1934 Scholz and Taussky [3681] computed some examples of principalization of ideals of imaginary quadratic fields in cubic extensions.

Iyanaga [1999, 2002] and Herbrand [1807] generalized the Principal Ideal Theorem to relative extensions. Some simplifications and additions were provided by Tannaka[58] [3993–3995].

It has been conjectured by Tannaka that if $\hat{K}$ is the Hilbert class-field of K, $K \subset L \subset \hat{K}$ and the extension $\hat{K}/L$ is cyclic, then every class of L/K invariant under its Galois group (*ambiguous class*) becomes principal in $\hat{K}$. This has been proved by Terada [4042] in 1950, and simpler proofs have been provided later by Tannaka [3996, 3999] and Terada [4043]. See also Tannaka [3997, 3998], Tannaka and Terada [4001], Terada [4044]. Later Terada [4045] showed that in this theorem the Hilbert class-field can be replaced by the genus field, and in 1977 Furuya [1361] established this for Abelian extensions of $\mathbf{Q}$.

For further generalizations see Thiebaud [4050], Zink [4475],

On a conference in Tokyo in 1955 Deuring proposed the following conjecture.

If $\hat{K}$ is the Hilbert class-field of K, then with every ideal I of K one can associate an element $\Phi(I)$ of $\hat{K}$ generating $I\mathbf{Z}_{\hat{K}}$ with the following property: for all ideals I, J the element

$$\frac{\Phi(I)\Phi(J)^{\sigma(I)}}{\Phi(IJ)},$$

(where $\sigma(I)$ is the Artin symbol for the extension $\hat{K}/K$) is a unit in K.

This conjecture has been proved in 1958 by Tannaka [4000] in a stronger form, replacing the Hilbert class-field by any ray class-field. This stronger result has been made explicit in the case

[55] Wilhelm Magnus (1907–1990), professor in Göttingen, Courant Institute and Polytechnic Institute of New York.

[56] Ernst Witt (1911–1991), professor in Hamburg. See [2139].

[57] Zenon Ivanovič Borevič (1922–1995), professor in Leningrad. See [3073].

[58] Tadao Tannaka (1908–1986), professor at the Tôhoku University. See [14].

$K = \mathbf{Q}$ by Takahashi [3980] in 1964 and for imaginary quadratic fields by Kempfert [2136] in 1966.

In the special case of an imaginary quadratic field $\mathbf{Q}(\sqrt{-d})$ the Principal Ideal Theorem has been proved in 1929 in the thesis of Schäfer [3591] in the cases $(d, 6) = 1$, and two years later Hasse [1673] showed how to modify Schäfer's proof to cover all cases.

9. Denote by Ω the field of all algebraic numbers and by Ω^+ the field of all real algebraic numbers. In a short paper [116] published in 1924 Artin showed that if K is a proper subfield of Ω, and the degree $[\Omega : K]$ is finite, then K is isomorphic to Ω^+. It follows that every finite subgroup of the Galois group of $\Omega/\mathbf{Q}$ has at most two elements, and all such subgroups are conjugated. Two years later Artin and Schreier [135, 136] constructed the theory of fields in which 0 is not a sum of non-zero squares (*formally real fields*) and used it to generalize the first part of the last result by showing ([136], Satz 4) that if a field L is algebraically closed and K is its proper subfield with finite $[L : K]$, then $[L : K] = 2$.

The theory of formally real fields permitted Artin [119] to show that if a rational function $f(X_1, \ldots, X_n)$ with real coefficients attains nonnegative values at every point of $\mathbf{R}^n$ at which it is defined, then it is a sum of squares of rational functions. This settled in the affirmative the 17th problem of Hilbert [1842]. In the case when f is a form attaining zero only trivially another proof has been given by Habicht [1578] in 1940.

A quantitative version of Artin's result on Hilbert's 17th problem has been established in 1967 by Pfister [3279]. In 1971 Pourchet [3337] proved that if K is an algebraic number field, then any polynomial $f(X) \in K[X]$ which attains with all conjugates nonnegative values at K is a sum of at most five squares, and for $K = \mathbf{Q}$ this is best possible. In 1974 Hsia and R.P. Johnson [1907] determined the best possible result of this question for all formally real fields.

Hsia studied in [1906] the representation of cyclotomic polynomials as sums of squares.

4.3.3 Hasse

1. The first result of Hasse is contained in a joint paper with Hensel [1799] dealing with norm residues, published in 1923. One year earlier Hensel [1795] introduced a $\mathfrak{p}$-adic approach to the theory of norm residues in cyclic extensions L/K of prime degree l, in the case when K contains ζ_l; hence $L = K(a^{1/l})$ with some $a \in K$, which is not an lth power in K. Let $\mathfrak{p}$ be a prime ideal of K, $\mathfrak{P}$ one of its prime ideal divisors in L, and let $K_\mathfrak{p}$, $L_\mathfrak{P}$ be the corresponding completions. An element a of $K_\mathfrak{p}$ has been called a norm residue if $a = N_{L_\mathfrak{P}/K_\mathfrak{p}}(b)$ for some $b \in L_\mathfrak{P}$, and Hensel noted that his definition is equivalent to one given earlier by Hilbert in [1836]. He gave a criterion for norm residues in all cases, except when $\mathfrak{p}$ divides l. For unramified prime ideals $\mathfrak{p}$ of L and a prime ideal $\mathfrak{P} \mid \mathfrak{p}$ Hensel described the group of norms in the local extension $L_\mathfrak{P}/K_\mathfrak{p}$ using the multiplicative form of elements of a $\mathfrak{p}$-adic field developed by him in [1787, 1794]. In the paper of Hasse and Hensel the case of prime ideal divisors of l has been treated.

Later Hensel [1796] extended his approach to arbitrary finite extensions L/K, showing that an element of K is for each $n \geq 1$ congruent mod $\mathfrak{p}^n$ to an element of $N_{L/K}(L)$ if and only if it lies in $N_{L_\mathfrak{P}/K_\mathfrak{p}}(L_\mathfrak{P})$ for all prime ideal divisors $\mathfrak{P}$ of $\mathfrak{p}$.

Certain results of [1799] were made more precise by Hasse in [1643] and [1646], who applied results of [1648] where a simple proof of Furtwängler's [1333] generalization of Theorem 150 in Hilbert's report [1836] (see Sect. 2.1.2) has been given.

2. The *local–global principle*, called also the *Hasse principle*, found its birth in 1923, when Hasse published two papers on quadratic forms. In the first [1640] he showed that if $f(X_1, \dots, X_n)$ is a quadratic form over $\mathbf{Q}$, then the equation $f(x_1, \dots, x_n) = r$ with $r \in \mathbf{Q}$ is solvable with rational x_j if and only if this equation is solvable in every completion of the rationals, i.e. in each p-adic field $\mathbf{Q}_p$ and in $\mathbf{R}$. The second paper [1641] brings a similar assertion for the equivalence of two quadratic forms. In the case of representations of zero by ternary quadratic forms over $\mathbf{Q}$ a simpler proof has been given by Hasse [1681] in 1934 (in the case of more variables this has been done in 1957 by Springer[59] [3880]). In the next year Hasse extended his results to all finite extensions of the rationals [1644, 1645].

In 1962 Hasse [1705] pointed out that he got the inspiration to these results from a letter of Hensel and an old result of Legendre (see pp. 512–513 of [2508], §294–298 of Gauss's [1394] and §156–157 of the second edition of Dirichlet's lectures [972]), giving a criterion for the non-trivial solvability of the equation $ax^2 + by^2 + cz^2 = 0$.

Another proof of the result in [1641] has been given by Siegel [3783] who showed also that if two quadratic forms with rational integral coefficients are equivalent over reals and over all $\mathbf{Z}_p$, then they are semi-equivalent; i.e. for every $m \in \mathbf{Z}$ they are equivalent over the ring of m-integral numbers. This assertion has been stated by Minkowski (on pp. 7–8 of [2871]), and in the case of three variables a proof has been given in 1867 by H.J.S. Smith [3831]. A method of checking the equivalence of two quadratic forms over $\mathbf{Z}_2$ has been given in 1944 by Jones [2064] (see also Jones and Durfee [2065]).

In [1642] showed Hasse that the local–global principle applies also to the solvability of the matrix equation

$$X^T AX = B$$

for matrices with rational entries.

3. Several papers of Hasse [1647, 1649–1651, 1661] were devoted to the search of explicit formulas for the product

$$\left(\frac{\alpha}{\beta}\right)_p \left(\frac{\beta}{\alpha}\right)_p,$$

[59] Tonny Albert Springer (1926–2011), professor at the Technical University Utrecht. See [4124].

where $\left(\frac{A}{B}\right)_p$ denotes the pth power residue symbol. Two papers on this subject, [133, 134], were written with Artin. A simpler proof of results of [1649] has been given in 1932 by Vennekohl [4211].

In 1995 Hill [1849] showed that the reciprocity law for prime exponents can be established using homotopy theory and homology groups.

In 1927 Hasse [1659] showed how the reciprocity law for mth power residues can be simply deduced from Artin's reciprocity law. He established also the product formula (2.4) for norm residues for extensions $k(a^{1/m})/k$, where k is a field containing ζ_m.

The norm residue symbol $\left(\frac{\beta, K/k}{\mathfrak{p}}\right)$ for arbitrary Abelian extensions K/k has been defined by Hasse [1666] in 1930, utilizing the Artin symbol $F_{K/k}(\mathfrak{p})$, extended by multiplicativity to all ideals. If $\mathfrak{f}$ is the conductor of the ideal group H, associated with K/k by class-field theory, and a prime ideal $\mathfrak{p}$ and $\beta \in K$ are given, then define $\mathfrak{f}_\mathfrak{p}$ by $\mathfrak{f} = \mathfrak{p}^\alpha \mathfrak{I}$ with $\mathfrak{p} \nmid \mathfrak{I}$, choose β_0 with

$$\beta_0 \equiv \beta \pmod{\mathfrak{p}^a}, \quad \beta_0 \equiv 1 \pmod{\mathfrak{I}},$$

write

$$\beta_0 \mathbf{Z}_K = \mathfrak{p}^a \mathfrak{B},$$

and put

$$\left(\frac{\beta, K/k}{\mathfrak{p}}\right) = F_{K/k}(\mathfrak{B}).$$

He established the analogue of the product formula (2.4)

$$\prod_{\mathfrak{p}} \left(\frac{\alpha, K}{\mathfrak{p}}\right) = 1, \tag{4.20}$$

($\mathfrak{p}$ running over all prime divisors, finite and infinite), proved that β is a norm residue for all powers of $\mathfrak{p}$ if and only if one has

$$\left(\frac{\beta, K/k}{\mathfrak{p}}\right) = 1,$$

and determined the set of values attained by the norm residue symbol for fixed $\mathfrak{p}$ and β running either through k^* or through all numbers of k prime to $\mathfrak{p}$. In the first case it is the splitting group and in the second case the inertia group of $\mathfrak{p}$.

The set of values of the norm residue symbol attained for β running through norm residues mod $\mathfrak{p}^r$ has been later determined by Iyanaga [2000].

A characterization of the norm residue symbol by its multiplicativity, the product formula (4.20) and $\mathfrak{p}$-continuity has been provided in 1932 by Grunwald[60] [1533], who generalized Hasse's [1654] result dealing with the case of the rational field.

For later development of the class-field theory see Sect. 5.2.

4.4 Class-Number and Class-Group

4.4.1 Quadratic Fields

1. In 1928 Rédei [3396] determined the value of

$$h(d) \bmod 2^{\omega(d)}$$

for negative $d \equiv 3 \bmod 4$. For any factorization

$$\sigma : -d = n_1 n_2 \cdots n_r$$

with $r = r(\sigma) \geq 2$ and $n_j \geq 2$ put

$$A_\sigma = \prod_{j=1}^{r-1} \left(\frac{n_j}{n_{j+1} \cdots n_r} \right) .$$

Then

$$h(d) \equiv 1 + \sum_\sigma \frac{(-1)^{r(\sigma)-1}}{r(\sigma)} A_\sigma \pmod{2^{\omega(d)}} .$$

In the case of real fields the divisibility of $h^*(pq)$ by 8 has been considered in 1934 by Rédei [3399] and Scholz [3672], who showed in particular that for primes $p = q = 1 \bmod 4$ the condition $8 | h^*(pq)$ is equivalent to $(p/q)_4 = (q/p)_4 = 1$ (an elementary proof has been provided much later by Nemenzo [3082]).

In 1969 Barrucand and Cohn [195] showed that for primes $p \equiv 1 \bmod 8$ the condition $8 \mid h(-p)$ is equivalent to the representability of p by the form $X^2 + 32Y^2$. Other proofs have been given by Hasse [1707, 1710] and K.S. Williams [4414]. Hasse gave in [1709] a similar result for the divisibility of $h(-2p)$ by 8 (another proof has been found by Brown [466]), in [1710] did this for $h(-pq)$ with distinct primes p, q, and in [1711] considered d positive and gave a criterion for $8 \mid h^*(pq)$. This has been extended to a condition for $2^n \mid h^*(pq)$ in 1971 by H. Bauer [217]. In 1972 Brown [465] used quadratic forms to prove for $p = a^2 + b^2 \equiv 1 \bmod 8$ (with $2 \nmid a$) the congruence

$$h(-p) \equiv \frac{p-1}{2} + b \pmod 8 ,$$

[60] Wilhelm Grunwald (1909–1989) worked in Göttingen.

and in 1973 jointly with C.J. Parry [469] characterized negative d with $\omega(d) = 3$ and $8 \mid h(d)$. In [467] Brown considered $h(-pq) \bmod 2^k$ for $k \leq 4$ in case $p \equiv q \bmod 4$ and in [468] studied divisibility of $h(pqr)$ by powers of 2.

Divisibility of $h^*(d)$ by 4 was considered by K.S. Williams [4419], and for divisibility by 8 see Reichardt [3414], A. Endô [1091], P. Kaplan [2110], Oriat [3190], K.S. Williams [4418, 4420] and K.S. Williams and Friesen [4421]. For divisibility by 16 see Oriat [3190], K.S. Williams [4417], P. Kaplan, K.S. Williams and K. Hardy [2114–2117]. In 2004 Basilla [208] gave elementary proofs of Kaplan's results.

2. The first result about divisibility of class-numbers of quadratic fields by arbitrary integers has been obtained in 1922 by Nagell [3014] who showed that there are infinitely many imaginary quadratic fields with class-number divisible by a given integer.

Other proofs were later given by P. Humbert [1923] in 1940, Ankeny and S. Chowla in 1955 [96] and Uehara [4099] in 1983. In 1964 S.-N. Kuroda [2385] proved that one can additionally demand the divisibility of the discriminant by a given number.

In 1970 Yamamoto [4450] established the analogue of Nagell's result for real quadratic fields (other proofs were given by Weinberger [4360], S. Nakano [3053] and Uehara [4100]) and showed also the existence of infinitely many imaginary quadratic fields whose class-group contains $C_n \oplus C_n$, with any given n. If $A_n(x)$, $B_n(x)$ denotes the number of $0 < d \leq x$ respectively $-x \leq d < 0$ such that n divides $h(d)$, then the bounds

$$A_n(x) \gg \begin{cases} x^{1/2n-\varepsilon} & \text{if } 4 \nmid n, \\ x^{1/4n-\varepsilon} & \text{if } 4|n, \end{cases} \tag{4.21}$$

and

$$B_n(x) \gg x^{1/2+1/n} \tag{4.22}$$

for every $\varepsilon > 0$ were established in 1999 by M.R. Murty [3000].

In the case of even n Soundararajan [3861] improved (4.22) to

$$B_n(x) \gg x^{1/2+c(n)-\varepsilon}$$

with

$$c(n) = \begin{cases} 2/n & \text{if } 4|n, \\ 3/(n+2) & \text{otherwise.} \end{cases}$$

The first improvement of (4.21) has been obtained in 2002 by G. Yu [4458] who for odd n got

$$A_n(x) \gg x^{1/n-\varepsilon},$$

and one year later Luca [2667] showed

$$A_n(x) \gg \frac{x^{1/n}}{\log x}$$

for n even. In 2008 Chakraborty, Luca and Mukhopadhyay [640] established for even n and large x the bound

$$A_n(x) \geq \frac{1}{5} x^{1/n} .$$

For $n = 5, 7$ Byeon [539] obtained $A_n(x) \gg x^{1/2}$ in 2006.

Nagell's result has been generalized to other classes of fields. In 1970 Madan[61] [2692] showed that if the sets of prime divisors of m and n coincide, then there exist infinitely many normal fields K of degree m with $n \mid h(K)$, and the same holds also for non-normal fields under the obvious assumption $m \neq 2$. Four years later Uchida [4095] proved Nagell's assertion for cyclic cubic fields, and the same has been done for pure cubic fields by S. Nakano [3053] in 1983.

Now one knows that there exist infinitely many fields of given degree and signature with the class-number divisible by a given integer. First Azuhata and Ichimura [162] established for every $n > 1$ the existence of infinitely many fields of given degree and signature (r_1, r_2) with $r_2 \geq 1$ with $H(K)$ containing $C_n^{r_2}$ as a subgroup, thus $n \mid h(K)$, and then S. Nakano [3054] removed the assumption $r_2 \geq 1$ with the stronger result: $C_n^{r_2+1} \subset H(K)$.

There is an old question whether every finite Abelian group is the class-group of an algebraic number field. In the case of Dedekind domains, where the class-group can be defined in the same way as in the case of number fields, the analogous question has been positively answered in 1966 by Claborn[62] [715]. (see also Chap. 3 in the book of Fossum [1220]). Another proof has been given in 1972 by Leedham-Green [2503], who showed that such Dedekind domains can be chosen to be quadratic extension of a principal ideal domain. It is not known whether in the case of finite groups always one can choose a ring of integers of an algebraic number field.

It has been shown in 1973 by Rosen [3512] that every finitely generated Abelian group can be realized as the class-group of a Dedekind domain of the form $\Omega[x, y]$, where $y^2 = 4x^3 - ax - b$ is an elliptic curve over Ω, the composite of all quadratic extensions of $\mathbf{Q}$. In 1976 he showed [3513] that the same holds for countable Abelian groups. In 2009 P.L. Clark [719] established that every Abelian group is the class-group of a certain overdomain of $\Omega[x, y]$.

In 1978 Yahagi [4448] proved that every finite Abelian p-group is the Sylow-p-subgroup of some $H(K)$. For elementary p-groups this has been earlier established by Gerth[63] [1417]. In 1979 Cornell [770] established that every finite Abelian group is a subgroup of the class-group of a cyclotomic field, and in 1999 Perret [3246] used this result to show that every finite Abelian group is the class-group of a ring of S-integers of an algebraic number field.

3. In 1924 Dickson [937] showed in an elementary way that if the class-number of quadratic forms of discriminant d is 1, then the ring of integers in $\mathbf{Q}\left(\sqrt{d}\right)$ is a unique factorization domain.

In 1928 Schaffstein [3592] presented tables of class-numbers of real quadratic fields with prime discriminant p in the ranges $p < 12\,000$, $100\,000 < p < 101\,000$ and $1\,000\,000 < p < 1\,001\,000$. He noted that for most primes the class-number equals 1. This holds in particular for all primes in the interval (3300, 3800).

In 1969 Shanks [3745] presented on a conference in Stony Brook an algorithm computing $h(d)$ with running time $O(d^{0.2})$ under the assumption of GRH (cf. H.W.Jr. Lenstra [2546] and Schoof [3685]. In 1998 Srinivasan [3882] described an unconditional algorithm for $h(d)$ with running time of order $O(d^c)$ for any $c > 0.2$.

Hendy [1760] and H.C. Williams and Broere [4404] computed $h(d)$ for $0 < d < 10^5$. Class-groups for real quadratic fields were determined by Oriat [3191] for $d(K) < 2572$. A survey on the computation of the class-number of real quadratic fields has been published in 1992 by Mollin and H.C. Williams [2914]. For more on computations of class-groups of quadratic fields see Sect. 5.3.1.

---

[61] Manohar Lal Madan (1935–2011), professor at Ohio State University.

[62] Luther Elic Claborn (1935–1967), professor at the University of Illinois in Urbana.

[63] Frank Gerth III (1945–2006), professor at the University of Texas in Austin.

4.4.2 Other Fields

1. A formula for the class-number of an imaginary quadratic extension of a real quadratic field has been given in 1921 by Hecke [1739], who also conjectured that if K is totally real and $L = K(\sqrt{\delta})$ is its totally imaginary extension, then the ratio $h(L)/h(K)$ can be expressed by an elementary arithmetical function of δ. He proved this in the case of real quadratic K.

A proof of the conjectured formula in the case of a totally real cubic K has been given by Reidemeister[64] [3424] in 1922.

In 1958 Leopoldt [2552] established Hecke's conjecture in the case of Abelian K, using the generalized Bernoulli numbers B_n^χ, attached to primitive characters χ mod f by formula

$$\sum_{j=0}^{\infty} \chi(j) \frac{te^{jt}}{e^{ft}-1} = \sum_{n=0}^{\infty} B_n^\chi \frac{t^n}{n!}. \tag{4.23}$$

He observed that one can define these numbers also by putting

$$B_n^\chi = -nL(1-n, \chi)$$

for positive integers n.

In case of real characters χ these numbers were considered already by A. Hurwitz [1926] in 1882, A. Berger [298] in 1891, as well by Ankeny, Artin and S. Chowla [91]) in 1952.

Arithmetical properties of B_χ^n were studied by Carlitz [579], Fresnel [1245] and Ernvall [1118, 1119]. Leopoldt applied these numbers in [2554] to describe the divisibility of the class-number of a real Abelian field of degree n by a prime not dividing $2n$.

In the general case it has been proved in 1973 by Goldstein [1469] that the ratio $h(L)/h(K)$ can be expressed in terms of periods of certain differential forms on manifolds (cf. also Goldstein and de la Torre [1471]).

The final step in the establishment of Hecke's conjecture has been done in 1976 by Shintani [3759].

2. In 1924 Čebotarev [619] considered subfields of the cyclotomic field $\mathbf{Q}(\zeta_{p^m})$ with prime p. He gave a new proof of Furtwängler's result stating that if $K \subset L \subset \mathbf{Q}(\zeta_{p^m})$, then $h(K)$ divides $h(L)$, and showed also that if $K \subset \mathbf{Q}(\zeta_{p^m})$ has degree n, the class-numbers of proper subfields of K are not divisible by p and the class-group $H(K)$ has a subgroup of index p, invariant under $Gal(K/\mathbf{Q})$, then $p \equiv 1 \bmod n$. In his next paper [624] he obtained a similar result for cyclic extensions of $\mathbf{Q}$.

A detailed analysis of these two papers and a survey of later generalizations have been presented by Metsänkylä [2843] in 2007.

3. A study of the p-part of the class-group of the fields $\mathbf{Q}(\zeta_p)$ and $\mathbf{Q}(\zeta_{p^2})$ in case of irregular primes p has been made in 1924 by Pollaczek [3330]. He established in particular the existence of a system $\{u_j\}$ of fundamental units of the maximal real subfield of $\mathbf{Q}(\zeta_p)$ such that if r is a primitive root mod p^2 and s is the automorphism

---

[64] Kurt Werner Reidemeister (1893–1971), professor in Königsberg, Marburg and Göttingen. See [141].

of $\mathbf{Q}(\zeta_p)$ with $s(\zeta_p) = \zeta_p^r$, then for $i = 1, 2, \ldots, (p-3)/2$ every unit $s(u_j)u_j^{-r^{2i}}$ is a pth power.

Another proof of this result has been given in 1979 by Washington [4303].

In the case of regular prime p Pollaczek showed that for any unit $u \in k = \mathbf{Q}(\zeta_p)$ the class-number of $k\left(u^{1/p}\right)$ is not divisible by p (a simpler proof of this result gave later Moriya [2954]).

4. In 1929 Vandiver [4159] proved that if $p \nmid h_p^+$, and for $j = 2, 4, \ldots, p-3$ the Bernoulli number B_{jp} is not divisible by p, then the second case of Fermat's theorem holds for the exponent p. Four years later he asserted in [4171] that $p \nmid h_p^+$ implies Fermat's theorem in the first case, but his proof is erroneous (see [3454], p. 188).

The assertion that the class-number h_p^+ is not divisible by p is called the *Vandiver conjecture*. It is still open.

A necessary condition for $p \mid h_p^+$, utilizing powers of the unit

$$\left(\frac{(1-\zeta_p^r)(1-\zeta_p^{-r})}{(1-\zeta_p)(1-\zeta_p^{-1})}\right)^{1/2}$$

(with r being a primitive root mod p) was given by Stafford and Vandiver in 1930 [3885]. Its sufficiency was established in the same year by Vandiver [4166]. The necessity in the case $p \| h_p^-$ was noted already by Kummer in 1857 [2366].

It was shown later by Sitaraman [3807] that Vandiver's argument in [4159] can be used to show that if $p^2 \nmid B_{p-3}$, and then Fermat's theorem holds for p in the first case (earlier a proof of this assertion was given by M. Kurihara [2381]. See also Mihailescu [2857].

A stronger conjecture asserting $h_p^+ < p$ was proposed in the seventies by Masley. In 1985 Cornell and Washington [771] showed that the converse inequality for $p = 11\,290\,018\,777$ follows from GRH, and this was established without extra assumptions by Seah, Washington and H.C. Williams [3722]. A few years later Schoof and Washington [3689] found a smaller counterexample, $p = 641\,492$. These examples were obtained by finding large class-numbers of cubic, respectively quintic Abelian fields with relatively small prime conductors, and used the fact that h_p^+ is divisible by the class-number of any real subfield of $\mathbf{Q}(\zeta_p)$. Several further examples were found by Louboutin [2646, 2648].

The truth of the Vandiver conjecture has been established for primes $p \leq 39 \cdot 2^{22}$ by Buhler and Harvey [508]. Earlier this has been done for $p < 10^6$ [507], $p < 4 \cdot 10^6$ [505] and $p < 12 \cdot 10^6$ [506]. Heuristical arguments presented in the book of Washington ([4304], pp. 159–160) indicate that it should be true for most primes p. Cf. Mihailescu [2856].

Various assertions equivalent to Vandiver's conjecture were given by Iwasawa [1993], Kersten and Michaliček [2140] and Anglés and Nuccio [85].

Vandiver conjecture is equivalent to the assertion that every real Abelian field K of prime conductor p has its class-number prime to p. This is true for quadratic K due to the inequality $h\left(Q(\sqrt{p})\right) < 2\sqrt{p}$ and has been established in 1981 for cubic K by Moser [2980] and for cyclic K of degrees 4 and 6 by Moser and Payan [2981].

In 1953 Dénes ([892] defined the irregularity index of $Q(\zeta_p)$ using divisibility by powers of p of Bernoulli numbers, and in [894] he deduced a formula for the maximal power of p dividing h_p^+ from the conjectural finiteness of that index. He promised to establish the truth of this conjecture and noted that it follows from the positive answer to the class-field tower problem. Dénes's conjecture has been established in 1979 by Washington [4303] as a consequence of the non-vanishing of the

p-adic regulator of Abelian fields established by Ax[65] [153] and Brumer [476]). A generalization to fields $Q(\zeta_{p^m})$ has been made by F. Kurihara [2380] in 1989.

4.5 Other Questions

4.5.1 *Galois Groups*

1. In 1929 Scholz [3665, 3666] applied class-field theory to discuss Galois extensions of algebraic number fields, having metabelian Galois groups and showed the existence of such extensions for a wide class of groups, including all metabelian groups of square-free degree. His papers contain one of the first steps towards the solution of the following embedding problem.

Let K/k be a Galois extension with group G_0, and let G be a group having a normal subgroup H with G/H isomorphic to G_0; thus the sequence

$$1 \longrightarrow H \longrightarrow G \longrightarrow G_0 \longrightarrow 1$$

of groups is exact. The embedding problem consists in finding necessary and sufficient conditions for the existence of an extension L/K such that L/k is Galois with Galois group G. In the case when G is the semi-direct product of G_0 and the kernel H one speaks about a split extension.

In [3666] Scholz defined a class of metabelian groups, called by him "Dispositionsgruppen", and showed that if G lies in this class, H is a normal Abelian subgroup of G, and the factor group G/H is the Galois group of an extension k/Q, then there exists an extension K/k with $Gal(K/\mathbf{Q}) = G$, solving the embedding problem in the case of split extensions with Abelian kernel.

A generalization of these groups has been presented in 1932 by Čebotarev [625] who applied them to study the embedding problem in the case of Abelian group H. He related this problem to the question of existence of certain prime ideals. For a class of imaginary quadratic fields k such prime ideals have been found by him in 1936 [629].

Later Scholz [3671, 3673] considered the case of metabelian groups of prime power orders and in [3675] showed that every group of odd prime power order occurs as a Galois group. The last result has been also obtained at the same time by Reichardt [3413].

A survey of various questions related to the problem of existence of extensions with given Galois group has been presented in 1934 by Čebotarev [626].

In 1941 Lednev [2498] applied results of A. Hurwitz [1930] on Riemann surfaces to show that the problem of existence of extensions with given Galois group of an algebraic number field k can be reduced to the question of the existence of k-points on a suitable algebraic surface.

A geometrical approach to Galois theory has been presented in 1940 by Delone [879], and this method has been later applied by him and Faddeev [883] to the extension problem, permitting to eliminate the use of class-field theory in some results of Reichardt [3413]. One finds in [883] a necessary condition for the existence of a solution of the embedding problem. This condition has been also found by Hasse [1692] (cf. [1689]), who conjectured that in the case of cyclic group H

[65] James Burton Ax (1937–2006), professor at Cornell University and in Stony Brook.

the condition is also sufficient. It has been later shown by Faddeev [1156] and Šafarevič [3548] (see also Beyer [334]) that it is not so, and later Beyer [333] established its sufficiency in the case when H is cyclic of an odd order. It has been shown later by Demuškin and Šafarevič [888] that in case of local fields this condition is sufficient for Abelian kernel H. A necessary condition for the solution of the embedding problem in the global case has been studied by them in [889]. For the embedding problem with Abelian kernel see also Hoechsmann [1876]. In 1972 Sonn [3854] showed that the embedding problem can be reduced to the cases of split extensions and extensions with solvable kernel.

In 1954 Šafarevič [3547] showed that if k is a finite extension of $\mathbf{Q}$, then every solvable group (note that in 1963 Feit[66] and J.G. Thompson [1170] showed that all groups of odd order are solvable) is the Galois group of an extension of k. His proof had a gap in the case of groups of even order, which has been filled by him in 1989 [3552]. A simpler proof has been given by Neukirch [3096] (see also Chap. IX of the book [3100] by Neukirch, A. Schmidt and Wingberg published in 2000). The local–global principle in the embedding problem has been studied by Neukirch [3092, 3096].

An important progress has been obtained with the use of the rigidity method, in which the theory of Riemann surfaces is applied to Noether's problem. It has been worked out by Shih [3753], M. Fried [1251], Belyĭ[67] [282], Matzat [2782, 2783] and J.G. Thompson [4055]. With its use Shih [3753] obtained in 1974 one of the first results for non-solvable linear groups. Using the theory of canonical systems of models developed by Shimura [3756, 3757] he showed the existence of polynomials with Galois group $PSL_2(\mathbf{F}_p)$ for prime $p \not\equiv \pm 1 \bmod 24$.

In 1978 E. Fried and Kollár [1249] proved that every finite group is the group of all automorphisms of infinitely many extensions of $\mathbf{Q}$. It turned out that there was an error in the proof, corrected later by M. Fried in Math. Reviews, who also made this result effective [1252]. A simpler proof has been given in 1983 by Geyer [1422]. The case of infinite groups has been considered in 1987 by Dugas and Göbel [1017].

In 1979 Belyĭ [282] succeeded in realization of linear groups as Galois groups over suitable Abelian fields, and in 1983 he established that every classical simple group is the Galois group of an extension of an Abelian field [283].

Realizability over $\mathbf{Q}$ or $\mathbf{Q}(t)$ of certain classes of simple orthogonal and symplectic groups has been established in 1992 by Malle and Matzat [2727] in 1985, Malle [2719–2724] in 1988–1993 and Häfner [1584]. For other classes of classical linear groups this has been done by Völklein [4239] in 1998, S. Reiter [3431] and Dentzer [895] in 1999, Dettweiler and S. Reiter [904, 905] in 1999–2000, Dettweiler, Kühn and S. Reiter [903] in 2001, Shiina [3754, 3755] in 2003–2004 and Dettweiler [902] in 2004.

The work of Matzat [2783, 2787], J.G. Thompson [4055], Hoyden-Siedersleben [1904], Matzat and Hoyden-Siedersleben [1905], Hunt [1924], Matzat and Zeh-Marschke [2788, 2789] and Pahlings [3207, 3208] showed that all sporadic simple groups are realizable over $\mathbf{Q}$. In [2783] it is also shown that all primitive non-solvable permutation groups of order ≤ 15 are realizable.

In 2000 Klüners and Malle [2177] showed that every transitive subgroup of S_n with $12 \leq n \leq 15$ is a Galois group over $\mathbf{Q}$.

Realization of certain other classes of groups was obtained by Feit [1169], Vila [4214, 4215], Kotlar, Schacher and Sonn [2230], Schacher and Sonn [3590], Sonn [3855], Epkenhans [1110, 1111] and Jehanne [2040].

Surveys have been presented by Matzat [2784, 2785] in 1985 and 1988 and by Serre [3738, 3739] in 1988 and 1992. The rigidity method has been exposed in the books by Völklein [4238] and by Malle and Matzat [2728].

In the book [2048] by C.U. Jensen, Ledet and Yui, published in 2002, a constructive approach to Noether's problem has been presented, giving in particular infinite families of polynomials with given Galois group of small cardinality, including all transitive subgroups of S_n with $n = 3, 4, 5, 7$ and 11.

---

[66] Walter Feit (1920–2004), professor at Yale. See [3719].

[67] Gennadiĭ Vladimirovič Belyĭ (1951–2001) worked in Vladimir University. See [377].

In 2002 Malinin [2718] proved that every finite group is the Galois group of an unramified extensions of some totally real field.

4.5.2 Algebraic Numbers in the Plane

1. In 1918 Schur [3698] generalized Kronecker's result about totally real numbers α with $\overline{|\alpha|} \leq 2$ showing that if I is an interval on the real axis of length <4, then there can be only finitely full sets of conjugates of algebraic integers (FCS) lying in I.

This has been extended in 1923 by Fekete[68] [1171], who proved that if A is a compact set on the complex plane having transfinite diameter <1, then there are only finitely many FCS in A. Let us recall that the *transfinite diameter* $\mathrm{trd}(A)$ of a compact subset A of the plane is defined by

$$\mathrm{trd}(A) = \lim_{n\to\infty} \max_{z_1,\dots,z_n\in A} \prod_{1\leq 1<j\leq n} |z_i - z_j|^{2/(n(n-1))} .$$

Much later, in 1955, Fekete and Szegő [1172] proved a kind of a converse theorem. They showed that if a subset A of the complex plane contains in its interior a subset E with $\mathrm{trd}(E) = 1$ which is closed under complex conjugation, then A contains infinitely many FCS. On the other hand R.M. Robinson[69] [3482] showed in 1964 that there exist real sets with arbitrary large transfinite diameter, consisting exclusively of transcendental numbers. He constructed compact nowhere dense sets A of large $\mathrm{trd}(A)$ and an uncountable set B such that the sets $\beta + A$ for $\beta \in B$ are pairwise disjoint. Then at least one of these sets consists of transcendental numbers. Moreover, it has been shown in [3482] with the use of Čebyšev polynomials that if A is a union of finitely many real intervals and $\mathrm{trd}(A) > 1$, then A contains infinitely many FCS. Earlier Robinson [3481] proved this for intervals. See also [3484], where he considered the same question for units. In [3483] one finds a description of algebraic integers of degrees ≤ 6 whose all conjugates lie in an interval of length <4, and this has been extended to degrees ≤ 14 by Capparelli, Del Fra and Sciò [564] in 2010, and next year Flammang, Rhin and Wu [1197] obtained this for the degree 15.

In the eighties the notion of the transfinite diameter has been generalized to adeles (for the definition of adeles see Sect. 6.3) by D.G. Cantor [558] and Rumely [3531]. Further development done by Rumely [3532] permitted him to show that if T is a finite set of primes and

$$\prod_{p\in T} p^{-1/(p-1)} > \frac{1}{r} ,$$

then the interval $[-2r, 2r]$ contains infinitely many FCS of algebraic integers, whose all conjugates are p-adic integers for $p \in T$ (see also Rumely [3533]).

The first result on full sets of conjugates of algebraic integers lying on a curve has been obtained in 1945 by Motzkin[70] [2987] who showed that there 11 families of straight lines on the plane containing FCS of integers.

In 1969 R.M. Robinson [3486] determined all circles with rational centre which contain infinitely many full sets of conjugated algebraic integers and conjectured that this is not true for circles with

[68] Michael Fekete (1886–1957), professor at the Hebrew University in Jerusalem. See [3492].

[69] Raphael Mitchell Robinson (1911–1995), professor at Berkeley. See [1761].

[70] Theodore Samuel Motzkin (1908–1970), professor at the Hebrew University in Jerusalem and the University of California in Los Angeles. See [9].

irrational real centres. This conjecture has been disproved in 1973 by Ennola [1103], who determined all such circles with irrational totally real centre. The case of circles having not totally real centres has been later studied by Ennola and Smyth [1106, 1107]. The analogue of these questions for other conics has been treated by Smyth [3838] (see also McKee [2802]). Algebraic integers with all conjugates lying on the circle $\{z: |z| = \sqrt{n}\}$ with $n \in \mathbf{Z}$ were called *Weil numbers* by Greaves[71] and Odoni[72] [1506], who showed that Weil numbers generate CM-fields, and described all primes p unramified in a CM-field K with $K = \mathbf{Q}(\alpha)$, where α is a Weil number with $|\alpha|^2 = p^j$ for some j. In 1991 Odoni obtained for fixed CM-field K asymptotics for the numbers of $n \le x$ such that there exists a Weil number α with $|\alpha|^2 = n$, generating K.

Algebraic integers whose conjugates lie in an ellipse were described in 2005 by Flammang and Rhin [1195].

2. Another result established in [3698] by Schur shows that there are only finitely many totally positive totally real integers α of degree n whose trace $T(\alpha)$ does not exceed cn for $c < \sqrt{e} = 1.6847\ldots$.

In 1945 Siegel [3788] proved a strengthening of the inequality between arithmetical and geometrical means, and this enabled him to obtain Schur's assertion for $c < 1.7336\ldots$. He noted that the example $\alpha = \cos(2\pi/p)$ (for prime $p \ge 3$) shows that one cannot have this result for $c > 2$, and showed that inequality $T(\alpha) \le 3n/2$ implies $\alpha \in \{1, (3 \pm \sqrt{5})/2\}$. The bound for c has been subsequently improved to $1.771\ldots$ (Smyth [3839]), $1.7783\ldots$ (McKee and Smyth [2804]), $1.78002\ldots$ (Aguirre, Bilbao and Peral [22]), $1.7836\ldots$ (Aguirre and Peral [23]), $1.7870\ldots$ (Flammang [1193]), $1.78839\ldots$ (McKee [2803]) and $1.7991\ldots$ (Y. Liang and Wu [2593]).

It is conjectured (the *Schur–Siegel–Smyth trace problem*) that one can take for c any number smaller than 2.

3. In 1928 Favard[73] [1163–1165] showed that if a is an algebraic integer of degree $n \ge 2$ and $a_1 = a, a_2, \ldots, a_n$ are its conjugates, then

$$\delta(a) = \max |a_i - a_j| \ge \sqrt{3/2}, \tag{4.24}$$

and if a is totally real, then $\delta(a) > 2$.

For totally real a R.M. Robinson [3486] established in 1969 $\delta(a) \ge \sqrt{5}$ and showed that for $n \ge 3$ one has $\delta(a) > 3$.

In 1984 Lloyd-Smith [2617] improved upon (4.24) by establishing $\delta(a) \ge 3/2$, and this has been replaced by $\delta(a) \ge \sqrt{2}$ by Langevin, Reyssat and Rhin [2467] in 1988. In the same year Langevin [2465] proved that for sufficiently large n one has $\delta(a) \ge c$ for any $c < 2$. The history of Favard's problem has been presented by Langevin [2465].

The dependence of $\delta(a)$ on the height of its minimal polynomial P has been considered in 1964 by Mahler [2706], who showed

$$\delta(a) \ge c(n)H(P)^{1-n},$$

where

$$c(n) = \sqrt{3}/(n+1)^{-n-1/2}.$$

In 2001 Collins [760] conjectured the bound

$$\delta(a) \ge n^{\alpha} H(P)^{-n/2}$$

---

[71] George Greaves (1941–2008), professor in Cardiff. See [1728].

[72] Robert Winston Keith Odoni (1947–2002), professor in Exeter and Glasgow. See [742].

[73] Jean Favard (1902–1965), professor in Grenoble and at the l'École Polytechnique in Paris.

with $\alpha = -n/4$, based on intensive computation. However it has been shown by Evertse [1143] and Schönhage [3684] that this fails already for $n = 3$. It has been shown in 2004 by Bugeaud and Mignotte [502] that one must have $\alpha \leq -n/2$, and this has been improved in 2011 by Bugeaud and Dujella [499] to

$$\alpha \leq -\left(\frac{n}{2} + \frac{n-2}{4n-4}\right) .$$

See also Beresnevich, Bernik, Götze [294], Dujella and Pejković [1020], Dubickas [1010] and Bugeaud and Dujella [500].

4.5.3 Infinite Extensions

1. Infinite extension of the rationals occurred for the first time in 1901 in a paper of Dedekind [854]. He observed on the example of the field generated by the set $\{\zeta_{p^n} : n = 1, 2, \dots\}$ for a fixed prime p that the existing in the finite case one-to-one correspondence between subgroups of the Galois group and the intermediate fields does not hold in the infinite case.

In 1913 Stiemke[74] [3935] proved that every additive group of algebraic integers is free, and in his dissertation [3936], prepared in 1914 and published in 1926, he deduced the existence of a denumerable integral basis and described ideal theory in the ring of integers of infinite extensions of $\mathbf{Q}$.

In 1964 C.U. Jensen [2047] showed that if K is an algebraic number field and the extension L/K is infinite, then the ring of algebraic integers of L is a free $\mathbf{Z}_K$-module. A simpler proof has been found in 1967 by Kulkarni [2346].

Some simple observations about factorizations in the field of all algebraic numbers were made in 1928 by E. Cahen [549].

2. An important step in the theory of infinite extensions was made by Krull in 1928 [2310], who extended Galois theory to the infinite case. Let L/K be a normal extension, and let G be its Galois group, i.e. the group of all automorphisms of L fixing elements of K. With every subgroup H of G Krull associated the H-invariant field

$$\mathfrak{K}(H) = \{a \in L : g(a) = a \text{ for } g \in H\} ,$$

and to every intermediate field M he associated the group

$$\Delta(M) = \{g \in G : g(a) = a \text{ for } a \in M\}.$$

He recalled a result of Dedekind [854] that one always has

$$\mathfrak{K}(\Delta(M)) = M ,$$

but the equality

[74]Erich Stiemke (1892–1915), killed in World War I.

$$\Delta(\Re(H)) = H \tag{4.25}$$

may fail for L/K infinite. Usually in this case one has only

$$H \subset \Delta(\Re(H)) \ .$$

To repair this situation Krull introduced a topology in the family

$$\Omega(L/K) = \{M : \ K \subset M \subset L\}$$

of all intermediate fields (*Krull topology*) and proved that the equality (4.25) holds for all closed subgroups H of G. In this topology the basis of open sets is formed by cosets of Galois groups of extensions L/M, where the extension M/K is Galois and finite In two later papers Krull [2311, 2312] extended Stiemke's theory, and this led to a description of ideals in the ring of integers in an infinite extension of an algebraic number field. In the related paper [2313] he observed that his assertion obtained in [2312] for these rings, stating that every invertible ideal is finitely generated is true for arbitrary rings with units.

Ideal theory in fields of infinite degree differs essentially from the theory for finite extensions. For example, there may exist prime ideals $\mathfrak{p}$ which are idempotent, i.e. $\mathfrak{p}^2 = \mathfrak{p}$ (see N. Nakano [3050–3052]).

4.5.4 Varia

1. In 1921 Hancock [1617] described the solutions of the equation $X^2 + Y^2 = Z^2$ in quadratic number fields. Certain classes of cubic and quartic equations in quadratic fields were studied in 1937 by Fogels[75] [1202].

In 1924–1925 Wahlin [4268, 4270] applied the theory of ideals in number fields to the solution of norm form equations.

2. In 1921 Skolem [3808] obtained a quantitative version of Hilbert's irreducibility theorem, showing that if $f(X, Y) \in \mathbf{Z}[X, Y]$ is irreducible over $\mathbf{Q}$, and $N_f(T)$ is the number of integers t with $|t| \le T$ for which $f(x, t)$ is reducible over $\mathbf{Q}$, then $N_f(T) = o(T)$. A similar assertion holds also for polynomials over algebraic number fields and in the case of several variables. For polynomials over $\mathbf{Z}$ this was later improved to $N_f(T) = o(T^c)$ with some $c < 1$ by Dörge [988, 989] (in the second paper also some bounds for the number of polynomials with small Galois group were given). Siegel [3777] deduced from Weil's results on algebraic curves [4348] a necessary and sufficient condition[76] for $\lim_{T\to\infty} N_f(T) = \infty$ (see Dörge [991] for the proof in a special case).

---

[75] Ernests Fogels (1910–1985) worked in Riga. See [2333].

[76] Dörge wrote in [991] that this condition has been also known to Weil.

In 1974 M. Fried [1250] established $N_f(T) = O(\sqrt{T})$. See also Neuhaus[77] [3088, 3089].

A survey of problems related to Hilbert's irreducibility theorem can be found in Sect. 4.4 of the book [3621] by Schinzel.

2. Zeros and, more generally, the sets of values of power series with coefficients in a finite extension of $\mathbf{Q}_p$ attained in an algebraic closure of $\mathbf{Q}_p$ have been studied by Strassmann [3940] in 1928. He showed that if R is the valuation ring of a field with a non-Archimedean valuation, and $f(X) = \sum_{n=0}^{\infty} a_n X^n$ is a non-zero power series with tending to zero coefficients $a_n \in R$, then the equation $f(x) = 0$ has only finitely many solutions in R. This result turned out later to be of importance in the theory of linear recurrent sequences (see, e.g. S. Chowla, Dunton, D.J. Lewis[78] [708] and van der Poorten,[79] Schlickewei [4130]).

A new proof of Strassmann's theorem has been given in 1972 by Lenskoĭ [2535].

4.5.5 Books

In 1923 Hecke published his book [1742] presenting algebraic number theory in a then modern fashion. In the first chapters he described tools from elementary number theory and group theory, presenting later fundamental notions of the theory of algebraic number fields, and proving the discriminant theorem and Dirichlet's unit theorem as well as the Ideal Theorem for ideal classes. This has been followed by a study of quadratic fields, and in the last chapter Gaussian sums in arbitrary number fields were introduced and used to the proof of the quadratic reciprocity law.

Hecke wrote in the introduction that his book was based on his lectures in Basel, Göttingen and Hamburg. His lectures presented at the Hamburg University in 1920 were published in 1987 [1745].

Four years later Landau published an introduction in number theory [2444] whose third volume contained a concise introduction in the theory of algebraic numbers, culminating in the proof of Kummer's theorem on the Fermat problem in the case of regular prime exponents. In the last chapter one finds a proof of the Eisenstein reciprocity law and the criteria for Fermat's Last Theorem of Furtwängler, Wieferich and Mirimanoff.

These two books served for years to come as the main introductions into the theory of algebraic numbers.

In 1924–1928 Fricke[80] published a three-volume algebra textbook[81] [1248] whose last volume was devoted to algebraic numbers.

[77]Friedrich Wilhelm Neuhaus (1899–1983), professor in Köln.

[78]Donald J. Lewis (1926–2015), professor at the Notre Dame University and the University of Michigan.

[79]Alfred Jacobus van der Poorten (1942–2010), professor at UNSW in Sydney and at the Macquarie University in North Ryde.

[80]Robert Fricke (1861–1930), professor in Braunschweig.

[81]He pointed out that it is based on Weber's book [4321].

A survey of the development of the theory of algebraic numbers has been prepared by Dickson,H.H. Mitchell, Vandiver and Wahlin. It appeared in 1923 [941], and in 1928 Vandiver and Wahlin prepared its continuation covering the years 1923–1927 [4189].

Chapter 5
The Thirties

5.1 Structure

5.1.1 *Ideal Theory*

1. In 1932 Prüfer [3343] constructed a general theory of divisibility in fields. Let K be a field, and let $R \subset K$ be a ring having K for its quotient field. An element b of K is said to be divisible by $a \in K$ if one has $b = ca$ with $c \in R$. If with every finite set $A = \{a_1, a_2, \dots, a_n\}$ a subset I_A of K is associated satisfying the following five conditions

(a) $A \subset I_A$,
(b) If $A \subset B$, then $I_A \subset I_B$,
(c) If $A = \{a\}$, then $I_A = \{ax : x \in R\}$,
(d) If $a \in I_A$, then for all $b \in K$ one has $ab \in \{ax : x \in I_A\}$,
(e) If $a, b \in K$, then $a + b \in I_{\{a+b\}}$,

then one speaks about an ideal system. Prüfer defined the product of his ideals by putting $I_A \cdot I_B = I_C$, where $C = \{ab : a \in A, b \in B\}$ and studied some additional properties which this multiplication may have. One of them stating that to every non-zero ideal I there exists J with $I \cdot J = I_{\{1\}}$ characterizes a class of domains called later *Prüfer domains*.[1] Prüfer illustrated his theory by three examples, one of them being the system of ideals as defined by Dedekind.

2. In 1928 Grell gave a new proof of the formula $N_{M/K}(I) = N_{L/K}(N_{M/L}(I)$ for fields $K \subset L \subset M$ and ideals $I \subset \mathbf{Z}_M$. Earlier proofs were based on the equivalence of the two definitions of the ideal norm.

In 1936 he showed [1517] that the integral closure of a Dedekind domain in an inseparable extension is Dedekind. He noted in a footnote on p. 505 that this result has been earlier established by F.K. Schmidt who presented it in his habilitation lecture.

[1]For the theory of Prüfer domains see the book [1207] by Fontana, Huckaba and Papick.

W. Narkiewicz, *The Story of Algebraic Numbers in the First Half of the 20th Century*, Springer Monographs in Mathematics,
https://doi.org/10.1007/978-3-030-03754-3_5

Another proof of Grell's result was given in 1955 by Northcott [3133].

It has been pointed out by Asano and Nakayama [146] in 1940 that if R is a Dedekind domain with field of fractions K and k is a subfield of K with finite $[K : k]$, then the intersection $R \cap k$ may be not a Dedekind domain. This happens, e.g. in the case when Ω is a field with $\mathrm{char}(\Omega) \neq 2$, $k = \Omega(X, Y)$, $K = \Omega(\sqrt{X} + Y)$ and $R = K[Y]$. The ring R is a Dedekind domain, but $R \cap k = \Omega[X, Y]$ is not. In the case when the extension K/k is Galois, they gave the following necessary and sufficient condition for $R \cap k$ to be Dedekind: the intersection of all domains conjugated to R should be a Dedekind domain.

3. An important presentation of the theory of ideals in commutative rings has been published in 1935 by Krull [2317], who later published a series of papers [2319–2325, 2327] in which he provided a more detailed exposition and generalizations of his theory. In particular the first paper of this series gave a theory of integrally closed domains, basing on two important assertions:

(a) A domain R is a valuation ring for some valuation if and only if for all $a, b \in R$ one has either $a \mid b$ or $b \mid a$,

(b) An integrally closed domain R with quotient field K is the intersection of all valuation rings D_v containing R for valuations v of the field K.

In 1939 Krull presented a short survey in the German mathematical encyclopedia [2326].

For later expositions of the ideal theory in commutative rings see the books by Zariski[2] and Samuel[3] [4466], Northcott [3132] and Gilmer [1434].

5.1.2 Integral Bases, Discriminants, Factorizations

1. In 1931 W.R. Thompson [4056] determined for given n and prime p all possible integers r such that there exists a field K of degree n with $p^r \parallel d(K)$ (see also Schur [3705]). In [4057] Thompson did this for relative extensions.

A description of the form of discriminants of solvable fields K of prime degree has been given in 1932 by Wegner [4340] (his proof contained an error corrected by Wegner in [4344]) and Porusch[4] ([3338], Satz VII). The paper of Wegner gives also formulas for the factorization of primes in such fields. A simpler proof gave Hasse [1686] in 1937 (see also Reichardt and Wegner [3416]).

Factorizations of prime ideals in extensions K/k of degree 60 with Galois group isomorphic with the icosahedral group in the case when $\zeta_5 \in k$ were studied by Gut [1552, 1553, 1555].

---

[2]Oscar Zariski (1899–1986), professor at the Johns Hopkins University and at Harvard. See [2996, 3218].

[3]Pierre Samuel (1921–2009), professor in Clermont-Ferrand and at the Université Paris-Sud.

[4]Israel Porush [Porusch-Mandel] (1907–1991), senior rabbi at the Great Synagoge in Sydney.

2. Explicit formulas for integral bases in Galois quartic fields were provided by Albert [43] in 1930. In [44] he made explicit the formulas for integral bases in cubic fields given in the book [3853] by Sommer, and in 1937 he considered a particular form of integral bases applying them in the case of cubic and quartic fields [49]. In the same year he discussed methods of constructing integral bases in fields of any degree n and gave applications for $n = 3, 4$ and 5 [50].

A method of constructing bases of ideals presented MacDuffee[5] [2678] in 1931.

In 1932 Nagell [3023] showed that computing an integral basis, discriminant, fundamental units and class-number of a given field K can be made in finite steps and showed how to check whether a given number α lies in a given field or belongs to a given ideal. Also methods for finding prime ideal factorization of a given prime, for equivalence of two ideals and for determining all ideals having a given norm were given. Another algorithm for integral bases has been proposed in 1935 by Petr [3273].

A table of totally real quartic fields of discriminants $\leq$8112 giving integral bases and Galois groups was presented in 1935 by B.N. Delone, Sominskiĭ and Billevič [884]. It was prepared with the use of a generalization of Klein's [2166] geometrical approach.

3. A construction of integral bases in orders has been given in 1931 by Nagell [3021] who also showed that the number of orders with a given discriminant in a fixed field is finite.

The maximal power of a prime dividing the discriminant of an order has been determined in 1933 by Akizuki[6] [39]. Later he presented with Y. Mori [41] a study of ideals in quadratic orders.

In 1936 Grell [1518] developed the ramification theory for orders,

4. In his book [1742] (Satz 176) Hecke utilized singular primary numbers, defined as integers α of K which are prime to 2, congruent to a square mod 4, and generating an ideal which is a square to prove that the ideal class of the different of a finite extension is a square. This implies immediately the same assertion for the class of the relative discriminant. A more precise version of Hecke's theorem in the case of metacyclic extensions has been established in 1932 by Herbrand [1805]. He showed that in this case one can write

$$d(L/K) = aI^2 ,$$

with an ideal I, of $\mathbf{Z}_K$ and $a \in \mathbf{Z}_K$ satisfying

$$a \equiv 1 \pmod{J} ,$$

---

[5]Cyrus Colton MacDuffee (1895–1961), professor at the Ohio State University and the University of Wisconsin.

[6]Yasuo Akizuki (1902–1984), professor in Kyoto.

where J is the maximal ideal divisor of $4\mathbf{Z}_K$ prime to the discriminant. He conjectured that the same holds for all normal extensions.

Herbrand's conjecture has been confirmed in 1960 by Fröhlich [1267].

A modern exposition of the part of [1742] dealing with singular primary numbers has been given in 2006 by Lemmermeyer [2529].

Another proof of Hecke's theorem has been given in 1967 by Armitage [105], who showed also that it holds also in the case of function fields. It has been shown in 1962 by Fröhlich, Serre and Tate [1285] that Hecke's assertion fails in certain Dedekind domains.

5. The history of the *normal basis theorem* begins in 1888 when Hensel [1765] showed that every finite field has a linear basis consisting of elements conjugated over the prime field. The assertion that if K is an infinite field and L/K is its separable Galois extension, then L has a K-basis consisting of conjugates of an element, called now the *normal basis theorem* appeared first in a letter of Dedekind to Frobenius [852] which has been quoted in Sect. 2.1.2. The first proof has been sketched in 1932 by E. Noether (Satz 3 and footnote 4 in [3125]). She utilized the notion of the group-ring $K[G]$ which she introduced in [3124], basing on the theory of "generalized groups" (the earliest name of G-modules) established earlier by Krull [2306, 2308]. Using this notion the normal basis theorem asserts the isomorphism of the $K[G]$-modules L and $K[G]$, and the existence of a normal integral basis for $K/\mathbf{Q}$ is equivalent with the existence of an isomorphism of $\mathbf{Z}_K$ and $\mathbf{Z}[G]$ as $\mathbf{Z}[G]$-modules, where G is the corresponding Galois group and $\mathbf{Z}[G]$ denotes the group-ring of G with rational integral coefficients.

There are several proofs of the normal basis theorem. In 1932 Deuring [907] provided a unified proof valid for finite and infinite base fields K (this paper contains a study of the influence of linear representations of the Galois group on properties of the extension, and this topic has been continued by Deuring in [911]). Chevalley presented in his thesis [677] a fresh proof due to Hasse.

Other proofs were provided by Fueter [1314], Artin [126], T.R. Berger, Reiner [300], Winter [4427], Waterhouse[7] [4307] and Blessenohl [370].

In 1981 Halter-Koch and Lorenz [1607] described normal fields having a normal basis consisting of units. For relative extensions this has been done in 1987 by Lorenz [2629].

An algorithm for explicit construction of a normal basis in the case when $\mathrm{char}(K) \nmid [L:K]$ was provided by Stauffer [3907] in 1936.

In 1941 Nakayama[8] [3057] extended Stauffer's method to the general case.

Another algorithm has been presented in 1999 by Girstmair [1438]. In the case of cyclic extensions of $\mathbf{Q}$ this has been done earlier by Schlickewei and Stepanov [3632] and Schwarz[9] [3709], and for Abelian extensions by Poli [3329] and Girstmair [1437].

If L/K is a given extension, then an element θ of L whose conjugates generate L as a linear K-space is called *regular*. If $\theta \in L$ is regular for all intermediate extensions L/M (with $K \subset M \subset L$), then it is called *completely regular*. In 1957 Faith[10] [1159] proved the existence of completely

---

[7] William Charles Waterhouse (1942–2016), professor at the Pennsylvania State University.

[8] Tadashi Nakayama (1912–1964), professor in Nagoya. See [10].

[9] Štefan Schwarz (1914–1996), professor in Bratislava. See [3464].

[10] Carl Faith (1927–2014), professor at the Rutgers University.

regular elements in every Galois extension of infinite fields and noted that this may fail in the case of finite fields, however later Blessenohl and Johnsen [371] proved their existence also for finite fields. Simpler proofs in the last case were later given by Blessenohl [369] and Hachenberger [1580]. For cyclic extensions L/K of prime power degree the regular elements (not using this name) of L were described in 1942 by S. Perlis[11] [3243].

An extension L/K is called completely regular if every its regular element is completely regular. Such Abelian extensions were described in 1991 by Blessenohl and Johnsen [372], and a simpler proof has been given by P. Meyer [2848].

6. In the study of the action of Galois group $G = Gal(L/K)$ on the additive group of a normal extension L/K an important role play the group algebras $\mathbf{Z}[G]$ and $\mathbf{Z}_K[G]$. One of the first papers on this subject has been written in 1932 by Fueter [1312], who applied results from Dickson's book [936] to study the structure of $\mathbf{Z}[G]$ in the case, when G is the Galois group of either $\mathbf{Q}(\sqrt{d}, \zeta_p)/\mathbf{Q}$ or $\mathbf{Q}(\sqrt{d}, \zeta_p)/\mathbf{Q}\left(\sqrt{d}\right)$ with square-free d and an odd prime p. The promise[12] to give arithmetical applications was never fulfilled.

In 1932 E. Noether [3125] showed that if L/K is Galois and $\mathfrak{p}$ is a prime ideal of $\mathbf{Z}_K$ not dividing $[L : K]$, then the corresponding local extension has a normal integral basis, i.e. a basis consisting of conjugate integers. This has been later extended to the assertion that a Galois extension of $\mathfrak{p}$-adic fields has a normal integral basis if and only if it is at most tamely ramified (see, e.g., Chap. 1 of Fröhlich's book [1282]).

Proofs were given also by Brinkhuis [458], Kawamoto [2132] and R.J. Chapman [654].

In the case of Abelian extensions $K/\mathbf{Q}$ the structure of $\mathbf{Z}_K$ as a G-module with G being the Galois group of $K/\mathbf{Q}$ has been determined by Leopoldt [2555] in 1959. He showed the existence of an order O in the group-ring $\mathbf{Q}[G]$, such that with an element $T_K \in K$ the G-module $\mathbf{Z}_K$ equals OT_K. An elementary proof of this result has been given by Lettl [2570] in 1990. He gave also a simple formula for the element T_K. In a series of papers Fröhlich [1274, 1275, 1278, 1279] presented a new approach to the study of Galois properties of algebraic number fields and summarized his results in a survey [1276] in 1974 and in his book [1282] published in 1983. In [1274] he showed that if $K/\mathbf{Q}$ is a normal octic extension with the quaternion Galois group, then it has a normal integral basis if and only if the root number $W(\chi)$ corresponding to the unique non-Abelian character χ equals 1. It has been shown by Armitage [106] that this condition implies $\zeta_K(1/2) \neq 0$.

Families of tame Abelian extensions L/K without normal integral bases were constructed by Kawamoto [2131], Brinkhuis [459] and Cougnard [780]. On the other hand it has been established by M.J. Taylor [4035] that $\mathbf{Z}_L$ is a free $\mathbf{Z}\Gamma$ module for tame extensions L/K with Abelian Galois group Γ.

It has been conjectured by Fröhlich (on p. 154 of [1280]) that if L/K is a tame extension with Galois group Γ and $(\mathbf{Z}_L)$ is the class of $\mathbf{Z}_L$ in the class-group of locally free $\mathbf{Z}\Gamma$-modules, then $(\mathbf{Z}_L)^2 = 1$. This conjecture has been established in 1981 by M.J. Taylor [4035] who also showed that $(\mathbf{Z}_L)$ depends only of $W(\chi)$ for symplectic characters of Γ. This implies in particular that every Galois extension of $\mathbf{Q}$ having odd degree has a normal integral basis.

A necessary and sufficient condition for the existence of a normal integral basis generated by a unit in a cyclic field of prime degree has been given in 1991 by Jakubec, Kostra and Nemoga [2030].

---

[11] Sam Perlis (1913–2009), professor at the Purdue University.

[12] '*Die zahlentheoretischen Anwendungen der Theorie ...hoffe ich in einer spätern Arbeit angeben zu können.*" ["*Number-theoretical applications of the theory ...I hope to present in a future paper.*"]

The books by Fröhlich [1282] and Snaith [3841] give expositions of the Galois module structure of rings of integers. The case of Abelian extensions of imaginary quadratic fields has been treated in the book [603] by Ph. Cassou-Noguès and M.J. Taylor.

Necessary and sufficient conditions for the existence of a normal integral basis in a tame Kummer extension were given in 1962 by Fröhlich [1271]. In the cyclic case of prime degree a necessary condition has been given in 1994 by Gómez-Ayala [1476]. The unramified case has been earlier treated by Childs [689].

Later Ichimura [1948] extended Gómez-Ayala's result to the case of a prime power degree. His assertion that his criterion works for arbitrary cyclic Kummer extensions was shown to be incorrect by del Corso and Rossi [870], who also provided a correct formulation. Later del Corso and Rossi [871] provided a criterion for a Kummerian tamely ramified extension L/K to have a normal integral basis .

5.1.3 Units

1. It has been known already to Kummer (he stated it without proof in the case of $\mathbf{Q}(\zeta_p)$ with prime p on p. 100 in [2354]) that cyclotomic fields have a real system of fundamental units. The first proof was given in 1857 by Kronecker [2282] (see also Theorem 127 in Hilbert's book [1836]). In 1934 Latimer [2480] proved the same assertion for all cyclic extensions of $\mathbf{Q}$. In the same paper he showed that if K is a cyclic extension of the rationals which is a composite of two fields K_1, K_2 with relatively prime degrees, then K has a system of fundamental units containing systems of fundamental units of K_1 and K_2. In 1939 Lednev [2497] showed that if K is totally real and $L = K(\zeta_{p^m})$ with an odd prime $p \nmid d(K)$, then L has a real system of fundamental units.

In 1952 appeared a posthumous paper of Remak [3443] in which it has been shown that if $K \subset L$, then the index $[U(L) : U(K)]$ is finite if and only if K is totally real, L is totally complex, and $[L : K] = 2$, i.e. L is a CM-field and $K = L^+$.

2. In 1933 Nehrkorn[13] [3079] gave a method of determination of fundamental units in Abelian fields of prime power degree using fundamental units of cyclic subfields.

In 1930 Herbrand [1800] generalized Minkowski's result on the existence of Minkowski units to relative Galois extensions. Another proof of this result, using Minkowski's approach, was found in 1932 by Artin [124].

A necessary and sufficient condition for the existence of a strong Minkowski unit in a cyclic extension $K/\mathbf{Q}$ has been given in 1934 by Latimer [2480]: a certain ideal in a matrix ring R associated with K should be principal (for the study of that ring see Latimer and MacDuffee [2482]). Latimer noted that in case of extensions of prime degree p the ring R is isomorphic with the ring of integers of $\mathbf{Q}(\zeta_p)$; hence $h(\mathbf{Q}(\zeta_p)) = 1$ is a sufficient condition for the existence of a strong Minkowski unit in a cyclic extension of degree p (it is now known that this happens only for $p \leq 19$ (Masley and Montgomery [2763])). In case $p = 3$ this assertion was known already

[13] Harald Nehrkorn (1910–2006), teacher in Schloss Bieberstein.

to Eisenstein [1071], Lehrsatz 4, and was rediscovered by B.N. Delone and Faddeev in 1940 [882]. Latimer's criterion has been generalized to arbitrary normal extensions by M.J. Weiss [4366] in 1936.

In 1969 Brumer [477] showed that if $K/\mathbf{Q}$ is cyclic of prime degree p and all ideals in $\mathbf{Q}(\zeta_p)$ of norm $h(K)$ are principal, then K has a strong Minkowski unit, and if none of such ideals has this property, then there is no strong Minkowski unit. A criterion in the case of tame extensions has been given by Marszałek [2744, 2745].

An analogue of Brumer's result for extensions of imaginary quadratic fields has been proved by Gillard [1431] in 1980. A criterion for the existence of a Minkowski unit for real cyclic fields of degree $pq < 39$ ($p \neq q$ prime) has been given by Marko [2739, 2740].

3. The norm of the fundamental unit ε_d in a real quadratic field K of discriminant d equals -1 if and only if the equation

$$X^2 - dY^2 = -4 \tag{5.1}$$

has integral solutions. This is equivalent to the assertion that the period of the continued fraction of $\sqrt{d}$ is odd.

It is clear that if d has a prime factor $p \equiv 3 \bmod 4$, then (5.1) has no solutions, and from results of Dirichlet [958] it follows that solutions exist either if $d = p_1 p_2$, both primes p_i are congruent to 1 mod 4, and p_1 is a quadratic non-residue mod 4, or if $d = 8p$ and either $p \equiv 5 \bmod 8$ or $p \equiv 9 \bmod 16$ and 2 is a quartic non-residue mod p. For $d = p_1 p_2 < 10\,000$ the sign of $N(\varepsilon_d)$ was determined by von Thielmann [4052] in 1926.

Some statistics for $d < 10\,000$ has been provided by Nagell [3024], who also proposed the following conjecture:

Let $A(x)$ be the number of square-free integers $0 < d \leq x$ with negative $N(\varepsilon_d)$, and let $B(x)$ be the number of square-free integers $0 \leq d \leq x$ having no prime factor congruent to 3 *mod* 4*. Then the limit* $\lim_{x\to\infty} A(x)/B(x)$ *exists and lies in the interval* $(0, 1)$.

Heuristical support for this conjecture in the more precise form

$$\lim_{x\to\infty} \frac{A(x)}{B(x)} = 1 - \prod_{j=1}^{\infty} \left(1 + 2^{-j}\right)^{-1} = 0.58057\ldots$$

has been given by Stevenhagen [3928, 3930], and numerical support has been provided by Bosma and Stevenhagen [407]. In 2010 Fouvry and Klüners [1223] established

$$0.41942\cdots + o(1) \leq \frac{A(x)}{B(x)} \leq \frac{2}{3} + o(1)\ ,$$

and in [1224] replaced $0.41942\ldots$ by $0.52475\ldots$.

Relations between the sign of $N(\varepsilon_d)$ and the continued fraction expansion of $\sqrt{d}$ were studied in 1934 by Epstein[14] [1113].

---

[14]Paul Epstein (1871–1939), professor in Frankfurt.

In 1934 Scholz [3672] applied class-field theory to study the sign of $N(\varepsilon_d)$ for odd discriminants and in several cases related it to quartic residuacity. Class-field theory has been also used by Rédei[15] to get necessary and sufficient conditions for $N(\varepsilon_d) = 1$ [3399, 3400, 3402, 3406].

In 1979 Morton [2972] presented a modern exposition of Rédei's results and extended them.

In 1980 Lagarias [2396] presented an algorithm determining the sign of $N(\varepsilon_d)$, essentially quicker than the usual approach using continued fractions. Another algorithm has been described by Golubeva [1475]. A criterion for $N(\varepsilon_d) = -1$ using Pythagorean triples has been given by Grytczuk, Luca and Wójtowicz [1535] in 2000.

4. In 1930 Nagell [3019] studied cubic extensions of the rationals, dealing in particular with orders and units. He showed that there exist only finitely many cubic units with given discriminant and norm (Satz 23), and the number of cubic units of negative norms lying in the interval $(1/x, 1)$ is for $x > 1$ equal to

$$\frac{8}{3}x^{3/2} + O(x),$$

(Satz 19). He proved also ([3018] and Satz 21 in [3019]) that for any cubic integer α the number $x + y\alpha$ with $x, y \in \mathbf{Z}$ can be a unit for at most two choices of x, y, except when the discriminant of α equals either -44 or -23, when there are two and four choices, respectively. The last result was earlier known in the case $\alpha = \sqrt[3]{D}$, in which case there is at most one choice for x, y (Delaunay [875, 878], Nagell [3015–3017]).

We noted already in Chap. 1 that Uspensky [4118] presented in 1931 an algorithm for determining fundamental units in orders of a cubic field.

An algorithm for fundamental units in totally real cubic fields has been found by Berwick in 1932 [328], who noted that his method works for all fields with unit rank two. One of the steps in his proof shows that in every such field there is a fundamental system of units $\varepsilon_1, \varepsilon_2$ with $\varepsilon_1 > 1$ and the remaining two conjugates of ε_1 lying inside the unit disk (i.e. ε_1 is a PV-number) (for the definition of PV-numbers see Sect. 5.4.4).

It has been shown in 1966 by Zlebov [4476] that the last assertion holds in every field having infinitely many units. A further generalization has been established by Brunotte and Halter-Koch [485].

Another approach, using appropriately defined reduced ideals, has been proposed by Berwick [329] two years later.

5. In 1937 Chabauty[16] [637] studied the structure of the set $U(V)$ of units of an algebraic number field which lie in a finite-dimensional algebraic variety V and gave a necessary condition for the set $U(V)$ to be infinite.

The more general question dealing with the structure of the intersection of an algebraic variety with a finitely generated group has been later studied by several authors. It is related to the conjecture

[15] László Rédei (1900–1980), professor in Szeged and Budapest. See [7, 2738].

[16] Claude Chabauty (1910–1990), professor in Grenoble.

of Lang (see Chabauty [639], Lang [2454, 2455, 2460, 2461], Liardet[17] [2594]), a consequence of Mordell's conjecture in algebraic geometry, which has been later proved by Faltings [1160] (see also Bombieri [389] and Vojta [4237]).

5.2 Class-Field Theory

5.2.1 *Hasse*

5.2.1.1 The Local Case

1. In 1930 Hasse [1667] applied the theory of norm residues to establish the main theorems of local class-field theory, at first for Abelian extensions K/k of $\mathfrak{p}$-adic fields, which are completions of number fields K_0 and k_0 such that the extension K_0/k_0 is Abelian. In Sect. VII of his paper he expressed his hope that this property is automatically satisfied for Abelian K/k, and soon this has been confirmed by F.K. Schmidt [3644].

An important role in the local class-field theory is taken by the group $H(K) \subset k^*$ consisting of all norms, i.e. $H(K) = N_{K/k}(K^*)$. If H is an arbitrary subgroup of k^* and r is the smallest positive integer (if it exists) with $1 + \mathfrak{p}^r \subset H$, then $\mathfrak{p}^r$ is called the *conductor* of H.

The following properties (i)–(iii) of Abelian K/k were proved by Hasse as consequences of the global class-field theory. He formulated also without proof the properties (iv) and (v).

(i) The Galois group of K/k is isomorphic to $k^*/H(K)$, the isomorphism explicitly given by the norm residue symbol,

(ii) This isomorphism agrees with the Galois correspondence between subgroups of the Galois group of K/k and fields M with $k \subset M \subset K$,

(iii) If K, L are Abelian extensions of k, then the relation $K \subset L$ is equivalent to $H(L) \subset H(K)$.

(iv) If K/k is not Abelian, then $[k^* : H(K)] < [K : k]$,

(v) If H is a subgroup of k^* containing non-units and having a conductor, then there exists an Abelian extension K/k with $H(K) = H$.

(It has been observed later that the assumption of (v) is equivalent to the finiteness of the index of H in k^*.)

In the same year 1930 F.K. Schmidt [3644] gave new proofs of (i)–(iii) and established also the properties (iv) and (v). He made the assertion of (iv) stronger, replacing the inequality by proper divisibility. In a footnote on p. 37 in [1665] Hasse mentions a talk of Schmidt in Halle, in which he presented the proof of the local class-field theory not utilizing the global theory. This proof has not been published.

[17] Pierre Liardet (1943–2014), professor at the Université de Provence. See [185, 4279]

In 2006 Pauli [3229] presented an algorithm producing the class-field associated with a given closed subgroup of finite index in K^*.

2. In 1935 Nakayama [3056] (see also [3058]) used factor sets to construct the isomorphism of the Galois group G of an Abelian extension K/k of local fields into the factor group $k^*/N_{K/k}(K^*)$. A factor set is a function $f : G^2 \longrightarrow K$ satisfying

$$f(g, hk)f(h, k) = f(gh, k)k(f(g, h)) \quad g, h, k \in G$$

and Nakayama showed that if $\Phi(g) = \prod_{h \in G} f(g, h)$, then the map

$$\phi : g \mapsto \Phi(g)N_{K/k}(K^*) \tag{5.2}$$

is a homomorphism of G/G' onto a subgroup of $k^*/N_{K/k}(K^*)$. If G is Abelian, then it follows from Theorem 0 in Chevalley's paper [678] that there exists a factor set such that ϕ is an isomorphism. One year later Akizuki [40] showed that ϕ is an isomorphism also in certain non-Abelian cases.

A modern proof of the result of Nakayama and Akizuki can be found in a footnote on p. 5 of the 2009 edition of the book [137] by Artin and Tate.

In 1938 Schilling [3608] considered the question of characterizing complete fields for which the main theorems of the local class-field theory hold and in [3609] (see also the book [3611]) presented a generalization of the local class-field theory to a class of fields complete under an archimedean valuation. For further results see Sect. 6.3.

A simpler way to prove the main theorems of the local class-field theory has been found by Hazewinkel [1725] in 1975. His method has been exposed in the book [1994] by Iwasawa. See also the book by Fesenko and Vostokov [1182]. Another book by Iwasawa [1994] on this subject has been based on the methods introduced by Lubin and Tate [2664] and Coleman[18] [759].

5.2.1.2 The Global Case

1. In 1931 Hasse [1671] showed that if K/k is a cyclic extension of global fields and non-zero $a \in K$ is a local norm everywhere, then a lies in $N_{K/k}(K^*)$. This has been earlier known for cyclic extensions of degree p of $\mathbf{Q}(\zeta_p)$ with regular prime p (Hilbert [1836], Theorem 167), for quadratic extensions (Hilbert, Satz 65 in [1838]) and for cyclic extensions of degree p of a field containing ζ_p (Furtwängler ([1333] in case $p \nmid h(k)$, [1343] in remaining cases).

On p. 38 of [1664] Hasse conjectured that the same assertion holds for all Abelian extensions, but in [1671] he observed that this fails for the extension $\mathbf{Q}(\sqrt{-3}, \sqrt{13})/\mathbf{Q}$, where 3 is a local norm everywhere, but not a global norm. One says that the *Hasse norm theorem* (HNT) holds for the extension K/k if all elements which are local norms everywhere are norms.

---

[18] Robert F. Coleman (1954–2014) professor at Berkeley. See [183].

In 1936 Scholz [3674] showed that if $K_i/\mathbf{Q}$ $(i = 1, 2)$ are cyclic of prime degree l with co-prime discriminants, and primes ramified in one of these extensions split in the second, then Hasse norm theorem does not hold in $K_1K_2/\mathbf{Q}$ and gave examples with $l = 2, 3$. He introduced also the *knot group* $\mathfrak{K}(K/k)$ as the quotient group of the group of all local norms by $N_{K/k}(K^*)$ as well as two related groups and continued the study of these groups in [3677–3679].

A cohomological reformulation of the Hasse norm theorem for Galois fields has been given in Tate's lecture at the Brighton meeting [4008]. Cohomological approach has been also used in 1970 by Tasaka [4006] to show the validity of HNT in all cubic extensions. Much later Bartels [198] proved that if the Galois group of the Galois closure of an extension K/k of degree m is dihedral of $2m$ elements, then in K/k HNT holds. In [199] he showed that HNT holds in all extensions of prime degree.

In 1975 Garbanati [1381] gave a criterion for the truth of HNT for Abelian extensions $k/\mathbf{Q}$: let g_k^+ be the degree of the maximal Abelian extension K/k unramified at finite prime ideals, and $K/\mathbf{Q}$ Abelian. Moreover let z_k^+ be the degree of the maximal Abelian extension L/k unramified at finite prime ideals such that $L/\mathbf{Q}$ is Galois, and the group $Gal(L/k)$ lies in the centre of $Gal(L/\mathbf{Q})$. The Hasse norm theorem holds for k if and only if $g_K^+ = z_K^+$. He deduced that if $p \equiv q \equiv 1 \bmod 4$ then HNT holds in $\mathbf{Q}(\sqrt{p}, \sqrt{q})$ if and only if p is a quadratic residue mod 4 and showed that HNT fails for $s \geq 4$ in the field $\mathbf{Q}(\sqrt{p_1}, \sqrt{p_2}, \dots, \sqrt{p_s})$. Later he characterized Abelian extensions $K/\mathbf{Q}$ of prime power degree which are composites of fields with prime power discriminants in which HNT holds [1382, 1385]. His results were extended in 1978 to relative extensions by Gurak [1543] who used Tate's reformulation of HNT to give some sufficient conditions for a Galois extension to satisfy HNT. In case of nilpotent Galois groups these conditions are also necessary.

In 1977 Gerth [1418] presented an algorithm to check whether for a given Abelian extension HNT holds, reducing the problem to the computation of ranks of a certain finite set of matrices. He applied it to give a list of cyclotomic extensions of $\mathbf{Q}$ satisfying HNT. The same result has been also obtained by Gurak [1543]. For the case of maximal real subfields of cyclotomic fields see Kagawa [2093].

Criteria for HNT in some classes of non-Abelian extensions were found by Gurak [1544, 1545] and Gerth [1419]. A method of finding the cardinality of $\mathfrak{K}(K/k)$ has been presented in 1977 by Razar [3395], who showed also that in an Abelian extension HNT holds if and only if it holds in its maximal subextensions of prime exponent (cf. Garbanati [1386]). He showed also that if all Sylow subgroups of $Gal(K/k)$ are cyclic, then HNT holds.

In 1979 Jehne [2044] presented a cohomological generalization of the knot groups introduced by Scholz and studied the question of the existence of extensions K/k with given $\mathfrak{K}(K/k)$. The case of Abelian K/k has been considered in 1982 by Steckel [3908]. For further study of the knot group see Opolka [3157–3162].

The case of extensions with Galois group C_p^n has been treated by M. Horie [1901, 1902]. Extensions K/k with $K = k(\zeta_q, \sqrt[q]{a})$ for prime power q were studied by Hamada [1611].

In 1979 Amano [60] applied cohomological methods to study HNT for non-Galois extensions. A generalization of HNT has been provided in 1980 by Lorenz [2627, 2628], and a weaker version of HNT has been studied by Leep and Wadsworth [2504] and Scharaschkin [3594].

A modern proof of HNT for cyclic extensions has been given by Gold [1455] in 1977.

2. In 1933 Grunwald [1534] studied the question whether there exists a cyclic extension L/K with prescribed local extensions of the completions $K_\mathfrak{p}$ at finitely many prime ideals $\mathfrak{p}$ and stated a sufficient condition for this to happen. Another proof has been given in 1942 by Whaples [4380], but it turned out in 1948 that these proofs had a lacuna, as S. Wang [4288] found counterexamples in the case when some of the chosen prime ideals were divisors of 2. In 1950 S. Wang [4289] and Hasse [1695] (for a correction see Morton [2976]) provided corrected versions. Hasse's paper contains also a generalization in which cyclic extensions are replaced by extensions with an arbitrary Abelian group, and the same has been also established by S. Wang in [4290]. This result is now called the *Grunwald–Hasse–Wang theorem.*

Simpler proofs were given by Artin and Tate (Theorem 4 of Chap. 10 in [137]) and Neukirch [3093]. An effective version has been provided by Song Wang [4291]. For a strengthening see Tomanov [4068]. Neukirch made a generalization to nilpotent groups of odd order [3092] and to solvable extensions of odd order [3094]. In [3096] he established an analogue in the case of pro-solvable profinite groups.

A generalization to abstract fields has been given in 1978 by Miki [2858] and Sueyoshi [3957] (see also Saltman [3566]).

In 1958 S. Lang and Tate [2463] conjectured a cohomological analogue of the Grunwald–Hasse–Wang theorem for Abelian varieties. A proof of it has been given in 2012 by Creutz [793].

An analogue of the Grunwald–Hasse–Wang theorem for the embedding problem has been considered in 1964 by Ikeda[19] [1958] and in 1970 by Adachi [21] for arbitrary cyclic H.

3. In 1930 Hasse [1663] utilized class-field theory to present the theory of non-cyclic cubic fields K. He parametrized such fields K by associating with K a pair (d, H), where d is the discriminant of the quadratic subfield L of the Galois closure M of K, and H is the subgroup of index 3 of the group of ideals of L, associated with the extension M/L by class-field theory. Hasse determined factorization of rational primes in K, and showed that the number of non-isomorphic cubic fields of given discriminant is unbounded. Another approach to the study of cubic fields has been presented by Reichardt [3411] in 1933.

Hasse's method has been generalized in 1933 by Porusch [3338] to fields whose Galois group G has the following property: its commutator subgroup G' is cyclic, it is the only Abelian normal subgroup $\neq 1$ of G, and G/G' is of prime order. A similar approach has been used later by Rosenblüth [3519] in the case of octic fields with quaternion group, by Dribin [997, 998] who considered fields with Galois group S_4, studying in particular their ramification groups, and by Bergström [302] for biquadratic fields.

In [1668] Hasse gave a purely algebraic proof of the conductor-discriminant formula for relative Abelian extensions. This formula, which generalizes the formula (1.35) for Abelian extensions of cyclotomic fields, has been already known as a consequence of Takagi's approach to the class-field theory (see Theorems 16 and 17 in Hasse's [1655, 1665] report), which used also analytical tools. In 1926 Sugawara [3959] presented a simpler proof in the case of cyclic extensions of prime power degree, and in 1929 Iyanaga [1998] found an arithmetical proof for all cyclic extensions. Hasse's proof has been later simplified by Vassiliou [4196], and in [1682] Hasse gave a still simpler proof.

In the last paper one finds also a proof of the important theorem concerning the sequence $G_0 \supset G_1 \supset \cdots \supset G_m \supset \dots$ of ramification groups in an Abelian extension of a local field. Define for real x

[19] Masatoshi Gündüz Ikeda (1926–2003), professor in Izmir and Ankara. See [2190].

$$G_x = G_{\lceil x \rceil} ,$$

and

$$\varphi(x) = \int_0^x \frac{dt}{[G_0 : G_t]} .$$

The theorem is now usually stated in the following form: *if for some i one has $G_i \neq G_{i+1}$, then $\varphi(i)$ is an integer* (see, e.g., Theorem 1 in §7 of Chap. 5 in Serre's book [3733]). Hasse established this result in the case of a finite residue field and used a more complicated, but equivalent formulation.

In 1939 Arf[20] [104] extended this theorem to the case when the residue field is not assumed to be finite (*Hasse-Arf theorem*).

4. In 1933 presented Hasse [1677] a new way to construct the theory of the norm residue symbol, utilizing results of the theory of cyclic algebras over algebraic number fields, developed in a series of papers by Albert and Hasse [53], R. Brauer [427], R. Brauer and Hasse [1670, 1672, 1676], Hasse and E. Noether [433].

A *cyclic algebra* is given by the following construction presented by Dickson [934] in 1914 (cf. also the German edition of his book [936]):

Let L/K be a cyclic extension of degree n, and let σ be the generator of its Galois group. For $a \in K^*$ let A be the algebra generated over L by an element u satisfying $u^n = a$ and $xu = ux^\sigma$ for all $x \in L$.

Hasse's paper [1677] contains a determination of the *Brauer group* of an algebraic number field, an object which later played an important role in the cohomological approach to class-field theory (in 1935 Chevalley and Nehrkorn [687] gave purely algebraic proofs of results presented in [1677]).

The Brauer group $Br(K)$ of a field K has been defined by R. Brauer [427] in the following way: it consists of division algebras of finite dimension over K, having K for their center, and the product of two division algebras A, B is defined as the division algebra C over which the algebra $A \otimes_K B$ is a matrix algebra (the existence of C follows from Wedderburn's[21] theorem [4336], established in 1908). Later this definition has been modified to the following equivalent form: one considers simple algebras over K for which K is their center, i.e., simple central algebras. Every such algebra A is the matrix algebra over a division algebra $D(A)$. The group $Br(K)$ consists of classes of simple central algebras, two algebras A, B lying in the same class if and only if the division algebras $D(A)$ and $D(B)$ are isomorphic over K. If C_1, C_2 are classes, then their product is the class containing $A_1 \otimes_K A_2$ for $A_i \in C_i$.

It turned out later that the group $Br(K)$ is isomorphic to the second cohomology group $H^2(Gal(K_s/K, K^*)$, where K_s is the separable closure of the field K (see, e.g., Serre [3735] and Sect. X.5 of [3733]).

Local class-field theory implies the isomorphism of $Br(K_v)$ with $\mathbf{Q}/\mathbf{Z}$ for every completion K_v of an algebraic number field K, as well as the exactness of the diagram

[20] Cahit Arf (1910–1997), professor in Istanbul. See [2631].

[21] Joseph Henry Maclagen Wedderburn (1882–1948), professor in Princeton. See [4034].

$$0 \longrightarrow Br(K) \longrightarrow \bigoplus_{v} Br(K_v) \longrightarrow \mathbf{Q}/\mathbf{Z} \longrightarrow 0 .$$

The inclusion in it is a consequence of the Brauer–Hasse–Noether theorem proved in 1932 [433], giving an explicit description of central simple algebras over an algebraic number field, which shows that they are cyclic.

In the case of division algebras the Brauer–Hasse–Noether theorem has been proved for degree 2 by Dickson ([936], p. 45 of the German edition), for degree 3 by Wedderburn [4337], for degree 4 by Albert [45] and in the general case by Albert and Hasse [53]. The history of this theorem has been described by Roquette [3505].

The theory of norm residues has been applied by Hasse [1680] in 1934 to an explicit construction of class-fields of a field K with cyclic class-group C_n in the case when K contains the nth roots of unity.

5. An exposition of the class-field theory has been given by Hasse [1678]. One finds there also the following result about roots of binomials:

Let $a_1, a_2, \ldots, a_r$ be elements of an algebraic number field K, put $\xi_j = a_j^{1/n_j}$ $(j = 1, 2, \ldots, r)$, and assume that an equality of the form

$$\prod_{j=1}^{r} \xi_j^{s_j} = a \in K$$

can hold with integral s_j only in the case $n_j \mid s_j$ for $j = 1, 2, \ldots, r$. If K contains all roots of unity of orders $n_1, n_2, \ldots, n_r$, the polynomial $P(X_1, \ldots, X_r)$ has coefficients in K, satisfies $\deg_{X_j} \leq n_j$ and $P(\xi_1, \ldots, \xi_r) = 0$, then $P = 0$.

A similar assertion in the case when $K = \mathbf{Q}$ and all ξ are real has been shown in 1940 by Besicovitch [330], and this has been generalized by Mordell [2934] to the case when K and all ξ's are real. See also Siegel [3796], Schinzel [3620] and Carr and O'Sullivan [583].

5.2.2 Chevalley

1. In one of his first papers Chevalley [674] presented a simple proof of the fact that Hensel's definition of norm residues given in [1795] is equivalent to the definition given by Hilbert in [1836]. In 1931 he gave with Herbrand [1808] a new proof of the following main part of the existence theorem in class-field theory:

If $\zeta_p \in K$ and H is a group of ideals in K of index p, then there exists a class-field corresponding to H.

In a short note published in the same year [675] he showed that if $k \subset K$ and K does not contain an unramified Abelian extension of k, then $h(k) \mid h(K)$. As a corollary he deduced that if K is a field with $h(K) = 1$, then there exists a sequence $\mathbf{Q} \subset K_1 \subset K_2 \subset \cdots \subset K_m = K$ such that there are no intermediate fields between K_i and K_{i+1}, and all fields K_1 have class-number one.

In 1933 Chevalley presented a new definition of the norm residue symbol, based on the theory of simple algebras, and this permitted him to prove the isomorphism theorem of the local class-field theory without using global class-field theory [678]. His paper [676] contained an important step toward arithmetization of class-field theory. He reduced there this aim to the case of cyclic extensions of prime degree. Results of these papers were incorporated in Chevalley's thesis [677], published in 1933, which gave a simplified approach to the class-field theory, reducing essentially the use of analytic tools in the global case, and eliminating them in the local case. His proof consisted of two parts: in the first he showed that if L/K is Abelian, then the decomposition of a prime ideal $\mathfrak{p}$ of K in this extension depends only on the class of $\mathfrak{p}$ in the factor group of a subgroup of the group of ideals of K by a suitable subgroup H. First he showed in a purely algebraic way how to associate such a group H to every Abelian L/K and then proved that the obtained group equals the group associated with L/K in Takagi's theory. In the last assertion he had to use analytic tools in the case when $L = K(\zeta)$ with a root of unity ζ. On the way he proved Artin's reciprocity law and then used it to establish the existence theorem.

2. Three years later presented Chevalley [681] a reformulation of the class-field theory based on the notion of ideles. If K is an algebraic number field, then the group I_K of ideles[22] is defined as the restricted product[23] of the multiplicative groups of all completions $K_\mathfrak{p}$ with respect to their unit groups $U_\mathfrak{p}$ (in case of infinite $\mathfrak{p}$ one defines $U_\mathfrak{p} = K_\mathfrak{p}^*$). Thus I_K is the group of all sequences $\{a_\mathfrak{p}\}$ with $a_\mathfrak{p} \in K_\mathfrak{p}^*$, with at most finitely many $a_\mathfrak{p}$ lying outside $U_\mathfrak{p}$. The multiplicative group of K forms a subgroup of I_K by the embedding

$$\xi \mapsto (a_\mathfrak{p}), \quad a_\mathfrak{p} = \xi ,$$

whose image is the group P_K of *principal ideles* of K. Another subgroup of I_K, denoted by U_K, consists of all $a \in I_K$ with $\mathfrak{p}$-integral $a_\mathfrak{p}$ for finite $\mathfrak{p}$, and $a_\mathfrak{p} > 0$ for infinite real $\mathfrak{p}$. The factor group I_K/U_K is isomorphic to $H^*(K)$; hence there is an epimorphism $\phi : I_K \longrightarrow H^*(K)$.

One defines congruences in I_K with respect to divisors $\mathfrak{m} = \prod_{i=1}^r \mathfrak{p}_i^{e_i}$ by putting

$$a \equiv b \pmod{\mathfrak{m}}$$

if for finite $\mathfrak{p}$ one has

$$\nu_\mathfrak{p}(a_\mathfrak{p} b_\mathfrak{p}^{-1}) \geq e_i ,$$

and for infinite real $\mathfrak{p}$ one has

$$a_\mathfrak{p} b_\mathfrak{p}^{-1} > 0 .$$

With every idele a one associates the divisor

[22]For the first time ideles, under the name 'éléments idéaux' occur in a letter of Chevalley to Hasse [679] written in June 1935.

[23]The definition of the restricted product in the general case occurs for the first time in the thesis of Braconnier [420, 421].

$$\phi(a) = \prod_{\mathfrak{p}} \mathfrak{p}^{e_{\mathfrak{p}}} ,$$

where $e_{\mathfrak{p}}$ is the exponent of $a_{\mathfrak{p}}$ in $K_{\mathfrak{p}}$ in case of finite $\mathfrak{p}$, and for infinite $\mathfrak{p}$ one puts

$$e_{\mathfrak{p}} = \begin{cases} 1 & \text{if } \mathfrak{p} \text{ is real and } a_{\mathfrak{p}} < 0, \\ 0 & \text{otherwise.} \end{cases}$$

Finally one defines the norm map $I_L \longrightarrow I_K$ for all finite extensions L/K by applying the norm componentwise.

If now L/K is Abelian, and H is a group of ideals of K, corresponding to L/K according to class-field theory, and $\mathfrak{f}$ is its conductor, then the set

$$H_{L/K} = \{a \in I_K : a \equiv 1 \pmod{\mathfrak{f}},\ \phi(a) \in H\}$$

is a subgroup of I_K. Chevalley proved that the group $H_{L/K}$ is generated by P_K and the group $N_{L/K}(J_L)$, and the extension L/K is determined by $H_{L/K}$. Moreover he established that the Galois group of L/K and the factor group $I_K/H_{L/K}$ are isomorphic, the isomorphism being defined in the following way:

For $a \in I_K$ write $a = ca_1$ with $c \in P_K$ and $a_1 \equiv 1 \pmod{\mathfrak{f}}$ and define the generalized Artin symbol with the use of the Artin symbol by

$$(a, L/K) = \left(\frac{L/K}{(a_1)}\right) .$$

Chevalley proved that the map

$$(a, L/K) : I_K/H_{L/K} \longrightarrow Gal(L/K)$$

is an isomorphism and showed also that a similar method leads to a generalization of the class-field theory to the case of infinite Abelian extensions.

It follows from a result of M. Bauer [220] that a normal extension of an algebraic number field is determined by its completions, and Chevalley asserted in the introduction to [681] that this holds also for Abelian infinite extensions. This has been shown to be false by Whaples [4381] in 1947.

Later ideles were also used outside the class-field theory. In 1960 Fröhlich [1267] used them to define $\partial(L/K)$ a new kind of discriminant for finite extensions of algebraic number fields. This discriminant is an element of the product

$$\prod_{v} K_v^*/U^2(K_v) ,$$

where v runs over all inequivalent valuations of K, K_v is the corresponding completion, and $U(K_v)$ is its group of units. This product is isomorphic to the factor group I_K/U_K^2.

Fröhlich used the idelic discriminant (see Sect. 3.2.3) to obtain a simple proof of Hecke's assertion (Theorem 177 in the book [1742]) that the ideal class of the discriminant $d(L/K)$ is always a square and to characterize extensions L/K having a relative integral basis as those with $\partial(L/K)$ being the image of a principal idele under the mapping $I_K \longrightarrow I_K/U_K^2$. The last result implies that any normal unramified extension of odd degree has a relative integral basis.

3. In 1940 Chevalley [682] made a further step in his simplification of the class-field theory. He stated in the introduction that this new method permits to cover also the case of infinite Abelian extensions and wrote: "*...elle*[24] *permet en effet d'éviter le maniement tojours un peu délicat des 'groupes de congruence', avec leurs multiples 'modules de définition'* ".

In the idele group I_K he introduced a topology defined by the family $I_K^n I_K^{E,n}$, where $n \geq 1$, E is a finite set of prime divisors $\mathfrak{p}$, $I_K^{E,n}$ is the group of ideles $(a_\mathfrak{p})$ with $a_\mathfrak{p} \in U_\mathfrak{p}$ for $\mathfrak{p} \notin E$, and for $\mathfrak{p} \in E$ $a_\mathfrak{p}$ is an nth power. He considered characters of the topological group I_K and showed that they all have finite order.

The main result of [682] asserts that if G_K is the Galois group of the maximal Abelian extension A_K/K of K with the Krull topology, then the topological groups of characters of G_K and I_K/P_K are isomorphic, the isomorphism being compatible with the norm map $N_{L/K}$. Characters of I_K, trivial on P_K, hence characters of I_K/P_K were called by Chevalley "differentials". He explained in a footnote on p. 403 of his paper, quoting an earlier paper of Weil [4349], that they played a role similar to the differentials in the theory of algebraic functions. He pointed also out that translating the last result into the language of ideals one recovers Artin's reciprocity law. His paper contains also (hidden in §3) the proof of the analogue of Dirichlet's unit theorem for S-units (*theorem of Dirichlet–Hasse–Chevalley*), following an idea of Hasse (let us recall that if S is a finite set of valuations of an algebraic number field K, then the ring

$$R = \{\alpha \in K : |v(a)| < 1 \text{ for all } v \in S\},$$

is called the ring of S-integers, and its invertible elements are called S-units).

5.3 Class-Number and Class-Group

5.3.1 Quadratic Fields

1. A relation between class-groups of distinct quadratic fields has been found in 1932 by Scholz [3669]. He applied class-field theory to show that if $K_1 = \mathbf{Q}\left(\sqrt{d}\right)$, $K_2 = \mathbf{Q}\left(\sqrt{-3d}\right)$, and r_i is the 3-rank of the class-groups of K_i, then

$$r_2 \leq r_1 \leq r_2 + 1.$$

In 1958 Leopoldt [2553] proved a generalization (Spiegelungssatz). For later developments see S.-N. Kuroda [2386], Oriat and Satgé [3193] and G. Gras ([1491] and §5 of Chap. 2 in [1492]). Conditions for the equality $r_1 = r_2$ in Scholz's theorem were given by Kishi [2159, 2160].

[24]...it permits in effect to avoid the somewhat delicate treatment of 'congruence groups' with their several 'modules of definition'.

2. Denote by $e_n(K)$ the number of invariants of $H^*(K)$ divisible by n. Gauss's theorem on the genera implies for quadratic fields the equality $e_2(K) = \omega(d(K)) - 1$, $\omega(d)$ denoting the number of distinct prime factor of d. The first result on $e_n(K)$ for quadratic fields beyond Gauss's theorem has been obtained in 1933 by Rédei and Reichardt [3407] who used class-field theory to show that $e_4(K)$ equals the number of decompositions $d(K) = d_1 d_2$, where d_1, d_2 are discriminants and the following conditions are satisfied:

$$\left(\frac{d_1}{p}\right) = 1 \quad \text{for } p \mid d_2, \quad \left(\frac{d_2}{p}\right) = 1 \quad \text{for } p \mid d_1 .$$

A simpler proof of this result has been given in 1934 by Rédei [3397] (see also Inaba [1960]), who used it to give an upper bound for $e_4(K)$ in [3398], and applied a similar approach to determine $e_8(K)$ in [3399]. This has been generalized by Reichardt [3410] who obtained a similar formula, involving factorizations of the discriminant, for $e_{2^n}(K)$ for arbitrary n. These results have been also established independently by Iyanaga [2003].

Other algorithms for determining $e_{2^n}(K)$ for imaginary quadratic K were given by Rédei [3406] in 1953 (a modern version of it has been presented in 1979 by Morton [2972]) and Shanks [3746] in 1971. For real quadratic fields this has been done by Hasse [1712] and A. Endô [1090] in 1973. In 2004 an algorithm leading to the structure of $H_2^*(K)$ for all quadratic K has been given by Basilla and Wada [209].

In 1936 Rédei [3401] determined the mean value of $e_4(d)$ and $e_8(d)$ for positive discriminants d without prime divisors congruent to 3 mod 4 and in 1939 showed the existence of infinitely many quadratic fields with given $e_2(d) \geq e_4(d) \geq e_8(d)$ [3402] (another proof has been given in 1982 by Morton [2974]).

In 1973 Waterhouse [4306] presented a way to compute $e_8(d)$. In [2112, 2113] P. Kaplan used the classical theory of quadratic forms to study the structure of $H_2^*(K)$.

In 1983 Cohn and Lagarias [752] formulated the following conjecture concerning $e_{2^k}(d)$:

For any integers $n \geq 1$ and $d \not\equiv 2 \bmod 4$ there exists an extension $K_n(d)/\mathbf{Q}$ such that if

$$F_{K_n(d)/\mathbf{Q}}(p) = F_{K_n(d)/\mathbf{Q}}(q) ,$$

(F denoting the Artin symbol) then for $j = 1, 2, \ldots, n$ one has

$$e_{2^j}(dp) = e_{2^j}(dq) .$$

They showed that if such field $K_n(d)$ exists then there is a unique such field of minimal degree, the *governing field*. Genus theory implies the truth of the conjecture for $n = 1$, its truth for $n = 2$ follows from results of Rédei and Reichardt [3407] using class-field theory, and for $n = 3$ it has been established in 1989 by Stevenhagen [3927]. Earlier Morton [2973–2975] showed its truth for $n = 3$ and infinitely many square-free d.

A criterion for $e_{16}(K) = 1$ has been given in 1993 by Costa [774].

It has been conjectured in 1962 by Šafarevič [3551] that the rank of the class-group of $\mathbf{Q}(\sqrt{p})$ for p prime is bounded, and it has been pointed out by Shanks and Weinberger [3749] that one expected this bound to be equal 2. They gave a counterexample to the last conjecture: $p = 188\,184\,253$ with 3-rank equal to 3. Later Quer [3354] found an example with 3-rank five. Šafarevič's conjecture is still open.

A table of the structure of $H(-d)$ for $0 < d < 24\,000$ has been published in 1970 by Wada [4259]. On the basis of his calculations he conjectured that if $H(-d) = \bigoplus_{i=1}^{m} C_{n_i}$ with $n_1 \mid n_2 \mid \cdots \mid n_t$, then at most two n_i's are not powers of two. This turned out to be rather over-optimistic, as it has been shown two years later by Shanks [3747] that for $d = 63\,199\,139$ one has $n_1 = 3, n_2 = 3^2, n_3 = 3^3, n_4 = 3^3 \times 116$. Later Craig [790] showed that $r_3(K) \geq 3$ happens for infinitely many imaginary quadratic fields, $e_p(K)$ denoting the number of invariants of the class-group of K, divisible by p.

Shanks and Serafin [3748] found in 1973 two examples ($d = 167 \times 12409 \times 42169$ and $d = 4 \times 83\,309\,629\,817$) with $e_3(K) = 4$, and Craig [791] showed in 1977 that this happens infinitely often. Another examples have been found by Neild and Shanks [3080] in 1974, Diaz y Diaz [921] in 1978, Diaz y Diaz, Shanks and H.C. Williams [928] in 1979. In 1977 Solderitsch [3851, 3852] gave examples with $e_5(K) = 3$ and $e_7(K) = 3$, and in 1983 Schoof [3686] found a field K with $e_5(K) = 4$. A field with $e_3(K) = 5$ has been found by Llorente and Quer [2616] in 1988, and the smallest discriminant with this property has been found by Belabas [275] in 2004. Fields with $e_3(K) = 6$ were found by Quer [3354]. In 2013 Dominguez, S.J. Miller and Wong [987] showed the existence of fields $\mathbf{Q}(\sqrt{-pq})$ with prime p, q with the class-group having an arbitrary large cyclic 2-Sylow subgroup.

In 1976 Buell [492, 493] computed the structure of $H(-d)$ first for $d < 4 \cdot 10^6$ and in 1987 extended his tables to $d < 25 \cdot 10^6$ [494] and $d < 2.2 \cdot 10^9$ [495]. An extension to $d < 10^{11}$ has been provided by M.J.Jr. Jacobson, Ramachandran and H.C. Williams [2014] in 2006, and a description of tables giving the class-group structure of imaginary quadratic fields with $\mid d(K) \mid < 2^{40} \approx 10^{12.04}$ has been given in 2016 by Mosunov and M.J.Jr. Jacobson [2984]. A method of computing this structure for very large discriminants has been proposed by Kleinjung [2169], who gave examples with discriminants having 100, 110, 120 and 130 decimal digits.

In the case of positive discriminants d it has been shown in 1978 by Diaz y Diaz [921] (see also Diaz y Diaz, Shanks and H.C. Williams [928]) the existence of infinitely many fields with $e_3(d) \geq 3$. In 1983 Mestre [2834] obtained infinitely many real quadratic fields with $e_p(d) \geq 2$ for $p = 5, 7$. Examples of such fields were earlier found by Buell [492].

3. An important step toward the solution of Gauss's class-number problem for imaginary quadratic fields has been done in 1933 by Deuring [908]. He showed that if the Riemann Hypothesis fails, then $h(d) = 1$ can happen only for finitely many $d < 0$.

He used the fact that for d negative the Dedekind zeta-function of the field $K = \mathbf{Q}\left(\sqrt{d}\right)$ with $h(K) = 1$ equals

$$\zeta_K(s) = \zeta(s) L(\chi_d, s) = \frac{1}{2} \sum_{0 \neq \alpha \in \mathbf{Z}_K} \frac{1}{N(\alpha)^s},$$

where $\chi_d(n) = \left(\frac{-4d}{n}\right)$, and the second factor on the right-hand side is the Epstein[25] zeta-function of the quadratic form

$$\mathbf{Q}(X, Y) = \begin{cases} X^2 + \frac{d}{4} Y^2 & \text{if } 4 \mid d, \\ X^2 + XY + \frac{1-d}{4} Y^2 & \text{if } d \equiv 1 \bmod 4. \end{cases}$$

[25]Epstein zeta-functions of quadratic forms were introduced in 1903 by Epstein [1112]. It has been found recently by Oswald and Steuding [3204] that these functions were earlier studied by A. Hurwitz.

He used this to deduce for $\Re s > 1/2$ and $\Im s > 1$ the equality

$$\zeta(s)L(\chi_d, s) = \zeta(2s) + d^{1/2-s}A(s) + O\left(\exp\left(-2d^{1/2}/\Im s\right)\right) ,$$

where

$$A(s) = \sqrt{\pi}\frac{\zeta(2s-1)\Gamma(s-1/2)}{\Gamma(s)} ,$$

and the implied constant does not depend on d.

If there are infinitely many d's with $h(-d) = 1$, say $d_1 < d_2 < \dots$, then this implies

$$\zeta(s) \lim_{i\to\infty} L(\chi_{d_i}, s) = \zeta(2s) \neq 0 ,$$

violating Riemann's Hypothesis.

He showed also that Landau's inequality (3.27) can be replaced by

$$\log(d_{i+1}) \geq c d_i^{1/4}$$

with some $c > 0$. Later S. Chowla [697] showed that the exponent $1/4$ can be replaced by $1/2$.

4. In the next year Mordell [2932] went one step further using a similar approach and deduced

$$\lim_{d\to\infty} h(-d) = \infty \tag{5.3}$$

from the falsity of Riemann Hypothesis.

In the same year Heilbronn[26] [1750] obtained a proof of (5.3) assuming the existence of a Dirichlet L-function with a real character having a zero in the half-plane $\Re s > 1/2$. In view of Hecke's result (3.24) this implied (5.3) unconditionally (some simplifications have been provided by S. Chowla [700, 701]).

The next step has been done by Heilbronn and Linfoot[27] [1756], who modified Deuring's approach to show that apart of the nine known discriminants there can be at most one more with $h(-d) = 1$. This has been extended in 1935 by Landau [2446], who gave a new proof of (5.3), and showed that there exists an absolute constant C such that for any h there can be at most one discriminant $-d$ with $d > Ch^8 \log^6(3h)$, satisfying $h(-d) = h$.

At that time it had been already known that $h(-d) > 1$ holds for $d \in [164, 5 \cdot 10^9]$ (D.H. Lehmer [2512]).

Another proof of the Heilbronn–Linfoot theorem has been given in 1967 by Ayoub [157].

Mordell's approach has been used by Mahler [2698] to establish a kind of counterpart to the inequality (3.24):

---

[26] Hans Arnold Heilbronn (1908–1975), professor in Bristol and Toronto. See [600].

[27] Edward Hubert Linfoot (1905–1982), lecturer at the University of Bristol. See [281].

If $c > 0$ and for a sufficiently large d the zeta-function of the imaginary quadratic field of discriminant $-d$ has a zero in the interval $(1 - c/\log d, 1)$, then

$$h(-d) \leq B(c)\frac{\sqrt{d}}{\log d}\sum_{j=1}^{h}\frac{1}{a_j},$$

where for $j = 1, 2, \ldots, h$ the forms $a_jX^2 + b_jXY + c_jY^2$ form a complete set of reduced forms of discriminant $-d$.

5. In 1935 Siegel [3778] established the important lower bound (*theorem of Siegel*)

$$L(1, \chi) > \frac{c(\varepsilon)}{k^\varepsilon} \tag{5.4}$$

for Dirichlet L-functions with real characters mod k. It led to the asymptotic relations for the class-number $h(d)$ for quadratic fields with discriminant d:

$$\log h(d) = (1/2 + o(1))\log(|d|) \tag{5.5}$$

for $d < 0$, and

$$\log\left(h(d)R(d)\right) = (1/2 + o(1))\log(|d|) \tag{5.6}$$

for $d > 0$, $R(d)$ being the regulator of $\mathbf{Q}\left(\sqrt{d}\right)$. Siegel noted that the method of his proof leads to (5.6) for solvable fields and conjectured that this formula should hold for all fields of a fixed degree (see Sect. 6.1.1).

The inequality (5.4) is not effective because of the possible existence of *Siegel zeros*, which are real zeros of $L(s, \chi)$ lying in the interval $(1 - c/\log d, 1)$ with some fixed $c > 0$, d denoting the conductor of χ.

It has been shown in 2000 by Granville and Stark [1487] that a version of the ABC-conjecture, due to Elkies [1080], implies the non-existence of Siegel zeros.

In 1967 Low [2656] established that if χ is an odd real character mod d with $d \leq 593\,060$ then the corresponding L-function does not have positive real zeros, except possibly for $d = 115\,147$, and in 2004 Watkins [4308] extended this for all $d \leq 3 \cdot 10^8$ with at most three exceptions. It is now known that for at least 20% of odd square-free d's the L-function corresponding to the character

$$\chi(n) = \left(\frac{-8d}{n}\right)$$

does not have zeros in $(1/2, 1)$ (Conrey and Soundararajan [766]).

For even real characters χ mod d the non-existence of positive real zeros of $L(s, \chi)$ has been established for $d \leq 227$ by Rosser [3523, 3524] in 1949–1950 and for $d \leq 200\,000$ by Chua [713] in 2005.

6. A simpler proof of (5.4) has been found in 1938 by Heilbronn [1751], who showed that an integral formula used by Siegel can be obtained as a special case of a general identity having a not difficult proof.

Later other proofs of Siegel's theorem were given by Čudakov[28] [796] in 1942, Estermann[29] [1124] in 1948, and S. Chowla [704], Linnik [2603] in 1950 and Tatuzawa [4013] in 1951. Linnik's paper contains also an elementary proof, which has been later simplified by Pintz [3295]. A very short proof presented Goldfeld [1460] in 1974. Further proofs were given by Ramachandra [3374] in 1980, and an elementary proof was found by Koukoulopoulos [2232] in 2013.

For later development in the class-number problem see Sect. 6.2.1.

7. Let $g(d)$ denote the number of genera of binary quadratic forms of discriminant d. Gauss's theorem on genera (see Sect. 1.1.2) gives for negative d the equality $g(d) = 2^{\omega(d)-1}$, where $\omega(d)$ is the number of prime divisors of d. It has been shown in 1934 by S. Chowla [698] that the ratio $h(d)/g(d)$ tends to infinity for $d \to -\infty$. This implied in particular that the number of imaginary quadratic fields with only one class in each genus, i.e., with the class-group of the form C_2^N, is finite.

Discriminants with $g(d) = h(d)$ are related to *idoneal numbers* introduced by Euler (see the surveys by Steinig [3912] and Kani [2107]), who found 65 such numbers, the biggest being 1848, and conjectured that hist list is complete. A numerical search has been performed by Dickson and Townes (chap. 5 in [938]) in the twenties.

It has been shown by Grube [1531] in 1874 that a number N is idoneal if and only if $g(-4N) = h(-4N)$ (a proof can be also found in Grosswald's[30] paper [1528] and in the book of Cox [786]). Grube gave also correct proofs of several Euler's assertions about idoneal numbers.

A set of necessary conditions for $g(d) = h(d)$ has been given in 1939 by N.A. Hall [1597].

The search for idoneal numbers has been extended to $|d| \leq 10^7$ by Swift [3967] in 1948, and it has been shown in 1954 by S. Chowla and Briggs [707] that GRH implies the completeness of Euler's list (another proof has been given in 1990 by Louboutin [2633]). They showed also that there is at most one such field discriminant d with $|d| > 10^{60}$. Grosswald quoted in [1528] a computation by Selfridge, Atkins and MacDonald going up to $10^{9.12919}$. In 2006 M.J.Jr. Jacobson, Ramachandran and H.C. Williams [2014] determined the structure of the class-group for $\mathbf{Q}\left(\sqrt{d}\right)$ with $0 > d > -10^{11}$, not finding any new discriminants with $g(d) = h(d)$. They applied a modification by Buchmann, M.J.Jr. Jacobson and Teske [490] of an algorithm of Shanks [3745]. Note that several errors in the earlier literature on idoneal numbers have been pointed out by Kani in [2107].

If one denotes by $Exp(K)$ the exponent of the class-group of K, then the equality $g(d) = h(d)$ is equivalent to $Exp(\mathbf{Q}(\sqrt{d})) \leq 2$. Boyd and Kisilevsky [419] and Weinberger [4361] showed that there are only finitely many imaginary quadratic fields $K_d = \mathbf{Q}\left(\sqrt{d}\right)$ with $Exp(K_d) = 3$, and if GRH holds for Dirichlet L-functions, then for $d < 0$ one has

$$Exp(K_d) \gg \frac{\log(|d|)}{\log\log(|d|)}. \tag{5.7}$$

In [419] it was shown that the implied constant can be equal to $1/2 - \varepsilon$ for $|d| > c(\varepsilon)$ for any $\varepsilon > 0$. In 1995 Pappalardi [3217] showed that the bound (5.7) holds unconditionally for almost all d. In 2003 Louboutin and Okazaki [2654] showed that (5.7) holds also for all CM-fields of fixed even degree.

---

[28] Nikolaĭ Grigorevič Čudakov (1904–1986), professor in Saratov, Moscow and Leningrad.

[29] Theodor Estermann (1902–1991), Reader at the University College, London.

[30] Emil Grosswald (1912–1989), professor at the Temple University.

In 1981 Earnest and Estes [1037] proved that $Exp(K_d) = 4$ can hold only for finitely many d, and one year later Earnest and Körner [1039] showed this for $Exp(K_d) = 2^s$ with fixed s. They showed, more generally, that if k is totally real, then there are only finitely may totally imaginary quadratic extensions K/k such that $Exp(K)$ is a fixed power of 2 (in [1037] this has been established for $Exp(K) = 4$ under the additional assumption $h(k) = 1$).

In 2008 Heath-Brown [1727] showed that $Exp(K_d) = 5$ and $Exp(K_d) = 2 \cdot 3^s$ can hold only for finitely many $d < 0$ and gave a new proof in the case of $Exp(K_d) = 2^s$. His results are non-effective.

It has been shown in 2008 by Chakraborty, Luca and Mukhopadhyay [641] that for given g and large x there are at least $x^{1/g}/5$ discriminants $0 < d \leq x$ with $Exp(d) = g$.

In 1990 K. Horie and M. Horie [1897] established that there are only finitely many imaginary Abelian fields K of 2-power degree with $Exp(K) = 2$.

In 2004 Ahn and Kwon [27] proved under GRH that there are only finitely many imaginary Abelian fields with Galois group and the class-group being both elementary 2-groups. They extended this in [28] to arbitrary imaginary Abelian fields of 2-power degree.

5.3.2 Other Fields

1. A cyclotomic field $\mathbf{Q}(\zeta_p)$ with prime p is called *properly irregular* if p divides h_p^- but does not divide h_p^+. All other irregular fields are called *improperly iregular*. Vandiver's conjecture asserts that such fields do not exist. For properly irregular fields Vandiver [4164, 4177, 4178] studied the structure of the p-Sylow subgroup H_p of the class-group, showed that its rank equals the number of the Bernoulli numbers B_n with $n \leq (p-3)/2$ which are divisible by p and described its generators. As a corollary he obtained that H_p is a cyclic G-module with G being the Galois group of the field.

Further properties of properly irregular fields were studied by Vandiver in [4180].

It was established in 1932 by Herbrand [1803] that H_p as a $\mathbf{Z}_p[G]$-module is equal to

$$\bigoplus_{i=0}^{p-2} A_i \ ,$$

where

$$A_i = \{a \in H_p : \ g(a) = \omega^i(g)a \text{ for } g \in G\} \ ,$$

ω being the Teichmüller character. Vandiver's conjecture is equivalent to $A_i = 0$ for even $i \leq p-3$. Herbrand proved also that if i is odd and $A_i \neq 0$, then p divides B_{p-i} (B_n denoting the nth Bernoulli number) and showed that the converse implication would follow from the truth of Vandiver's conjecture.

In 1976 Ribet [3458] established this implication unconditionally.

In the same paper [1803] Herbrand presented a construction of all unramified extensions of $\mathbf{Q}(\zeta_p)$ in the case when $p \nmid h^+(p)$.

In 1992 M. Kurihara [2381] obtained $A_{p-3} = 0$ using bounds for the fourth algebraic K-group of $\mathbf{Z}$, and in 1999 Soulé [3860] applied K-theory to show that if n is odd, and p is very large in comparison to n, say

$$\log p > n^{224n^4},$$

then $A_{p-n} = 0$. An exposition has been presented by Ghate [1423].

It should be noted that although an analogue of the theorems of Herbrand and Ribet has been established for function fields by Taelman [3972], the analogue of Vandiver conjecture fails in that case, as shown by Anglés and Taelman [86, 87] in 2013.

2. Let L/K be a cyclic extension of prime degree p, and denote by $Am(L/K)$ the group of ideal classes invariant under the action of the Galois group. In 1933 Moriya [2955] showed that if L/K is unramified, then the ratio $|Am(L/K)|/h(K)$ is a power of p, whereas in the unramified case one has $|Am(L/K)| = h(K)/p$. He established also that in case $K = \mathbf{Q}$ the rank of the p-part of the class-group does not exceed $(p-1)(\omega(d(L)) - 1)$.

5.4 Other Questions

5.4.1 Additive Problems

1. In 1935–1937 Siegel published three papers dealing with the analytical theory of quadratic forms [3779, 3781, 3782]. In the first two he studied quadratic forms in n variables over the ring of rational integers, whereas in the last paper he generalized his results to the case of forms with integral coefficients in an algebraic number field K (cf. [3785]). We shall not describe his main results, as they belong properly to the theory of quadratic forms, and restrict our attention to their consequences for sums of squares. Two quadratic forms in n variables over $\mathbf{Z}_K$ having the same discriminant are said to lie in the same genus if they are equivalent in all localizations. Denote by $\gamma_n(K)$ the number of classes of forms lying in the genus of the form $\sum_{j=1}^n X_j^2$. It has been noted by Siegel at the end of [3782] that if K is totally real and $\gamma_4(K) = 1$, then one can obtain a formula for the number of representation of $a \in \mathbf{Z}_K$ as a sum of four squares, analogous to the classical formula[31] of Jacobi in the case of $K = \mathbf{Q}$. He showed also that this can happen only for finitely many totally real fields K, providing a bound for their discriminant and class-number, and pointed out that the only such fields known are $\mathbf{Q}$, $\mathbf{Q}(\sqrt{2})$, $\mathbf{Q}\left(\sqrt{5}\right)$, without however conjecturing that they are the only ones.

In 1960 Dzewas [1034] showed that the observation of Siegel is true for every n: if K is totally real and $\gamma_n(K) = 1$, then one can obtain an explicit formula for the number of representations of $a \in \mathbf{Z}_K$ as a sum of n squares. Dzewas characterized real quadratic fields K having $\gamma_n(K) = 1$, showed that there are none with $n > 4$ and proved that $\gamma_3(K) = 1$ holds only for $K = \mathbf{Q}\left(\sqrt{d}\right)$

[31] A simple proof of Jacobi's formula has been given in 2000 by Spearman and K.S. Williams [3866].

with $d = 2, 5, 17$ and $\gamma_4(K)$ holds for $d = 2, 5$. He gave also a list of real quadratic fields with $\gamma_2(K) = 1$ for $d \le 100$. He proved moreover that there are no totally real fields with $\gamma_n = 1$ with $n \ge 9$, $\gamma_8(K) = 1$ holds only for the rational field, $\gamma_n(K) = 1$ with $m = 2, 3$ can hold only for finitely many totally real fields of given degree, and for $n \ge 4$ there are only finitely such fields, all having degree ≤ 11 and $d(K) \le 62\,122\,500$. Later Barner [189] proved that $\mathbf{Q}$, $\mathbf{Q}(\sqrt{2})$, $\mathbf{Q}\left(\sqrt{5}\right)$ are the only fields with $\gamma_4(K) = 1$.

In 1971 Pfeuffer [3276] showed that if $\gamma_3(K) = 1$, then deg $K \le 20$, and $d(K)$ is bounded by $\left(2^{32}3^{38}\pi^{80}\right)^{1/3}$. He proved also that the only cubic fields with $\gamma_3(K) = 1$ have discriminants 49 and 148, and there are no such quartic fields . He conjectured that there are also no such fields of degree > 4, and this has been established in 1977 by Peters [3264], who also proved that there are no new fields with $\gamma_m(K) = 1$ for $m \ge 3$. Using GRH he was able to show that there are only finitely many fields with $\gamma_2(K) = 1$.

2. In 1932 van der Waerden [4135] published an exercise, whose solution implied that if in a field K the element -1 is a sum of squares, then the minimal number $s(K)$ of needed summands either exceeds 15 or equals $1, 2, 4$ or 8. A proof has been later given by H. Kneser[32] [2182], and with the use of Siegel's result in [3769] this led to the assertion that for totally complex algebraic number fields one has $s(K) \in \{1, 2, 4\}$. Obviously $s(K) = 1$ is equivalent to $i \in K$; hence Siegel's bound leaves open only the question for which K one has $s(K) = 2$.

The number $s(K)$ has been later called the *Stufe* of K and denoted $s(K)$; now it is usually called the *level*. In the same way one defines the level of an integral domain.

In 1962 Nagell [3031] proved that if $n \equiv 3 \bmod 8$, then $s(\mathbf{Q}(\zeta_n)) = 2$ (see also P. Chowla [694]).

It has been shown in 1965 by Pfister [3278] that for every not formally real field the level is a power of 2.

In 1969 P. Chowla [695] showed that if for a prime $p \equiv 3 \bmod 8$ the order of 2 mod p is even, then for the ring R_p of integers of the field $\mathbf{Q}(\zeta_p)$ with prime $p \equiv 3 \bmod 8$ one has $s(R_p) = 2$, and next year she established with her father S. Chowla [696] the equality $s(R_p) = 4$ for $p \equiv 7 \bmod 8$. They noted in that paper that in unpublished papers by J.H. Smith and S. Chowla it has been shown that in case $p \equiv 5 \bmod 8$ one has $s(R_p) = 2$.

In 1970 Moser [2978, 2979] determined $s(K)$ for imaginary quadratic and cyclotomic fields, Abelian extensions of p-adic fields, as well as for rings of integers of imaginary quadratic fields. Next year a determination of the level for all algebraic number fields has been achieved by Fein, Gordon and J.H. Smith [1167], who applied the theory of linear algebras. They showed that $s(K) = 2$ holds if and only if for all prime ideals $\mathfrak{p}$ dividing 2 the local degree $K_\mathfrak{p}/\mathbf{Q}_2$ is odd. For quadratic $K = \mathbf{Q}(\sqrt{m})$ with m square-free this implies that $s(K) = 2$ holds if and only if $m \not\equiv 7 \bmod 8$, and for $K = \mathbf{Q}(\zeta_n)$ one has $s(K) = 2$ if and only if $2 \nmid n$ and 2 has even order mod n. A proof based on Hasse's local–global principle for quadratic forms has been given in 1972 by F.W. Barnes [193].

All solutions of the equation $-1 = x^2 + y^2$ in imaginary quadratic fields have been described in 1972 by Nagell [3043]. It has observed by Risman [3476] in 1974 that the determination of $s(K)$ for quadratic fields can be used to a proof of the three squares theorem, and Rajwade [3373] and Small [3830] did this in a completely elementary way. An elementary approach in the case of quartic fields has been made by Parnami, Agrawal and Rajwade [3224]. The level of pure extensions of the rationals has been determined by Nassirou [3077] in 1999.

The level of orders in imaginary quadratic fields has been determined by Peters [3261] in 1971, who also observed that the value of $s(K)$ for algebraic number fields follows from Theorem 4 in a paper of Hasse [1644] published in 1924.

---

[32] Hellmuth Kneser (1898–1973), professor in Greifswald and Tübingen, son of Adolf Kneser, father of Martin Kneser. See [4398].

The level of arbitrary commutative rings has been treated by Dai, Lam and Peng [809]. For the case of Dedekind domains see Baeza [174] and Arason, Baeza [103].

The generalization of the level $s(K)$ to the case of fourth powers for imaginary quadratic fields has been considered by Parnami, Agrawal, Rajwade [3223] in 1981.

3. In 1936 Siegel [3780] invented a method of obtaining asymptotical formulas for sums of the form

$$\sum_{(a,a')\in G} f(a),$$

where f is a function defined on integers of a real quadratic field, G is a region in the plane, and a' is the conjugate of a. He obtained this using an integral formula involving Euler's Beta-function. He pointed out that the same approach can be applied in the case of arbitrary totally real fields.

In the sixties Schaal [3588, 3589] applied this method in the case when $f(a)$ is the number of representations of a as a sum of two squares, first in real quadratic fields, and then in arbitrary totally real fields. His results have been improved later by Rausch [3387] who in [3389, 3390] applied this method to the study of the analogue of the problem of ellipsoids in algebraic number fields.

Grotz [1529, 1530] showed how Siegel's method can be extended to arbitrary algebraic number fields and applied it to obtained asymptotic formulas for the mean values of various arithmetical functions.

5.4.2 *Galois Groups*

1. In 1935 Rosenblüth [3519] gave a necessary condition for embedding of a biquadratic field in a field with quaternion Galois group H_8, and Richter [3460] and Reichardt [3412] showed later that it is also sufficient. A construction of all extensions with group H_8 of an arbitrary field has been described by Witt [4431] in 1936.

Conditions for embedding a quadratic field in a generalized quaternion algebra can be deduced from Hasse's results on splitting of algebras, proved in [1677]. An elementary proof has been given in 1936 by Latimer [2481].

In [3460] Richter solved also the problem of embedding a biquadratic field into an octic dihedral field . He used this result as well as a theorem of Hasse [1677] asserting that in this question the Hasse principle holds and gave a sufficient condition for the solvability of the embedding problem in some cases. For embedding of a field with group $C_2 \oplus C_2$ in a octic field with the dihedral or quaternion group he gave a necessary and sufficient condition.

For embedding of biquadratic fields in dihedral fields of order 2^n see Halter-Koch [1598].

In a later paper gave Richter [3461] necessary and sufficient conditions for the embedding of a field with group C_{p^m} (p prime) into a field with group C_{p^n} with $n > m$ and noted that here the Hasse principle is not always applicable.

In 1958 Šafarevič [3550] proved that for every finite extension of the rationals the embedding problem is solvable in the split case with nilpotent kernel (his proof has been later simplified by Išhanov [1968]). Earlier he did this in the split case with the kernel being a p-group [3549]. It has been pointed out by Sonn in his review of Neukirch's paper [3095] in Math. Reviews (MR 53#8013) that the result in [3550] lead to a simpler proof of Šafarevič's theorem [3547] on the realizability of every finite solvable group as the Galois group of an extension of any algebraic number field.

5.4.3 Euclidean Algorithm

1. In the German translation of Dickson's book on algebras ([936], pp. 150–151) it has been shown that the only norm-Euclidean imaginary quadratic fields are these listed by Dedekind in §159 of [848]. Dickson's assertion that the only norm-Euclidean real quadratic fields K are those with $d(K) \in \{5, 8, 12, 13\}$ turned out to be incorrect when in 1932 Perron [3255] found six other such fields with $d(K) = 17, 21, 24, 28, 29, 44$. A shorter proof of Perron's list has been provided in 1934 by Oppenheim[33] [3163], who also showed that the discriminants 33, 37 and 41 should be added to it (the same has been obtained independently by Remak [3442], and Heinhold [1757] gave another proof in 1939). In 1935 E. Berg [295] added 76 to this list (this has been later repeated by Behrbohm and Rédei [274]) completing the list of all real quadratic norm-Euclidean fields with $4 \mid d(K)$. In the same year Hofreiter [1881] showed that $\mathbf{Q}\left(\sqrt{57}\right)$ is also norm-Euclidean. Various necessary conditions for the existence of the Euclidean algorithm in real quadratic fields were given by Hofreiter [1880, 1881] and Behrbohm and Rédei [274].

In 1938 Erdős and Chao Ko[34] [1116] proved that there are only finitely many real quadratic norm-Euclidean fields with prime discriminant, and Heilbronn [1752] used their method to establish this also in the case when $d(K)$ is the product of two primes. Since a real quadratic field $\mathbf{Q}\left(\sqrt{d}\right)$ with odd d can have class-number one only if d is either a prime or is a product of two primes congruent to 3 mod 4 this showed that there are only finitely many quadratic norm-Euclidean fields. For later results on Euclidean fields see Sect. 6.4.

5.4.4 Algebraic Numbers on the Plane

1. In 1936 Pisot[35] [3301] considered algebraic integers $\alpha > 1$ whose all remaining conjugates lie inside the unit circle. Such integers are now called the *numbers of Pisot–Vijayaraghavan* (or PV-numbers). since they were also studied by Vija-

[33] Alexander Victor Oppenheim (1903–1997), professor in Singapore and Kuala Lumpur.

[34] Chao Ko (1910–2002), professor at Sichuan University.

[35] Charles Pisot (1910–1984), professor in Bordeaux and Paris. See [62].

yaraghavan[36] [4212] a few years later. Actually PV-numbers occurred already in a paper of Thue [4059] published in 1912 and also in a 1919 paper of G.H. Hardy [1624].

Pisot gave several characterizations of PV-numbers. In 1936 he showed [3301] that $\alpha > 1$ is a PV-number if and only if for some $0 < \theta < 1$ one has

$$\| \alpha^n \| = O(\theta^n)$$

(where $\| x \|$ denotes the distance of the real number x from the nearest rational integer), noted that every algebraic number field contains PV-numbers and deduced a necessary and sufficient condition for a complex number α to be algebraic. In [3302] he established that α is a PV-number if and only if for some $\lambda \neq 0$ the series

$$\sum_{n=1}^{\infty} \| \lambda\alpha^n \|^2$$

converges, and in [3303] he showed that another equivalent conditions is the convergence of the series

$$\sum_{n=1}^{\infty} \sin^2 \left(\pi\alpha^n\right).$$

See also Pisot [3305].

In 1941 Vijayaraghavan [4212] showed that an irrational algebraic number α is a PV-number of and only if the set of limit points of the sequence $\{\alpha^n\}$ is finite. He conjectured that it is not necessary to assume that α is algebraic and showed in [4213] that there can exist at most countably many such transcendent numbers. He conjectured also that the set of all PV-numbers is closed and nowhere dense and this has been established in 1944 by Salem[37] [3559]. In the same year Siegel [3787] found two smallest PV-numbers and conjectured that $(1 + \sqrt{5})/2$ is the smallest limit point of the set of PV-numbers. This has been shown to be true by Dufresnoy and Pisot [1013], who later determined several small PV-numbers [1014–1016]. For later work on determination of PV-numbers see Boyd [409, 413, 415] and the survey by Bertin and Pathiaux-Delefosse [323].

The question posed by Kuba [2330] whether there exist irrational algebraic PV-numbers α with $\lim_{n\to\infty}\{\alpha^n\} = 0$ got a negative answer in 2000 by Luca [2666] and Dubickas [1006]. Dubickas characterized also algebraic numbers α satisfying $\lim_{n\to\infty}\{\alpha^n\} = 1$.

A survey of the theory of PV-numbers was written by Bertin, Decomps-Gilloux, Grandet-Hugot, Pathiaux-Delefosse and Schreiber [322] in 1992.

Algebraic integers $\alpha > 1$ which are not PV-numbers but whose remaining conjugates lie in the closed unit disc (*Salem numbers*) were studied in 1945 by Salem [3561], who showed that their set is closed, and every PV-number is a limit point of a sequence of Salem numbers. Small Salem numbers were studied by Boyd [408] and Flammang, Grandcolas and Rhin [1194] (see also [323]). It has been shown in 2005 by McKee and Smyth [2805] that for any given $n \in \mathbf{Z}$ there are Salem numbers of trace n.

---

[36] Tirukannapuram Vijayaraghavan (1902–1955), professor at the Andhra University and the Ramanujan Institute for Mathematics in Madras. See [824].

[37] Raphael Salem (1898–1963), professor at MIT, in Caen and Paris.

It has been observed by Salem [3558] that PV-numbers and Salem numbers are of importance in the theory of harmonic analysis. See the papers of Salem [3560, 3562] and Salem and Zygmund[38] [3564], as well the books of Salem [3563] and Y. Meyer [2849]. These numbers appear also in various algebraic, combinatorial and geometrical contexts. See the papers of Bartholdi and Ceccherini-Silberstein [200], Chinburg [690], McMullen [2808], Lakatos [2400, 2401], W.Parry[39] [3227] and Sury [3964]. A survey has been presented by Ghate and Hironaka [1424] in 2001.

Various generalizations of PV-numbers and/or Salem numbers were considered by D.G. Cantor [557], Garth [1387], Kelly [2134] and Samet [3568, 3569].

2. It is now customary to formulate Kronecker's theorem on monic polynomials with all roots in the unit disk using the *house* $\overline{|a|}$ of an algebraic integer a of degree n, defined as the maximal absolute value of conjugates of a. Thus Kronecker's first result states that if $0 < \overline{|a|} \leq 1$, then a is a root of unity, and the second result describes the totally real integers with $\overline{|a|} \leq 2$.

Related to the house $\overline{|a|}$ is the *Mahler measure* $M(f)$ of monic polynomials $f \in \mathbf{Z}[X]$, defined by

$$M(f) = \prod_{i=1}^{n} \max\{|a_i|, 1\}, \tag{5.8}$$

where $a_1, a_2, \dots, a_n$ are the roots of f. One speaks also about Mahler measure $M(a)$ of an algebraic integer a, which equals $M(f)$ with f being the minimal polynomial of a. Obviously one has

$$\overline{|a|} \leq M(a) \leq \left(\overline{|a|}\right)^n .$$

For non-monic f with leading coefficient A one defines its Mahler measure by

$$M(f) = |A| \prod_{i=1}^{n} \max\{|a_i|, 1\} ,$$

a_i being the roots of f.

In 1960 Mahler [2705] observed that Jensen's formula[40] (Jensen[41] [2050]) implies the equality

$$M(f) = \exp\left(\int_0^1 \log|f(e^{2\pi i t})| dt\right)$$

[38]Antoni Szczepan Zygmund (1900–1992), professor in Wilno and Chicago. See [1166].

[39]William Parry (1934–2006), professor in Warwick.

[40]It has been noted in 1914 by Landau [2425] that Jensen's formula in the case of polynomials occurs already in a paper of Jacobi [2006], published in 1827.

[41]Johan Ludwig William Valdemar Jensen (1859–1925), worked in a telephone company. See [3129].

The *absolute logarithmic height* $h(\alpha)$ whose definition goes back to Weil [4352] (who modified[42] the definition of the height given earlier by Northcott[43] [3130, 3131]) is defined for algebraic numbers α of degree N by the equality

$$h(\alpha) = \frac{1}{N}\left(\sum_{\mathfrak{p}}[K_{\mathfrak{p}} : \mathbf{Q}_p]\log^+(|\alpha|_{\mathfrak{p}}) + \sum_{\sigma}\epsilon(\sigma)\log^+(|\sigma(\alpha)|)\right), \tag{5.9}$$

where $K = \mathbf{Q}(\alpha)$, $\mathfrak{p}$ runs over all prime ideals of K, $K_{\mathfrak{p}}$ denotes the completion of K at $\mathfrak{p}$, σ runs over all non-conjugated embeddings of K in the complex field, $\log^+ x = \max\{\log x, 0\}$ and

$$\epsilon(\sigma) = \begin{cases} 1 & \text{if } \sigma \text{ is real,} \\ 2 & \text{if } \sigma \text{ is complex}. \end{cases}$$

It is related to Mahler measure by the equality

$$h(\alpha) = \frac{1}{N}\log M(\alpha).$$

3. Let $a_1, a_2, \ldots, a_n$ be roots of a monic polynomial $f \in \mathbf{Z}[X]$. In 1933 D.H. Lehmer [2513] studied the sequence

$$\Delta_k(f) = \prod_{j=1}^{n}\left(a_j^k - 1\right) \quad (k = 1, 2, \ldots)$$

introduced in 1916 by Pierce[44] [3288], who showed i.a. that for prime p the congruence $f(x) \equiv 0 \bmod p$ is solvable if and only if p divides $\Delta_{p-1}(f)$. In Sect. 13 of his paper D.H. Lehmer looked for polynomials f with $1 < M(f) < 1 + \varepsilon$ for a given ε, $M(f)$ being the Mahler measure, defined by (5.8) (he denoted Mahler measure by $\Omega(f)$), determined the minimal value of $M(f)$ for $\deg f \leq 4$, made computations revealing the rather small measure $M(f) = 1.1762\ldots$ for the polynomial

$$f(X) = X^{10} - X^9 - X^7 - X^6 - X^5 - X^4 - X^3 + X + 1,$$

and wrote "*Whether or not the problem has a solution for* $\varepsilon < 0.176$ *we do not know*". Later it has been customary to call *Lehmer's conjecture* the assertion that there exists $\varepsilon > 0$ such that for all irreducible non-cyclotomic monic $f \in \mathbf{Z}[X]$ one has $M(f) \geq 1 + \varepsilon$. Because of the inequality

---

[42] Weil pointed out that the first definition of the height for algebraic numbers appeared in the paper of Siegel [3777]. The name "height" ("Höhe" in German) has been introduced by Hasse in [1687].

[43] Douglas Geoffrey Northcott (1916–2005), professor at Sheffield University. See [3751].

[44] Tracy Augustus Pierce (1891–1945), professor at the University of Nebraska.

$$\overline{|\alpha|} \geq M(\alpha)^{1/n}$$

with $n = \deg \alpha$, every inequality of the form $M(\alpha) \geq 1 + c$ implies $\overline{|\alpha|} \geq 1 + 2c$ for non-rational α; hence the truth of Lehmer's conjecture would imply

$$\overline{|\alpha|} \geq \exp(\varepsilon/2n) \geq 1 + \frac{\varepsilon}{2n} \tag{5.10}$$

for all algebraic integers which are not roots of unity with an absolute constant $\varepsilon > 0$

Using the absolute logarithmic height $h(\alpha)$ (defined by (5.9)) Lehmer's conjecture may be written as

$$h(\alpha) \geq \frac{c}{\deg \alpha}, \tag{5.11}$$

with an absolute positive constant c.

The first result on Lehmer's conjecture has been obtained in 1951 by Breusch[45] [448] who showed that if the minimal polynomial f of an algebraic integer α is non-reciprocal, then $M(\alpha) \geq 1.179$. This has been later rediscovered twenty years later by Smyth [3836]. If f is reciprocal, then

$$M(\alpha) \leq (\overline{|\alpha|})^{n/2} ,$$

thus a bound $M(\alpha) \geq 1 + c$ implies

$$\overline{|\alpha|} \geq 1 + \frac{2c}{n} .$$

The results of Breusch [448] and Smyth [3836] imply that for non-reciprocal α of degree n one has

$$\overline{|\alpha|} \geq 1 + \frac{C}{n} \tag{5.12}$$

with $C = \log \theta$, where $\theta^3 - \theta - 1 = 0$. In 1980 Boyd [410] conjectured that in this case one has $C = 3 \log \theta / 2n = 0.4217\ldots$, and Dubickas [1004] showed that for large n one can take for C any number smaller that 0.3096.

The first step toward (5.10) covering also the reciprocal case has been made in 1965 by Schinzel and Zassenhaus [3625]. They showed that if an algebraic integer a is not a root of unity and has $2s$ non-real conjugates, then

$$\overline{|a|} > 1 + 4^{-s-2} ,$$

and if a is a totally real integer of degree n, $a \neq \cos(2\pi r)$ with rational r, then

$$\overline{|a|} > 1 + 4^{-2n-3}. \tag{5.13}$$

They conjectured the existence of an absolute constant $C > 0$ such that one would have (5.12) for all algebraic integers α, not roots of unity. In view of (5.10) their conjecture would follow from the truth of the Lehmer conjecture.

In the same year R.M. Robinson [3485] showed that if $\mathbf{Q}(a)/\mathbf{Q}$ is Abelian, and a is not a root of unity, then $\overline{|a|} \geq \sqrt{2}$. Much later it has been established by Callahan, Newman and Sheingorn [554] that the same holds if a lies in a normal extension of the rationals having the property that the complex conjugation lies in the centre of its Galois group.

---

[45] Robert Hermann Breusch (1907–1995), professor at the Amherst College.

In 1966 Cassels[46] [595] showed that if for some a the inequality (5.12) fails with $C = 1/10$, then a and $1/a$ are conjugated (for a refinement see Blanksby [367]). Schinzel [3616] showed that a small change in the proof permits to replace $1/10$ by $1/5$.

In 1971 Blanksby and Montgomery [368] used Fourier series to show that for integers of degree n which are not roots of unity one has

$$M(a) > 1 + \frac{1}{52n\log(6n)}$$

and

$$\overline{|a|} > 1 + \frac{\log(6n)}{30n^2},$$

and in 1978 C.L. Stewart [3931] obtained a similar bound applying an auxiliary function having many zeros.

In the same year Dobrowolski [978] gave a short proof of

$$\overline{|a|} \geq 1 + \frac{\log n}{6n^2},$$

and in the next year he established in [979] the inequality

$$M(a) \geq 1 + A\left(\frac{\log\log n}{\log n}\right)^3$$

with $A = 1/1200$ (for sufficiently large n one can take $A = 1 - e$ for any $\varepsilon > 0$). This implied

$$\overline{|a|} \geq 1 + \frac{B}{n}\left(\frac{\log\log n}{\log n}\right)^3$$

with $B = 2 + o(1)$.

Other proofs of these results were given by Rausch [3386] in 1985 and Amoroso [65] in 1998. The values of the constants appearing here were improved by Rausch [3386], D.G. Cantor[47] and Straus[48] [559], and the best known results are due to R. Louboutin [2632] ($A = 2.45 + o(1)$) and Dubickas [1002] ($B = 64/\pi^2 + o(1)$). See also Matveev [2780] and Voutier [4251].

For numerical results see Boyd [410, 414, 418], Mossinghoff [2982], Rhin and Sac-Épée [3450], Flammang, Rhin and Sac-Épée [1196], Mossinghoff, Rhin and Wu [2983].

In 1979 Mignotte [2851] gave a lower bound for $|a - 1|$ for an algebraic number with $M(a) \leq 2$, and this has been strengthened by Mignotte and Waldschmidt [2855], who gave a lower bound depending on $M(a)$. Their result implied in particular the inequality

$$\log M(a) \geq \frac{1}{500n\log n}$$

for algebraic numbers a of degree n, not roots of unity. For further results on $|a - 1|$ see Amoroso [64], Bugeaud, Mignotte and Normandin [503] and Dubickas [1003, 1005].

In 2007 the inequality (5.12) has been established by Rhin and Wu [3451] for $\deg a \leq 28$ with $C = 3\log\theta/2$, θ being the real zero of the polynomial $X^3 - X - 1$.

The range of $M(a)$ has been studied by Boyd [412, 416, 417], Flammang [1191], Dixon and Dubickas [976], Dubickas [1008, 1009], Schinzel [3622] and Drungilas and Dubickas [1000].

---

[46] John William Scott Cassels (1922–2015), professor in Cambridge.

[47] David Geoffrey Cantor (1935–2012), professor at the University of Washington and the University of California at Los Angeles.

[48] Ernst Gabor Straus (1922–1983), professor at the University of California at Los Angeles. See [1115, 1458].

Stronger lower bounds for $M(a)$ and $\overline{|a|}$ were obtained for elements of certain classes of fields. In 1973 Schinzel [3619] showed that if $|a| \neq 1$ and a generates either a CM-field or a totally real field, then

$$M(a)^{2/d} \geq \frac{1+\sqrt{5}}{2}.$$

In the totally real case a very simple proof has been found in 1993 by Höhn and Skoruppa [1888]. For improvements see Flammang [1192].

See Rhin [3449], Garza [1391] and Höhn [1887] for generalizations to arbitrary fields with the bound depending on the number of real conjugates.

In 1995 Dubickas [1003] essentially improved the bound (5.13), establishing

$$\overline{|a|} > 2 + 3.8\frac{(\log\log n)^3}{d\log^4 n}$$

for totally real integers a of degree n, not of the form $2\cos\pi r$ with rational r.

In the case when the extension $\mathbf{Q}(a)/\mathbf{Q}$ is Galois the conjecture (5.11) has been established by Amoroso and S. David [66] in 1999. Later Amoroso and Dvornicich [69] showed that if a generates an Abelian field of degree n, then for the logarithmic Weil height (see (5.9)) one has

$$h(a) \geq \frac{\log 15}{12n},$$

implying a corresponding lower bound for $M(a)$ (see also Amoroso, Zannier [71, 72], Amoroso, S. David and Zannier [68]). Quite recently Amoroso and Masser [70] showed that if $\mathbf{Q}(a)$ is a Galois extension, then for every $\varepsilon > 0$ one has

$$h(a) \geq \frac{c(\varepsilon)}{n^\varepsilon}$$

with $c(\varepsilon) > 0$.

In 2007 Borwein, Dobrowolski and Mossinghoff [406] established Lehmer's conjecture for polynomials with all coefficients congruent to unity mod n for some $n > 1$ (for strengthenings see Dubickas and Mossinghoff [1011], Samuels [3571], and Garza, Ishak, Mossinghoff, Pinner and B. Wiles [1392]).

The first bounds for $M(f)$ depending on the number $k(f)$ of non-vanishing coefficients of polynomials $f \in \mathbf{Z}[X]$ were obtained in 1983 by Dobrowolski, Lawton and Schinzel [983]. Their result has been later improved by Dobrowolski [980, 981].

In 2008 Dobrowolski [982] showed that if f is a non-cyclotomic monic, irreducible and reciprocal polynomial over $\mathbf{Z}$ with zeros $\alpha_1, \alpha_1^{-1}, \dots, \alpha_n, \alpha_n^{-1}$, and the polynomial

$$\prod_{i=1}^{n}(X - \alpha_i - \alpha_i^{-1})$$

is the characteristic polynomial of a symmetric matrix, then $M(f) > 1.043$. This bound has been improved in 2012 by McKee and Smyth [2806] and generalized by G. Taylor [4032]. See also Greaves and G. Taylor [1507].

A generalization of Lehmer's problem to several variables has been formulated by Boyd [411] (see also Boyd [412] and Smyth [3837]. Strong lower bounds for Mahler measure in this case were given in 2000 by Amoroso and S. David [67].

A survey was presented by Smyth [3840] in 2008.

5.4.5 *Infinite Extensions*

1. In the early thirties Herbrand devoted three papers [1801, 1804, 1806] to the study of extensions (finite and infinite) of algebraic number fields of infinite degree. In [1804] he described the behavior of primary ideals of a field k of infinite degree under finite extensions K/k, extended Hilbert's theory of ramification groups to infinite Galois extensions and presented a theory of the different for these extensions (a lemma in [1804] has been corrected by Moriya [2958] in 1936). In [1806] these questions were considered for infinite extensions of fields of infinite degree.

In his last paper Scholz [3680] completed the results of Krull [2311, 2312] and Herbrand [1804, 1806] concerning the factorization of prime ideals in algebraic number fields of infinite degree. He showed in particular that if such extension is Galois, then all prime ideals dividing a prime number are conjugated, a fact overlooked by Krull and Herbrand.

Ramification groups of infinite extensions of $\mathfrak{p}$-adic fields were later studied by Satake [3582] and Kawada [2128]. The case of Abelian extensions has been treated by Marshall [2743] in 1971, Gordeev [1477] in 1977 and Laubie [2484] in 1981. For $\mathbf{Z}_p$-extensions the ramification groups were studied by G. Gras [1489].

2. In 1935 Moriya [2956, 2957] presented an exposition of the theory of number fields of infinite degree and studied their completions. In the next year he developed both local and global class-field theory for infinite extensions [2959–2961]. Let $K/\mathbf{Q}$ be such extension, and let Ω be the family of all subfields k of K with $[k:\mathbf{Q}] < \infty$. For prime p and $k \in \Omega$ let $p^{\alpha(p,k)} \parallel [k:\mathbf{Q}]$, define $0 \leq a(p) \leq \infty$ as the least upper bound of $\alpha(p,k)$ for $k \in \Omega$, and let $N_\infty(K)$ be the product of primes p with $a(p) = \infty$. Finally let $I(K)$ be the group of invertible ideals in the ring of algebraic integers in K. Moriya showed that for finite Abelian extensions L/K one can construct a subgroup H of $I(K)$ such that $Gal(L/K) = I(K)/H$ if and only if the degree of L/K is prime to $N_\infty(K)$. In the local case he proceeded in a similar way. This theory has been further developed by Schilling [3606, 3607].

In 1951 Kawada [2127] showed that Moriya's class-field theory can be also interpreted in the language of ideles.

Fresh proofs of the local class-field theory for infinite extensions were given in 1937 by Moriya and Schilling [2971].

3. In 1937 Gut considered the case of infinite extensions $K/\mathbf{Q}$ in which all prime ideals have bounded degree and ramification indices. For this class, which encompasses, e.g. the composite of all cyclic extensions of a fixed prime degree, the general theory becomes much simpler. In his next papers [1557, 1558] he showed that for such fields K there exists a sequence $k_1 \subset k_2 \subset \cdots \subset K$ of fields with

$$\bigcup_{i=1}^{\infty} k_i = K ,$$

such that for $\Re s > 1$ there exists the limit

$$\lim_{i\to\infty} \zeta_{k_j}^{1/n_j}(s)$$

(with $[k_i : \mathbf{Q}] = n_i$).

4. In 1939 Mac Lane and Schilling [2691] described fields K in which every valuation in K is discrete, and for non-zero $a \in K$ only finite many valuations attain a non-trivial value. They showed in particular that there are infinite extensions of the rationals having these properties.

5.4.6 Local Fields

1. In 1933 Hasse and F.K. Schmidt [1714] determined the structure of fields complete under a discrete valuation, showing that they are finite extensions either of a p-adic field or of the field of formal power series over a finite field. Other proofs were given later by Witt [4432], Teichmüller[49] [4040, 4041] and Mac Lane [2688] who earlier pointed out [2687] certain inaccuracies in the arguments of [1714]. In [2688, 2689] Mac Lane studied automorphisms and isomorphisms of fields with valuations.

A broad presentation of the proof of the Hasse-Schmidt theorem theorem has been given in Chap. 2 of Hasse's book [1694].

2. In 1933 Mahler [2695] generalized the Thue–Siegel theorem to local fields, and in the next year later [2696] introduced continued fractions in $\mathbf{Q}_p$ and used them to study the approximations of irrational p-adic numbers. In [2697] he proved the p-adic analogue of Minkowski's theorem on linear forms and in the next year he obtained [2699] a p-adic analogue of Gelfond's [1403] theorem on the transcendence or rationality of $\log\alpha/\log\beta$ for algebraic $\alpha, \beta \neq 0, 1$, which solved the seventh problem of Hilbert. Later [2704] he described a geometrical way of presenting p-adic numbers.

A generalization of the Thue–Siegel theorem for the case of approximations of elements of a local field by algebraic numbers of fixed degree has been given by C.J. Parry [3225] in 1950. In 1958 Ridout [3462] established the analogue of the Thue–Siegel–Roth theorem for p-adic numbers. A far-reaching generalization of the Thue–Siegel–Roth theorem for $\mathfrak{p}$-adic fields was established in 1977 by Schlickewei [3628]. See Sect. 9.3 of the book [497] by Bugeaud for a survey of these questions.

A kind of continued fractions in $\mathbf{Q}_p$ has been introduced in the sixties by Schneider [3660]. Their properties were later investigated by Bundschuh [513], de Weger [920], P.-G. Becker [263] and Hirsh and Washington [1870]. Another types of continued fractions in local fields were considered by Ruban [3527], Browkin[50] [461, 462] and L.X. Wang [4286, 4287].

3. In 1933 Schur [3706] associated with each sequence $\{a_n\}$ and $p \neq 0$ a kind of derivative defined by

---

[49] Oswald Teichmüller (1913–1943), worked in Berlin. See [1719].

[50] Jerzy Browkin (1934–2015), professor in Warsaw. See [3624].

$$\Delta a_n = \frac{a_{n+1} - a_n}{p^n} ,$$

and generalized the small theorem of Fermat by showing that if p is prime, then for the kth iteration Δ^k of Δ one has

$$\Delta^k a^{p^m} \in \mathbf{Z}$$

for $a \in \mathbf{Z}$. A simpler proof has been provided in 1937 by A. Brauer [425].

In 1937 Zorn[51] [4480] considered a p-adic interpretation of Schur's theorem and applied his result in a study of several p-adic functions, including the exponential function and the logarithm.

In 1953 Carlitz [575] extended the results of Schur and Zorn to algebraic numbers (see also Carlitz [574, 576] and Overholtzer [3205]).

4. The group $U_1(L)$ of principal units in a tame Galois extension L/K with group G as a module over $\mathbf{Z}_p[G]$ has been studied by Krasner [2240, 2248] who determined its rank. He showed also that if $\mathbf{Q}_p \subset K \subset L, \zeta_p \notin L$ and L/K is a tame Galois extension with group G, then the $\mathbf{Z}_p[G]$-module $U_1(L)$ is the direct sum of a finite cyclic group and a free $\mathbf{Z}_p[G]$-module; i.e., there exist $\varepsilon_1, \dots, \varepsilon_N \in U_1(L)$ with $N = [L : \mathbf{Q}_p]$ such that every $\varepsilon \in U_1(L)$ has a unique representation in the form

$$\varepsilon = \zeta^m \prod_{i=1}^{N} \varepsilon_i^{\gamma_i} ,$$

where ζ is the generator of the group of roots of unity in $U_1(L)$, and γ_i are elements of the group-ring $\mathbf{Z}_p[G]$. One says that the 1-units of L have a normal basis.

In 1942 Gilbarg [1428] proved the converse by showing that if $\mathbf{Q}_p \subset K \subset L, \zeta_p \notin L$, and L/K is not tame, then it has no normal basis for 1-units.

Krasner showed also in [2248] that a normal basis for $U_1(L)$ exists also in the case when L/K is cyclic of degree not divisible by p.

A simpler proof has been later found by Gilbarg [1428], who also showed by the simple example $Q_2(\sqrt{2})/Q_2$ that the assumption about the degree cannot be dropped.

A characterization of all extensions L/K with a normal basis for 1-units has been given in 1965 by Borevič and Skopin [404].

The structure of the group of 1-units as a Galois module in the tamely ramified case has been considered by Iwasawa [1982] in 1955, who in [1988] determined that structure for the extensions $\mathbf{Q}_p(\zeta_{p^m})/\mathbf{Q}_p$. Later Pieper [3286] considered arbitrary tame extensions.

In the case when $\mathbf{Q}_p \subset K \subset L$, the extension L/K is cyclic and $\zeta_p \notin L$ (in such case one says that the field L is *regular*) the structure of the $Z_p[G]$-module $U_1(L)$ has been determined by Borevič [398, 402] and Arutjunjan [140]. For a class of regular Abelian p-extensions this has been done in 1978 by P. David [828].

---

[51] Max Zorn (1906–1993), professor at Yale, University of California at Los Angeles and Indiana University.

In the case of cyclic irregular p-extensions the structure has been described by Borevič [399, 401] and Borevič, Gerlovin [403]. In 1965 Borevič [400] did this for L regular in the case when the Galois group of the extension L/K can be generated by $\leq [K : \mathbf{Q}_p]$ elements. More general cyclic extensions in the irregular case were treated by Rosenbaum [3516–3518].

5. Krasner [2241, 2244] studied in 1938 the number of extensions of a given degree of a given $\mathfrak{p}$-adic field and in [2243] has been concerned with primitive extensions. He was also concerned with a generalization of ramification groups to non-Galois extensions [2242, 2249, 2250, 2267].

The number of extensions with various given properties of $\mathfrak{p}$-adic fields has been later studied by Krasner in a series of papers [2260–2265]. Later Serre [3737] considered the case of totally ramified extensions (cf. Krasner [2266]). The number of Abelian extensions has been studied by Travesa [4074].

A method of listing all extensions of $\mathbf{Q}_p$ of given degree n has been proposed in 2001 by Pauli and Roblot [3230]. It has been later applied by Jones and Roberts [2067, 2068, 2070] and Awtrey [152].

6. In 1938 Turkstra [4084] considered Diophantine approximations in p-adic fields and introduced a measure for open-closed subsets of $\mathbf{Q}_p$. Later Monna [2915] showed that one can obtain this measure as the image of Lebesgue measure in [0, 1] under a one-to-one mapping (see also Monna [2916, 2917]).

5.4.7 Algebraic Numbers and Matrices

1. In 1933 Latimer and MacDuffee [2482] showed that if $f(X)$ is the minimal polynomial of a generator of the field K of degree n, then the class-number of K equals the number of classes of $n \times n$ matrices A with integral rational coefficients satisfying $f(A) = 0$ under conjugation by matrices from $GL_n(\mathbf{Z})$ (the case of quadratic fields has been earlier treated by MacDuffee [2676]).

A simpler proof has been given by Taussky [4022] in 1949, who later [4023, 4025, 4026] developed properties of the correspondence between classes of ideals and classes of matrices introduced in [2482] (see also Bender [284] and Buccino [486]). Another proof has been given by Wallace [4280] in 1984. For a generalization to general commutative rings see Estes and Guralnick [1127].

In [4027–4030] Taussky studied matrices transforming an integral basis into a basis of an ideal (*ideal matrices*). Such matrices were earlier considered by Poincaré [3323] and MacDuffee [2677]. Generalizations to relative extensions were made by Bhandari and Nanda [335] and to arbitrary Dedekind domains by G.B. Wagner [4261].

In 1942 MacDuffee [2679] modified Poincaré's approach, which permitted him to interpret the multiplication of ideals in the language of matrices.

If $\omega_1, \ldots, \omega_n$ is a basis of a field K and $\beta \in K$, then with some matrix $A = [\alpha_{ij}]$ with rational entries one has

$$\beta\omega_i = \sum_{j=1}^{n} \alpha_{ij}\omega_j \ .$$

The matrices A obtained in this way were characterized in 1940 by Taussky and J. Todd[52] [4031].

[52] John Todd (1911–2007), husband of Olga Taussky-Todd, professor at CalTech. See [215].

In 2002 Behn and van der Merwe [269] gave an effective version of the Latimer–MacDuffee correspondence in case $n = 2$.

In 1937 Albert [49] described totally real fields as those generated by roots of the characteristic polynomial of products of diagonal and symmetric matrices.

See also the later papers of Gorškov [1478], Faddeev [1155], Krakowski [2238], Estes [1126], Estes and Guralnick [1128], Bass, Estes and Guralnick [211] and Fitzgerald [1190], where relations between algebraic numbers and symmetric matrices were studied.

5.4.8 *Varia*

1. In 1935 Skolem [3810] applied $\mathfrak{p}$-adic power series to exponential equations. One of his tools was a p-adic version of the Vorbereitungssatz of Weierstrass[53] (fifth paper in [4347]; cf. Cartan[54] [584]).

The method used by Skolem has been later applied by Chabauty [638] to show that the following result of Mahler [2695] can be obtained without using Diophantine approximations:

If α, β, γ are different integers of K and P is a finite set of prime ideals of K, then there are only finitely many co-prime pairs x, y of rational integers such that all prime divisors of the product $(x - \alpha y)(x - \beta y)(x - \gamma y)$ belong to P.

Certain classes of cubic and quartic equations in quadratic fields were studied in 1937 by Fogels[55] [1202]. He showed in particular that the equation $X^4 + Y^4 = 8Z^4$ has non-trivial solutions in quadratic integers only in $\mathbf{Q}(\sqrt{d})$ with $d = \pm 2$, and for prime $p \equiv 3 \bmod 4$ the equation $X^4 + Y^4 = 2p^2Z^4$ has quadratic integral solutions only in $\mathbf{Q}(\sqrt{d})$ with $d = \pm p$.

2. Various elementary congruences in $\mathbf{Z}$, due to M. Bauer [218], Lubelski [2658], Ore [3171] and Rados [3371], were generalized to algebraic number fields by M. Bauer [248], Lubelski [2659], Vandiver [4141] and Zányi [4464, 4465].

A way of determining the greatest common divisor of two algebraic integers in the case when it exists has been proposed in 1935 by MacDuffee and Jenkins [2680]. In the same year Venkatachaliengar [4206] gave an algorithm leading to the determination of the structure of the class-group.

Let L/K be a cyclic extension, and let s be the generator of the Galois group. In 1936 Carlitz [571] gave a necessary and sufficient condition for the solvability of the equations $s(x) = ax$ and $s(x) = ax + b$ in non-zero integers a, b of L.

Vandiver [4173] presented a way to construct the splitting field of a polynomial and in [4174] gave a constructive method for ordering the set of all real algebraic numbers.

In 1937 Bilharz [354] reformulated Artin's conjecture on primitive roots of rational primes asserting that if an integer $a \neq -1$ which is not a square, then a is a

---

[53] Karl Weierstrass (1815–1897), professor in Berlin. See [2411, 3325].

[54] Henri Cartan (1904–2008), professor in Paris. See [3740].

[55] Ernests Fogels (1910–1985), worked in Riga. See [2333].

primitive root for infinitely many primes p, and the set of such primes has a positive density. He proved that it is equivalent with the following statement: *there are infinitely many primes p which do not split in the field* $\mathbf{Q}(\sqrt[q]{1}, \sqrt[q]{a})$ *for any prime* $q \neq p$. Using the analogue of this in the case of fields of algebraic functions over finite fields he was able to establish the analogue of Artin's conjecture in this case.

This reformulation has been used by Hooley [1892], who showed in 1967 that Artin's conjecture is a consequence of the Generalized Riemann Hypothesis.

For generalizations of Artin's conjecture to algebraic number fields see Weinberger [4360], Lenstra [2543], Egami [1053] and Hinz [1859, 1863].

3. In 1931–1932 two volumes of the book [1619] by Hancock appeared[56] presenting a broad introduction to the theory of algebraic number fields.

[56] See Ore [3186] for a review of it.

Chapter 6
The Forties

6.1 Analytic Methods

6.1.1 *General Results*

1. In 1947 R. Brauer [428] found a proof of Artin's conjecture on the divisibility of Dedekind zeta-functions for Galois extensions,[1] showing first that in Artin's theorem about linear combinations of characters induced by cyclic subgroups the rational coefficients may be taken to be nonnegative. As a corollary he obtained for normal extensions L/K a representation of $(\zeta_L(s)/\zeta_K(s))^n$ as a product of Abelian L-functions. He pointed out that from the truth of Artin's conjecture on the integrality of Artin L-functions this corollary would hold also for non-normal extensions.

In 1975 Uchida [4096] and van der Waall [4140] showed that if L/K is normal with solvable Galois group, and $K \subset M \subset L$, then the ratio $\zeta_M(s)/\zeta_L(s)$ is entire.

In 1989 Foote and V.K. Murty [1209] found another proof of the Aramata–Brauer theorem and showed also that if L/K is Galois, then for any corresponding Artin L-function $L(s, \chi)$ the only pole of the ratio $\zeta_K(s)/L(s, \chi)$ occurs at $s = 1$ (cf. Rhoades [3452]).

In another paper of 1947 R. Brauer [429] showed that in the above theorem one may replace the rational coefficients by integral numbers, not necessarily nonnegative, and this implied in particular that Artin L-functions are meromorphic on the whole plane.

In 1973 R. Brauer [432] made a generalization to the case of several fields, proving in particular the following result:

If L_1, L_2 are normal fields, $k = L_1 \cap L_2$ and $K = L_1 L_2$, then the ratio

$$\frac{\zeta_k(s)\zeta_K(s)}{\zeta_{L_1}(s)\zeta_{L_2}(s)}$$

is entire.

---

[1] In a footnote on p. 243 Brauer acknowledged the priority of Aramata, adding "I publish my proof …since it seems to be somewhat simpler".

W. Narkiewicz, *The Story of Algebraic Numbers in the First Half of the 20th Century*, Springer Monographs in Mathematics,
https://doi.org/10.1007/978-3-030-03754-3_6

This result has been later extended to other classes of fields L_1, L_2 by Sato [3583–3587] and van der Waall and Sato [4132].

2. The second theorem in Brauer's paper [428] generalized Siegel's results (5.5), (5.6) to arbitrary fields, showing that for all fields K of a fixed degree one has

$$\log\left(h(K)R(K)\right) = \left(\frac{1}{2} + o(1)\right)\log(|d(K)|). \tag{6.1}$$

Simpler proofs have been given later by Pintz [3296] and Louboutin [2647, 2649].

Three years later R. Brauer [430] that (6.1) holds also for every sequence of fields K_j, normal over the rationals, provided it satisfies

$$\lim_{j\to\infty} \frac{n(K_j)}{\log(|d(K_j)|)} = 0, \tag{6.2}$$

$n(K)$ being the degree of K. This result is usually called the *Brauer–Siegel theorem*. To prove it R. Brauer established for all normal fields K the inequality

$$h(K)\kappa(K) > \frac{1}{c(\varepsilon)^{n(K)}|d(K)|^{\varepsilon}}$$

valid for all $\varepsilon > 0$ with some ineffective $c(\varepsilon)$. If K satisfies (6.2), then this gives

$$h(K)\kappa(K) \gg_{\varepsilon} \frac{1}{|d(K)|^{\varepsilon}}. \tag{6.3}$$

For non-normal fields $K \neq \mathbf{Q}$ he showed that if $n^*(K)$ is the degree of the minimal normal field containing K, then one has

$$\frac{1}{n^*(K)}\left|\frac{\log(h(K)R(K))}{\log(|d(K)|)} - \frac{1}{2}\right| = O(1)\ ,$$

and the only limit point of the left-hand side is 0.

In 1956 Ankeny, R. Brauer and S. Chowla [92] used (6.1) to show that for every $\varepsilon > 0$ there are infinitely many fields K of a given signature with

$$h(K) > |d(K)|^{1/2-\varepsilon}\ ,$$

thus Landau's bound (3.26) is close to the best possible. In 1977 Montgomery and Weinberger [2921] improved upon this in the case of real quadratic fields by showing

$$h(d) \gg \frac{\sqrt{d}\log\log d}{\log d}$$

for infinitely many d.

Brauer's result (6.1) has been made more precise by A.I. Vinogradov [4217, 4218] in 1962, who proved the formula

$$\log(h(K)R(K)) = \log\left(\sqrt{|d(K)|}\right) + \log(1-\beta) + O(\log\log(|d(K)|), \tag{6.4}$$

valid for fields K of fixed degree. Here β denoted the largest positive zero of $\zeta_K(s)$ if it exists. He showed also in [4219] that for almost all fields of fixed degree the term $\log(1-\beta)$ may be omitted.

A simpler proof of (6.4) has been given in 1966 by Levin and Tuljaganova [2582].

In 1970 Lavrik [2489] obtained the simply looking bound

$$h(K)R(K) < w\sqrt{D}\log^{n-1} D$$

where $D = |d(K)| > 5$, n is the degree of K, and w is the number of roots of unity in K. For improvements see Lavrik and Edgorov [2491] and Louboutin [2642–2644].

The Brauer–Siegel theorem has been used in 1971 by Uchida [4090, 4091, 4094] to show that there are only finitely many imaginary Abelian fields having a given class-number. In the first paper he proved that if K is normal and imaginary and its maximal real subfield K^+ is also normal, then the ratio $h^-(K) = h(K)/h(K^+)$ becomes arbitrary large if $[K : \mathbf{Q}]/\log d(K)$ is sufficiently small, and gave an elementary argument to show that for Abelian K the last condition is satisfied for almost all fields. In the second paper he applied the estimate of $L(1, \chi)$, due to Tatuzawa [4013], to show that there is an effective upper bound for the conductors of imaginary Abelian fields with given class-number except for quadratic and biquadratic fields. As a corollary he deduced the bound 2400 for the maximal prime p with $h_p^- = 1$. This has been strengthened in [4092] where Uchida proved that for primes $p \geq 23$ one has $h_p^- > 1$. In [4093] he got bounds for conductors of imaginary cyclic fields of degree 4 and 6 having $h = 1$, and determined imaginary Abelian fields K with $Gal(K/\mathbf{Q}) = C_2^N$ and $h(K) = 1$ with one possible exception. Imaginary Abelian fields of two-power degree and $h = 1$ were determined by Uchida [4098] in 1986.

All cyclotomic fields with class-number one were found in 1976 by Masley and Montgomery [2763], Masley [2760] showed that $h = 2$ occurs for these fields only in two cases ($\mathbf{Q}(\zeta_m)$ with $m = 39, 56$), and in [2761] he found all cyclotomic fields with $h \leq 10$ (there 44 such fields).

In 1974 Brown and C.J.Parry [470] determined all 47 imaginary biquadratic fields with $h = 1$, and in 1980 Setzer [3742] found all seven imaginary cyclic quartic fields with $h = 1$. All imaginary cyclic quartic fields with $h = 2$ were determined by K. Hardy, Hudson,[2] Richman and K.S. Williams [1631] in 1989. The maximal conductor equals to 119.

In 1992 Yamamura [4452, 4453] determined all imaginary Abelian fields with class-number one. There are 172 such fields. Their maximal degree, 24, is attained by five fields: $\mathbf{Q}(\zeta_m)$ with $m = 35, 45, 81$, $\mathbf{Q}(\zeta_{21}, \sqrt{5})$ and $\mathbf{Q}(\zeta_{15}.\cos(2\pi/7)$. The field with maximal conductor is the biquadratic field $\mathbf{Q}(\sqrt{-67}, \sqrt{-163})$ of conductor 10 921.

In 1974 Stark [3896] obtained an effective version of (6.3). He considered the class $\mathcal{N}$ of fields K for which there exists a sequence $\mathbf{Q} = k_0 \subset k_1 \subset \cdots \subset k_m = K$ such that each extension k_{i+1}/k_i is normal, and proved that for $K \in \mathcal{N}$ one has

$$h(K)\kappa(K) > \frac{c}{n|d(K)|^{1/n}}$$

with an effective $c > 0$. For other fields this inequality holds with the factor n in the denominator replaced by $nn!$. If K does not contain a quadratic subfield, then the denominator can be replaced by $\log(|d(K)|)$.

In the case when K is a normal CM-field and $f = |d(K)|/d(K^+)^2$, then this implies for $0 < \varepsilon < 1/2$ the inequality

$$h(K) > \frac{c(\varepsilon)^n}{n}\frac{(|d(K)|f)^{1/2-1/n}}{|d(K)|^\varepsilon},$$

[2]Richard Howard Hudson (1945–2016), professor at the University of South Carolina.

with effective $c(\varepsilon)$. For non-normal CM-fields the factor $n!$ should be inserted in the denominator.

This implies that there are only finitely many CM-fields of fixed degree $n \geq 6$ with a given class-number, and Stark conjectured that the assumption about the degree can be waived. This conjecture has been established in an effective form by Odlyzko [3137] in 1975 for CM-fields K of degrees ≥ 6 with $K^+ \subset \mathcal{N}$ as a corollary of his estimates of discriminants. In 1979 Hoffstein [1877] gave lower bounds for the class-number of CM-fields with degree ≥ 20 and showed in particular that the degree of a normal CM-field with class-number 1 does not exceed 436. This bound has been reduced to 266 by Bessassi [332] in 2003 and to 216 by G.-N. Lee and Kwon [2499] in 2006. Under GRH Bessassi [332] obtained the bound 164, and G.-N. Lee and Kwon [2499] replaced it by 96.

In 2001 V.K. Murty [3005] proved Stark's conjecture for fields of degree ≥ 6 having a solvable normal closure.

All non-Abelian normal CM-fields with $h = 1$ and degree <48 are known, as well as all such dihedral fields of arbitrary degree. This has been done by the work of Louboutin and his collaborators Lefeuvre, Lemmermeyer, Okazaki, Olivier and Y.-H. Park [2505, 2506, 2534, 2635, 2637, 2638, 2640, 2651–2653, 2655, 3222]. In 2003 Chang and Kwon [652] showed that there is only one non-Abelian normal CM-field of degree 48 having a normal CM-subfield of degree 16. In 2007 S.-M. Park and Kwon [3221] showed that the not yet known normal CM-fields with $h = 1$ have to be of degree 64 or 96.

In the case of non-Abelian non-normal CM-fields with $h = 1$ only partial results are known, obtained by Hoffstein and Jochnowitz [1878, 1879], Chang and Kwon [651], Louboutin, G.-N. Lee and Kwon [2500], Kwon, Louboutin and S.-M. Park [2391] and Ahn, Boutteaux, Kwon and Louboutin [26]

In 1994 Louboutin [2634] gave a bound for the discriminants of non-normal quartic CM-fields with $h = 1$, and a list of all such fields was given by Louboutin and Okazaki [2651].

Assuming GRH and the Artin conjecture for the ratio $\zeta_K(s)/\zeta(s)$ Duke [1021] proved in 2003 the bound

$$h(K) \gg_n \sqrt{d} \left(\frac{\log\log |d(K)|}{\log(|d(K)|)} \right)^{n-1}$$

for infinitely many totally real fields of fixed degree n and Galois group S_n, and in [1022] obtained this unconditionally in the case of Abelian cubic fields. For non-Abelian cubics this has been shown in 2006 by Daileda [810]. Conditional results in some other cases were obtained later by Cho and H.H. Kim [692, 693] and Cho [691].

3. An elementary proof of the Prime Ideal Theorem using the following generalization of Selberg's[3] identity in the rational case [3725] has been given in 1949 by Shapiro[4] [3750]:

$$\sum_{N(\mathfrak{p}) \leq x} \log^2 N(\mathfrak{p}) + \sum_{N(\mathfrak{p}\mathfrak{q}) \leq x} \log N(\mathfrak{p}) \log N(\mathfrak{q}) = 2x \log x + O(x) ,$$

$\mathfrak{p}$, $\mathfrak{q}$ running over prime ideals.

Other proofs of this identity were found by Ayoub[5] [156] and Rieger [3467, 3468, 3471].

In 1954 Forman and Shapiro [1218] presented an elementary method of establishing asymptotics for generalized primes, obtaining in a particular case a proof of the Prime Ideal Theorem. For further development of this approach see Amitsur [63]. A similar method has been applied by Bredikhin [438–440].

[3] Atle Selberg (1917–2007), professor in Princeton. See [163].

[4] Harold Nathaniel Shapiro (1922–2013), professor at the New York University.

[5] Raymond Ayoub (1923–2013), professor at the Penn State University.

A proof based on Wiener's approach has been found by Gál [1379] in 1963. An elementary proof of the Prime Ideal Theorem leading to the error term $O(x\log^{-c}x)$ with some $c>0$ has been given in 1967 by Eda and Nakagoshi [1042], who used a method applied by Kuhn [2344] for the elementary proof of the Prime Number Theorem.

A proof based on Daboussi's [808] approach to the Prime Number Theorem was provided by Touibi and Zargouni [4072], and in the next year Touibi [4071] gave a proof applying the method used in the rational case by Hildebrand [1848] and based on the large sieve. In 1993 Hinz [1866] presented another proof based on the Selberg sieve.

4. A new way of establishing functional equations for a class of L- and zeta-functions, encompassing Dedekind zeta-function and L-functions of Hecke characters, has been showed in 1950 by Tate in his thesis[6] [4007]. He considered *quasi-characters* χ of the group I_K of ideles of K defined as continuous homomorphisms of I_K into the multiplicative group of the complex field. The quasi-characters of the idele class-group $C(K)=I_K/P_K$ are related to characters of $C(K)$ in the following way. Define the *volume* $V(x)$ of an idele $x=\langle x_v\rangle$ by

$$V(x)=\prod_v v(x_v)\,,$$

where the valuations v are normalized so that for $a\in K^*$ they satisfy

$$v(a)=N(\mathfrak{p})^{-\nu_{\mathfrak{p}}(a)}$$

if v corresponds to the prime ideal $\mathfrak{p}$ of K and $\nu_{\mathfrak{p}}(a)$ is the corresponding exponent, $v(a)=|F(a)|$ if v corresponds to a real embedding F of K and $v(a)=|F(a)^2$ if v corresponds to a complex embedding F of K. If J_K denotes the kernel of V, then $P_K\subset J_K$, and hence $V(x)$ depends only of the class of x in I_K/P_K.

If now q is a quasi-character of I_K/P_K, then with a complex number s one has

$$q(xP_K)=\chi(xP_K)V(x)^s, \tag{6.5}$$

where χ is a character of $C(K)$. The number $\exp(q)=\Re s$ is called the *exponent of* q.

Every character χ of C_K can be written in the form

$$\chi(x)=\prod_v \chi_v(x_v)$$

for $x=\langle x_v\rangle$, χ_v being a character of K_v^*. If S is a finite set of valuations containing all infinite v as well as those for which χ_v is non-trivial on the set of units of K_v, then for $v\notin S$ the value $\chi(x_v)$ depends only on $v(x_v)$. For such v one can consider χ_v as a character of the group of fractional ideals of K_v, and this permits to define a multiplicative function $X(I)$ by putting for prime ideals $\mathfrak{p}$

[6] Tate's thesis has been published only after 17 years.

$$X(\mathfrak{p}) = \chi_v(\mathfrak{p}_v)$$

if $v \notin S$ and $\mathfrak{p}$ induces the ideal $\mathfrak{p}_v$ in K_v, and $X(\mathfrak{p}) = 0$ otherwise. The set of all functions $X(I)$ coincides with the set of Hecke characters of K.

For complex-valued functions $f(x)$ on I_K Tate considered its Mellin transform, defined for $q = \chi \cdot V^s$ by

$$\hat{f}(q) = \int_{I_K} f(x)q(x)dm_I(x) ,$$

m_I being a suitably normalized Haar measure on I_K. If this integral is well-defined for all q with $\exp(q) > 1$, then for fixed χ it defines a function $Z(f, s, \chi)$ of the complex variable s satisfying $\Re s > 1$. Under certain assumptions on f this function is regular in $\Re s > 1$, and Tate proved that it can be continued either to a meromorphic function with two poles (at $s = 0$ and $s = 1$) if $\chi = 1$, or to an entire function.

By choosing appropriately the function f one obtains in this way the continuation and functional equation for various zeta-functions and L-functions, including $\zeta_K(s)$.

For expositions of Tate's thesis see the books [1829, 2457, 3065, 3378]. A similar approach has been later used by Ono [3155] in the case of another class of zeta-functions.

6.1.2 Additive Problems

6.1.2.1 Sums of Squares

1. In 1940 Niven [3113] showed that an integer $a+bi \in Q[i]$ is a sum of two squares of integers if and only if b and at least one of the numbers $a, b/2$ are even (simpler proofs were given later by Leahey [2492], Mordell [2937], K.S. Williams [4407, 4411] and Joly [2062]). It has been observed later by Siegel [3789] that Niven's assertion about sums of three squares in $\mathbf{Q}(\sqrt{-m})$ for $0 < m \equiv 3 \bmod 4$ is incorrect. Niven utilized results on representations of quadratic forms as sums of squares of linear forms obtained by Mordell [2930] (for further development of this subject see Mordell [2931, 2933], Braun[7] [434], J. Hardy [1630]).

In the same paper [3789] Siegel proved that every totally positive integer of a field K is a sum of squares if and only if either K is not totally real and $2 \nmid d(K)$, or $K = \mathbf{Q}, \mathbf{Q}\left(\sqrt{5}\right)$.

Denote for shortness the field $\mathbf{Q}(\sqrt{m})$ by K_m, and let $r_{m,n}(a)$ be the number of representations of $a \in K_m$ as a sum of n squares of integers in K_m.

In 1951 Pall [3216] gave formulas for $r_{m,2}(a)$ for $m = -1, 2, 5$ (for the case $m = -1$ another proof has been found by K.S. Williams [4407, 4409]; cf. [4410]). Pall's method has been later applied by J. Hardy [1629] for $m = 3, 7, 13$ and 37. In the case $m < 0$ J. Hardy partitioned

[7] Hel Braun (1914–1986), professor in Göttingen and Hamburg. See [3938].

representations into a finite number of classes and gave formulas for the number of these classes for $m = -2$ and $m = -p$ with prime p.

Integers which are sums of two integral squares in K_m in the case $h(K_{-|m|}) = 1$ were described in 1953 by Nagell [3026], who later dealt also with $m = \pm 5, \pm 13$ [3030] and $m = \pm 37$ [3032]. A generalization of results in [3026] to representations by other binary quadratic forms has been obtained in the dissertation of Christofferson [709]. Nagell showed also in [3031] that in an arbitrary algebraic number field K the number of representations of its integers a sum of two squares is finite if and only if either K is totally real or $K = \mathbf{Q}(i)$ (this paper contains also several corrections to [3026]).

In 1959 Cohn [746] made a numerical study of sums of squares in a family of quadratic fields and conjectured that if $m \not\equiv 1 \bmod 8$ and $a \in K_m$ is a sum of squares, then it is a sum of at most 5 squares, and if moreover its norm is sufficiently large, then already three squares are sufficient. Next year [747] he showed that all integers of K_2 are sums of two integral squares, obtained an analogue of Jacobi's formula for $r_{2,4}(a)$, provided also a formula for $r_{3,4}(a)$, involving a certain error function $L(a)$, studied later in [750] and proved that all integers of K_3 of sufficiently large norm are sums of 4 squares. He conjectured that in both fields three squares should be sufficient. In [749] he related the class-number of imaginary quadratic extensions of K_m with $m = 2, 3, 5$ to the values of $r_{m,4}$. Elementary proofs of formulas for $r_{2,4}$ and $r_{5,4}$ were given later by Cohn and Pall [754]. New proofs of the four squares theorem for K_2 and K_5 were found by Deutsch [918, 919].

In 1974 Peters [3263] showed that in a not totally real field every integer which is a sum of integral squares is a sum of five squares. In totally real fields the situation is different, as Scharlau [3595] showed that there are such fields with arbitrary large Waring constant for exponent 2.

In 1983 Estes and Hsia [1129] determined imaginary quadratic fields in which every integer is a sum of three integral squares (see also Ji, Y. Wang and Xu [2054]), and Ji [2053] did this for $\mathbf{Q}(\zeta_m)$ with odd m.

A positive-definite quadratic form is said to be universal in a ring $\mathbf{Z}_K$ if it represents all totally positive integers in K; thus, Maass's result in [2673] shows that $X^2 + Y^2 + Z^2$ is universal in K_5. It has been established in 1996 by Chan, M.-H. Kim and Raghavan[8] [646] that a universal ternary quadratic form exists in K_m with $m > 0$ only for $m = 2, 3$ and 5. It has been shown in 1997 by Earnest and Khosravani [1038] that if K is totally real of odd degree, then there can be only finitely many inequivalent positive quaternary universal forms in K, and their discriminants do not exceed 1073/4. In 2002 Iwabuchi [1978] showed that this bound is best possible. All 58 universal quaternary quadratic forms in K_5 were listed in 2008 by Y.M. Lee [2502], and H. Sasaki [3578] showed in 2009 that there are only two such forms in K_{13}.

Sums of two squares in quadratic extensions of cyclotomic fields were studied by Asimi and Lbekkouri [148] in 2005. The Waring constants for squares in orders of real quadratic fields were determined by Peters [3262] in 1973, and in 2007 Ji and Wei [2055] showed that this constant for cyclotomic fields $\mathbf{Q}(\zeta_n)$ ($n \not\equiv 2 \bmod 4$) equals to 3 except when $2 \nmid n \cdot \mathrm{ord}_n(2)$, in which case it equals to 4.

In 2011 B.M. Kim and P.-S. Park [2147] showed that if an integer of norm exceeding $160 + 4m + m^2$ in a real quadratic field K_m is a sum of four integral squares, then these squares can be chosen to be distinct. This has been made more explicit for $m = 5$ by P.-S. Park [3220] and for $m = 2, 3$ and 6 by J.Y. Kim and Y.M. Lee [2148].

A description of integers in $K_{\pm p}$ (with prime p) which are sums of two integral squares has been given by Wei [4345], who later [4346] did this for fields K_{-2p}.

Integers of quadratic fields which are sums of two squares are of importance in K-theory. See Qin [3355–3357].

[8] Srinivasacharya Raghavan (1934–2014), professor at Tata Institute of Fundamental Research.

6.1.2.2 Waring Problem

1. A description of the set of integers of a quadratic field which can be written as sums of kth powers of elements of K has been given in 1941 by Niven [3114] in the case when either k is odd or the field is imaginary. He also determined quadratic fields in which every integer is a sum of kth powers. In [3115] he showed that for $K = \mathbf{Q}(i)$ the number $a + bi$ is a sum of fourth powers if and only if $24 \mid b$, and every such element is a sum of 18 fourth powers.

Much later Revoy [3448] replaced 18 by 12 and showed also that every integer in $\mathbf{Q}(\sqrt{-3})$ is also a sum of 12 fourth powers.

2. At the end of [3775], where the results of [3774] were exposed, Siegel promised to study the analogue of Waring problem in algebraic number fields. He fulfilled his promise after 22 years[9] in [3784].

For an algebraic number field K denote by $J_K(k)$ the subring of $\mathbf{Z}_K$, generated by kth powers, and let $G(k, n)$ be the *large Waring constant*, i.e. the smallest integer s such that if $\deg K = n$, then all totally positive elements of $J_K(k)$ with sufficiently large norm are sums of s such powers. The main result of [3784] gives the following asymptotic formula for the number $A_{k,s}(\alpha)$ of representations of a totally positive integer α of a field K of degree n as a sum of $s > (2^{k-1} + n)kn$ kth powers of totally positive integers $x_j \in K$, satisfying

$$| \varphi_j(x_i) |^k \leq | \varphi_j(\alpha) | ,$$

φ_j running over the non-conjugated non-real embeddings of K into the complex field:

$$A_{k,s}(\alpha) = \sigma_0 \sigma(\alpha) \, |d(K)|^{(1-s)/2} \, N(\alpha)^{s/k-1} + o\left(N(\alpha)^{s/k-1}\right) ,$$

where $\sigma_0 > 0$ depends only on n, r_1, s and k, and $\sigma(\alpha)$ is the *singular series* which for $\alpha \in J_K(k)$ satisfies $0 < c_1 \leq s(\alpha) \leq c_2$ for some c_1, c_2 independent on α, and vanishes for α outside $J_K(k)$.

This formula implies the bound

$$G(k, n) \leq 1 + kn \left(2^{k-1} + n\right) . \tag{6.6}$$

Siegel presented the proof only in the case of a totally real field K and wrote on p. 124: "*the proof in the general case proceeds on the same lines, the formulae being somewhat more cumbersome ...*". He conjectured that $G(k, n)$ can be made to be independent on the degree n of K.

In 1958 Tatuzawa [4016] obtained an improvement of (6.6) proving

[9] In 1936 Rademacher [3366] announced some results on the Waring problem in totally real fields, but no details were published.

$$G(k,n) \le 8(k+n)kn \ ,$$

and in 1961 Körner [2218] replaced the factor $k+n$ in Tatuzawa's result essentially by $3\log k + c\log n$ for $c > 6$ and large k (his bound is actually more precise).

Siegel's conjecture has been established by Birch. He considered first the case of $k = p$, a prime, obtaining in [357]

$$G(p) := \max_n G(p,n) \le 1 + 2^p \ ,$$

and proving in 1964 the bound $G(k) \le \max\{2^k+1, k^{16k^2}\}$ for arbitrary k [358]. This was improved by Ramanujam[10] [3380] to $G(k) \le \max\{2^k+1, 8k^5\}$. Both these results depended on the progress in the evaluation of the Waring constant in local fields.

Tatuzawa's bound has been improved for large k by Eda [1040] 1971, who showed

$$G(k,n) \le 2kn(\log k + 3\log\log k + 2\log n + \log\log n)$$

for $k > c(K)$, and later [1041] replaced here the term $2kn\log\log n$ by $28kn + 1$. At the same time Tatuzawa [4018, 4019] showed $G(k,n) \le 2n(2k + G(k,1)) + 1$. Using Wooley's [4445] bound for $G(k,1)$ this implies

$$G(k,n) \le 2kn(\log k + \log\log k + O(1)) \ .$$

The best known bound is due to Davidson [829], who in 1999 obtained

$$G(k,n) \le kn(\log k + \log\log k + 6) + O\left(\frac{kn\log\log k}{\log k}\right) \ .$$

A characterization of fields with $J_K(k) = \mathbf{Z}_K$ for k prime has been given in 1962 by Bateman and Stemmler[11] [214], and in the case of composite k this has been done by Bhaskaran [344]. A simpler proof has been given in 1970 by Körner [2222].

An elementary approach to the Waring problem in algebraic number fields has been presented in 1962–1963 by Rieger [3472, 3473].

In 1991 Y. Wang published a book in which the generalization of Waring's problem to number fields has been exposed [4294]. Surveys on Waring's problem and its generalizations have been given by Ellison [1083] in 1971 and Vaughan and Wooley [4200] in 2000. The problem of Waring in arbitrary fields and/or rings has been studied by Joly [2059–2061], Ellison [1083–1085], E. Becker [262] and Vaserstein [4194, 4195].

6.2 The Class-Number

6.2.1 Class-Number of Quadratic Fields

1. In 1948 Gelfond[12] and Linnik [1408] showed that the truth of the following assertion in the case $n = 3$ would imply an effective bound for discriminants of imaginary quadratic fields with given class-number:

---

[10] Chidambaran Padmanabhan Ramanujam (1938–1973), professor at Tata Institute of Fundamental Research. See [3379].

[11] Rosemarie S. Stemmler (1930–2011), professor at the Purdue University.

[12] Aleksandr Osipovič Gelfond (1906–1968), professor in Moscow. See [2581].

If for every $\varepsilon > 0$ *there exists an effective constant* $C(\varepsilon)$ *such that if* α_i $(i = 1, 2, \ldots, n)$ *are non-zero algebraic numbers whose logarithms are linearly independent over* $\mathbf{Q}$, x_i $(i = 1, 2, \ldots, n)$ *are rational integers, and* $x = \max |x_i|$, *then one has*

$$\left|\sum_{i=1}^{n} x_i \log \alpha_i \geq C(\varepsilon) \exp(-\varepsilon x)\right|. \tag{6.7}$$

At that time such bound was known only in the case of $n = 2$, proved by Gelfond [1404] in 1939.

In 1950 Ankeny and S. Chowla [94] showed that if $p \equiv 3 \bmod 4$ is a prime, then a strong upper bound for the number of primes $\leq x$ in residue classes mod p implies an explicit lower bound for $h(-p)$.

In 1952 Heegner[13] [1747] published a proof of the assertion, going back to Gauss, that the only imaginary quadratic number fields with class-number 1 are the fields $\mathbf{Q}\left(\sqrt{-d}\right)$ with $d = 3, 4, 7, 8, 11, 19, 43, 67$ and 163. His proof utilized certain assertions of H. Weber concerning modular forms, and since it was not clear whether all of them have valid proofs, Heegner's result was regarded as incomplete.

In 1966 Stark [3889] improved upon D.H. Lehmer's [2512] bound, showing that the hypothetic tenth discriminant satisfies $|d| > \exp(2.2 \cdot 10^7)$, and soon was able to show that this hypothetic discriminant does not exist [3890, 3891]. In [3893] he discussed connections of his proof with Heegner's method.

At the same time A. Baker [177] established for all n a stronger version of the bound (6.7), and noted that "*...it follows from work of Gelfond and Linnik ...that the theorem suffices, at least in principle, to settle the celebrated conjecture, dating back to Gauss ...*". The details have been worked out by Bundschuh and Hock [514] in 1969 and led to the bound $|d| \leq \exp(1.6 \cdot 10^5)$.

A proof based on Gelfond's result in the case $n = 2$ of (6.7) [1404] has been sketched in 1969 by Čudakov [797], and details were given by Feldman[14] and Čudakov [1173] in 1972, leading to the bound $|d| < 10^{40}$. In 1987 Cherubini and Wallisser [673] used a strengthening of Gelfond's result due to Mignotte and Waldschmidt [2853] to get the bound $|d| \leq 10^{34}$ and later improvements (Mignotte, Waldschmidt [2854]) led to $d| < 3 \cdot 10^{16}$.

In 1968 Deuring [916], Stark [3892] and Birch [359] analysed Heegner's proof and showed that it is essentially correct. This has been also done by C. Meyer [2847] in 1970. For that reason the result is often called the *Heegner–Stark–Baker theorem*.

A discussion of Weber's results used by Heegner can be found in Schertz [3603] (see also Schertz [3604]), and a detailed exposition of Heegner's method has been given in 2011 by Birch [359].

Another proof of the Heegner–Stark–Baker theorem using modular forms has been given by Siegel [3793] in 1968 (a simplification has been provided in 1999 by I. Chen [668]). A short proof based on Kronecker's limit formula has been given in 1970 by S. Chowla [706].

Imaginary quadratic fields $\mathbf{Q}\left(\sqrt{-d}\right)$ with $d \leq 6000$ and $h = 2$ were listed in 1952 by Iseki[15] [1966], who in [1967] showed that there can be only one such field with $d > 90\,000$. In the case when $d \not\equiv 3 \bmod 8$ A. Baker [178] proved in 1969 that $h(-d) = 2$ implies $d \leq 10^{500}$, and two years later Ellison, Pesek, Stall and Lunnon [1087] as well as by Kenku [2138] reduced this bound to $d \leq 58$.

Methods for effective determination of all imaginary quadratic fields with $h = 2$ were given by A. Baker [179] and Stark [3894] with the use of Baker's method of lower bounds for linear

---

[13] Kurt Heegner (1893–1965). See [3228].

[14] Naum Ilič Feldman (1918–1994), professor in Moscow. See [2224].

[15] Kiyoshi Iseki (1919–2011), professor in Osaka.

combinations of logarithms of algebraic numbers (see also the book [181] by A. Baker). Stark's paper contains the upper bound $d < 10^{1100}$, and he replaced it by $d < 10^{1030}$ in [3895]. This bound has been used by him [3897] and also by Montgomery and Weinberger [2920] to determine all imaginary quadratic fields with $h = 2$. This implied the non-existence of the exceptional field in the list of Iseki. The same assertion has been obtained in 1974 by Abraškin [19] who used Heegner's method.

In 1976 Goldfeld [1461] showed that if there exists an elliptic curve E, defined over $\mathbf{Q}$ with conductor N, having complex multiplication, whose L-function has a zero of order g at $s = 1$, then for $d > f(N, g)$ (with effective $f(N, g)$) and any real primitive character $\chi \bmod d$ with $(d, N) = 1$ one has

$$L(1, \chi) \geq \frac{c}{g^{4g} N^{13}} \frac{\log^{g-\mu-1} d}{\sqrt{d}} \exp\left(-21\sqrt{g \log\log d}\right)$$

with $\mu \in \{1, 2\}$, so that

$$\chi(-N) = (-1)^{g-\mu}$$

holds. The constant c is effective.

In view of Dirichlet's class-number formula the existence of such curve would lead to an effective lower bound for the class-number of imaginary quadratic fields, which tends to infinity with $|d(K)|$. The first such curve (with $g = 3$)

$$-139y^2 = x^3 + 10x^2 - 20x + 8$$

has been found by Gross and Zagier [1526, 1527] in 1983. This led to the inequality

$$h(-d) > \kappa(\varepsilon) \log^{1-\varepsilon} d$$

for every $\varepsilon > 0$ with effective $\kappa(\varepsilon)$.

Goldfeld's proof has been simplified by Oesterlé [3144], who also showed that one has

$$h(-d) \geq c \log d \prod_{p|d} \left(1 - \frac{\sqrt{p}}{p+1}\right) \tag{6.8}$$

with $c = 1/7000$, and if $5077 \nmid d$, then $c = 1/55$. Later Mestre [2835] showed that for every d one can take $c = 1/55$. An exposition has been given in the Bourbaki talk of Coates [721]. This approach has been illustrated in 2004 by Goldfeld [1463] in the example of fields with $h = 1$.

In 2002 Conrey and Iwaniec [765] showed that if there are many small gaps between zeros of Hecke L-functions associated with class-group characters, then for some constant $A > 0$ one has

$$h(-d) \gg \frac{\sqrt{d}}{\log^A d}.$$

Surveys on the development of the Gauss class-number problem were given by A. Baker [180], Goldfeld [1462] in 1985 and Oesterlé [3145] in 1988.

By now all imaginary quadratic fields with class-numbers ≤ 100 are known. See Oesterlé [3144] ($h = 3$), Arno [111] in 1992 ($h = 4$), C. Wagner [4260] in 1996 ($h = 5, 6, 7$), Arno, M.L. Robinson and Wheeler [112] in 1998 (odd $h \leq 23$) and Watkins [4309] in 2004 (the remaining $h \leq 100$).

It has been shown in 2007 by Soundararajan [3862] that if $F(n)$ denotes the number of imaginary quadratic fields with class-number n, then

$$\sum_{n \leq x} F(n) = 3\frac{\zeta(2)}{\zeta(3)} x^2 + O\left(\frac{x^2}{\log^c x}\right)$$

holds for every $c < 1/2$, and recently Lamzouri [2413] improved the error term to

$$O(x^2 (\log\log x)^3 / \log x).$$

2. In Chap. 1 we mentioned the conjecture of Gauss on the mean value of class-numbers of quadratic forms, which can be also interpreted in the language of quadratic fields. An important improvement in the evaluation of the error term $R(x)$ in the Gaussian equality (1.3) has been made in 1949 by I.M. Vinogradov who established

$$R(x) = O(x^{a+\varepsilon}), \tag{6.9}$$

for $a = 113/462 = 0.6975\ldots$ and all $\varepsilon > 0$ [4229], and later showed this with $a = 11/16 = 0.6875$ [4230] and $a = 19/28 = 0.6785\ldots$ [4232].

In 1962 J.R. Chen [669] got (6.9) $a = 35/52 = 0.6730\ldots$, and next year I.M. Vinogradov and Chen [670, 671, 4233] showed independently that $a = 2/3$ suffices. In the same year Vinogradov made this more precise, obtaining the bound $R(x) = O(x^{2/3}\log^b x)$ with a certain b [4234]. The next improvement came from Chamizo and Iwaniec [643, 644] who in 1998 showed that (6.9) holds with $a = 29/44 = 0.6590\ldots$. A further progress in the sphere problem obtained by Heath-Brown [1726] led to $a = 21/32 = 0.65625$. In 2012 Chamizo and Cristóbal [642] analysed the dependence of the influence on the error term of certain hypotheses concerning L-functions.

The function $\Phi(x)$ in (1.3) counts the values of the Gaussian class-number $H(D)$. Its analogue for $h^*(d)$ has the form

$$\Psi(x) = \sum_{d\le x} h^*(-d) = \frac{\pi}{18\zeta(3)} x^{3/2} + R(x)\ ,$$

with $R(x) = O(x\log x)$, established by Siegel (formula (22) in [3786]) in 1944.

In 1981 A.I. Vinogradov and Tahtadžyan [4223]) made the error term more precise by showing

$$R(x) = -\frac{3}{2\pi^2}x + O\left(x^{3/4}\log^3 x\right)\ .$$

The moments of class-numbers of positive-definite quadratic forms[16] were evaluated for $k = 2, 3$ by Lavrik [2488] in 1959 and for arbitrary k in 1962 by Barban[17] [186], who for every $\alpha < 1/2$ showed

$$\sum_{d=1}^{N} h^k(-d) = c(k)x^{1+k/2}(1 + O(r(x))) \tag{6.10}$$

with $r(x) = \exp(-\log^\alpha x)$.

For certain small values of k the error term has been improved by Saparnijazov [3574] in 1965, and in the next year Barban and Gordover [188] got $r(x) = x^{-1/5k^2}$. In 1969 Wolke proved $r(x) = x^{-1/k}$, Warlimont [4300] obtained $r(x) = x^{-2/(k-2)}$, and Wolke obtained $r(x) = x^{-c/k}$) for any $c < 8$ in 1971 ([4442] and $r(x) = x^{-1/4}$) [4443] in 1972.

[16] This problem is closely related to the problem of moments of the values at $s = 1$ of Dirichlet L-functions associated with real characters $\chi_k(x) = \left(\frac{-k}{x}\right)$.

[17] Mark Borisovič Barban (1935–1968). See [4221].

The case when d runs through a given arithmetical progression has been considered in 1971 by Lavrik [2490], and studied later by Saparnijazov and Faĭnleĭb [3575] in 1975 and by Stankus [3886] in 1976.

Moments of class-numbers of imaginary quadratic fields were evaluated in 1973 by Jutila [2084], who got a formula similar to (6.10) with the error term $O\left(x^{k/2-1/2}\log^{\beta(k)}\right)$.

3. The analogue for indefinite quadratic forms of the equality (1.4) has been proved in 1944 by Siegel [3786] in the following form:

$$\sum_{d\leq x} h^*(4d)\log(\varepsilon_{4d}) \approx \frac{4\pi^2}{21\zeta(3)}x^{3/2}, \tag{6.11}$$

where $\varepsilon_m = (T + U\sqrt{m})/2$ and the pair T, U is the smallest positive solution of the equation $T^2 - dU^2 = 4$.

He established also the asymptotic formula

$$\sum_{d\leq x} h^*(d)\log(\varepsilon_d) = \frac{\pi^2}{18\zeta(3)}x^{3/2} + r(x), \tag{6.12}$$

with $r(x) = O(x\log x)$, giving two proofs, one of them elementary, utilizing only the Pólya–Vinogradov inequality and Dirichlet's formula for the class-number.

In 1975 Shintani[18] [3758] improved (6.12) showing

$$r(x) = -\frac{3}{\pi^2}x\log x - cx + O\left(x^{3/4+\varepsilon}\right)$$

for every $\varepsilon > 0$ and

$$c = \frac{3}{\pi^2}(\log(2\pi) - \frac{\zeta'(2)}{\zeta(2)} - 1)\,.$$

Much later, in 2006, Chamizo and Ubis [645] made (6.2.1) more precise, replacing the exponent in the error term by $21/32 + \varepsilon$ and providing in (6.11) the error term equal to

$$-\frac{3}{\pi^2}x\log x + C_1 x + O\left(x^{21/32+\varepsilon}\right)$$

for all $\varepsilon > 0$.

Lower bounds for the error terms in all these formulas are yet unknown, but Kühleitner [2343] obtained such bound in the related asymptotical formula for the sum of class-numbers number of all (not necessarily primitive) quadratic forms of given negative determinant.

In 1982 Sarnak [3577] used the Selberg trace formula to obtain asymptotics for the mean value of the class-number of orders of real quadratic fields arranged according to the value of the regulator.

For extension of Sarnak's results to totally complex cubic and quartic fields see Deitmar [859], Deitmar, Hoffmann [861], Deitmar, Pavey [862]. The case of real fields was treated by Deitmar in [860]. These results were obtained using the prime geodesics theorem for manifolds.

The mean value of the class-number of quadratic fields was determined by Goldfeld and Hoffstein [1464] and Datskovsky [814]. Both sums

$$\sum_{d\leq x} h(-d),\quad \sum_{d\leq x} h(d)R(d)$$

[18] Takuro Shintani (1943–1980), professor in Tokyo. See [1956].

(with $R(d)$ denoting the regulator of $\mathbf{Q}(\sqrt{d})$) are asymptotically equal to $cx^{3/2}$ with

$$c = \frac{\pi}{18}\prod_p\left(1 - \frac{1}{p^2} - \frac{1}{p^3} + \frac{1}{p^4}\right).$$

For generalizations and variants see Taniguchi [3989], Raulf [3385], Hashimoto [1639].

From the class-number formulas (1.33),(1.34) it follows that the mean value of class-numbers of quadratic fields is related to the sum

$$\sum_{d\le x} L(1, \chi_d),$$

where χ_d is the primitive real character associated with the discriminant d. The more general sums

$$\sum_{d\le x} L(s, \chi_d),$$

with complex s were later studied by Goldfeld and Viola [1465], A.I. Vinogradov, Tahtadžyan [4222], Jutila [2085], Stankus [3887], Mai [2707]. Jutila's result implied that for infinitely many primes one has $L(1/2, \chi_p) \ne 0$. For the case of arbitrary s see Goldfeld, Hoffstein [1464] and Stankus [3888].

6.2.2 Class-Number of Cyclotomic Fields

1. Kummer's conjecture which we mentioned in Sect. 1.2.2 states that for prime p the first class-number number h_p^- of the field $\mathbf{Q}(\zeta_p)$ is asymptotically equal to

$$L(p) = p^{(p+3)/4}/c_p\ ,$$

with $c_p = 2^{(p-3)/2}\pi^{(p-1)/2}$.

Write for shortness $A(p) = h_p^-/L(p)$. The first result about Kummer's conjecture was obtained in 1949 by Ankeny and S. Chowla [93, 95] who used the Brun–Titchmarsh inequality (Titchmarsh[19] [4065]) and Walfisz's bound in the Prime Number Theorem [4277] to prove[20]

$$\log A(p) = o(\log p), \tag{6.13}$$

and showed that the Generalized Riemann Hypothesis implies the bound

$$\log A(p) = O\left(\log^{1/2} p \log\log p\right).$$

They observed that (6.13) implies that h_p^- is increasing for sufficiently large p. They showed moreover that the truth of Kummer's conjecture would imply

[19] Edward Charles Titchmarsh (1899–1963), professor in Liverpool and Oxford. See [592].

[20] The later paper [3792] of Siegel contains a weaker result.

$$\sum_q \frac{\varepsilon(q)}{q} = O\left(p^{-1}\right),$$

where q runs over all primes and

$$\varepsilon(q) = \begin{cases} 1 & \text{if } q \equiv 1 \bmod p, \\ -1 & \text{if } q \equiv -1 \bmod p, \\ 0 & \text{in other cases.} \end{cases}$$

In 1953 Tatuzawa [4014] obtained the stronger inequalities

$$\frac{c(\varepsilon)}{p^{\varepsilon}} < A(p) < \log^a p$$

for certain positive $a, c(\varepsilon)$ and arbitrary $\varepsilon > 0$. The first effective bounds for $A(p)$ were given in 1974 by Lepistö [2560], who applied his earlier results about zero-free region of Dirichlet L-functions [2559] to prove that for $p > 200$ one has

$$|\log A(p)| \leq 7.26 \log\log p \text{ if } p \equiv 1 \bmod 4,$$

and

$$-1.76 \log p - 6 \log(10) \leq \log A(p) \leq 14.76 \log\log p \text{ if } p \equiv 3 \bmod 4.$$

He showed also that the Generalized Riemann Hypothesis implies

$$\log A(p) = \theta(p) \left(\log\log p + 3.46 + \frac{912 \log p}{p^2} \right)$$

with $|\theta(p)| \leq 1$. His results imply that for $p > 224$ one has $h_p^- \geq 3.1 \cdot 10^{46}$. The constant 7.26 was later reduced by Lu and W. Zhang [2657] to 7/6.

In 1977 Pajunen [3209, 3210] checked that h_p^- increases in the interval [19, 1097]. His conjecture that this happens for all $p \geq 19$ is still unproved, but Lepistö [2560] showed that under Generalized Riemann Hypothesis h_p^- increases for $p \geq p_0$ with $p_0 \leq 2 \cdot 10^{13}$. An effective value of p_0 is obtainable from the Birch–Swinnerton-Dyer conjecture in the theory of elliptic curves (see Fung, Granville and H.C. Williams [1325]).

In 1990 Granville [1485] shed some doubts about Kummer's conjecture by showing that it is incompatible with the conjunction of two known conjectures in the prime number theory: the conjecture of G.H. Hardy and Littlewood [1627] asserting that the number of primes $p \leq x$ such that $2p+1$ is a prime exceeds $cx/\log^2 x$ for some positive c, and the conjecture of Elliott and Halberstam [1082] about the mean value of the error term in the prime number theorem for progressions. He showed also that for any $c > 1$ one has

$$\frac{1}{c} \leq A(p) \leq c \tag{6.14}$$

for a set of primes of positive density $\varrho(c)$ with $\lim_{c\to\infty} \varrho(c) = 1$. Later, in 2001 M.R. Murty and Petridis [3003] proved the existence of $c > 1$ such that (6.14) holds for almost all primes, and if the Elliott–Halberstam conjecture is true, then it holds for all $c > 1$ (see also Croot, Granville [794]).

In 2000 Puchta [3344] proved

$$\log A(p) = \log(1 - \beta) + O(\log\log p), \tag{6.15}$$

where β is a real zero of some $L(s, \chi)$ with odd character χ of conductor p, and in 2014 Debaene [834] made the error term in 6.15 more precise and proved that for $p > 9649$ one has

$$\log A(p) \leq (5 + o(1)) \log\log p ,$$

and

$$h_p^- < 2p\left(\frac{p}{31}\right)^{(p-1)/4} .$$

Results analogous to (6.13) for arbitrary fields $\mathbf{Q}(\zeta_n)$ were obtained by Lepistö, for prime powers n in 1964 [2557, 2558]. In 1967 Hyyrö [1946] established

$$h_n^- \leq 2^{c_n} ,$$

with $c_n = (n-6)2^{n-3} + n$. This has been much later rediscovered by Feng [1175, 1176] (cf. Metsänkylä [2842]).

It has been shown by Cornell and Washington [771] in 1985 that for infinitely many n one has

$$h_n^+ \geq B(c)n^c$$

for every $c < 3/2$ with some not effective $B(c) > 0$. A weaker but effective result obtained Louboutin [2645] in 2004, who showed that for infinitely many n one has $h_m^+ \geq n$.

In 2014 J.C. Miller [2861] computed h_m^+ for all m with $\varphi(m) \leq 116$ and found that except for $m = 136, 145$ and 212 these numbers are equal to 1.

A new way of studying the question of divisibility of h_{p^n} by primes has been found by Iwasawa who presented it at a conference in 1956 [1985]. Let K be a fixed field, and let L/K be its infinite Galois extension with Galois group isomorphic to Γ, the additive group $\mathbf{Z}_p^+$ of p-adic integers (Iwasawa called them Γ-extensions, and now they are called $\mathbf{Z}_p$-extensions). A $\mathbf{Z}_p$-extension L/K is called *cyclotomic* if L is generated over K be the set $\{\zeta_p.\zeta_{p^2}, \dots, \zeta_{p^n} \dots\}$.

The main theorem of [1985] shows that if L/K is a $\mathbf{Z}_p$-extension, L_n is the unique subfield of L with $[L_n : K) = p^n$ and $p^{e_n} \parallel h(L_n)$, then with certain λ, μ, ν depending on L/K and p one has

$$e_n = \lambda_p n + \mu_p p^n + \nu_p \tag{6.16}$$

for sufficiently large n.

This result applied to the cyclotomic $\mathbf{Z}_p$-extension of $\mathbf{Q}(\zeta_p)$ if p is odd and of $\mathbf{Q}(i)$ if $p = 2$ gives information about the power of p dividing the class-number of $\mathbf{Q}(\zeta_{p^n})$ [1987]. A necessary and sufficient condition for the vanishing of μ_p in this case has been given by Iwasawa in [1984] (for regular primes p one has $\lambda_p = \mu_p = \nu_p = 0$ by the result of Furtwängler [1341]). Another criterion was given by K. Shiratani [3761], and a cohomological criterion has been provided in 1959 by Iwasawa in [1985].

Iwasawa obtained his results by a study of $\mathbf{Z}_p[\Gamma]$-modules, which he continued in [1986]. Important simplifications were provided later by Serre [3732] who based his approach on the study of modules over the ring $\mathbf{Z}_p[[T]]$ of power series in one variable over $\mathbf{Z}_p$.

It has been conjectured in 1970 by Iwasawa [1991] that the coefficient μ_p vanishes for all $\mathbf{Z}_p$-extensions of any field, but in 1973 he found examples of non-cyclotomic extensions with arbitrary large μ_p [1992], so he restricted his conjecture to the case of cyclotomic $\mathbf{Z}_p$-extensions and in a joint paper with C. Sims [1997] he showed that this holds in the case $K = \mathbf{Q}$ for all primes below 4001. This was extended to $p < 8000$ by W. Johnson [2057]. Later computations by W. Johnson [2058], Wagstaff [4262], Ernvall and Metsänkylä [1121, 1122], Buhler, Crandall, Ernvall and Metsänkylä [505], Buhler, Crandall, Ernvall, Metsänkylä and Shokrollahi [506], and Buhler and Harvey [508] lifted this bound up to $163 \cdot 10^6$.

It has been observed by Wagstaff [4262] that for the cyclotomic $\mathbf{Z}_p$-extension of $\mathbf{Q}(\zeta_p)$ (with $p > 2$) the coefficient λ_p equals to the index of irregularity of p. This has been confirmed for all $p < 163 \cdot 10^6$ by the computations quoted above.

Iwasawa's conjecture has been established for all Abelian K in 1979 by Ferrero and Washington [1181] (for expositions see Oesterlé [3143] and Chap. 11 of the book [2459] by Lang). Another proof has been given in 1984 by Sinnott [3805]. Earlier this was known for $p = 2, 3$ (Ferrero [1180]).

For primes $q \neq p$ the q-primary part of the class-group in layers of a cyclotomic $\mathbf{Z}_p$-extension of an Abelian field has bounded cardinality; in fact, it is constant from some point on. This was established in 1978 by Washington [4302]. Another proof was provided later by Sinnott [3806]. See Friedman [1255] for a generalization to $\mathbf{Z}_{p_1} \times \cdots \mathbf{Z}_{p_s}$-extensions (cf. Sinnott [3806]).

Iwasawa's formula (6.16) has been generalized in 1981 to the case of $\mathbf{Z}_p^d$-extensions by Cuoco and Monsky [800].

6.3 Class-Field Theory

1. A generalization of the local class-field theory to fields which are complete under a discrete valuation and whose residue class-field has exactly one extension of given degree and this extension is separable has been studied by Moriya and Nakayama in a series of papers [2962–2966, 2968, 2969, 3058–3060] (cf. Schilling [3612, 3613]). Another generalization has been done by Whaples[21] [4382–4388] (see also Chap. 13 of the book [3733] by Serre).

From the local class-field theory it follows (see, e.g., Theorem 2 in Sect. 14.6 of Serre's book [3733]) that the maximal Abelian extension of $\mathbf{Q}_p$ is generated by all roots of unity. This is an analogue of the Kronecker–Weber theorem. It has been shown later by Lubin and Tate [2664] that if K is a local field with prime element π and residue class-field of q elements, then the maximal Abelian extension of K equals to $K_u K_\pi$, where K_u is the maximal unramified extension of K and K_π is generated by zeros of all iterations of the polynomial $X^q + \pi x$. Other proofs were given by Lubin [2663], Gold [1456] in 1981 and Kozuka [2235] in 1990. In the case of $\mathfrak{p}$-adic K a simple proof was provided by Rosen [3515].

In a series of papers Krasner presented an attempt to generalize class-field theory to non-Abelian extensions, both local [2245–2247, 2251–2253] and global [2254–2259], using the notion of a hypergroup introduced by Marty [2757, 2758] in 1934; see also Dresher and Ore [993].

Later an extension of the local class-field theory to metabelian extensions has been constructed by Koch and de Shalit [2198, 2199], and to arbitrary non-Abelian extensions by Laubie [2485].

2. In 1950 Hochschild[22] [1874] applied cohomology of groups, developed by Eilenberg[23] and Mac Lane [1062–1064], to prove the main results of the local class-field theory. His paper starts with a simple proof of $H^1(G, L^*) = 1$ for L/K being a Galois extension with Galois group G of an arbitrary field, a result generalizing

[21] George William Whaples (1914–1981), professor at the University of Massachusetts in Amherst and Indiana University. See [1025, 1026].

[22] Gerhard Hochschild (1915–2010), professor at the University of Illinois at Urbana and the University of California at Berkeley. See [3573].

[23] Samuel Eilenberg (1913–1998), professor at the University of Michigan, University of Indiana and the Columbia University. See [210].

Hilbert's theorem 90, which has been established in 1919 by Speiser [3872] in an elementary fashion. In Theorem 7.1 he showed that the isomorphism

$$Gal(L/K) \longrightarrow L^*/N_{L/K}(L^*)$$

can be obtained as the image of a certain element of group $H^2(Gal(L/K), L^*)$ under the map ϕ defined in (5.2). He pointed out that his arguments follow Chevalley's approach presented in [682], the theory of algebras being replaced by group cohomology.

In his next paper [1875] Hochschild showed how the cohomological approach can be used to deduce the global class-field theory from the local. This led in particular to a new proof of Artin's reciprocity law.

In 1951 Iyanaga[24] and Tamagawa [2005] gave an explicit description of the class-field theory for Abelian extensions of **Q**.

3. In 1943 Gut [1559, 1560] described the maximal Abelian subfield of the Hilbert class-field of a composite of cyclic extensions $K/\mathbf{Q}$ of given prime degree p (such subfield is now called the *genus field* of K. Gut's paper contains also a description of the ray class-fields of imaginary quadratic fields.

An approach to the theory of genera in quadratic number fields based on class-field theory has been presented in 1951 by Hasse [1698]. He defined the *genus field* of a quadratic number field K as the maximal unramified extension of K which is Abelian over **Q** and showed that it coincides with the field L in (2.1). In the same year the principal genus and the genus group have been defined by Iyanaga and Tamagawa [2005] in the case of cyclic extensions of the rationals in terms of ideles.

Hasse's method has been extended in 1953 to the case when K is an arbitrary Abelian field by Leopoldt[25] [2549], who extended the definition of the genus field to these extensions, and described this field in terms of characters of the Galois group of $K/\mathbf{Q}$. Later Hasse [1708] showed how the classes in the principal genus can be characterized using the norm residue symbol (cf. Gold [1453] and Queen [3352]).

In 1959 Fröhlich [1265, 1266] defined the genus field[26] $\bar{K}$ for arbitrary normal extensions $K/\mathbf{Q}$ as the maximal unramified extension of K having the form $K\Omega/K$ with Abelian $\Omega/\mathbf{Q}$, the principal genus being the subgroup of $H(K)$ corresponding to $K\Omega$ by class-field theory. The degree $g(K)$ of the extension $\bar{K}/K$ is called the *genus number of* K. In 1976 Gold [1454] showed that also in this case the principal genus can described by the norm residue symbol (see also Gurak[27] [1542]).

In 1967 Furuta [1359] extended Fröhlich's definition to normal relative extensions K/k, defining the genus field K^* of K over k as the maximal unramified extension of K of the form $K^* = kL$, where L/k is Abelian. He called the degree $g(K/k) = [K^* : K]$ the genus number of K/k and proved a formula for its value. This has been generalized to non-Galois extensions by Halter-Koch [1601].

The case of K/k quadratic has been studied in 1972 by Goldstein [1468], who gave an explicit construction of the genus field in the case when k is totally real and $h^*(k) = 1$.

A characterization of the principal genus by the norm residue symbol was given by Gold [1454] in 1976 (see also Gurak [1542]).

---

[24] Shokichi Iyanaga (1906–2006), professor in Tokyo and the Gakushuin University.

[25] Heinrich-Wolfgang Leopoldt (1927–2011), professor in Karlsruhe. See [3510].

[26] These fields occur already in a paper of Scholz [3679] published in 1940.

[27] Stanley Gurak (1949–2010), professor at the University of San Diego.

Constructions of genus fields for various classes of fields, including Eisenstein fields, fields of prime degree and pure fields were given by Ishida [1970–1974].

In 1981 Hamamura [1612] proved that there are only finitely many imaginary Abelian fields K with $g(K) = h(K)$ (later a proof has been given also by Louboutin [2636]). All such non-quadratic fields are now known due to the work of Miyada [2896], Loboutin [2639, 2641] and Chang and Kwon [650].

4. In 1946 Artin and Whaples [138] presented an axiomatic approach to the theory of algebraic fields. They showed that the following axioms for a field K characterize finite extensions of $\mathbf{Q}$ and $k(X)$, where k is a finite field.

($\mathbf{A}_1$): *There is a set* $\mathfrak{M} = \{v_\mathfrak{p}\}$ *of valuations of* K *such that for non-zero* $\alpha \in K$ *one has* $v_\mathfrak{p}(\alpha) = 1$ *for all except finitely many* $\mathfrak{p}$, *and the equality*

$$\prod_{\mathfrak{p}} v_\mathfrak{p}(\alpha) = 1 \tag{6.17}$$

holds for all non-zero $\alpha \in K$.

($\mathbf{A}_2$): *At least one valuation* $v_\mathfrak{p} \in \mathfrak{M}$ *is either discrete with finite field of residue classes or archimedean whose completion equals to either* $\mathbf{R}$ or $\mathbf{C}$.

It turned out that the Dirichlet–Hasse–Chevalley unit theorem as well as the finiteness of the class-number can be deduced directly from these axioms, omitting Minkowski's theory of lattice points and the ideal theory, used in previous proofs. One of the tools was the following approximation theorem for valuations:

If $v_1, v_2, \ldots, v_n$ *are non-trivial inequivalent valuations of a field* K *and the elements* $\alpha_1, \alpha_2, \ldots, \alpha_n$ *of* K *are given, then for every* $\varepsilon > 0$ *there exists* $\alpha \in K$ *with*

$$v(\alpha - \alpha_i) < \varepsilon \quad (i = 1, 2, \ldots, n)\,.$$

They wrote on p. 469 of [138]: "*This shows that the theorems of class-field theory are consequences of two simple axioms concerning the valuations, and suggests the possibility of deriving these theorems directly from our axioms*".

In this paper they considered *valuation vectors* of a field K of finite degree over the rationals as elements of the restricted product A_K of the rings K_v (v ranging over all valuations of K) with respect to the valuation rings of v. Later the elements of A_K have been called *adeles*. The group I_K of ideles of K coincides with the multiplicative group of A_K. In the case of function fields adeles were introduced already in 1938 by Weil [4349].

In 1953 Iwasawa [1980] characterized the adele rings of global fields as locally compact, neither compact nor discrete, semi-simple commutative rings R with unit which have a discrete subfield R with compact residue space R/K.

In their next paper [139] Artin and Whaples showed that without changing essentially the assertions the axiom A_1 may be weakened to the form

(A_1^*): *There is a set* $\mathfrak{M} = \{v_\mathfrak{p}\}$ *of valuations of* K *such that for all non-zero* $\alpha \in K$ *the equality* (6.17) *holds, the product being absolutely convergent; i.e. the series* $\sum_\mathfrak{p} \log(v_\mathfrak{p}(\alpha))$ *converges absolutely to* 0 .

A characterization of algebraic (finite and infinite) extensions of fields satisfying the axioms of Artin and Whaples has been given in 1953 by Jaffard [2022].

5. In 1949 Šafarevič [3544] announced an explicit formula for the qth power norm residue symbol

$$\left(\frac{\alpha, \beta}{\mathfrak{p}}\right)_q ,$$

where q is a power of a prime p and $\mathfrak{p}$ is a prime ideal dividing p. Next year he published a proof in the case $q = p \neq 2$ [3545]. An exposition has been given by Hasse [1699] (cf. M.Kneser [2183]).

A similar formula has been obtained in 1951 by Mills [2865, 2866] for arbitrary prime powers. For further developments see Brückner [473, 474], Henniart [1762], M. Kurihara [2382], Sen [3729, 3730] and Vostokov [4250].

6. There are several books dealing with class-field theory. In 1951/1952 Artin and Tate had a seminar in Princeton in which the cohomological approach to class-field theory has been exposed. Notes from that seminar were published in 1968 [137]. They based their presentation on the notion of *class formations* $(G, \{G_F\}, A)$, where G is a group, $\{G_F\}_{F\in\Sigma}$ is a family of finite index subgroups of G closed under intersection and conjugation and satisfying $\bigcap_{F\in\Sigma} G_F = \{1\}$, and A is a G-module whose each element is left-fixed by some group G_F. There are moreover two axioms dealing with the first two cohomological groups $H^i(G_{K/F}, A_K)$, where $K, F \in \Sigma$, $G_K \subset G_F$, $G_{K/F} = G_F/G_K$, and $A_K = A^{G_K}$. The authors show first that the family of idele classes of global fields form a class formation, and this leads to the reciprocity law. In the next chapters they prove the existence theorem in global class-field theory, prove the Grunwald–Wang theorem and treat the explicit reciprocity laws.

In 1954 Chevalley [683] presented an exposition of the cohomological approach to class-field theory. It is a bit difficult to read it, as several notions are denoted by Japanese characters.

An introduction to Galois cohomology presented Serre [3735] in 1964, and a book on applications of cohomology to algebraic number fields has been written by Neukirch, A. Schmidt and Wingberg [3100].

In 1965 a conference in Brighton [599] was devoted to a survey of algebraic number theory, with detailed expositions of the local (Serre [3736]) and global class-field theory (Tate [4008]), based on cohomology.

Cohomological approach to the class-field theory has been also presented in the book of Neukirch [3091], published in 1967. Later he presented [3097, 3098] an axiomatic approach to the class-field theory, both local and global, in which the use of the second cohomology group has been eliminated.

Introductions to class-field theory were presented by Cohn [751] in 1978 and Childress [688] in 2009. Two expositions of the local class-field theory have been written by Iwasawa [1994, 1995]. Several applications of the class-field theory were given in the monograph [1492] by G. Gras, published in 2003.

6.4 Euclidean Algorithm

1. In 1942 Rédei [3404] showed that the field $\mathbf{Q}\left(\sqrt{d}\right)$ with $d = 73$ is norm-Euclidean, and asserted that the same happens for $d = 97$, but in 1952 E.S. Barnes[28] and Swinnerton-Dyer [192] showed that this assertion is not correct.

In 1944 Hua [1909] made effective the result of Erdős, Chao Ko [1116] and Heilbronn [1752] by showing that the discriminants of real quadratic norm-Euclidean fields are bounded by e^{250}. The work of Schuster [3708], A. Brauer [426], Chatland [665], Chatland and Davenport [666], Hua and Min [1912], Hua and Shih [1913], Inkeri [1963], Min [2868, 2869] and Rédei [3403, 3404] completed the list of norm-Euclidean real quadratic fields.

In 1951 Davenport [823] reduced Hua's bound to 2^{14} giving a new proof of the completeness of the list of real quadratic Euclidean fields. Uniform proofs for Euclidicity of those fields have been given by Varnavides [4193] in 1952, Ennola [1102] in 1958 and Eggleton, Lacampagne and Selfridge [1055] in 1992.

2. A criterion for the existence of an Euclidean algorithm in an integral domain has been presented in 1949 by Motzkin [2989], who used it to show the existence of a principal ideal ring without any Euclidean algorithm.

It has been shown in 1965 by O'Meara [3154] that if S is a sufficiently large finite set of valuations of K containing all Archimedean valuations, then the ring of S-integers is norm-Euclidean (see also Queen [3349]).

3. In 1949 Davenport proved that there are only finitely many norm-Euclidean not totally real cubic fields [820, 821], and in the next year he did the same for totally complex quartic fields [822]. For cyclic cubic fields this has been established in the same year by Heilbronn [1753], who in the next year proved this for cyclic fields of fixed degree having a prime power discriminant [1754]. He conjectured in [1753] that there are infinitely many totally real cubic non-Galois norm-Euclidean fields (several such fields were found in 1967 by Godwin [1449] who in [1448] did this in the case of quartic and quintic fields).

Several norm-Euclidean cyclic cubic fields with small discriminants were listed in 1969 by J.R. Smith [3835], and McGown [2800] found all of them with discriminants below 10^{10} and in [2801] showed that under GRH this list is complete. In 2017 Lezowski and McGown [2590] determined all cyclic quintic and septic norm-Euclidean fields.

---

[28] Eric Stephen Barnes (1924–2000), professor in Adelaide.

Later the finiteness of norm-Euclidean fields has been established for various classes of fields. Egami [1052] did this for pure quartic fields $\mathbf{Q}(\sqrt[4]{m})$ with $m \neq 2p^2$ with prime $p \equiv 3 \bmod 8$, and in 1984 he showed this for cyclic fields of any fixed degree in which every ramified prime fully ramifies [1054]. This covers in particular cyclic fields of a fixed prime degree. In 1989 Lemmermeyer [2523] did this for pure quintics.

In certain cases one can effectively list all norm-Euclidean fields or at least give a bound for their discriminants. In 1952 Cassels [594] gave such a bound in the case of totally complex quartic fields. In 1984 van der Linden noted in his thesis [4126] that Cassels' argument needs a correction, and finally arrived at the huge bound 230 202 117.

In 1979 Cioffari showed that $\mathbf{Q}(\sqrt[3]{d})$ is norm-Euclidean only for $d = 2, 3, 10$, and $\mathbf{Q}(\sqrt[4]{d})$ with $d < 0$ only for $d = -2, -3, -7$ and perhaps also for $d = -12, -44$ and -67, and van der Linden [4127] showed that the only cyclic quartic totally complex norm-Euclidean fields are the unique quartic subfields of $\mathbf{Q}(\zeta_p)$ with $p = 5, 13$.

In 1976 H.W.Jr. Lenstra [2542] gave the following sufficient condition for a field K to be norm-Euclidean:

Let $L(K)$ be the maximal cardinality of a subset $\{u_1 = 0, u_2 = 1, \dots, u_k\}$ of K such that all non-zero differences $u_i - u_j$ are units. If

$$L(K) > \frac{n!}{n^n}\left(\frac{4}{\pi}\right)^{r_2(K)} |d(K)|^{1/2},$$

where $n = [K : \mathbf{Q}]$, then K is norm-Euclidean.

This permitted him to find 132 new norm-Euclidean fields of degrees ≤ 8.

To determine the value of $L(K)$, called now the *Lenstra constant* one has to know the exceptional units of K (see Sect. 4.1.3).

Lenstra's criterion has been used to find new norm-Euclidean fields by Mestre [2833], Leutbecher and Martinet [2575], Leutbecher [2574], Leutbecher and Niklasch [2576], and Houriet [1903]. An algorithm constructed by Quême [3353] in 1998 produced 1200 new norm-Euclidean fields.

For a field K put

$$M(K) = \sup_{x \in K} \inf_{y \in \mathbf{Z}_K} |\ N(x - y)\ | \ .$$

Norm-Euclidean fields are characterized by the inequality $M(K) < 1$. It follows from the conjecture (1.24), which is known to be correct for $n \leq 8$ (see Sect. 1.3.2) that if K is totally real of degree n, then one has

$$M(K) \leq \frac{\sqrt{|d(K)|}}{2^n}. \tag{6.18}$$

In 2006 Bayer-Fluckiger [257] proved

$$M(K) \leq \frac{|d(K)|}{2^n}$$

for arbitrary fields and established (6.18) for all cyclotomic fields and also for maximal real subfields of $\mathbf{Q}(\zeta_q)$, where q is an odd prime power. The truth of (6.18) has been extended to the maximal real subfields of $\mathbf{Q}(\zeta_{4m})$ for odd $m > 1$ (Bayer-Fluckiger, Suarez [260]), and to all Abelian fields of prime power conductor (Bayer-Fluckiger, Maciak [258]). The last paper contains also the proof of the inequality (6.18) for all Abelian fields of prime power conductor.

An algorithm for determining $M(K)$ for totally real fields has been given by Cerri [636] in 2007, and four years later Lezowski [2589] extended it to arbitrary fields.

A survey on Euclidean fields was prepared by Lemmermeyer [2526] in 1998, and evaluations of $M(K)$ were surveyed by Bayer-Fluckiger and Maciak [259].

In 1973 Weinberger [4360] proved that if GRH is true, then a field with class-number 1 and infinitely many units is Euclidean under a specific norm, and in 1977 H.W.Jr. Lenstra [2543] extended this to rings K_S of S-integers in the case when S is a finite set of prime ideals, containing S_∞, $|S| \geq 2$, and K_S is a unique factorization domain.

In 1987 R. Gupta, M.R. Murty and V.K. Murty [1540] showed that this assertion holds unconditionally for a field K of degree n provided $|S| \geq \max\{2, 2n-3\}$ and either K is not totally complex or $\zeta_g \in K$, where

$$g = GCD\{N(\mathfrak{p}) - 1 : \mathfrak{p} \in S \setminus S_\infty\}.$$

The first example, $\mathbf{Q}(\sqrt{69})$, of an Euclidean but not norm-Euclidean quadratic field has been provided in 1994 by D.A. Clark [716] (cf. Niklasch [3110]), and in 1995 D.A. Clark and M.R. Murty Murty, M.R. [718] showed that a totally real, normal, quartic field with class-number one in which there exists a prime ideal $\mathfrak{p}$ such that every invertible residue class mod $\mathfrak{p}^2$ contains a unit is Euclidean, and provided a generalization to the case of totally real normal extensions of arbitrary degree. In 1996 D.A. Clark [717] found two cubic fields which are Euclidean, but not norm-Euclidean, and in 2000 Cavallar and Lemmermeyer [613] gave more such examples.

The ring $\mathbf{Z}[\sqrt{14}]$ attracted particular attention. The question, whether it is Euclidean, has been asked in 1971 by Samuel [3570]. First Bedocchi [264, 265] deduced its Euclidicity assuming a certain unproved condition, and in 1988 M. Nagata [3012] introduced a generalization of the Euclidean algorithm and showed that it works in $\mathbf{Z}[\sqrt{14})$ (see Leu [2573] for an exposition). Finally in 2004 a positive answer to Samuel's question has been given unconditionally by Harper [1633]. In 1997 Cardon [566] showed that the ring of two integers of $\mathbf{Q}(\sqrt{14})$ is norm-Euclidean.

In 2004 Harper and M.R. Murty [1634] established Weinberger's assertion unconditionally for Galois fields having unit rank ≥ 4. It has been observed by Narkiewicz [3070]) that a small change of the proof in [1634] shows that this holds also for real quadratic fields with at most two exceptions, and for cyclic cubic fields with at most one exception. Recently M.R. Murty and K.L. Petersen [3002] showed that if K has unit rank ≥ 4, and for some of its subfield M the extension K/M is Galois and $[K : M] \geq 4$, then its class-number equals to 1 if and only if K is Euclidean.

In 1976 Cooke[29] [767, 768] introduced for $k = 1, 2, \ldots, \omega$ the k-stage Euclidean domains, satisfying a weakening of the norm-Euclidean property.

It has been conjectured that real quadratic fields K with $h(K) = 1$ are a two-stage Euclidean, and this has been checked for fields with $d(K) \leq 8000$ (Guitart and Masdeu [1538]).

6.5 Other Topics

1. In 1940 Hasse [1688] used Gaussian sums for characters of multiplicative groups of finite fields to modify Dirichlet's formula for the class-number of $\mathbf{Q}(\sqrt{p})$ for prime $p \equiv 1 \bmod 4$. His method has been generalized in 1944 to the case of arbitrary real quadratic fields by Bergström [304] and to cyclic cubic and biquadratic fields in 1948 by Hasse [1690]. In his book [1700] Hasse generalized this method to all Abelian extensions of the rationals.

In 1948 Kiselev [2157] considered the class-number of $\mathbf{Q}(\sqrt{p})$ with prime $p \equiv 1 \bmod 4$ and established a relation of it with Bernoulli numbers, by showing

$$h(p)U \equiv TB_{(p-1)/2} \pmod{p}, \tag{6.19}$$

where $(T + U\sqrt{p})/2 > 1$ is the fundamental unit.

---

[29] George Erskine Cooke (1942–1976), professor at the Cornell University and the University of Maryland.

A generalization of the congruence (6.19) to quadratic fields of arbitrary discriminant $d > 0$ and primes $3 < p \mid d$ has been proved by Ankeny, Artin and S. Chowla [90, 91] in 1952:

$$-2h\frac{U}{T} \equiv B_{(p-1)/2,\chi} \pmod{p}, \tag{6.20}$$

where $B_{n,\chi}$ is defined by (4.23).

They obtained also for primes $p \mid d$ the congruence

$$\frac{2hU}{T} \equiv -\frac{1}{m}\sum_{j=1}^{d-1}\frac{1}{j}\left(\frac{d}{j}\right)\left[\frac{j}{p}\right] \pmod{p},$$

with $m = d/p > 1$.

They conjectured (*the Ankeny–Artin–Chowla conjecture*) that in the case of prime discriminant $p \equiv 1 \bmod 4$ one has $p \nmid U$, and checked it for primes $p \equiv 5 \bmod 8$ below 2000.

In 1960 Mordell [2935] established the conjecture for regular primes and noted that it has been checked for all primes $p < 100\,000$. Later Beach, H.C. Williams and Zarnke [261] extended the check to $p < 6\,270\,714$, and Stephens and H.C. Williams [3924] did this for $p < 10^9$ (they quote an unpublished computation performed in 1986 by Soleng for $p < 100\,028\,010$). It is now known (van der Poorten, te Riele and H.C. Williams [4131]) that the conjecture is true for all primes $p \equiv 1 \bmod 4$ below $2 \cdot 10^{11}$.

In 1959 the Ankeny–Artin–Chowla conjecture has been extended by Kiselev and Slavutskiĭ [2158] to fields with discriminant $4p$ for primes $p \equiv 3 \bmod 4$. They checked it for $p < 2000$.

In 1961 Mordell [2936] showed that for $p \equiv 3 \bmod 4$ the divisibility of U by p is equivalent to $p \mid E_{(p-3)/4}$, where E_n is the sequence of Euler numbers (see (1.20)).

In 1965 Slavutskiĭ [3822, 3823] used Hua's [1908] inequality

$$L(1,\chi) < 1 + \frac{\log d}{2}$$

for

$$\chi(n) = \left(\frac{n}{d}\right)$$

to deduce the bound

$$h(d) < d$$

for positive d. This allows in view of (6.19) to state the Ankeny–Artin–Chowla conjecture in the form

$$p \nmid B_{(p-1)/2}\,.$$

In the announcement [90] of their results Ankeny, Artin and S. Chowla stated that if $p \equiv 1 \bmod 4$ is a prime, then

$$2h(p)\frac{U(p)}{T(p)} \equiv \frac{A+B}{p} \pmod{p},$$

where A is the product of quadratic residues mod p and B is the product of quadratic non-residues. A proof has been given by Carlitz [573] (see also [578]).

Similar congruences modulo powers of p have been later obtained by Slavutskiĭ [3821, 3824, 3826].

Several kinds of congruences for class-numbers of quadratic fields and their linear combinations were discussed in the book [4109] by Urbanowicz[30] and K.S. Williams.

An analogue of the congruence (6.20) for pure cubic fields has been obtained in 1984 by Ito [1976], for pure quartic and sextic fields this has been done in 1987 by Kamei [2095], and the case of cyclic fields of prime degree has been considered by Jakubec [2026–2028].

An elementary method of computing $h(d)$ for $d > 0$ has been proposed by Slavutskiĭ in [3823].

[30] Jerzy Urbanowicz (1951–2012), professor in Warsaw. See [3623].

2. The structure of the p-part of the class-group of a field which is a cyclic extension L/K of degree p of some its subfield K has been studied in 1940 by Inaba [1960], who later looked more closely at the case when $K = \mathbf{Q}$ [1961], extending to odd p the results obtained in case of $p = 2$ by Reichardt [3410] and Iyanaga [2003].

A generalization the case of cyclic extensions of prime power degree has been given in 1954 by Fröhlich [1264].

In 1944 Rédei [3405] (see also [3406]) presented methods of determining the invariants of the class-group of any algebraic number field, extending his earlier results in the quadratic case [3402].

It has been shown in 1949 by Litver [2611] that if K has class-number one, p is a prime and $L = K(\alpha_1^{1/p}, \dots, \alpha_n^{1/p})$, where the numbers $\alpha_i \in \mathbf{Z}_K$ are p-independent; i.e. an equality

$$\prod_{i=1}^{n} \alpha_i^{x_i} = \beta^p$$

with $x_i \in \mathbf{Z}$ and $\beta \in \mathbf{Z}_K$ can hold only if all exponents x_i are divisible by p, then with some integer A one has

$$h(L) = p^A \prod_{j=1}^{m} h(K_j) ,$$

where $K_1, K_2, \dots, K_m$ are subfields of L of the form $K(\gamma^{1/p})$. For $p = 2$ this has been earlier established by Herglotz [1810] in 1922 (see Sect. 1.1.3).

3. In 1940 Gelfond [1405] considered the difference between powers of algebraic numbers and showed that if $\alpha \neq \beta$ are such numbers lying in a field K with $|\alpha|, |\beta| \neq 0, 1$, the ratio $\log\alpha/\log\beta$ is irrational, and $\mathfrak{p}$ is a prime ideal of K, then for every $\varepsilon > 0$ there exists $n_0(\varepsilon)$ such that if $n \geq m > 0$ and $n > n_0(\varepsilon)$, then the difference $\alpha^n - \beta^m$ cannot be divisible by a power of $\mathfrak{p}$ exceeding $\log^{3+\varepsilon} n$. As an application he showed that if at least one of the numbers $\alpha, \beta, \gamma \in K$ is not a unit and one has $|\alpha|, |\beta|, |\gamma| \neq 0, 1$, then the equation

$$\alpha^x + \beta^y = \gamma^z$$

can have at most finitely many solutions in rational integers, except when $\alpha, \beta, \gamma \in \{\pm 1, \pm 2, \dots, \pm 2^n, \dots\}$.

Quadratic equations in two variables over quadratic fields were studied in 1942 by Niven [3116]. He showed in particular (Theorem 4) that the Pell equation

$$X^2 - \alpha Y^2 = 1 \tag{6.21}$$

with $0 \neq \alpha \in \mathbf{Z}_K$ has infinitely many solutions in $\mathbf{Z}_K$ if and only if either K is real and α is not totally imaginary, or K is complex and α is not a square. In this case he found the general solution of (6.21), similar to that known in the case of rational integers, except in the case when K is real and α is totally positive non-square integer [3117].

The Pell equation in totally real fields and imaginary quadratic fields has been studied by Skolem ([3812, 3813], cf. [3811]), who proved an analogue of Störmer's[31] (see [3937]) theorem which states that if $d > 0$ is not a square then the equation $x^2 - dy^2$ can have only one solution with all prime divisors of y dividing d. Skolem considered also several other equations in number fields [3814, 3815].

Methods to find roots of a p-adic polynomial were described in 1943 by Thurston [4061].

Various cubic equations in real quadratic fields were studied by Ljunggren[32] [2613].

4. Let K be a finite extension of $\mathbf{Q}_l$, and let $p \neq l$ be a prime. In 1940 Schilling [3610] described the Galois group $G_p(K)$ of the maximal p-extension of K. If K does not contain ζ_p, then $G_p(K) = \mathbf{Z}_p$, and if $\zeta_p \in K$, then $G_p(K)$ is a pro-p-group with two generators and one relation. The case $p = l$ has been treated in 1947 by Šafarevič [3543], who showed that if $\zeta_p \notin K$, then $G_p(K)$ is the free pro-p-group with $[K : \mathbf{Q}_p] + 1$ generators.

Let us recall that a pro-p-group G is defined as the inverse limit of a sequence G_n of finite p-groups. To define the free pro-p-group $F_{p,I}$ with generators $\{g_i\}_{i \in I}$ let F_I be the free group with these generators, $\mathfrak{N}$, and let be the set of all normal subgroups H of F_I having p-power index and containing all elements g_i except finitely many. Then

$$F_{p,I} = \lim_{\leftarrow} F_I/H \ ,$$

where H runs over $\mathfrak{N}$.

In the case when $p = l$ and $\zeta_p \in K$ it has been shown by Kawada [2129] in 1954 that $G_p(K)$ is a pro-p-group with $[K : \mathbf{Q}_p] + 2$ generators and one relation. Another proof has been given by Faddeev and Skopin [1158] in 1959. In 1961 Demuškin [886] made this theorem more explicit for odd p and two years later did this for $p = 2$ [887], except for one case, which has been completed by Labute [2392] two years later. Later Labute introduced a class of pro-p-groups (*Demuškin groups*), containing all groups $G_p(K)$ for local fields K, and classified them in [2393–2395]. Demuškin groups which appear as $G_2(K)$ for some K were determined in 1992 by Mináč and Ware [2870].

Groups which are Galois groups of maximal Abelian extensions of fields were described in 1994 by C.U. Jensen and Prestel [2049], and in 1997 Efrat [1051] described finitely generated pro-p-groups of the form $G_p(K)$ for K being an algebraic extension of $\mathbf{Q}$.

The more general problem of description of $\Omega(K)$, the Galois group of the algebraic closure of a local field K, has been considered in 1955 by Iwasawa [1982]. He described the groups $\Omega(V)$ and $Gal(V/K)$ as profinite group, where V denotes the maximal tamely ramified extension on K. Some information about the extension of $\Omega(V)$ by $Gal(V/K)$ was also obtained. A more precise result has been achieved by Koch [2191] in 1961. In 1968 Jakovlev [2023] proved that in the case when the degree m of $K/\mathbf{Q}_p$ is odd, then $\Omega(K)$ is a profinite group with $m + 3$ generators with explicitly described generators (there were some inaccuracies in case $p = 2$). A cohomological description of $\Omega(K)$ has been given in 1978 by Koch [2196].

---

[31] Carl Störmer (1874–1957), professor in Kristiania (Oslo). See [482].

[32] Wilhelm Ljunggren (1905–1973), professor in Oslo.

In the eighties Jannsen and Wingberg [2032, 2033, 4426] obtained a very precise description of $\Omega(K)$ as a profinite group with $[K : \mathbf{Q}_p] + 3$ generators and explicit relations in the case when K is an extension of $\mathbf{Q}_p$ with odd p. In the case $p = 2$ the group $\Omega(K)$ has been in certain cases described by Diekert [942].

5. In 1943 Fogels [1203] showed that in the field $\mathbf{Q}(\sqrt{-5})$ almost all integers as well as almost all natural numbers have non-unique factorization into irreducibles. His method works also for all imaginary quadratic fields with one class per genus and non-trivial class-numbers.

In 1960 Carlitz [580] showed that all factorizations of an integer of K have the same length if and only if $h(K) \leq 2$. It has been later established by Rush [3536] and Kaczorowski [2091] (see also Halter-Koch [1602] and Geroldinger [1414]) that the class-group of a field can be defined with the use of factorization properties of its integers.

Similar descriptions of fields having a given small class-number or a prescribed type of the class-group were given by S.T. Chapman and W.W. Smith [655], Czogala [807], Di Franco and Pace [947], Feng [1177], Geroldinger [1412], Kaczorowski [2089], Krause [2272] and Salce and Zanardo [3557].

It has been shown in the sixties that if K is an algebraic number field with $h(K) \geq 2$, then for every $k \geq 2$ almost all integers of K have at least k different factorizations into irreducibles, and the same happens for natural numbers in the case, when K is a Galois field. Similar results hold also for numbers having at least k factorizations of distinct lengths, provided $h(K) \geq 3$ (Narkiewicz [3061, 3062]).

Asymptotics for the number of non-associated irreducible elements $\alpha \in \mathbf{Z}_K$ with $|N(\alpha)| \leq x$, as well as for the mean value of the number of distinct factorizations in $\mathbf{Z}_K$ of elements of $\mathbf{Z}_K$ and natural numbers has been determined in 1966 by Rémond [3445]. A more precise formula appears in Kaczorowski [2090] and Halter-Koch [1606].

Asymptotics for the number of non-associated integers α of K with $|N(\alpha)| \leq x$, resp. of natural numbers $n \leq x$ having at most k distinct factorizations into irreducibles in K, as well as for similar counting functions of numbers having at most factorizations of distinct lengths functions, were obtained by Geroldinger [1411, 1413], Kaczorowski [2090], Narkiewicz [3064], Odoni [3141], Śliwa [3828] (see also Chap. 9 of [3065] and Chaps. 8 and 9 of [1415]). Another way of counting numbers with particular factorization properties was used by Helmut Weber [4332].

Similar functions defined in more general situations were treated by Halter-Koch and Müller [1608], Halter-Koch [1605] in 1992, and Geroldinger, Halter-Koch and Kaczorowski [1416] in 1995.

The asymptotic formulas for these functions have usually the form

$$(C + o(1))\frac{x(\log\log x)^a}{\log^b x} ,$$

with $C > 0$ and the exponents a, b depending on the class-group of the field. It has been observed by Narkiewicz and Śliwa [3067, 3072] that in some cases these exponents have a combinatorial interpretation (see the book [1415] by Geroldinger and Halter-Koch and the large literature quoted there for a treatment of this approach in a more general setting.)

6. If L/K is an extension and I is an ideal in $\mathbf{Z}_L$, then an additive homomorphism $d : \mathbf{Z}_L \longrightarrow \mathbf{Z}_L/I$ is called an I-derivation over K if for all $x, y \in \mathbf{Z}_L$ one has

$$d(xy) = xd(y) + yd(x) ,$$

and $d(\mathbf{Z}_K) = 0$. It has been observed by Weil [4350] that an I-derivation whose image contains at least one non-zero-divisor exists if and only if I divides the different $D_{L/K}$. The first proof of this assertion has been published in 1951 by Kawada [2126], who showed how the theory of the different can be based on a study of derivations.

Two years later Moriya generalized Kawada's results to arbitrary Dedekind domains [2970]. See also Kinohara [2155]. In 1967 Neukirch [3090] presented the theory of the different based on Weil's definition. Other proofs of Weil's assertion were given later by Kinohara [2154] and Narkiewicz [3063].

7. If the ideal generated by an algebraic integer α has all its prime ideal divisors principal, then α has only one factorization into irreducibles. In the case of quadratic fields an elementary proof not using ideals has been given in 1945 by Pall [3215].

The number of representations of an integer of a quadratic field K as product of two elements of given norms has been later determined by Butts and Pall [534, 535]. They treated also the analogous question for ideals, also in orders of K.

Factorization of prime ideals in an extension with Galois group isomorphic to the unique simple group of order 168 has been described in the thesis of Büsser [530] in 1944.

In 1946 Châtelet [664] described cyclic cubic extensions $K/\mathbf{Q}$, proved by explicit construction the existence of a normal basis in K in the case when the conductor of K is square-free, noted that its discriminant is of the form $9^a m^2$ with $a = 0, 1$ and m being a product of distinct primes congruent to 1 mod 3, and showed that if N has this form, then there are $2^{\omega(N)-1}$ cyclic cubic fields of discriminant N.

The case of fields with cyclic Galois group of prime power order has been later dealt with by Oriat [3189].

A simple proof of the fact that in every algebraic number field there are infinitely many prime ideals of degree one has been given in 1950 by Moriya [2967].

Later other proofs of this old result were given by Nakayama and Tuzuku [3013] in 1953, Duičev [1019] in 1956 and Voloch [4240] in 2000. Certain generalizations had been presented by M. Nagata in 1953 and Dress [994] in 1964.

8. A characterization of algebraic numbers using Diophantine approximations has been presented in 1942 by Pisot [3304].

It has been shown in 1947 by Skolem [3816] that for any finite extension $K/\mathbf{Q}$ the factor group $K^*/U(K)$ is free Abelian. Later Abelian groups G with torsion subgroup T were called *regular* if the factor group G/T is free Abelian (Iwasawa [1981]).

Skolem's theorem has been also proved in 1953 by Iwasawa [1981], who showed moreover that if K is an algebraic number field of finite degree and L/K is an infinite Abelian extension, then the multiplicative group L^* is regular. In 1964 Schenkman [3599] proved that every group generated by algebraic numbers of bounded degree is regular, and in 1990 K. Horie [1894] applied Iwasawa's method to establish the regularity of the multiplicative group of infinite algebraic number fields whose Galois closure contains only finitely many roots of unity. Regularity of multiplicative groups of a class of non-algebraic extensions has been established in 1972 by Jehne [2041]. Since a subgroup of a regular group is also regular in all these cases the group of units is regular.

9. Tame extensions of $\mathfrak{p}$-adic fields were described in 1940 by Albert [51], who presented their construction, gave a necessary and sufficient condition for their normality, and described the resulting Galois group.

It has been established in 1947 by Motzkin [2988] that if there are given $n-1$ distinct points $z_1, z_2, \dots, z_{n-1}$ on the complex plane, symmetric to the real plane, then for every $\varepsilon > 0$ there is an algebraic integer α of degree n with conjugates $a_1 = a, a_2, \dots, a_n$ such that for $i = 1, 2, \dots, n-1$ one has

$$|a_i - z_i| < \varepsilon .$$

An effective proof of this theorem has been provided in 2004 by Dubickas [1007].

In 1948 Hasse [1690] gave a way to find the fundamental units in cyclic cubic and biquadratic fields, and in [1691] announced results about the structure of the unit group as a Galois module in these cases as well as for the Galois closure of a real non-cyclic cubic field. He presented the proof in [1696].

6.6 Books

1. In 1940 Weyl published his lectures on the theory of algebraic numbers, based on a modern form of Kronecker's approach [4378], introducing divisors axiomatically. A presentation of that theory based on valuation theory, written by Artin [128], appeared in 1951.

A geometrical theory of algebraic number fields based on an extension of the methods of Minkowski [2880], developed later for quadratic fields by Klein [2167] and for cubic fields by Furtwängler [1328], has been presented in 1940 by B.N. Delone and Faddeev [882], who used it to give a detailed treatment of cubic fields. They presented Voronoĭ's algorithms [4243] to construct an integral basis and fundamental units. They gave also a table of fundamental units in cubic fields K with $-379 \leq d(K) \leq 0)$ and in pure cubic fields $\mathbf{Q}(\sqrt[3]{a})$, and presented a study of integers represented by binary cubic forms.

An introduction to the theory of algebraic numbers presented Pollard [3332] in 1950, and in 1955 the book [2734] by Mann appeared, giving an introduction to algebraic numbers in the classical way. In the sixties several books dealt with that subject. The books by E. Weiss [4365] in 1963 and S. Lang [2456] in 1964 gave introductions to that theory, and the book [4355] by Weil presented local and global class-field theory based on the theory of simple algebras. The book [1059] by Eichler, published in 1963, dealt with both number fields and function fields. In 1970 S. Lang [2457] presented an expanded version of [2456], exposing the class-field theory and modern analytical methods. Later several books presented the fundamentals of the theory of algebraic numbers: Ribenboim [3453] in 1972, and [3457] in 2001, Janusz [2034] in 1973, Narkiewicz [3065] in 1974, Long [2626] in 1977, I. Stewart and Tall [3932] in 1979, Ono [3156] in 1990, and Fröhlich and M.J. Taylor [1286] in 1993.

Computational aspects of the theory of algebraic numbers were treated by Zimmer [4473] in 1972, Pohst and Zassenhaus [3320] in 1989, H. Cohen [732, 733] in 1993 and 2000, and Pohst [3317] in 1993.

Bibliography

1. —: Carl Herz (1930–1995). Notices Amer. Math. Soc. **43**, 768–771 (1996)
2. —: Correspondance mathématique entre Legendre et Jacobi. J. Reine Angew. Math. **80**, 205–279 (1875). [Reprint: [3287].]
3. —: Festschrift Heinrich Weber zu seinem siebzigsten Geburtstag am 5. März 1912. Teubner (1912). [Reprint: Chelsea, 1912.]
4. —: Festschrift zur Feier des 100. Geburtstages Eduard Kummers. Teubner (1910)
5. —: Heinrich Grell (3.2.1903–21.8.1974). Math. Nachr. **65**, 5–6 (1975)
6. —: Nikolaĭ Mikhaĭlovič Korobov (November 23, 1917 – October 25, 2004). Čebyshevskiĭ. Sb. 6, 224–230 (2005)
7. —: László Rédei (1900–1980). Acta Sci. Math. (Szeged) **43**, 3–4 (1981)
8. —: Leonhard Euler: 1707–1783. Beiträge zu Leben und Werk. Birkhäuser (1983)
9. —: Theodore Samuel Motzkin: Professor of mathematics 1908–1970. J. Comb. Th. A **14**, 271–272 (1973)
10. —: Obituary: Tadasi Nakayama. Nagoya Math. J. **27**, i–vii (1966)
11. —: Oystein Ore (1899–1968). J. Combin. Theory **8**, i–iii (1970)
12. —: Paul T. Bateman — biography. Internat. J. Number Th. **11(5)**, xv–xviii (2015)
13. —: Séminaire Bourbaki. W.A. Benjamin, New York (1966)
14. —: Tadao Tannaka: 1908–1986. Tôhoku Math. J. **39**, i–iv (1987)
15. Abel, N.H.: Beweis der Unmöglichkeit, algebraische Gleichungen von höheren Graden als dem vierten allgemein aufzulösen. J. Reine Angew. Math. **1**, 65–84 (1826) [French translation: [18] (The pages of Abel's papers in [18] refer to the first edition.) **1**, 5–24.]
16. Abel, N.H.: Recherches sur les fonctions elliptiques. J. Reine Angew. Math. **2**, 101–181 (1827); **3**, 160–190 (1828) [[18], **1**, 141–249.]
17. Abel, N.H.: Mémoire sur une classe particulière d'équations résolubles algébriquement. J. Reine Angew. Math. **4**, 131–156 (1829) [[18], **1**, 114–140.]
18. Abel, N.H.: Oeuvres complétes, 1-2. Chr. Gröndahl, 1839. [2nd ed. 1881; reprint: J. Gabay, 1992.]
19. Abraškin, V.A.: Determination of the imaginary quadratic fields of class number two with even discriminant by the method of Heegner. Mat. Zametki **15**, 241–246 (1974) (Russian)
20. Adachi, N.: On the class number of an absolutely cyclic number field of prime degree. Proc. Japan Acad. Sci. **45**, 647–650 (1969)
21. Adachi, N.: On the relatively cyclic imbedding problem with given local behavior. J. Math. Soc. Japan **22**, 298–307 (1970)

W. Narkiewicz, *The Story of Algebraic Numbers in the First Half of the 20th Century*, Springer Monographs in Mathematics,
https://doi.org/10.1007/978-3-030-03754-3

22. Aguirre, J., Bilbao, M., Peral, J.C.: The trace of totally positive algebraic integers. Math. Comp. **75**, 385–393 (2006)
23. Aguirre, J., Peral, J.C.: The integer Chebyshev constant of Farey intervals. Publicacions Mat. Volume Extra, 11–27 (2007)
24. Ahern, P.R.: The asymptotic distribution of prime ideals in ideal classes. Indag. Math. **26**, 10–14 (1964)
25. Ahmad, H., Hajja, M., Kang, M.-C.: Rationality of some projective linear actions. J. Algebra **228**, 643–658 (2000)
26. Ahn, J.-H., Boutteaux, G., Kwon, S.-H., Louboutin, S.: The class number one problem for some non-normal CM-fields of degree $2p$. J. Number Theory **132**, 1793–1806 (2012)
27. Ahn, J.-H., Kwon, S.-H.: The class groups of the imaginary abelian number fields with Galois group $(Z/2Z)^n$. Bull. Aust. Math. Soc. **70**, 267–277 (2004)
28. Ahn, J.-H., Kwon, S.-H.: The imaginary abelian number fields of 2-power degrees with ideal class groups of exponent ≤ 2. Math. Comp. **81**, 533–554 (2012)
29. Ahn, J.-H., Kwon, S.-H.: Some explicit zero-free regions for Hecke L-functions. J. Number Theory **145**, 433–473 (2014)
30. Aigner, A.: Über die Möglichkeit von $x^4 + y^4 = z^4$ in quadratischen Körpern. Jahresber. Deutsch. Math.-Verein. **43**, 226–228 (1934)
31. Aigner, A:, Kriterium zu 8. and 16. Potenzcharakter der Reste 2 und -2. Deutsche Math. **4**, 44–52 (1939)
32. Aigner, A.: Weitere Ergebnisse über $x^3 + y^3 = z^3$ in quadratischen Körpern. Monatsh. Math. **56**, 240–252 (1952)
33. Aigner, A.: Ein zweiter Fall der Unmöglichkeit von $x^3 + y^3 = z^3$ in quadratischen Körpern mit durch 3 teilbarer Klassenzahl. Monatsh. Math. **56**, 335–338 (1952)
34. Aigner, A.: Zur einfachen Bestimmung der Klassengruppe eines imaginär quadratischen Körpers. Arch. Math. (Basel) **4**, 408–411 (1953)
35. Aigner, A.: Unmöglichkeitskernzahlen der kubischen Fermatgleichung mit Primfaktoren der Art $3n + 1$. J. Reine Angew. Math. **195**, 175–179 (1955)
36. Aigner, A.: Die kubische Fermatgleichung in quadratischen Körpern. J. Reine Angew. Math. **195**, 3–17 (1956)
37. Aigner, A.: Die Unmöglichkeit von $x^6 + y^6 = z^6$ und $x^9 + y^9 = z^9$ in quadratischen Körpern. Monatsh. Math. **61**, 147–150 (1957)
38. Akizuki, Y.: Bemerkungen über den Aufbau des Nullideals. Proc. Phys.-Math. Soc. Japan (3) **14**, 253–262 (1932)
39. Akizuki, Y.: Über die Konstruktionen der Zahlringe vom endlichen Grad und ihre Diskriminantenteiler. Tôhoku Math. J. **37**, 1–16 (1933)
40. Akizuki, Y.: Eine homomorphe Zuordnung der Elemente der galoisschen Gruppe zu den Elementen einer Untergruppe der Normklassengruppe. Math. Ann. **112**, 566–571 (1936)
41. Akizuki, Y., Mori, Y.: Ideale in quadratischen Zahlringen. Jpn. J. Math. **11**, 35–57 (1934)
42. Alaca, Ş., Williams, K.S.: On Voronoi's method for finding an integral basis of a cubic field. Utilitas Math. **65**, 163–166 (2004)
43. Albert, A.A.: The integers of normal quartic fields. Ann. Math. (2) **31**, 381–418 (1930)
44. Albert, A.A.: A determination of the integers of all cubic fields. Ann. Math. (2) **31**, 550–566 (1930)
45. Albert, A.A.: Normal division algebras of degree four over an algebraic field. Trans. Amer. Math. Soc. **34**, 363–372 (1932)
46. Albert, A.A.: Cyclic fields of degree eight. Trans. Amer. Math. Soc. **35**, 949–964 (1933)
47. Albert, A.A.: On normal Kummer fields over a non-modular field. Trans. Amer. Math. Soc. **36**, 885–892 (1934)
48. Albert, A.A.: On cyclic fields. Trans. Amer. Math. Soc. **37**, 454–462 (1935)
49. Albert, A.A.: A note on matrices defining total real fields. Bull. Amer. Math. Soc. **43**, 242–244 (1937)
50. Albert, A.A.: Normalized integral bases of algebraic number fields. Ann. Math. (2) **38**, 923–957 (1937)

51. Albert, A.A.: On p-adic fields and rational division algebras. Ann. Math. (2) **41**, 674–693 (1940)
52. Albert, A.A.: Leonard Eugene Dickson. Bull. Amer. Math. Soc. **61**, 331–345 (1955)
53. Albert, A.A., Hasse, H.: A determination of all normal division algebras over an algebraic number field. Trans. Amer. Math. Soc. **34**, 722–726 (1932)
54. Albis González, V.S.: A remark on primitive roots and ramification. Rev. Colombiana Mat. **7**, 93–98 (1973)
55. Albu, T.: On a paper of Uchida concerning simple finite extensions of Dedekind domains. Osaka J. Math. **16**, 65–69 (1979)
56. Albu, T., Nicolae, F.: Heckesche Systeme idealer Zahlen und Knesersche Körpererweiterungen. Acta Arith. **73**, 43–50 (1995)
57. Aleksandrov, A.D. et al.: Dmitriĭ Konstantinovič Faddeev (on the occasion of his eightieth birthday). Uspekhi Mat. Nauk **44**(3), 187–193 (1989) (Russian)
58. Aleksandrov, P.S. (ed.): Hilbert's Problems. Nauka (1969) (Russian) [German translation: Geest–Portig, 1971; 2nd ed. 1979.]
59. Alexanderson, G.L., Lange, L.H.: Obituary: George Pólya. Bull. Lond. Math. Soc. **19**, 559–608 (1987)
60. Amano, K.: On the Hasse norm principle for a separable extension. Bull. Fac. Gen. Ed. Gifu Univ. **15**, 10–13 (1979)
61. Amberg, E.J.: Über die Körper, dessen Zahlen sich aus zwei Quadratwurzeln zusammensetzen. Diss., Zürich (1897)
62. Amice, Y. et al.: Charles Pisot. Acta Arith. **51**, 1–4 (1988)
63. Amitsur, S.A.: Arithmetic linear transformations and abstract prime number theorems. Canad. J. Math. **13**, 83–109 (1961); Corr. **21**, 1–5 (1969)
64. Amoroso, F.: Algebraic numbers close to 1 and variants of Mahler's measure. J. Number Theory **60**, 80–96 (1996)
65. Amoroso, F.: Upper bound for the resultant and Diophantine applications. In: Number Theory (Eger 1996), 23–36, de Gruyter (1998)
66. Amoroso, F., David, S.: Le problème de Lehmer en dimension supérieure. J. Reine Angew. Math. **513**, 145–179 (1999)
67. Amoroso, F., David, S.: Minoration de la hauteur normalisée des hypersurfaces. Acta Arith. **92**, 339–366 (2000)
68. Amoroso, F., David, S., Zannier, U.: On fields with Property (B). Proc. Amer. Math. Soc. **142**, 1893–1910 (2014)
69. Amoroso, F., Dvornicich, R.: A lower bound for the height in abelian extensions. J. Number Theory **80**, 260–272 (2000)
70. Amoroso, F., Masser, D.: Lower bounds for the height in Galois extensions. Bull. Lond. Math. Soc. **48**, 1008–1012 (2016)
71. Amoroso, F., Zannier, U.: A relative Dobrowolski lower bound over abelian extensions. Ann. Scuola Norm. Sup. Pisa, Cl. Sci. (4) **29**, 711–727 (2000)
72. Amoroso, F., Zannier, U.: A uniform relative Dobrowolski's lower bound over abelian extensions. Bull. Lond. Math. Soc. **42**, 489–498 (2010)
73. Anderson, D.D.: Multiplication ideals, multiplication rings, and the ring $R(X)$. Canad. J. Math. **28**, 760–768 (1976)
74. Anderson, D.D.: Some remarks on multiplication ideals. Math. Japon. **25**, 463–469 (1980)
75. Anderson, D.D.: π-domains without identity. Lecture Notes Pure Appl. Math. **205**, 25–30 (1999)
76. Anderson, D.D., Johnson, E.W.: A new characterization of Dedekind domains. Glasg. Math. J. **28**, 237–239 (1986)
77. Anderson, D.D., Matijevic, J.: Graded π-rings. Canad. J. Math. **31**, 449–457 (1979)
78. Anderson, D.F., Kim, H., Park, J.: Factorable domains. Comm. Algebra **30**, 4113–4120 (2002)
79. Andrews, G.E. et al. (ed.): The Rademacher Legacy to Mathematics. Contemp. Math. **166** (1994)

80. Andrews, G.E. et al. (ed.): Ramanujan Revisited (Proceedings of the Centenary Conference). Academic Press (1988)
81. Andrianov, A.N. et al.: Boris F. Skubenko. An essay on his life and work. Zap. Naučn. Sem. POMI **212**, 5–14 (1995) (Russian)
82. Angell, I.O.: A table of complex cubic fields. Bull. Lond. Math. Soc. **5**, 37–38 (1973)
83. Angell, I.O.: A table of totally real cubic fields. Math. Comp. **30**, 184–187 (1976)
84. Anglès, B.: Norm residue symbol and the first case of Fermat's equation. J. Number Theory **91**, 297–311 (2001)
85. Anglès, B., Nuccio, F.A.E.: On Jacobi sums in $\mathbf{Q}(\zeta_p)$. Acta Arith. **142**, 199–218 (2010)
86. Anglès, B., Taelman, L.: On a problem à la Kummer-Vandiver for function fields. J. Number Theory **133**, 830–841 (2013)
87. Anglès, B., Taelman, L.: The Spiegelungssatz for the Carlitz module; Addendum to "On a problem à la Kummer-Vandiver for function fields". J. Number Theory **133**, 2139–2142 (2013)
88. Ankeny, N.C.: A generalization of a theorem of Suetuna on Dirichlet series. Proc. Japan Acad. Sci. **28**, 389–395 (1952)
89. Ankeny, N.C.: Representations of primes by quadratic forms. Amer. J. Math. **74**, 913–919 (1952)
90. Ankeny, N.C., Artin, E., Chowla, S.: The class-number of real quadratic fields. Proc. Nat. Acad. Sci. U.S.A. **37**, 524–525 (1951)
91. Ankeny, N.C., Artin, E., Chowla, S.: The class number of real quadratic fields. Ann. Math. (2) **56**, 479–493 (1952)
92. Ankeny, N.C., Brauer, R., Chowla, S.: A note on the class-numbers of algebraic number fields. Amer. J. Math. **78**, 51–61 (1956)
93. Ankeny, N.C., Chowla, S.: The class number of the cyclotomic field. Proc. Nat. Acad. Sci. U.S.A. **35**, 529–532 (1949)
94. Ankeny, N.C., Chowla, S.: The relation between the class number and the distribution of primes. Proc. Amer. Math. Soc. **1**, 775–776 (1950)
95. Ankeny, N.C., Chowla, S.: The class number of the cyclotomic field. Canad. J. Math. **3**, 486–491 (1951)
96. Ankeny, N.C., Chowla, S.: On the divisibility of the class number of quadratic fields. Pacific J. Math. **5**, 321–324 (1955)
97. Ankeny, N.C., Chowla, S., Hasse, H.: On the class-number of the maximal real subfield of a cyclotomic field. J. Reine Angew. Math. **217**, 217–220 (1965)
98. Apostol, T.: The resultant of the cyclotomic polynomials $F_m(ax)$ and $F_n(bx)$. Math. Comp. **29**, 1–6 (1975)
99. Aramata, H.: Über die Teilbarkeit der Zetafunktionen gewisser algebraischer Zahlkörper. Proc. Imp. Acad. Tokyo **7**, 334–336 (1931)
100. Aramata, H.: Über die Teilbarkeit der Dedekindschen Zetafunktionen. Proc. Imp. Acad. Tokyo **9**, 31–34 (1933)
101. Aramata, H.: Über die Eindeutigkeit der Artinschen L-Funktionen. Proc. Imp. Acad. Tokyo **15**, 124–126 (1939)
102. Aranés, M., Arenas, A.: On the defining polynomials of maximal real cyclotomic extensions. Rev. R. Acad. Cienc. Exactas Fís. Nat. Ser. A Math. **102**, 183–191 (2008)
103. Arason, J.K., Baeza, R.: On the level of principal ideal domains. Arch. Math. (Basel) **96**, 519–524 (2011)
104. Arf, C.: Untersuchungen über reinverzweigte Erweiterungen diskret bewerteter perfekter Körper. J. Reine Angew. Math. **181**, 1–44 (1939)
105. Armitage, J.V.: On a theorem of Hecke in number fields and function fields. Invent. Math. **2**, 238–246 (1967)
106. Armitage, J.V.: Zeta functions with a zero at $s = 1/2$. Invent. Math. **15**, 199–205 (1972)
107. Armitage, J.V., Fröhlich, A.: Classnumbers and unit signatures. Mathematika **14**, 94–98 (1967)

108. Arndt, F.: Zur Theorie der binären kubischen Formen. J. Reine Angew. Math. **53**, 309–321 (1857)
109. Arndt, F., Ueber die Anzahl der Genera der quadratischen Formen. J. Reine Angew. Math. **56**, 72–78 (1858)
110. Arndt, F.: Einfacher Beweis der Irreductibilität einer Gleichung in der Kreistheilung. J. Reine Angew. Math. **56**, 178–181 (1858)
111. Arno, S.: The imaginary quadratic fields of class number 4. Acta Arith. **60**, 321–334 (1992)
112. Arno, S., Robinson, M.L., Wheeler, F.S.: Imaginary quadratic fields with small odd class number. Acta Arith. **83**, 295–330 (1998)
113. Arnold, A., Monagan, M.: Calculating cyclotomic polynomials. Math. Comp. **80**, 2359–2379 (2011)
114. Arthur, J. et al.: Armand Borel (1923–2003). Notices Amer. Math. Soc. **51**, 498–524 (2004)
115. Artin, E.: Über die Zetafunktionen gewisser algebraischer Zahlkörper. Math. Ann. **89**, 147–156 (1923) [[134], 95–104.]
116. Artin, E.: Quadratische Körper im Gebiet der höheren Kongruenzen, I. Math. Z. **19**, 153–206 (1924) [[132], 1–54.]
117. Artin, E.: Quadratische Körper im Gebiet der höheren Kongruenzen, II. Math. Z. **19**, 207–246 (1924) [[132], 55–94.]
118. Artin, E.: Über eine neue Art von L-Reihen. Abh. Math. Semin. Univ. Hambg **3**, 89–108 (1924) [[132], 105–124.]
119. Artin, E.: Über die Zerlegung definiter Funktionen in Quadrate. Abh. Math. Semin. Univ. Hambg **5**, 100–115 (1927) [[132], 273–288.]
120. Artin, E.: Beweis des allgemeinen Reziprozitätsgesetzes, Abh. Math. Semin. Univ. Hambg **5**, 353–363 (1927) [[132], 131–141.]
121. Artin, E.: Idealklassen in Oberkörpern und allgemeines Reziprozitätsgesetz. Abh. Math. Semin. Univ. Hambg **7**, 46–51 (1929) [[132], 159–164.]
122. Artin, E.: Zur Theorie der L-Reihen mit allgemeinen Gruppencharakteren. Abh. Math. Semin. Univ. Hambg **8**, 292–306 (1930) [[132], 165–179.]
123. Artin, E.: Die gruppentheoretische Struktur der Diskriminanten algebraischer Zahlkörper. J. Reine Angew. Math. **164**, 1–11 (1931) [[132], 180–194.]
124. Artin, E.: Über Einheiten relativ galoischer Zahlkörper. J. Reine Angew. Math. **167**, 153–156 (1932) [[132], 195–198.]
125. Artin, E.: Über die Bewertungen algebraischer Zahlkörper. J. Reine Angew. Math. **167**, 157–159 (1932) [[132], 199–201.]
126. Artin, E.: Linear mappings and the existence of a normal basis. In: Studies and Essays presented to R. Courant on his 60th Birthday, 1–5. Interscience Publ., (1948) [[132], 376–379.]
127. Artin, E.: Questions de base minimale dans la théorie des nombres algébriques. In: Algébre et Théorie des Nombres, Colloques du CNRS **24**, 19–20 (1950) [[132], 229–231.]
128. Artin, E.: Algebraic numbers and algebraic functions, I. New York University (1951)
129. Artin, E.: Theory of Algebraic Numbers. George Strike (1959)
130. Artin, E.: Algebraic Numbers and Algebraic Functions. Gordon and Breach (1967) [Reprint: Amer. Math. Soc. 2006.]
131. Artin, E.: Quadratische Körper über Polynombereichen Galois'scher Felder und ihre Zetafunktionen. Abh. Math. Semin. Univ. Hambg **70**, 3–30 (2000)
132. Artin, E.: The Collected Papers of Emil Artin. Addison-Wesley (1965) [2nd ed. Springer (1982).]
133. Artin, E., Hasse, H.: Über den zweiten Ergänzungssatz zum Reziprozitätsgesetz der l-ten Potenzreste im Körper k_ζ der l-ten Einheitswurzeln und in Oberkörpern von k_ζ. J. Reine Angew. Math. **154**, 143–148 (1925) [[132], 125–130; [1713], **1**, 247–252.]
134. Artin, E., Hasse, H.: Die beiden Ergänzungssätze zum Reziprozitätsgesetz der l^n-ten Potenzreste im Körper der l^nten Einheitswurzeln. Abh. Math. Semin. Univ. Hambg **6**, 146–162 (1928) [[132], 142–158; [1713], **1**, 326–342.]

135. Artin, E., Schreier, O.: Algebraische Konstruktion reeller Körper. Abh. Math. Semin. Univ. Hambg **5**, 85–99 (1926) [[132], 258–272.]
136. Artin, E., Schreier, O.: Eine Kennzeichnung der reell abgeschlossenen Körper. Abh. Math. Semin. Univ. Hambg **5**, 225–231 (1927). [[132], 289–295.]
137. Artin, E., Tate, J.: Class Field Theory. W.A. Benjamin (1967) [2nd ed. Addison-Wesley (1990); reprint: Amer. Math. Soc., Chelsea (2009).]
138. Artin, E., Whaples, G.: Axiomatic characterization of fields by the product formula for valuations. Bull. Amer. Math. Soc. **51**, 469–492 (1945) [[132], 202–225.]
139. Artin, E., Whaples, G.: A note on axiomatic characterization of fields. Bull. Amer. Math. Soc. **52**, 245–247 (1946) [[132], 226–228.]
140. Arutjunjan, L.Z.: Generators of the group of principal units of a cyclic p-extension of a regular local field. Zap. Naučn. Sem. LOMI **71**, 16–23 (1977) (Russian)
141. Artzy, R.: Kurt Reidemeister (13.10.1893–8.7.1971). Jahresber. Deutsch. Math.-Verein. **74**, 96–104 (1972/73)
142. Arwin, A.: Einige periodische Kettenbruchentwicklungen. J. Reine Angew. Math. **155**, 111–128 (1926)
143. Arwin, A.: Weitere periodische Formationen. J. Reine Angew. Math. **159**, 180–196 (1928)
144. Asano, K.: Über Moduln und Elementarteilertheorie im Körper, in dem Arithmetik definiert ist. Jpn. J. Math. **20**, 55–71 (1950)
145. Asano, K.: Über kommutative Ringe, in denen jedes Ideal als Produkt von Primidealen darstellbar ist. J. Math. Soc. Japan **3**, 82–90 (1951)
146. Asano, K., Nakayama, T.: A remark on the arithmetic in a subfield. Proc. Japan Acad. Sci. **16**, 529–531 (1940)
147. Ashrafi, N., Vámos, P.: On the unit sum number of some rings. Quart. J. Math., Oxford ser. **56**, 1–12 (2005)
148. Asimi, A., Lbekkouri, A.: Sum of two squares in a cyclotomi-quadratic field. Algebras Groups Geom. **22**, 201–213 (2005)
149. Askey, R., Nevai, P.: Gabor Szegő: 1895–1985. Math. Intelligencer **18/2**, 10–22 (1996)
150. Atiyah, M.F.: Obituary: John Arthur Todd. Bull. Lond. Math. Soc. **30**, 305–316 (1998)
151. Auslander, L., Tolimieri, R., Winograd, S.: Hecke's theorem in quadratic reciprocity, finite nilpotent groups and the Cooley-Tukey algorithm. Adv. Math. **43**, 122–172 (1982)
152. Awtrey, C.: Dodecic 3-adic fields. Internat. J. Number Th. **8**, 933–944 (2012)
153. Ax, J.: On the units of an algebraic number field. Illinois J. Math. **9**, 584–589 (1965)
154. Axer, A.: Zahlentheoretische Funktionen und deren asymptotische Werte im Gebiete der aus den dritten Einheitswurzeln gebildeten ganzen komplexen Zahlen. Monatsh. Math. Phys. **15**, 239–291 (1904)
155. Ayoub, R.G.: On the Waring-Siegel theorem. Canad. J. Math. **5**, 439–450 (1953)
156. Ayoub, R.G.: On Selberg's lemma for algebraic fields. Canad. J. Math. **7**, 138–143 (1955)
157. Ayoub, R.G.: A note on the class number of imaginary quadratic fields. Math. Comp. **21**, 442–445 (1967)
158. Ayoub, R.G.: On the coefficients of the zeta function of an imaginary quadratic field. Acta Arith. **13**, 375–381 (1967/68)
159. Ayoub, R.G.: Sarvadaman Chowla. J. Number Theory **11**, 286–301 (1979); Corr. **12**, p. 139 (1980)
160. Ayoub, R.[G.]: The lemniscate and Fagnano's contributions to elliptic integrals. Arch. Hist. Exact Sci. **29**, 131–149 (1984)
161. Ayoub, R.G., Chowla, S.: On Euler's polynomial. J. Number Theory **13**, 443–445 (1981)
162. Azuhata, T., Ichimura, H.: On the divisibility problem of the class numbers of algebraic number fields. J. Fac. Sci. Univ. Tokyo **30**, 579–585 (1984)
163. Baas, N.A., Skau, F.C.: The lord of the number, Atle Selberg. On his life and mathematics. Bull. Amer. Math. Soc. (N.S.) **45**, 617–649 (2008)
164. Babai, L., Pomerance, C., Vértesi, P.: The mathematics of Paul Erdős. Notices Amer. Math. Soc. **45**, 19–31 (1998)
165. Babai, L., Spencer, J.: Paul Erdős (1913–1996). Notices Amer. Math. Soc. **45**, 64–73 (1998)

166. Bachman, G.: Flat cyclotomic polynomials of order three. Bull. Lond. Math. Soc. **38**, 53–60 (2006)
167. Bachmann, P.: De unitatum complexarum theoria. Habilitation thesis, Schade (1864)
168. Bachmann, P.: Zur Theorie der complexen Zahlen. J. Reine Angew. Math. **67**, 200–204 (1867)
169. Bachmann, P.: Die Lehre von der Kreistheilung und ihre Beziehung zur Zahlentheorie. Teubner (1872)
170. Bachmann, P.: Zur Theorie von Jacobi's Kettenbruch-Algorithmen. J. Reine Angew. Math. **75**, 25–35 (1873)
171. Bachmann, P.: Ergänzung einer Untersuchung von Dirichlet. Math. Ann. **16**, 537–550 (1880)
172. Bachmann, P.: Zahlentheorie. Versuch einer Gesamtdarstellung dieser Wissenschaft in ihren Hauptteilen. V. Allgemeine Arithmetik der Zahlenkörper. Teubner (1905)
173. Bachmann, P.: Das Fermatproblem in seiner bisherigen Entwicklung. De Gruyter (1919)
174. Baeza, R.: Über die Stufe von Dedekind Ringen. Arch. Math. (Basel) **33**, 226–231 (1979/80)
175. Baily, A.M.: On the density of discriminants of quartic fields. J. Reine Angew. Math. **315**, 190–210 (1980)
176. Baily, A.M., On octic fields of exponent 2. J. Reine Angew. Math. **328**, 33–38 (1981)
177. Baker, A.: Linear forms in the logarithms of algebraic numbers. Mathematika **13**, 204–216 (1966)
178. Baker, A.: A remark on the class number of quadratic fields. Bull. Lond. Math. Soc. **1**, 98–102 (1969)
179. Baker, A.: Imaginary quadratic fields with class number 2. Ann. Math. (2) **94**, 139–152 (1971)
180. Baker, A.: On the class number of imaginary quadratic fields. Bull. Amer. Math. Soc. **77**, 678–684 (1971)
181. Baker, A.: Transcendental Number Theory. Cambridge University Press (1975); 2nd ed. (1990)
182. Baker, H.F.: Francis Sowerby Macaulay. J. Lond. Math. Soc. **13**, 157–190 (1938)
183. Baker, M., Mazur, B., Ribet, K.: Robert F. Coleman 1954–2014. Res. Math. Sci. **2/21** 1–15 (2015)
184. Bambah, R.P., Woods, A.C.: Minkowski's conjecture for $n = 5$; a theorem of Skubenko. J. Number Theory **12**, 27–48 (1980)
185. Barat, G., Grabner, P.J., Hellekalek, P.: Pierre Liardet (1943–2014) in memoriam. Eur. Math. Soc. Newsl. **97**, 52–58 (2015)
186. Barban, M.B.: The "large sieve" of Ju.V.Linnik and limit theorem for the class-number of an imaginary quadratic field. Izv. Akad. Nauk SSSR, Ser. Mat. **26**, 573–580 (1962) (Russian)
187. Barban, M.B.: On a theorem of P. Bateman, S. Chowla and P. Erdös. Magyar Tud. Akad. Mat. Kutató Int. Kőzl. **9**, 429–435 (1965) (Russian)
188. Barban, M.B., Gordover, G.: On the moments of the class-numbers of purely radical quadratic forms of negative determinant. Dokl. Akad. Nauk SSSR **167**, 267–269 (1966) (Russian)
189. Barner, K.: Über die quaternäre Einheitsform in total reellen algebraischen Zahlkörpern. J. Reine Angew. Math. **229**, 194–208 (1968)
190. Barner, K.: Über die Werte der Ringklassen-L-Funktionen reellquadratischer Zahlkörper an natürlichen Argumentstellen. J. Number Theory **1**, 28–64 (1969)
191. Barner, K.: Paul Wolfskehl and the Wolfskehl prize. Notices Amer. Math. Soc. **44**, 1294–1303 (1997)
192. Barnes, E.S., Swinnerton-Dyer, H.P.F.: The inhomogeneous minima of binary quadratic forms, I. Acta Math. **87**, 259–323 (1952)
193. Barnes, F.W.: On the stufe of an algebraic number field. J. Number Theory **4**, 474–476 (1972)
194. Barroero, F., Frei, C., Tichy, R.F.: Additive unit representations in rings over global fields – a survey. Publ. Math. Debrecen **79**, 291–307 (2011)

195. Barrucand, P., Cohn, H.: Note on primes of type $x^2 + 32y^2$, class number, and residuacity. J. Reine Angew. Math. **238**, 67–70 (1969)
196. Barrucand, P., Cohn, H.: A rational genus, class number divisibility, and unit theory for pure cubic fields. J. Number Theory **2**, 7–21 (1970)
197. Barrucand, P., Laubie, F.: Sur les symboles des restes quadratiques des discriminants. Acta Arith. **48**, 81–88 (1987)
198. Bartels, H.J.: Zur Arithmetik in Diedergruppenerweiterungen. Math. Ann. **256**, 465–473 (1981)
199. Bartels, H.J.: Zur Arithmetik von Konjugationsklassen in algebraischen Gruppen. J. Algebra **70**, 179–199 (1981)
200. Bartholdi, L., Ceccherini-Silberstein, T.G.: Salem numbers and growth series of some hyperbolic graphs. Geom. Dedicata **90**, 107–114 (2002)
201. Bartz, K.[M.]: On a theorem of A.V. Sokolovskiĭ. Acta Arith. **34**, 113–126 (1977/78)
202. Bartz, K.[M.]: On zero-free regions for the Hecke-Landau zeta functions. Funct. Approx. Comment. Math. **14**, 101–107 (1984)
203. Bartz, K.M.: Some remarks on zero-free regions for Hecke-Landau zeta functions. Discuss. Math. **7**, 113–117 (1985)
204. Bartz, K.M.: An effective order of Hecke–Landau zeta functions near the line $\sigma = 1$, I. Acta Arith. **50**, 183–193 (1988); corr. **58**, p. 211 (1991)
205. Bartz, K.M.: An effective order of Hecke-Landau zeta functions near the line $\sigma = 1$. (Some applications). Acta Arith. **52**, 163–170 (1989)
206. Bartz, K.M., Fryska, T.L.: An effective estimate for the density of zeros of Hecke-Landau zeta-functions. Acta Arith. **52**, 339–352 (1989)
207. Bartz, K.[M.], Staś, W.: On the order of Hecke-Landau zeta functions near the line $\sigma = 1$. Funct. Approx. Comment. Math. **15**, 1986, 131–137 (1986)
208. Basilla, J.M.: The quadratic fields with discriminant divisible by exactly two primes and with "narrow" class-number divisible by 8. Proc. Japan Acad. Sci. **80**, 187–190 (2004)
209. Basilla, J.M., Wada, H.: On efficient computation of the 2-parts of ideal class groups of quadratic fields. Proc. Japan Acad. Sci. **80**, 191–193 (2004)
210. Bass, H. et al.: Samuel Eilenberg (1913–1998). Notices Amer. Math. Soc. **45**, 1344–1352 (1998)
211. Bass, H., Estes, D.R., Guralnick, R.M.: Eigenvalues of symmetric matrices and graphs. J. Algebra **168**, 536–567 (1994)
212. Bass, H., Lam, T.Y.: Irving Kaplansky 1917–2006. Notices Amer. Math. Soc. **54**, 1477–1493 (2007)
213. Bateman, P.T., Chowla, S., Erdös, P.: Remarks on the size of $L(1, \chi)$. Publ. Math. Debrecen **1**, 165–182 (1950)
214. Bateman, P.T., Stemmler, R.M.: Waring's problem for algebraic number fields and primes of the form $(p^r - 1)/(p^d - 1)$. Illinois J. Math. **6**, 142–156 (1962)
215. Bauer, F.L.: In memoriam John Todd (1911–2007). Numer. Math. **108**, 1–6 (2007)
216. Bauer, H.: Numerische Bestimmung von Klassenzahlen reeller zyklischer Zahlkörper. J. Number Theory **1**, 161–162 (1969)
217. Bauer, H.: Zur Berechnung der 2-Klassenzahl der quadratischen Zahlkörper mit genau zwei verschiedenen Diskriminantenprimteilern. J. Reine Angew. Math. **248**, 42–46 (1971)
218. Bauer, M.: Sur les congruences identiques. Nouv. Ann. Math. (4) **2**, 256–264 (1902)
219. Bauer, M.: Über einen Satz von Kronecker. Arch. Math. Phys. (3) **6**, 218–219 (1903)
220. Bauer, M.: Über Kreisteilungsgleichungen. Arch. Math. Phys. (3) **6**, p. 220 (1903)
221. Bauer, M.: Über zusammengesetzte Körper. Arch. Math. Phys. (3) **6**, 221–222 (1903)
222. Bauer, M.: Beitrag zur Theorie der irreduziblen Gleichungen. J. Reine Angew. Math. **128**, 298–301 (1905)
223. Bauer, M.: Zur allgemeinen Theorie der algebraischen Grössen. J. Reine Angew. Math. **132**, 21–32 (1907)
224. Bauer, M.: Über Gleichungen ohne Affekt. J. Reine Angew. Math. **132**, 33–35 (1907)
225. Bauer, M.: Ganzzahlige Gleichungen ohne Affekt. Math. Ann. **64**, 325–327 (1907)

226. Bauer, M.: Über die ausserwesentlichen Diskriminantenteiler einer Gattung. Math. Ann. **64**, 573–576 (1907)
227. Bauer, M.: Zur Theorie der arithmetischen Progression. Arch. Math. Phys. (3) **25**, 131–134 (1916)
228. Bauer, M.: Zur Theorie der algebraischen Zahlkörper. Math. Ann. **77**, 353–356 (1916)
229. Bauer, M.: Über zusammengesetzte Zahlkörper. Math. Ann. **77**, 357–361 (1916)
230. Bauer, M.: Eine algebraische Behauptung von Gauss. Jahresber. Deutsch. Math.-Verein. **26**, 348–349 (1918)
231. Bauer, M.: Bemerkungen über die Differente des algebraischen Zahlkörpers. Math. Ann. **79**, 321–322 (1919)
232. Bauer, M.: Zur Theorie der Fundamentalgleichung. J. Reine Angew. Math. **149**, 89–96 (1919)
233. Bauer, M.: Bemerkungen über die Zusammensetzung der algebraischen Zahlkörper. J. Reine Angew. Math. **150**, 185–188 (1919/20)
234. Bauer, M.: Über relativ Galoissche Zahlkörper. Math. Ann. **83**, 70–73 (1921)
235. Bauer, M.: Über die Differente eines algebraischen Zahlkörpers. Math. Ann. **83**, 74–76 (1921)
236. Bauer, M.: Beweis von einigen bekannten Sätzen über zusammengesetzte Körper ohne Anwendung der Idealtheorie. Jahresber. Deutsch. Math.-Verein. **30**, 186–188 (1921)
237. Bauer, M.: Über ein Problem von Dedekind. Acta Litt. Sci. Reg. Univ. Hungar. Franc.-Jos., sect. Math. **1**, 14–17 (1922/23)
238. Bauer, M.: Die Theorie der p-adischen bzw. $\mathfrak{P}$-adischen Zahlen und die gewöhnlichen algebraischen Zahlkörper. Math. Z. **14**, 244–249 (1922)
239. Bauer, M.: Verschiedene Bemerkungen über die Differente und die Diskriminante eines algebraischen Zahlkörpers. Math. Z. **16**, 1–12 (1923)
240. Bauer, M.: Ganzzahlige Gleichungen ohne Affekt. Math. Z. **16**, 318–319 (1923)
241. Bauer, M.: Über die Erweiterung eines algebraischen Zahlkörpers durch Henselsche Grenzwerte. Acta Litt. Sci. Reg. Univ. Hungar. Franc.-Jos., sect. Math. **1**, 74–79 (1922/23)
242. Bauer, M.: Bemerkungen zur Theorie der Differente. Acta Litt. Sci. Reg. Univ. Hungar. Franc.-Jos., sect. Math. **1**, 195–198 (1922/23)
243. Bauer, M.: Über die Erweiterung des Körpers der p-adischen Zahlen zu einem algebraisch abgeschlossenen Körper. Math. Z. **19**, 308–312 (1924)
244. Bauer, M.: Die Theorie der p-adischen bzw. $\mathfrak{P}$-adischen Zahlen und die gewöhnlichen algebraischen Zahlkörper. II. Math. Z. **20**, 95–97 (1924)
245. Bauer, M.: Zur Theorie der algebraischen Körper. Acta Litt. Sci. Reg. Univ. Hungar. Franc.-Jos., sect. Math. **2**, 69–71 (1924/26)
246. Bauer, M.: Bemerkung zur Algebra. Acta Litt. Sci. Reg. Univ. Hungar. Franc.-Jos., sect. Math. **4**, 244–245 (1929)
247. Bauer, M.: Über einen Takagischen Satz. J. Reine Angew. Math. **163**, 249–250 (1930)
248. Bauer, M.: Elementare Bemerkung über identische Kongruenzen. Acta Litt. Sci. Reg. Univ. Hungar. Franc.-Jos., sect. Math. **6**, 46–48 (1932/34)
249. Bauer, M.: Bemerkungen zum Hensel-Oreschen Hauptsätze. Acta Litt. Sci. Reg. Univ. Hungar. Franc.-Jos., sect. Math. **8**, 64–67 (1936/37)
250. Bauer, M.: Über den Führer eines Ringes in algebraischen Zahlkörpern. Duke Math. J. **2**, 578–580 (1936)
251. Bauer, M.: Bemerkungen über die Galoissche Gruppe einer Gleichung. Math. Ann. **114**, 352–354 (1937)
252. Bauer, M.: Zur Theorie der Kreiskörper. Acta Litt. Sci. Reg. Univ. Hungar. Franc.-Jos., sect. Math. **9**, 110–112 (1938/40)
253. Bauer, M.: Über zusammengesetzte relativ Galois'sche Zahlkörper. Acta Litt. Sci. Reg. Univ. Hungar. Franc.-Jos., sect. Math. **9**, 206–211 (1938/40)
254. Bauer, M.: Über die Zusammensetzung algebraischer Zahlkörper. Acta Litt. Sci. Reg. Univ. Hungar. Franc.-Jos., sect. Math. **9**, 212–217 (1938/40)

255. Bauer, M., Tschebotaröw, N. [Čebotarev, N.G.]: p-adischer Beweis des zweiten Hauptsatzes von Herrn Ore. Acta Litt. Sci. Reg. Univ. Hungar. Franc.-Jos., sect. Math. **4**, 56–57 (1928/29)
256. Bayad, A.: Loi de réciprocité quadratique dans les corps quadratiques imaginaires. Ann. Inst. Fourier (Grenoble) **45**, 1223–1237 (1995)
257. Bayer-Fluckiger, E.: Upper bounds for Euclidean minima of algebraic number fields. J. Number Theory **121**, 305–323 (2006)
258. Bayer-Fluckiger, E. Maciak, P.: Upper bounds for the Euclidean minima of abelian fields of odd prime power conductor. Math. Ann. **357**, 1071–1089 (2013)
259. Bayer-Fluckiger, E., Maciak, P.: Upper bounds for the Euclidean minima of abelian fields. J. Théor. Nombres Bordeaux **27**, 689–697 (2015)
260. Bayer-Fluckiger, E., Suarez, I.: Ideal lattices over totally real number fields and Euclidean minima. Arch. Math. (Basel) **86**, 217–225 (2006)
261. Beach, B.D., Williams, H.C., Zarnke, C.R.: Some computer results on units in quadratic and cubic fields. Proceedings of the Twenty-Fifth Summer Meeting of the Canadian Math. Congress, 609–648. Lakehead University (1971)
262. Becker, E.: Summen n-ter Potenzen in Körpern. J. Reine Angew. Math. **307/308**, 8–30 (1979)
263. Becker, P.-G..: Periodizitätseigenschaften p-adischer Kettenbrüche. Elem. Math. **45**, 1–8 (1990)
264. Bedocchi, E.: L'anneau $Z[\sqrt{14}]$ et l'algorithme euclidien. Manuscripta Math. **53**, 199–216 (1985)
265. Bedocchi, E.: On the second minimum of a quadratic form and its applications. Riv. Mat. Univ. Parma (4) **15**, 175–190 (1989)
266. Beeger, N.G.W.H.: Über die Teilkörper des Kreiskörpers $K(e^{2\pi i/l^h})$. Proc. Nederl. Akad. Wet. Amsterdam **21**, 454–465 (1919) 758–773, 774–779
267. Beeger, N.G.W.H.: Bestimmung der Klassenzahl der Ideale aller Unterkörper des Kreiskörpers der m-ten Einheitswurzeln, wo m durch mehr als eine Primzahl teilbar ist. Proc. Nederl. Akad. Wet. Amsterdam **22**, 331–350, 395–414 (1920); corr. **23**, 1399–1401 (1922)
268. Beeger, N.G.W.H.: On a problem of the theory of numbers and its history. Nieuw Arch. Wiskd. (2) **22**, 306–309 (1948)
269. Behn, A., van der Merwe, A.B.: An algorithmic version of the theorem by Latimer and Mac Duffee for 2×2 integral matrices. Linear Algebra Appl. **346**, 1–14 (2002)
270. Behnke, H.: Über die Verteilung von Irrationalitäten mod. 1. Abh. Math. Semin. Univ. Hambg **1**, 252–267 (1922)
271. Behnke, H.: Über analytische Funktionen und algebraische Zahlen. Abh. Math. Semin. Univ. Hambg **2**, 81–111 (1923)
272. Behnke, H.: Paul Sengenhorst (1894–1968). Math.-Phys. Semesterberichte **15**, 235–239 (1968)
273. Behnke, H., Köthe, G.: Heinz Prüfer. Jahresber. Deutsch. Math.-Verein. **45**, 32–40 (1935)
274. Behrbohm, H., Rédei, L.: Der Euklidische Algorithmus in quadratischen Körpern. J. Reine Angew. Math. **174**, 192–205 (1936)
275. Belabas, K.: On quadratic fields with large 3-rank. Math. Comp. **73**, 2061–2074 (2004)
276. Belabas, K., Bhargava, M., Pomerance, C.: Error estimates for the Davenport-Heilbronn theorems. Duke Math. J. **153**, 173–210 (2010)
277. Belabas, K., Fouvry, É.: Discriminants cubiques et progressions arithmétiques. Internat. J. Number Th. **6**, 1491–1529 (2010)
278. Belcher, P.: Integers expressible as sums of distinct units. Bull. Lond. Math. Soc. **6**, 66–68 (1974)
279. Belcher, P.: A test for integers being sums of distinct units applied to cubic fields. J. Lond. Math. Soc. (2) **12**, 141–148 (1975/76)
280. Belhoste, B.: Cauchy, 1789–1857. Librairie Classique Eugéne Belin (1985) [English translation: Augustin-Louis Cauchy. A biography, Springer (1991)]
281. Bell, J.L.: Obituary: Edward Hubert Linfoot. Bull. Lond. Math. Soc. **16**, 52–58 (1934)

282. Belyĭ, G.V.: Galois extensions of a maximal cyclotomic field. Izv. Akad. Nauk SSSR, Ser. Mat. **43**, 267–276 (1979) (Russian)
283. Belyĭ, G.V.: On extensions of the maximal cyclotomic field having a given classical Galois group. J. Reine Angew. Math. **341**, 147–156 (1983)
284. Bender, E.A.: Classes of matrices and quadratic fields. Linear Algebra Appl. **1**, 15–201 (1968)
285. Benjamin, E.: On imaginary quadratic number fields with 2-class group of rank 4 and infinite 2-class field tower. Pacific J. Math. **201**, 257–266 (2001)
286. Benjamin, E., Lemmermeyer, F., Snyder, C.: Imaginary quadratic fields k with cyclic $Cl_2(k^1)$. J. Number Theory **67**, 229–245 (1997)
287. Benjamin, E., Lemmermeyer, F., Snyder, C.: Imaginary quadratic fields k with $Cl_2(k) \cong (2, 2m)$ and rank $Cl_2(k^1) = 2$. Pacific J. Math. **198**, 15–31 (2001)
288. Benjamin, E., Lemmermeyer, F., Snyder, C.: Imaginary quadratic fields with $Cl_2(k) \cong (2, 2, 2)$, J. Number Theory **103**, 38–70 (2003)
289. Benkart, K., Kaplansky, I., McCrimmon, K., Saltman, D.J., Seligman, G.B.: Nathan Jacobson (1910–1999). Notices Amer. Math. Soc. **47**, 1061–1071 (2000)
290. Bennett, E.R.: Factoring in a domain of rationality. Amer. Math. Monthly **15**, 222–226 (1908)
291. Bensebaa, B., Movahhedi, A., Salinier, A.: The Galois group of $X^p + aX^s + a$. Acta Arith. **134**, 55–65 (2008)
292. Bensebaa, B., Movahhedi, A., Salinier, A.: The Galois group of $X^p + aX^{p-1} + a$. J. Number Theory **129**, 824–830 (2009)
293. Bérczes, A., Evertse, J.-H., Győry, K.: On the number of equivalence classes of binary forms of given degree and given discriminant. Acta Arith. **113**, 363–399 (2004)
294. Beresnevich, V., Bernik, V.[I.], Götze, F.: The distribution of close conjugate algebraic numbers. Compositio Math. **146**, 1165–1179 (2010)
295. Berg, E.: Über die Existenz eines Euklidischen Algorithmus in quadratischen Zahlkörpern. Fysiograf. Sällsk. Lund Forhandl. **5(5)** 1–6 (1935)
296. Berg, M.C.: The Fourier-analytic Proof of Quadratic Reciprocity. Wiley-Interscience (2000)
297. Bergé, A.-M., Martinet, J., Olivier, M.: The computation of sextic fields with a quadratic subfield. Math. Comp. **54**, 869–884 (1990)
298. Berger, A.: Recherches sur les nombres et les fonctions de Bernoulli. Acta Math. **14**, 249–304 (1891)
299. Berger, R.I.: Hasse's class number product formula for generalized Dirichlet fields and other types of number fields. Manuscripta Math. **76**, 397–406 (1992)
300. Berger, T.R., Reiner, I.: A proof of the normal basis theorem. Amer. Math. Monthly **82**, 915–918 (1975)
301. Bergström, H.: Über die Methode von Woronoj zur Berechnung einer Basis eines kubischen Zahlkörpers. Ark. Mat. **25(26)**, 1–8 (1937)
302. Bergström, H.: Zur Theorie der biquadratischen Zahlkörper. Die Arithmetik auf klassenkörpertheoretischer Grundlage. Nova Acta Soc. Sci. Upsal. (4) **10(8)**, 1–56 (1937)
303. Bergström, H.: Vereinfachter Beweis des Hauptidealsatzes der Klassenkörpertheorie. Ark. Mat. **29B(6)**, 1–6 (1943)
304. Bergström, H.: Die Klassenzahlformel für reelle quadratische Zahlkörper mit zusammengesetzter Diskriminante als Produkt verallgemeinerter Gaussscher Summen. J. Reine Angew. Math. **186**, 91–115 (1944)
305. Bernays, P.: Über die Darstellung von positiven, ganzen Zahlen durch die primitiven, binären quadratischen Formen einer nicht-quadratischen Diskriminante. Dissertation, Univ. Göttingen (1912)
306. Berndt, B.C.: On the average order of ideal functions and other arithmetical functions. Bull. Amer. Math. Soc. **76**, 1270–1274 (1970)
307. Berndt, B.C.: On the average order of a class of arithmetical functions. I. J. Number Theory **3**, 184–203 (1971)

308. Berndt, B.C.: On the average order of a class of arithmetical functions. II. J. Number Theory **3**, 288–305 (1971)
309. Bernoulli, N.: Regula generalis inveniendi aequationes, per quam alia quaepiam data, modo reducibilis sit, dividi potest. In: Leibniz, G.W., Mathematische Schriften, **3**, 827–835. H.W. Schmidt (1856)
310. Bernstein, F.: Über den Klassenkörper eines algebraischen Zahlkörpers. Nachr. Ges. Wiss. Göttingen **1903**, 46–58
311. Bernstein, F.: Über den Klassenkörper eines algebraischen Zahlkörpers. II, Nachr. Ges. Wiss. Göttingen **1903**, 304–311
312. Bernstein, F.: Über unverzweigte Abelsche Körper (Klassenkörper) in einem imaginären Grundbereich. Jahresber. Deutsch. Math.-Verein. **13**, 116–119 (1904)
313. Bernstein, F.: Über den letzten Fermatschen Lehrsatz. Nachr. Ges. Wiss. Göttingen **1910**, 482–488
314. Bernstein, F.: Über den zweiten Fall des letzten Fermatschen Lehrsatzes. Nachr. Ges. Wiss. Göttingen **1910**, 1910, 507–516.
315. Bernstein, L.: Periodical continued fractions for irrationals of degree n by Jacobi's algorithm. J. Reine Angew. Math. **213**, 31–38 (1963/64)
316. Bernstein, L.: Periodicity of Jacobi's algorithm for a special type of cubic irrationals. J. Reine Angew. Math. **213**, 137–146 (1963/64)
317. Bernstein, L.: Representation of $(D^n - d)^{1/n}$ as a periodic continued fraction by Jacobi's algorithm. Math. Nachr. **29**, 179–200 (1965)
318. Bernstein, L.: The Jacobi-Perron Algorithm, its Theory and Applications. Lecture Notes Math. 207 (1971)
319. Bernstein, L.: Units and periodic Jacobi-Perron algorithms in real algebraic number fields of degree 3. Trans. Amer. Math. Soc. **212**, 295–306 (1975)
320. Bernstein, L.: Gaining units from units. Canad. J. Math. **29**, 93–106 (1977)
321. Bernstein, L. Hasse, H.: An explicit formula for the units of an algebraic number field of degree $n \geq 2$. Pacific J. Math. **30**, 293–365 (1969)
322. Bertin, M.-J., Decomps-Guilloux, A., Grandet-Hugot, M., Pathiaux-Delefosse, M., Schreiber, J.-P.: Pisot and Salem Numbers. Birkhäuser (1992)
323. Bertin, M.-J., Pathiaux-Delefosse, M.: Conjecture de Lehmer et petits nombres de Salem. Queen's Papers in Pure and Applied Mathematics **81**. Queen's University (1989)
324. Berwick, W.E.H.: The classification of ideal numbers that depend on a cubic irrationality. Proc. London Math. Soc. (2) **12**, 393–429 (1913)
325. W.E.H.B. [Berwick, W.E.H.]: George Ballard Mathews. Proc. London Math. Soc. (2) **21**, xlvi–l (1923)
326. Berwick, W.E.H.: On cubic fields with a given discriminant. Proc. London Math. Soc. (2) **23**, 359–378 (1925)
327. Berwick, W.E.H.: Integral Bases. Cambridge University Press (1927). [Reprint: Stechert-Hafner (1964)]
328. Berwick, W.E.H.: Algebraic number-fields with two independent units. Proc. London Math. Soc. (2) **34**, 360–379 (1932)
329. Berwick, W.E.H.: The classification of ideal numbers in a cubic field. Proc. London Math. Soc. (2) **38**, 217–240 (1935)
330. Besicovitch, A.S.: On the linear independence of fractional powers of integers. J. Lond. Math. Soc. **15**, 3–6 (1940)
331. Besov, O.V. et al.: Sergeĭ Borisovič Stečkin. Uspekhi Mat. Nauk **51(6)**, 3–10 (1996) (Russian)
332. Bessassi, S.: Bounds for the degrees of CM-fields of class number one. Acta Arith. **106**, 213–245 (2003)
333. Beyer, G.: Über relativ-zyklische Erweiterungen galoisscher Körper. J. Reine Angew. Math. **196** , 34–58 (1956)
334. Beyer, G.: Über eine Vermutung von Hasse zum Erweiterungsproblem galoisscher Zahlkörper. J. Reine Angew. Math. **196**, 205–212 (1956)

335. Bhandari, S.K., Nanda V.C.: Ideal matrices for relative extensions. Abh. Math. Semin. Univ. Hambg **49**, 3–17 (1979)
336. Bhargava, M.: Higher composition laws, I. A new view on Gauss composition, and quadratic generalizations. Ann. Math. (2) **159**, 217–250 (2004)
337. Bhargava, M.: Higher composition laws, II. Ann. Math. (2) **159**, 865–886 (2004)
338. Bhargava, M.: Higher composition laws, III. The parametrization of quartic rings, Ann. Math. (2) **159**, 1329–1360 (2004)
339. Bhargava, M.: The density of discriminants of quartic rings and fields. Ann. Math. (2) **162**, 1031–1063 (2005)
340. Bhargava, M.: The density of discriminants of quintic rings and fields. Ann. Math. (2) **172**, 1559–1591 (2010)
341. Bhargava, M., Shankar, A., Tsimerman, J.: On the Davenport-Heilbronn theorems and second order terms. Invent. Math. **193**, 439–499 (2013)
342. Bhargava, M., Shankar, A., Wang, X.: Squarefree values of polynomial discriminants, I. arXiv:1611.09806.
343. Bhargava, M., Wood, M.M.: The density of discriminants of S_3-sextic number fields. Proc. Amer. Math. Soc. **136**, 1581–1587 (2008)
344. Bhaskaran, M.: Sums of mth powers in algebraic and Abelian number fields. Arch. Math. (Basel) **17**, 497–504 (1966); corr. **22**, 370–371 (1971)
345. Bianchi, L.: Sulle forme quadratiche a coefficienti e a indeterminate complesse. Rend. Accad. Lincei (4) **5(1)** 589–599 (1889)
346. Bianchi, L.: Sui gruppi di sostituzioni lineari e sulle forme quadratiche di Dirichlet e di Hermite. Rend. Accad. Lincei (4) **7(2)** 3–11 (1891)
347. Bianchi, L.: Geometrische Darstellung der Gruppen linearer Substitutionen mit ganzen complexen Coefficienten nebst Anwendungen auf die Zahlentheorie. Math. Ann. **38**, 313–333 (1891)
348. Bianchi, L.: Sui gruppi di sostituzioni lineari con coefficienti appartenenti a corpi quadratici immaginari. Math. Ann. **40**, 332–412 (1892)
349. Bickmore, C.E.: Tables connected with the Pellian equation. British Association Report **53**, 73–120 (1893)
350. Bickmore, C.E., Western, O.: A table of complex prime factors in the field of 8th roots of unity. Messenger of Math. (2) **41**, 52–64 (1911)
351. Biermann, K.-R.: Gotthold Eisenstein. Die wichtigsten Daten seines Lebens und Wirkens. J. Reine Angew. Math. **214/215**, 19–30 (1964)
352. Bilhan, M.: Théorème de Bauer dans les corps globaux. Bull. Sci. Math. (2) **105**, 299–303 (1981)
353. Bilhan, M.: Tchebotarev's density theorem in global fields. Doğa Mat. **13**, 1–8 (1989)
354. Bilharz, H.: Primdivisoren mit vorgegebener Primitivwurzel. Math. Ann. **114**, 476–492 (1937)
355. Bilu, Y., Gaál, I., Győry, K.: Index form equations in sextic fields: a hard computation. Acta Arith. **115**, 85–96 (2004)
356. Bindschedler, C.: Die Teilungskörper der elliptischen Funktionen im Bereich der dritten Einheitswurzel. J. Reine Angew. Math. **152**, 49–75 (1923)
357. Birch, B.J.: Waring's problem in algebraic number fields. Proc. Cambridge Philos. Soc. **57**, 449–459 (1961)
358. Birch, B.J.: Waring's problem for p-adic number fields. Acta Arith. **9**, 169–176 (1964)
359. Birch, B.J.: Weber's class invariants. Mathematika **16**, 283–294 (1969)
360. Birch, B.J.: Heegner's proof. In: Arithmetic of L-functions. 281–291, Amer. Math. Soc. (2011)
361. Birch, B.J., Merriman, J.R.: Finiteness theorems for binary forms with given discriminant. Proc. London Math. Soc. (3) **24**, 385–394 (1972)
362. Birch, B.J., Taylor, M.J.: Albrecht Fröhlich, 22 May 1916 — 8 November 2001. Biogr. Mems. Fell. Roy. Soc. **51**, 149–168 (2005)

363. Bird, R.F., Parry, C.J.: Integral bases for bicyclic biquadratic fields over quadratic subfields. Pacific J. Math. **66**, 29–36 (1976)
364. Birkhoff, G.D.: Note on certain quadratic number systems for which factorization is unique. Amer. Math. Monthly **13**, 156–159 (1906)
365. Bishnoi, A., Khanduja, S.K.: A class of trinomials with Galois group S_n. Algebra Colloq. **19**, Special issue, nr. 1, 905–911 (2012)
366. Bjerknes, V.: Axel Thue. Norsk Mat. Tidsskr. **4**, 33–46 (1922)
367. Blanksby, P.E.: A note on algebraic integers. J. Number Theory **1**, 155–160 (1969)
368. Blanksby, P.E., Montgomery, H.L.: Algebraic integers near the unit circle. Acta Arith. **18**, 355–369 (1971)
369. Blessenohl, D.: Supplement to: "A sharpening of the normal basis theorem". J. Algebra **132**, 154–159 (1990)
370. Blessenohl, D.: On the normal basis theorem. Note Mat. **27**, 5–10 (2007)
371. Blessenohl, D., Johnsen, K.: Eine Verschärfung des Satzes von der Normalbasis. J. Algebra **103**, 141–159 (1986)
372. Blessenohl, D., Johnsen, K.: Stabile Teilkörper galoisscher Erweiterungen und ein Problem von C.Faith. Arch. Math. (Basel) **56**, 245–253 (1991)
373. Bley, W., Boltje, R.: Cohomological Mackey functors in number theory. J. Number Theory **105**, 1–37 (2004)
374. Blichfeldt, H.F.: A new principle in the geometry of numbers, with some applications. Trans. Amer. Math. Soc. **15**, 227–235 (1914)
375. Blichfeldt, H.F.: Note on the minimum value of the discriminant of an algebraic field. Monatsh. Math. Phys. **48**, 531–533 (1939)
376. Bogomolov, F.A.: The Brauer group of quotient spaces of linear representations. Izv. Akad. Nauk SSSR, Ser. Mat. **51**, 485–516 (1987) (Russian)
377. Bogomolov, F.A. et al.: Gennadiĭ Vladimirovič Belyĭ. Uspekhi Mat. Nauk **57(5)**, 139–140 (2002) (Russian)
378. Bochniček, S. [Bohnicek, S.]: Zur Theorie des relativbiquadratischen Zahlkörpers. Math. Ann. **63**, 85–144 (1907)
379. Bohnicek, S.: Zur Theorie der achten Einheitswurzeln. SBer. Kais. Akad. Wissensch. Wien **120**, 25–47 (1911)
380. Bohnicek, S.: Bemerkungen zur Kreisteilung. SBer. Kais. Akad. Wissensch. Wien **121**, 719–725 (1912)
381. Bohnicek, S.: Über die Unmöglichkeit der diophantischen Gleichung $\alpha^{2^{n-1}} + \beta^{2^{n-1}} + \gamma^{2^{n-1}} = 0$ im Kreiskörper der 2^n-ten Einheitswurzeln, wenn n grösser als 2 ist. SBer. Kais. Akad. Wissensch. Wien **121**, 727–742 (1912)
382. Bohr, H., Landau, E.: Über das Verhalten von $\zeta(s)$ und $\zeta_k(s)$ in der Nähe der Geraden $\sigma = 1$. Nachr. Ges. Wiss. Göttingen **1910**, 303–330
383. Bölling, R.: Kummer vor der Erfindung der "idealen complexen Zahlen": Das Jahr 1844. Acta Hist. Leopold. **27**, 145–157 (1997)
384. Bölling, R.: From reciprocity laws to ideal numbers: an (un)known manuscript by E.E. Kummer. In: [1467], 271–290
385. Bölling, R.: Jacobi als Wegbereiter für Kummers Idee der idealen Zahlen. Mitt. Dtsch. Math.-Ver. **16**, 274–281 (2008)
386. Boltje, R.: Class group relations from Burnside ring idempotents. J. Number Theory **66**, 291–305 (1997)
387. Bombieri, E.: Counting points on curves over finite fields (d'aprés S.A. Stepanov). Lecture Notes Math. **383**, 234–241 (1974)
388. Bombieri, E.: Hilbert's 8th problem: an analogue. Proc. Symposia Pure Math. **28**, 269–274 (1976)
389. Bombieri, E.: The Mordell conjecture revisited. Ann. Scuola Norm. Sup. Pisa, Cl. Sci. (4) **17**, 615–640 (1990)
390. Bonaventura, P.: Sul teoremo di reciprocità delle teorie dei residui quadratici nei numeri interi del campo $(1, i\sqrt{2})$. Giornale di Mat. **30**, 221–234 (1892)

391. Bonaventura, P.: Il teorema di reciprocità pei numeri interi complessi e le funzioni lemniscatiche. Giornale di Mat. **30**, 300–310 (1892)
392. Bonciocat, N.C.: Schönemann-Eisenstein-Dumas-type irreducibility conditions that use arbitrarily many prime numbers. Comm. Algebra **43**, 3102–3122 (2015)
393. Booker, A.R.: Poles of Artin L-functions and the strong Artin conjecture. Ann. Math. (2) **158**, 1089–1098 (2003)
394. Bopp, K.: Leo Koenigsberger als Historiker der mathematischen Wissenschaften. Jahresber. Deutsch. Math.-Verein. **33**, 104–112 (1924)
395. Bordellés, O.: On the ideal theorem for number fields. Funct. Approx. Comment. Math. **53**, 31–45 (2015)
396. Borel, A., Chowla, S., Herz, C.S., Iwasawa, K., Serre, J.-P.: Seminar on Complex Multiplication. Lecture Notes Math. **21** (1966)
397. Borevič, Z.I.: On the proof of the principal ideal theorem. Vestnik Leningrad. Univ. Ser. Mat. Mekh. Astronom. **12(13)**, 5–8 (1957) (Russian)
398. Borevič, Z.I.: Multiplicative group of a regular local field with a cyclic operator group. Izv. Akad. Nauk SSSR, Ser. Mat. **28**, 707–712 (1964) (Russian)
399. Borevič, Z.I.: On the multiplicative group of cyclic p-extensions of a local field. Trudy Mat. Inst. im. Steklova **80**, 16–29 (1965) (Russian)
400. Borevič, Z.I.: On the multiplicative group of cyclic p-extensions of a local field. On the group of principal units of a normal p-extension of a regular local field. Trudy Mat. Inst. im. Steklova **80**, 30–44 (1965) (Russian)
401. Borevič, Z.I.: On the multiplicative group of cyclic p-extensions of a local field, II. Vestnik Leningrad. Univ. Ser. Mat. Mekh. Astronom. **20(13)**, 5–12 (1965) (Russian)
402. Borevič, Z.I.: On groups of principal units of p-extensions of a local field. Dokl. Akad. Nauk SSSR **173**, 253–255 (1967) (Russian)
403. Borevič, Z.I., Gerlovin, E.I., Structure of the group of principal units of a cyclic p-extension of a local field. Zap. Naučn. Sem. LOMI **57**, 51–63 (1976) (Russian)
404. Borevič, Z.I., Skopin, A.I.: Extensions of a local field with a normal basis for the principal units. Trudy Mat. Inst. im. Steklova **80**, 45–50 (1965) (Russian)
405. Borwein, J.M., Bugeaud, Y., Coons, M.: The legacy of Kurt Mahler. Notices Amer. Math. Soc. **62**, 526–531 (2015)
406. Borwein, P., Dobrowolski, E., Mossinghoff, M.J.: Lehmer's problem for polynomials with odd coefficients. Ann. Math. (2) **166**, 347–366 (2007)
407. Bosma, W., Stevenhagen, P.: Density computations for real quadratic unit. Math. Comp. **65**, 1327–1337 (1996)
408. Boyd, D.W.: Small Salem numbers. Duke Math. J. **44**, 315–328 (1977)
409. Boyd, D.W.: Pisot and Salem numbers in intervals of the real line. Math. Comp. **32**, 1244–1260 (1978)
410. Boyd, D.W.: Reciprocal polynomials having small measure. Math. Comp. **35**, 1361–1377 (1980)
411. Boyd, D.W.: Kronecker's theorem and Lehmer's problem for polynomials in several variables. J. Number Theory **13**, 116–121 (1981)
412. Boyd, D.W.: Speculations concerning the range of Mahler's measure. Canad. Math. Bull. **24**, 453–469 (1981)
413. Boyd, D.W.: Pisot numbers in the neighborhood of a limit point, II. Math. Comp. **43**, 596–602 (1984)
414. Boyd, D.W.: The maximal modulus of an algebraic integer. Math. Comp. **45**, 243–249, S17–S20 (1985)
415. Boyd, D.W.: Pisot numbers in the neighbourhood of a limit point, I. J. Number Theory **21**, 17–43 (1985)
416. Boyd, D.W.: Perron units which are not Mahler measures. Ergodic Theory Dynam. Systems **6**, 485–488 (1986)
417. Boyd, D.W.: Inverse problems for Mahler's measure. In: Diophantine Analysis (Kensington 1985), 147–158, London Math. Soc. Lecture Note Ser. **109** (1986)

418. Boyd, D.W.: Reciprocal polynomials having small measure, II. Math. Comp. **53**, 355–357, S1–S5 (1989)
419. Boyd, D.W., Kisilevsky, H.: On the exponent of the ideal class groups of complex quadratic fields. Proc. Amer. Math. Soc. **31**, 433–436 (1972)
420. Braconnier, J.: Sur les groupes topologiques localement compacts. Dissertation, Univ. de Nancy (1945)
421. Braconnier, J.: Sur les groupes topologiques localement compacts. J. Math. Pures Appl. (9) **27**, 1–85 (1948)
422. Brahana, H.R.: George Abram Miller (1863–1951). Bull. Amer. Math. Soc. **57** 377–382 (1951)
423. Branchini, G.: Algoritmo del massimo comun divisore nel corpo delle radici quinte dell' unità. Atti Accad. Naz. Lincei (5) **32(1)** 68–72 (1923)
424. Brauer, A.: Über diophantische Gleichungen mit endlich vielen Lösungen. J. Reine Angew. Math. **160**, 70–99 (1929)
425. Brauer, A.: Über eine Erweiterung des kleinen Fermatschen Satzes. Math. Z. **42**, 255–262 (1937)
426. Brauer, A.: On the non-existence of the Euclidean algorithm in certain quadratic number fields. Amer. J. Math. **62**, 697–716 (1940)
427. Brauer, R.: Über die algebraische Struktur von Schiefkörpern. J. Reine Angew. Math. **166**, 241–252 (1932)
428. Brauer, R.: On the zeta-functions of algebraic number fields. Amer. J. Math. **69**, 243–250 (1947)
429. Brauer, R.: On Artin's L-series with general group characters. Ann. Math. (2) **48**, 502–514 (1947)
430. Brauer, R.: On the zeta-functions of algebraic number fields, II. Amer. J. Math. **72**, 739–746 (1950)
431. Brauer, R.: Beziehungen zwischen Klassenzahlen von Teilkörpern eines galoisschen Körpers. Math. Nachr. **4**, 158–174 (1951)
432. Brauer, R.: A note on zeta-functions of algebraic number fields. Acta Arith. **24**, 325–327 (1973)
433. Brauer, R., Hasse, H., Noether, E.: Beweis eines Hauptsatzes in der Theorie der Algebren. J. Reine Angew. Math. **167**, 399–404 (1932)
434. Braun, H.: Über die Zerlegung quadratischer Formen in Quadrate. J. Reine Angew. Math. **178**, 34–64 (1937)
435. Braun, H.: Geschlechter quadratischer Formen. J. Reine Angew. Math. **182**, 32–49 (1940)
436. Bravais, A.: Mémoire sur les systèmes formés par des points distribués régulièrement sur un plan ou dans l'espace. J. École Polytech. **19**, 1–128 (1850)
437. Brawley, J.V.: Dedicated to Leonard Carlitz: the man and his work. Finite Fields Appl. **1**, 135–151 (1995)
438. Bredikhin, B.M.,: Free numerical semigroups with power densities. Mat. Sb. **46**, 143–158 (1958) (Russian)
439. Bredikhin, B.M.: Elementary solutions of inverse problems on bases of free semigroups. Mat. Sb. **50**, 221–232 (1960) (Russian)
440. Bredikhin, B.M.: The remainder term in the asymptotic formula for $\nu_G(x)$. Izv. Vysš. Učeb. Zaved. Mat. **1960(6)**, 40–49. (Russian)
441. Bremner, A.: On power bases in cyclotomic fields. J. Number Theory **28**, 288–298 (1988)
442. Bresslau, H.: Dirichlets Satz von der arithmetischen Reihe für den Körper der dritten Einheitswurzeln. Dissertation, Univ. Strassburg (1907)
443. Breuer, S.: Zyklische Gleichungen 6. Grades und Minimalbasis. Math. Ann. **86**, 108–113 (1922)
444. Breuer, S.: Zur Bestimmung der metazyklischen Minimalbasis von Primzahlgrad. Math. Ann. **92**, 126–144 (1924)
445. Breuer, S.: Zur Theorie der metazyklischen Gleichungen von Primzahlgrad. SBer. Heidelberg. Akad. Wiss. **1925(5)**, 7–11.

446. Breuer, S.: Metazyklische Minimalbasis und komplexe Primzahlen. J. Reine Angew. Math. **156**, 13–42 (1926)
447. Breuer, S.: Zyklische Minimalbasis zusammengesetzten Grades. J. Reine Angew. Math. **166**, 54–58 (1932)
448. Breusch, R.: On the distribution of the roots of a polynomial with integral coefficients. Proc. Amer. Math. Soc. **2**, 939–941 (1951)
449. Bricard, R.: Georges Fontené. Nouv. Ann. Math. (5) **1**, 361–363 (1923)
450. Briggs, W.E.: An elementary proof of a theorem about the representation of primes by quadratic forms. Canad. J. Math. **6**, 353–363 (1954)
451. Brill, A.: Ueber die Discriminante. Math. Ann. **12**, 87–89 (1877)
452. Brill, A.: Max Noether, Jahresber. Deutsch. Math.-Verein. **32**, 211–233 (1924)
453. Brill, A., Noether, M.: Die Entwickelung der Theorie der algebraischen Functionen in älterer und neuerer Zeit. Jahresber. Deutsch. Math.-Verein. **3**, 107–566 (1892/93)
454. Brillhart, J.: Derrick Henry Lehmer. Acta Arith. **62**, 207–213 (1992)
455. Brillhart, J.: Emma Lehmer 1906–2007. Notices Amer. Math. Soc. **54**, 1500–1501 (2007)
456. Brindza, B., Győry, K.: On unit equations with rational coefficients. Acta Arith. **53**, 367–388 (1990)
457. Brink, D.: Remark on infinite unramified extensions of number fields with class number one. J. Number Theory **130**, 304–306 (2010)
458. Brinkhuis, J.: Galois modules and embedding problems. J. Reine Angew. Math. **346**, 141–165 (1984)
459. Brinkhuis, J.: Normal integral bases and complex conjugation. J. Reine Angew. Math. **375/376**, 157–166 (1987)
460. Browder, F.E. (ed.): Mathematical Developments Arising from Hilbert Problems. Proc. Symposia Pure Math. **28** (1976)
461. Browkin, J.: Continued fractions in local fields, I. Demonstratio Math. **11**, 67–82 (1978)
462. Browkin, J.: Continued fractions in local fields, II. Math. Comp. **70**, 1281–1292 (2001)
463. Browkin, J., Brzeziński, J., Xu, K.: On exceptions in the Brauer-Kuroda relations. Bull. Acad. Pol. Sci., sér. sci. math. astr. phys. **59**, 207–214 (2011)
464. Browkin, J., Xu, K.: On exceptional pq-groups. Sci. China Math. **55**, 2081–2093 (2012)
465. Brown, E.: The class number of $Q(\sqrt{-p})$, for $p \equiv 1 \bmod 8$, a prime. Proc. Amer. Math. Soc. **31**, 381–383 (1972)
466. Brown, E.: The power of 2 dividing the class-number of a binary quadratic discriminant. J. Number Theory **5**, 413–419 (1973)
467. Brown, E.: Class number of complex quadratic fields. J. Number Theory **6**, 185–191 (1974)
468. Brown, E.: Class number of real quadratic fields. Trans. Amer. Math. Soc. **90**, 99–107 (1974)
469. Brown, E., Parry, C.J.: Class numbers of imaginary quadratic fields having exactly three discriminantal divisors. J. Reine Angew. Math. **260**, 31–34 (1973)
470. Brown, E., Parry, C.J.: The imaginary bicyclic biquadratic fields with class number 1, I. J. Reine Angew. Math. **266**, 118–120 (1974)
471. Bruche, C.: Classes de Steinitz d'extensions non abéliennes de degré p^3. Acta Arith. **137**, 177–191 (2009)
472. Bruche, C., Sodaïgui, B.: On realizable Galois module classes and Steinitz classes of non-abelian extensions. J. Number Theory **128**, 954–978 (2008)
473. Brückner, H.: Eine explizite Formel zum Reziprozitätsgesetz für Primzahlexponenten p. Algebraische Zahlentheorie (Ber. Tagung Math. Forschungsinst. Oberwolfach, 1964), 31–39, Bibliographisches Institut (1967)
474. Brückner, H.: Explizites Reziprozitätsgesetz und Anwendungen. Vorlesungen aus dem Fachbereich Mathematik der Universität Essen, 1–83 (1979)
475. Brumer, A.: Ramification and class towers of number fields. Michigan Math. J. **12**, 129–131 (1965)
476. Brumer, A.: On the units of algebraic number fields. Mathematika **14**, 121–124 (1967)
477. Brumer, A.: On the group of units of an absolutely cyclic number field of prime degree, J. Math. Soc. Japan **21**, 357–358 (1969)

478. Brumer, A., Rosen, M.: Class number and ramification in number fields. Nagoya Math. J. **23**, 97–101 (1963)
479. Brun, V.: Le crible d'Eratosthène et le théorème de Goldbach. C. R. Acad. Sci. Paris **168**, 544–546 (1919)
480. Brun, V.: En generalisation av kjedebröken, I. Skrifter Videnskaps. Kristiania **1919(6)**, 1–29 (1919)
481. Brun, V.: En generalisation av kjedebröken, II. Skrifter Videnskaps. Kristiania **1920(2)**, 1–24 (1920)
482. Brun, V.: Carl Störmer in memoriam. Acta Math. **100**, I-VII (1958)
483. Brun, V.: Axel Thue. In: [4060], xv–xxv
484. Brunotte, H., Halter-Koch, F.: Metrische Kennzeichnung von Erzeugenden für Einheitengruppen vom Rang 1 oder 2 in algebraischen Zahlkörpern. J. Number Theory **13**, 320–333 (1981)
485. Brunotte, H., Halter-Koch, F.: Grundeinheitensysteme algebraischer Zahlkörper mit vorgegebener Verteilung der Konjugiertenbeträge. Arch. Math. (Basel) **37**, 512–513 (1981)
486. Buccino, A.: Matrix classes and ideal classes Illinois J. Math. **13**, 188–191 (1969)
487. Buchmann, J.: A generalization of Voronoĭ's unit algorithm, I. J. Number Theory **20**, 177–191 (1985)
488. Buchmann, J.: A generalization of Voronoĭ's unit algorithm, II. **20**, 192–209 (1985)
489. Buchmann, J.: The computation of the fundamental unit of totally complex quartic orders. Math. Comp. **48**, 39–54 (1987)
490. Buchmann, J., Jacobson, M.J.Jr., Teske, E.: On some computational problems in finite abelian groups. Math. Comp. **66**, 1663–1687 (1997)
491. Bucht, G.: Über einige algebraische Körper achten Grades. Ark. Mat. **6(30)**, 1–30 (1911)
492. Buell, D.A.: Class groups of quadratic fields. Math. Comp. **30**, 610–623 (1976)
493. Buell, D.A.: Small class numbers and extreme values of L-functions of quadratic fields. Math. Comp. **31**, 786–796 (1977)
494. Buell, D.A.: Class groups of quadratic fields, II. Math. Comp. **48**, 85–93 (1987)
495. Buell, D.A.: The last exhaustive computation of class groups of complex quadratic number fields. In: Number Theory (Ottawa, ON, 1996), 35–53, CRM Proc. Lecture Notes **19**, Amer. Math. Soc. (1999)
496. Bugeaud, Y.: Bornes effectives pour les solutions des équations en S-unités et des équations de Thue-Mahler. J. Number Theory **71**, 227–244 (1998)
497. Bugeaud, Y.: Approximation by Algebraic Numbers. Cambridge University Press, Cambridge (2004)
498. Bugeaud, Y.: Quantitative versions of the subspace theorem and applications. J. Théor. Nombres Bordeaux **23**, 35–57 (2011)
499. Bugeaud, Y., Dujella, A.: Root separation for irreducible integer polynomials. Bull. Lond. Math. Soc. **43**, 1239–1244 (2011)
500. Bugeaud, Y., Dujella, A.: Root separation for reducible integer polynomials. Acta Arith. **162**, 393–403 (2014)
501. Bugeaud, Y., Győry, K.: Bounds for the solutions of unit equations. Acta Arith. **74**, 67–80 (1996)
502. Bugeaud, Y., Mignotte, M.: On the distance between roots of integer polynomials. Proc. Edinb. Math. Soc. (2) **47**, 553–556 (2004)
503. Bugeaud, Y., Mignotte, M., Normandin, F.: Nombres algébriques de petite mesure et formes linéaires en un logarithme. C. R. Acad. Sci. Paris **321**, 517–522 (1995)
504. Buhler, J.P.: Icosahedral Galois Representations. Lecture Notes Math. **654** (1978)
505. Buhler, J.[P.], Crandall, R., Ernvall, R., Metsänkylä, T.: Irregular primes and cyclotomic invariants to four million. Math. Comp. **61**, 151–153 (1993)
506. Buhler, J.[P.], Crandall, R., Ernvall, R., Metsänkylä, T., Shokrollahi, M.A.: Irregular primes and cyclotomic invariants to 12 million. J. Symbolic Comput. **31**, 89–96 (2001)
507. Buhler, J.[P.], Crandall, R., Sompolski, R.W.: Irregular primes to one million. Math. Comp. **59**, 717–722 (1992)

508. Buhler, J.P., Harvey, D.: Irregular primes to 163 million. Math. Comp. **80**, 2435–2444 (2011)
509. Bullig, G.: Die Berechnung der Grundeinheit in den kubischen Körpern mit negativer Diskriminante. Math. Ann. **112**, 325–394 (1936)
510. Bullig, G.: Ein periodisches Verfahren zur Berechnung eines Systems von Grundeinheiten in den total reellen kubischen Körpern. Abh. Math. Sem. Hansischen Univ. **12**, 369–414 (1938)
511. Bullig, G.: Zur Zahlengeometrie in den total reellen kubischen Körpern. Math. Z. **45**, 511–532 (1939)
512. Bulota, K.: On Hecke Z-functions and the distribution of the prime numbers of an imaginary quadratic field. Lit. Mat. Sb. **4**, 309–328 (1964) (Russian)
513. Bundschuh, P.: p-adische Kettenbrüche und Irrationalität p-adischer Zahlen. Elem. Math. **32**, 36–40 (1977)
514. Bundschuh, P., Hock, A.: Bestimmung aller imaginär-quadratischen Zahlkörper der Klassenzahl Eins mit Hilfe eines Satzes von Baker. Math. Z. **111**, 191–204 (1969)
515. Bungers, R.: Über Zahlkörper mit gemeinsamen ausserwesentlichen Diskriminantenteilern. Jahresber. Deutsch. Math.-Verein. **46**, 93–96 (1936); corr. **47**, p. 56 (1937)
516. Burde, K.: Ein rationales biquadratisches Reziprozitätsgesetz. J. Reine Angew. Math. **235**, 175–184 (1969)
517. Burde, G., Schwarz, W.: Wolfgang Franz zum Gedächtniss. Jahresber. Deutsch. Math.-Verein. **100**, 284–292 (1998)
518. Burgess, D.A.: On character sums and primitive roots. Proc. London Math. Soc. (3) **12**, 179–192 (1962)
519. Burkill, J.C.: Charles-Joseph de la Vallée Poussin. J. Lond. Math. Soc. **39**, 165–175 (1964)
520. Burkill, J.C.: John Edensor Littlewood. Bull. Lond. Math. Soc. **11**, 59–103 (1979)
521. Burnside, W.: Theory of Groups of Finite Order. London (1897); 2nd ed. (1911) [Reprint: Dover (1955)]
522. Burnside, W.: On the rational solutions of the equation $x^3 + y^3 + z^3 = 0$ in quadratic fields. Proc. London Math. Soc. (2) **14**, 1–4 (1915)
523. Burnside, W.: On cyclotomic quinquisection. Proc. London Math. Soc. (2) **14**, 251–259 (1915)
524. Busam, R., Freitag, E.: Hans Maaß. Jahresber. Deutsch. Math.-Verein. **101**, 135–150 (1999)
525. Busche, E.: Arithmetischer Beweis des Reciprocitätsgesetzes für die biquadratischen Reste. J. Reine Angew. Math. **99**, 261–274 (1886)
526. Busche, E.: Beweis des quadratischen Reciprocitätsgesetzes in der aus den vierten Wurzeln der Einheit gebildeten complexen Zahlen. Mitt. Math. Ges. Hamburg **2**, 80–92 (1890)
527. Busche, E.: Ueber eine reale Darstellung der imaginären Gebilde einer reellen Ebene und einige Anwendungen davon auf die Zahlentheorie. J. Reine Angew. Math. **122**, 227–262 (1900)
528. Busche, E.: Über die Theorie der biquadratischer Reste. J. Reine Angew. Math. **141**, 146–162 (1912)
529. Bush, M.R.: Computation of Galois groups associated to the 2-class towers of some quadratic fields. J. Number Theory **100**, 313–325 (2003)
530. Büsser, A.H.: Über die Primidealzerlegung in Relativkörpern mit der Relativgruppe $\mathfrak{G}_{168}$. Dissertation, Univ. Zürich (1944)
531. Butts, H.S.: Unique factorization of ideals into nonfactorable ideals. Proc. Amer. Math. Soc. **15**, 21 (1964)
532. Butts, H.S., Estes, D.: Modules and binary quadratic forms over integral domains. Linear Algebra Appl. **1**, 153–180 (1968)
533. Butts, H.S., Gilmer, R.W.Jr.: Primary ideals and prime power ideals. Canad. J. Math. **18**, 1183–1195 (1966)
534. Butts, H.S., Pall, G.: Factorization in quadratic rings. Duke Math. J. **34**, 139–146 (1967)
535. Butts, H.S., Pall, G.: Ideals not prime to the conductor in quadratic orders. Acta Arith. **21**, 261–270 (1972)

536. Butts, H.S., Wade, L.I.: Two criteria for Dedekind domains. Amer. Math. Monthly **73**, 14–21 (1966)
537. Buzzard, K., Dickinson, M., Shepherd-Barron, N., Taylor, R.: On icosahedral Artin representations. Duke Math. J. **109**, 283–318 (2001)
538. Buzzard, K., Stein, W.A.: A mod five approach to modularity of icosahedral Galois representations. Pacific J. Math. **203**, 265–282 (2002)
539. Byeon, D.: Real quadratic fields with class number divisible by 5 or 7. Manuscripta Math. **120**, 211–215 (2006)
540. Byeon, D., Lee, J.: A complete determination of Rabinowitsch polynomials. J. Number Theory **131**, 1513–1529 (2011)
541. Byeon, D., Stark, H.M.: On the finiteness of certain Rabinowitsch polynomials. J. Number Theory **94**, 177–180 (2002)
542. Byeon, D., Stark, H.M.: On the finiteness of certain Rabinowitsch polynomials, II. J. Number Theory **99**, 219–221 (2003)
543. Byott, N.P., Greither, C., Sodaïgui, B.: Classes réalisables d'extensions non abéliennes. J. Reine Angew. Math. **601**, 1–27 (2006)
544. Byott, N.P., Sodaïgui, B.: Galois module structure for dihedral extensions of degree 8: realizable classes over the group ring. J. Number Theory **112**, 1–19 (2005)
545. Byott, N.P., Sodaïgui, B.: Realizable Galois module classes for tetrahedral extensions. Compositio Math. **141**, 573–582 (2005)
546. Byott, N.P., Sodaïgui, B.: Realizable Galois module classes over the group ring for non abelian extensions. Ann. Inst. Fourier (Grenoble) **63**, 303–371 (2013)
547. Cahen, E.: Sur la fonction $\zeta(s)$ de Riemann et sur des fonctions analogues. Ann. Sci. Éc. Norm. Supér. (3) **11**, 75–164 (1894)
548. Cahen, E.: Sur une note de M. Fontené relative aux entiers algébriques de la forme $x + y\sqrt{-5}$. Nouv. Ann. Math. (4) **3**, 444–447 (1903)
549. Cahen, E.: Sur l'arithmétique du corps de tous les nombres algébriques. Bull. Soc. Math. France **56**, 7–17 (1928)
550. Cahen, P.-J., Chabert, J.-L.: Integer-valued Polynomials. Amer. Math. Soc. (1997)
551. Cahen, P.-J., Chabert, J.-L.: What you should know about integer-valued polynomials. Amer. Math. Monthly **123**, 311–337 (2016)
552. Cajori, F.: Pierre Laurent Wantzel. Bull. Amer. Math. Soc. **24**, 339–347 (1918)
553. Calegari, F.: The Artin conjecture for some S_5-extensions. Math. Ann. 356, 2013, 191–207. (2016)
554. Callahan, T., Newman, M., Sheingorn, M.: Fields with large Kronecker constants. J. Number Theory **9**, 182–186 (1977)
555. Calloway, J.: On the discriminant of arbitrary algebraic number fields. Proc. Amer. Math. Soc. **6**, 482–489 (1955)
556. Canci, J.K.: Rational periodic points for quadratic maps. Ann. Inst. Fourier (Grenoble) **60**, 953–985. (2010)
557. Cantor, D.G.: On sets of algebraic integers whose remaining conjugates lie in the unit circle. Trans. Amer. Math. Soc. **105**, 391–406 (1962)
558. Cantor, D.G.: On an extension of the definition of transfinite diameter and some applications. J. Reine Angew. Math. **316**, 160–207 (1980)
559. Cantor, D.G., Straus, E.G.: On a conjecture of D.H. Lehmer. Acta Arith. **42**, 97–100 (1988); corr. p. 325
560. Cantor, G.: Ueber die Ausdehnung eines Satzes aus der Theorie der trigonometrischen Reihen. Math. Ann. **5**, 123–133 (1872)
561. Capelli, A.: Sulla riduttibilità delle equazioni algebriche. Rend. Accad. Sci. fis. mat. Napoli (3) **3**, 243–252 (1897)
562. Capelli, A.: Sulla riduttibilità delle equazioni algebriche, II. Rend. Accad. Sci. fis. mat. Napoli (3) **4**, 84–90, 243–247 (1898)
563. Capelli, A.: Sulla riduttibilità delle funzione $x^n - A$ in un campo qualunque di razionalitá. Math. Ann. **54**, 602–603 (1901)

564. Capparelli, S., Del Fra, A., Sciò, C.: On the span of polynomials with integer coefficients. Math. Comp. **79**, 967–981 (2010)
565. Caratheodory, C. et al. (ed.): Mathematische Abhandlungen Hermann Amandus Schwarz zu seinem fünfzigjährigen Doktorjubiläum am 6. Aug. 1914 gewidmet von Freunden und Schülern. Springer (1914) [Reprint: Chelsea (2003)]
566. Cardon, D.A.: A Euclidean ring containing $Z[\sqrt{14}]$. C.R. Math. Acad. Sci. Soc. R. Canada **19**, 28–32 (1997)
567. Carey, F.S.: Notes on the division of the circle. Quart. J. Math., Oxford ser. **26**, 322–371 (1893)
568. Carleman, T: L.E.Phragmén in memoriam, Acta Math. 69, XXXI–XXXIII (1938)
569. Carlitz, L.: On a function connected with a cubic field. Bull. Amer. Math. Soc. **37**, 73–75 (1931)
570. Carlitz, L.: On abelian fields. Trans. Amer. Math. Soc. **35**, 122–136 (1933)
571. Carlitz, L.: On certain equations in relative-cyclic fields. Duke Math. J. **2**, 650–659 (1936)
572. Carlitz, L.: A theorem of Stickelberger. Math. Scand. **1**, 82–84 (1953)
573. Carlitz, L.: Note on the class number of real quadratic fields. Proc. Amer. Math. Soc. **4**, 535–537 (1953)
574. Carlitz, L.: The Schur derivative of a polynomial. Proc. Glasgow Math. Assoc. **1**, 159–163 (1953)
575. Carlitz, L.: Some theorems on the Schur derivative. Pacific J. Math. **3**, 321–332 (1953)
576. Carlitz, L.: Kummer congruences and the Schur derivative. Amer. J. Math. **75**, 699–706 (1953)
577. Carlitz, L.: Note on irregular primes. Proc. Amer. Math. Soc. **5**, 329–331 (1954)
578. Carlitz, L.: A note on power residues. Duke Math. J. **22**, 583–587 (1955)
579. Carlitz, L.: Arithmetic properties of generalized Bernoulli numbers. J. Reine Angew. Math. **202**, 174–182 (1959)
580. Carlitz, L.: A characterization of algebraic number fields with class number two. Proc. Amer. Math. Soc. **11**, 391–392 (1960)
581. Carlitz, L.: A generalization of Maillet's determinant and a bound for the first factor of the class number. Proc. Amer. Math. Soc. **12**, 256–261 (1961)
582. Carlitz, L. Olson, F.R.: Maillet's determinant. Proc. Amer. Math. Soc. **6**, 265–269 (1955)
583. Carr, R., O'Sullivan, C.: On the linear independence of roots. Internat. J. Number Th. **5**, 161–171 (2009)
584. Cartan, H.: Sur le théorème de préparation de Weierstrass. In: Festschrift Gedächtnisfeier K.Weierstrass, 155–168. Westdeutscher Verlag (1966)
585. Carter, J.E.: Steinitz classes of a nonabelian extension of degree p^3. Colloq. Math. **71**, 297–303 (1996)
586. Carter, J.E.: Steinitz classes of nonabelian extensions of degree p^3. Acta Arith. **78**, 297–303 (1997)
587. Carter, J.E.: Module structure of integers in metacyclic extensions. Colloq. Math. **76**, 191–199 (1998)
588. Carter, J.E.: A generalization of a result on integers in metacyclic extensions. Colloq. Math. **81**, 153–156 (1999)
589. Carter, J.E.: Normal integral bases in quadratic and cyclic cubic extensions of quadratic fields. Arch. Math. (Basel) **81**, 266–271 (2003); corr. **83**, vi–vii (2004)
590. Carter, J.E., Sodaïgui, B.: Classes de Steinitz d'extensions quaternioniennes généralisées de degré $4p^r$. J. Lond. Math. Soc. (2) **76**, 331–344 (2007)
591. Cartier, P.: Über einige Integralformeln in der Theorie der quadratischen Formen. Math. Z. **84**, 93–100 (1964)
592. Cartwright, M.L.: Edward Charles Titchmarsh. J. Lond. Math. Soc. **39**, 544–565 (1964)
593. Carver, W.B.: Ideals of a quadratic number field in canonic form. Amer. Math. Monthly **18**, 81–87 (1911)
594. Cassels, J.W.S.: The inhomogeneous minimum of binary quadratic, ternary cubic and quaternary quartic forms. Proc. Cambridge Philos. Soc. **48**, 72–86 (1952); add.: 519–520

595. Cassels, J.W.S.: On a problem of Schinzel and Zassenhaus. J. Math. Sci. (N.Y.) **1**, 1–8 (1966)
596. Cassels, J.W.S.: L.J.Mordell. Bull. Lond. Math. Soc. **6**, 69–96 (1974)
597. Cassels, J.W.S.: Trygve Nagell. Acta Arith. **55**, 109–118 (1991)
598. Cassels, J.W.S.: Obituary of Kurt Mahler. Acta Arith. **58**, 215–228 (1991)
599. Cassels, J.W.S., Fröhlich, A. (ed.): Algebraic Number Theory (Proceedings of an Instructional Conference, Brighton, 1965). Thompson (1967)
600. Cassels, J.W.S., Fröhlich, A.: Hans Arnold Heilbronn. Bull. Lond. Math. Soc. **9**, 219–232 (1977)
601. Cassels, J.W.S., Vaughan, R.C.: Obituary: Ivan Matveevich Vinogradov. Bull. Lond. Math. Soc. **17**, 584–600 (1985)
602. Cassou-Nogués, Ph., Taylor, M.J.: Fonctions elliptiques et génération d'anneaux d'entiers. Astérisque **147/148**, 49–70 (1987)
603. Cassou-Nogués, Ph., Taylor, M.J.: Elliptic Functions and Rings of Integers. Progr. Math. **66**, Birkhäuser (1987)
604. Cassou-Nogués, Ph., Taylor, M.J.: Unités modulaires et monogénéité d'anneaux d'entiers. In: Séminaire de Théorie des Nombres, Paris 1986/87, 35–64. Birkhäuser (1987)
605. Cassou-Nogués, Ph., Taylor, M.J.: A note on elliptic curves and the monogeneity of rings of integers. J. Lond. Math. Soc. (2) **37**, 63–72 (1988)
606. Castela, C.: Nombre de classes d'idéaux d'une extension diédrale d'un corps de nombres. C. R. Acad. Sci. Paris **287**, A483–A486 (1978)
607. Castillo, A., Dietmann, R.: On Hilbert's irreducibility theorem. Acta Arith. **180**, 1–14 (2017)
608. Cauchy, A.: Rapport sur un Mémoire de M. Lamé, relatif au dernier théorème de Fermat. C. R. Acad. Sci. Paris **9**, 359–363 (1839) [J. Math. Pures Appl. (1) **5**, 211–215 (1840) [[612] 1_4, 499–504.]
609. Cauchy, A.: Sur les fonctions alternées et sur divers formulaes d'Analyse. C. R. Acad. Sci. Paris **10**, 178–181 (1840) [[612] 1_5, 81–85.]
610. Cauchy, A.: Mémoires sur la théorie des nombres. Mém. Acad. Sci. Paris **17**, 249–769 (1840) [[612] 1_3, 5–450.]
611. Cauchy, A.: Mémoire sur de nouvelles formules à la théorie des pôlynomes radicaux, et sur la dernier théorème de Fermat. C. R. Acad. Sci. Paris **24**, 469–481, 516–528, 578–584, 633–636, 661–666 (1847) [[612] 1_{10}, 254–292.]
612. Cauchy, A.: Oeuvres complètes. Gauthier-Villars (1882–1938) [Reprint: Cambridge University Press (2009)]
613. Cavallar, S., Lemmermeyer, F.: Euclidean windows. LMS J. Comput. Math. **3**, 336–355 (2000)
614. Cayley, A.: On the binomial equation $x^p - 1 = 0$; Trisection and quartisection. Proc. London Math. Soc. **11**, 4–17 (1880/81)
615. Cayley, A.: On the binomial equation $x^p - 1 = 0$; quinquisection. Proc. London Math. Soc. **12**, 15–16 (1881/82)
616. Cayley, A.: On the binomial equation $x^p - 1 = 0$; quinquisection; second note. Proc. London Math. Soc. **16**, 61–63 (1884/85)
617. Čebotarev, N.G.: Determination of the density of the set of primes corresponding to a given class of permutations. Izv. Akad. Nauk SSSR, Ser. Mat. **17**, 205–250 (1923) (Russian) [[631] **1**, 27–65.]
618. Čebotarev, N.G.: Proof of the Kronecker-Weber theorem on Abelian fields. Mat. Sb. **31**, 302–309 (1923) (Russian) [[631] **1**, 18–26.]
619. Čebotarev, N.G.: A generalization of Minkowski's theorem with application to the study of ideal classes of a field. Žurnal Naučn.-issled. kafedr v Odesse **4**, 1–4 (1924) (Russian) [[631] **1**, 66–70.]
620. Tschebotareff, N.G. [Čebotarev, N.G.]: Die Bestimmung der Dichtigkeit einer Menge von Primzahlen, welche zu einer gegebener Substitutionsklasse gehören. Math. Ann. **95**, 191–228 (1926)

621. Čebotarev, N.G.: On the problem of finding algebraic equations with a given group. Bull. Soc. Math.-Phys. Kazan (3) **1**, 26–32 (1926) (Russian) [[631] **1**, 87–94.]
622. Čebotarev, N.G.: Studies of prime number densities, I. Bull. Soc. Math.-Phys. Kazan (3) **2**, 14–20 (1927) (Russian) [[631] **1**, 95–101.]
623. Čebotarev, N.G.: Studies of prime number densities, II. Bull. Soc. Math.-Phys. Kazan (3) **3**, 1–17 (1927) (Russian) [[631] **1**, 102–118.]
624. Čebotarev, N.G. [Tschebotaröw, N.]: Zur Gruppentheorie des Klassenkörpers. J. Reine Angew. Math. **161**, 179–193 (1929); corr. **164**. p. 196 (1931) [[631] **1**, 121–140. (Russian)]
625. Čebotarev, N.G. [Tschebotaröw, N.]: Untersuchungen über relativ Abelsche Zahlkörper. J. Reine Angew. Math. **167**, 98–121 (1932) [[631] **1**, 141–171. (Russian)]
626. Čebotarev, N.G. [Tschebotaröw, N.]: Die Probleme der modernen Galoisschen Theorie. Comment. Math. Helv. **6**, 1934, 235–283.
627. Čebotarev, N.G.: The Principles of Galois Theory. 1–2, ONTI, Leningrad (1934–1937) (Russian)
628. Čebotarev, N.G. [Tschebotaröw, N.]: Kurzer Beweis des Diskriminantensatzes. Acta Arith. **1**, 78–82 (1935) [[631] **1**, 222–225. (Russian)]
629. Čebotarev, N.G. [Tschebotaröw, N.]: Eine Aufgabe aus der algebraischen Zahlentheorie. Acta Arith. **2**, 221–229 (1936) [[631] **1**, 226–234. (Russian)]
630. Čebotarev, N.G.: On the development of the theory of ideals according to Zolotarev. Uspekhi Mat. Nauk **2(6)**, 52–67 (1947) (Russian) [[631] **1**, 71–86.]
631. Čebotarev, N.G.: Collected Works. **1,2**, Izdat. AN SSSR (1949) (Russian)
632. Čebyšev, P.L.: Sur la fonction qui détermine la totalité des nombres premiers inférieurs à une limite donnée. Mémoires des savants étrangers de l'Acad. Sci. St. Pétersbourg **6**, 1–19 (1848) [J. Math. Pures Appl. **17**, 341–365 (1852); [631], **1**, 27–48.]
633. Čebyšev, P.L.: Mémoire sur nombres premiers. Mémoires des savants étrangers de l'Acad. Sci. St. Pétersbourg **7**, 17–33 (1850) [J. Math. Pures Appl. **17**, 366–390 (1852); [631], **1**, 49–70.]
634. Čebyšev, P.L.: Oeuvres. St. Pétersbourg (1899) [Reprint: Chelsea (1962)]
635. Cerri, J.-P.: De l'euclidianité de $Q(\sqrt{2+\sqrt{2+\sqrt{2}}})$ et $Q(\sqrt{2+\sqrt{2}})$ pour la norme. J. Théor. Nombres Bordeaux **12**, 103–126 (2000)
636. Cerri, J.-P.: Euclidean minima of totally real number fields: algorithmic determination. Math. Comp. **76**, 1547–1575 (2007)
637. Chabauty, C.: Sur les équations diophantiennes liées aux unités d'un corps de nombres algébriques fini. Ann. Mat. Pura Appl. **17**, 127–168 (1938)
638. Chabauty, C.: Démonstration nouvelle d'un théoréme de Thue et Mahler sur les formes binaires. Bull. Sci. Math. (2) **65**, 112–130 (1941)
639. Chabauty, C.: Sur les points rationnels des courbes algébriques de genre supérieur á l'unité. C. R. Acad. Sci. Paris **212**, 882–885 (1941)
640. Chakraborty, K., Luca, F., Mukhopadhyay, A.: Class numbers with many prime factors. J. Number Theory **128**, 2559–2572 (2008)
641. Chakraborty, K., Luca, F., Mukhopadhyay, A.: Exponents of class groups of real quadratic fields. Internat. J. Number Th. **4**, 597–611 (2008)
642. Chamizo, F., Cristóbal, E.: The sphere problem and the L-functions. Acta Math. Hungar. **135**, 97–115 (2012)
643. Chamizo, F., Iwaniec, H.: On the sphere problem. Rev. Mat. Iberoamericana **11**, 417–429 (1995)
644. Chamizo, F., Iwaniec, H.: On the Gauss mean-value theorem for class number. Nagoya Math. J. **151**, 199–208. (1998)
645. Chamizo, F., Ubis, A.: An average formula for the class number. Acta Arith. **122**, 75–90 (2006)
646. Chan, W.-K., Kim, M.-H., Raghavan, S.: Ternary universal integral quadratic forms over real quadratic fields. Jpn. J. Math. (N.S.) **22**, 263–273 (1996)
647. Chandrasekharan, K., Narasimhan, R.: Hecke's functional equation and the average order of arithmetical functions. Acta Arith. **6**, 487–503 (1960/61)

648. Chandrasekharan, K., Narasimhan, R.: Functional equations with multiple gamma factors and the average order of arithmetical functions. Ann. Math. (2) **76**, 93–136 (1962)
649. Chandrasekharan, K., Narasimhan, R.: On the mean value of the error term for a class of arithmetical functions. Acta Math. **112**, 41–67 (1964)
650. Chang, K.-Y., Kwon, S.-H.: The imaginary abelian number fields with class numbers equal to their genus class numbers. J. Théor. Nombres Bordeaux **12**, 349–365 (2000)
651. Chang, K.-Y., Kwon, S.-H.: The non-abelian normal CM-fields of degree 36 with class number one. Acta Arith. **101**, 53–61 (2002)
652. Chang, K.-Y., Kwon, S.-H.: The class number one problem for some non-abelian normal $CM-$fields of degree 48. Math. Comp. **72**, 1003–1017 (2003)
653. Changa, M.E. et al.: Scientific achievements of Anatoliĭ Alekseevič Karatsuba. Trudy Mat. Inst. im. Steklova **280**, Supp. 2, 1–22 (2013)
654. Chapman, R.J.: A simple proof of Noether's theorem. Glasg. Math. J. **38**, 49–51 (1996)
655. Chapman, S.T., Smith, W.W.: On a characterization of algebraic number fields with class number less than three. J. Algebra **135**, 381–387 (1990)
656. Charkani, M.E., Deajim, A.: Generating a power basis over a Dedekind ring. J. Number Theory **132**, 2267–2276 (2012)
657. Charve, L.: De la réduction des formes quadratiques ternaires positives et de son application aux irrationnelles du troisième degré. Ann. Sci. Éc. Norm. Supér. (2) **9**, Suppl., 3–156 (1880)
658. Châtelet, A.: Sur les corps abéliens du troisiéme degré. C. R. Acad. Sci. Paris **152**, 1290–1292 (1911)
659. Châtelet, A.: Sur certains ensembles de tableaux et leur application à la théorie des nombres. Ann. Sci. Éc. Norm. Supér. (3) **28**, 105–202 (1911)
660. Châtelet, A.: Sur une représentation des idéaux. C. R. Acad. Sci. Paris **154**, 502–504 (1912)
661. Châtelet, A.: Leçons sur la théorie des nombres: Modules, entiers algébriques, réduction continuelle. Gauthier-Villars (1913)
662. Châtelet, A.: Sur les corps abéliens de degré premier. C. R. Acad. Sci. Paris **170**, 651–653 (1920)
663. Châtelet, A.: Énumération et constitution des corps abéliens quélconques. C. R. Acad. Sci. Paris **171**, 658–661 (1920)
664. Châtelet, A.: Arithmétique des corps abéliens du troisième degré. Ann. Sci. Éc. Norm. Supér. (3) **63**, 109–160 (1946)
665. Chatland, H.: On the Euclidean algorithm in quadratic number fields. Bull. Amer. Math. Soc. **55**, 948–953 (1949)
666. Chatland, H., Davenport, H.: Euclid's algorithm in real quadratic fields. Canad. J. Math. **2**, 289–296 (1950)
667. Chella, T.: Dimostrazione dell' esistenza di un algoritmo delle divisioni successive per alcuni corpi circolari. Ann. Mat. Pura Appl. (4) **1**, 199–218 (1924)
668. Chen, I.: On Siegel's modular curve of level 5 and the class number one problem. J. Number Theory **74**, 278–297 (1999)
669. Chen, J.-R.: The number of lattice points in a given region. Chinese Math. **3**, 439–452 (1962)
670. Chen, J.-R.: Improvements on asymptotic formulas for the number of lattice points in a region of three dimensions. Sci. Sinica **12**, 151–161 (1963)
671. Chen, J.-R.: Improvements on asymptotic formulas for the number of lattice points in a region of three dimensions, II. Sci. Sinica **12**, 751–764 (1963)
672. Chen, J.-R.: On the representation of a larger even integer as the sum of a prime and the product of at most two primes,. Sci. Sinica **16**, 157–176 (1973)
673. Cherubini, J.M., Wallisser, R.V.: On the computation of all imaginary quadratic fields of class number one. Math. Comp. **49**, 295–299 (1987)
674. Chevalley, C.: Sur la théorie des restes normiques. C. R. Acad. Sci. Paris **191**, 426–428 (1930)
675. Chevalley, C.: Rélation entre le nombre de classes d'un sous-corps et celui d'un surcorps. C. R. Acad. Sci. Paris **192**, 257–258 (1931)

676. Chevalley, C.: La structure de la théorie du corps de classes. C. R. Acad. Sci. Paris **194**, 766–769 (1932)
677. Chevalley, C.: Sur la théorie du corps de classes dans les corps finis et les corps locaux. J. Fac. Sci. Univ. Tokyo **2**, 365–476 (1933)
678. Chevalley, C.: La théorie du symbole de restes normiques. J. Reine Angew. Math. **169**, 140–157 (1933)
679. Chevalley, C.: Letter to Hasse of June 20th, 1935. In: [1026], 25–29
680. Chevalley, C.: L'arithmétique dans les algèbres de matrices. Hermann (1936)
681. Chevalley, C.: Généralisation de la théorie du corps de classes pour les extensions infinies. J. Math. Pures Appl. (9) **15**, 359–371 (1936)
682. Chevalley, C.: La théorie du corps de classes. Ann. Math. (2) **41**, 394–418 (1940)
683. Chevalley, C.: Class Field Theory. University of Nagoya (1954)
684. Chevalley, C.: Invariants of finite groups generated by reflections. Amer. J. Math. **77**, 778–782 (1955)
685. Chevalley, C.: Emil Artin (1898–1962). Bull. Soc. Math. France **92**, 1–10 (1964)
686. Chevalley, C., Weil, A.: Hermann Weyl (1885–1955). Enseign. Math. (2) **3**, 157–187 (1957)
687. Chevalley, C., Nehrkorn, H.: Sur les démonstrations arithmétiques dans la théorie du corps de classes. Math. Ann. **111**, 364–371 (1935)
688. Childress, N.: Class Field Theory. Springer (2009)
689. Childs, L.N.: The group of unramified Kummer extensions of prime degree. Proc. London Math. Soc. **35**, 1977, 407–422 (1977)
690. Chinburg, T.: Salem numbers and L-functions. J. Number Theory **18**, 213–214 (1984)
691. Cho, P.J.: The strong Artin conjecture and large class numbers. Quart. J. Math., Oxford ser. **65**, 101–111 (2014)
692. Cho, P.J., Kim, H.H.: Dihedral and cyclic extensions with large class numbers. J. Théor. Nombres Bordeaux **24**, 583–603 (2012)
693. Cho, P.J., Kim, H.H.: Application of the strong Artin conjecture to the class number problem. Canad. J. Math. **65**, 1201–1216 (2013)
694. Chowla, P.: On the representation of -1 as a sum of squares in a cyclotomic field. J. Number Theory **1**, 208–210 (1969)
695. Chowla, P.: On the representation of -1 as a sum of two squares of cyclotomic integers. Norske Vid. Selsk. Forh., Trondheim **42**, 51–52 (1969)
696. Chowla, P., Chowla, S.: Determination of the Stufe of certain cyclotomic fields. J. Number Theory **2**, 271–272. (1970)
697. Chowla, S.: The class-number of binary quadratic form. Quart. J. Math., Oxford ser. **5**, 302–303. (1934)
698. Chowla, S.: An extension of Heilbronn's class-number theorem. Quart. J. Math., Oxford ser. **5**, 304–307 (1934)
699. Chowla, S.: On the k-analogue of a result in the theory of the Riemann zeta function. Math. Z. **38**, 483–487 (1934)
700. Chowla, S.: Heilbronn's class-number theorem. J. Indian Math. Soc. **2**, 66–68 (1934)
701. Chowla, S.: Heilbronn's class-number theorem. II. Proc. Indian Acad. Sci., Math. Sci. **1**, 145–146 (1934)
702. Chowla, S.: On the class-number of the corpus $P(\sqrt{-k})$. Proc. Nat. Inst. Sci. India **13**, 197–200 (1947)
703. Chowla, S.: Improvement of a theorem of Linnik and Walfisz. Proc. London Math. Soc. (2) **50**, 423–429 (1949)
704. Chowla, S.: A new proof of a theorem of Siegel. Ann. Math. (2) **51**, 120–122 (1950)
705. Chowla, S.: Proof of a conjecture of Julia Robinson. Norske Vid. Selsk. Forh., Trondheim **34**, 100–101 (1961)
706. Chowla, S.: The Heegner-Stark-Baker-Deuring-Siegel theorem. J. Reine Angew. Math. **241**, 47–48 (1970)
707. Chowla, S., Briggs, W.E.: On discriminants of binary quadratic forms with a single class in each genus. Canad. J. Math. **6**, 463–470 (1954)

708. Chowla, S., Dunton, M., Lewis, D.J.: Linear recurrences of order two. Pacific J. Math. **11**, 833–845 (1961)
709. Christofferson, S.: On representation of integers by binary quadratic forms in algebraic number fields. Dissertation, University of Uppsala (1962)
710. Chu, H., Hu, S.-J., Kang, M.-C.: Noether's problem for dihedral 2-groups. Comment. Math. Helv. **79**, 147–159 (2004)
711. Chu, H., Hu, S.-J., Kang, M.-C., Prokhorov, Y.G.: Noether's problem for groups of order 32. J. Algebra **320**, 3022–3025 (2008)
712. Chu, H., Kang, M.-C.: Rationality of p-group actions. J. Algebra **237**, 673–690 (2001)
713. Chua, K.S.: Real zeros of Dedekind zeta functions of real quadratic fields. Math. Comp. **74**, 1457–1470 (2005)
714. Cioffari, V.G.: The Euclidean condition in pure cubic and complex quartic fields. Math. Comp. **33**, 389–398 (1979)
715. Claborn, L.: Every abelian group is a class group. Pacific J. Math. **18**, 219–222 (1966)
716. Clark, D.A.: A quadratic field which is Euclidean but not norm-Euclidean. Manuscripta Math. **83**, 327–330 (1994)
717. Clark, D.A.: Non-Galois cubic fields which are Euclidean but not norm-Euclidean. Math. Comp. **65**, 1675–1679 (1996)
718. Clark, D.A., Murty, M.R.: The Euclidean algorithm for Galois extensions of Q. J. Reine Angew. Math. **459**, 151–162 (1995)
719. Clark, P.L.: Elliptic Dedekind domains revisited. Enseign. Math. (2) **55**, 213–225 (2009)
720. Clement Fernández, R., Echarri Hernández, J.M., Gómez Ayala, E.J.: A geometric proof of Kummer's reciprocity law for seventh powers. Acta Arith. **146**, 299–318 (2011)
721. Coates, J.: The work of Gross and Zagier on Heegner points and the derivatives of L-series. Astérisque **133/134**, 57–72 (1986)
722. Coates, J.: Kenkichi Iwasawa (1917–1998). Notices Amer. Math. Soc. **46**, 1221–1225 (1999)
723. Coates, J., Sinnott, W.: On p-adic L-functions over real quadratic fields. Invent. Math. **25**, 253–279 (1974)
724. Coates, J., Sinnott, W.: Integrality properties of the values of partial zeta functions. Proc. London Math. Soc. (3) **34**, 365–384 (1977)
725. Coates, J.H., van der Poorten, A.J.: Kurt Mahler, 1903–1988. Number Theory Research Reports, 92–118, Macquarie University (1992)
726. Cobbe, A.: Steinitz classes of tamely ramified Galois extensions of algebraic number fields. J. Number Theory **130**, 1129–1154 (2010)
727. Cobbe, A.: Steinitz classes of some abelian and nonabelian extensions of even degree. J. Théor. Nombres Bordeaux **22**, 607–628 (2010)
728. Cobbe, A.: Steinitz classes of tamely ramified nonabelian extensions of odd prime power order. Acta Arith. **149**, 347–359 (2011)
729. Cochrane, T., Mitchell, P.: Small solutions of the Legendre equation. J. Number Theory **70**, 62–66 (1998)
730. Cogdell, J. et al.: Ilya Piatetski-Shapiro, in memoriam. Notices Amer. Math. Soc. **57**, 1260–1275 (2010)
731. Cohen, H.: Variations sur un thème de Siegel–Hecke. Sém. Théor. Nombres Bordeaux **1973/74**, exp. 18, 1–45.
732. Cohen, H.: A Course in Computational Algebraic Number Theory. Springer (1993)
733. Cohen, H.: Advanced Topics in Computational Number Theory. Springer (2000)
734. Cohen, H., Diaz y Diaz, F., Olivier, M.: Tables of octic fields with a quartic subfield. Math. Comp. **68**, 1701–1716 (1999)
735. Cohen, H., Diaz y Diaz, F., Olivier, M.: Enumerating quartic dihedral extensions of Q. Compositio Math. **133**, 65–93 (2002)
736. Cohen, I.S.: Commutative rings with restricted minimum condition. Duke Math. J. **17**, 27–42 (1950)
737. Cohen, J., Sonn, J.: On the Ono invariants of imaginary quadratic fields. J. Number Theory **95**, 259–267 (2002)

738. Cohen, S.D.: The distribution of the Galois groups of integral polynomials. Illinois J. Math. **23**, 135–152 (1979)
739. Cohen, S.D.: The Galois group of a polynomial with two indeterminate coefficients. Pacific J. Math. **90**, 63–76 (1980); corr. **97**, 483–486 (1981)
740. Cohen, S.D.: The distribution of Galois groups and Hilbert's irreducibility theorem. Proc. London Math. Soc. (3) **43**, 227–250 (1981)
741. Cohen, S.D.: Galois groups of trinomials. Acta Arith. **54**, 43–49 (1989); corr. **70**, p. 193 (1995)
742. Cohen, S.D.: Obituary: Robert Winston Keith Odoni (1947–2002). Glasg. Math. J. **45**, 569–575 (2003)
743. Cohen, S.D., Movahhedi, A., Salinier, A.: Double transitivity of Galois groups of trinomials. Acta Arith. **82**, 1–15 (1997)
744. Cohen, S.D., Movahhedi, A., Salinier, A.: Galois grops of trinomials. J. Algebra **222**, 561–573 (1999)
745. Cohn, H.: The density of abelian cubic fields. Proc. Amer. Math. Soc. **5**, 476–477 (1954)
746. Cohn, H.: Numerical study of the representation of a totally positive quadratic integer as the sum of quadratic integral squares. Numer. Math. **1**, 121–134 (1959)
747. Cohn, H.: Decomposition into four integral squares in the fields of $2^{1/2}$ and $3^{1/2}$. Amer. J. Math. **82**, 301–322 (1960)
748. Cohn, H.: A numerical study of Weber's real class number calculation, I. Numer. Math. **2**, 347–362 (1960)
749. Cohn, H.: Calculation of class numbers by decomposition into three integral squares in the field of $2^{1/2}$ and $3^{1/2}$. Amer. J. Math. **83**, 33–56 (1961)
750. Cohn, H.: Cusp forms arising from Hilbert's modular functions for the field of $3^{1/2}$. Amer. J. Math. **84**, 283–305 (1962)
751. Cohn, H.: A Classical Invitation to Algebraic Numbers and Class Fields. Springer (1978)
752. Cohn, H., Lagarias, J.C.: On the existence of fields governing the 2-invariants of the class-group of $Q(\sqrt{dp})$ as p varies. Math. Comp. **41**, 711–730 (1983)
753. Cohn, H., Gorn, S.: A computation of cyclic cubic units. J. Res. Nat. Bur. Standards **59**, 155–168 (1957)
754. Cohn, H., Pall, G.: Sums of four squares in a quadratic ring. Trans. Amer. Math. Soc. **105**, 536–556 (1962)
755. Coleman, J.B.: A test for the type of irrationality represented by a periodic ternary continued fraction. Amer. J. Math. **52**, 835–842 (1930)
756. Coleman, J.B.: The Jacobian algorithm for periodic continued fractions as defining a cubic irrationality. Amer. J. Math. **55**, 585–592 (1934)
757. Coleman, M.D.: A zero-free region for the Hecke L-functions. Mathematika **37**, 287–304 (1990)
758. Coleman, M.D.: The Rosser-Iwaniec sieve in number fields, with an application. Acta Arith. **65**, 53–83 (1993)
759. Coleman, R.F.: Division values in local fields. Invent. Math. **53**, 91–116 (1979)
760. Collins, G.E.: Polynomial minimum root separation. J. Symbolic Comput. **32**, 467–473 (2001)
761. Connell, I.G.: On algebraic number fields with unique factorization. Canad. Math. Bull. **5**, 151–156 (1962)
762. Connell, I.G., Sussman, D.: The p-dimension of class-groups of number fields. J. Lond. Math. Soc. (2) **2**, 525–529 (1970)
763. Conrad, K.: The origin of representation theory. Enseign. Math. (2) **44**, 361–392 (1998)
764. Conrad, M.: Construction of bases for the group of cyclotomic units. J. Number Theory **81**, 1–15 (2000)
765. Conrey, [J.]B., Iwaniec, H.: Spacing of zeros of Hecke L-functions and the class number problem. Acta Arith. **103**, 259–312 (2002)
766. Conrey, J.B., Soundararajan, K.: Real zeros of quadratic Dirichlet L-functions. Invent. Math. **150**, 1–44 (2002)

767. Cooke, G.E.: A weakening of the Euclidean property for integral domains and applications to algebraic number theory, I. J. Reine Angew. Math. **282**, 133–156 (1976)
768. Cooke, G.E.: A weakening of the Euclidean property for integral domains and applications to algebraic number theory, II. J. Reine Angew. Math. **283/284**, 71–85 (1976)
769. Coray, D.F.: Cubic hypersurfaces and a result of Hermite. Duke Math. J. **54**, 657–670 (1987)
770. Cornell, G.: Abhyankar's lemma and the class group. In: Number theory, Carbondale 1979, Lecture Notes Math. **751**, 82–88 (1979)
771. Cornell, G., Washington, L.: Class numbers of cyclotomic fields. J. Number Theory **21**, 260–274 (1985)
772. Corry, L.: Fermat comes to America: Harry Schultz Vandiver and FLT (1914–1963). Math. Intelligencer **29**, 30–40 (2007)
773. Corry, L.: Number crunching vs. number theory: computers and FLT, from Kummer to SWAC (1850–1960), and beyond. Arch. Hist. Exact Sci. **62**, 393–455 (2008)
774. Costa, A.: A generalization of a result of Barrucand and Cohn on class numbers. J. Number Theory **45**, 254–260 (1993)
775. Cougnard, J.: Monogénéité de l'anneau des entiers des extensions cycliques imaginaires de degré $2l$ (l premier ≥ 5). J. Reine Angew. Math. **375/376**, 42–46 (1987)
776. Cougnard, J.: Conditions necessaires de monogénéité. Application aux extensions cycliques de degré premier $l \geq 5$ d'un corps quadratique imaginaire. J. Lond. Math. Soc. (2) **37**, 73–87 (1988)
777. Cougnard, J.: Modéle de Legendre d'une courbe elliptique à multiplication complexe et monogénéité d'anneaux d'entiers, I. Acta Arith. **54**, 191–212 (1990)
778. Cougnard, J.: Modéle de Legendre d'une courbe elliptique à multiplication complexe et monogénéité d'anneaux d'entiers, II. Acta Arith. **55**, 75–81 (1990)
779. Cougnard, J.: Construction de base normale pour les extensions de Q à groupe D_4. J. Théor. Nombres Bordeaux **12**, 399–409 (2000)
780. Cougnard, J.: Nouveaux exemples d'extension relatives sans base normale. Ann. Fac. Sci. Toulouse, Math. (6) **10**, 493–505 (2001)
781. Cougnard, J.: Normal integral bases for A_4 extensions of the rationals. Math. Comp. **75**, 485–496 (2006)
782. Cougnard, J., Fleckinger, V.: Sur la monogénéité de l'anneau des entiers de certains corps de rayon. Manuscripta Math. **63**, 365–376 (1989)
783. Cougnard, J., Queyrut, J.: Construction de bases normales pour les extensions galoisiennes absolues à groupe de Galois quaternionien d'ordre 12. J. Théor. Nombres Bordeaux **14**, 87–102 (2002)
784. Courant, R.: Felix Klein. Jahresber. Deutsch. Math.-Verein. **34**, 197–213 (1926)
785. Courant, R.: Carl Runge als Mathematiker. Naturwissenschaften **15**, 229–231 (1927)
786. Cox, D.A.: Primes of the form $x^2 + ny^2$. J. Wiley (1989); 2nd ed. (2013)
787. Cox, D.A., Hyde, T.: The Galois theory of the lemniscate. J. Number Theory **135**, 43–59 (2014)
788. Cox, D.A., Shurman, J.: Geometry and number theory on clovers. Amer. Math. Monthly **112**, 682–704 (2005)
789. Coykendall, J.: A remark on arithmetic equivalence and the normset. Acta Arith. **92**, 105–108 (2000)
790. Craig, M.: A type of class group for imaginary quadratic fields. Acta Arith. **22**, 449–459 (1973)
791. Craig, M.: A construction for irregular discriminants. Osaka J. Math. **14**, 365–402 (1977); corr. **15**, p. 461 (1978)
792. Cresse, G.H.: Arithmetical deduction of Kronecker's class-number relations. Ann. Math. (2) **23**, 271–279 (1922)
793. Creutz, B.: A Grunwald-Wang type theorem for abelian varieties. Acta Arith. **154**, 353–370 (2012)
794. Croot, E.S.III, Granville, A.: Unit fractions and the class number of a cyclotomic field. J. Lond. Math. Soc. (2) **66**, 579–591 (2002)

795. Čudakov, N.G. [Tchudakoff, N.]: On zeros of Dirichlet's L-functions. Mat. Sb. (N.S.) **1**, 591–602 (1936)
796. Čudakov, N.G.: On Siegel's theorem. Izv. Akad. Nauk SSSR, Ser. Mat. **6**, 135–142 (1942) (Russian)
797. Čudakov, N.G.: The upper bound for the discriminant of the tenth imaginary quadratic field with class number one. Studies in Number Theory (Saratov Univ.) **3**, 75–77 (1969) (Russian)
798. Cunningham, A.J.C.: On 2 as a 16-ic residue. Proc. London Math. Soc. **27**, 85–122 (1895)
799. Cunningham, A.J.C.: Theory of numbers tables — errata. Messenger of Math. **46**, 49–69 (1916)
800. Cuoco, A.A., Monsky, P.: Class numbers in Z_p^d-extensions. Math. Ann. **255**, 235–258 (1981)
801. Čupr, K.: Prof. Matyáš Lerch. Časopis mat. fys. **52**, 301–313 (1923)
802. Cusick, T.W.: Finding fundamental units in cubic fields. Math. Proc. Cambridge Philos. Soc. **92**, 385–389 (1982)
803. Cusick, T.W.: Finding fundamental units in totally real fields. Math. Proc. Cambridge Philos. Soc. **96**, 191–194 (1984)
804. Cusick, T.W., Schoenfeld, L.: A table of fundamental pairs of units in totally real cubic fields. Math. Comp. **48**, 147–158 (1987)
805. Cvetkov, V.M.: On a theorem of Stickelberger and Voronoĭ. Zap. Naučn. Sem. LOMI **121**, 171–175 (1983) (Russian)
806. Cvetkov, V.M.: A simple example of extensions without an integral basis. Zap. Naučn. Sem. POMI **272**, 341–344 (2000) (Russian)
807. Czogała, A.: Arithmetic characterization of algebraic number fields with small class numbers. Math. Z. **176**, 247–253 (1981)
808. Daboussi, H.: Sur le théorème des nombres premiers. C. R. Acad. Sci. Paris **298**, 161–164 (1984)
809. Dai, Z.D., Lam, T.Y., Peng, C.K.: Levels in algebra and topology. Bull. Amer. Math. Soc. (N.S.) **3**, 845–848 (1980)
810. Daileda, R.C.: Non-abelian number fields with very large class numbers. Acta Arith. **125**, 215–255 (2006)
811. Dalawat, C.S.: Local discriminants, Kummerian extensions, and elliptic curves. J. Ramanujan Math. Soc. **25**, 25–80 (2010)
812. Dalen, K.: On a theorem of Stickelberger. Math. Scand. **3**, 124–126 (1955)
813. Dantscher, V.: Bemerkungen zum analytischen Beweise des cubischen Reciprocitätsgesetzes. Math. Ann. **12**, 241–254 (1877)
814. Datskovsky, B.A.: A mean-value theorem for class numbers of quadratic extensions. Contemp. Math. **143**, 179–242 (1993)
815. Datskovsky, B.[A.], Wright, D.J.: Density of discriminants of cubic extensions. J. Reine Angew. Math. **386**, 116–138 (1988)
816. Daus, P.H.: Normal ternary continued fraction expansions for the cube roots of integers. Amer. J. Math. **44**, 279–296 (1922)
817. Daus, P.H.: Normal ternary continued fraction expansions for cubic irrationalities. Amer. J. Math. **51**, 67–98 (1929)
818. Daus, P.H.: Ternary continued fractions in a cubic field. Tôhoku Math. J. **41**, 337–348 (1935)
819. Davenport, H.: W.E.H. Berwick. J. Lond. Math. Soc. **21**, 74–80 (1946)
820. Davenport, H.: Sur les corps cubiques à discriminants négatifs. C. R. Acad. Sci. Paris **228**, 883–885 (1949)
821. Davenport, H.: Euclid's algorithm in cubic fields of negative discriminant. Acta Math. **84**, 159–179 (1950)
822. Davenport, H.: Euclid's algorithm in certain quartic fields. Trans. Amer. Math. Soc. **68**, 508–532 (1950)
823. Davenport, H.: Indefinite binary quadratic forms, and Euclid's algorithm in real quadratic fields. Proc. London Math. Soc. (2) **53**, 65–82 (1951)
824. Davenport, H.: T.Vijayaraghavan. J. Lond. Math. Soc. **33**, 252–255 (1958)
825. Davenport, H.: L.J. Mordell. Acta Arith. **9**, 3–22 (1964)

826. Davenport, H., Heilbronn, H.: On the density of discriminants of cubic fields. Bull. Lond. Math. Soc. **1**, 345–348 (1969) [[1755], 528–531.]
827. Davenport, H., Heilbronn, H.: On the density of discriminants of cubic fields, II. Proc. Royal Soc. London, ser. A **322**, 405–420 (1971) [[1755], 532–547.]
828. David, P.: Détermination de la structure du groupe des unités principales d'une p-extension abélienne K d'un corps local régulier, considéré comme un $Z_p(G(K/k))$-module. C. R. Acad. Sci. Paris **286**, A985–A986 (1978)
829. Davidson, M.: On Waring's problem in number fields. J. Lond. Math. Soc. (2) **59**, 435–447 (1999)
830. Davis, D.: Computing the number of totally positive circular units which are squares. J. Number Theory **10**, 1–9 (1978)
831. Davis, R.W.: Class number formulae for imaginary quadratic fields. J. Reine Angew. Math. **286/287**, 369–379 (1976)
832. Davis, R.W.: Class number formulae for imaginary quadratic fields, II. J. Reine Angew. Math. **299/300**, 247–255 (1978)
833. Davis, S., Duke, W., Sun, X.: Probabilistic Galois theory of reciprocal polynomials. Expo. Math. **16**, 263–270 (1998)
834. Debaene, K.: The first factor of the class number of the pth cyclotomic field. Arch. Math. (Basel) **102**, 237–244 (2014)
835. Debarre, O., Klassen, M.J.: Points of low degree on smooth plane curves. J. Reine Angew. Math. **446**, 81–87 (1994)
836. Dedekind, I., Dugac, P., Geyer, W.-D., Scharlau, W.: Richard Dedekind, 1831–1981. Vieweg (1981)
837. Dedekind, R.: Beweis für die Irreductibilität der Kreisteilungsgleichungen. J. Reine Angew. Math. **54**, 27–30 (1857) [[856], **1**, 68–71.]
838. Dedekind, R.: Anzeige der zweiten Auflage von Dirichlets Vorlesungen über Zahlentheorie. Göttingische gelehrte Anzeigen **1871**, 1481–1494. [[856], **3**, 399–407.]
839. Dedekind, R.: X Supplement, Über die Komposition der binären quadratischen Formen. In: [972], 2nd ed., 423–462. [[856], **3**, 223–261.]
840. Dedekind, R.: Anzeige von P. Bachmann, Die Lehre von der Kreistheilung und ihre Beziehung zur Zahlentheorie. Zeitschr. Math. Phys. **18**, 14–24 (1873) [[856], **3**, 409–420.]
841. Dedekind, R.: Sur la théorie des nombres entiers algébriques. Bull. Sci. Math. **11**, 278–288 (1876); errata p. 310; Bull. Sci. Math. (2) **1**, 17–41, 69–92, 114–164 (1877) [Gauthier-Villars, 1877; [856], **3**, 262–296; English translation: Theory of Algebraic Integers, Cambridge University Press (1996)]
842. Dedekind, R.: Über die Anzahl der Ideal-Klassen in den verschiedenen Ordnungen eines endlichen Körpers. In: Festschrift der Technischen Hochschule in Braunschweig zur Säkularfeier des Geburtstages von C. F. Gauß, 1–55, Vieweg (1877) [[856], **1**, 105–158.]
843. Dedekind, R.: Über den Zusammenhang zwischen der Theorie der Ideale und der Theorie der höheren Kongruenzen. Abhandl. Kgl. Ges. Wiss. Göttingen **23**, 1–23 (1878) [[856], **1**, 202–232.]
844. Dedekind, R.: XI Supplement, Über die Theorie der ganzen algebraischen Zahlen. In: [972], 3rd. ed., 515–530. [[856], **3**, 1–222.]
845. Dedekind, R.: Über die Diskriminanten endlicher Körper. Abhandl. Kgl. Ges. Wiss. Göttingen **29**, 1–56 (1882) [[856], **1**, 351–397.]
846. Dedekind, R.: Grundideale in Kreiskörpern. In: [856], **2**, 401–409.
847. Dedekind, R.: Aus Briefen an Frobenius. Letter of June 8th, 1882. In: [856], **2**, 414–415.
848. Dedekind, R.: XI Supplement, Über die Theorie der ganzen algebraischen Zahlen. In: [972], 4th ed., 434–657. [[856], **3**, 297–314.]
849. Dedekind, R.: Zur Theorie der Ideale. Nachr. Ges. Wiss. Göttingen **1894**, 272–277. [[856], **2**, 43–49.]
850. Dedekind, R.: Über die Begründung der Idealtheorie. Nachr. Ges. Wiss. Göttingen **1895**, 106–113. [[856], **2**, 50–58.]

851. Dedekind, R.: Über eine Erweiterung des Symbols $(\mathfrak{a},\mathfrak{b})$ in der Theorie der Moduln. Nachr. Ges. Wiss. Göttingen **1895**, 183–188. [[856], **2**, 59–86.]
852. Dedekind, R.: Aus Briefen an Frobenius. Letter of July 8th, 1896. In: [856], **2**, 433–437.
853. Dedekind, R.: Über die Anzahl von Idealklassen in reinen kubischen Zahlkörpern. J. Reine Angew. Math. **121**, 40–123 (1900) [[856], **2**, 148–233.]
854. Dedekind, R.: Über die Permutationen des Körpers aller algebraischen Zahlen. Abhandl. Kgl. Ges. Wiss. Göttingen **1901**, 1–17. [[856], **2**, 272–292.]
855. Dedekind, R.: Charakteristische Eigenschaft einklassiger Körper Ω. In: [856], **2**, 373–375.]
856. Dedekind, R.: Gesammelte mathematische Werke. **1–3**, Vieweg (1930) [Reprint: Chelsea (1968)]
857. Dedekind, R., Weber, H.: Theorie der algebraischen Funktionen einer Veränderlichen. J. Reine Angew. Math. **92**, 181–290 (1882) [[856], **1**, 238–249.]
858. Degen, C.F.: Canon Pellianus. Bonnier (1817) [Reprint: Nabu Press (2012)]
859. Deitmar, A.: Class numbers of orders in cubic fields. J. Number Theory **95**, 150–166 (2002)
860. Deitmar, A.: A prime geodesic theorem for higher rank spaces. Geom. Funct. Anal. **14**, 1238–1266 (2004)
861. Deitmar, A., Hoffmann, W.: Asymptotics of class numbers. Invent. Math. **160**, 647–675 (2005)
862. Deitmar, A., Pavey, M.: Class numbers of orders in complex quartic fields. Math. Ann. **338**, 767–799 (2007)
863. de la Maza, A.-C.: Bounds for the smallest norm in an ideal class. Math. Comp. **71**, 1745–1758 (2002)
864. Del Corso, I.: On Kronecker's use of indeterminate coefficients. Rend. Sem. Mat. Univ. Politec. Torino **53**, 261–275 (1995)
865. Del Corso, I.: Kronecker's method of indeterminate coefficients. Comm. Algebra **28**, 4749–4765 (2000)
866. Del Corso, I., Dvornicich, R.: Number fields with the same index. Acta Arith. **102**, 323–337 (2002)
867. Del Corso, I., Dvornicich, R.: An equivalence between local fields. J. Number Theory **115**, 230–248 (2005)
868. Del Corso, I., Dvornicich, R.: On Ore's conjecture and its developments. Trans. Amer. Math. Soc. **357**, 3813–3829 (2005)
869. Del Corso, I., Dvornicich, R.: Non-invariance of the index in wildly ramified extensions. Internat. J. Number Th. **6**, 1855–1868 (2010)
870. Del Corso, I., Rossi, L.P.: Normal integral bases for cyclic Kummer extensions. J. Pure Appl. Algebra **214**, 385–391 (2010)
871. Del Corso, I., Rossi, L.P.: Normal integral bases and tameness conditions for Kummer extensions. Acta Arith. **160**, 1–23 (2013)
872. Deligne, P.: Les constantes des équations fonctionnelles des fonctions L. Lecture Notes Math. **349**, 501–597 (1973); corr. Lecture Notes Math. **476**, p. 149 (1975)
873. Deligne, P., Serre, J.-P.: Formes modulaires de poids 1. Ann. Sci. Éc. Norm. Supér. (4) **7**, 507–530 (1974)
874. Delone, B. [B.N.]: On defining algebraic realms by congruences (with applications to Abelian equations). Soobšč. Kharkov Mat. Obšč. **14**, 271–274 (1915) (Russian)
875. Delone, B.N. [Delaunay, B.]: La solution générale de l'équation $X^3\varrho + Y^3 = 1$. C. R. Acad. Sci. Paris **162**, 150–151 (1916)
876. Delone, B.N. [Delaunay, B.]: Zur Bestimmung algebraischer Zahlkörper durch Kongruenzen; eine Anwendung auf die Abelschen Gleichungen. J. Reine Angew. Math. **152**, 120–123 (1923)
877. Delone, B.N. [Delaunay, B.]: Interprétation géométrique de la généralisation de l'algorithme des fractions continues donnée par Voronoi. C. R. Acad. Sci. Paris **176**, 554–556 (1923)
878. Delone, B.N. [Delaunay, B.]: Vollständige Lösung der unbestimmten Gleichung $X^3q + Y^3 = 1$ in ganzen Zahlen. Math. Z. **28**, 1–9 (1928)

879. Delone, B.N.: On the geometry of Galois theory. In: Sbornik posviaščennyĭ pamĭati akademika D.A. Grave, 52–62. Moskva (1940) (Russian)
880. Delone, B.N.: Dmitriĭ Aleksandrovič Grave (1863–1939). Izv. Akad. Nauk SSSR, Ser. Mat. **4**, 349–356 (1940) (Russian)
881. Delone, B.N.: Nikolaĭ Grigor'evič Čebotarev, 1894–1947. Izv. Akad. Nauk SSSR, Ser. Mat. **12**, 337–340 (1948) (Russian)
882. Delone, B.N., Faddeev, D.K.: The theory of irrationalities of third degree. Trudy Mat. Inst. im. Steklova **11**, 1–340 (1940) (Russian) . [English translation: Amer. Math. Soc. (1964)]
883. Delone, B.N., Faddeev, D.K.: Investigations in the geometry of Galois theory. Mat. Sb. **15**, 243–284 (1944) (Russian)
884. Delone, B.N., Sominskiĭ, J., Billevič, K.: A table of totally real fields of degree four. Izv. Akad. Nauk SSSR, Ser. Mat. (7) **1935**, 1267–1297 (Russian)
885. Demchenko, O., Gurevich, A.: Reciprocity laws through formal groups. Proc. Amer. Math. Soc. **141**, 1591–1596 (2013)
886. Demuškin, S.P.: The group of a maximal p-extension of a local field. Izv. Akad. Nauk SSSR, Ser. Mat. **25**, 329–346 (1961) (Russian)
887. Demuškin, S.P.: On 2-extensions of a local field. Sibir. Mat. Zh. **4**, 951–955 (1963) (Russian)
888. Demuškin, S.P., Šafarevič, I.R.: The imbedding problem for local fields. Izv. Akad. Nauk SSSR, Ser. Mat. **23**, 823–840 (1959) (Russian)
889. Demuškin, S.P., Šafarevič, I.R.: The second obstruction for the imbedding problem for the field of algebraic numbers. Izv. Akad. Nauk SSSR, Ser. Mat. **26**, 911–924 (1962) (Russian)
890. Dénes, P.: Über den ersten Fall des letzten Fermatschen Satzes. Monatsh. Math. **54**, 161–174 (1950)
891. Dénes, P.: Proof of a conjecture of Kummer. Publ. Math. Debrecen **2**, 206–214 (1952)
892. Dénes, P.: Über irreguläre Kreiskörper. Publ. Math. Debrecen **3**, 17–23 (1953)
893. Dénes, P.: Über die Kummerschen logarithmischen Hilfsfunktionen. Acta Sci. Math. (Szeged) **15**, 115–125 (1954)
894. Dénes, P.: Über den zweiten Faktor der Klassenzahl und den Irregularitätsgrad der irregulären Kreiskörper. Publ. Math. Debrecen **4**, 163–170 (1956)
895. Dentzer, R.: Projektive symplektische Gruppen $PSp_4(p)$ als Galoisgruppen über $Q(t)$. Arch. Math. (Basel) **53**, 337–346 (1989)
896. de Rham, G.: Gustave Dumas. Elem. Math. **10**, 121–122 (1955)
897. de Roton, A.: On the mean square of the error term for an extended Selberg class. Acta Arith. **126**, 27–55 (2007)
898. Descombes, R., Poitou, G.: Sur l'approximation dans $R(i\sqrt{11})$. C. R. Acad. Sci. Paris **231**, 264–266 (1950)
899. de Smit, B.: Generating arithmetically equivalent number fields with elliptic curves. Lecture Notes in Computer Science **1423**, 392–399 (1998)
900. de Smit, B.: On arithmetically equivalent fields with distinct p-class numbers. J. Algebra **272**, 417–424 (2004)
901. de Smit, B., Perlis, R.: Zeta functions do not determine class numbers. Bull. Amer. Math. Soc. (N.S.) **31** 213–215 (1994)
902. Dettweiler, M.: Middle convolution and Galois realizations. Develop. Math. **11**, 143–158 (2004)
903. Dettweiler, M., Kühn, U., Reiter, S.: On Galois representations via Siegel modular forms of genus two. Math. Res. Lett. **8**, 577–588 (2001)
904. Dettweiler, M., Reiter, S.: On rigid tuples in linear groups of odd dimension. J. Amer. Math. Soc. **222**, 550–560 (1999)
905. Dettweiler, M., Reiter, S.: An algorithm of Katz and its application to the inverse Galois problem. J. Symbolic Comput. **30**, 761–798 (2000)
906. Deuring, M.: Verzweigungstheorie bewerteter Körper. Math. Ann. **105**, 277–307 (1931)
907. Deuring, M.: Galoissche Theorie und Darstellungstheorie. Math. Ann. **107**, 140–144 (1932)
908. Deuring, M.: Imaginäre quadratische Zahlkörper mit der Klassenzahl 1. Math. Z. **37**, 405–415 (1933)

909. Deuring, M.: Über den Tschebotareffschen Dichtigkeitssatz. Math. Ann. **110**, 414–415 (1934)
910. Deuring, M.: Neuer Beweis des Bauerschen Satzes. J. Reine Angew. Math. **173**, 1–4 (1935)
911. Deuring, M.: Anwendungen der Darstellungen von Gruppen durch lineare Substitutionen auf die galoissche Theorie. Math. Ann. **113**, 40–47 (1936)
912. Deuring, M.: Die Typen der Multiplikatorenringe elliptischer Funktionenkörper. Abh. Math. Sem. Hansischen Univ. **14**, 197–272 (1941)
913. Deuring, M.: Algebraische Begründung der komplexen Multiplikation. Abh. Math. Semin. Univ. Hambg **16**, 32–47 (1949)
914. Deuring, M.: Die Struktur der elliptischen Funktionen-Körper und die Klassenkörper der imaginären quadratischen Zahlkörper. Math. Ann. **124**, 393–426 (1952)
915. Deuring, M.: Die Klassenkörper der komplexen Multiplikation. Enzyklopädie der Mathematischen Wissenschaften, Bd. 1_2, Heft 10_2, Teubner (1958)
916. Deuring, M.: Imaginäre quadratische Zahlkörper mit der Klassenzahl Eins. Invent. Math. **5**, 169–179 (1968)
917. Deuring, M.: Carl Ludwig Siegel, December 12, 1896–April 4, 1981. Acta Arith. 45, 93–113 (1986)
918. Deutsch, J.I.: Geometry of numbers proof of Götzky's four-squares theorem. J. Number Theory **96**, 417–431 (2002)
919. Deutsch, J.I.: An alternate proof of Cohn's four squares theorem. J. Number Theory **104**, 263–278 (2004)
920. de Weger, B.M.M.: Periodicity of p-adic continued fractions. Elem. Math. **43**, 112–116 (1988)
921. Diaz y Diaz, F.: On some families of imaginary quadratic fields. Math. Comp. **32**, 637–650 (1978)
922. Diaz y Diaz, F.: Sur le 3-rang des corps quadratiques. Publ. Math. d'Orsay **78(11)**, 1–93 (1978)
923. Diaz y Diaz, F.: Valeur minima du discriminant des corps de degré 7 ayant une place réelle. C. R. Acad. Sci. Paris **296**, 137–139 (1983)
924. Diaz y Diaz, F.: Valeur minima de discriminant pour certains types de corps de degré 7. Ann. Inst. Fourier (Grenoble) **34(3)**, 29–38 (1984)
925. Diaz y Diaz, F.: Petits discriminants des corps de nombres totalement imaginaires de degré 8. J. Number Theory **25**, 34–52 (1987)
926. Diaz y Diaz, F.: Discriminant minimal et petit discriminants des corps de nombres de degré 7 avec cinq places réels. J. Lond. Math. Soc. (2) **38**, 33–46 (1988)
927. Diaz y Diaz, F., Olivier, M.: Imprimitive ninth-degree number fields with small discriminants. Math. Comp. **64**, 305–321 (1995)
928. Diaz y Diaz, F., Shanks, D., Williams, H.C.: Quadratic fields with 3-rank equal to 4. Math. Comp. **33**, 836–840 (1979)
929. Dick, A.: Emmy Noether: 1882–1935. Elem. Math. Beiheft no. **13**, 1–72 (1970) [English translation: Birkhäuser, 1981.]
930. Dick, A.: Franz Mertens, 1840–1927. Forschungszentrum Graz (1981)
931. Dickson, L.E.: The inscription of regular polygon. Amer. Math. Monthly **1**, 299–301, 342–345, 376–377, 423–425 (1894); **2**, 7–9, 38–40 (1895)
932. Dickson, L.E.: On the factorization of integral functions with p-adic coefficients. Bull. Amer. Math. Soc. (2) **17**, 19–23 (1910)
933. Dickson, L.E.: On the negative discriminants for which there is a single class of positive primitive binary quadratic forms. Bull. Amer. Math. Soc. (2) **17**, 534–537 (1911)
934. Dickson, L.E.: Linear associative algebras and Abelian equations. Trans. Amer. Math. Soc. **15**, 31–46 (1914)
935. Dickson, L.E.: History of the Theory of Numbers. **1–3**, Carnegie Institute (1919–1923) [Reprints: Stechert (1934); Chelsea (1952, 1966)]
936. Dickson, L.E.: Algebras and their Arithmetics. University of Chicago Press (1923) [Reprint: Dover (1960); German translation: Algebren und ihre Zahlentheorie, Orell Füssli (1927)]

937. Dickson, L.E.: Quadratic fields in which factorization is always unique. Bull. Amer. Math. Soc. **30**, 328–334 (1924)
938. Dickson, L.E.: Introduction to the theory of numbers. University of Chicago Press (1929) [German translation: Einführung in die Zahlentheorie, Teubner (1931)]
939. Dickson, L.E.: Cyclotomy, higher congruences, and Waring's problem. Amer. J. Math. **57**, 391–424 (1935)
940. Dickson, L.E.: Hans Frederik Blichfeldt. 1873–1945. Bull. Amer. Math. Soc. **53**, 882–883 (1947)
941. Dickson, L.E., Mitchell, H.H., Vandiver, H.S., Wahlin, G.E.: Algebraic numbers. Report of the Committee on algebraic numbers, **1**. Bull. National Research Council, **5**, part 3, 1–96 (1923) [Reprint: Chelsea (1967)]
942. Diekert, V.: Über die absolute Galoisgruppe dyadischer Zahlkörper. J. Reine Angew. Math. **350**, 152–172 (1984)
943. Dieudonné, J.: Jacques Herbrand et la théorie des nombres. Stud. Logic Found. Math. **107**, 3–7 (1982)
944. Dieudonné, J., Tits, J.: Claude Chevalley (1909–1984). Bull. Amer. Math. Soc. (N.S.) **17**, 1–7 (1987)
945. Dietmann, R.: Probabilistic Galois theory for quartic polynomials. Glasg. Math. J. **48**, 553–556 (2006)
946. Dietmann, R.: Probabilistic Galois theory. Bull. Lond. Math. Soc. **45**, 453–462 (2013)
947. Di Franco, F., Pace, F.: Arithmetical characterization of rings of algebraic integers with class number three and four. Boll. Unione Mat. Ital. (6) **4**, 63–69 (1985)
948. Dinghas, A.: Erhard Schmidt (Erinnerungen und Werk). Jahresber. Deutsch. Math.-Verein. **72**, 3–17 (1970/71)
949. Dintzl, E.: Ueber den zweiten Ergänzungssatz des biquadratischen Reciprocitätsgesetzes. Monatsh. Math. Phys. **10**, 88–96 (1899)
950. Dintzl, E.: Der zweite Ergänzungssatz des kubischen Reciprocitätsgesetzes. Monatsh. Math. Phys. **10**, 303–306 (1899)
951. Dintzl, E.: Über die Legendreschen Symbole für quadratische Reste in einem imaginären quadratischen Zahlkörper mit der Klassenanzahl 1. SBer. Kais. Akad. Wissensch. Wien **116**, 785–800 (1907)
952. Dintzl, E.: Über die Zahlen im Körper $k(\sqrt{-2})$, welche den Bernoullischen Zahlen analog sind. SBer. Kais. Akad. Wissensch. Wien **118**, 173–201 (1909)
953. Dintzl, E.: Über die Entwicklungskoeffizienten der elliptischen Funktionen, insbesondere im Falle singulärer Module. Monatsh. Math. **25**, 125–151 (1914)
954. Dirichlet, P.G.L.: Recherches sur les diviseurs premiers d'une classe de formules du quatrième degré. J. Reine Angew. Math. **3**, 35–69 (1828) [[973], **1**, 63–98.]
955. Dirichlet, P.G.L.: Mémoire sur l'impossibilité de quelques équations indéterminées du cinquième degré. J. Reine Angew. Math. **3**, 354–375 (1828) [[973], **1**, 21–46.]
956. Dirichlet, P.G.L.: Démonstration d'une propriété analogue à la loi de réciprocité qui existe entre deux nombres premiers quelconques. J. Reine Angew. Math. **9**, 379–389 (1832) [[973], **1**, 173–188.]
957. Dirichlet, P.G.L.: Démonstration du théorème de Fermat pour le cas de 14iémes puissances. J. Reine Angew. Math. **9**, 390–393 (1832) [[973], **1**, 181–194.]
958. Dirichlet, P.G.L.: Einige neue Sätze über unbestimmte Gleichungen. Abh. Kgl. Preuss. Akad. Wiss. Berlin **1834**, 649–664. [[973], **1**, 219–236.]
959. Dirichlet, P.G.L.: Beweis eines Satzes ueber die arithmetische Progression. Ber. Verhandl. Kgl. Preuß. Akad. Wiss. Berlin **1837**, 108–110. [[973], **1**, 307–312.]
960. Dirichlet, P.G.L.: Beweis des Satzes, dass jede unbegrenzte arithmetische Progression, deren erstes Glied und Differenz ganze Zahlen ohne gemeinschaftlichen Factor sind, unendlich viele Primzahlen enthält. Abh. Kgl. Preuss. Akad. Wiss. Berlin **1837**, 45–81. [[973], **1**, 313–342; French translation: J. Math. Pures Appl. **4**, 393–422 (1839)]
961. Dirichlet, P.G.L.: Sur la manière de résoudre l'équation $t^2 - pu^2 = 1$ au moyen des fonctions circulaires. J. Reine Angew. Math. **17**, 286–290 (1837) [[973], **1**, 343–350.]

962. Dirichlet, P.G.L.: Sur l'usage des séries infinies dans la théorie des nombres. J. Reine Angew. Math. **18**, 259–274 (1838) [[973], **1**, 357–374.]
963. Dirichlet, P.G.L.: Recherches sur diverses applications de l'analyse infinitésimale à la théorie des nombres. J. Reine Angew. Math. **19**, 324–369 (1839) [[973], **1**, 411–461.]
964. Dirichlet, P.G.L.: Recherches sur diverses applications de l'analyse infinitésimale à la théorie des nombres, II, J. Reine Angew. Math. **21**, 1–12, 134–155 (1840) [[973], **1**, 461–472, 473–496]
965. Dirichlet, P.G.L.: Sur la théorie des nombres. C. R. Acad. Sci. Paris **10**, 285–288 (1840) ["Extrait d'une lettre de M. Lejeune-Dirichlet à M. Liouville", J. Math. Pures Appl. (1) **5**, 72–74 (1840); [973], **1**, 619–623.]
966. Dirichlet, P.G.L.: Einige Resultate von Untersuchungen über eine Classe homogener Functionen des dritten und höheren Grades. Verhandl. Kgl. Preuss. Akad. Wiss. Berlin **1841**, 280–285 [[973], **1**, 625–632.]
967. Dirichlet, P.G.L.: Untersuchungen über die Theorie der complexen Zahlen. J. Reine Angew. Math. **22**, 375–378 (1841) [[973], **1**, 503–508.]
968. Dirichlet, P.G.L.: Untersuchungen über die Theorie der complexen Zahlen. Abh. Kgl. Preuss. Akad. Wiss. Berlin **1841**, 141–161 [[973], **1**, 509–532; French translation: J. Math. Pures Appl. (1) **9**, 245–269 (1844)]
969. Dirichlet, P.G.L.: Recherches sur les formes quadratiques à coefficients et à indéterminées complexes. J. Reine Angew. Math. **24**, 291–371 (1842) [[973], **1**, 533–618.]
970. Dirichlet, P.G.L.: Verallgemeinerung eines Satzes aus der Lehre von den Kettenbrüchen nebst einigen Anwendungen auf die Theorie der Zahlen. Monatsber. Preuss. Akad. Wiss. **1842**, 93–95. [[973], **1**, 633–638.]
971. Dirichlet, P.G.L.: Zur Theorie der complexen Einheiten. Verhandl. Kgl. Preuss. Akad. Wiss. Berlin **1846**, 103–107. [[973], **1**, 639–644.]
972. Dirichlet, P.G.L., Vorlesungen über Zahlentheorie. Vieweg (1863); 2nd ed. (1871); 3rd ed. (1879); 4th ed. (1894) [Reprint of the 4th edition: Chelsea (1968); English translation of the first edition: Lectures on Number Theory, Amer. Math. Soc. & London Math. Soc. (1999)]
973. Dirichlet, P.G.L.: Werke. **1, 2**, G. Reimer (1889,1897) [Reprint: Chelsea (1969)]
974. Disse, A.: Über die Beziehungen zwischen Logarithmus und Numerus in einem $\mathfrak{p}$-adischen algebraischen Körper. J. Reine Angew. Math. **154** 178–198 (1925)
975. Disse, A.: Das Fundamentalsystem für die Logarithmen eines $\mathfrak{p}$-adischen algebraischen Körpers und sein Regulator. J. Reine Angew. Math. **155**, 225–250 (1926)
976. Dixon, J.D., Dubickas, A.: The values of Mahler measures. Mathematika **51**, 131–148 (2004)
977. Dobrovolskiĭ, V.A.: Dmitriĭ Aleksandrovič Grave (1863–1939). Nauka **1968**
978. Dobrowolski, E.: On the maximal modulus of conjugates of an algebraic integer. Bull. Acad. Pol. Sci., sér. sci. math. astr. phys. **26**, 291–292 (1978)
979. Dobrowolski, E.: On a question of Lehmer and the number of irreducible factors of a polynomial. Acta Arith. **34**, 391–401 (1979)
980. Dobrowolski, E.: Mahler's measure of a polynomial in function of the number of its coefficients. Canad. Math. Bull. **34**, 186–195 (1991)
981. Dobrowolski, E.: Mahler's measure of a polynomial in terms of the number of its monomials. Acta Arith. **123**, 201–231 (2006)
982. Dobrowolski, E.: A note on integer symmetric matrices and Mahler's measure. Canad. Math. Bull. **51**, 57–59 (2008)
983. Dobrowolski, E., Lawton, W., Schinzel, A.: On a problem of Lehmer. In: Studies in pure mathematics, 135–144, Birkhäuser (1983)
984. Dohmae, K.: Demjanenko matrix for imaginary abelian fields of odd conductors. Proc. Japan Acad. Sci. **70**, 292–294 (1994)
985. Dombek, D., Hajdu, L., Pethő, A.: Representing algebraic integers as linear combinations of units. Period. Math. Hung. **68**, 135–142 (2014)
986. Dombek, D., Masáková, Z., Ziegler, V.: On distinct unit generated fields that are totally complex. J. Number Theory **148**, 311–327 (2015)

987. Dominguez, C., Miller, S.J., Wong, S.: Quadratic fields with cyclic 2-class groups. J. Number Theory **133**, 926–939 (2013)
988. Dörge, K.: Zum Hilbertschen Irreduzibilitätssatz. Math. Ann. **95**, 84–97 (1926)
989. Dörge, K.: Über die Seltenheit der reduziblen Polynomen und der Normalgleichungen. Math. Ann. **95**, 247–256 (1926); corr. **96**, p. 182 (1927)
990. Dörge, K.: Einfacher Beweis des Hilbertschen Irreduzibilitätssatzes. Math. Ann. **96**, 176–182 (1927)
991. Dörge, K.: Bemerkung zum Hilbertschen Irreduzibilitätssatz. Math. Ann. **102**, 521–530 (1929)
992. Dörrie, H.: Das quadratische Reciprocitätsgesetz im quadratischen Zahlkörper mit der Klassenzahl 1. Dissertation. Univ. Göttingen (1898)
993. Dresher, M., Ore, O.: Theory of multigroups. Amer. J. Math. **60**, 705–733 (1938)
994. Dress, A.[W.M.]: Zu einem Satz aus der Theorie der algebraischen Zahlen. J. Reine Angew. Math. **216**, 218–219 (1964)
995. Dress, A.W.M.: Notes on the theory of representations of finite groups. Part I: The Burnside ring of a finite group and some AGN-applications. Univ. Bielefeld (1971)
996. Dress, A.W.M.: Contributions to the theory of induced representations. Lecture Notes Math. **342**, 183–240 (1973)
997. Dribin, D.M.: Quartic fields with the symmetric group. Ann. Math. (2) **38**, 739–749 (1937)
998. Dribin, D.M.: Normal extensions of quartic fields with the symmetric group. Ann. Math. (2) **39**, 341–349 (1938)
999. Driver, E.D., Jones, J.W.: Minimum discriminants of imprimitive decic fields. Experiment. Math. **19**, 475–479 (2010)
1000. Drungilas, P., Dubickas, A.: Every real algebraic integer is a difference of two Mahler measures. Canad. Math. Bull. **50**, 191–195 (2007)
1001. Duarte, F.J.: Sur l'équation $x^3 + y^3 = z^3$. Enseign. Math. **32**, 68–74 (1933)
1002. Dubickas, A.: On a conjecture of A. Schinzel and H. Zassenhaus. Acta Arith. **63**, 15–20 (1993)
1003. Dubickas, A.: On algebraic numbers of small measure. Liet. Mat. Rink. **35**, 421–431 (1995) [Lith. Math. J. **35**, 333–342 (1995)]
1004. Dubickas, A.: The maximal conjugate of a non-reciprocal algebraic integer. Liet. Mat. Rink. **37**, 168–174 (1997) [Lith. Math. J. **37**, 129–133 (1997)]
1005. Dubickas, A.: On algebraic numbers close to 1. Bull. Aust. Math. Soc. **58**, 423–434 (1998)
1006. Dubickas, A.: A note on powers of Pisot numbers. Publ. Math. Debrecen **56**, 141–144 (2000)
1007. Dubickas, A.: Conjugate algebraic numbers that are close to a symmetric set. Algebra i Analiz **16**, 123–127 (2004) (Russian)
1008. Dubickas, A.: Mahler measures in a cubic field. Czechoslovak Math. J. **56**, 949–956 (2006)
1009. Dubickas, A.: Mahler measures in a field are dense modulo 1. Arch. Math. (Basel) **88**, 29–34 (2007)
1010. Dubickas, A.: Polynomial root separation in terms of the Remak height. Turkish J. Math. **37**, 747–761 (2013)
1011. Dubickas, A., Mossinghoff, M.J.: Auxiliary polynomials for some problems regarding Mahler's measure. Acta Arith. **119**, 65–79 (2005)
1012. Dubois, E., Rhin, G.: Sur la majoration de formes linéaires à coefficients algébriques réels et p-adiques. Démonstration d'une conjecture de K. Mahler. C. R. Acad. Sci. Paris **282**, A1211–A1214 (1976)
1013. Dufresnoy, J., Pisot, C.: Sur les dérivés successifs d'un ensemble fermé d'entiers algébriques. Bull. Sci. Math. (2) **77**, 129–136 (1953)
1014. Dufresnoy, J., Pisot, C.: Sur un ensemble fermé d'entiers algébriques. Ann. Sci. Éc. Norm. Supér. (3) **70**, 105–133 (1953)
1015. Dufresnoy, J., Pisot, C.: Sur les petits éléments d'un ensemble remarquable d'entiers algébriques. C. R. Acad. Sci. Paris **238**, 1551–1553 (1954)

1016. Dufresnoy, J., Pisot, C.: Étude de certaines fonctions méromorphes bornées sur le cercle unité. Application à un ensemble fermé d'entiers algébriques. Ann. Sci. Éc. Norm. Supér. (3) **72**, 69–92 (1955)
1017. Dugas, M., Göbel, R.: All infinite groups are Galois groups over any field. Trans. Amer. Math. Soc. **304**, 355–384 (1987)
1018. Duhem, P.: Emile Mathieu, his life and work. Bull. Amer. Math. Soc. **1**, 156–168 (1892)
1019. Duĭčev, J.: On prime ideals of degree one. Acta Math. Hungar. **7**, 71–73 (1956) (Russian)
1020. Dujella, A., Pejković, T.: Root separation for reducible monic quartics. Rend. Semin. Mat. Univ. Padova **126**, 63–72 (2011)
1021. Duke, W.: Extreme values of Artin L-functions and class numbers. Compositio Math. **136**, 103–115 (2003)
1022. Duke, W.: Number fields with large class group. In: Number theory. CRM Proc. Lecture Notes **36**, 117–126 (2004)
1023. Dulin, B.J., Butts, H.S.: Composition of binary quadratic forms over integral domains. Acta Arith. **20**, 223–251 (1972)
1024. Dumas, G.: Sur quelques cases d'irréductibilité des polynomes à coefficients rationnels. J. Math. Pures Appl. (6) **2**, 191–258 (1906)
1025. Dumbaugh, D., Schwermer, J.: The collaboration of Emil Artin and George Whaples: Artin's mathematical circle extends to America. Arch. Hist. Exact Sci. **66**, 465–484 (2012)
1026. Dumbaugh, D., Schwermer, J.: Emil Artin and Beyond — Class Field Theory and L-functions. European Math. Soc. (2015)
1027. Dummit, D.S., Kisilevsky, H.: Indices in cubic fields. In: Number Theory and Algebra, 29–42. Academic Press (1977)
1028. Dunnington, G.W.: Carl Friedrich Gauss: Titan of Science. A study of his life and work. Exposition Press (1955) [Reprint: Math. Assoc. of America (2004)]
1029. Dupré, A.: Sur le nombre des divisions à effectuer pour obtenir le plus grand commun diviseur entre deux nombres entiers. J. Math. Pures Appl. **11** 41–64 (1846)
1030. Dupré, A.: Sur le nombre de divisions à effectuer pour trouver le plus grand commun diviseur entre deux nombres complexes de la forme $a + b\sqrt{-1}$, oú a et b sont entiers. J. Math. Pures Appl. **13**, 333–343 (1848)
1031. Dwork, B.: On the Artin root number. Amer. J. Math. **78**, 444–472 (1956)
1032. Dyson, F.J.: The approximation to algebraic numbers by rationals. Acta Math. **79**, 225–240 (1947)
1033. Dyson, F.J.: On the product of four non-homogeneous linear forms. Ann. Math. (2) **49**, 82–109 (1948)
1034. Dzewas, J.: Quadratsummen in reellquadratischen Zahlkörpern. Math. Nachr. **21**, 233–284 (1960)
1035. Earnest, A.G.: Binary quadratic forms over rings of algebraic integers: a survey of recent results. In: Théorie des nombres (Quebec, 1987), 133–159. de Gruyter (1989)
1036. Earnest, A.G., Estes, D.R.: Class groups in the genus and spinor genus of binary quadratic lattices. Proc. London Math. Soc. (3) **40**, 40–52 (1980)
1037. Earnest, A.G., Estes, D.R.: An algebraic approach to the growth of class numbers of binary quadratic lattices. Mathematika **29**, 160–168 (1981)
1038. Earnest, A.G., Khosravani, A.: Universal positive quaternary quadratic lattices over totally real number fields. Mathematika **44**, 342–347 (1997)
1039. Earnest, A.G., Körner, O.H.: On ideal class groups of 2-power exponent. Proc. Amer. Math. Soc. **86**, 196–198 (1982)
1040. Eda, Y.: On the Waring problem in an algebraic number field. In: Seminar on Modern Methods in Number Theory, nr. 10, 1–11, Inst. Statist. Math. Tokyo (1971)
1041. Eda, Y.: On the Waring's problem in algebraic number fields. Rev. Colomb. Mat. **9**, 29–73 (1975)
1042. Eda, Y., Nakagoshi, N.: An elementary proof of the prime ideal theorem with remainder term. Sci. Rep. Kanazawa Univ. **12**, 1–12 (1967)

1043. Edwards, H.M.: The background of Kummer's proof of Fermat's last theorem for regular primes. Arch. Hist. Exact Sci. **14**, 219–236 (1975)
1044. Edwards, H.M.: Postscript to: "The background of Kummer's proof of Fermat's last theorem for regular primes". Arch. Hist. Exact Sci. **17**, 381–394 (1977)
1045. Edwards, H.M.: Fermat's Last Theorem, A Genetic Introduction to Algebraic Number Theory. Springer (1977); 2nd ed. (1996)
1046. Edwards, H.M.: The genesis of ideal theory. Arch. Hist. Exact Sci. **23**, 321–378 (1980)
1047. Edwards, H.M.: Dedekind's invention of ideals. Bull. Amer. Math. Soc. **15**, 8–17 (1983)
1048. Edwards, H.M.: Divisor Theory. Birkhäuser (1990)
1049. Edwards, H.M.: Composition of binary quadratic forms and the foundations of mathematics. In: [1467], 129–144.
1050. Edwards, H.[M.], Neumann, O., Purkert, W.: Dedekinds "Bunte Bemerkungen" zu Kroneckers "Grundzüge". Arch. Hist. Exact Sci. **27**, 49–85 (1982)
1051. Efrat, I.: Pro-p Galois groups of algebraic extensions of Q. J. Number Theory **64**, 84–99 (1997)
1052. Egami, S.: Euclid's algorithm in pure quartic fields. Tokyo J. Math. **2**, 379–385 (1979)
1053. Egami, S.: Average version of Artin's conjecture in an algebraic number field. Tokyo J. Math. **4**, 203–212 (1981)
1054. Egami, S.: On finiteness of the numbers of Euclidean fields in some classes of number fields. Tokyo J. Math. **7**, 183–196 (1984)
1055. Eggleton, R.B., Lacampagne, C.B., Selfridge, J.L.: Euclidean quadratic fields. Amer. Math. Monthly **99**, 829–837 (1992)
1056. Ehlich, H.: Ein elementarer Beweis des Primzahlsatzes für binäre quadratische Formen. J. Reine Angew. Math. **201**, 1–36 (1959)
1057. Eichler, M.: Zum Hilbertschen Irreduzibilitätssatz. Math. Ann. **116**, 742–748 (1939)
1058. Eichler, M.: Der Hilbertsche Klassenkörper eines imaginärquadratischen Zahlkörpers. Math. Z. **64**, 229–242 (1956); corr. **65**, p. 214 (1956)
1059. Eichler, M.: Einführung in die Theorie der algebraischen Zahlen und Funktionen. Birkhäuser (1963) [English translation: Introduction to the Theory of Algebraic Numbers and Functions, Academic Press (1966)]
1060. Eichler, M.: Das wissenschaftliche Werk von Max Deuring. Acta Arith. **47**, 187–192 (1986)
1061. Eichler, M.: Alexander Ostrowski. Über sein Leben und Werk. Acta Arith. **51**, 295–298 (1988)
1062. Eilenberg, S., Mac Lane, S.: Cohomology theory in abstract groups. I. Ann. Math. (2) **48**, 51–78 (1947)
1063. Eilenberg, S., Mac Lane, S.: Cohomology theory in abstract groups. II, Group extensions with a non-Abelian kernel. Ann. Math. (2) **48**, 326–341 (1947)
1064. Eilenberg, S., Mac Lane, S.: Cohomology and Galois theory. I. Normality of algebras and Teichmüller's cocycle. Trans. Amer. Math. Soc. **64**, 1–20 (1948)
1065. Eisenstein, G.: Beweis des Reciprocitätssatzes für die cubische Reste in der Theorie der aus dritten Wurzel der Einheit zusammengesetzten complexen Zahlen. J. Reine Angew. Math. **27**, 289–310 (1844) [[1079], **1**, 59–80.]
1066. Eisenstein, G.: Über die Anzahl der quadratischen Formen, welche in der Theorie der complexen Zahlen zu einer reellen Determinante gehören. J. Reine Angew. Math. **27**, p. 80 (1844) [[1079], **1**, p. 6.]
1067. Eisenstein, G.: Über die Anzahl der quadratischen Formen in verschiedenen complexen Theorien. J. Reine Angew. Math. **27**, 311–316 (1844) [[1079], **1**, 89–94.]
1068. Eisenstein, G.: Nachtrag zum cubischen Reciprocitätssatze für die aus dritten Wurzel der Einheit zusammengesetzten complexen Zahlen. Criterien des cubischen Characters der Zahl 3 und ihrer Theiler. J. Reine Angew. Math. **28**, 28–35 (1844) [[1079], **1**, 81–88.]
1069. Eisenstein, G.: Lois de réciprocité. J. Reine Angew. Math. **28**, 53–67 (1844) [[1079], **1**, 126–140.]
1070. Eisenstein, G.: Einfacher Beweis und Verallgemeinerung des Fundamentaltheorems für biquadratische Reste. J. Reine Angew. Math. **28**, 223–245 (1844) [[1079], **1**, 141–163.]

1071. Eisenstein, G.: Allgemeine Untersuchungen über die Formen dritten Grades mit drei Variabeln, welche der Kreistheilung ihre Entstehung verdanken. J. Reine Angew. Math. **28**, 289–374 (1844); **29**, 19–53 (1845) [[1079], **1**, 167–286.]
1072. Eisenstein, G.: Applications de l'algébre à l'arithmétique transcendante. J. Reine Angew. Math. **29**, 177–184 (1845) [[1079], **1**, 291–298.]
1073. Eisenstein, G.: Beiträge zur Theorie der elliptischen Functionen. I. Ableitung des biquadratischen Fundamentaltheorems aus der Theorie der Lemniscatenfunctionen, nebst Bemerkungen zu den Multiplications- und Transformationsformeln. J. Reine Angew. Math. **30**, 185–210 (1846) [[1079], **1**, 299–324.]
1074. Eisenstein, G.: Zur Theorie der quadratischer Zerfällung der Primzahlen $8n + 3$, $7n + 2$, und $7n + 4$. J. Reine Angew. Math. **37**, 97–125 (1848) [[1079], **2**, 506–535.]
1075. Eisenstein, G.: Über die Irreductibilität und einige andere Eigenschaften der Gleichung, von welcher die Theilung der ganzen Lemniscate abhängt. J. Reine Angew. Math. **39**, 160–179, 224–287 (1850) [[1079], **2**, 536–619.]
1076. Eisenstein, G.: Lehrsätze. J. Reine Angew. Math. **39**, 180–182 (1850) [[1079], **2**, 620–622.]
1077. Eisenstein, G.: Über ein einfaches Mittel zur Auffindung der höheren Reciprocitätsgesetze und der mit ihnen zu verbindenden Ergänzungssätze. J. Reine Angew. Math. **39**, 351–364 (1850) [[1079], **2**, 623–636.]
1078. Eisenstein, G.: Beweis der allgemeinsten Reciprocitätsgesetze zwischen reellen und complexen Zahlen. Ber. Verhandl. Kgl. Preuß. Akad. Wiss. Berlin **1850**, 189–198. [[1079], **2**, 712–721.]
1079. Eisenstein, G.: Mathematische Werke. **1**,**2**. Chelsea (1975); 2nd ed. (1989)
1080. Elkies, N.D.: ABC implies Mordell. Int. Math. Res. Not. **7**, 99–109 (1991)
1081. Ellenberg, J.S., Venkatesh, A.: The number of extensions of a number field with fixed degree and bounded discriminant. Ann. Math. (2) **163**, 723–741 (2006)
1082. Elliott, P.D.T.A., Halberstam, H.: A conjecture in prime number theory. Symposia Math. **4**, 59–72 (1970)
1083. Ellison, W.J.: Waring's problem. Amer. Math. Monthly **78**, 10–36 (1971)
1084. Ellison, W.J.: Waring's problem for fields. Sém. Théor. Nombres Bordeaux **1970/71**, exp. 10, 1–8
1085. Ellison, W.J.: Waring's problem for fields. Publ. Math. Univ. Bordeaux **1973/74(1)**, 23–33
1086. Ellison, W.[J.]: Waring's problem for fields. Acta Arith. **159**, 315–330 (2013)
1087. Ellison, W., Pesek, J., Stall, D.S., Lunnon, W.F.: A postscript to a paper of A. Baker. Bull. Lond. Math. Soc. **3**, 75–78 (1971)
1088. Elstrodt, J., Grunewald, F., Mennicke, J.: On unramified A_m-extensions of quadratic number fields. Glasg. Math. J. **27**, 31–37 (1985)
1089. Endler, O.: Valuation Theory. Springer (1972)
1090. Endô, A.: On the 2-rank of the ideal class groups of quadratic number fields. Mem. Fac. Sci. Kyushu Univ. **27**, 7–12 (1973)
1091. Endô, A.: On the 2-class number of certain quadratic number fields. Mem. Fac. Sci. Kyushu Univ. **27**, 111–120 (1973)
1092. Endô, A.: The relative class number of certain imaginary abelian fields. Abh. Math. Semin. Univ. Hambg **58**, 237–243 (1988)
1093. Endô, A.: On the Stickelberger ideal of $(2, \dots, 2)$-extensions of a cyclotomic number field. Manuscripta Math. **69**, 107–132 (1990)
1094. Endô, A.: The relative class numbers of certain imaginary abelian number fields and determinants. J. Number Theory **34**, 13–20 (1990)
1095. Endô, A.: A generalization of the Maillet determinant and the Demyanenko matrix. Abh. Math. Semin. Univ. Hambg **73**, 181–193 (2003)
1096. Endô, A.: On the Stickelberger ideal of a quadratic twist of a cyclotomic field. J. Number Theory **151**, 1–6 (2015)
1097. Endô,S., Miyata, T.: Invariants of finite abelian groups. J. Math. Soc. Japan **25**, 7–26 (1973)
1098. Engler, A.J., Prestel, A.: Valued Fields. Springer (2005)

1099. Engstrom, H.T.: On the common index divisors of an algebraic field. Trans. Amer. Math. Soc. **32**, 223–237 (1930)
1100. Engstrom, H.T.: The theorem of Dedekind in the ideal theory of Zolotarev. Trans. Amer. Math. Soc. **32**, 879–887 (1930)
1101. Engstrom, H.T.: An example of the ideal theory of Zolotarev. Amer. Math. Monthly **37**, 128–129 (1930)
1102. Ennola, V.: On the first inhomogeneous minimum of indefinite binary quadratic forms and Euclid's algorithm in real quadratic fields. Ann. Univ. Turku **A I 28**, 1–58 (1958)
1103. Ennola, V.: Conjugate algebraic integers on a circle with irrational center. Math. Z. **134**, 337–350 (1973)
1104. Ennola, V.: On a conjecture of H.J. Godwin on cubic units. Ann. Acad. Sci. Fenn. Math., Ser. A I **12**, 319–328 (1987)
1105. Ennola, V.: Cubic number fields with exceptional units. In: Computational number theory (Debrecen, 1989), 103–128, de Gruyter (1991)
1106. Ennola, V., Smyth, C.J.: Conjugate algebraic numbers on a circle. Ann. Acad. Sci. Fenn. Math., Ser. A I **582**, 1–31 (1974)
1107. Ennola, V., Smyth, C.J.: Conjugate algebraic numbers on circles. Acta Arith. **29**, 147–157 (1976)
1108. Ennola, V., Turunen, R.: On totally real cubic fields. Math. Comp. **44**, 495–518 (1985)
1109. Epelbaum, B.: The construction of a basis of G.F.Voronoĭ's type for a field of algebraic numbers. Dokl. Akad. Nauk SSSR **64**, 637–640 (1949) (Russian)
1110. Epkenhans, M.: On double covers of the generalized symmetric group $Z_d \wr \mathfrak{S}_n$ as Galois groups over algebraic number fields K with $\mu_d \subset K$. J. Algebra **163**, 404–423 (1994)
1111. Epkenhans, M.: On double covers of the generalized alternating group $Z_d \wr \mathfrak{A}$ as Galois groups over algebraic number fields. Acta Arith. **82**, 129–145 (1997)
1112. Epstein, P.: Zur Theorie allgemeiner Zetafunctionen. Math. Ann. **56**, 615–644 (1903)
1113. Epstein, P.: Zur Auflösbarkeit der Gleichung $x^2 - Dy^2 = -1$. J. Reine Angew. Math. **171**, 243–252 (1934)
1114. Erdős, P.: Some personal reminiscences of the mathematical work of Paul Turán. Acta Arith. **37**, 3–8 (1980)
1115. Erdős, P.: E. Straus (1921–1983). Rocky Mountain J. Math. **15**, 331–341 (1985)
1116. Erdős, P., Chao Ko: Note on the Euclidean algorithm. J. Lond. Math. Soc. **13**, 3–8 (1938)
1117. Erdős, P., Vaughan, R.C.: Bounds for the r-th coefficients of cyclotomic polynomials. J. Lond. Math. Soc. (2) **8**, 393–400 (1974)
1118. Ernvall, R.: Generalized Bernoulli numbers, generalized irregular primes, and class number. Ann. Univ. Turku **A I 178**, 1–72 (1979)
1119. Ernvall, R.: Generalized irregular primes. Mathematika **30**, 67–73 (1983)
1120. Ernvall, R., Metsänkylä, T.: Cyclotomic invariants and E-irregular primes. Math. Comp. **32**, 617–629 (1978); corr. **33**, p. 433 (1979)
1121. Ernvall, R., Metsänkylä, T.: Cyclotomic invariants for primes between 125000 and 150000. Math. Comp. **56**, 851–858 (1991)
1122. Ernvall, R., Metsänkylä, T.: Cyclotomic invariants for primes to one million. Math. Comp. **59**, 249–250 (1992)
1123. Eršov, Yu.L.: The Dedekind criterion for arbitrary valuation rings. Dokl. Akad. Nauk SSSR **410**, 158–160 (2006) (Russian)
1124. Estermann, T.: On Dirichlet's L-functions. J. Lond. Math. Soc. **23**, 275–279 (1948)
1125. Estes, D.R.: On the parity of the class number of the field of qth roots of unity. Rocky Mountain J. Math. **19**, 675–682 (1989)
1126. Estes, D.R.: Eigenvalues of symmetric integer matrices. J. Number Theory **42**, 292–296 (1992)
1127. Estes, D.R., Guralnick, R.M.: Representations under ring extensions: Latimer-MacDuffee and Taussky correspondences. Adv. Math. **54**, 302–313 (1984)
1128. Estes, D.R., Guralnick, R.M.: Minimal polynomials of integral symmetric matrices. Linear Algebra Appl. **192**, 83–99 (1993)

1129. Estes, D.R., Hsia, J.S.: Sums of three integer squares in complex quadratic fields. Proc. Amer. Math. Soc. **89**, 211–214 (1983)
1130. Euler, L.: Institutiones calculi differentialis cum eius usu in analysi finitorum ac doctrina serierum. Acad. Sci. Petropol., (1755) [[1137], I_{10}.]
1131. Euler, L.: De usu novi algorithmi in problemate Pelliano solvendo. Novi Comment. Acad. Sci. Petropol. **11**, 24–66 (1767) [[1137], I_3, 73–111.]
1132. Euler, L.: Extrait d'un lettre de M. Euler le pére à M. Bernoulli concernant le memoire imprimé parmi ceux de 1771. p. 318. Nouv. Mém. Acad. Berlin **1772**, 35–36. [[1137], I_3, 335–337.]
1133. Euler, L.: De casibus quibus hanc formulam $x^4 + kxxyy + y^4$ ad quadratum reducere licet. Nova Acta Acad. Sci. Petropol. **10**, 27–40 (1797) [[1137], I_4, 235–244.]
1134. Euler, L.: Regula facilis problemata Diophantea per numeros integros expedite resolvendi. Mem. Acad. Sci. St.-Petersbourg **4**, 3–17 (1813) [[1137], I_4, 406–417.]
1135. Euler, L.: Resolutio facilis quaestionis difficillimae, qua haec formula maxime generalis $vvzz(axx + byy)^2 + \Delta xxyy(avv + bzz)^2$ ad quadratum reduci postulatur. Mem. Acad. Sci. St.-Petersbourg **9**, 14–19 (1824) [[1137], I_5, 71–76.]
1136. Euler, L.: Vollständige Einleitung zur Algebra. St. Petersburg (1771) [[1137], I_1, 1–651; Reprint: Cambridge University Press (2009); English translation: Longman, (1822, 1840); Reprints: Springer (1984), Cambridge University Press (2009).]
1137. Euler, L.: Opera Omnia. Teubner, continued by Societas Scientiarum Naturalium Helveticae and Springer (1910– ...)
1138. Evans, R.J.: Generalized cyclotomic periods. Proc. Amer. Math. Soc. **81**, 207–212 (1981)
1139. Evans, R.J.: The octic periodic polynomial. Proc. Amer. Math. Soc. **87**, 389–393 (1983)
1140. Evertse, J.-H.: On equations in S-units and the Thue-Mahler equation. Invent. Math. **75**, 561–584 (1984)
1141. Evertse, J.-H.: On sums of S-units and linear recurrences. Compositio Math. **53**, 225–244 (1984)
1142. Evertse, J.-H.: Estimates for reduced binary forms. J. Reine Angew. Math. **434**, 159–190 (1993)
1143. Evertse, J.-H.: Distances between the conjugates of an algebraic number. Publ. Math. Debrecen **65**, 323–340 (2004)
1144. Evertse, J.-H., Ferretti, R.G.: A further improvement of the quantitative subspace theorem. Ann. Math. (2) **177**, 513–590 (2013)
1145. Evertse, J.-H., Győry, K.: On unit equations and decomposable form equations. J. Reine Angew. Math. **358**, 6–19 (1985)
1146. Evertse, J.-H., Győry, K.: On the numbers of solutions of weighted unit equations. Compositio Math. **66**, 329–354 (1988)
1147. Evertse, J.-H., Győry, K.: Effective finiteness results for binary forms with given discriminant. Compositio Math. **79**, 169–204 (1991)
1148. Evertse, J.-H., Győry, K.: Effective results for unit equations over finitely generated integral domains. Math. Proc. Cambridge Philos. Soc. **154**, 351–380 (2013)
1149. Evertse, J.-H., Győry, K.: Unit Equations in Diophantine Number Theory. Cambridge University Press (2015)
1150. Evertse, J.-H., Győry, K.: Discriminant Equations in Diophantine Number Theory. Cambridge University Press (2017)
1151. Evertse, J.-H., Győry, K., Stewart, C.L., Tijdeman, R.: S-unit equations and their applications. In: New advances in transcendence theory (Durham, 1986), 110–174, Cambridge University Press (1988)
1152. Evertse, J.-H., Győry, K., Stewart, C.L., Tijdeman, R.: On S-unit equations in two unknowns. Invent. Math. **92**, 461–477 (1988)
1153. Evertse, J.-H., Schlickewei, H.P.: A quantitative version of the absolute subspace theorem. J. Reine Angew. Math. **548**, 21–127 (2002)
1154. Evertse, J.-H., Schlickewei, H.P., Schmidt, W.M.: Linear equations in variables which lie in a multiplicative group. Ann. Math. (2) **155**, 807–836 (2002)

1155. Faddeev, D.K.: On the characteristic equations of rational symmetric matrices. Dokl. Akad. Nauk SSSR **58**, 753–754 (1947) (Russian)
1156. Faddeev, D.K.: On a hypothesis of Hasse. Dokl. Akad. Nauk SSSR **94**, 1013–1016 (1954) (Russian)
1157. Faddeev, D.K.: Group of divisor classes on the curve defined by the equation $x^4 + y^4 = 1$. Dokl. Akad. Nauk SSSR **134**, 776–777 (1960) (Russian)
1158. Faddeev, D.K., Skopin, A.I.: Proof of a theorem of Kawada. Dokl. Akad. Nauk SSSR **127**, 529–530 (1959) (Russian)
1159. Faith, C.C.: Extensions of normal bases and completely basic fields. Trans. Amer. Math. Soc. **85**, 406–427 (1957)
1160. Faltings, G.: Endlichkeitssätze für abelsche Varietäten über Zahlkörpern. Invent. Math. **73**, 349–366 (1983); corr. **75**, p. 381 (1984)
1161. Fanta, E.: Beweis, dass jede lineare Funktion, deren Koeffizienten dem kubischen Kreisteilungskörper entnommene ganze teilerfremde Zahlen sind, unendlich viele Primzahlen dieses Körpers darstellt. Monatsh. Math. Phys. **12**, 1–44 (1901)
1162. Farhat, M., Sodaïgui, B.: On Steinitz classes of nonabelian Galois extensions and p-ary cyclic Hamming codes. J. Number Theory **134**, 93–108 (2014)
1163. Favard, J.: Sur les nombres algébriques. C. R. Acad. Sci. Paris **186**, 1181–1182 (1928)
1164. Favard, J.: Sur les formes décomposables et les nombres algébriques. Bull. Soc. Math. France **57**, 50–71 (1929)
1165. Favard, J.: Sur les nombres algébriques. Mathematica (Cluj) **4**, 109–113 (1930)
1166. Fefferman, C., Kahane, J.-P., Stein, E.M.: The scientific achievements of Antoni Zygmund. Wiadomości matematyczne (2) **19**, 91–126 (1976) (Polish)
1167. Fein, B., Gordon, B., Smith, J.H.: On the representation of -1 as a sum of two squares in an algebraic number field. J. Number Theory **3**, 310–315 (1971)
1168. Feit, W.: Richard D. Brauer, Bull. Amer. Math. Soc. (2) **1**, 1–20 (1979)
1169. Feit, W.: $\tilde{A}_5$ and $\tilde{A}_7$ are Galois groups over number fields. J. Algebra **104**, 231–260 (1986)
1170. Feit, W., Thompson, J.G.: Solvability of groups of odd order. Pacific J. Math. **13**, 775–1029 (1963)
1171. Fekete, M., Über die Verteilung der Wurzeln bei gewisser algebraischen Gleichungen mit ganzzahligen Koeffizienten. Math. Z. **17**, 228–249 (1923)
1172. Fekete, M., Szegö, G.: On algebraic equations with integral coefficients whose roots belong to a given point set. Math. Z. **63**, 228–249 (1955)
1173. Feldman, N.I., Čudakov, N.G.: On a theorem of Stark. Mat. Zametki **11**, 329–340 (1972) (Russian)
1174. Fellman, E.A.: Leonhard Euler. Rowohlt (1995) [English translation: Birkhäuser (2007)]
1175. Feng, K.Q.: On the first factor of the class number of a cyclotomic field. Proc. Amer. Math. Soc. **84**, 479–482 (1982)
1176. Feng, K.Q.: Upper bound for first factor of class number of cyclotomic fields. Sci. Sinica **27**, 1121–1128 (1984)
1177. Feng K.Q.: Arithmetic characterization of algebraic number fields with given class group. Acta Math. Sin. **1**, 47–54 (1985)
1178. Fenster, D.D.: Leonard Eugene Dickson (1874–1954): an American legacy in mathematics. Math. Intelligencer **21(4)**, 54–59 (1999)
1179. Fenster, D.D., Schwermer, J.: Composition of quadratic forms: an algebraic perspective. in: [1467], 145–158.
1180. Ferrero, B.: Iwasawa invariants of Abelian number fields. Math. Ann. **234**, 9–24 (1978)
1181. Ferrero, B., Washington, L.C.: The Iwasawa invariant μ_p vanishes for abelian number fields. Ann. Math. (2) **109**, 377–395 (1979)
1182. Fesenko, I.B., Vostokov, S.V.: Local fields and their extensions. A constructive approach. Amer. Math. Soc. (1993); 2nd ed. (2002)
1183. von Fiecker, H.: A.A. Friedmann †, Zeitschr. f. angew. Math. **5**, 526–527 (1925)
1184. Fieker, C., Klüners, J.: Minimal discriminants for fields with small Frobenius groups as Galois groups. J. Number Theory **99**, 318–337 (2003)

1185. Fili, P., Miner, Z.: A generalization of Dirichlet's unit theorem. Acta Arith. **162**, 355–368 (2014)
1186. Filipin, A., Tichy, R., Ziegler, V.: On the quantitative unit sum number problem — an application of the subspace theorem. Acta Arith. **133**, 297–308 (2008)
1187. Finsterwalder, S.: Alexander v.Brill, Ein Lebensbild. Math. Ann. **112**, 653–663 (1936)
1188. Fischer, E.: Die Isomorphie der Invariantenkörper der endlicher Abelschen Gruppen linearer Transformationen. Nachr. Ges. Wiss. Göttingen **1915**, 77–80
1189. Fischer, E.: Zur Theorie der endlichen Abelschen Gruppen. Math. Ann. **77**, 81–88 (1915)
1190. Fitzgerald, R.W.: Characteristic polynomials of symmetric matrices. Linear Algebra Appl. **36**, 233–237 (1994)
1191. Flammang, V.: Two new points in the spectrum of the absolute Mahler measure of totally positive algebraic integers. Math. Comp. **65**, 307–311 (1996)
1192. Flammang, V.: Inégalités sur la mesure de Mahler d'un polynôme. J. Théor. Nombres Bordeaux **9**, 69–74 (1997)
1193. Flammang, V.: Trace of totally positive algebraic integers and integer transfinite diameter. Math. Comp. **78**, 1119–1125 (2009)
1194. Flammang, V., Grandcolas, M., Rhin, G.: Small Salem numbers, in: Number Theory in Progress, **1**, 165–168, de Gruyter (1999)
1195. Flammang, V., Rhin, G.: Algebraic integers whose conjugates all lie in an ellipse. Math. Comp. **74**, 2007–2015 (2005)
1196. Flammang, V., Rhin, G., Sac-Épée, J.-M.: Integer transfinite diameter and polynomials with small Mahler measure. Math. Comp. **76**, 1527–1540 (2006)
1197. Flammang, V., Rhin, G., Wu, Q.: The totally real algebraic integers with diameter less than 4. Moscow J. Comb. Number Th. **1**, 17–25 (2011)
1198. Flatto, L.: Invariants of finite reflection groups. Enseign. Math. (2) **24**, 237–292 (1978)
1199. Fleckenstein, J.O., van der Waerden, B.L.: Zum Gedenken an Andreas Speiser. Elem. Math. **26**, 97–102 (1971)
1200. Fleckinger, V.: Monogénéité de l'anneau des entiers de certains corps de classes de rayon. Ann. Inst. Fourier (Grenoble) **38(1)**, 17–57 (1988)
1201. Fleckinger, V.: Génération de bases d'entiers a partir de la courbe $y^2 = 4x^3 + 1$. Publ. Math. Fac. Sci. Besançon **1986/87–1987/88(1)** 1–22
1202. Fogels, E.: Über die Möglichkeit einiger diophantischer Gleichungen 3. und 4. Grades in quadratischen Körpern. Comment. Math. Helv. **10** 263–269 (1937)
1203. Fogels, E.: Zur Arithmetik quadratischer Zahlenkörper. Wiss. Abh. Univ. Riga, Kl. Math. **1** 23–47 (1943)
1204. Fogels, E.: On the zeros of Hecke's L-functions, I. Acta Arith. **7**, 87–106 (1961/62)
1205. Fogels, E.: On the zeros of Hecke's L-functions, II. Acta Arith. **7**, 131–147 (1961/62)
1206. Fogels, E.: Über die Ausnahmenullstelle der Heckeschen L-Funktionen. Acta Arith. **8**, 307–309 (1962/63)
1207. Fontana, M., Huckaba, J.A., Papick, I.J.: Prüfer Domains. Marcel Dekker (1997)
1208. Fontené, G.: Sur les entiers algébriques de la forme $x + y\sqrt{-5}$. Nouv. Ann. Math. (4) **3**, 209–214 (1903)
1209. Foote, R., Murty, V.K.: Zeros and poles of Artin L-series. Math. Proc. Cambridge Philos. Soc. **105**, 5–11 (1989)
1210. Ford, D.: Minimum discriminants of primitive sextic fields. Lecture Notes in Computer Science **122**, 141–143 (1996)
1211. Ford, D., Pohst, M.: The totally real A_5 extension of degree 6 with minimum discriminant. Experiment. Math. **1**, 231–235 (1992)
1212. Ford, D., Pohst, M.: The totally real A_6 extension of degree 6 with minimum discriminant. Experiment. Math. **2**, 231–232 (1993)
1213. Ford, D., Pohst, M., Daberkow, M., Haddad, N.: The S_3 extensions of degree 6 with minimum discriminant. Experiment. Math. **7**, 121–124 (1998)
1214. Ford, K., Wooley, T.D.: On Vinogradov's mean value theorem: strongly diagonal behaviour via efficient congruencing. Ann. Math. (2) **213**, 199–236. (2014)

1215. Ford, L.R.: Sur l'approximation des irrationnelles complexes. C. R. Acad. Sci. Paris **162**, 459–461 (1916)
1216. Ford, L.R.: Rational approximations to irrational complex numbers. Trans. Amer. Math. Soc. **19**, 1–42 (1918)
1217. Ford, L.R.: On the closeness of approach of complex rational fractions to a complex irrational number. Trans. Amer. Math. Soc. **27**, 146–154 (1925)
1218. Forman, W., Shapiro, H.N.: Abstract prime number theorems. Comm. Pure Appl. Math. **7**, 587–619 (1954)
1219. Forsyth, A.R.: William Burnside. J. Lond. Math. Soc. **2**, 64–80 (1928)
1220. Fossum, R.M.: The Divisor Class Group of a Krull Domain. Springer (1973)
1221. Fouvry, É.: On the size of the fundamental solution of the Pell equation. J. Reine Angew. Math. **717**, 1–33 (2016)
1222. Fouvry, É. Jouve, F.: A positive density of fundamental discriminants with large regulator. Pacific J. Math. **262**, 81–107 (2013)
1223. Fouvry, É. Klüners, J.: On the negative Pell equation. Ann. Math. (2) **17**, 2035–2104 (2010)
1224. Fouvry, É., Klüners, J.: The parity of the period of the continued fraction of $\sqrt{d}$. Proc. London Math. Soc. (3) **101**, 337–391 (2010)
1225. Fraenkel, A.: Axiomatische Begründung von Hensel's p-adischen Zahlen. J. Reine Angew. Math. **141**, 43–76 (1912)
1226. Fraenkel, A.: Über die Teiler der Null und die Zerlegung von Ringen. J. Reine Angew. Math. **145**, 139–176 (1915)
1227. Fraenkel, A.: Über gewisse Teilbereiche und Erweiterungen von Ringen. Teubner (1916)
1228. Fraenkel, A.: Über einfache Erweiterungen zerlegbarer Ringe. J. Reine Angew. Math. **151**, 121–166 (1921)
1229. Fraenkel, A.: Alfred Loewy (1873–1935). Scripta Math. **5**, 17–22 (1938)
1230. Frank, E.: Oskar Perron (1880–1975). J. Number Theory **14**, 281–291 (1982)
1231. Franz, W.: Untersuchungen zum Hilbertschen Irreduzibilitätssatz. Math. Z. **33**, 275–293 (1931)
1232. Franz, W.: Zur vorstehenden Arbeit von Herrn A.Korselt. J. Reine Angew. Math. **164**, p.63 (1931)
1233. Franz, W.: Elementarteilertheorie in algebraischen Zahlkörpern. J. Reine Angew. Math. **171**, 149–161 (1934)
1234. Franz, W.: Die Teilwerte der Weberschen Tau-Funktion. J. Reine Angew. Math. **173**, 60–64 (1935)
1235. Franz, W.: Torsionsideale, Torsionsklassen und Torsion. J. Reine Angew. Math. **176**, 113–124 (1936)
1236. Frei, C.: On rings of integers generated by their units. Bull. Lond. Math. Soc. **44**, 167–182 (2012)
1237. Frei, C., Tichy, R., Ziegler, V.: On sums of S-integers of bounded norm. Monatsh. Math. **175**, 241–247 (2014)
1238. Frei, G.: Helmut Hasse (1898–1979). Expo. Math. **3**, 55–69 (1985)
1239. Frei, G.: Heinrich Weber and the emergence of class field theory. In: The History of Modern Mathematics, **I**, 425–450. Academic Press (1989)
1240. Frei, G.: Zum Gedenken an Bartel Leendert van der Waerden (2.2.1903–12.1.1996). Elem. Math. **53**, 133–138 (1998)
1241. Frei, G.: On the history of the Artin reciprocity law in abelian extensions of algebraic number fields: how Artin was led to his reciprocity law. In: The legacy of Niels Henrik Abel, 267–294. Springer (2004)
1242. Frei, G.: The unpublished section eight: on the way to functions fields over a finite field. In: [1467] 159–198
1243. Freitas, N., Siksek, S.: The asymptotic Fermat's Last Theorem for five-sixth of real quadratic fields. Compositio Math. **151**, 1395–1415 (2015)
1244. Freitas, N., Siksek, S.: Fermat's Last Theorem for some small real quadratic fields. Algebra & Number Th. **9**, 875–895 (2015)

1245. Fresnel, J.: Nombres de Bernoulli et fonctions L p-adiques. Ann. Inst. Fourier (Grenoble) **17(2)**, 281–333 (1967)
1246. Frewer, M.: Felix Bernstein. Jahresber. Deutsch. Math.-Verein. **83**, 84–95 (1981)
1247. Frey, G. (ed.): On Artin's conjecture for odd 2-dimensional representations. Lecture Notes Math. **1575** (1994)
1248. Fricke, R.: Lehrbuch der Algebra, I–III. Vieweg (1924–1928)
1249. Fried, E., Kollár, J., Automorphism groups of algebraic number fields. Math. Z. **163**, 121–123 (1978)
1250. Fried, M.: On Hilbert's Irreducibility Theorem. J. Number Theory **6**, 211–231 (1974)
1251. Fried, M.: Fields of definition of function fields and Hurwitz families — groups as Galois groups. Comm. Algebra **5**, 17–82 (1977)
1252. Fried, M.: A note on automorphism groups of algebraic number fields. Proc. Amer. Math. Soc. **80**, 386–388 (1980)
1253. Fried, M.: The nonregular analogue of Tchebotarev's theorem. Pacific J. Math. **112**, 303–311 (1984)
1254. Friedlander, J.B., Iwaniec, H.: Summation formulae for coefficients of L-functions. Canad. J. Math. **57**, 494–505 (2005)
1255. Friedman, E.C.: Ideal class groups in basic $Z_{p_1} \times \ldots Z_{p_s}$-extensions of abelian number fields. Invent. Math. **65**, 425–440 (1981/82)
1256. Friedmann, A., Tamarkine, J.: Quelques formules concernant la théorie de la fonction $[x]$ et des nombres de Bernoulli. J. Reine Angew. Math. **135**, 145–156 (1908)
1257. Friesen, C., Spearman, B.K., Williams, K.S.: Another proof of Eisenstein's law of cubic reciprocity and its supplement. Rocky Mountain J. Math. **16**, 395–402 (1986)
1258. Frobenius, G.: Gedächtnisrede auf Leopold Kronecker. Abh. Kgl. Preuss. Akad. Wiss. Berlin **1893**, 1–22. [[1262], 3, 7705–724.]
1259. Frobenius, G.: Ueber Beziehungen zwischen den Primidealen eines algebraischen Körpers und den Substitutionen seiner Gruppe. SBer. Kgl. Preuß. Akad. Wiss. Berlin **1896**, 689–703. [[1262], 2, 719–733.]
1260. Frobenius, G.: Gegenseitige Reduktion algebraischer Körper. Math. Ann. **70**, 457–458 (1911) [[1262], 3, 491–492.]
1261. Frobenius, G.: Über quadratische Formen, die viele Primzahlen darstellen. SBer. Kgl. Preuß. Akad. Wiss. Berlin **1912**, 966–980. [[1262], 3, 573–587.]
1262. Frobenius, G.: Gesammelte Abhandlungen. **1–3**. Springer (1968)
1263. Frobenius, G., Stickelberger, L.: Ueber Gruppen von vertauschbaren Elementen. J. Reine Angew. Math. **88**, 217–261 (1879) [[1262], 1, 545–590.]
1264. Fröhlich, A.: The generalization of a theorem of L.Rédei's. Quart. J. Math., Oxford ser. (2) **5**, 130–140 (1954)
1265. Fröhlich, A.: The genus field and genus group in finite number fields. Mathematika **16**, 40–46 (1959)
1266. Fröhlich, A.: The genus field and genus group in finite number fields, II. Mathematika **16**, 142–146 (1959)
1267. Fröhlich, A.: Discriminants of algebraic number fields. Math. Z. **74**, 18–28 (1960)
1268. Fröhlich, A.: Ideals in an extension field as modules over the algebraic integers in a finite number field. Math. Z. Math. Z. **74**, 29–38 (1960)
1269. Fröhlich, A.: The discriminants of relative extensions and the existence of integral bases. Mathematika **7**, 15–22 (1960)
1270. Fröhlich, A.: Discriminants and module invariants over a Dedekind domain. Mathematika **7**, 15–22 (1960)
1271. Fröhlich, A.: The module structure of Kummer extensions over Dedekind domains. J. Reine Angew. Math. **209**, 39–53 (1962)
1272. Fröhlich, A.: A remark on the class field tower of the field $P(\sqrt[l]{m})$. J. Lond. Math. Soc. **37**, 193–194 (1962)
1273. Fröhlich, A.: On non-ramified extensions with prescribed Galois group. Mathematika **9**, 133–134 (1962)

1274. Fröhlich, A.: Artin root numbers and normal integral bases for quaternion fields. Invent. Math. **17**, 143–166 (1972)
1275. Fröhlich, A.: Module invariants and root numbers for quaternion fields of degree $4l^\gamma$. Proc. Cambridge Philos. Soc. **76**, 393–399 (1974)
1276. Fröhlich, A.: Galois module structure and Artin L-functions. In: Proceedings of the International Congress of Mathematicians (Vancouver, B.C., 1974), **1**, 351–356, Canad. Math. Congress, Montreal (1975)
1277. Fröhlich, A.: Stickelberger without Gauss sums. In: [1281], 589–607.
1278. Fröhlich, A.: A normal integral basis theorem. J. Algebra **39**, 131–137 (1976)
1279. Fröhlich, A.: Arithmetic and Galois module structure for tame extensions. J. Reine Angew. Math. **286/287**, 380–440 (1976)
1280. Fröhlich, A.: Galois module structure. In: [1281], 133–191
1281. Fröhlich, A. (ed.): Algebraic Number Fields (L-functions and Galois properties). Academic Press (1977)
1282. Fröhlich, A.: Galois Module Structure of Algebraic Integers. Springer, Berlin (1983)
1283. Fröhlich, A., Keating, M.E., Wilson, S.M.J.: The class groups of quaternion and dihedral 2-groups. Mathematika **21**, 64–71 (1974)
1284. Fröhlich, A., Queyrut, J.: On the functional equation of the Artin L-function for characters of real representations. Invent. Math. **20**, 125–138 (1973)
1285. Fröhlich, A., Serre, J.-P., Tate, J.: A different with an odd class. J. Reine Angew. Math. **209**, 6–7 (1962)
1286. Fröhlich, A., Taylor, M.J.: Algebraic Number Theory. Cambridge University Press (1993)
1287. Fubini, G.: Luigi Bianchi e la sua opera scientifica. Ann. Mat. Pura Appl. (4) **6**, 45–83 (1928/29)
1288. Fuchs, C., Tichy, R., Ziegler, V.: On quantitative aspects of the unit sum number problem. Arch. Math. (Basel) **93**, 259–268 (2009)
1289. Fuchs, L.: Ueber die Perioden, welche aus den Wurzeln der Gleichung $w^n = 1$ gebildet sind, wenn n eine zusammengesetzte Zahl ist. J. Reine Angew. Math. **61**, 374–386 (1863) [[1291], I, 53–66.]
1290. Fuchs, L.: Ueber die aus Einheitswurzeln gebildeten complexen Zahlen von periodischen Verhalten, insbesondere die Bestimmung der Klassenzahl derselben. J. Reine Angew. Math. **65**, 74–111 (1866) [[1291], I, 69–109.]
1291. Fuchs, L.: Gesammelte mathematische Werke. Mayer & Müller (1904–1909)
1292. Fuchs, L.: A theorem on the relative norm of an ideal. Comment. Math. Helv. **21**, 29–43 (1948)
1293. Fuchs, L.: A simple proof of the basis theorem for finite abelian groups. Norske Vid. Selsk. Forh., Trondheim **25**, 117–118 (1953)
1294. Fuchs, L.: On the fundamental theorem of commutative ideal theory. Acta Math. Hungar. **5**, 95–99 (1954)
1295. Fuchs, L., Göbel, R.: Friedrich Wilhelm Levi, 1888–1966. In: Abelian groups (Curaçao, 1991), 1–14, Marcel Dekker (1993)
1296. Fueter, R.: Der Klassenkörper der quadratischer Körper und die komplexe Multiplikation. Dissertation, Univ. Göttingen (1903)
1297. Fueter, R.: Die Theorie der Zahlstrahlen. J. Reine Angew. Math. **130**, 197–237 (1905)
1298. Fueter, R.: Die Theorie der Zahlstrahlen, II. J. Reine Angew. Math. **132**, 255–269 (1907)
1299. Fueter, R.: Die Klassenanzahl der Körper der komplexen Multiplikation. Nachr. Ges. Wiss. Göttingen **1907**, 288–298
1300. Fueter, R.: Die verallgemeinerte Kroneckersche Grenzformel und ihre Anwendung auf die Berechnung der Klassenzahl. Rend. Circ. Mat. Palermo **29**, 380–395 (1910)
1301. Fueter, R.: Die Klassenkörper der komplexen Multiplikation und ihr Einfluss auf die Entwicklung der Zahlentheorie. Jahresber. Deutsch. Math.-Verein. **20**, 1–47 (1911)
1302. Fueter, R.: Die diophantische Gleichung $\xi^3 + \eta^3 + \zeta^3 = 0$. SBer. Heidelberg. Akad. Wiss. **1913**, 1–25.

1303. Fueter, R.: Abelsche Gleichungen in quadratisch-imaginären Körpern. Math. Ann. **75**, 177–255 (1914)
1304. Fueter, R.: Die Klassenzahl zyklischer Körper vom Primzahlgrad, deren Diskriminante nur eine Primzahl enthält. J. Reine Angew. Math. **147**, 174–183 (1917)
1305. Fueter, R.: Synthetische Zahlentheorie. Göschen (1917); 2nd ed. de Gruyter (1925); 3rd ed. de Gruyter (1950)
1306. Fueter, R.: Vorlesungen über die singulären Moduln und die komplexe Multiplikation der elliptischen Funktionen, I–II. Teubner (1924–1927)
1307. Fueter, R.: Die Diskriminanten der Körper der singulären Moduln und der Teilungskörper der elliptischen Funktionen. Acta Math. **48**, 43–89 (1926)
1308. Fueter, R.: Reziprozitätsgesetze in quadratisch-imaginären Körpern, I. Nachr. Ges. Wiss. Göttingen **1927**, 336–346
1309. Fueter, R.: Reziprozitätsgesetze in quadratisch-imaginären Körpern, II. Nachr. Ges. Wiss. Göttingen **1927**, 427–445
1310. Fueter, R.: Les lois de réciprocité dans un corps quadratique imaginaire. Enseign. Math. **26**, 316–317 (1928)
1311. Fueter, R.: Über kubische diophantische Gleichungen. Comment. Math. Helv. **2**, 69–89 (1930)
1312. Fueter, R.: Über eine spezielle Algebra. J. Reine Angew. Math. **167**, 52–61 (1932)
1313. Fueter, R.: Ein Satz über die Ring- und Strahlklassenzahlen in algebraischen Zahlkörpern. Comment. Math. Helv. **5**, 319–322 (1933)
1314. Fueter, R.: Über die Normalbasis in einem absolut Abelschen Zahlkörper. In: Festschrift zum 60. Geburtstag von Prof. Dr. Andreas Speiser, 141–152, Füssli (1945)
1315. Fujisaki, G.: On an example of an unramified Galois extension. Sûgaku **9**, 97–99 (1957/58) (Japanese)
1316. Fujisaki, G.: Note on a paper of E. Artin. Sci. Papers College Gen. Ed. Univ. Tokyo **24**, 93–98 (1974)
1317. Fujisaki, G.: A generalization of Carlitz's determinant. Sci. Papers College Arts Sci. Univ. Tokyo **40**, 63–68 (1990)
1318. Fujita, H.: The minimum discriminant of totally real algebraic number fields of degree 9 with cubic subfields. Math. Comp. **60**, 801–810 (1993)
1319. Fukuda, T., Komatsu, K.: Weber's class number problem in the cyclotomic Z_2-extension of Q. Experiment. Math. **18**, 213–222 (2009)
1320. Fukuda, T., Komatsu, K.: Weber's class number problem in the cyclotomic Z_2-extension of Q. II, J. Théor. Nombres Bordeaux **22**, 359–368 (2010)
1321. Fukuda, T., Komatsu, K.: Weber's class number problem in the cyclotomic Z_2-extension of Q. III, Internat. J. Number Th. **7**, 1627–1635 (2011)
1322. Fukuda, T., Komatsu, K., Morisawa, T.: Weber's class number one problem. In: Iwasawa theory 2012, Contributions in Math. Comput. Sci. **7**, 221–226 (2014)
1323. Funakura, T.: On integral bases for pure quartic fields. Math. J. Okayama Univ. **26**, 27–41 (1984)
1324. Funakura, T.: On coefficients of Artin L functions as Dirichlet series. Canad. Math. Bull. **33**, 50–54 (1990)
1325. Fung, G.[W.], Granville, A., Williams, H.C.: Computation of the first factor of the class number of cyclotomic fields. J. Number Theory **42**, 297–312
1326. Fung, G.W., Williams, H.C.: On the computation of a table of complex cubic fields with discriminant $D< -10^6$. Math. Comp. **55**, 313–325 (1990); corr. **63**, p. 433 (1994)
1327. Furtwängler, Ph.: Zur Begründung der Idealtheorie. Nachr. Ges. Wiss. Göttingen **1895**, 381–384
1328. Furtwängler, Ph.: Zur Theorie der in Linearfactoren zerlegbaren, ganzzahligen ternären kubischen Formen. Dissertation, Univ. Göttingen (1896)
1329. Furtwängler, Ph.: Über das Reziprozitätsgesetz der l-ten Potenzreste in algebraischen Zahlkörpern, wenn l eine ungerade Primzahl bedeutet. Abhandl. Kgl. Ges. Wiss. Göttingen **2**, 3–82 (1902)

1330. Furtwängler, Ph.: Die Konstruktion des Klassenkörpers für solche algebraischen Zahlkörper, die eine l-te Einheitswurzel enthalten, und deren Idealklassen eine zyklische Gruppe vom Grade l^h bilden. Nachr. Ges. Wiss. Göttingen **1903**, 203–217
1331. Furtwängler, Ph.: Über die Konstruktion des Klassenkörpers für beliebige algebraische Zahlkörper, die eine l-te Einheitswurzel enthalten. Nachr. Ges. Wiss. Göttingen **1903**, 282–303
1332. Furtwängler, Ph.: Die Konstruktion des Klassenkörpers für beliebige algebraische Zahlkörper. Nachr. Ges. Wiss. Göttingen **1904**, 173–194
1333. Furtwängler, Ph.: Über die Reziprozitätsgesetze zwischen l-ten Potenzresten in algebraischen Zahlkörpern, wenn l eine ungerade Primzahl bedeutet. Math. Ann. **58**, 1–50 (1904)
1334. Furtwängler, Ph.: Eine charakteristische Eigenschaft des Klassenkörpers, I. Nachr. Ges. Wiss. Göttingen **1906**, 417–434
1335. Furtwängler, Ph.: Eine charakteristische Eigenschaft des Klassenkörpers, II. Nachr. Ges. Wiss. Göttingen **1907**, 1–24
1336. Furtwängler, Ph.: Allgemeiner Existenzbeweis für den Klassenkörper eines beliebiges algebraischen Zahlkörpers. Math. Ann. **63**, 1–37 (1907)
1337. Furtwängler, Ph.: Über die Klassenzahlen Abelscher Zahlkörper. J. Reine Angew. Math. **134**, 91–94 (1908)
1338. Furtwängler, Ph.: Die Reziprozitätsgesetze für Potenzreste mit Primzahlexponenten in algebraischen Zahlkörpern, I. Math. Ann. **67**, 1–31 (1909)
1339. Furtwängler, Ph.: Untersuchungen über die Kreisteilungskörper und den letzten Fermatschen Satz, I. Nachr. Ges. Wiss. Göttingen **1910**, 554–562
1340. Furtwängler, Ph.: Allgemeiner Beweis des Zerlegungssatzes für den Klassenkörper. Nachr. Ges. Wiss. Göttingen **1911**, 293–317
1341. Furtwängler, Ph.: Über die Klassenzahlen der Kreisteilungskörper. J. Reine Angew. Math. **140**, 29–32 (1911)
1342. Furtwängler, Ph.: Letzter Fermatscher Satz und Eisensteinsches Reziprozitätsprinzip. SBer. Kais. Akad. Wissensch. Wien **121**, 589–592 (1912)
1343. Furtwängler, Ph.: Die Reziprozitätsgesetze für Potenzreste mit Primzahlexponenten in algebraischen Zahlkörpern, II. Math. Ann. **72**, 346–386 (1912)
1344. Furtwängler, Ph.: Die Reziprozitätsgesetze für Potenzreste mit Primzahlexponenten in algebraischen Zahlkörpern, III. Math. Ann. **74**, 413–429 (1913)
1345. Furtwängler, Ph.: Über das Verhalten der Ideale des Grundkörpers im Klassenkörper. Monatsh. Math. Phys. **27**, 1–15 (1916)
1346. Furtwängler, Ph.: Über Kriterien für die algebraischen Zahlen. SBer. Kais. Akad. Wissensch. Wien **126**, 299–309 (1917)
1347. Furtwängler, Ph.: Über die Führer von Zahlringen. SBer. Kais. Akad. Wissensch. Wien **128**, 239–245 (1919)
1348. Furtwängler, Ph.: Über die Ringklassenkörper für imaginäre quadratische Körper, I. SBer. Kais. Akad. Wissensch. Wien **128**, 247–280 (1919)
1349. Furtwängler, Ph.: Über die Ringklassenkörper für imaginäre quadratische Körper, II. SBer. Kais. Akad. Wissensch. Wien **129**, 161–200 (1920)
1350. Furtwängler, Ph.: Punktgitter und Idealtheorie. Math. Ann. **82**, 256–279 (1921)
1351. Furtwängler, Ph.: Über Kriterien für irreduzible und für primitive Gleichungen und über die Aufstellung affektfreier Gleichungen. Math. Ann. **85**, 34–40 (1922)
1352. Furtwängler, Ph.: Über Minimalbasen für Körper rationaler Funktionen. SBer. Kais. Akad. Wissensch. Wien **134**, 69–80 (1925)
1353. Furtwängler, Ph.: Über die Reziprozitätsgesetze für Primzahlpotenzexponenten. J. Reine Angew. Math. **157**, 15–25 (1926)
1354. Furtwängler, Ph.: Über die Reziprozitätsgesetze für ungerade Primzahlexponenten. Math. Ann. **98**, 539–543 (1927)
1355. Furtwängler, Ph.: Über affektfreie Gleichungen. Monatsh. Math. Phys. **36**, 89–96 (1929)
1356. Furtwängler, Ph.: Beweis des Hauptidealsatzes für die Klassenkörper algebraischer Zahlkörper. Abh. Math. Semin. Univ. Hambg **7**, 14–36 (1929)

1357. Furtwängler, Ph.: Über die Verschärfung des Hauptidealsatzes für algebraische Zahlkörper. J. Reine Angew. Math. **167**, 279–387 (1932)
1358. Furtwängler, Ph.: Über einen Determinantensatz. SBer. Kais. Akad. Wissensch. Wien **145**, 527–528 (1936)
1359. Furuta, Y.: The genus field and genus number in algebraic number fields. Nagoya Math. J. **29**, 281–285 (1967)
1360. Furuta, Y.: On class field towers and the rank of ideal class groups. Nagoya Math. J. **48**, 147–157 (1972)
1361. Furuya, H.: Principal ideal theorems in the genus field for absolutely Abelian extensions. J. Number Theory **9**, 4–15 (1977)
1362. Gaál, I.: Power integral bases in orders of families of quartic fields. Publ. Math. Debrecen **42**, 253–263 (1993)
1363. Gaál, I.: Computing all power integral bases in orders of totally real cyclic sextic number fields. Math. Comp. **65**, 801–822 (1996)
1364. Gaál, I.: Computing power integral bases in algebraic number fields. In: Number Theory (Eger 1996), 243–254. de Gruyter (1998)
1365. Gaál, I.: Computing power integral bases in algebraic number fields, II. In: Algebraic Number Theory and Diophantine Analysis (Graz 1998), 153–161. de Gruyter (2000)
1366. Gaál, I.: Solving index form equations in fields of degree 9 with cubic subfields. J. Symbolic Comput. **30**, 181–193 (2000)
1367. Gaál, I.: Power integral bases in cubic relative extensions. Experiment. Math. **10**, 133–139 (2001)
1368. Gaál, I.: Diophantine Equations and Power Integral Bases. Birkhäuser (2002)
1369. Gaál, I., Győry, K.: Index form equations in quintic fields. Acta Arith. **89**, 379–396 (1999)
1370. Gaál, I., Nyul, G.: Computing all monogeneous mixed dihedral quartic extensions of a quadratic field. J. Théor. Nombres Bordeaux **13**, 137–142 (2001)
1371. Gaál, I., Pethő, A., Pohst, M.: On the resolution of index form equations in biquadratic number fields, I. J. Number Theory **38**, 18–34 (1991)
1372. Gaál, I., Pethő, A., Pohst, M.: On the resolution of index form equations in biquadratic number fields, II. J. Number Theory **38**, 35–51 (1991)
1373. Gaál, I., Pethő, A., Pohst, M.: On the indices of biquadratic number fields having Galois group V_4. Arch. Math. (Basel) **57**, 357–361 (1991)
1374. Gaál, I., Pethő, A., Pohst, M.: On the resolution of index form equations in quartic number fields. J. Symbolic Comput. **16**, 563–584 (1993)
1375. Gaál, I., Pethő, A., Pohst, M.: On the resolution of index form equations in dihedral quartic number fields. Experiment. Math. **3**, 245–254 (1994)
1376. Gaál, I., Pethő, A., Pohst, M.: On the resolution of index form equations in biquadratic number fields, III. The bicyclic biquadratic case. J. Number Theory **53**, 100–114 (1995)
1377. Gaál, I., Robertson, L.: Power integral bases in prime-power cyclotomic fields. J. Number Theory **120**, 372–384 (2006)
1378. Gaál, I., Schulte, N.: Computing all power integral bases of cubic fields. Math. Comp. **53**, 689–696 (1989)
1379. Gál, I.S.: The asymptotic distribution of primes. Indag. Math. **25**, 282–294 (1963)
1380. Gallagher, P.X.: The large sieve and probabilistic Galois theory. Proc. Symposia Pure Math. **24**, 91–101 (1973)
1381. Garbanati, D.[A.]: Extensions of the Hasse norm theorem. Bull. Amer. Math. Soc. **81**, 583–586 (1975)
1382. Garbanati, D.A.: Extensions of the Hasse norm theorem, II. Linear Multilinear Algebra **3**, 143–145 (1975/76)
1383. Garbanati, D.A.: Unit signatures, and even class numbers, and relative class numbers. J. Reine Angew. Math. **274/275**, 376–384 (1975)
1384. Garbanati, D.A.: Units with norm -1 and signatures of units. J. Reine Angew. Math. **283/284**, 164–175 (1976)

1385. Garbanati, D.A.: The Hasse norm theorem for l-extensions of the rationals. In: Number theory and algebra, 77–90. Academic Press (1977)
1386. Garbanati, D.A.: The Hasse norm theorem for non-cyclic extensions of the rationals. Proc. London Math. Soc. (3) **37**, 143–164 (1978)
1387. Garth, D.: On limits of PV k-tuples. Acta Arith. **90**, 291–299 (1999)
1388. Garver, R.: Quartic equations with certain groups. Ann. Math. (2) **30**, 47–51 (1928)
1389. Garver, R.: Two notes on cyclic cubics. Amer. Math. Monthly **35**, 435–436 (1928)
1390. Garver, R.: Quartic equations with the alternating group. Tôhoku Math. J. **32**, 306–311 (1930)
1391. Garza, J.: On the height of algebraic numbers with real conjugates. Acta Arith. **128**, 385–389 (2007)
1392. Garza, J., Ishak, M.I.M., Mossinghoff, M.J., Pinner, C.G., Wiles, B.: Heights of roots of polynomials with odd coefficients. J. Théor. Nombres Bordeaux **22**, 369–381 (2010)
1393. Gassmann, F.: Bemerkungen zu der bevorstehender Arbeit von Hurwitz. Math. Z. **25**, 665–675 (1926)
1394. Gauss, C.F.: Disquisitiones Arithmeticae. Gottingae (1801) [[1397], **1**, 1–474; German translation: Untersuchungen über höhere Arithmetik, Springer (1889); reprint: Chelsea (1965); English translations: Yale University Press (1966), Springer (1986)]
1395. Gauss, C.F.: Theoria residuorum biquadraticorum. Comm. Soc. Reg. Sci. Gottingensis, **6**, (1828); **7**, (1832) [[1397], **2**, 65–92, 93–148.]
1396. Gauss, C.F.: De nexu inter multitudinem classium, in quas formae binariae secundi gradus distribunter, earumque determinantem. In: [1397], **2**, 269–291.
1397. Gauss, C.F.: Werke. **1–12**, Königliche Gesellschaft der Wissenschaften zu Göttingen, Teubner, Springer, (1863–1929) [Reprint: G. Olms (1973)]
1398. Gegenbauer, L.: Ueber das cubische Reciprocitätsgesetz. SBer. Kais. Akad. Wissensch. Wien **81**, 1–5 (1880)
1399. Gegenbauer, L.: Zur Theorie der aus den vierten Einheitswurzeln gebildeten complexen Zahlen. Denkschr. Kais. Akad. Wiss., Math.-Nat. Cl., **50**, I. Abt., 153–184 (1885)
1400. Gegenbauer, L.: Ueber die ganzen complexen Zahlen von der Form $a + bi$. SBer. Kais. Akad. Wissensch. Wien **91**, 1047–1058 (1885)
1401. Gegenbauer, L.: Wahrscheinlichkeiten im Gebiet der aus den vierten Einheitswurzeln gebildeten complexen Zahlen. SBer. Kais. Akad. Wissensch. Wien **98**, 635–646 (1888)
1402. Gegenbauer, L.: Einige arithmetische Sätze. Monatsh. Math. **1**, 39–46 (1890)
1403. Gelfond, A.O.: Sur la septième probléme de Hilbert. Izv. Akad. Nauk SSSR, Ser. Mat. **1934**, 623–634 [[1407], 51–56. (Russian)]
1404. Gelfond, A.O.: On the approximation of the ratio of logarithms of two algebraic numbers by algebraic numbers. Izv. Akad. Nauk SSSR, Ser. Mat. **3**, 509–518 (1939) (Russian)
1405. Gelfond, A.O.: Sur la divisibilité de la différence des puissances de deux nombres entiers par une puissance d'un idéal premier. Mat. Sb. (N.S.) **7**, 7–25 (1940)
1406. Gelfond, A.O.: Approximation of algebraic irrationalities and their logarithms. Vestnik Moskov. Univ. Ser. I. Mat. Mekh. **9**, 3–25 (1948) (Russian) [[1407], 129–150.]
1407. Gelfond, A.O.: Izbrannye trudy. Nauka (1973) (Russian)
1408. Gelfond, A.O., Linnik, Yu.V.: On Thue's method in the problem of effectiveness in quadratic fields. Dokl. Akad. Nauk SSSR **61**, 773–776 (1948) (Russian) [[1407], 151–154.]
1409. Genocchi, A.: Intorni ad un teorema di Cauchy. Ann. Mat. Pura Appl. (2) **2**, 216–219 (1868/69)
1410. Gerboud, G.: Construction, sur un anneau de Dedekind, d'une base réguliere de polynômes à valeurs entiéres. Manuscripta Math. **65**, 167–179 (1989)
1411. Geroldinger, A.: Ein quantitatives Resultat über Faktorisierungen verschiedener Länge in algebraischen Zahlkörpern. Math. Z. **205**, 159–162 (1990)
1412. Geroldinger, A.: Arithmetical characterizations of divisor class group. Arch. Math. (Basel) **54**, 455–464 (1990)
1413. Geroldinger, A.: Factorization of natural numbers in algebraic number fields. Acta Arith. **57**, 365–373 (1991)

1414. Geroldinger, A.: Arithmetical characterizations of divisor class group, II. Acta Math. Univ. Comenian., (N.S.) **61**, 193–208 (1992)
1415. Geroldinger, A., Halter-Koch, F.: Non-unique Factorizations, Algebraic Combinatorics and Analytic Theory. Chapman & Hall/CRC (2006)
1416. Geroldinger, A., Halter-Koch, F., Kaczorowski, J.: Non-unique factorizations in orders of global fields. J. Reine Angew. Math. **459**, 89–118 (1995)
1417. Gerth, F.III: Number fields with prescribed l-class groups. Proc. Amer. Math. Soc. **49**, 284–288 (1975)
1418. Gerth, F.III: The Hasse norm principle for abelian extensions of number fields. Bull. Amer. Math. Soc. **83**, 264–266 (1977)
1419. Gerth, F.III: The Hasse norm principle in metacyclic extensions of number fields. J. Lond. Math. Soc. (2) **16**, 203–208 (1977)
1420. Gerth, F.III: Imaginary quadratic fields with 2-class rank equal to 4 and infinite Hilbert 2-class field towers. JP J. Algebra, Number Theory and Appl. **4**, 391–405 (2004)
1421. Gerth, F.III: Densities for some real quadratic fields with infinite Hilbert 2-class field towers. J. Number Theory **118**, 90–97 (2006)
1422. Geyer, W.-D.: Jede endliche Gruppe ist Automorphismengruppe einer endlichen Erweiterung K/Q. Arch. Math. (Basel) **41**, 139–142 (1983)
1423. Ghate, E.: Vandiver's conjecture via K-theory. In: Cyclotomic Fields and Related Topics (Pune 1999), 285–298. Bhaskaracharya Pratishthana (2000)
1424. Ghate, E., Hironaka, E.: The arithmetic and geometry of Salem numbers. Bull. Amer. Math. Soc. (N.S.) **38**, 293–314 (2001)
1425. Gierster, J.: Ueber Relationen zwischen Classenanzahlen binärer quadratischer Formen von negativer Determinante. Math. Ann. **17**, 71–84 (1880)
1426. Gierster, J.: Ueber Relationen zwischen Klassenzahlen binärer quadratischer Formen von negativer Determinante. Math. Ann. **21**, 1–51 (1882)
1427. Gierster, J.: Ueber Relationen zwischen Klassenzahlen binärer quadratischer Formen von negativer Determinante. Math. Ann. **22**, 190–211 (1883)
1428. Gilbarg, D.: The structure of the group of $\mathfrak{P}$-adic units. Duke Math. J. **9**, 262–271 (1942)
1429. Giles, J.R., Wallis, J.S.: George Szekeres. With affection and respect. J. Aus. Math. Soc. A **21**, 385–392 (1976)
1430. Gillard, R.: Remarques sur les unités cyclotomiques et les unités elliptiques. J. Number Theory **11**, 21–48 (1979)
1431. Gillard, R.: Unités elliptiques et unités de Minkowski. J. Math. Soc. Japan **32**, 697–701 (1980)
1432. Gilmer, R.W.[Jr.]: On a classical theorem of Noether in ideal theory. Pacific J. Math. **13**, 579–583 (1963)
1433. Gilmer, R.[W.Jr.]: A note on two criteria for Dedekind domains. Enseign. Math. **13**, 253–256 (1967)
1434. Gilmer, R.W.Jr.: Multiplicative Ideal Theory. Queen's University (1968) [2nd. ed. Marcel Dekker (1972); reprint: Queen's University (1992)]
1435. Gilmer, R.W.Jr., Mott, J.L.: Multiplication rings as rings in which ideals with prime radical are primary. Trans. Amer. Math. Soc. **114**, 40–52 (1965)
1436. Gilmore, P.C., Robinson, A.: Metamathematical considerations on the relative irreducibility of polynomials. Canad. J. Math. **7**, 483–489 (1955)
1437. Girstmair, K.: A remark on normal bases. J. Number Theory **58**, 64–65 (1996)
1438. Girstmair, K.: An algorithm for the construction of a normal basis. J. Number Theory **78**, 36–45 (1999)
1439. Girstmair, K., Kühleitner, M., Müller, W., Nowak, W.G.: The Piltz divisor problem in number fields: an improved lower bound by Soundararajan's method. Acta Arith. **117**, 187–206 (2005)
1440. Glaisher, J.W.L.: On the expressions for the number of classes of a negative determinant, and on the numbers of positives in the octants of P. Quart. J. Pure Appl. Math. **34**, 178–204 (1903)

1441. Gmeiner, J.A.: Die Ergänzungssätze zum bikubischen Reciprocitätsgesetze. SBer. Kais. Akad. Wissensch. Wien **100**, 1330–1361 (1890)
1442. Gmeiner, J.A.: Das allgemeine bikubische Reciprocitätsgesetz. SBer. Kais. Akad. Wissensch. Wien **101**, 563–584 (1892)
1443. Gmeiner, J.A.: Die bikubische Reciprocität zwischen einer reellen und einer zweigliedrigen regulären Zahl. Monatsh. Math. Phys. **3**, 179–192, 199–210 (1892)
1444. Gmeiner, J.A.: Ueber die ganzen Zahlen im Rationalitätsgebiete der fünften Einheitswurzeln. Programm des Staats-Gymnasium in Pola **6**, 1–24 (1896)
1445. Gmeiner, J.A.: Die Einheiten im Rationalitätsgebiete der fünften Einheitswurzeln. Monatsh. Math. Phys. **9**, 184–206 (1898)
1446. Gmeiner, J.A.: Ueber die Primzahlen und Primideale im Rationalitätsgebiete der fünften Einheitswurzeln. Monatsh. Math. Phys. **11**, 1–27 (1900)
1447. Godwin, H.J.: The determination of units in totally real cubic fields. Proc. Cambridge Philos. Soc. **56**, 318–321 (1960)
1448. Godwin, H.J.: On Euclid's algorithm in some quartic and quintic fields. J. Lond. Math. Soc. **40**, 699–704 (1965)
1449. Godwin, H.J.: On Euclid's algorithm in some cubic fields with signature one. Quart. J. Math., Oxford ser. (2) **18**, 333–338 (1967)
1450. Godwin, H.J.: A note on Cusick's theorem on units in totally real cubic fields. Math. Proc. Cambridge Philos. Soc. **95**, 1–2 (1984)
1451. Godwin, H.J., Samet, P.A.: A table of real cubic fields. J. Lond. Math. Soc. **34**, 108–110 (1959); corr. (2) **9**, p. 624 (1974/75)
1452. Gogia, S.K., Luthar, I.S.: Real characters of the ideal class-group and the narrow ideal class-group of $Q(\sqrt{d})$. Colloq. Math. **41**, 153–159 (1979/80)
1453. Gold, R.: Genera in abelian extensions. Proc. Amer. Math. Soc. **47**, 25–28 (1975)
1454. Gold, R.: Genera in normal extensions. Pacific J. Math. **63**, 397–400 (1976)
1455. Gold, R.: The principal genus and Hasse's norm theorem. Indiana Univ. Math. J. **26**, 183–189 (1977)
1456. Gold, R.: Local class field theory via Lubin-Tate groups. Indiana Univ. Math. J. **30**, 795–798 (1981)
1457. Gold, R., Kim, J.M.: Bases for cyclotomic units. Compositio Math. **71**, 13–27 (1989)
1458. Goldberg, M.: Ernst G.Straus (1922–1983). Linear Algebra Appl. **64**, 1–19 (1985)
1459. Goldberg, M.: Morris Newman — a discrete mathematician for all seasons. Linear Algebra Appl. **254**, 7–18 (1997)
1460. Goldfeld, D.M.: A simple proof of Siegel's theorem. Proc. Nat. Acad. Sci. U.S.A. **71**, p. 1055 (1974)
1461. Goldfeld, D.: The class number of quadratic fields and the conjectures of Birch and Swinnerton-Dyer. Ann. Scuola Norm. Sup. Pisa, Cl. Sci. (4) **3**, 623–663 (1976)
1462. Goldfeld, D.: Gauss's class number problem for imaginary quadratic fields. Bull. Amer. Math. Soc. (N.S.) **13**, 23–37 (1985)
1463. Goldfeld, D.: The Gauss class number problem for imaginary quadratic fields. In: Heegner points and Rankin L-series, 25–36. Cambridge University Press (2004)
1464. Goldfeld, D., Hoffstein, J.: Eisenstein series of $1/2$-integral weight and the mean value of real Dirichlet L-series. Invent. Math. 80, 1985, 185–208 (1985)
1465. Goldfeld, D., Viola, C.: Mean values of L-functions associated to elliptic, Fermat and other curves at the centre of the critical strip. J. Number Theory 11, 1979, 305–320 (1979)
1466. Goldscheider, F.: Das Reciprocitätsgesetz der achten Potenzreste. Programm des Berliner Luisenstadt Realgymnasium, 1889, 1–29 (1889)
1467. Goldstein, C., Schappacher, N., Schwermer, J. (ed.): The Shaping of Arithmetic after C.F. Gauss's Disquisitiones Arithmeticae. Springer (2007)
1468. Goldstein, L.J.: On prime discriminants. Nagoya Math. J. **45**, 119–127 (1972)
1469. Goldstein, L.J.: On a conjecture of Hecke concerning elementary class number formulas. Manuscripta Math. **9**, 245–305 (1973)
1470. Goldstein, L.J.: On a formula of Hecke. Israel J. Math. **17**, 283–301 (1974)

1471. Goldstein, L.[J.], de la Torre, P.: On a function analogous to log $\eta(\tau)$. Nagoya Math. J. **59**, 169–198 (1975)
1472. Golod, E.S., Šafarevič, I.R.: On the class-field tower. Izv. Akad. Nauk SSSR, Ser. Mat. **29**, 261–272 (1964) (Russian)
1473. Golomb, S., Harris, T., Seberry, J.: Albert Leon Whiteman (1915–1995). Notices Amer. Math. Soc. **44**, 217–219 (1997)
1474. Golubeva, E.P.: On the Pell equation. Zap. Naučn. Sem. POMI **286**, Anal. Teor. Čisel i Teor. Funk. **18**, 36–39 (2002) (Russian)
1475. Golubeva, E.P.: On the negative Pell equation. Zap. Naučn. Sem. POMI **383**, Anal. Teor. Čisel i Teor. Funk. **125**, 53–62 (2010) (Russian)
1476. Gómez-Ayala, E.J.: Bases normales d'entiers dans les extensions de Kummer de degré premier. J. Théor. Nombres Bordeaux **6**, 95–116 (1994)
1477. Gordeev, N.L.: Ramification groups of infinite p-extensions of a local field. Zap. Naučn. Sem. LOMI **71**, 96–132 (1977) (Russian)
1478. Gorškov, D.S.: Cubic fields and symmetric matrices. Dokl. Akad. Nauk SSSR **31**, 842–844 (1941) (Russian)
1479. Götzky, F.: Über eine zahlentheoretische Anwendung von Modulfunktionen zweier Veränderlicher. Math. Ann. **100**, 411–437 (1928)
1480. Grandjot, K.: Über die Irreduzibilität der Kreisteilungsgleichung. Math. Z. **19**, 128–129 (1924)
1481. Grant, D.: Sequences of fields with many solutions to the unit equation. Rocky Mountain J. Math. **26**, 1017–1029 (1996)
1482. Grant, D.: A proof of quintic reciprocity using the arithmetic of $y^2 = x^5 + 1/4$. Acta Arith. **75**, 321–327 (1996)
1483. Grant, D.: Geometric proofs of reciprocity laws. J. Reine Angew. Math. **586**, 91–124 (2005)
1484. Grant, H.S.: Concerning powers of certain classes of ideals in a cyclotomic realm which give the principal class. Ann. Math. (2) **35**, 220–238 (1934)
1485. Granville, A.: On the size of the first factor of the class number of a cyclotomic field. Invent. Math. **100**, 321–338 (1990)
1486. Granville, A., Mollin, R.A.: Rabinowitsch revisited. Acta Arith. **96**, 139–153 (2000)
1487. Granville, A.J., Stark, H.M.: ABC implies no "Siegel zero" for L-functions of characters with negative discriminant. Invent. Math. **139**, 509–523 (2000)
1488. Gras, G.: Critère de parité du nombre de classes des extensions abéliennes réelles de Q de degré impair. Bull. Soc. Math. France **103**, 177–190 (1975)
1489. Gras, G.: Decomposition and inertia groups in Z_p-extensions. Tokyo J. Math. **9**, 41–51 (1986)
1490. Gras, G.: Non monogénéité d'anneaux d'entiers. Ann. Sci. Univ. Besançon Math. Théorie des nombres **1**, 1–43 (1986/87-87/88)
1491. Gras, G.: Théorèmes de réflexion. J. Théor. Nombres Bordeaux **10**, 399–499 (1998)
1492. Gras, G.: Class Field Theory, From Theory to Practice. Springer (2003)
1493. Gras, G., Gras, M.-N.: Signature des unités cyclotomiques et parité du nombre de classes des extensions cycliques de Q de degré premier impair. Ann. Inst. Fourier (Grenoble) **25(1)**, 1–22 (1975)
1494. Gras, M.-N.: Sur les corps cubiques cycliques dont l'anneau des entiers est monogène. Ann. Sci. Univ. Besançon Math. **3(6)** 1–26 (1973)
1495. Gras, M.-N.: Sur les corps cubiques cycliques dont l'anneau des entiers est monogène. C. R. Acad. Sci. Paris **278**, 59–62 (1974)
1496. Gras, M.-N.: Méthodes et algorithmes pour le calcul numérique du nombre de classes et des unités des extensions cubiques cycliques de Q. J. Reine Angew. Math. **277**, 89–116 (1975)
1497. Gras, M.-N.: Note à propos d'une conjecture de H.J. Godwin sur les unités des corps cubiques. Ann. Inst. Fourier (Grenoble) **30(4)**, 1–6 (1980)
1498. Gras, M.-N.: Z-bases d'entiers $1, \theta, \theta^2, \theta^3$ dans les extensions cycliques de degré 4 de $\mathbf{Q}$. Publ. Math. Fac. Sci. Besançon **1980/81**, 1–14

1499. Gras, M.-N.: Non monogénéité de l'anneau des entiers de certaines extensions abéliennes de Q. Publ. Math. Fac. Sci. Besançon **1983/84(5)**, 1–25
1500. Gras, M.-N.: Non monogénéité de l'anneau des entiers des extensions cycliques de Q de degré premier $l \geq 5$. J. Number Theory **23**, 347–353 (1986)
1501. Gras, M.-N., Tanoé, F.: Corps biquadratiques monogènes. Manuscripta Math. **86**, 63–79 (1995)
1502. Grauert, H., Remmert, R.: In memoriam Heinrich Behnke. Math. Ann. **255**, 1–4 (1981)
1503. Grave, D.A.: On the main theorems of the theory of ideal numbers. Mat. Sb. **32**, 135–151 (1924/25) (Russian)
1504. Grave, D.A.: On decomposition of prime numbers into ideal factors. Mat. Sb. **32**, 542–561 (1924/25) (Russian)
1505. Gray, J.: König, Hadamard and Kürschák, and abstract algebra. Math. Intelligencer **19**, 61–64 (1997)
1506. Greaves, A., Odoni, R.W.K.: Weil-numbers and CM-fields. I. J. Reine Angew. Math. **391**, 198–212 (1988)
1507. Greaves, G., Taylor, G.: Lehmer's conjecture for Hermitian matrices over the Eisenstein and Gaussian integers. Electron. J. Combin. **20(1/42)**, 1–22 (2013)
1508. Green, J.A.: Richard Dagobert Brauer. Bull. Lond. Math. Soc. **10**, 317–342 (1978)
1509. Greenberg, M.J.: An elementary proof of the Kronecker-Weber theorem. Amer. Math. Monthly **1**, 601–607 (1974); corr. **82**, p. 803 (1975)
1510. Greenberg, R.: A generalization of Kummer's criterion. Invent. Math. **21**, 247–254 (1973)
1511. Greenberg, R.: On p-adic L-functions and cyclotomic fields, II. Nagoya Math. J. **67**, 139–158 (1977)
1512. Greiter, G.: A simple proof for a theorem of Kronecker. Amer. Math. Monthly **85**, 756–757(1978)
1513. Greither, C., Johnston, H.: On totally real Hilbert-Speiser fields of type C_p. Acta Arith. **138**, 329–336 (2009)
1514. Greither, C., Replogle, D.R., Rubin, K., Srivastav, A.: Swan modules and Hilbert-Speiser number fields. J. Number Theory **79**, 164–173 (1999)
1515. Grell, H.: Beziehungen zwischen den Idealen verschiedener Ringe. Math. Ann. **97**, 490–523 (1927)
1516. Grell, H.: Zur Theorie der Ordnungen in algebraischen Zahl- und Funktionenkörpern. Math. Ann. **97**, 524–558 (1927)
1517. Grell, H.: Über die Gültigkeit der gewöhnlichen Idealtheorie in endlichen algebraischen Erweiterungen erster und zweiter Art. Math. Z. **40**, 503–505 (1936)
1518. Grell, H.: Verzweigungstheorie in allgemeinen Ordnungen algebraischer Zahlkörper. Math. Z. **40**, 629–657 (1936)
1519. Grenié, L., Molteni, G.: Explicit versions of the prime ideal theorem for Dedekind zeta functions under GRH. Math. Comp. **85**, 889–906 (2016)
1520. Griffin, M.: Multiplication rings via their total quotient rings. Canad. J. Math. **26**, 430–449 (1974)
1521. Gröbner, W.: Über Minimalbasen für die Invariantenkörper zyklischer und metazyklischer Permutationsgruppen. Anzeiger Akad. Wiss. Wien **69**, 43–44 (1932)
1522. Gröbner, W.: Minimalbasis der Quaternionengruppe. Monatsh. Math. Phys. **41**, 78–84 (1934)
1523. Grodzenskiĭ, S.Ya.: Andreĭ Andreevič Markov. 1856–1922. Nauka (1987)
1524. Gronwall, T.H.: Sur les séries de Dirichlet correspondant à des caractéres complexes. Rend. Circ. Mat. Palermo **35**, 145–159 (1913)
1525. Gross, B.H., Rohrlich, D.E.: Some results on the Mordell-Weil group of the Jacobian of the Fermat curve. Invent. Math. **44**, 201–224 (19378)
1526. Gross, B.H., Zagier, D.: Points de Heegner et dérivées de fonctions L. C. R. Acad. Sci. Paris **297**, 85–87 (1983)
1527. Gross, B.H., Zagier, D.: Heegner points and derivatives of L-series. Invent. Math. **84**, 225–320 (1986)

1528. Grosswald, E.: Negative discriminants of binary quadratic forms with one class in each genus. Acta Arith. **8**, 295–306 (1962/63)
1529. Grotz, W.: Mittelwert der Eulerschen φ-Funktion und des Quadrates der Dirichletschen Teilerfunktion in algebraischen Zahlkörpern. Monatsh. Math. **88**, 219–228 (1979)
1530. Grotz, W.: Einige Anwendungen der Siegelschen Summenformel. Acta Arith. **38**, 69–95 (1980/81)
1531. Grube, F.: Ueber einige Euler'sche Sätze aus der Theorie der quadratischen Formen. Zeitschr. Math. Phys. (5) **19**, 492–519 (1874)
1532. Gruenberg, K.W.: Cohomological topics in group theory. Lecture Notes Math. **143** (1970)
1533. Grunwald, W.: Charakterisierung des Normenrestsymbols durch die $\mathfrak{p}$-Stetigkeit, den vorderen Zerlegungssatz und die Produktformel. Math. Ann. **107**, 145–164 (1932)
1534. Grunwald, W.: Ein allgemeines Existenztheorem für algebraische Zahlkörper. J. Reine Angew. Math. **169**, 103–107 (1933)
1535. Grytczuk, A., Luca, F., Wójtowicz, M.: The negative Pell equation and Pythagorean triples. Proc. Japan Acad. Sci. **76**, 91–94 (2000)
1536. Gu, H., Gu, D., Liu, Y.: Ono invariants of imaginary quadratic fields with class number three. J. Number Theory **127**, 262–271 (2007)
1537. Gudermann, C.: Theorie der Modular-Functionen und der Modular-Integrale. J. Reine Angew. Math. **18**, 1–54 (1838)
1538. Guitart, X., Masdeu, M.: Continued fractions in 2-stage Euclidean quadratic fields. Math. Comp. **82**, 1223–1233 (2013)
1539. Guo, X., Qin, H.: Imaginary quadratic fields with Ono number 3. Comm. Algebra **38**, 230–232 (2010)
1540. Gupta, R., Murty M.R., Murty, V.K.: The Euclidean algorithm for S-integers. In: Number Theory (Montreal, 1985), 189–201, Amer. Math. Soc. (1987)
1541. Gupta, S., Zagier, D.: On the coefficients of the minimal polynomials of Gaussian periods. Math. Comp. **60**, 385–398 (1993)
1542. Gurak, S.J.: Ideal-theoretic characterization of the relative genus field. J. Reine Angew. Math. **296**, 119–124 (1977)
1543. Gurak, S.: On the Hasse norm principle. J. Reine Angew. Math. **299/300**, 16–27 (1978)
1544. Gurak, S.: The Hasse norm principle in non-abelian extensions. J. Reine Angew. Math. **303/304**, 314–318 (1978)
1545. Gurak, S.: The Hasse norm principle in a compositum of radical extensions. J. Lond. Math. Soc. (2) **22**, 385–397 (1980)
1546. Gurak, S.: Minimal polynomials for Gauss circulants and cyclotomic units. Pacific J. Math. **102**, 347–353 (1982)
1547. Gurak, S.: Minimal polynomials for circular numbers. Pacific J. Math. **112**, 313–331 (1984)
1548. Gurak, S.: Minimal polynomials for Gauss periods with $f = 2$. Acta Arith. **121**, 233–257 (2006)
1549. Gurak, S.: On the minimal polynomials of Gauss periods for prime powers. Math. Comp. **75**, 2021–2035 (2006)
1550. Gut, M.: Über die Klassenanzahl der quadratischen Körper. Vierteljahrsschr. Naturforsch. Ges. Zürich **72**, 197–204 (1927)
1551. Gut, M.: Die Zetafunktion, die Klassenzahl und die Kroneckersche Grenzformel eines beliebigen Kreiskörpers. Comment. Math. Helv. **1**, 160–226 (1929)
1552. Gut, M.: Über die Primidealzerlegung in gewissen relativ–ikosaedrischen Zahlkörpern. Comment. Math. Helv. **4**, 219–229 (1932)
1553. Gut, M.: Weitere Untersuchungen über die Primidealzerlegung in gewissen relativ-ikosaedrischen Zahlkörpern. Comment. Math. Helv. **6**, 47–75 (1933)
1554. Gut, M.: Über die Primideale im Wurzelkörper einer Gleichung. Comment. Math. Helv. **6**, 185–191 (1934)
1555. Gut, M.: Über die Gradteilerzerlegung in gewissen relativ-ikosaedrischen Zahlkörpern. Comment. Math. Helv. **7**, 103–130 (1934)

1556. Gut, M.: Über Erweiterungen von unendlichen algebraischen Zahlkörpern. Comment. Math. Helv. **9**, 136–155 (1937)
1557. Gut, M.: Folgen von Dedekindschen Zetafunktionen. Monatsh. Math. Phys. **48**, 153–160 (1939)
1558. Gut, M.: Mittel aus Dirichlet-Reihen mit reellen Restcharakteren. Vierteljahr. Naturforsch. Ges. Zürich **85**, Beiblatt, 214–224 (1940)
1559. Gut, M.: Zur Theorie der Kreiskörper, insbesondere der Strahlklassenkörper der quadratisch imaginären Zahlkörper. Comment. Math. Helv. **15**, 81–119 (1943)
1560. Gut, M.: Zur Theorie der Strahlklassenkörper der quadratisch reellen Zahlkörper. Comment. Math. Helv. **16**, 37–59 (1944)
1561. Gut, M.: Eulersche Zahlen und grosser Fermat'scher Satz. Comment. Math. Helv. **24**, 73–99 (1950)
1562. Gut, M., Stünzi, M.: Kongruenzen zwischen Koeffizienten trigonometrischer Reihen und Klassenzahlen quadratisch imaginärer Körper. Comment. Math. Helv. **41**, 287–302 (1966/67)
1563. Gutiérrez, C., Gutiérrez, F.: Carlos Grandjot, tres décadas de matemáticas in Chile (1930–1960), Bol. Asoz. Mat. Venezolana **11**, 55–84 (2004)
1564. Győry, K.: Représentation des nombres par des formes decomposables, I. Publ. Math. Debrecen **16**, 253–263 (1969)
1565. Győry, K.: Sur les polynômes à coefficients entiers et de discriminant donné. Acta Arith. **23**, 419–426 (1973)
1566. Győry, K.: Sur les polynômes à coefficients entiers et de discriminant donné, II. Publ. Math. Debrecen **21**, 125–144 (193674)
1567. Győry, K.: Sur les polynômes à coefficients entiers et de discriminant donné, III. Publ. Math. Debrecen **23**, 141–165 (1976)
1568. Győry, K.: Sur les polynômes à coefficients entiers et de discriminant donné, IV. Publ. Math. Debrecen **25**, 155–167 (1978)
1569. Győry, K.: On the number of solutions of linear equations in units of an algebraic number field. Comment. Math. Helv. **54**, 583–600 (1979)
1570. Győry, K.: Corps de nombres algébriques d'anneau d'entiers monogène. Sém. Delange–Pisot–Poitou **20**, exp. 26, 1–7 (1978/79)
1571. Győry, K.: On S-integral solutions of norm form, discriminant form and index form equations. Studia Sci. Math. Hungar. **16**, 149–161 (1981)
1572. Győry, K.: On discriminants and indices of integers of an algebraic number field. J. Reine Angew. Math. **324**, 114–126 (1981)
1573. Győry, K.: Some recent applications of S-unit equations. Astérisque **209**, 17–38 (1992)
1574. Győry, K.: Upper bounds for the numbers of solutions of unit equations in two unknowns. Liet. Mat. Rink. **32**, 53–58 (1992) [Lith. Math. J. 32, 1992, 40–44 (1992)]
1575. Győry, K.: Polynomials and binary forms with given discriminant. Publ. Math. Debrecen **69**, 473–499 (2006)
1576. Győry, K., Papp, Z.Z.: On discriminant form and index form equations. Studia Sci. Math. Hungar. **12**, 47–60 (1977)
1577. Győry, K., Yu, K.: Bounds for the solutions of S-unit equations and decomposable form equations. Acta Arith. **123**, 9–41 (2006)
1578. Habicht, W.: Über die Zerlegung strikte definiter Formen in Quadrate. Comment. Math. Helv. **12**, 317–322 (1940)
1579. Habicht, W.: Ein elementarer Beweis des kubischen Reziprozitätsgesetzes. Math. Ann. **139**, 343–365 (1960)
1580. Hachenberger, D.: On completely free elements in finite fields. Des. Codes Cryptogr. **4**, 129–143 (1994)
1581. Hadamard, J.: Sur la distribution des zéros de la fonction $\zeta(s)$ et ses conséquences arithmétiques. Bull. Soc. Math. France **24**, 199–220 (1896) [J. Hadamard, Selecta, 111–132, Gauthier-Villars (1935)]

1582. Haeuslein, G.K.: On the invariants of finite groups having an abelian normal subgroup of prime index. J. Lond. Math. Soc. (2) **3**, 355–360 (1971)
1583. Haffner, E.: Strategical use(s) of arithmetic in Richard Dedekind and Heinrich Weber's *Theorie der algebraischen Funktionen einer Veränderlichen*. Historia Math. **44**, 31–69 (2017)
1584. Häfner, F.: Einige orthogonale und symplektische Gruppen als Galoisgruppen über Q. Math. Ann. **292**, 587–618 (1992)
1585. Hafner, J.L.: On the average order of a class of arithmetical functions. J. Number Theory **15**, 36–76 (1982)
1586. Hafner, J.L.: The distribution and average order of the coefficients of Dedekind ζ functions. J. Number Theory **17**, 183–190 (1983)
1587. Hajdu, L.: Arithmetic progressions in linear combinations of S-units. Period. Math. Hung. **54**, 175–181 (2007)
1588. Hajdu, L., Ziegler, V.: Distinct unit generated totally complex quartic fields. Math. Comp. **83**, 1495–1512 (2014)
1589. Hajir, F.: On a theorem of Koch. Pacific J. Math. **176**, 15–18 (1996)
1590. Hajir, F., Maire, C.: Tamely ramified towers and discriminant bounds for number fields. Compositio Math. **128**, 35–53 (2001)
1591. Hajir, F., Maire, C.: Tamely ramified towers and discriminant bounds for number fields, II. J. Symbolic Comput. **33**, 415–423 (2002)
1592. Hajja, M.: Rational invariants of meta-abelian groups of linear automorphisms. J. Algebra **80**, 295–305 (1983)
1593. Halász, G.: The number-theoretic work of Paul Turán. Acta Arith. **37**, 9–19 (1980)
1594. Halász, G., Lovász, L., Simonovits,M., Sós,V.T. (ed.): Paul Erdős and his Mathematics, **1**,**2**. Springer (2002)
1595. Halberstam, H., Richert, H.-E.: Sieve Methods. Academic Press (1974)
1596. Hall, M.: Indices in cubic fields. Bull. Amer. Math. Soc. **43**, 104–108 (1937)
1597. Hall, N.A.: Binary quadratic discriminants with a single class of forms in each genus. Math. Z. **44**, 85–90 (1939)
1598. Halter-Koch, F.: Arithmetische Theorie der Normalkörper von 2-Potenzgrad mit Diedergruppe. J. Number Theory **3**, 412–443 (1971)
1599. Halter-Koch, F.: Ein Satz über die Geschlechter relativ-zyklischer Zahlkörper von Primzahlgrad und seine Anwendung auf biquadratisch-bizyklische Körper. J. Number Theory **4**, 144–156 (1972)
1600. Halter-Koch, F.: Einheiten und Divisorenklassen in Galois'schen algebraischen Zahlkörpern mit Diedergruppe der Ordnung $2l$ für eine ungerade Primzahl l. Acta Arith. **33**, 353–364 (1977)
1601. Halter-Koch, F.: Zur Geschlechtertheorie algebraischer Zahlkörper. Arch. Math. (Basel) **31**, 137–142 (1978/79)
1602. Halter-Koch, F.: Factorization of algebraic integers, Ber. Math. Stat. Sektion, Graz **191**, 1–24 (1983)
1603. Halter-Koch, F.: Prime-producing quadratic polynomials and class numbers of quadratic orders. In: Computational Number Theory, 73–82. de Gruyter (1991)
1604. Halter-Koch, F.: Der Čebotarev'sche Dichtigkeitssatz und ein Analogon zum Dirichlet'schen Primzahlsatz für algebraische Funktionenkörper. Manuscripta Math. **72**, 205–211 (1991)
1605. Halter-Koch, F.: Chebotarev formations and quantitative aspects of non-unique factorizations. Acta Arith. **62**, 173–206 (1992)
1606. Halter-Koch, F.: A generalization of Davenport's constant and its arithmetical applications. Colloq. Math. **63**, 203–210 (1992)
1607. Halter-Koch, F., Lorenz, F.: Ein Normalbasissatz für Einheiten algebraischer Zahlkörper. Math. Ann. **257**, 335–339 (1981)
1608. Halter-Koch, F., Müller, W.: Quantitative aspects of non-unique factorization: a general theory with applications to algebraic functions fields. J. Reine Angew. Math. **421**, 159–188 (1991)

1609. Halter-Koch, F., Narkiewicz, W.: Polynomial cycles and dynamical units. In: Proceedings of a Conference on Analytic and Elementary Number Theory (Wien 1996), 70–80, Wien (1997)
1610. Halter-Koch, F., Narkiewicz, W.: Scarcity of finite polynomial orbits. Publ. Math. Debrecen **56**, 405–414 (2000)
1611. Hamada, S.: Norm theorem on splitting fields of some binomial polynomials. Kodai Math. Rep. **6**, 47–50 (1983)
1612. Hamamura, M.: On absolute class fields of certain algebraic number fields. Natur. Sci. Rep. Ochanomizu Univ. **32**, 23–34 (1981)
1613. Hamburger, M.: Gedächtnisrede auf Immanuel Lazarus Fuchs. Arch. Math. Phys. (3) **3**, 177–186 (1902)
1614. Hammerstein, A.: Zwei Beiträge zur Zahlentheorie, Dissertation, Univ. Göttingen (1919)
1615. Hancock, H.: Méthode de décomposition des polynômes entiers à plusieurs variables en facteurs irréductibles. Ann. Sci. Éc. Norm. Supér. (3) **17**, 89–102 (1900)
1616. Hancock, H.: Sur les systémes modulaires de Kronecker. Ann. Sci. Éc. Norm. Supér. (3) **18** Suppl., 3–115 (1901)
1617. Hancock, H.: Integral solutions of the equation $\xi^2 + \eta^2 = \zeta^2$ in the quadratic realms of rationality. J. Math. Pures Appl. (8) **4**, 327–341 (1921)
1618. Hancock, H.: Trigonometric realms of rationality. Rend. Circ. Mat. Palermo **49**, 263–276 (1925)
1619. Hancock, H.: Foundations of the Theory of Algebraic Numbers, **1**,**2**. MacMillan (1931–1932) [Reprint: Dover (1964)]
1620. Hanlon, P.: To the Latimer-MacDuffee theorem and beyond! Linear Algebra Appl. **280**, 21–37 (1998)
1621. Hans-Gill, R.J., Madhu, R., Ranjeet, S.: On conjectures of Minkowski and Woods for $n = 7$. J. Number Theory **129**, 1011–1033 (2009)
1622. Hans-Gill, R.J., Madhu, R., Ranjeet, S.: On conjectures of Minkowski and Woods for $n = 8$. Acta Arith. **147**, 337–385 (2011)
1623. Hao, F.H., Parry, C.J.: The Fermat equation over quadratic fields. J. Number Theory **19**, 115–130 (1984)
1624. Hardy, G.H.: A problem of diophantine approximation. J. Indian Math. Soc. **11**, 162–166 (1919)
1625. Hardy, G.H.: Ramanujan. Cambridge University Press (1940) [Reprint: Chelsea (1959)]
1626. Hardy, G.H., Heilbronn, H.: Edmund Landau. J. Lond. Math. Soc. **13**, 302–310 (1938) [[2447], **1**, 15–23; [1755], 351–359.]
1627. Hardy, G.H., Littlewood, J.E.: Some problems of 'Partitio Numerorum': III. On the expression of a number as a sum of primes. Acta Math. **44**, 1–70 (1923)
1628. Hardy, G.H., Ramanujan, S.: Asymptotic formulae in combinatory analysis. Proc. London Math. Soc. **17**, 75–115 (1918)
1629. Hardy, J.: Sums of two squares in a quadratic ring. Acta Arith. **14**, 357–369 (1967/8)
1630. Hardy, J.: A note on the representability of binary quadratic forms with Gaussian integer coefficients as sums of squares of two linear forms. Acta Arith. **15**, 77–84 (1968)
1631. Hardy, K. Hudson, R.H., Richman,D., Williams, K.S.: Determination of all imaginary cyclic quartic fields with class number 2. Trans. Amer. Math. Soc. **311**, 1–55 (1989)
1632. Harman, G., Lewis, P.: Gaussian primes in narrow sectors. Mathematika **48**, 119–135 (2001)
1633. Harper, M.: $Z[\sqrt{14}]$ is Euclidean. Canad. J. Math. **56**, 55–70 (2004)
1634. Harper, M., Murty, M.R.: Euclidean rings of algebraic integers. Canad. J. Math. **56**, 71–76 (2004)
1635. Hart, W., Harvey, D., Ong, W.: Irregular primes to two billion. Math. Comp. **86**, 3031–3049 (2017)
1636. Hartman, P.: Aurel Wintner. J. Lond. Math. Soc. **37**, 483–503 (1962)
1637. Hashimoto, K.-I., Hoshi, A.: Families of cyclic polynomials obtained from geometric generalization of Gaussian period relations. Math. Comp. **74**, 1519–1530 (2005)

1638. Hashimoto, K.-I., Hoshi, A.: Geometric generalization of Gaussian period relations with application to Noether's problem for meta-cyclic groups. Tokyo J. Math. **28**, 13–32 (2005)
1639. Hashimoto, Y.: Asymptotic formulas for class number sums of indefinite binary quadratic forms on arithmetic progressions. Internat. J. Number Th. **9**, 27–51 (2013)
1640. Hasse, H.: Über die Darstellbarkeit von Zahlen durch quadratische Formen im Körper der rationalen Zahlen. J. Reine Angew. Math. **152**, 129–148 (1923) [[1713], **1**, 3–22.]
1641. Hasse, H.: Über die Äquivalenz quadratischer Formen im Körper der rationalen Zahlen. J. Reine Angew. Math. **152**, 205–224 (1923) [[1713], **1**, 23–42.]
1642. Hasse, H.: Symmetrische Matrizen im Körper der rationalen Zahlen. J. Reine Angew. Math. **153**, 12–43 (1924) [[1713], **1**, 43–74.]
1643. Hasse, H.: Zur Theorie des quadratischen Hilbertschen Normenrestsymbols in algebraischen Körpern. J. Reine Angew. Math. **153**, 76–93 (1924)
1644. Hasse, H.: Darstellbarkeit von Zahlen durch quadratische Formen in einem beliebigen algebraischen Zahlkörper. J. Reine Angew. Math. **153**, 113–130 (1924) [[1713], **1**, 75–92.]
1645. Hasse, H.: Äquivalenz quadratischer Formen in einem beliebigen algebraischen Zahlkörper. J. Reine Angew. Math. **153**, 158–162 (1924) [[1713], **1**, 93–97.]
1646. Hasse, H.: Zur Theorie des Hilbertschen Normenrestsymbols in algebraischen Zahlkörpern. II. Fall eines ungeraden l. J. Reine Angew. Math. **153**, 184–191 (1924)
1647. Hasse, H.: Das allgemeine Reziprozitätsgesetz und seine Ergänzungssätze in beliebigen algebraischen Zahlkörpern für gewisse, nicht-primäre Zahlen. J. Reine Angew. Math. **153**, 192–207 (1924) [[1713], **1**, 217–232.]
1648. Hasse, H.: Direkter Beweis des Zerlegungs- und Vertauschungssatzes für das Hilbertsche Normenrestsymbol in einem algebraischen Zahlkörper im Falle eines Primteilers $\mathfrak{l}$ des Relativgrades l. J. Reine Angew. Math. **154**, 20–35 (1925) [[1713], **1**, 118–133.]
1649. Hasse, H.: Über das allgemeine Reziprozitätsgesetz der l-ten Potenzreste im Körper k_ζ der l-ten Einheitswurzeln und in Oberkörpern von k_ζ. J. Reine Angew. Math. **154**, 96–109 (19235 [[1713], **1**, 233–246.]
1650. Hasse, H.: Das allgemeine Reziprozitätsgesetz der l-ten Potenzreste für beliebige, zu l prime Zahlen in gewissen Oberkörpern des Körpers der l-ten Einheitswurzeln. J. Reine Angew. Math. **154**, 1925, 199–214 (1925) [[1713], **1**, 253–268.]
1651. Hasse, H.: Der zweite Ergänzungssatz zum Reziprozitätsgesetz der l-ten Potenzreste für beliebige zu l prime Zahlen in gewissen Oberkörpern des Körpers der l-ten Einheitswurzeln. J. Reine Angew. Math. **154**, 215–218 (1925)
1652. Hasse, H.: Zwei Existenztheoreme über algebraische Zahlkörper. Math. Ann. **95**, 229–238 (1926)
1653. Hasse, H.: Ein weiteres Existenztheorem in der Theorie der algebraischen Zahlkörper. Math. Z. **24**, 149–160 (1926)
1654. Hasse, H.: Über die Einzigkeit der beiden Fundamentalsätze der elementaren Zahlentheorie. J. Reine Angew. Math. **155**, 199–220 (1926) [[1713], **3**, 425–446.]
1655. Hasse, H.: Bericht über neuere Untersuchungen und Probleme aus der Theorie der algebraischen Zahlkörper, I, Klassenkörpertheorie. Jahresber. Deutsch. Math.-Verein. **35**, 1–55 (1926) [[1665].]
1656. Hasse, H.: Bericht über neuere Untersuchungen und Probleme aus der Theorie der algebraischen Zahlkörper, Ia, Beweise zu I. Jahresber. Deutsch. Math.-Verein. **36**, 233–311 (1927) [[1665].]
1657. Hasse, H.: Neue Begründung der komplexen Multiplikation. I. Einordnung in die allgemeine Klassenkörpertheorie. J. Reine Angew. Math. **157**, 115–139 (1927) [[1713], **2**, 3–27.]
1658. Hasse, H.: Das Eisensteinsche Reziprozitätsgesetz der n-ten Potenzreste. Math. Ann. **97**, 599–623 (1927) [[1713], **1**, 269–293.]
1659. Hasse, H.: Existenz gewisser algebraischer Zahlkörper. SBer. Preuß. Akad. Wiss. Berlin **1927**, 229–234.
1660. Hasse, H.: Über eindeutige Zerlegung in Primelemente oder in Primhauptideale in Integritätsbereichen. J. Reine Angew. Math. **159**, 3–12 (1928)

1661. Hasse, H.: Zum expliziten Reziprozitätsgesetz. Abh. Math. Semin. Univ. Hambg **7**, 52–63 (1929) [[1713], **1**, 343–354.]
1662. Hasse, H.: Ein Satz über relativ-Galoische Zahlkörper und seine Anwendung auf relativ-Abelsche Zahlkörper. Math. Z. **31**, 559–564 (1930) [[1713], **1**, 417–422.]
1663. Hasse, H.: Arithmetische Theorie der kubischen Zahlkörper auf klassenkörpertheoretischer Grundlage. Math. Z. **31**, 565–582 (1930); corr. p.799. [[1713], **1**, 423–440.]
1664. Hasse, H.: Bericht über neuere Untersuchungen und Probleme aus der Theorie der algebraischen Zahlkörper, II, Reziprozitätsgesetz. Jahresber. Deutsch. Math.-Verein. **VI** Erg.Bd. (1930) [[1665]]
1665. Hasse, H.: Bericht über neuere Untersuchungen und Probleme aus der Theorie der algebraischen Zahlkörper. Teubner (1930) [2nd ed. Physica-Verlag (1965); reprint: Physica-Verlag (1970)]
1666. Hasse, H.: Neue Begründung und Verallgemeinerung der Theorie des Normenrestsymbols. J. Reine Angew. Math. **162**, 134–144 (1930) [[1713], **1**, 134–144.]
1667. Hasse, H.: Die Normenresttheorie relativ-Abelscher Zahlkörper als Klassenkörpertheorie im Kleinen. J. Reine Angew. Math. **162**, 145–154 (1930) [[1713], **1**, 145–154.]
1668. Hasse, H.: Führer, Diskriminante und Verzweigungskörper relativ-Abelscher Zahlkörper. J. Reine Angew. Math. **162**, 169–184 (1930) [[1713], **1**, 161–176.]
1669. Hasse, H.: Neue Begründung der komplexen Multiplikation, II. J. Reine Angew. Math. **165**, 64–88 (1931) [[1713], **2**, 28–52.]
1670. Hasse, H.: Über $\mathfrak{p}$-adische Schiefkörper und ihre Bedeutung für die Arithmetik hyperkomplexer Zahlsysteme. Math. Ann. **104**, 495–534 (1931) [[1713], **1**, 455–494.]
1671. Hasse, H.: Beweis eines Satzes und Widerlegung einer Vermutung über das allgemeine Normenrestsymbol. Nachr. Ges. Wiss. Göttingen **1931**, 64–69 [[1713], **1**, 155–160.]
1672. Hasse, H.: Theorie der zyklischen Algebren über einem algebraischen Zahlkörper. Nachr. Ges. Wiss. Göttingen **1931**, 70–79.
1673. Hasse, H.: Zum Hauptidealsatz der komplexen Multiplikation. Monatsh. Math. Phys. **38**, 315-322 (1931)
1674. Hasse, H.: Ein Satz über die Ringklassenkörper der komplexen Multiplikation. Monatsh. Math. Phys. **38**, 323–330 (1931)
1675. Hasse, H.: Das Zerlegungsgesetz für die Teiler des Moduls in den Ringklassenkörpern der komplexen Multiplikation. Monatsh. Math. Phys. **38**, 331–344 (1931)
1676. Hasse, H.: Theory of cyclic algebras over an algebraic number field. Trans. Amer. Math. Soc. **34**, 171–214 (1932); add. 727–730
1677. Hasse, H.: Die Struktur der R.Brauerschen Algebrenklassengruppe über einem algebraischen Zahlkörper. Insbesondere Begründung der Theorie des Normenrestsymbols und Herleitung des Reziprozitätsgesetzes mit nichtkommutativen Hilfsmitteln. Math. Ann. **107**, 731–760 (1933) [[1713], **1**, 501–530.]
1678. Hasse, H.: Vorlesungen über Klassenkörpertheorie, preprint (1933). [In a book form: Physica-Verlag (1967)]
1679. Hasse, H.: Abstrakte Begründung der komplexen Multiplikation und Riemannsche Vermutung in Funktionenkörper. Abh. Math. Semin. Univ. Hambg **10**, 325–348 (1934) [[1713], **2**, 109–132.]
1680. Hasse, H.: Explizite Konstruktion zyklischer Klassenkörper. Math. Ann. **109**, 191–195 (1934) [[1713], **1**, 441–445.]
1681. Hasse, H.: Elementarer Beweis des Hauptsatzes über ternäre quadratische Formen mit rationalen Koeffizienten. J. Reine Angew. Math. **172**, 129–132 (1934)
1682. Hasse, H.: Normenresttheorie galoisscher Zahlkörper mit Anwendungen auf Führer und Diskriminante abelscher Zahlkörper. J. Fac. Sci. Univ. Tokyo **2**, 477–498 (1934) [[1713], **1**, 182–203.]
1683. Hasse, H.: Zur Theorie der abstrakten elliptischen Funktionenkörper, I, Die Struktur der Gruppe der Divisorenklassen endlicher Ordnung. J. Reine Angew. Math. **175**, 55–62 (1936) [[1713], **2**, 223–230.]

1684. Hasse, H.: Zur Theorie der abstrakten elliptischen Funktionenkörper, II, Automorphismen und Meromorphismen. J. Reine Angew. Math. 175, 1936, 69–88 (1923) [[1713], **2**, 231–250.]
1685. Hasse, H.: Zur Theorie der abstrakten elliptischen Funktionenkörper, III, Die Struktur des Meromorphismenringes. Die Riemannsche Vermutung. J. Reine Angew. Math. **175**, 193–208 (1936) [[1713], **2**, 251–266.]
1686. Hasse, H.: Über die Diskriminante auflösbarer Körper von Primzahlgrad. J. Reine Angew. Math. **176**, 12–17 (1937)
1687. Hasse, H.: Simultane Approximation algebraischer Zahlen durch algebraische Zahlen. Monatsh. Math. Phys. **48**, 205–225 (1939)
1688. Hasse, H.: Produktformeln für verallgemeinerte Gausssche Summen und ihre Anwendung auf die Klassenzahlformel für reelle quadratische Zahlkörper. Math. Z. **46**, 303–314 (1940) [[1713], **3**, 3–14.]
1689. Hasse, H.: Invariante Kennzeichnung relativ-Abelscher Zahlkörper mit vorgegebener Galoisgruppe über einem Teilkörper des Grundkörpers. Abh. Deutsch. Akad. Wiss., Berlin, Kl. Math. Nat. **1947(8)**, 1–56 (1947)
1690. Hasse, H.: Arithmetische Bestimmung von Grundeinheit und Klassenzahl in zyklischen kubischen und biquadratischen Zahlkörpern. Abh. Deutsch. Akad. Wiss., Berlin, Kl. Math. Nat. 1948, nr. 2, 1–95 (1923)
1691. Hasse, H.: Die Einheitengruppe in einem total-reellen nichtzyklischen kubischen Zahlkörper und in zugehörigen bikubischen Normalkörper. Arch. Math. (Basel) **1**, 42–46 (1948)
1692. Hasse, H.: Existenz und Mannigfaltigkeit abelscher Algebren mit vorgegebener Galoisgruppe über einem Teilkörper des Grundkörpers, I. Math. Nachr. **1**, 40–61 (1948)
1693. Hasse, H.: Kurt Hensel zum Gedächtnis. J. Reine Angew. Math. **187**, 1–13 (1949)
1694. Hasse, H.: Zahlentheorie. Akademie-Verlag (1949); 2nd ed. (1963); 3rd ed. (1969) [English translation: Number Theory, Akademie-Verlag (1979), Springer (1980)]
1695. Hasse, H.: Zum Existenzsatz von Grunwald in der Klassenkörpertheorie. J. Reine Angew. Math. **188**, 40–64 (1950)
1696. Hasse, H.: Die Einheitengruppe in einem totalreellen nicht-zyklischen kubischen Zahlkörper und im zugehörigen bikubischen Normalkörper. Misc. Acad. Berolinensia **1**, 1950, 1–24 (1950)
1697. Hasse, H.: Allgemeine Theorie der Gaußschen Summen in algebraischen Zahlkörpern. Abh. Deutsch. Akad. Wiss., Berlin, Kl. Math. Nat. **1951(1)** 1–23. [[1713], **3**, 15–34.]
1698. Hasse, H.: Zur Geschlechtertheorie in quadratischen Zahlkörpern. J. Math. Soc. Japan **3**, 45–51 (1951)
1699. Hasse, H.: Zur Arbeit von Šafarevič über das allgemeine Reziprozitätsgesetz. Math. Nachr. **5**, 301–327 (1951)
1700. Hasse, H.: Über die Klassenzahl abelscher Zahlkörper. Akademie-Verlag (1952) [Reprint: Akademie-Verlag (1985), Springer (1985)]
1701. Hasse, H.: Gausssche Summen zu Normalkörpern über endlich-algebraischen Zahlkörpern. Abh. Deutsch. Akad. Wiss., Berlin, Kl. Math. Nat. **1952(1)**, 1–19
1702. Hasse, H.: Artinsche Führer, Artinsche L-Funktionen und Gausssche Summen über endlich-algebraischen Zahlkörpern. Acta Salman. Ci. **1954(4)** 1–113
1703. Hasse, H.: Die dyadische Einheitsoperatorengruppe der 2^n-ten Einheitswurzeln nebst Anwendung auf die Klassenzahl seines grössten reellen Teilkörpers. Rev. Fac. Sci. Univ. Istanbul A **20**, 7–16 (1955)
1704. Hasse, H.: Der 2^n-te Potenzcharakter von 2 im Körper der 2^n-ten Einheitswurzeln. Rend. Circ. Mat. Palermo (2) **7**, 185–244 (1958)
1705. Hasse, H.: Kurt Hensels entscheidender Anstoss zur Entdeckung des Lokal-Global-Prinzips. J. Reine Angew. Math. **209**, 3–4 (1962)
1706. Hasse, H.: Vandiver's congruence for the relative class number of the pth cyclotomic field. J. Math. Anal. Appl. **15**, 87–90 (1966)
1707. Hasse, H.: Über die Klassenzahl des Körpers $P(\sqrt{-p})$ mit einer Primzahl $p \equiv 1 \bmod 2^3$. Aequationes Math. **3**, 165–169 (1969)

1708. Hasse, H.: A supplement to Leopoldt's theory of genera in abelian number fields. J. Number Theory **1**, 4–7 (1969)
1709. Hasse, H.: Über die Klassenzahl des Körpers $P(\sqrt{-2p})$ mit einer Primzahl $p \neq 2$. J. Number Theory **1**, 231–234 (1969)
1710. Hasse, H.: Über die Teilbarkeit durch 2^3 der Klassenzahl imaginärquadratischer Zahlkörper mit genau zwei verschiedenen Diskriminantenprimteilern. J. Reine Angew. Math. **241**, 1–6 (1970)
1711. Hasse, H.: Über Teilbarkeit durch 2^3 der Klassenzahl quadratischer Zahlkörper mit genau zwei verschiedenen Diskriminantenprimteilern. Math. Nachr. **46**, 61–70 (1970)
1712. Hasse, H.: An algorithm for determining the structure of the 2-Sylow-subgroups of the divisor class group of a quadratic number field. Symposia Math. **15**, 341–352 (1975)
1713. Hasse, H.: Mathematische Abhandlungen, **1–3**. de Gruyter (1975)
1714. Hasse, H., Schmidt, F.K.: Die Struktur diskret bewerteter Körper. J. Reine Angew. Math. **170**, 4–63 (1933) [[1713], **3**, 460–519.]
1715. Hasse, H., Scholz, A.: Zur Klassenkörpertheorie auf Takagischer Grundlage. Math. Z. **29**, 60–69 (1928)
1716. Hasse, H., Suetuna, Z.: Ein allgemeines Teilerproblem der Idealtheorie. J. Fac. Sci. Univ. Tokyo **2**, 133–154 (1931)
1717. Hathaway, A.S.: A memoir in the theory of numbers. Amer. J. Math. **9**, 162–179 (1887)
1718. Haubrich, R.: Zur Entstehung der algebraischen Zahlentheorie Richard Dedekinds. Dissertation, Univ. Göttingen (1992)
1719. Hauser, K. et al.: Oswald Teichmüller — Leben und Werk. Jahresber. Deutsch. Math.-Verein. **94**, 1–39 (1992)
1720. Hausmann, B.A.: A new simplification of Kronecker's method of factorization of polynomials. Amer. Math. Monthly **44**, 574–576 (1937)
1721. Hawkins, T.: The Mathematics of Frobenius in Context. Springer (2013)
1722. Hayashi, H.: On a simple proof of Eisenstein's reciprocity law. Mem. Fac. Sci. Kyushu Univ. **28**, 93–99 (1974)
1723. Hayashi, H.: Note on a product formula for the Bayad function and a law of quadratic reciprocity. Acta Arith. **149**, 321–336 (2011)
1724. Hays, J.H.: Reductions of ideals in commutative rings. Trans. Amer. Math. Soc. **177**, 51–63 (1973)
1725. Hazewinkel, M.: Local class field theory is easy. Adv. Math. **18**, 148–181 (1975)
1726. Heath-Brown, D.R.: Lattice points in the sphere. In: Number Theory in Progress, **2**, 883–892. de Gruyter (1999)
1727. Heath-Brown, D.R.: Imaginary quadratic fields with class group exponent 5. Forum Math. **20**, 275–283 (2008)
1728. Heath-Brown, D.R., Huxley, M.N.: George Greaves, 1941–2008. Bull. Lond. Math. Soc. 43, 2011, 203–205 (1992)
1729. Hecke, E.: Über die Konstruktion der Klassenkörper reeller quadratischer Körper mit Hilfe von automorphen Funktionen. Nachr. Ges. Wiss. Göttingen **1910**, 619–623 [[1746], 64–68.]
1730. Hecke, E.: Höhere Modulfunktionen und ihre Anwendung auf die Zahlentheorie. Math. Ann. **71**, 1–37 (1912) [[1746], 21–57.]
1731. Hecke, E.: Über die Konstruktion relativ-abelscher Zahlkörper durch Modulfunktionen von zwei Variablen. Math. Ann. **74**, 465–510 (1913) [[1746], 69–114.]
1732. Hecke, E.: Über die Zetafunktion beliebiger algebraischer Zahlkörper. Nachr. Ges. Wiss. Göttingen **1917**, 77–89 [[1746], 159–171.]
1733. Hecke, E.: Über eine neue Anwendung der Zetafunktionen auf die Arithmetik der Zahlkörper. Nachr. Ges. Wiss. Göttingen **1917**, 90–95 [[1746], 215–234.]
1734. Hecke, E.: Über die L-Funktionen und den Dirichletschen Primzahlsatz für einen beliebigen Zahlkörper. Nachr. Ges. Wiss. Göttingen **1917**, 299–318 [[1746], 178–197.]
1735. Hecke, E.: Über die Kroneckersche Grenzformel für reelle quadratische Körper und die Klassenzahl relativ-Abelscher Körper. Verhandl. Naturf. Ges. Basel **28**, 363–372 (1917) [[1746], 198–207.]

1736. Hecke, E.: Eine neue Art von Zetafunktionen und ihre Beziehungen zur Verteilung der Primzahlen. Math. Z. **1**, 357–376 (1918) [[1746], 215–234.]
1737. Hecke, E.: Reziprozitätsgesetz und Gausssche Summen in quadratischen Zahlkörpern. Nachr. Ges. Wiss. Göttingen **1919**, 265–278. [[1746], 235–248.]
1738. Hecke, E.: Eine neue Art von Zetafunktionen und ihre Beziehungen zur Verteilung der Primzahlen, II. Math. Z. **6**, 11–51 (1920) [[1746], 249–289.]
1739. Hecke, E.: Bestimmung der Klassenzahl einer neuen Reihe von algebraischen Zahlkörpern. Nachr. Ges. Wiss. Göttingen **1921**, 1–13. [[1746], 290–312.]
1740. Hecke, E.: Über analytische Funktionen und die Verteilung von Zahlen mod. eins. Abh. Math. Semin. Univ. Hambg **1**, 54–76 (1922) [[1746], 381–404.]
1741. Hecke, E.: Analytische Funktionen und algebraische Zahlen, I. Abh. Math. Semin. Univ. Hambg **1**, 102–126 (1922) [[1746], 336–360.]
1742. Hecke, E.: Vorlesungen über die Theorie der algebraischen Zahlen. Akademische Verlagsgesellschaft (1923) [2nd ed. Gerst & Portig K.-G., (1954); Reprint: Chelsea (1970); English translation: Lectures on the Theory of Algebraic Numbers, Springer (1981)]
1743. Hecke, E.: Analytische Funktionen und algebraische Zahlen, II. Abh. Math. Semin. Univ. Hambg **3**, 213–236 (1924) [[1746], 381–404.]
1744. Hecke, E.: Darstellung von Klassenzahlen als Perioden von Integralen 3. Gattung aus dem Gebiet der elliptischen Modulfunktionen. Abh. Math. Semin. Univ. Hambg **4**, 211–223 (1925) [[1746], 381–404.]
1745. Hecke, E.: Analysis und Zahlentheorie, DMG, Vieweg (1987)
1746. Hecke, E.: Mathematische Werke. Vandenhoeck & Ruprecht (1959); 2nd ed. (1970); 3rd ed. (1983)
1747. Heegner, K.: Diophantische Analysis und Modulfunktionen. Math. Z. **56**, 227–253 (1952)
1748. Heffter, L.: Ludwig Stickelberger. Jahresber. Deutsch. Math.-Verein. **47**, 79–86 (1937)
1749. Heidaryan, B., Rajaei, A.: Biquadratic Pólya fields with only one quadratic Pólya subfield. J. Number Theory **143**, 279–285 (2014)
1750. Heilbronn, H.: On the class–number in imaginary quadratic fields. Quart. J. Math., Oxford ser. **5**, 150–160 (1934) [[1755], 177–187.]
1751. Heilbronn, H.: On Dirichlet series which satisfy a certain functional equation. Quart. J. Math., Oxford ser. **9**, 194–195 (1938) [[1755], 343–344.]
1752. Heilbronn, H.: On Euclid's algorithm in real quadratic fields. Proc. Cambridge Philos. Soc. **34**, 521–526 (1938) [[1755], 345–350.]
1753. Heilbronn, H.: On Euclid's algorithm in cubic self-conjugated fields. Proc. Cambridge Philos. Soc. **46**, 377–382 (1950) [[1755], 414–419.]
1754. Heilbronn, H.: On Euclid's algorithm in cyclic fields. Canad. J. Math. **3**, 257–268 (1951) [[1755], 428–438.]
1755. Heilbronn, H.: Collected Papers. J. Wiley (1988)
1756. Heilbronn, H., Linfoot, E.H.: On the imaginary quadratic corpora of class-number one. Quart. J. Math., Oxford ser. **5**, 293–301 (1934) [[1755], 188–196.]
1757. Heinhold, J.: Verallgemeinerung und Verschärfung eines Minkowskischen Satzes. Math. Z. **44**, 659–688 (1939)
1758. Heinhold, J.: Oskar Perron. Jahresber. Deutsch. Math.-Verein. **90**, 184–199 (1988)
1759. Hendy, M.D.: Prime quadratics associated with complex quadratic fields of class number two. Proc. Amer. Math. Soc. **43**, 253–260 (1974)
1760. Hendy, M.D.: The distribution of ideal class numbers of real quadratic fields. Math. Comp. **29**, 1129–1134 (1975); corr. **30**, p. 679 (1976)
1761. Henkin, L.: In memoriam: Raphael Mitchell Robinson. Bull. Symbolic Logic **1**, 340–343 (1995)
1762. Henniart, G.: Sur les lois de réciprocité explicites, I. J. Reine Angew. Math. **329**, 177–203 (1981)
1763. Hensel, K.: Arithmetische Untersuchungen über Discriminanten und ihre ausserwesentlichen Theiler, Dissertation, Univ. Berlin (1884)

1764. Hensel, K.: Untersuchung der ganzen algebraischen Zahlen eines gegebenen Gattungsbereiches für einen beliebigen algebraischen Primdivisor. J. Reine Angew. Math. **101**, 99–141 (1887)
1765. Hensel, K.: Ueber die Darstellung der Zahlen eines Gattungsbereiches für einen beliebigen Primdivisor. J. Reine Angew. Math. **103**, 230–237 (1888)
1766. Hensel, K.: Ueber Gattungen, welche durch Composition aus zwei anderen Gattungen entstehen. J. Reine Angew. Math. **105**, 329–344 (1889)
1767. Hensel, K.: Ueber die Darstellung der Determinante eines Systems, welches aus zwei andern componirt ist. Acta Math. **14**, 317–319 (1891)
1768. Hensel, K.: Untersuchung der Fundamentalgleichung einer Gattung für eine reelle Primzahl als Modul und Bestimmung der Theiler ihrer Discriminante. J. Reine Angew. Math. **113**, 61–83 (1894)
1769. Hensel, K.: Arithmetische Untersuchungen über die gemeinsamen Discriminantentheiler einer Gattung. J. Reine Angew. Math. **113**, 128–160 (1894)
1770. Hensel, K.: Ueber die Bestimmung der Discriminante eines algebraischen Körpers. Nachr. Ges. Wiss. Göttingen **1897**, 247–253
1771. Hensel, K.: Ueber die Fundamentalgleichung und die ausserwesentliche Discriminantentheiler eines algebraischen Körpers. Nachr. Ges. Wiss. Göttingen **1897**, 254–260
1772. Hensel, K.: Ueber eine neue Begründung der Theorie der algebraischen Zahlen. Jahresber. Deutsch. Math.-Verein. **6**, 83–88 (1899)
1773. Hensel, K.: Ueber diejenigen algebraischen Körper, welche aus zwei anderen componirt sind. J. Reine Angew. Math. **120**, 99–108 (1899)
1774. Hensel, K.: Über die Entwickelung der algebraischen Zahlen in Potenzreihen. Math. Ann. **55**, 301–336 (1902)
1775. Hensel, K.: Neue Grundlagen der Arithmetik. J. Reine Angew. Math. **127**, 51–84 (1904)
1776. Hensel, K.: Über eine neue Begründung der Theorie der algebraischen Zahlen. J. Reine Angew. Math. **128**, 1–32 (1904)
1777. Hensel, K.: Über die zu einem algebraischen Körper gehörigen Invarianten. J. Reine Angew. Math. **129**, 68–85 (1905)
1778. Hensel, K.: Über die arithmetischen Eigenschaften der algebraischen und transzendenten Zahlen. Jahresber. Deutsch. Math.-Verein. **14**, 545–558 (1905)
1779. Hensel, K.: Über die arithmetischen Eigenschaften der Zahlen. Jahresber. Deutsch. Math.-Verein. **16**, 299–319, 388–393, 473–496 (1907)
1780. Hensel, K.: Theorie der algebraischen Zahlen. Bd.I, Teubner (1908)
1781. Hensel, K.: Über die zu einer algebraischer Gleichung gehörigen Auflösungskörper. J. Reine Angew. Math. **136**, 183–209 (1909)
1782. Hensel, K.: Gedächtnisrede auf Ernst Eduard Kummer. In: [4], 1–37. [[2377],**1**, 31–69.]
1783. Hensel, K.: Zahlentheorie, Göschen (1913)
1784. Hensel, K.: Über die Grundlagen einer neuen Theorie der quadratischer Zahlkörper. J. Reine Angew. Math. **144**, 57–70 (1914)
1785. Hensel, K.: Die Exponentialdarstellung der Zahlen eines algebraischen Zahlkörpers für den Bereich eines Primdivisors. In: [565], 61–75
1786. Hensel, K.: Untersuchungen der Zahlen eines algebraischen Körpers für den Bereich eines beliebigen Primteilers. J. Reine Angew. Math. **145**, 92–113 (1915)
1787. Hensel, K.: Die multiplikative Darstellung der algebraischen Zahlen für den Bereich eines beliebigen Primteilers. J. Reine Angew. Math. **146**, 189–215 (1916)
1788. Hensel, K.: Untersuchung der Zahlen eines algebraischen Körpers für eine beliebige Primteilerpotenz als Modul. J. Reine Angew. Math. **146**, 216–228 (1916)
1789. Hensel, K.: Allgemeine Theorie der Kongruenzklassgruppen und ihrer Invarianten in algebraischen Körpern. J. Reine Angew. Math. **147**, 1–15 (1917)
1790. Hensel, K.: Die Verallgemeinerung des Legendreschen Symboles für allgemeine algebraische Körper. J. Reine Angew. Math. **147**, 233–248 (1917)
1791. Hensel, K.: Eine neue Theorie der algebraischen Zahlen. Math. Z. **2**, 433–452 (1918)

1792. Hensel, K.: Über die Zerlegung der Primteiler in relativ zyklischen Körpern, nebst einer Anwendung auf die Kummerschen Körper. J. Reine Angew. Math. **151**, 112–120 (1921)
1793. Hensel, K.: Die Zerlegung der Primteiler eines beliebigen Zahlkörpers in einem auflösbaren Oberkörper. J. Reine Angew. Math. **151**, 200–209 (1921)
1794. Hensel, K.: Zur multiplikativen Darstellung der algebraischen Zahlen für den Bereich eines Primteilers. J. Reine Angew. Math. **151**, 210–212 (1921)
1795. Hensel, K.: Über die Normenreste und Nichtreste in den allgemeinsten relativ Abelschen Zahlkörpern. Math. Ann. **85**, 1–10 (1922)
1796. Hensel, K.: Über ein neues Normenrestsymbol und seine Anwendung auf die Theorie der Normenreste in allgemeinen algebraischen Körpern. J. Reine Angew. Math. **152**, 225–234 (1923)
1797. Hensel, K.: Paul Bachmann und sein Lebenswerk. Jahresber. Deutsch. Math.-Verein. **36**, 31–73 (1927)
1798. Hensel, K.: Über eindeutige Zerlegung in Primelemente. J. Reine Angew. Math. **158**, 195–198 (1927)
1799. Hensel, K., Hasse, H.: Über die Normenreste eines relativ-zyklischen Körpers vom Primzahlgrad l nach einem Primteiler $\mathfrak{l}$ von l. Math. Ann. **90**, 262–278 (1923)
1800. Herbrand, J.: Nouvelle démonstration et généralisation d'un théorème de Minkowski. C. R. Acad. Sci. Paris **191**, 1282–1285 (1930)
1801. Herbrand, J.: Sur la théorie des corps des nombres de degré infini. C. R. Acad. Sci. Paris **193**, 504–506 (1931)
1802. Herbrand, J.: Sur la théorie des groupes de décomposition, d'inertie et de ramification. J. Math. Pures Appl. (9) **10**, 481–498 (1931)
1803. Herbrand, J.: Sur les classes des corps circulaires. J. Math. Pures Appl. (9) **11**, 417–441 (1932)
1804. Herbrand, J.: Théorie arithmétique des corps des nombres de degré infini, Extensions algébriques finies de corps infinis. Math. Ann. **106**, 473-501 (1932)
1805. Herbrand, J.: Une propriété du discriminant des corps algébriques. Ann. Sci. Éc. Norm. Supér. (3) **49**, 105–112 (1932)
1806. Herbrand, J.: Théorie arithmétique des corps des nombres de degré infini, II, Extensions algébriques de degré infini. Math. Ann. **108**, 699–717 (1933)
1807. Herbrand, J.: Sur les théorèmes du genre principal et des idéaux principaux. Abh. Math. Semin. Univ. Hambg **9**, 84–92 (1933)
1808. Herbrand, J., Chevalley, C.: Nouvelle démonstration du théorème d'existence en théorie du corps de classes. C. R. Acad. Sci. Paris **193**, 814–815 (1931)
1809. Herglotz, G.: Über das quadratische Reziprozitätsgesetz in imaginären quadratischen Zahlkörpern. Ber. Verhandl. Sächsische Akad. Wiss. Leipzig, Math.-Nat. Kl. **73**, 303-310 (1921)
1810. Herglotz, G.: Über einen Dirichletschen Satz. Math. Z. **12**, 255–261 (1922)
1811. Herglotz, G.: Über die Kroneckersche Grenzformel für reelle, quadratische Körper. Ber. Verhandl. Sächsische Akad. Wiss. Leipzig, Math.-Nat. Kl. **75**, 3–14, 31–37 (1923)
1812. Hermes, J.: Ueber die Teilung des Kreises in 65537 gleiche Teile. Nachr. Ges. Wiss. Göttingen **1894**, 170–186
1813. Hermite, C.: Note sur la réduction des fonctions homogènes à coefficients entiers et à deux indéterminées. J. Reine Angew. Math. **36**, 357–364 (1848) [[1826], **1**, 84–93.]
1814. Hermite, C.: Extraits de lettres de M.Ch.Hermite à M. Jacobi sur différents objects de la théorie des nombres, Premiére lettre. J. Reine Angew. Math. **40**, 261–277 (1850) [[1826], **1**, 100–121.]
1815. Hermite, C.: Extraits de lettres de M.Ch. Hermite à M. Jacobi sur différents objects de la théorie des nombres, Deuxiéme lettre. J. Reine Angew. Math. **40**, 279–290 (1850) [[1826], **1**, 122–135.]
1816. Hermite, C.: Sur l'introduction des variables continues dans la théorie des nombres. J. Reine Angew. Math. **41**, 191–216 (1851) [[1826], **1**, 164–192.]

1817. Hermite, C.: Sur la théorie des formes quadratiques, I. J. Reine Angew. Math. **47**, 313–342 (1854) [[1826], **1**, 200–233.]
1818. Hermite, C.: Sur la théorie des formes quadratiques, II. J. Reine Angew. Math. **47**, 343–368 (1854) [[1826], **1**, 234–263.]
1819. Hermite, C.: Extrait d'une lettre de M.C.Hermite à M.Borchardt sur le nombre limité d'irrationalités auxquelles se réduisent les racines des équations à coefficients entiers complexes d'un degré et d'un discriminant donnés. J. Reine Angew. Math. **53**, 182–192 (1857) [[1826], **1**, 414–428.]
1820. Hermite, C.: Sur l'invariant du 18e ordre des formes du cinquième degré et sur le role qu'il joue dans la résolution de l'équation du cinquiéme degré. J. Reine Angew. Math. **59**, 304–305 (1861) [[1826], **2**, 107–108.]
1821. Hermite, C.: Sur la théorie des formes quadratiques. C. R. Acad. Sci. Paris **55**, 684–692 (1862) [[1826], **2**, 255–263.]
1822. Hermite, C.: Sur la théorie des fonctions elliptiques et ses applications à l'arithmétique. J. Math. Pures Appl. (2) **7**, 25–40 (1862) [[1826], **2**, 108–124.]
1823. Hermite, C.: Sur les théorèmes de M. Kronecker relatifs aux formes quadratiques. J. Math. Pures Appl. (2) **9**, 149–159 (1864) [[1826], **2**, 241–254.]
1824. Hermite, C.: Sur quelques conséquences arithmétiques des formules de la théorie des fonctions elliptiques. Bull. Acad. Sci. St.-Petersbourg **29**, 325–352 (1864) [Acta Math. **5**, 297–330 (1884); [1826], **4**, 138–168]
1825. Hermite, C.: Remarques sur les formes quadratiques de déterminant négatif. Bull. Sci. Math. (2) **10**, 23–30 (1886) [[1826], **4**, 215–222.]
1826. Hermite, C.: Oeuvres. **–4**. Gauthier-Villars (1905–1917)
1827. Herrmann, O.: Über Hilbertsche Modulfunktionen und die Dirichletschen Reihen mit Eulerscher Produktentwicklung. Math. Ann. **127**, 357–400 (1954)
1828. Hida, H.: On the values of Hecke's L-functions at non-positive integers. J. Math. Soc. Japan **30**, 249–278 (1978)
1829. Hida, H.,: Elementary Theory of L-functions and Eisenstein Series. Cambridge University Press (1993)
1830. Hilbert, D.: Ueber die Irreducibilität ganzer rationaler Functionen mit ganzzahligen Coefficienten. J. Reine Angew. Math. **110**, 104–129 (1892) [[1846], **2**, 264–286.]
1831. Hilbert, D.: Ueber die Zerlegung der Ideale eines Zahlkörpers in Primideale. Math. Ann. **44**, 1–8 (1894) [[1846], **1**, 6–12.]
1832. Hilbert, D.: Grundzüge einer Theorie des Galois'schen Zahlkörpers. Nachr. Ges. Wiss. Göttingen **1894**, 224–236 [[1846], **1**, 13–23.]
1833. Hilbert, D.: Ueber den Dirichlet'schen biquadratischen Zahlkörper. Math. Ann. **45**, 309–340 (1894) [[1846], **1**, 24–52.]
1834. Hilbert, D.: Ein neuer Beweis des Kronecker'schen Fundamentalsatzes über Abel'sche Zahlkörper. Nachr. Ges. Wiss. Göttingen **1896**, 29–39 [[1846], **1**, 53–62.]
1835. Hilbert, D.: Ueber Diophantische Gleichungen. Nachr. Ges. Wiss. Göttingen **1897**, 48–54 [[1846], **2**, 384–389.]
1836. Hilbert, D.: Die Theorie der algebraischen Zahlkörper. Jahresber. Deutsch. Math.-Verein. **4**, 1897, 175–546. [[1846], **1**, 63–363; French translation (To this translation Humbert and Got appended several comments in Ann. Fac. Sci. Toulouse, Math. (3) **3**, 1–62 (1911)): Théorie des corps de nombres algébriques. Ann. Fac. Sci. Toulouse, Math. (3) **1**, 257–328 (1909) and (3) **2**, 225–456 (1910); reprint: Jacques Gabay (1991); English translation: The Theory of Algebraic Number Fields. Springer (1998)]
1837. Hilbert, D.: Ueber die Theorie der relativ-Abel'schen Zahlkörper. Nachr. Ges. Wiss. Göttingen **1898**, 370–399.
1838. Hilbert, D.: Ueber die Theorie des relativquadratischen Zahlkörpers. Jahresber. Deutsch. Math.-Verein. **6**, 88–94 (1899) [[1846], **1**, 364–369.]
1839. Hilbert, D.: Ueber die Theorie des relativquadratischen Zahlkörpers. Math. Ann. **51**, 1–127 (1899) [[1846], **1**, 370–482.]

1840. Hilbert, D.: Theorie der algebraischen Zahlkörper. Encyklop. Math. Wiss. **1**, 675–714, Teubner (1900)
1841. Hilbert, D.: Grundlagen der Geometrie. Teubner (1899); 2nd ed. (1903); 3rd ed. (1909); ..., 13th ed. (1987) [English translation: The Foundations of Geometry, The Open Court Publishing Co., (1902); reprint (1959)]
1842. Hilbert, D.: Mathematische Probleme. Nachr. Ges. Wiss. Göttingen **1900**, 253–297 [Arch. Math. Phys. (3) **1**, 44–63, 213–237 (1901); [1846], **3**, 290–329; English translation: Bull. Amer. Math. Soc. **8**, 437–479 (1902); reprint: Bull. Amer. Math. Soc. (N.S.) **37**, 407–436 (2000)]
1843. Hilbert, D.: Über die Theorie der relativ-Abelschen Zahlkörper. Acta Math. **26**, 99–132 (1902) [[1846], **1**, 483–509.]
1844. Hilbert, D.: Herrmann Minkowski. Nachr. Ges. Wiss. Göttingen **1909**, 72–101. [Math. Ann. **68**, 445–471 (1910); [1846], **3**, 339–364; [2882], 197–223.]
1845. Hilbert, D.: Adolf Hurwitz. Math. Ann. **83**, 161–172 (1921) [[1846], **3**, 370–377.]
1846. Hilbert, D.: Gesammelte Abhandlungen. **1–3**. Springer (1932–1935); 2nd ed. (1970) [Reprint: Chelsea (1965)]
1847. Hilbert, K.S.: Das allgemeine quadratische Reciprocitätsgesetz in ausgewählten Kreiskörpern der 2^n-ten Einheitswurzeln, Dissertation, Univ. Göttingen (1900)
1848. Hildebrand, A.: The prime number theorem via the large sieve. Mathematika **33**, 23–30 (1986)
1849. Hill, R.: A geometric proof of a reciprocity law. Nagoya Math. J. **137**, 77–144 (1995)
1850. Hille, E.: Thomas Hakon Gronwall — in memoriam. Amer. Math. Monthly **3**, 775-786 (1932)
1851. Hille, E.: Jacob David Tamarkin — his life and work. Bull. Amer. Math. Soc. **53**, 440–457 (1947)
1852. H.H. [Hilton, H.]: Camille Jordan. Proc. London Math. Soc. **21**, xliii–xliv (1923)
1853. Hinz, J.G.: Régions libres de zéros des fonctions dzétas de Hecke et la distribution des idéaux premiers. C. R. Acad. Sci. Paris **283**, 919–920 (1976)
1854. Hinz, J.G.: Eine Erweiterung des nullstellenfreien Bereiches der Heckeschen Zetafunktion und Primideale in Idealklassen. Acta Arith. **38**, 209–254 (1980/81)
1855. Hinz, J.G.: On the representation of even integers as sums of two almost-primes in algebraic number fields. Mathematika **29**, 93–108 (1982)
1856. Hinz, J.: Character sums in algebraic number fields. J. Number Theory **17**, 52–70 (1983)
1857. Hinz, J.: Character sums and primitive roots in algebraic number fields. Monatsh. Math. **95**, 275–286 (1983)
1858. Hinz, J.: The average order of magnitude of least primitive roots in algebraic number fields. Mathematika **30**, 11–25 (1983)
1859. Hinz, J.: Some applications of sieve methods in algebraic number fields. Manuscripta Math. **48**, 117–137 (1984)
1860. Hinz, J.: Über die Verteilung von primen primitiven Wurzeln in algebraischen Zahlkörpern. Monatsh. Math. **100**, 259–275 (1985)
1861. Hinz, J.: Cubefree ideal modulus character sums. Arch. Math. (Basel) **46**, 307–314 (1986)
1862. Hinz, J.: Least primitive roots modulo the square of a prime ideal. Mathematika **33**, 244–251 (1986)
1863. Hinz, J.: A note on Artin's conjecture in algebraic number fields. J. Number Theory **22**, 334–349 (1986)
1864. Hinz, J.: A generalization of Bombieri's prime number theorem to algebraic number fields. Acta Arith. **51**, 173–193 (1988)
1865. Hinz, J.G.: Chen's theorem in totally real algebraic number fields. Acta Arith. **58**, 335–361 (1991)
1866. Hinz, J.: On the prime ideal theorem. J. Indian Math. Soc. (N.S.) **59**, 243–260 (1993)
1867. Hinz, J., Lodemann, M.: On Siegel zeros of Hecke-Landau zeta-functions. Monatsh. Math. **118**, 231–248 (1994)

1868. Hirabayashi, M.: Inkeri's determinant for an imaginary abelian number field. Arch. Math. (Basel) **79**, 175–181 (2002)
1869. Hirabayashi, M.: A generalization of Newman's formula. Arch. Math. (Basel) **90**, 223–229 (2008)
1870. Hirsh, J., Washington, L.C.: p-adic continued fractions. The Ramanujan J. **25**, 389–403 (2011)
1871. Hlawka, E.: Carl Ludwig Siegel (31/12/1896–4/4/1981). J. Number Theory **20**, 373–404 (1985)
1872. Hlawka, E.: Nachruf auf Nikolaus Hofreiter. Monatsh. Math. **116**, 1993, 263–273 (1993)
1873. Hlawka, E.: Olga Taussky-Todd, 1906–1995. Monatsh. Math. **123**, 189–201 (1997)
1874. Hochschild, G.: Local class field theory. Ann. Math. (2) **51**, 331–347 (1950)
1875. Hochschild, G.: Note on Artin's Reciprocity Law. Ann. Math. (2) **52**, 694–701 (1985)
1876. Hoechsmann, K.: Zum Einbettungsproblem. J. Reine Angew. Math. **229**, 81–106 (1968)
1877. Hoffstein, J.: Some analytic bounds for zeta functions and class numbers. Invent. Math. **55**, 37–47 (1979)
1878. Hoffstein, J., Jochnowitz, N.: On Artin's conjecture and the class number of certain CM fields, I. Duke Math. J. **59**, 553–563 (1989)
1879. Hoffstein, J., Jochnowitz, N.: On Artin's conjecture and the class number of certain CM fields, II. Duke Math. J. **59**, 565–584 (1989)
1880. Hofreiter, N.: Quadratische Zahlkörper ohne Euklidischen Algorithmus. Math. Ann. **110**, 195–196 (1934)
1881. Hofreiter, N.: Quadratische Körper mit und ohne euklidischen Algorithmus. Monatsh. Math. Phys. **42**, 397–400 (1935)
1882. Hofreiter, N.: Über die Approximation von komplexen Zahlen. Monatsh. Math. Phys. **42**, 401–416 (1935)
1883. Hofreiter, N.: Diophantische Approximationen in imaginär quadratischen Zahlkörpern. Monatsh. Math. Phys. **45**, 175–190 (1937)
1884. Hofreiter, N.: Über die Kettenbruchentwicklung komplexer Zahlen und Anwendungen auf diophantische Approximationen. Monatsh. Math. Phys. **46**, 379–383 (1938)
1885. Hofreiter, N.: Nachruf auf Philipp Furtwängler. Monatsh. Math. Phys. **49**, 219–227 (1940)
1886. Hofreiter, N.: Diophantische Approximationen komplexer Zahlen. Monatsh. Math. Phys. **49**, 299–302 (1940)
1887. Höhn, G.: On a theorem of Garza regarding algebraic numbers with real conjugates. Internat. J. Number Th. **7**, 943–945 (2011)
1888. Höhn, G., Skoruppa, N.-P.: Un résultat de Schinzel. J. Théor. Nombres Bordeaux **5**, p. 185 (1993)
1889. Holzer, L.: Takagische Klassenkörpertheorie, Hassesche Reziprozitätsformel und Fermatsche Vermutung. J. Reine Angew. Math. **173**, 114–124 (1935)
1890. Holzer, L.: Minimal solutions of Diophantine equations. Canad. J. Math. **2**, 238–244 (1950)
1891. Honda, K.: Teiji Takagi: a biography — on the 100th anniversary of his birth. Comment. Math. Univ. St. Pauli **24**, 141–167 (1975/76)
1892. Hooley, C.: On Artin's conjecture. J. Reine Angew. Math. **225**, 209–220 (1967)
1893. Hooley, C.: On the Pellian equation and the class number of indefinite binary quadratic forms. J. Reine Angew. Math. **353**, 98–131 (1984)
1894. Horie, K.: Multiplicative groups in some infinite algebraic number fields. Arch. Math. (Basel) **54**, 32–35 (1990)
1895. Horie, K.: Ideal class groups of Iwasawa-theoretical abelian extensions over the rational field. J. Lond. Math. Soc. (2) **66**, 257–275 (2002)
1896. Horie, K.: Primary components of the ideal class group of the Z_p-extension over Q for typical inert primes. Proc. Japan Acad. Sci. **81**, 40–43 (2005)
1897. Horie, K., Horie, M.: CM-fields and exponents of their ideal class groups. Acta Arith. **55**, 157–170 (1990)
1898. Horie, K., Horie, M.: The narrow class groups of some Z_p-extensions over the rationals. Acta Arith. **135**, 159–180 (2008)

1899. Horie, K., Horie, M.: The ideal class group of the Z_{23}-extension over the rational field. Proc. Japan Acad. Sci. **85**, 155–159 (2009)
1900. Horie, K., Horie, M.: The narrow class groups of the Z_{17} and Z_{19}-extensions over the rational field. Abh. Math. Semin. Univ. Hambg **80**, 47–57 (2010)
1901. Horie, M.: The Hasse principle for elementary abelian fields. Mem. Fac. Sci. Kyushu Univ. **45**, 41–54 (1991)
1902. Horie, M.: The Hasse norm principle for elementary abelian extensions. Proc. Amer. Math. Soc. **118**, 47–56 (1993)
1903. Houriet, J.: Exceptional units and Euclidean number fields. Arch. Math. (Basel) **88**, 425–433 (2007)
1904. Hoyden-Siedersleben, G.: Realisierung der Jankogruppen J_1 und J_2 als Galoisgruppen über Q. J. Algebra **97**, 17–22 (1985)
1905. Hoyden-Siedersleben, G., Matzat, B.H.: Realisierung sporadischer einfacher Gruppen als Galoisgruppen über Kreisteilungskörpern. J. Algebra **101**, 273–286 (1986)
1906. Hsia, J.S.: On the representation of cyclotomic polynomials as sums of squares. Acta Arith. **25**, 115–120 (1973/74)
1907. Hsia, J.S., Johnson, R.P.: On the representation in sums of squares for definite functions in one variable over an algebraic number field. Amer. J. Math. **96**, 448–453 (1974)
1908. Hua, L.-K.: On the least solution of Pell's equation. Bull. Amer. Math. Soc. **48**, 731–735 (1942) [[1911], 119–123.]
1909. Hua, L.-K.: On the distribution of quadratic nonresidues and the Euclidean algorithm in real quadratic fields, I. Trans. Amer. Math. Soc. **56**, 537–546 (1944) [[1911], 137–146.]
1910. Hua, L.-K.: Additive prime number theory. Trudy Mat. Inst. im. Steklova **22**, (1947). (Russian) [German translation: Additive Primzahltheorie. Teubner (1959); English translation: Additive Theory of Prime Numbers. Amer. Math. Soc. (1965)]
1911. Hua, L.-K., Selected Papers. Springer (1983) [Reprint: Springer (2015)]
1912. Hua, L.-K., Min, S.-H.: On the distribution of quadratic non-residues and the Euclidean algorithm in real quadratic fields, II. Trans. Amer. Math. Soc. **56**, 547–569 (1944) [[1911], 147–169.]
1913. Hua, L.-K., Shih, W.T.: On the lack of an Euclidean algorithm in $R(\sqrt{61})$. Amer. J. Math. **67**, 209–211 (1945)
1914. Huard, J.G., Spearman, B.K., Williams, K.S.: Integral bases for quartic fields with quadratic subfields. J. Number Theory **51**, 87–102 (1995)
1915. Huber, A.: Philipp Furtwängler. Jahresber. Deutsch. Math.-Verein. **50**, 167–178 (1940)
1916. Huckaba, J.A., Papick, I.J.: A localization of $R[x]$. Canad. J. Math. **33**, 103–115 (1981)
1917. Hudson, R.H., Markham, T.L.: Alfred T. Brauer as a mathematician and teacher. Linear Algebra Appl. **59**, 1–17 (1984)
1918. Hughes, I., Mollin, R.: Totally positive units and squares. Proc. Amer. Math. Soc. **87**, 613–616 (1983)
1919. Hull, R.: A determination of all cyclotomic quintic fields. Ann. Math. (2) **36**, 366–372 (1935)
1920. Humbert, G.: Sur quelques conséquences arithmétiques de la théorie des fonctions abéliennes. C. R. Acad. Sci. Paris **142**, 537–541 (1906)
1921. Humbert, G.: Sur les fonctions abéliennes singuliéres d'invariants huit, douze et cinq. J. Math. Pures Appl. (6) **2**, 329–355 (1906)
1922. Humbert, G.: Formules relatives aux nombres de classes des formes quadratiques binaires et positives, J. Math. Pures Appl. (6) **3**, 337–449 (1907)
1923. Humbert, P.: Sur les nombres de classes de certains corps quadratiques. Comment. Math. Helv. **12**, 233–245 (1939/40); add.: **13**, p. 67 (1940/41)
1924. Hunt, D.C.: Rational rigidity and the sporadic groups. J. Algebra **99**, 577–592 (1986)
1925. Hunter, J.: The minimum discriminants of quintic fields. Glasg. Math. J. **3**, 1957, 57–67 (1957)
1926. Hurwitz, A.: Einige Eigenschaften der Dirichlet'schen Functionen $F(s) = \sum \left(\frac{D}{n}\right) \cdot \frac{1}{n^s}$, die bei der Bestimmung der Classenanzahlen binärer quadratischer Formen auftreten. Zeitschr. Math. Phys. **27**, 86–101 (1882) [[1939], **1**, 72–88.]

1927. Hurwitz, A.: Ueber Relationen zwischen Klassenanzahlen binärer quadratischer Formen von negativer Determinante. Math. Ann. **25**, 157–196 (1884) [[1939], **2**, 8–50.]
1928. Hurwitz, A.: Ueber die Anzahl der Klassen quadratischer Formen von negativer Determinante. J. Reine Angew. Math. **99**, 165–168 (1885) [[1939], **2**, 68–71]
1929. Hurwitz, A.: Über die Entwicklung complexer Grössen in Kettenbrüche. Acta Math. **11**, 187–200 (1887/88) [[1939], **2**, 72–83.]
1930. Hurwitz, A.: Ueber die angenäherte Darstellung der Irrationalzahlen durch rationale Brüche. Math. Ann. **39**, 279–284 (1891) [[1939], **2**, 122–128.]
1931. Hurwitz, A.: Ueber die Theorie der Ideale. Nachr. Ges. Wiss. Göttingen **1894**, 291–298 [[1939], **2**, 191–197.]
1932. Hurwitz, A.: Zur Theorie der algebraischen Zahlen. Nachr. Ges. Wiss. Göttingen **1895**, 324–331 [[1939], **2**, 236–243]
1933. Hurwitz, A.: Die unimodularen Substitutionen in einem algebraischen Zahlenkörper. Nachr. Ges. Wiss. Göttingen **1895**, 332–356 [[1939], **2**, 244–268.]
1934. Hurwitz, A.: Ueber die Anzahl der Klassen binärer quadratischer Formen von negativer Determinante. Acta Math. **19**, 351–384 (1895) [[1939], **2**, 208–235.]
1935. Hurwitz, A.: Ueber die Entwickelungscoefficienten der lemniskatischen Functionen. Math. Ann. **51**, 196–226 (1899) [[1939], **2**, 342–373.]
1936. Hurwitz, A.: Über eine Darstellung der Klassenzahl binärer quadratischer Formen durch unendliche Reihen. J. Reine Angew. Math. **129**, 187–213 (1905) [[1939], **2**, 385–409.]
1937. Hurwitz, A.: Der Euklidische Divisionssatz in einem endlichen algebraischen Zahlkörper. Math. Z. **3**, 123–126 (1919) [[1939], **2**, 471–474.]
1938. Hurwitz, A.: Über Beziehungen zwischen den Primidealen eines algebraischen Körpers und den Substitutionen seiner Gruppe. Math. Z. **25**, 661–665 (1926) [[1939], **2**, 733–737.]
1939. Hurwitz, A.: Mathematische Werke. **1**,**2**, Birkhäuser (1932–1933) [Reprint: Birkhäuser (1962–1963)]
1940. Hurwitz, A., Rudio, F. (ed.): Briefe von G. Eisenstein an M.A. Stern. Zeitschr. Math. Phys. **40**, Suppl., 169–203 (1895) [[1079], **2**, 791–823.]
1941. Hurwitz, J.: Ueber eine besondere Art der Kettenbruch-Entwickelung complexer Grössen, Dissertation, Univ. Halle (1895)
1942. Hurwitz, J.: Über die Reduktion der binären quadratischen Formen mit komplexen Koeffizienten und Variabeln. Acta Math. **25**, 231–290 (1902)
1943. Hykšová, M.: Karel Rychlìk (1885–1968). Prometheus (2003)
1944. Hymo, J.A., Parry, C.J.: On relative integral bases for cyclic quartic fields. J. Number Theory **34**, 189–197 (1990)
1945. Hymo, J.A., Parry, C.J.: On relative integral bases for pure quartic fields. Indian J. Pure Appl. Math. **23**, 359–376 (1992)
1946. Hyyrö, S.: Über eine Determinantenidentität und den ersten Faktor der Klassenzahl des Kreiskörpers. Ann. Acad. Sci. Fenn. Math., Ser. A I **398**, 1–7 (1967)
1947. Ibragimov, I.A. et al.: Juriĭ Vladimirovič Linnik: obituary. Uspekhi Mat. Nauk **28(2)**, 197–213 (1973) (Russian)
1948. Ichimura, H.: On the ring of integers of a tame Kummer extension over a number field. J. Pure Appl. Algebra **187**, 169–182 (2004)
1949. Ichimura, H.: A class number formula for the p-cyclotomic field. Arch. Math. (Basel) **87**, 539–545 (2006)
1950. Ichimura, H.: Note on Hilbert-Speiser number fields at a prime p. Yokohama Math. J. **54**, 45–53 (2007)
1951. Ichimura, H.: Hilbert-Speiser number fields for a prime p inside the p-cyclotomic field. J. Number Theory **128**, 858–864 (2008)
1952. Ichimura, H.: On the parity of the class number of the 7^nth cyclotomic field. Math. Slovaca **59**, 357–364 (2009)
1953. Ichimura, H.: Hilbert-Speiser number fields and the complex conjugation. J. Math. Soc. Japan **62**, 83–94 (2010)

1954. Ichimura, H., Sumida-Takahashi, H.: Imaginary quadratic fields satisfying the Hilbert-Speiser type condition for a small prime p. Acta Arith. **127**, 179–191 (2007)
1955. Ichimura, H., Sumida-Takahashi, H.: On Hilbert-Speiser type imaginary quadratic fields. Acta Arith. **136**, 385–389 (2009)
1956. Ihara, T.: Takuro Shintani (1943–1980). J. Fac. Sci. Univ. Tokyo **28(3)**, iii–vi (1981)
1957. Iimura, K.: A note on the Stickelberger ideal of conductor level. Arch. Math. (Basel) **36**, 45–52 (1981)
1958. Ikeda, M.: Zum Existenzsatz von Grunwald. J. Reine Angew. Math. **216**, 12–24 (1964)
1959. Ikehara, S.: An extension of Landau's theorem in the analytical theory of numbers. J. Math. Phys. **10**, 1–12 (1931)
1960. Inaba, E.: Über die Struktur der l-Klassengruppe zyklischer Zahlkörper vom Primzahlgrad l. J. Fac. Sci. Univ. Tokyo **4**, 61–115 (1940)
1961. Inaba, E.: Klassenkörpertheoretische Deutung der Struktur der Klassengruppe des zyklischen Zahlkörpers. Proc. Imp. Acad. Tokyo **17**, 125–128 (1941)
1962. Inaba, E.: Über den Hilbertschen Irreduzibilitätssatz. Jpn. J. Math. **19**, 1–25 (1944)
1963. Inkeri, K.: Über den Euklidischen Algorithmus in quadratischen Zahlkörpern. Ann. Acad. Sci. Fenn. Math., Ser. A I **41**, 1–35 (1947)
1964. Inkeri, K.: Über die Klassenzahl des Kreiskörpers der $l$ten Einheitswurzeln. Ann. Acad. Sci. Fenn. Math., Ser. A I **199**, 1–12 (1955)
1965. Ireland, K., Rosen, M.: A Classical Introduction to Modern Number Theory. Springer (1972)
1966. Iseki, K.: Über die negativen Fundamentaldiskriminanten mit der Klassenzahl zwei. Natur. Sci. Rep. Ochanomizu Univ. **3**, 23–29 (1952)
1967. Iseki, K.: On the imaginary quadratic fields of class-number one or two. Jpn. J. Math. **21**, 145–162 (1952)
1968. Išhanov, V.V.: The semidirect imbedding problem with a nilpotent kernel. Izv. Akad. Nauk SSSR, Ser. Mat. **40**, 3–25 (1976) (Russian)
1969. Ishibashi, M.: Effective version of the Tschebotareff density theorem in function fields over finite fields. Bull. Lond. Math. Soc. **24**, 52–56 (1992)
1970. Ishida, M.: On the genus field of an algebraic number field of odd prime degree. J. Math. Soc. Japan **27**, 289–293 (1975)
1971. Ishida, M.: The genus field of algebraic number fields. Lecture Notes Math. **555** (1976)
1972. Ishida, M.: An algorithm for constructing the genus field of an algebraic number field of odd prime degree. J. Fac. Sci. Univ. Tokyo **24**, 61–75 (1977)
1973. Ishida, M.: On the genus fields of pure number fields. Tokyo J. Math. **3**, 163–171 (1980)
1974. Ishida, M.: On the genus fields of pure number fields, II. Tokyo J. Math. **4**, 213–220 (1981)
1975. Ishikawa, T.: On Dedekind rings. J. Math. Soc. Japan **11**, 83–84 (1959)
1976. Ito, H.: Congruence relations of Ankeny-Artin-Chowla type for pure cubic fields. Nagoya Math. J. **96**, 95–112 (1984)
1977. Ivanov, I.I.: Integral complex numbers. St. Petersburg (1891) (Russian)
1978. Iwabuchi, H.: The universal quaternary quadratic form with the maximal discriminant. Mathematika **49**, 197–199 (2002)
1979. Iwasaki, K.: Simple proof of a theorem of Ankeny on Dirichlet series. Proc. Japan Acad. Sci. **28**, 555–557 (1952)
1980. Iwasawa, K.: On the rings of valuation vectors. Ann. Math. (2) **57**, 351–356 (1953)
1981. Iwasawa, K.: A note on Kummer extensions. J. Math. Soc. Japan **5**, 253–262 (1953)
1982. Iwasawa, K.: On Galois groups of local fields. Trans. Amer. Math. Soc. **80**, 448–469 (1955)
1983. Iwasawa, K.: A note on class numbers of algebraic number fields. Abh. Math. Semin. Univ. Hambg **20**, 257–258 (1956)
1984. Iwasawa, K.: On some invariants of cyclotomic fields. Amer. J. Math. **80**, 773–783 (1956); corr.: **81**, p. 280 (1959)
1985. Iwasawa, K.: On Γ-extensions of algebraic number fields. Bull. Amer. Math. Soc. **65**, 183–226 (1959)
1986. Iwasawa, K.: On some properties of Γ-finite modules. Ann. of Math. (2) **70**, 291–312 (1959)
1987. Iwasawa, K.: On the theory of cyclotomic fields. Ann. Math. (2) **70**, 530–561 (1959)

1988. Iwasawa, K.: On local cyclotomic fields. J. Math. Soc. Japan **12**, 16–21 (1960)
1989. Iwasawa, K.: A class number formula for cyclotomic fields. Ann. Math. (2) **76**, 171–179 (1962)
1990. Iwasawa, K.: A note on ideal class groups. Nagoya Math. J. **27**, 239–247 (1966)
1991. Iwasawa, K.: On some infinite Abelian extensions of algebraic number fields. In: Actes du Congrés International des Mathématiciens (Nice, 1970), **1**, 391–394. Gauthier-Villars (1971)
1992. Iwasawa, K.: On the μ-invariants of Z_l-extensions. In: Number theory, algebraic geometry and commutative algebra, in honor of Yasuo Akizuki, 1–11. Kinokuniya (1973)
1993. Iwasawa, K.: A note on Jacobi sums. Symposia Math. **15**, (Convegno di Strutture in Corpi Algebrici, INDAM, Rome, 1973), 447–459. Academic Press (1975)
1994. Iwasawa, K.: Local Class field Theory. Iwanami Shoten (1980) (Japanese) [Russian translation: Mir (1983)]
1995. Iwasawa, K.: Local Class Field Theory. The Clarendon Press, Oxford University Press (1986)
1996. Iwasawa, K.: On papers of Takagi in number theory. In: [3979], 342–351
1997. Iwasawa, K., Sims, C.: Computation of invariants in the theory of cyclotomic fields. J. Math. Soc. Japan **18**, 86–96 (1966)
1998. Iyanaga, S.: Über den Führer eines relativ zyklischen Zahlkörpers. Proc. Imp. Acad. Tokyo **5**, 108–110 (1929)
1999. Iyanaga, S.: Über den allgemeinen Hauptidealsatz. Jpn. J. Math. **7**, 315–333 (1930)
2000. Iyanaga, S.: Über den Wertvorrat des Normenrestsymbols. Abh. Math. Semin. Univ. Hambg **9**, 159–165 (1933)
2001. Iyanaga, S.: Zum Beweis des Hauptidealsatzes. Abh. Math. Semin. Univ. Hambg **10**, 349–357 (1934)
2002. Iyanaga, S.: Zur Theorie der Geschlechtermoduln. J. Reine Angew. Math. **171**, 12–18 (1934)
2003. Iyanaga, S.: Sur les classes d'idéaux dans les corps quadratiques. Actual. Sci. Industr. **197**, 1–15 (1935)
2004. Iyanaga, S.: Travaux de Claude Chevalley sur la théorie du corps de classes: introduction. Jpn. J. Math. **1**, 25–85 (2006)
2005. Iyanaga, S., Tamagawa, T.: Sur la théorie du corps de classes sur le corps des nombres rationnels. J. Math. Soc. Japan **3**, 220–227 (1951)
2006. Jacobi, C.G.J.: Ueber den Ausdruck der verschiedenen Wurzeln einer Gleichung durch bestimmte Integrale. J. Reine Angew. Math. **2**, 1–8 (1827) [[2012], **6**, 12–20.]
2007. Jacobi, C.G.J.: De residuis cubicis commentatio numerosa. J. Reine Angew. Math. **2**, 66–69 (1827) [[2012], **6**, 233–237.]
2008. Jacobi, C.G.J.: Vorlesungen über Zahlentheorie, Wintersemester 1836/37. E.-Rauner Verlag (2007)
2009. Jacobi, C.G.J.: Ueber die complexen Primzahlen, welche in der Theorie der Reste der 5ten, 8ten und 12ten Potenzen zu betrachten sind. J. Reine Angew. Math. **19**, 314–318 (1839) [[2012], **6**, 275–280; French translation: J. Math. Pures Appl. (1) **8**, 268–272 (1843)]
2010. Jacobi, C.G.J.: Über die Kreistheilung und ihre Anwendung auf die Zahlentheorie. J. Reine Angew. Math. **30**, 166–182 (1846) [[2012], **6**, 254–274.]
2011. Jacobi, C.G.J.: Allgemeine Theorie der kettenbruchähnlichen Algorithmen, in welchen jede Zahl aus drei vorhergehenden gebildet wird. J. Reine Angew. Math. **69**, 29–64 (1868) [[2012], **6**, 385–426.]
2012. Jacobi, C.G.J.: Gesammelte Werke. G. Reimer (1881–1891)
2013. Jacobson, B.: Sums of distinct divisors and sums of distinct units. Proc. Amer. Math. Soc. **15**, 179–183 (1964)
2014. Jacobson, M.J.Jr., Ramachandran, S., Williams, H.C.: Numerical results on class groups of imaginary quadratic fields. Lecture Notes in Computer Science **4076**, 87–101 (2006)
2015. Jacobson, N.: Abstract derivation and Lie algebras. Trans. Amer. Math. Soc. **42**, 206–224 (1937)

2016. Jacobson, N.: The Theory of Rings. Amer. Math. Soc. (1943) [4th printing: Amer. Math. Soc. (1968)]
2017. Jacobson, N.: Abraham Adrian Albert (1905–1972). Bull. Amer. Math. Soc. **80**, 1075–1100 (1974)
2018. Jacobsthal, E.: Diophantische Gleichungen im Bereich aller ganzen algebraischen Zahlen. Math. Ann. **74**, 31–65 (1913)
2019. Jacquet, H., Piatetski-Shapiro, I.J., Shalika, J.: Automorphic forms on $GL(3)$, I. Ann. Math. (2) **109**, 169–212 (1979)
2020. Jacquet, H., Piatetski-Shapiro, I.J., Shalika, J.: Automorphic forms on $GL(3)$, II. Ann. Math. (2) **109**, 213–258 (1979)
2021. Jaeger, Ch.G.: A character symbol for primes relatives to a cubic field. Amer. J. Math. **52**, 85–96 (1930)
2022. Jaffard, P.: Extensions algébriques infinies de PF-corps. Ann. Sci. Éc. Norm. Supér. (3) **70**, 181–198 (1953)
2023. Jakovlev, A.V.: The Galois group of the algebraic closure of a local field. Izv. Akad. Nauk SSSR, Ser. Mat. **32**, 1283–1322 (1968) (Russian)
2024. Jakubec, S.: On the divisibility of h^+ by the prime 3. Rocky Mountain J. Math. **24** 1467–1473 (1994)
2025. Jakubec, S.: On divisibility of h^+ by the prime 5. In: Number Theory (Račkova dolina, 1993). Math. Slovaca **44**, 651–661 (1994)
2026. Jakubec, S.: Congruence of Ankeny-Artin-Chowla type for cyclic fields of prime degree l. Math. Proc. Cambridge Philos. Soc. **119**, 17–23 (1996)
2027. Jakubec, S.: Congruence of Ankeny-Artin-Chowla type modulo p^2 for cyclic fields of prime degree l. Acta Arith. **74**, 293–310 (1996)
2028. Jakubec, S.: Note on the congruence of Ankeny-Artin-Chowla type modulo p^2. Acta Arith. **85**, 377–388 (1998)
2029. Jakubec, S.: On divisibility of the class number h^+ of the real cyclotomic fields of prime degree l. Math. Comp. **67**, 369–398 (1998)
2030. Jakubec, S., Kostra, J., Nemoga, K.: On the existence of an integral normal basis generated by a unit in prime extensions of rational numbers. Math. Comp. **56**, 809–815 (1991)
2031. Jakubec, S., Trojovský, P.: On divisibility of the class number h^+ of the real cyclotomic fields $Q(\zeta_p + \zeta_p^{-1})$ by primes $q \leq 5000$. Abh. Math. Semin. Univ. Hambg **67**, 269–280 (1997)
2032. Jannsen, U., Wingberg, K.: Einbettungsprobleme und Galoisstruktur lokaler Körper. J. Reine Angew. Math. **319**, 196–212 (1980)
2033. Jannsen, U., Wingberg, K.: Die Struktur der absoluten Galoisgruppe p-adischer Zahlkörper. Invent. Math. **70**, 70–98 (1982/83)
2034. Janusz, G.J.: Algebraic Number Fields. Academic Press (1973) [2nd ed. Amer. Math. Soc. (1996)]
2035. Janusz, G.J.: Irving Reiner 1924–1986. Illinois J. Math. **32**, 315–328 (1988)
2036. Járási, I.: Power integral bases in sextic fields with a cubic subfield. Acta Sci. Math. (Szeged) **69**, 3–15 (2003)
2037. Jarden, M.: The Čebotarev density theorem for function fields: an elementary approach. Math. Ann. **261**, 467–475 (1982)
2038. Jarden, M., Narkiewicz, W.: On sums of units. Monatsh. Math. **150**, 327–332 (2007)
2039. Jarvis, F., Meekin, P.: The Fermat equation over $\mathbf{Q}(\sqrt{2})$. J. Number Theory **109**, 182–196 (2004)
2040. Jehanne, A.: Realization over Q of the groups $\tilde{A}_5$ and $\hat{A}_5$. J. Number Theory **89**, 340–368 (2001)
2041. Jehne, W.: Über die Struktur der Multiplikationsgruppe von Körpern. J. Reine Angew. Math. **256**, 190–209 (1972)
2042. Jehne, W.: Kronecker classes of algebraic number fields. J. Number Theory **9**, 279–320 (1977)

2043. Jehne, W.: On Kronecker classes of atomic extensions. Proc. London Math. Soc. (3) **34**, 32–64 (1977)
2044. Jehne, W.: On knots in algebraic number theory. J. Reine Angew. Math. **311/312**, 215–254 (1979)
2045. Jeltsch-Fricker, R.: In memoriam: Alexander M. Ostrowski (1893 bis 1986). Elem. Math. **43**, 33–38 (1988)
2046. Jensen, C.U.: On characterization of Prüfer rings. Math. Scand. **13**, 90–98 (1963)
2047. Jensen, C.U.: A remark on relative integral bases for infinite extensions of finite number fields. Mathematika **11**, 64–66 (1964)
2048. Jensen, C.U., Ledet, A., Yui, N.: Generic Polynomials. Constructive aspects of the inverse Galois problem. Cambridge University Press (2002)
2049. Jensen, C.U., Prestel, A.: Realization of finitely generated profinite groups by maximal abelian extensions of fields. J. Reine Angew. Math. **447**, 201–218 (1994)
2050. Jensen, J.L.W.V.: Om Raekkers Konvergens. Math. Tidsskr. (5) **2**, 63–72 (1884)
2051. Jensen, K.L.: Om talteoriske Egenskaber ved de Bernouliiske Tal. Nyt Tidsskr. Math. B **3**, 73–83 (1915)
2052. Jha, V.: Faster computation of the first factor of the class number of $Q(\zeta_p)$. Math. Comp. **64**, 1705–1710 (1995)
2053. Ji, C.-G.: Sums of three integral squares in cyclotomic fields. Bull. Aust. Math. Soc. **68**, 101–106 (2003)
2054. Ji, C.-G., Wang, Y., Xu, F.: Sums of three squares over imaginary quadratic fields. Forum Math. **18**, 585–601 (2006)
2055. Ji, C.-G., Wei, D.: Sums of integral squares in cyclotomic fields. C. R. Acad. Sci. Paris **344**, 413–416 (2007)
2056. Johnson, E.W.: Almost-Dedekind rings. Glasg. Math. J. **36**, 131–134 (1994)
2057. Johnson, W.: On the vanishing of the Iwasawa invariant μ_p for $p<8000$. Math. Comp. **27**, 387–396 (1973)
2058. Johnson, W.: Irregular primes and cyclotomic invariants. Math. Comp. **29**, 113–120 (1975)
2059. Joly, J.-R.: Constantes de Waring des corps commutatifs. C. R. Acad. Sci. Paris **266**, A516–A518 (1968)
2060. Joly, J.-R.: Sommes de puissances $m^{iémes}$ dans les anneaux p-adiques et les anneaux d'entiers algébriques. Enseign. Math. (2) **14**, 197–204 (1968)
2061. Joly, J.-R.: Sommes de puissances $d$iémes dans un anneau commutatif. Acta Arith. **17**, 37–114 (1970)
2062. Joly, J.-R.: Sommes des carrés dans certains anneaux principaux. Bull. Sci. Math. (2) **94**, 85–95 (1970)
2063. Joly, J.-R.: Démonstration cyclotomique de la loi de réciprocité cubique. Bull. Sci. Math. (2) **96**, 273–278 (1972)
2064. Jones, B.W.: A canonical quadratic form for the ring of 2-adic integers. Duke Math. J. **11**, 715–727 (1944)
2065. Jones, B.W., Durfee, W.H.: A theorem on quadratic forms over the ring 2-adic integers. Bull. Amer. Math. Soc. **55**, 758–762 (1949)
2066. Jones, J.W.: Minimal solvable nonic fields. LMS J. Comput. Math. **16**, 130–138 (2013)
2067. Jones, J.W., Roberts, D.P.: Nonic 3-adic fields. Lecture Notes in Computer Science **3076**, 293–308 (2004)
2068. Jones, J.W., Roberts, D.P.: A database of local fields. J. Symbolic Comput. **41**, 80–97 (2006)
2069. Jones, J.W., Roberts, D.P.: : Galois number fields with small root discriminant. J. Number Theory **122**, 379–407 (2004)
2070. Jones, J.W., Roberts, D.P.: Octic 2-adic fields. J. Number Theory **128**, 1410–1429 (2008)
2071. Jones, J.W., Wallington, R.: Number fields with solvable Galois groups and small Galois root discriminants. Math. Comp. **81**, 555–567 (2012)
2072. Jones, M., Rouse, J.: Solutions of the cubic Fermat equation in quadratic fields. Internat. J. Number Th. **9**, 1579–1591 (2013)

2073. Jordan, C.: Traité des substitutions et des équations algébriques. Gauthier-Villars (1870) [Reprints: A. Blanchard (1957); J. Gabay (1989)]
2074. Jordan, C., Bonnett, O., Faye, H., Renan ,E.: Discours prononces aux funérailles de M.Serret le jeudi 5 mars 1885. Bull. Sci. Math. (2) **9**, 123–132 (1885)
2075. Jorgenson, J., Krantz, S.G.: Serge Lang, 1927–2005. Notices Amer. Math. Soc. **53**, 536–553 (2006)
2076. Joris, H.: Un Ω-théorème pour la fonction des idéaux d'un corps de nombres algébriques. C. R. Acad. Sci. Paris **270**, p. A1713 (1970)
2077. Joris, H.: Ω-Sätze für zwei arithmetische Funktionen. Comment. Math. Helv. **47**, 220–248 (1972)
2078. Joris, H.: Ω-Sätze für gewisse multiplikative arithmetische Funktionen. Comment. Math. Helv. **48**, 409–435 (1973)
2079. Joseph , A., Melnikov, A., Rentschler, R. (ed.): Studies in Memory of Issai Schur. Progr. Math. **210** (2003)
2080. Joubert, P.: Sur la théorie des fonctions elliptiques et son application à la théorie des nombres. C. R. Acad. Sci. Paris **50**, 774–779, 832–837, 907–912, 1040–1045 (1860)
2081. Joubert, P.: Sur l'équation du sixième degré. C. R. Acad. Sci. Paris **64**, 1025–1029 (1867)
2082. Julia, G.: La vie et l'oeuvre de J.-L. Lagrange. Enseign. Math. **39**, 9–21 (1942–1950)
2083. Jung, H.Y., Ahn, J.[-H.]: Group determinant formulas and class numbers of cyclotomic fields. J. Korean Math. Soc. **44**, 499–509 (2007)
2084. Jutila, M.: On character sums and class numbers. J. Number Theory **5**, 203–314 (1973)
2085. Jutila, M.: On the mean value of $L(1/2, \chi)$ for real characters. Analysis **1**, 149–161 (1981)
2086. Kable, A.C.: Power bases in dihedral quartic fields. J. Number Theory **76**, 120–129 (1999)
2087. Kable, A.C., Yukie, A.: A construction of quintic rings. Nagoya Math. J. **173**, 163–203 (2004)
2088. Kable, A.C., Yukie, A.: On the number of quintic fields. Invent. Math. **160**, 217–259 (2005)
2089. Kaczorowski, J.: A pure arithmetical characterization for certain fields with a given class group. Colloq. Math. **45**, 327–330 (1981)
2090. Kaczorowski, J.: Some remarks on factorization in algebraic number fields. Acta Arith. **43**, 53–68 (1983)
2091. Kaczorowski, J.: A pure arithmetical definition of the classgroup. Colloq. Math. **48**, 265–267 (1984)
2092. Kadiri, H.: Explicit zero-free regions for Dedekind zeta functions. Internat. J. Number Th. **8**, 125–147 (2012)
2093. Kagawa, T.: The Hasse norm principle for the maximal real subfields of cyclotomic fields. Tokyo J. Math. **18**, 221–229 (1995)
2094. Kalužnin, L.A. et al.: Vladimir Petrovič Vel'min. Obituary. Uspekhi Mat. Nauk **31(1)**, 229–235 (1976) (Russian)
2095. Kamei, M.: Congruences of Ankeny-Artin-Chowla type for pure quartic and sectic fields. Nagoya Math. J. **108**, 131–144 (1987)
2096. Kaminski, M.: Cyclotomic polynomials and units in cyclotomic number fields. J. Number Theory **28**, 283–287 (1988)
2097. Kamke, E.: Verallgemeinerungen des Waring-Hilbertschen Satzes. Math. Ann. **83**, 85–112 (1921)
2098. Kamke, E.: Über die Zerfällung rationaler Zahlen in rationale Polynomwerte. Math. Z. **12**, 323–328 (1922)
2099. Kamke, E.: Zum Waringschen Problem für rationale Zahlen und Polynome. Math. Ann. **87**, 238–245 (1922)
2100. Kanemitsu, S., Kuzumaki, T.: On a generalization of the Maillet determinant. In: Number theory (Eger, 1996), 271–287. de Gruyter (1998)
2101. Kanemitsu, S., Kuzumaki, T.: On a generalization of the Maillet determinant, II. Acta Arith. **99**, 343–361 (2001)
2102. Kang, M.-C.: Noether's problem for dihedral 2-groups, II. Pacific J. Math. **222**, 301–316 (2005)

2103. Kang, M.-C.: Noether's problem for metacyclic p-groups. Adv. Math. **203**, 554–567 (2005)
2104. Kang, M.-C.: Rationality problem for some meta-abelian groups. J. Algebra **322**, 1214–1219 (2009)
2105. Kang, M.-C., Michailov, I.M., Zhou, J.: Noether's problem for the groups with a cyclic subgroup of index 4. Transform. Groups **17**, 1037–1058 (2012)
2106. Kang, M.-C., Zhou, J.: The rationality problem for finite subgroups of $GL_4(Q)$. J. Algebra **368**, 53–69 (2012)
2107. Kani, E.: Idoneal numbers and some generalizations. Ann. Sci. Math. Québec **35**, 197–227 (2011)
2108. Kaplan, N.: Flat cyclotomic polynomials of order three. J. Number Theory **127**, 118–126 (2007)
2109. Kaplan, N.: Flat cyclotomic polynomials of order four and higher. Integers **10**, 357–363 (2010)
2110. Kaplan, P.: Divisibilité par 8 du nombre des classes des corps quadratiques dont le 2−groupe des classes est cyclique et réciprocité quadratique. J. Math. Soc. Japan **25**, 596–608 (1973)
2111. Kaplan, P.: Comparaison des 2-groupes des classes d'idéaux au sens large et au sens étroit d'un corps quadratique réel. Proc. Japan Acad. Sci. **50**, 688–693 (1974)
2112. Kaplan, P.: Sur le 2-groupe des classes d'idéaux des corps quadratiques. J. Reine Angew. Math. **283/284**, 313–363 (1976)
2113. Kaplan, P.: Cycles d'ordre au moins 16 dans le 2-groupe des classes d'idéaux de certains corps quadratiques. Bull. Soc. Math. France Mém. **49/50**, 113–124 (1977)
2114. Kaplan, P., Williams, K.S.: On the class numbers of $Q\left(\sqrt{\pm 2p}\right)$ modulo 16, for $p \equiv 1 \pmod 8$, a prime. Acta Arith. **40**, 289–296 (1982)
2115. Kaplan, P., Williams, K.S.: Congruences modulo 16 for the class numbers of the quadratic fields $Q\left(\sqrt{\pm p}\right)$ and $Q\left(\sqrt{\pm 2p}\right)$ for p a prime congruent to 5 modulo 8. Acta Arith. **40**, 375–397 (1982)
2116. Kaplan, P., Williams, K.S.: On the strict class number of $Q\left(\sqrt{2p}\right)$ modulo 16, $p \equiv 1 \pmod 8$ prime. Osaka J. Math. **21**, 23–29 (1984)
2117. Kaplan, P., Williams, K.S., Hardy, K.: Divisibilité par 16 des classes au sens strict des corps quadratiques réels dont le deux-groupe des classes est cyclique. Osaka J. Math. **23**, 479–489 (1986)
2118. Kaplansky, I.: Modules over Dedekind rings and valuation rings. Trans. Amer. Math. Soc. **72**, 327–340 (1952)
2119. Kaplansky, I.: Composition of binary quadratic forms. Scripta Math. **31**, 523–530 (1968)
2120. Kaplansky, I.: Abraham Adrian Albert (1905–1972). Biographical Mem. Nat. Acad. Sci. **51**, 3–22 (1980)
2121. Karatsuba, A.A.: The mean value of the modulus of a trigonometric sum. Izv. Akad. Nauk SSSR, Ser. Mat. **37**, 1203–1227 (1973)
2122. Katz, N.M.: The congruences of Clausen-von Staudt and Kummer for Bernoulli-Hurwitz numbers. Math. Ann. **216**, 1–4 (1975)
2123. Katz, N.M., Tate, J.: Bernard Dwork (1923–1998). Notices Amer. Math. Soc. **46**, 338–343 (1999)
2124. Kaufmann-Bühler, W.: Gauss. A Biographical Study. Springer (1981)
2125. Kaur, G.: The minimum discriminant of sixth degree totally real algebraic number fields. J. Indian Math. Soc. (N.S.) **34**, 123–134 (1970)
2126. Kawada, Y.: On the derivations in number fields. Ann. Math. (2) **54**, 302–314 (1951)
2127. Kawada, Y.: On the class field theory on algebraic number fields with infinite degree. J. Math. Soc. Japan **3**, 104–115 (1951)
2128. Kawada, Y.: On the ramification theory of infinite algebraic extensions. Ann. Math. (2) **58**, 24–47 (1953)
2129. Kawada, Y.: On the structure of the Galois group of some infinite extensions. I. J. Fac. Sci. Univ. Tokyo **7**, 1–18 (1954)
2130. Kawada, Y.: A remark on the principal ideal theorem. J. Math. Soc. Japan **20**, 166–169 (1968)

2131. Kawamoto, F.: On normal integral bases. Tokyo J. Math. **7**, 221–231 (1984)
2132. Kawamoto, F.: On normal integral bases in local fields. J. Algebra **98**, 197–199 (1986)
2133. Kedlaya, K.S.: A construction of polynomials with squarefree discriminants. Proc. Amer. Math. Soc. **140**, 3025–3033 (2012)
2134. Kelly, J.B.: A closed set of algebraic integers. Amer. J. Math. **72**, 565–572 (1950)
2135. Kelly, J.B.: On factorization of polynomials. Amer. Math. Monthly **60**, 375–379 (1953)
2136. Kempfert, H.: Zum allgemeinen Hauptidealsatz, II. J. Reine Angew. Math. **223**, 28–55 (1966)
2137. Kempfert, H.: On the factorization of polynomials. J. Number Theory **1**, 116–120 (1969)
2138. Kenku, M.A.: Determination of the even discriminants of complex quadratic fields of class-number 2. Proc. London Math. Soc. (3) **22**, 731–746 (1971)
2139. Kersten, I.: Ernst Witt 1911–1991. Jahresber. Deutsch. Math.-Verein. **95**, 166–180 (1993)
2140. Kersten, I., Michaliček, J.: On Vandiver's conjecture and $\mathbf{Z}_p$-extensions of $\mathbf{Q}(\zeta_p)$. J. Number Theory **32**, 371–386 (1989)
2141. Kervaire, M.: Fractions rationnelles invariantes (d'aprés H.W.Lenstra). Lecture Notes Math. **431**, 170–189 (1975)
2142. Khanduja, S.K., Kawada, Y.: On a theorem of Dedekind. Internat. J. Number Th. **4**, 1019–1025 (2008)
2143. Khanduja, S.K., Kumar, M.: On Dedekind criterion and simple extensions of valuation rings. Comm. Algebra **38**, 684–696 (2010)
2144. Khare, C., Wintenberger, J.-P.: Serre's modularity conjecture, I. Invent. Math. **178**, 485–504 (2009)
2145. Khare, C., Wintenberger, J.-P.: Serre's modularity conjecture, II. Invent. Math. **178**, 505–586 (2009)
2146. Kiepert, L.: Siebzehntheilung des Lemniscatenumfangs durch alleinige Anwendung von Lineal und Cirkel. J. Reine Angew. Math. **75**, 255–263 (1872): add.: p. 348.
2147. Kim, B.M., Park, P.-S.: Sums of distinct integral squares in real quadratic fields. C. R. Acad. Sci. Paris **349**, 497–500 (2011)
2148. Kim, J.Y., Lee, Y.M.: Sums of distinct integral squares in $Q(\sqrt{2})$, $Q(\sqrt{3})$ and $Q(\sqrt{6})$. Bull. Aust. Math. Soc. **85**, 1–10 (2012)
2149. Kim, M.-H., Lim, S.-G.: Square classes of totally positive units. J. Number Theory **125**, 1–6 (2007)
2150. Kiming, I.: On the experimental verification of the Artin conjecture for 2-dimensional odd Galois representations over Q. Liftings of 2-dimensional projective Galois representations over Q. In: [1247], 1–36 (1994)
2151. Kiming, I.: Applications of massive computations: the Artin conjecture. In: Analysis, Algebra, and Computers in Mathematical Research (Luleåa, 1992). Lecture Notes Pure Appl. Math. **156**, 185–199 (1994)
2152. Kiming, I., Wang, X.D.: Examples of 2-dimensional, odd Galois representations of A_5-type over Q satisfying the Artin conjecture. In: [1247], 109–121 (1994)
2153. Kinkelin, H.: Allgemeine Theorie der harmonischen Reihen, mit Anwendungen auf die Zahlentheorie. Programm der Gewerbeschule Basel **1862**, 1–32
2154. Kinohara, A.: On the derivations and the relative differents in algebraic number fields. J. Sci. Hiroshima Univ. **A 16**, 261–266 (1952)
2155. Kinohara, A.: On the derivations and the relative differents in commutative fields. J. Sci. Hiroshima Univ. **A 16**, 441–456 (1953)
2156. Kirmse, J.: Zur Darstellung total positiver Zahlen als Summen von vier Quadraten. Math. Z. **21**, 195–202 (1924)
2157. Kiselev, A.A.: Expression of the class-number of ideals of real quadratic fields by means of Bernoulli numbers. Dokl. Akad. Nauk SSSR **61**, 777–779 (1948) (Russian)
2158. Kiselev, A.A., Slavutskiĭ, I.Š.: On the number of classes of ideals of a quadratic field and its rings. Dokl. Akad. Nauk SSSR **126**, 1191–1194 (1959) (Russian)
2159. Kishi, Y.: A constructive approach to Spiegelung relations between 3-ranks of absolute ideal class groups and congruent ones modulo $(3)^2$ in quadratic fields. J. Number Theory **83**, 1–49 (2000)

2160. Kishi, Y.: On the 3-rank of the ideal class group of quadratic fields. Kodai Math. Rep. **36**, 275–283 (2013)
2161. Kisilevsky, H.: Number fields with class number congruent to 4 mod 8 and Hilbert's theorem 94. J. Number Theory **8**, 271–279 (1976)
2162. Kisilevsky, H.: Olga Taussky–Todd's work in class field theory. Pacific J. Math. Special Issue: Olga Taussky-Todd: in memoriam, 219–224 (1997)
2163. Kisin, M.: Modularity of 2-adic Barsotti-Tate representations. Invent. Math. **178**, 587–634 (2009)
2164. Klassen, M.[J.], Tzermias, P.: Algebraic points of low degree on the Fermat quintic. Acta Arith. **82**, 393–401 (1997)
2165. Klein, F.: Ueber die Transformation der elliptischen Functionen und die Auflösung der Gleichungen fünften Grades. Math. Ann. **14**, 111–114 (1879)
2166. Klein, F.: Ueber die Composition der binären quadratischen Formen. Nachr. Ges. Wiss. Göttingen **1893**, 106–109.
2167. Klein, F.: Ausgewählte Kapitel der Zahlentheorie, II. Göttingen (1897)
2168. Klein, F.: Ernst Schering. Jahresber. Deutsch. Math.-Verein. **6**, 25–27 (1899)
2169. Kleinjung, T.: Quadratic sieving. Math. Comp. **85**, 1861–1873 (2016)
2170. Klingen, H.: Über die Werte der Dedekindschen Zetafunktion. Math. Ann. **145**, 265–272 (1961/62)
2171. Klingen, N.: Atomare Kronecker-Klassen mit speziellen Galoisgruppen. Abh. Math. Semin. Univ. Hambg **48**, 42–53 (1979)
2172. Klingen, N.: Allgemeine Primstellen und Kroneckerklassen unendlichen Körpererweiterungen. Resultate Math. **6**, 183–193 (1983)
2173. Klingen, N.: Arithmetical Similarities. Oxford University Press (1998)
2174. Kloosterman, H.D.: Thetareihen in total-reellen algebraischen Zahlkörpern. Math. Ann. **103**, 279–299 (1930)
2175. Klüners, J.: A counterexample to Malle's conjecture on the asymptotics of discriminants. C. R. Acad. Sci. Paris **340**, 411–414 (2005)
2176. Klüners, J.: Asymptotics of number fields and the Cohen-Lenstra heuristics. J. Théor. Nombres Bordeaux **18**, 607–615 (2006)
2177. Klüners, J., Malle, G.: Explicit Galois realization of transitive groups of degree up to 15. J. Symbolic Comput. **30**, 675–716 (2000)
2178. Klüners, J., Malle, G.: Counting nilpotent Galois extensions. J. Reine Angew. Math. **572**, 1–26 (2004)
2179. Knapowski, S.: On a theorem of Hecke. J. Number Theory **1**, 235–251 (1969)
2180. Kneser, A.: Ueber die Gattung niedrigster Ordnung, unter welcher gegebene Gattungen algebraischer Grössen enthalten sind. Math. Ann. **30**, 179–202 (1987)
2181. Kneser, A.: Leopold Kronecker. Jahresber. Deutsch. Math.-Verein. **33**, 210–228 (1925)
2182. Kneser, H.: Verschwindente Quadratsummen in Körpern. Jahresber. Deutsch. Math.-Verein. **44**, 143–146 (1934)
2183. Kneser, M.: Zum expliziten Reziprozitätsgesetz von I.R.Šafarevič. Math. Nachr. **6**, 89–96 (1951)
2184. Kneser, M.: Kleine Lösungen der diophantischen Gleichung $ax^2 + by^2 = cz^2$. Abh. Math. Semin. Univ. Hambg **23**, 163–173 (1959)
2185. Kneser, M.: Composition of binary quadratic forms. J. Number Theory **15**, 406–413 (1982)
2186. Kneser, M.: Martin Eichler (1912–1992). Acta Arith. **65**, 293–296 (1993)
2187. Knopfmacher, A., Knopfmacher, J.: The number of steps in the Euclidean algorithm over complex quadratic fields. BIT **31**, 286–292 (1991)
2188. Knopp, K.: Edmund Landau. Jahresber. Deutsch. Math.-Verein. **54**, 55–62 (1951
2189. Kobald, E., Gmeiner, J.A., Stolz, O.: Leopold Gegenbauer. Monatsh. Math. Phys. **15**, 3–10, 129–136 (1904)
2190. Koç, C.: Professor Masatoshi Gündüz Ikeda: a life devoted to mathematics. Turkish J. Math. **27**, 461–471 (2003)

2191. Koch, H.: The Galois group of a local field. Dokl. Akad. Nauk SSSR **137**, 1291–1294 (1961) (Russian)
2192. Koch, H.: Über den 2-Klassenkörperturm eines quadratischen Zahlkörpers, I. J. Reine Angew. Math. **214/215**, 201–206 (1964)
2193. Koch, H.: Zum Satz von Golod-Schafarewitsch. Math. Nachr. **42**, 321–333 (1969)
2194. Koch, H.: Galoissche Theorie der p-Erweiterungen. Springer, VEB Deutscher Verlag der Wissenschaften (1970)
2195. Koch, H.: Zum Satz von Golod-Schafarewitsch. J. Reine Angew. Math. **274/275**, 240–243 (1975)
2196. Koch, H.: Galois group of a $p-$closed extension of a local field. Dokl. Akad. Nauk SSSR **238**, 19–22 (1978) (Russian)
2197. Koch, H.: Nachruf auf Hans Reichardt. Jahresber. Deutsch. Math.-Verein. **95**, 135–140 (1993)
2198. Koch, H.: Local class field theory for metabelian extensions. In: Proceedings of the 2nd Gauss Symposium (Munich, 1993), 287–300. de Gruyter (1995)
2199. Koch, H., de Shalit, E.: Metabelian local class field theory. J. Reine Angew. Math. **478**, 85–106 (1996)
2200. Kohl, E.: Ueber die Lemniskatenteilung. SBer. Kais. Akad. Wissensch. Wien **98**, 364–387 (1888)
2201. Kolyvagin, V.A.: The Fermat equations over cyclotomic fields. Trudy Mat. Inst. im. Steklova **208**, 163–185 (1995) (Russian)
2202. Kolyvagin, V.A.: The first case of Fermat's theorem for cyclotomic fields. Izv. Ross. Akad. Nauk, Ser. Mat. **63**, 147–158 (1999) (Russian)
2203. Kolyvagin, V.A.: On the first case of the Fermat theorem for cyclotomic fields. J. Math. Sci. (N.Y.) **106**, 3302–3311 (2001)
2204. Kolyvagin, V.A.: The Fermat equation over a tower of cyclotomic fields. Izv. Ross. Akad. Nauk, Ser. Mat. **65**, 85–122 (2001) (Russian)
2205. Komatsu, K.: On the Galois group of $x^p + ax + a = 0$. Tokyo J. Math. **14**, 227–229 (1991)
2206. König, J.: Einleitung in die allgemeine Theorie der algebraischen Grössen. Teubner (1903)
2207. König, R.: Über die quadratischen Formen mit rationalen Funktionen als Koeffizienten. Monatsh. Math. Phys. **23**, 321–346 (1912)
2208. König, R.: Über quadratische Formen und Zahlkörper sowie zwei Gruppensätze. Jahresber. Deutsch. Math.-Verein. **22**, 239–254 (1913)
2209. Königsberger, L. [Koenigsberger, L.]: Ueber den Eisenstein'schen Satz von der Irreductibilität algebraischer Gleichungen. J. Reine Angew. Math. **115**, 53–78 (1895)
2210. Koenigsberger, L.: Carl Gustav Jacob Jacobi. Jahresber. Deutsch. Math.-Verein. **13**, 405–435 (1904)
2211. Koenigsberger, L.: Carl Gustav Jacob Jacobi. Teubner (1904)
2212. Konyagin, S., Maier, H., Wirsing, E.: Cyclotomic polynomials with many primes dividing their orders. Period. Math. Hung. **49**, 99–106 (2004)
2213. Konyagin, S., Soundararajan, K.: Two S-unit equations with many solutions. J. Number Theory **124**, 193–199 (2007)
2214. Koppenhöfer, D.: Determining the monogeneity of a quartic number field. Math. Nachr. **172**, 191–198 (1995)
2215. Körner, O.: Übertragung des Goldbach-Vinogradovschen Satzes auf reell-quadratische Zahlkörper. Math. Ann. **141**, 343–366 (1960)
2216. Körner, O.: Erweiterter Goldbach-Vinogradovscher Satz in beliebigen algebraischen Zahlkörpern. Math. Ann. **143**, 344–378 (1961)
2217. Körner, O.: Zur additiven Primzahltheorie algebraischer Zahlkörper. Math. Ann. **144**, 97–109 (1961)
2218. Körner, O.: Über das Waringsche Problem in algebraischen Zahlkörpern. Math. Ann. **144**, 224–238 (1961)
2219. Körner, O.: Über Mittelwerte trigonometrischer Summen und ihre Anwendung in algebraischen Zahlkörpern. Math. Ann. **147**, 205–239 (1962); Corr.: **149**, p. 462 (1963)

2220. Körner, O.: Ganze algebraische Zahlen als Summen von Polynomwerten. Math. Ann. **149**, 97–104 (1962/63)
2221. Körner, O.: Darstellung ganzer Grössen durch Primzahlpotenzen in algebraischen Zahlkörpern. Math. Ann. **155**, 204–245 (1964)
2222. Körner, O.: Über durch Potenzen erzeugte Ringe und Gruppen in algebraischen Zahlkörpern. Manuscripta Math. **3**, 157–174 (1970)
2223. Korobov, N.M.: Estimates of exponential sums and their applications. Uspekhi Mat. Nauk **13(4)**, 185–192 (1958) (Russian)
2224. Korobov, N.M., Nesterenko, Yu.V., Šidlovskiĭ, A.N.: Naum Il'ič Fel'dman. Uspekhi Mat. Nauk **50(6)**, 157–162 (1995) (Russian)
2225. Kortum, H.: Rudolf Lipschitz. Jahresber. Deutsch. Math.-Verein. **15**, 56–59 (1906)
2226. Korselt, A.: Einfacher Beweis eines Satzes von Kronecker. J. Reine Angew. Math. **164**, 61–62 (1931)
2227. Koschmieder, L.: Beweis des kubischen Reziprozitätsgesetzes mit Hilfe der elliptischen Funktionen. Math. Ann. **83**, 280–285 (1921)
2228. Koschmieder, L.: Adolf Kneser. SBer. Berliner Math. Ges. **29**, 78–102 (1930)
2229. Kostrikin, A.I.: On defining groups by means of generators and defining relations. Izv. Akad. Nauk SSSR, Ser. Mat. **29**, 1119–1122 (1965) (Russian)
2230. Kotlar, D., Schacher, M., Sonn, J.: Central extensions of symmetric groups as Galois groups. J. Algebra **124**, 183–198 (1989)
2231. Kotov, S.V., Trelina, L.A.: S-ganze Punkte auf elliptischen Kurven. J. Reine Angew. Math. **306**, 28–41 (1979)
2232. Koukoulopoulos, D.: Pretentious multiplicative functions and the prime number theorem for arithmetic progressions. Compositio Math. **149**, 1129–1149 (2013)
2233. Koutský, K.: Památce prof. dr. Karla Petra. Časopis mat. fys. **75**, D341–D345 (1950)
2234. Kovalčik, F.B.: Density theorems for sectors and progressions. Lit. Mat. Sb. **15**, 133–151 (1975) (Russian)
2235. Kozuka, K.: On the local Kronecker-Weber theorem. Arch. Math. (Basel) **54**, 162–163 (1990)
2236. Koyama, T., Nishi, M., Yanagihara, H.: On characterizations of Dedekind domains. J. Sci. Hiroshima Univ. **4**, 71–74 (1974)
2237. Kraft, H.: A result of Hermite and equations of degree 5 and 6. J. Algebra **297**, 234–253 (2006)
2238. Krakowski, F.: Eigenwerte und Minimalpolynome symmetrischer Matrizen in kommutativen Körpern. Comment. Math. Helv. **32**, 224–240 (1958)
2239. Kramer, D.: On the values at integers of the Dedekind zeta function of a real quadratic field. Trans. Amer. Math. Soc. **229**, 59–79 (1987)
2240. Krasner, M.: Sur la représentation multiplicative dans les corps de nombres p-adiques relativement galoisiens. C. R. Acad. Sci. Paris **203**, 907–908 (1936)
2241. Krasner, M.: Le nombre des surcorps d'un degré donné d'un corps de nombres $\mathfrak{g}$-adiques. C. R. Acad. Sci. Paris **205**, 1026–1028 (1937); Corr.: p. 1267
2242. Krasner, M.: Sur la théorie de la ramification des idéaux de corps nongaloisiens de nombres algébriques, Mém. Acad. Belg. **11(8)**, 1–110 (1937)
2243. Krasner, M.: Sur la primitivité des corps $\mathfrak{P}$-adiques. Mathematica (Cluj) **13**, 72–191 (1937)
2244. Krasner, M.: Le nombre des surcorps primitifs d'un degré donné et le nombre des surcorps métagaloisiens d'un degré donné d'un corps de nombres p-adiques. C. R. Acad. Sci. Paris **206**, 876–878 (1938); Corr.: p. 1152
2245. Krasner, M.: Une généralisation de la théorie locale des corps de classes. Généralisation du symbole de Hasse et la loi d'isomorphisme pour les extensions galoisiennes. Analogue local de la loi de densites de Tschebotaröw. C. R. Acad. Sci. Paris **206**, 1940–1942 (1938)
2246. Krasner, M.: Une généralisation de la théorie locale des corps de classes. Conducteur, loi d'unicité, loi d'ordination, loi d'existence. C. R. Acad. Sci. Paris **206**, 1534–1536 (1938)

2247. Krasner, M.: Une généralisation de la thèorie locale des corps de classes. Valeur de conducteur. Interpretation d'une formule de M. Artin. Loi de limitation pour les extensions galoisiennes. Structure des $s(f', e')K/k$ et sa liaison avec la théorie de la ramification. C. R. Acad. Sci. Paris **206**, 1696–1699 (1938)
2248. Krasner, M.: Sur la représentation exponentielle dans les corps relativement galoisiens de nombres $\mathfrak{p}$-adiques. Acta Arith. **3**, 133–173 (1939)
2249. Krasner, M.: La loi de Jordan-Hölder dans les hypergroupes et les suites génératrices des corps de nombres P-adiques. Duke Math. J. **6**, 120–140 (1940)
2250. Krasner, M.: La loi de Jordan-Hölder dans les hypergroupes et les suites génératrices des corps de nombres P-adiques. Chapitre II. Suites génératrices des corps de nombres P-adiques. Duke Math. J. **7**, 121–135 (1940)
2251. Krasner, M.: Théorie non abélienne des corps de classes pour les extensions finies et séparables des corps valués complets: principes fondamentaux; espaces de polynomes et transformation T; lois d'unicité, d'ordination et d'existence. C. R. Acad. Sci. Paris **222**, 626–628 (1946)
2252. Krasner, M.: Théorie non-abélienne des corps de classes pour les extensions finies et séparables des corps valués complets: conducteur, théorie de l'irrégularité. C. R. Acad. Sci. Paris **222**, 984–986 (1946)
2253. Krasner, M.: Théorie non-abélienne des corps de classes pour les extensions finies et séparables des corps valués complets: relations avec la théorie de la ramification; loi de limitation pour les extensions galoisiennes. C. R. Acad. Sci. Paris **222**, 1370–1372 (1946)
2254. Krasner, M.: Théorie non-abélienne des corps de classes pour les extensions galoisiennes des corps de nombres algébriques: bimatrices; représentations bimatricielles des semi-groupes abéliens libres. C. R. Acad. Sci. Paris **225**, 785–787 (1947)
2255. Krasner, M.: Théorie non abélienne des corps de classes pour les extensions galoisiennes des corps de nombres algébriques: anneau des représentations d'un groupe; représentations associées du groupe de Galois et du semi-groupe des idéaux. C. R. Acad. Sci. Paris **225**, 973–975 (1947)
2256. Krasner, M.: Théorie non abélienne des corps de classes pour les extensions galoisiennes des corps de nombres algébriques: anneau principal; lois d'unicité, d'ordination, d'existence (forme provisoire), d'isomorphisme et de décomposition; loi de monodromie. C. R. Acad. Sci. Paris **225**, 1113–1115 (1947)
2257. Krasner, M.: Théorie non abélienne des corps de classes pour les extensions galoisiennes des corps de nombres algébriques: conséquences de la loi de monodromie; résumé de la théorie locale. C. R. Acad. Sci. Paris **226**, 535–537 (1948)
2258. Krasner, M.: Théorie non abélienne des corps de classes pour les extensions galoisiennes des corps de nombres algébriques: f-extensions; conducteur. C. R. Acad. Sci. Paris **226**, 1231–1233 (1948)
2259. Krasner, M.: Théorie non abélienne des corps de classes pour les extensions galoisiennes des corps de nombres algébriques: forme définitive de la loi d'existence. C. R. Acad. Sci. Paris **226**, 1656–1658 (1948)
2260. Krasner, M.: Nombre des extensions d'un degré donné d'un corps $\wp$-adique: énoncé des résultats et préliminaires de la démonstration (espace des polynomes, transformation T). C. R. Acad. Sci. Paris **254**, 3470–3472 (1962)
2261. Krasner, M.: Nombre des extensions d'un degré donné d'un corps $\wp$-adique: suite de la démonstration. C. R. Acad. Sci. Paris **255**, 224–226 (1962)
2262. Krasner, M.: Nombre des extensions de degré donné d'un corps $\wp$-adique: compléments au théorème 1 dans le cas non p-adique; démonstration du théorème 2. C. R. Acad. Sci. Paris **255**, 3095–3097 (1962)
2263. Krasner, M.: Nombre des extensions de degré donné d'un corps de nombres p-adiques: les conditions d'Ore et la caractérisation de $E_{k,j}^{(n)}$; préliminaires du calcul de $N_{k,j,s}^{(n)}$. C. R. Acad. Sci. Paris **255**, 1682–1684 (1962)
2264. Krasner, M.: Nombre des extensions de degré donné d'un corps p-adique: calcul de $N_{k,j,s}^{(n)}$; démonstration du théorème 1. C. R. Acad. Sci. Paris **255**, 2342–2344(1962)

2265. Krasner, M.: Nombre des extensions d'un degré donné d'un corps p-adique. In: Les Tendances Géometriques en Algébre et Théorie des Nombres, 143–169. CNRS (1966)
2266. Krasner, M.: Remarques au sujet d'une note de J.-P. Serre: "Une 'formule de masse' pour les extensions totalement ramifiées de degré donné d'un corps local": une démonstration de la formule de M.Serre á partir de mon théorème sur le nombre des extensions séparables d'un corps valué localement compact, qui sont d'un degré et d'une différente donnés. C. R. Acad. Sci. Paris **288**, A863–A865 (1979)
2267. Krasner, M., Kuntzmann, J.: Remarques sur les hypergroupes. C. R. Acad. Sci. Paris **224**, 525–527 (1947)
2268. Krasner, M., Ranulac, B.: Sur une propriété des polynomes de la division du cercle. C. R. Acad. Sci. Paris **204**, 397–399 (1937)
2269. Krätzel, E., Lamm, C.: Von Wiesbaden nach Tiflis. Die wechselvolle Lebensgeschichte des Zahlentheoretikes Arnold Walfisz. Mitt. Dtsch. Math.-Ver. **21**, 42–51 (2013)
2270. Kraus, A.: Remarques sur le premier cas du théorème de Fermat sur les corps de nombres. Acta Arith. **167**, 133–141 (2015)
2271. Kraus, A.: Équation de Fermat et nombres premiers inertes. Internat. J. Number Th. **11**, 2341–2351 (2015)
2272. Krause, U.: A characterization of algebraic number fields with cyclic class groups of prime power order. Math. Z. **186**, 143–148 (1984)
2273. Kravchenko, R.V., Mazur, M., Petrenko, B.V.: On the smallest number of generators and the probability of generating an algebra. Algebra & Number Th. **6**, 243–291 (2012)
2274. Kravchenko, R.V., Mazur, M., Petrenko, B.V.: Generators of maximal orders. J. Algebra **426**, 32–50 (2015)
2275. Krazer, A.: Lehrbuch der Thetafunktionen. Teubner (1903)
2276. Kronecker, L.: De unitatibus complexis. Dissertation, Univ. Berlin (1845) [J. Reine Angew. Math. **93**, 1–52 (1882); [2300], **1**, 5–73.]
2277. Kronecker, L.: Über die algebraisch auflösbaren Gleichungen. Monatsber. Preuss. Akad. Wiss. **1853**, 365–374; **1856**, 203–215. [[2300], **4**, 1–11, 25–37.]
2278. Kronecker, L.: Mémoire sur les facteurs irréductiblés de l'expression $x^n - 1$. J. Math. Pures Appl. **19**, 177–192 (1854) [[2300], **1**, 75–92.]
2279. Kronecker, L.: Démonstration d'un théorème de M. Kummer. J. Math. Pures Appl. (2) **1**, 396–398 (1856) [[2300], **1**, 93–97.]
2280. Kronecker, L.: Über die algebraisch auflösbare Gleichungen. Monatsber. Preuss. Akad. Wiss. **1856**, 203–215. [[2300], **4**, 27–37.]
2281. Kronecker, L.: Zwei Sätze über Gleichungen mit ganzzahligen Coefficienten. J. Reine Angew. Math. **53**, 173–175 (1857) [[2300], **1**, 103–108.]
2282. Kronecker, L.: Über complexe Einheiten. J. Reine Angew. Math. **53**, 176–181 (1857) [[2300], **1**, 109–118.]
2283. Kronecker, L.: Über die elliptischen Functionen für welche complexe Multiplication stattfindet. Monatsber. Preuss. Akad. Wiss. **1857**, 455–460. [[2300], **4**, 177–183.]
2284. Kronecker, L.: Ueber die Anzahl der verschiedenen Classen quadratischer Formen von negativer Determinante. J. Reine Angew. Math. **57**, 248–255 (1860) [[2300], **4** 185–195.]
2285. Kronecker, L.: Über die complexe Multiplication der elliptischen Functionen. Monatsber. Preuss. Akad. Wiss. **1862**, 363–372. [[2300], **4**, 207–217.]
2286. Kronecker, L.: Bemerkung über die Klassenzahl der aus Wurzeln der Einheit gebildeten complexen Zahlen. Monatsber. Preuss. Akad. Wiss. **1863**, 340–341. [[2300], **1**, 123–131.]
2287. Kronecker, L.: Über den Gebrauch der Dirichletschen Methoden in der Theorie der quadratischen Formen. Monatsber. Preuss. Akad. Wiss. **1864**, 285–303. [[2300], **4**, 227–244.]
2288. Kronecker, L.: Auseinandersetzung einiger Eigenschaften der Klassenzahl idealer complexer Zahlen. Monatsber. Preuss. Akad. Wiss. **1870**, 881–889. [[2300], **1**, 271–282.]
2289. Kronecker, L.: Ueber quadratische Formen von negativer Determinante. Monatsber. Preuss. Akad. Wiss. **1875**, 233–236. [[2300], **4**, 247–259.]

2290. Kronecker, L.: Bemerkungen über das Werk des Herrn Reuschle. Monatsber. Preuss. Akad. Wiss. **1875**, 236–238. [[2300], **5**, 449–452.]
2291. Kronecker, L.: Über Abelsche Gleichungen. Monatsber. Preuss. Akad. Wiss. **1877**, 845–851. [[2300], **4**, 63–71.]
2292. Kronecker, L.: Auszug aus einem Briefe von L. Kronecker an R. Dedekind vom 15. März 1880. SBer. Kgl. Preuß. Akad. Wiss. Berlin **1895**, 115–117. [[2300], **5**, 455–457.]
2293. Kronecker, L.: Ueber die Irreductibilität von Gleichungen. Monatsber. Preuss. Akad. Wiss. **1880**, 155–162. [[2300], **2**, 85–93.]
2294. Kronecker, L.: Grundzüge einer arithmetischen Theorie der algebraischen Grössen. J. Reine Angew. Math. **92**, 1–122 (1882). [[2300], **2**, 237–387.]
2295. Kronecker, L.: Die kubischen Abelschen Gleichungen des Bereichs $\sqrt{-31}$. SBer. Kgl. Preuß. Akad. Wiss. Berlin **1882**, 1151–1154. [[2300], **4**, 123–129.]
2296. Kronecker, L.: Die Zerlegung der ganzen Grössen eines natürlichen Rationalitäts-Bereichs in ihre irreductiblen Factoren. J. Reine Angew. Math. **94**, 344–349 (1883) [[2300], **2**, 409–416.]
2297. Kronecker, L.: Sur les unités complexes. C. R. Acad. Sci. Paris **9**, 93–98, 148–152, 216–222 (1883) [[2300], 3_1, 1–20.]
2298. Kronecker, L.: Ueber bilineare Formen mit vier Variabeln. Abh. Kgl. Preuss. Akad. Wiss. Berlin **1883**, II, nr. 2, 1–60. [[2300], **2**, 425–495
2299. Kronecker, L.: Zur Theorie der elliptischen Funktionen. SBer. Kgl. Preuß. Akad. Wiss. Berlin **1885**, 761–784. [[2300], **4**, 363–389.]
2300. Kronecker, L.: Werke **1–5**. Teubner (1895–1930) [Reprint: Chelsea (1968)]
2301. Krull, W.: Algebraische Theorie der Ringe, I. Math. Ann. **88**, 80–122 (1922)
2302. Krull, W.: Ein neuer Beweis für die Hauptsätze der allgemeinen Idealtheorie. Math. Ann. **90**, 55–64 (1923)
2303. Krull, W.: Axiomatische Begründung der allgemeinen Idealtheorie. SBer. Erlangen **56**, 47–63 (1924)
2304. Krull, W.: Algebraische Theorie der Ringe. II. Math. Ann. **91**, 1–46 (1924)
2305. Krull, W.: Algebraische Theorie der zerlegbaren Ringe (Algebraische Theorie der Ringe. III). Math. Ann. **92**, 183–213 (1924)
2306. Krull, W.: Über verallgemeinerte endliche Abelsche Gruppen. Math. Z. **23**, 161–196 (1925)
2307. Krull, W.: Über Multiplikationsringe. SBer. Heidelberg. Akad. Wiss. **1925**, 13–18.
2308. Krull, W.: Theorie und Anwendung der verallgemeinerten Abelschen Gruppen. SBer. Heidelberg. Akad. Wiss. **1926(1)**, 1–32 (1926)
2309. Krull, W.: Zur Theorie der allgemeinen Zahlringe. Math. Ann. **99**, 51–70 (1928)
2310. Krull, W.: Galoissche Theorie der unendlichen algebraischen Erweiterungen. Math. Ann. **100**, 687–698 (1928)
2311. Krull, W.: Idealtheorie in unendlichen algebraischen Zahlkörpern. Math. Z. **29**, 42–54 (1928)
2312. Krull, W.: Idealtheorie in unendlichen algebraischen Zahlkörpern, II. Math. Z. **31**, 527–557 (1930)
2313. Krull, W.: Ein Hauptsatz über umkehrbare Ideale. Math. Z. **31**, p. 558 (1930)
2314. Krull, W.: Galoissche Theorie bewerteter Körper. SBer. Bayer. Akad. Wiss. **1930(2)**, 225–238
2315. Krull, W.: Matrizen, Moduln und verallgemeinerte Abelsche Gruppen im Bereich der ganzen algebraischen Zahlen. SBer. Heidelberg. Akad. Wiss. **1932(2)**, 13–38
2316. Krull, W.: Allgemeine Bewertungstheorie. J. Reine Angew. Math. **167**, 160–196 (1932)
2317. Krull, W.: Idealtheorie. Springer (1935); 2nd ed. (1968)
2318. Krull, W.: Über allgemeine Multiplikationsringe. Tôhoku Math. J. **41**, 320–326 (1936)
2319. Krull, W.: Beiträge zur Arithmetik kommutativer Integritätsbereiche. Math. Z. **41**, 545–577 (1936)
2320. Krull, W.: Beiträge zur Arithmetik kommutativer Integritätsbereiche, II. v-Ideale und vollständig ganz abgeschlossene Integritätsbereiche. Math. Z. **41**, 665–679 (1936)
2321. Krull, W.: Beiträge zur Arithmetik kommutativer Integritätsbereiche, III. Zum Dimensionsbegriff der Idealtheorie. Math. Z. **42**, 745–766. (1937); Add.: **43**, p. 767 (1937)

2322. Krull, W.: Beiträge zur Arithmetik kommutativer Integritätsbereiche, IV. Unendliche algebraische Erweiterungen endlicher diskreter Hauptordnungen. Math. Z. **42**, 767–773 (1937)
2323. Krull, W.: Beiträge zur Arithmetik kommutativer Integritätsbereiche, V. Potenzreihenringe. Math. Z. **43**, 768–782 (1937)
2324. Krull, W.: Beiträge zur Arithmetik kommutativer Integritätsbereiche, VI. Der allgemeine Diskriminantensatz. Unverzweigte Ringerweiterungen. Math. Z. **45**, 1–19 (1939)
2325. Krull, W.: Beiträge zur Arithmetik kommutativer Integritätsbereiche, VII. Inseparable Grundkörpererweiterung. Bemerkungen zur Körpertheorie. Math. Z. **45**, 319–334 (1939)
2326. Krull, W.: Allgemeine Modul-, Ring- und Idealtheorie. Encyklop. Math. Wiss. **I**, 1, 11, 1–54. Teubner (1939)
2327. Krull, W.: Beiträge zur Arithmetik kommutativer Integritätsbereiche, Eine Bemerkung zu den Beiträgen VI und VII. Math. Z. **48**, 530–531 (1942)
2328. Krull, W.: Die Verzweigungsgruppen in der Galoisschen Theorie beliebiger arithmetischer Körper. Math. Ann. **121**, 446–466 (1950)
2329. Krull, W.: Zur Galoisschen Theorie der arithmetischen Körper. Math. Ann. **126**, 239–252 (1953)
2330. Kuba, G.: The number of lattice points below a logarithmic curve. Arch. Math. (Basel) **69**, 156–163 (1997)
2331. Kubilius, J.: Distribution of prime numbers of the Gaussian field in sectors and contours. Uč. Zap. LGU **137**, 40–52 (1950) (Russian)
2332. Kubilius, J.: On certain problems of the geometry of prime numbers. Mat. Sb. **31**, 507–542 (1952) (Russian)
2333. Kubilius, J., Reiziņš,L., Riekstiņš, E.: Ernests Fogels (1910–1985). Acta Arith. **57**, 178–187 (1991)
2334. Kubo, K.: Über die Noetherschen fünf Axiome in kommutativen Ringen. J. Sci. Hiroshima Univ. **A 10**, 77–84 (1940)
2335. Kubota, T.: Über die Beziehung der Klassenzahlen der Unterkörper des bizyklischen biquadratischen Zahlkörpers. Nagoya Math. J. **6**, 119–127 (1953)
2336. Kubota, T.: Über den bizyklischen biquadratischen Zahlkörper. Nagoya Math. J. **10**, 65–85 (1956)
2337. Kubota, T.: Anwendung Jacobischer Thetafunktionen auf die Potenzreste. Nagoya Math. J. **19**, 1–13 (1961)
2338. Kubota, T.: Some arithmetical applications of an elliptic function. J. Reine Angew. Math. **214/215**, 141–145 (1964)
2339. Kučera, R.: The basis of the Stickelberger ideal and a system of principal circular units of a cyclotomic field. Zap. Naučn. Sem. LOMI **175**, 69–74 (1989) (Russian)
2340. Kučera, R.: Circular units and class groups of abelian fields. Ann. Sci. Math. Québec **28**, 121–136 (2004)
2341. Kučera, R.: The circular units and the Stickelberger ideal of a cyclotomic field revisited. Acta Arith. **174**, 217–238 (2016)
2342. Kučera, R.: Formulae for the relative class number of an imaginary abelian field in the form of a determinant. Nagoya Math. J. **163**, 167–191 (2001)
2343. Kühleitner, M.: On the class number of binary quadratic forms: an omega estimate for the error term. Math. Pannon. **13**, 63–78 (2002)
2344. Kuhn, P.: Eine Verbesserung des Restgliedes beim elementaren Beweis des Primzahlsatzes. Math. Scand. **3**, 75–89 (1955)
2345. Kühnová, J.: Maillet's determinant $D_{p^{n+1}}$. Arch. Math. (Brno) **15**, 209–212 (1979)
2346. Kulkarni, R.S.: On a theorem of Jensen. Amer. Math. Monthly **74**, 960–961 (1967)
2347. Kummer, E.E.: De numeris complexis qui radicibus unitatis et numeris integris realibus constant. Dissertation, Univ. Breslau (1844) [J. Math. Pures Appl. (1) **12**, 185–212 (1847); [2377], **1**, 165–192.]
2348. Kummer, E.E.: Über die Divisoren gewisser Formen der Zahlen, welche aus der Theorie der Kreistheilung entstehen. J. Reine Angew. Math. **29**, 107–116 (1845) [[2377], **1**, 193–202.]

2349. Kummer, E.E.: Extrait d'une lettre de M. Kummer a M. Liouville. J. Math. Pures Appl. (1) **12**, p. 136 (1847) [[2377], **1**, p. 298.]
2350. Kummer, E.E.: Zur Theorie der complexen Zahlen. J. Reine Angew. Math. **35**, 319–325 (1847) [[2377], **1**, 203–210.]
2351. Kummer, E.E.: Über die Zerlegung der aus Wurzeln der Einheit gebildeten complexen Zahlen in ihre Primfaktoren. J. Reine Angew. Math. **35**, 327–367 (1847) [[2377], **1**, 211–251.]
2352. Kummer, E.E.: Beweis des Fermatschen Satzes der Unmöglichkeit von $x^\lambda + y^\lambda = z^\lambda$ für eine unendliche Anzahl von Primzahlen λ. Monatsber. Preuss. Akad. Wiss. **1847**, 132–141, 305–319 [[2377], **1**, 274–319.]
2353. Kummer, E.E.: Letter to Kronecker of December, 28th, 1849. In: [4], 84–87. [[2377], **1**, 114–117.]
2354. Kummer, E.E.: Bestimmung der Anzahl nicht äquivalenter Classen für die aus λ-ten Wurzeln der Einheit gebildeten complexen Zahlen und die ideale Faktoren derselben. J. Reine Angew. Math. **40**, 93–116 (1850) [[2377], **1**, 299–322.]
2355. Kummer, E.E.: Zwei besondere Untersuchungen über die Classen Anzahl und über die Einheiten der aus den λ-ten Wurzeln der Einheit gebildeten complexen Zahlen. J. Reine Angew. Math. **40**, 117–129 (1850) [[2377], **1**, 323 –335.]
2356. Kummer, E.E.: Allgemeiner Beweis des Fermatschen Satzes, dass die Gleichung $x^\lambda + y^\lambda = z^\lambda$ durch ganze Zahlen unlösbar ist, für alle diejenige Potenz-Exponenten λ, welche ungerade Primzahlen sind und in den Zählern der ersten $\frac{1}{2}(\lambda - 3)$ Bernoulli'schen Zahlen nicht vorkommen. J. Reine Angew. Math. **40**, 130–138 (1850) [[2377], **1**, 336–344.]
2357. Kummer, E.E.: Allgemeine Reciprocitätsgesetze für beliebig hohe Potenzreste. Monatsber. Preuss. Akad. Wiss. **1850**, 154–165 [[2377], **1**, 345–357.]
2358. Kummer, E.E.: Mémoire sur la théorie des nombres complexes composés de racines de l'unité et des nombres entiers. J. Math. Pures Appl. **16**, 377–498 (1851) [[2377], **1**, 363–384.]
2359. Kummer, E.E.: Über die Ergänzungssätze zu den allgemeinen Reciprocitätsgesetzen. J. Reine Angew. Math. **44**, 93–146 (1852) [[2377], **1**, 485–538.]
2360. Kummer, E.E.: Über die Irregularität der Discriminanten. Monatsber. Preuss. Akad. Wiss. **1853**, 194–200. [[2377], **1**, 539–545.]
2361. Kummer, E.E.: Letter to Kronecker of March, 12th, 1853. In: [4], 92–93. [[2377], **1**, 122–123.]
2362. Kummer, E.E.: Letter to Kronecker of April, 24th, 1853. In: [4], 93–94. [[2377], **1**, 123–124.]
2363. Kummer, E.E.: Über eine besondere Art, aus complexen Einheiten gebildeter Ausdrücke. J. Reine Angew. Math. **50**, 212–232 (1855) [[2377], **1**, 552–572.]
2364. Kummer, E.E.: Theorie der idealen Primfaktoren der complexen Zahlen, welche aus den Wurzeln der Gleichung $\omega^n = 1$ gebildet sind, wenn n eine zusammengesetzte Zahl ist. Abh. Kgl. Preuss. Akad. Wiss. Berlin **1856**, 1–47. [[2377], **1**, 583–629.]
2365. Kummer, E.E.: Über die Gaussische Perioden der Kreistheilung entsprechenden Congruenzwurzeln. J. Reine Angew. Math. **53**, 142–148 (1857) [[2377], **1**, 573–580.]
2366. Kummer, E.E.: Einige Sätze über die aus der Wurzeln der Gleichung $\alpha^\lambda = 1$ gebildeten complexen Zahlen für den Fall, dass die Classenzahl durch λ teilbar ist, nebst Anwendung derselben auf einen weiteren Beweis des letzten Fermatschen Satzes. Abh. Kgl. Preuss. Akad. Wiss. Berlin **1857**, 41–74. [[2377], **1**, 639–672.]
2367. Kummer, E.E.: Über die allgemeinen Reciprocitätsgesetze der Potenzreste. Monatsber. Preuss. Akad. Wiss. **1858**, 158–171. [[2377], **1**, 673–687.]
2368. Kummer, E.E.: Über die Ergänzungssätze zu den allgemeinen Reciprocitätsgesetzen. J. Reine Angew. Math. **56**, 270–279 (1859) [[2377], **1**, 688–697.]
2369. Kummer, E.E.: Über die allgemeinen Reziprozitätsgesetze unter den Resten und Nichtresten der Potenzen, deren Grad eine Primzahl ist. Abh. Kgl. Preuss. Akad. Wiss. Berlin **1859**, 19–159. [[2377], **1**, 699–839.]
2370. Kummer, E.E.: Zwei neue Beweise der allgemeinen Reziprozitätsgesetze unter den Resten und Nichtresten der Potenzen, deren Grad eine Primzahl ist. Abh. Kgl. Preuss. Akad. Wiss. Berlin **1861**, 81–122. [J. Reine Angew. Math. **100**, 10–50 (1887); [2377], **1**, 842–882.]

2371. Kummer, E.E.: Über die Classenanzahl der aus n-ten Einheitswurzeln gebildeten complexen Zahlen. Monatsber. Preuss. Akad. Wiss. **1861**, 1051–1053. [[2377], **1**, 883–885.]
2372. Kummer, E.E.: Über die Classenanzahl der aus zusammengesetzten Einheitswurzeln gebildeten idealen complexen Zahlen. Monatsber. Preuss. Akad. Wiss. **1863**, 21–28. [[2377], **1**, 887–894.]
2373. Kummer, E.: Über die aus 31sten Wurzeln der Einheit gebildeten Zahlen. Monatsber. Preuss. Akad. Wiss. **1870**, 755–769. [[2377], **1**, 907–918.]
2374. Kummer, E.E.: Ueber eine Eigenschaft der Einheiten der aus den Wurzeln der Gleichung $\alpha^\lambda = 1$ gebildeten complexen Zahlen und über den zweiten Factor der Klassenzahl. Monatsber. Preuss. Akad. Wiss. **1870**, 855–880. [[2377], **1**, 919–944.]
2375. Kummer, E.E.: Ueber diejenigen Primzahlen λ, für welche die Klassenzahl der aus λ^{ten} Einheitswurzeln gebildeten complexen Zahlen durch λ theilbar ist. SBer. Kgl. Preuß. Akad. Wiss. Berlin **1874**, 239–248. [[2377], **1**, 945–954.]
2376. Kummer, E.E.: Zwei neue Beweise der allgemeinen Reciprocitätsgesetze unter den Resten und Nichtresten der Potenzen, deren Grad eine Primzahl ist. J. Reine Angew. Math. **100**, 10–50 (1887) [[2377], **1**, 842–882.]
2377. Kummer, E.E.: Collected Papers, **1**,**2**. Springer (1975)
2378. Kunert, D.: Ein neuer Beweis für die Reziprozitätsformel der Gaussschen Summen in beliebigen algebraischen Zahlkörpern. Math. Z. **40**, 326–347 (1936)
2379. Kunz, E., Nastold, H.-J.: In memoriam Friedrich Karl Schmidt. Jahresber. Deutsch. Math.-Verein. **83**, 169–181 (1981)
2380. Kurihara, M.: On the p-adic expansion of units of cyclotomic fields. J. Number Theory **32**, 226–253 (1989)
2381. Kurihara, M.: Some remarks on conjectures about cyclotomic fields and K-groups of Z. Compositio Math. **81**, 223–236 (1992)
2382. Kurihara, M.: The exponential homomorphisms for the Milnor K-groups and an explicit reciprocity law. J. Reine Angew. Math. **498**, 201–221 (1998)
2383. Kuroda, S.: Über den Dirichletschen Körper. J. Fac. Sci. Univ. Tokyo **4**, 383–406 (1943)
2384. Kuroda, S.: Über die Klassenzahlen algebraischer Zahlkörper. Nagoya Math. J. **1**, 1–10 (1950)
2385. Kuroda, S.-N.: On the class number of imaginary quadratic number fields. Proc. Japan Acad. Sci. **40**, 365–367 (1964)
2386. Kuroda, S.-N.: Über den allgemeinen Spiegelungssatz für Galoissche Zahlkörper. J. Number Theory **2**, 282–297 (1970)
2387. Kürschak, J.: Über Limesbildung und allgemeine Körpertheorie. J. Reine Angew. Math. **142**, 211–253 (1931)
2388. Kutsuna, M.: On a criterion for the class number of a quadratic field to be one. Nagoya Math. J. **79**, 123–129 (1980)
2389. Kuzmin, R.O.: Ivan Ivanovič Ivanov (1862–1939). Izv. Akad. Nauk SSSR, Ser. Mat. **4**, 357–362 (1940) (Russian)
2390. Kuzmin, R.O.: The life and scientific activity of Egor Ivanovič Zolotarev. Uspekhi Mat. Nauk **2(6)**, 22–51 (1947) (Russian)
2391. Kwon, S.-H., Louboutin, S., Park, S.-M.: Nonabelian normal CM-fields of degree $2pq$. J. Aus. Math. Soc. **87**, 129–144 (2009)
2392. Labute, J.[P.]: Classification des groupes de Demuškin. C. R. Acad. Sci. Paris **260**, 1043–1046 (1965)
2393. Labute, J.P.: Les groupes de Demuškin de rang dénombrable. C. R. Acad. Sci. Paris **262**, A4–A7 (1966)
2394. Labute, J.P.: Demuškin groups of rank $\aleph_0$. Bull. Soc. Math. France **94**, 211–244(1966)
2395. Labute, J.P.: Classification of Demushkin groups. Canad. J. Math. **19**, 106–132 (1967)
2396. Lagarias, J.C.: On the computational complexity of determining the solvability or unsolvability of the equation $X^2 - DY^2 = -1$. Trans. Amer. Math. Soc. **260**, 485–508 (1980)
2397. Lagarias, J.C., Montgomery, H.L., Odlyzko, A.M.: A bound for the least prime ideal in the Chebotarev density theorem. Invent. Math. **54**, 271–296 (1979)

2398. Lagarias, J.C., Odlyzko, A.M.: Effective versions of the Chebotarev density theorem. In: Algebraic Number Fields. 409–464. Academic Press (1977)
2399. Lagrange, J.L.: Recherches d'arithmétique, Nouv. Mém. Acad. Roy. Sci. et Belles-Lettres, Berlin **1773**, 265–312; **1775**, 325–356 [Lagrange, J.L., Oeuvres. **3**, 695–795. Gauthier-Villars (1869)]
2400. Lakatos, P.: Salem numbers. PV numbers and spectral radii of Coxeter transformations. C.R. Math. Acad. Sci. Soc. R. Canada **23**, 71–77 (2001)
2401. Lakatos, P.: Salem numbers defined by Coxeter transformation. Linear Algebra Appl. **432**, 144–154 (2010)
2402. Lakein, R.B.: Euclid's algorithm in complex quartic fields. Acta Arith. **20**, 393–400 (1972)
2403. Lakkis, K.: Die verallgemeinerten Gaußschen Summen. Arch. Math. (Basel) **17**, 505–509 (1966)
2404. Lakkis, K.: Die galoisschen Gauss'schen Summen von Hasse. Bull. Soc. Math. Grèce **7**, 183–371 (1966)
2405. Lakkis, K.: Die lokalen verallgemeinerten Gauss'schen Summen. Bull. Soc. Math. Grèce **8**, 143–150 (1967)
2406. Lam, T.Y., Leroy, A.: Hilbert 90 theorems over division rings. Trans. Amer. Math. Soc. **345**, 595–622 (1994)
2407. Lamé, G.: Mémoire d'analyse indetérminée, démontrant que l'équation $x^7 + y^7 = z^7$ est impossible en nombres entiers. J. Math. Pures Appl. (1) **5**, 195–210 (1840)
2408. Lamé, G.: Démonstration générale du thèoréme de Fermat. C. R. Acad. Sci. Paris **24**, 310–315 (1847)
2409. Lamé, G.: Mémoire sur la résolution, en nombres complexes, de l'équation $A^n + B^n + C^n = 0$. J. Math. Pures Appl. (1) **12**, 172–184 (1847)
2410. Lampe, E.: Nachruf für Ernst Eduard Kummer. Jahresber. Deutsch. Math.-Verein. **2**, 13–28 (1892/93)
2411. Lampe, E.: Karl Weierstrass. Jahresber. Deutsch. Math.-Verein. **6**, 27–44 (1899)
2412. Lamprecht, E.: Allgemeine Theorie der Gaussschen Summen in endlichen kommutativen Ringen. Math. Nachr. **9**, 149–196 (1953)
2413. Lamzouri, Y.: On the average of the number of imaginary quadratic fields with a given class number. The Ramanujan J. **44**, 411–416 (2017)
2414. Landau, E.: Ueber die zu einem algebraischen Zahlkörper gehörige Zetafunction und die Ausdehnung der Tschebyscheffschen Primzahlentheorie auf das Problem der Verteilung der Primideale. J. Reine Angew. Math. **125**, 64–188 (1903) [[2447], **1**, 201–325.]
2415. Landau, E.: Neuer Beweis des Primzahlsatzes und Beweis des Primidealsatzes. Math. Ann. **56**, 645–670 (1903) [[2447], **1**, 327–352.]
2416. Landau, E.: Über die Klassenzahl der binären quadratischen Formen von negativer Diskriminante. Math. Ann. **56**, 671–676 (1903) [[2447], **1**, 354–359.]
2417. Landau, E.: Über die Darstellung der Anzahl der Idealklassen eines algebraischen Körpers durch eine unendliche Reihe. J. Reine Angew. Math. **127**, 167–174 (1904) [[2447], **2**, 145–174.]
2418. Landau, E.: Über den Zusammenhang einiger neuerer Sätze der analytischen Zahlentheorie. SBer. Kais. Akad. Wissensch. Wien **115**, 589–632 (1906) [[2447], **2**, 346–388.]
2419. Landau, E.: Über die Verteilung der Primideale in den Idealklassen eines algebraischen Zahlkörpers. Math. Ann. **63**, 145–204 (1907) [[2447], **3**, 181–240.]
2420. Landau, E.: Über die Primzahlen in einer arithmetischen Progression und die Primideale in einer Idealklasse. SBer. Kais. Akad. Wissensch. Wien **117**, 1095–1107 (1908) [2447], **4**, 73–85.]
2421. Landau, E.: Die Bedeutung der Pfeifferschen Methode für die analytische Zahlentheorie. SBer. Kais. Akad. Wissensch. Wien **121**, 2195–2332 (1912) [[2447], **5**, 284–419.]
2422. Landau, E.: Über eine idealtheoretische Funktion. Trans. Amer. Math. Soc. **13**, 1912, 1–21 (1912) [[2447], **5**, 107–127.]
2423. Landau, E.: Über die Anzahl der Gitterpunkte in gewissen Bereichen. Nachr. Ges. Wiss. Göttingen **1912**, 687–770 [[2447], **5**, 156–239.]

2424. Landau, E.: Über die Primzahlen in definiten quadratischen Formen und die Zetafunktion reiner kubischer Körper. In: [565], 244–273 (1914) [[2447], **6**, 105–134.]
2425. Landau, E.: Über eine Aufgabe der Funktionentheorie. Tôhoku Math. J. **5**, 97–116 (1914) [[2447]. 6, 167–186.]
2426. Landau, E.: Neuer Beweis eines analytischen Satzes des Herrn de la Vallée Poussin. Jahresber. Deutsch. Math.-Verein. **24**, 250–278 (1915) [[2447], **6**, 308–342.]
2427. Landau, E.: Über die Anzahl der Gitterpunkte in gewissen Bereichen, II. Nachr. Ges. Wiss. Göttingen **1915**, 209–243 [[2447], **6**, 308–342.]
2428. Landau, E.: Über die Heckesche Funktionalgleichung. Nachr. Ges. Wiss. Göttingen **1917**, 102–111 [[2447], **6**, 439–448.]
2429. Landau, E.: Richard Dedekind. Gedächtnisrede. Nachr. Ges. Wiss. Göttingen **1917**, 50–70 [[2447], **4**, 449–465.]
2430. Landau, E.: Über die Anzahl der Gitterpunkte in gewissen Bereichen, III. Nachr. Ges. Wiss. Göttingen **1917**, 96–101. [[2447], **6**, 433–438.]
2431. Landau, E.: Einführung in die elementare und analytische Theorie der algebraischen Zahlen und der Ideale. Teubner (1918), 2nd ed. (1927); [Reprint: Chelsea (1949)]
2432. Landau, E.: Über Ideale und Primideale in Idealklassen. Math. Z. **2**, 52–154 (1918) [[2447], **7**, 11–113.]
2433. Landau, E.: Abschätzungen von Charaktersummen, Einheiten und Klassenzahlen. Nachr. Ges. Wiss. Göttingen **1918**, 79–97. [[2447], **7**, 114–132.]
2434. Landau, E.: Über imaginär-quadratische Zahlkörper mit gleicher Klassenzahl. Nachr. Ges. Wiss. Göttingen **1918**, 277–284. [[2447], **7**, 142–149.]
2435. Landau, E.: Über die Klassenzahl imaginär-quadratischer Zahlkörper. Nachr. Ges. Wiss. Göttingen **1918**, 285–295. [[2447], **7**, 150–160.]
2436. Landau, E.: Verallgemeinerung eines Pólyaschen Satzes auf algebraische Zahlkörper. Nachr. Ges. Wiss. Göttingen **1918**, 478–488. [[2447], **7**, 161–171.]
2437. Landau, E.: Zur Theorie der Heckeschen Zetafunktionen, welche komplexen Charakteren entsprechen. Math. Z. **4**, 1919, 152–162 [[2447], **7**, 176–186.]
2438. Landau, E.: Über die Wurzeln der Zetafunktion eines algebraischen Zahlkörpers. Math. Ann. **79**, 1919, 388–401 (1919) [[2447], **7**, 187–200.]
2439. Landau, E.: Über die Zerlegung total positiver Zahlen in Quadrate. Nachr. Ges. Wiss. Göttingen **1919**, 392–396 [[2447], **7**, 201–204.]
2440. Landau, E.: Der Minkowskische Satz über die Körperdiskriminante. Nachr. Ges. Wiss. Göttingen **1922**, 80–82 [[2447], **7**, 424–426.]
2441. Landau, E.: Über die Anzahl der Gitterpunkte in gewissen Bereichen, IV. Nachr. Ges. Wiss. Göttingen **1924**, 137–150 [[2447], **8**, 145–158.]
2442. Landau, E.: Bemerkungen zu der Arbeit des Herrn Walfisz: Über das Piltzsche Teilerproblem in algebraischen Zahlkörpern. Math. Z. **22**, 189–205 (1925); add.: **24**, 1925, p. 192 (1925) [[2447], **8**, 211–227.]
2443. Landau, E.: Über Dirichletsche Reihen mit komplexen Charakteren. J. Reine Angew. Math. **157**, 26–32 (1927) [[2447], **8**, 377–383.]
2444. Landau, E.: Vorlesungen über Zahlentheorie. **1–3**. S. Hirzel (1927)
2445. Landau, E.: Über die Irreduzibilität der Kreisteilungsgleichung. Math. Z. **29**, p. 462 (1929) [[2447], **9**, p. 59.]
2446. Landau, E.: Bemerkungen zum Heilbronnschen Satz. Acta Arith. **1**, 1–18 (1935) [[2447], **9**, p. 59.]
2447. Landau, E.: Collected Works. **1–9**. Thales Verlag (1985–1987)
2448. Landsberg, G.: Ueber das Fundamentalsystem und die Diskriminante der Gattungen algebraischer Zahlen welche aus Wurzelgrössen gebildet sind. J. Reine Angew. Math. **117**, 140–147 (1897)
2449. Landsberg, G.: Ueber Modulsysteme zweiter Stufe und Zahlenringe. Nachr. Ges. Wiss. Göttingen **1897**, 277–303.
2450. Landsberg, G.: Über Reduktion von Gleichungen durch Adjunktion. J. Reine Angew. Math. **132**, 1–20 (1907)

2451. Lang, A.: Arnold Meyer. Jahresber. Deutsch. Math.-Verein. **5**, 18–20 (1897)
2452. Lang, H.: Über eine Gattung elementar-arithmetischer Klasseninvarianten reell-quadratischer Zahlkörper. J. Reine Angew. Math. **233**, 123–175 (1968)
2453. Lang, H.: Über Anwendungen höherer Dedekindscher Summen auf die Struktur elementar-arithmetischer Klasseninvarianten reell-quadratischer Zahlkörper. J. Reine Angew. Math. **254**, 17–32 (1972)
2454. Lang, S.: Integral points on curves. Publ. Math. Inst. Hautes Études Sci. **6**, 27–43 (1960)
2455. Lang, S.: Diophantine geometry. John Wiley & Sons (1962)
2456. Lang, S.: Algebraic numbers. Addison-Wesley (1964)
2457. Lang, S.: Algebraic number theory. Addison-Wesley (1970) [2nd ed. Springer (1994)]
2458. Lang, S.: Elliptic Functions. Addison-Wesley (1973) [2nd ed. Springer (1987)]
2459. Lang, S.: Cyclotomic Fields. **1**, **2**. Springer (1978–1980); 2nd ed. (1990)
2460. Lang, S.: Fundamentals of Diophantine geometry. Springer (1983)
2461. Lang, S.: Number theory. III, Diophantine Geometry. Encyclopaedia of Mathematical Sciences. **60**. Springer (1991)
2462. Lang, S.: Le théorème d'irréductibilité de Hilbert. Sém. Bourbaki **5**, 451–463, Soc. Math. France (1995)
2463. Lang, S., Tate, J.: Principal homogeneous spaces over abelian varieties. Amer. J. Math. **80**, 659–684 (1958)
2464. Lang, S.D.: Note on the class-number of the maximal real subfield of a cyclotomic field. J. Reine Angew. Math. **290**, 70–72 (1977)
2465. Langevin, M.: Solution des problèmes de Favard. Ann. Inst. Fourier (Grenoble) **38(2)**, 1–10 (1988)
2466. Langevin, M.: Solution et histoire d'un problème de Favard. Progr. Math. **75**, 221–269 (1988)
2467. Langevin, M., Reyssat, E., Rhin, G.: Diamètres transfinis et problème de Favard. Ann. Inst. Fourier (Grenoble) **38(1)**, 1–16 (1988)
2468. Langlands, R.P.: On Artin's L-functions. Rice Univ. Studies **56**, 23–28 (1970)
2469. Langlands, R.P.: Some contemporary problems with origins in the Jugendtraum. In: [460], 401–418
2470. Langlands, R.P.: Base change for $GL(2)$. Ann. Math. Stud. **96**, 1–237 (1980)
2471. Lao, H.: On the distribution of integral ideals and Hecke Grössencharacters. Chin. Ann. Math. **31**, 385–392 (2010)
2472. Larson, E., Rolen, L.: Progress towards counting D_5 quintic fields. Involve **5**, 91–97 (2012)
2473. Lasker, E.: Zur Theorie der Moduln und Ideale. Math. Ann. **60**, 20–115 (1905); Corr.: **60**, 607–608 (1905)
2474. Lasker, E.: Über eine Eigenschaft der Diskriminante. SBer. Berliner Math. Ges. **15**, 176–178 (1916)
2475. Latimer, C.G.: On the prime ideals of a general cubic Galois field. Amer. J. Math. **51**, 295–304 (1929)
2476. Latimer, C.G.: On the class number of cubic cyclotomic fields. Ann. Math. (2) **31**, 347–354 (1930)
2477. Latimer, C.G.: A generalization of Eisenstein's canonical cubic and associated forms. Amer. J. Math. **52**, 127–136 (1930)
2478. Latimer, C.G.: On the class number of a cyclic field. Trans. Amer. Math. Soc. **35**, 411–417 (1933)
2479. Latimer, C.G.: On the class number of a cyclic field and a subfield. Bull. Amer. Math. Soc. **39**, 115–118 (1933)
2480. Latimer, C.G.: On the units in a cyclic field. Amer. J. Math. **56**, 69–74 (1934)
2481. Latimer, C.G.: The quadratic subfields of a generalized quaternion algebra. Duke Math. J. **2**, 681–684 (1936)
2482. Latimer, C.G., MacDuffee, C.C.: A correspondence between classes of ideals and classes of matrices. Ann. Math. (2) **34**, 313–316 (1933)

2483. Lau, Y.-K.: On the mean square formula of the error term for a class of arithmetical functions. Monatsh. Math. **128**, 111–129 (1999)
2484. Laubie, F.: Groupes de ramification et corps résiduels. Bull. Sci. Math. (2) **105**, 309–320 (1981)
2485. Laubie, F.: Une théorie du corps de classes local non abélien. Compositio Math. **143**, 339–362 (2007)
2486. Laudal, O.A., Piene, R. (ed.): The legacy of Niels Henrik Abel. Papers from the Abel Bicentennial Conference held at the University of Oslo, Oslo, June 3–8, 2002. Springer (2004)
2487. Laugwitz, D.: Bernhard Riemann 1826–1866. Wendepunkte in der Auffassung der Mathematik. Birkhäuser (1996) [English translation: Bernhard Riemann 1826–1866. Turning points in the conception of mathematics. Birkhäuser (1999); reprint: (2008)]
2488. Lavrik, A.F.: On the problem of distribution of the values of class-numbers of purely radical quadratic forms with negative determinant. Izv. Akad. Nauk Uzb. SSR., ser. Fiz.-Mat. Nauk **1959(1)**, 81–90 (Russian)
2489. Lavrik, A.F.: A note on the Siegel-Brauer theorem concerning parameters of algebraic number fields. Mat. Zametki **8**, 259–263 (1970) (Russian)
2490. Lavrik, A.F.: A method of evaluation of double sums with a real quadratic character and its applications. Izv. Akad. Nauk SSSR, Ser. Mat. **35**, 1189–1207 (1971) (Russian)
2491. Lavrik, A.F., Edgorov, Ž.: The product of the number of divisor classes by the regulator of algebraic number fields. Izv. Akad. Nauk Uzb. SSR., ser. Fiz.-Mat. Nauk **1975(2)**, 77–79 (Russian)
2492. Leahey, W.J.: A note on a theorem of I.Niven. Proc. Amer. Math. Soc. **16**, 1130–1131 (1965)
2493. Lebesgue, H.: L'oeuvre mathématique de Georges Humbert, quelques mots sur Camille Jordan. Bull. Sci. Math. (2) **46**, 220–233 (1922)
2494. Lebesgue, H.: Notice sur la vie et les travaux de Camille Jordan (1838–1922). Mém. Acad. Sci. Paris (2) $\mathbf{58}_1$, XXXIX–LXVI (1926)
2495. Lebesgue, V.A.: Démonstration de l'impossibilité de résoudre l'équation $x^7 + y^7 + z^7 = 0$ en nombres entiers. J. Math. Pures Appl. (1) **5**, 276–279, 348–349 (1840)
2496. Lebesgue, V.A.: Démonstration de l'irréductibilité de l'équation aux racines primitives de l'unité. J. Math. Pures Appl. (2) **4**, 105–110 (1859)
2497. Lednev, N.A.: On units of relatively cyclic algebraic fields. Mat. Sb. **6**, 227–261 (1939) (Russian)
2498. Lednev, N.A.: On the inverse problem of Galois theory. Mat. Sb. **9**, 137–164 (1941) (Russian)
2499. Lee, G.-N., Kwon, S.-H.: CM-fields with relative class number one. Math. Comp. **75**, 997–1013 (2006)
2500. Lee, G.-N., Kwon, S.-H.: The zeros of Dedekind zeta functions and class numbers of CM-fields. Math. Comp. **77**, 2437–2445 (2008)
2501. Lee, K.-C.: On the average order of characters in totally real algebraic number fields. Chin. J. Math. **7**, 77–90 (1979)
2502. Lee, Y.M.: Universal forms over $Q(\sqrt{5})$. The Ramanujan J. **16**, 97–104 (2008)
2503. Leedham-Green, C.R.: The classgroup of Dedekind domains. Trans. Amer. Math. Soc. **163**, 493–500 (1972)
2504. Leep, D.B., Wadsworth, A.R.: The Hasse norm theorem mod squares. J. Number Theory **42**, 337–348 (1992)
2505. Lefeuvre, Y.: Corps diédraux à multiplication complexe principaux. Ann. Inst. Fourier (Grenoble) **50**, 67–103 (2000)
2506. Lefeuvre, Y., Louboutin, S.: The class number one problem for the dihedral CM-fields. In: Algebraic Number Theory and Diophantine Analysis (Graz, 1998), 249–275. de Gruyter (2000)
2507. Lefton, P.: On the Galois groups of cubics and trinomials. Acta Arith. **35**, 239–246 (1979)
2508. Legendre, A.-M.: Recherches d'analyse indéterminée. Histoire de l'Academie Royale des Sciences de Paris **1785**, 465–559

2509. Legendre, A.-M.: Essai sur la théorie des nombres. Duprat, An 6 (1797/1798). [2nd ed.: Courcier (1808); reprint: Cambridge University Press (2009); 3rd ed. "Théorie des nombres", Firmin Didot (1830); reprint: Hermann (1899); German translation: Teubner (1886,1894)]
2510. Legendre, A.-M.: Sur quelques objets d'analyse indéterminée et particulièrment sur le théorème de Fermat. Mém. Acad. Roy. Sci. **6**, 1–60 (1823) [Reprint: Sphinx-Oedipe **4**, 97–128 (1909); err. **5**, p. 112 (1910)]
2511. Lehmer, D.H.: A list of errors in tables of the Pell equation. Bull. Amer. Math. Soc. **32**, 545–550 (1926)
2512. Lehmer, D.H.: On imaginary quadratic fields whose class-number is unity. Amer. Math. Monthly **39**, p. 360 (1933)
2513. Lehmer, D.H.: Factorization of certain cyclotomic functions. Ann. Math. (2) **34**, 461–479 (1933)
2514. Lehmer, D.H.: The mathematical work of Morgan Ward. Math. Comp. **61**, 307–311 (1993)
2515. Lehmer, D.H., Lehmer, E.: Cyclotomy with short periods. Math. Comp. **41**, 743–758 (1983)
2516. Lehmer, D.H., Lehmer, E.: The sextic period polynomial. Pacific J. Math. **111**, 341–355 (1984)
2517. Lehmer, D.H., Lehmer, E., Vandiver, H.S.: An application of high-speed computing to Fermat's last theorem. Proc. Nat. Acad. Sci. U.S.A. **40**, 25–33 (1954)
2518. Lehmer, D.H., Masley, J.M.: Table of the cyclotomic class numbers $h^*(p)$ and their factors for $200<p<521$. Math. Comp. **32**, 577–582 (1978)
2519. Lehmer, D.N.: On Jacobi's extension of the continued fraction algorithm. Proc. Nat. Acad. Sci. U.S.A. **4**, 360–364 (1918)
2520. Lehmer, D.N.: On ternary continued fractions. Tôhoku Math. J. **37**, 436–445 (1933)
2521. Lehmer, E.: A numerical function applied to cyclotomy. Amer. Math. Monthly **36**, 291–298 (1930)
2522. Lehmer, E.: On congruences involving Bernoulli numbers and the quotients of Fermat and Wilson. Ann. Math. (2) **39**, 350–360 (1938)
2523. Lemmermeyer, F.: Euklidische Ringe, Diplomarbeit, Univ. Heidelberg (1989)
2524. Lemmermeyer, F.: Kuroda's class number formula. Acta Arith. **66**, 245–260 (1994)
2525. Lemmermeyer, F.: On 2-class field towers of imaginary quadratic number fields. J. Théor. Nombres Bordeaux **6**, 261–272 (1994)
2526. Lemmermeyer, F.: The Euclidean algorithm in algebraic number fields. Expo. Math. **13**, 385–416 (1995)
2527. Lemmermeyer, F.: Reciprocity Laws. Springer (2000)
2528. Lemmermeyer, F.: Kronecker-Weber via Stickelberger. J. Théor. Nombres Bordeaux **17**, 555–558 (2005)
2529. Lemmermeyer, F.: Selmer groups and quadratic reciprocity. Abh. Math. Semin. Univ. Hambg **76**, 279–293 (2006)
2530. Lemmermeyer, F.: The development of the principal genus theorem. In: [1467], 529–561
2531. Lemmermeyer, F.: Jacobi and Kummer's ideal numbers. Abh. Math. Semin. Univ. Hambg **79**, 165–187 (2009)
2532. Lemmermeyer, F.: Dedekind's Arbeit über rein kubische Zahlkörper. Schr. Gesch. Math. Didaktik 3, 102–133 (2017)
2533. Lemmermeyer, F.: David Hilbert: Die Theorie der algebraischen Zahlkörper. Jahresber. Deutsch. Math.-Verein. 120, 41–79 (2018)
2534. Lemmermeyer, F., Louboutin, S., Okazaki, R.: The class number one problem for some non-abelian normal CM-fields of degree 24. J. Théor. Nombres Bordeaux **11**, 387–406 (1999)
2535. Lenskoĭ, D.N.: A new proof of Strassmann's theorem. In: Studies in number theory (Saratov) **4**, 68–72 (1972) (Russian)
2536. Lenstra, A.K.: Factoring multivariate integral polynomials. Lecture Notes in Computer Science **154**, 458–465 (1983)

2537. Lenstra, A.K.: Factoring polynomials over algebraic number fields. Lecture Notes in Computer Science **162**, 245–254 (1983)
2538. Lenstra, A.K.: Factoring multivariate polynomials over algebraic number fields. Lecture Notes in Computer Science **176**, 389–396 (1984)
2539. Lenstra, A.K., Lenstra, H.W.Jr., Lovász, L.: Factoring polynomials with rational coefficients. Math. Ann. **261**, 515–534 (1982)
2540. Lenstra, H.W.Jr.: Rational functions invariant under a finite abelian group. Invent. Math. **25**, 299–325 (1974)
2541. Lenstra, H.W.Jr.: Euclid's algorithm in cyclotomic fields. J. Lond. Math. Soc. (2) **10**, 457–465 (1975)
2542. Lenstra, H.W.Jr.: Euclidean number fields of large degree. Invent. Math. **38**, 237–254 (1976/77)
2543. Lenstra, H.W.Jr.: On Artin's conjecture and Euclid's algorithm in global fields. Invent. Math. **42**, 201–224 (1977)
2544. Lenstra, H.W.Jr.: Euclidean number fields, Math. Intelligencer **2**, 6–15 (1980)
2545. Lenstra, H.W.Jr.: Rational functions invariant under a cyclic group. In: Proceedings of the Queen's Number Theory Conference, 1979. Queen's Papers in Pure and Appl. Math. **54**, 91–99 (1980)
2546. Lenstra, H.W.Jr.: On the calculation of regulators and class numbers of quadratic fields. In: Number Theory Days 1980 (Exeter, 1980), 123–150. Cambridge University Press (1982)
2547. Lenstra, H.W.Jr., Stevenhagen, P.: Primes of degree one and algebraic cases of Čebotarev's theorem. Enseign. Math. (2) **37**, 17–30 (1991)
2548. Lenz, H., Aigner, M., Deubner, W.: Richard Rado 1906–1989. Jahresber. Deutsch. Math.-Verein. **93**, 127–145 (1991)
2549. Leopoldt, H.-W.: Zur Geschlechtertheorie in abelschen Zahlkörpern. Math. Nachr. **9**, 350–362 (1953)
2550. Leopoldt, H.-W.: Über Einheitengruppe und Klassenzahl reeller abelscher Zahlkörper. Abh. Deutsch. Akad. Wiss., Berlin, Kl. Math. Nat. **1953(2)**, 1–48
2551. Leopoldt, H.-W.: Über ein Fundamentalproblem der Theorie der Einheiten algebraischer Zahlkörper. SBer. Bayer. Akad. Wiss. **1956**, 41–48
2552. Leopoldt, H.-W.: Eine Verallgemeinerung der Bernoullischen Zahlen. Abh. Math. Semin. Univ. Hambg **22**, 131–140 (1958)
2553. Leopoldt, H.-W.: Zur Struktur der l-Klassengruppe galoisscher Zahlkörper. J. Reine Angew. Math. **199**, 165–174 (1968)
2554. Leopoldt, H.-W.: Über Klassenzahlprimteiler reeller abelscher Zahlkörper als Primteiler verallgemeinerter Bernoullischer Zahlen. Abh. Math. Semin. Univ. Hambg **23**, 36–47 (1959)
2555. Leopoldt, H.-W.: Über die Hauptordnung der ganzen Elemente eines abelschen Zahlkörpers. J. Reine Angew. Math. **201**, 119–149 (1959)
2556. Leopoldt, H.-W.: Obituary: Sigekatu Kuroda (November 11, 1905 – November 3, 1972). J. Number Theory **7**, 1–4 (1975)
2557. Lepistö, T.: The first factor of the class number of the cyclotomic field $k(\exp(2\pi i/p^n))$. Ann. Univ. Turku **70**, 1–7 (1964)
2558. Lepistö, T.: On the first factor of the class-number of the cyclotomic field and Dirichlet's L-functions. Ann. Acad. Sci. Fenn. Math., Ser. A I **387**, 1–53 (1966)
2559. Lepistö, T.: A zero-free region for certain L-functions. Ann. Acad. Sci. Fenn. Math., Ser. A I **576**, 1–13 (1974)
2560. Lepistö, T.: On the growth of the first factor of the class number of the prime cyclotomic field. Ann. Acad. Sci. Fenn. Math., Ser. A I **577**, 1–21 (1974)
2561. Lequain, Y.: A local characterization of Noetherian and Dedekind rings. Proc. Amer. Math. Soc. **94**, 369–370 (1985)
2562. Lerch, M.: Über die arithmetische Gleichung $Cl(-\Delta) = 1$. Math. Ann. **57**, 568–571 (1903)
2563. Lerch, M.: Sur le nombre des classes de formes quadratiques binaires d'un discriminant positif fondamental. J. Math. Pures Appl. (5) **9**, 377–401 (1903)

2564. Lerch, M.: Essais sur le calcul du nombre des classes de formes quadratiques binaires aux coefficients entiers. Acta Math. **29**, 333–424 (1905)
2565. Lerch, M.: Essais sur le calcul du nombre des classes de formes quadratiques binaires aux coefficients entiers, II. Acta Math. **30**, 203–293 (1906)
2566. Lerch, M.: Zjednodušeni Lejeune-Dirichletova postupu při odvozeni vzorců pro počet tříd kvadratickich forem záporného diskriminantu. Časopis mat. fys. **40**, 425–446 (1911)
2567. Leriche, A.: Pólya fields, Pólya groups and Pólya extensions: a question of capitulation. J. Théor. Nombres Bordeaux **23**, 235–249 (2011)
2568. Leriche, A.: Cubic, quartic and sextic Pólya fields. J. Number Theory **133—**, 59–71 (2013)
2569. Leriche, A.: About the embedding of a number field in a Pólya field. J. Number Theory **145**, 210–229 (2014)
2570. Lettl, G.: The ring of integers of an abelian field. J. Reine Angew. Math. **404**, 162–170 (1990)
2571. Lettl, G.: A note on Thaine's circular units. J. Number Theory **35**, 224–226 (1990)
2572. Lettl, G., Prabpayak, C.: Conductor ideals of orders in algebraic number fields. Arch. Math. (Basel) **103**, 133–138 (2014)
2573. Leu, M.-G.: The restricted Nagata's pairwise algorithm and the Euclidean algorithm. Osaka J. Math. **15**, 807–818 (2008)
2574. Leutbecher, A.: Euclidean fields having a large Lenstra constant. Ann. Inst. Fourier (Grenoble) **35(2)**, 83–106 (1985)
2575. Leutbecher, A., Martinet, J.: Lenstra's constant and Euclidean number fields. Astérisque **94**, 87–131 (1982)
2576. Leutbecher, A., Niklasch, G.: On cliques of exceptional units and Lenstra's construction of Euclidean fields. Lecture Notes Math. **1380**, 150–178 (1989)
2577. LeVeque, W.J.: Topics in Number Theory. Addison-Wesley (1956) [Reprint: Dover (2002)]
2578. LeVeque, W.J.: Rational points on curves of genus greater than 1. J. Reine Angew. Math. **206**, 45–52 (1961)
2579. Levi, F.: Kubische Zahlkörper und binäre kubische Formenklassen. Ber. Verhandl. Sächsische Akad. Wiss. Leipzig, Math.-Nat. Kl. **66**, 26–37 (1914)
2580. Levi, F.: Zur Irreduzibilität der Kreisteilungspolynome. Compositio Math. **1**, 303–304 (1934)
2581. Levin, B.V., Fel'dman, N.I., Šidlovski, A.B.: Alexander O. Gelfond. Acta Arith. **17**, 315–336 (1970/71)
2582. Levin, B.V., Tuljaganova, M.I.: The sieve of A. Selberg in algebraic number fields. Lit. Mat. Sb. **6**, 59–73 (1966) (Russian)
2583. Levitz, K.B.: A characterization of general Z.P.I.-rings. Proc. Amer. Math. Soc. **32**, 376–380 (1972)
2584. Levitz, K.B.: A characterization of general Z.P.I.-rings. Pacific J. Math. **42**, 147–151 (1972)
2585. Lévy, P.: L'arithmétique des lois de probabilité. C. R. Acad. Sci. Paris **204**, 80–82 (1937)
2586. Lévy, P., Mandelbrojt, S., Malgrange, B., Malliavin, P.: La vie et l'oeuvre de Jacques Hadamard (1865–1963). Inst. Mat. Univ. Genève (1967)
2587. Lewis, D.J., Mahler, K.: On the representation of integers by binary forms. Acta Arith. **6**, 333–363 (1960/61)
2588. Lewis, D.J., Schinzel, A., Zassenhaus, H.: An extension of the theorem of Bauer and polynomials of certain special type. Acta Arith. **11**, 345–352 (1966)
2589. Lezowski, P.: Computation of the Euclidean minimum of algebraic number fields. Math. Comp. **83**, 1397–1426 (2014)
2590. Lezowski, P., McGown, K.J.: The Euclidean algorithm in quintic and septic cyclic fields. Math. Comp. **86**, 2535–2549 (2017)
2591. Liang, J.J.: On the integral basis of the maximal real subfield of a cyclotomic field. J. Reine Angew. Math. **286/287**, 223–226 (1976)
2592. Liang, J.[J.], Zassenhaus, H.: The minimum discriminant of sixth degree totally complex algebraic number fields. J. Number Theory **9**, 16–35 (1977)

2593. Liang, Y., Wu, Q.: The trace problem for totally positive algebraic integers. J. Aus. Math. Soc. **90**, 341–354 (2011)
2594. Liardet, P.: Sur une conjecture de Serge Lang. C. R. Acad. Sci. Paris **A279**, 435–437 (1974)
2595. Lidl, R., Niederreiter, H.: Finite Fields. Addison-Wesley (1983) [2nd ed. Cambridge University Press (1997)]
2596. Lietzmann, W.: Über das biquadratische Reziprozitätsgesetz in algebraischen Zahlkörpern. Dissertation, Univ. Göttingen (1904)
2597. Lietzmann, W.: Zur Theorie der n-ten Potenzreste in algebraischen Zahlenkörpern. Math. Ann. **60**, 263–284 (1905)
2598. Lietzmann, W.: Zur Theorie der n-ten Potenzreste in algebraischen Zahlenkörpern, II. Math. Ann. **61**, 372–391 (1905)
2599. Lietzmann, W.: Das spezielle Reziprozitätsgesetz im relativ-biquadratischen Zahlkörper. Math. Ann. **68**, 119–124 (1910)
2600. Linnik, Yu.V.: On a conditional theorem of J.E. Littlewood. Dokl. Akad. Nauk SSSR **37**, 142–144 (1942) (Russian)
2601. Linnik, Yu.V.: On the least prime in an arithmetic progression, I. The basic theorem. Mat. Sb. **15**, 139–178 (1944) (Russian)
2602. Linnik, Yu.V.: On the least prime in an arithmetic progression, II. The Deuring-Heilbronn phenomenon. Mat. Sb. **15**, 347–368 (1944) (Russian)
2603. Linnik, Yu.V.: An elementary proof of Siegel's theorem based on the method of I.M.Vinogradov. Izv. Akad. Nauk SSSR, Ser. Mat. **14**, 327–342 (1944) (Russian)
2604. Liouville, J.: Sur la division du périmetre de la lemniscate, le diviseur étant un nombre entier réel ou complexe quelconque. J. Math. Pures Appl. **8**, 507–512 (1843)
2605. Liouville, J.: Sur de classes très eténdues de quantités dont la valeur n'est ni algébrique, ni même reductible à des irrationelles algébriques. J. Math. Pures Appl. **16**, 133–142 (1851)
2606. Liouville, J.: Réponse de M. Liouville. J. Math. Pures Appl. (2) **7**, 41–43 (1862)
2607. Liouville, J.: Note de M. Liouville. J. Math. Pures Appl. (2) **7**, 44–48 (1862)
2608. Lipschitz, R.: Einige Sätze aus der Theorie der quadratischen Formen. J. Reine Angew. Math. **53**, 238–259 (1857)
2609. Lipschitz, R.: Ueber die asymptotischen Gesetze von gewisser Gattungen zahlentheoretischer Funktionen. Monatsber. Preuss. Akad. Wiss. **1865**, 174–185
2610. Littlewood, J.E.: On the class-number of the corpus $P(\sqrt{-k})$. Proc. London Math. Soc. (2) **27**, 358–372 (1928)
2611. Litver, E.L.: On the number of ideal classes in certain special fields. Dokl. Akad. Nauk SSSR **66**, 335–338 (1949) (Russian)
2612. Liu, J., Ye, Y.: Perron's formula and the prime number theorem for automorphic L-functions. Pure Appl. Math. Q. **3**, 481–497 (2007)
2613. Ljunggren, W.: Sur la résolution de quelques équations diophantiennes cubiques dans certains corps quadratiques. Pure Appl. Math. Q. **1943(14)**, 1–23
2614. Llorente, P., Nart, E.: Effective determination of the decomposition of the rational primes in a cubic field. Proc. Amer. Math. Soc. **87**, 579–585 (1983)
2615. Llorente, P., Oneto, A.V.: On the real cubic fields. Math. Comp. **39**, 689–692 (1982)
2616. Llorente, P., Quer, J.: On the 3-Sylow subgroup of the class group of quadratic fields. Math. Comp. **50**, 321–333 (1988)
2617. Lloyd-Smith, C.W.: On a problem of Favard concerning algebraic integers. Bull. Aust. Math. Soc. **30**, 111–121 (1984)
2618. Lochter, M.: New characterization of Kronecker equivalence. J. Number Theory **53**, 115–136 (1995)
2619. Loewy, A.: Eine algebraische Behauptung von Gauß. Jahresber. Deutsch. Math.-Verein. **26**, 100–109 (1918)
2620. Loewy, A.: Eine algebraische Behauptung von Gauß, II. Jahresber. Deutsch. Math.-Verein. **30**, 155–158 (1921)
2621. Loewy, A.: Über die Reduktion algebraischer Gleichungen durch Adjunktion insbesondere reeller Radikale. Math. Z. **15**, 261–273 (1922)

2622. Löffler, E.: Alexander von Brill. Jahresber. Deutsch. Math.-Verein. **53**, 82–89 (1943)
2623. Long, R.L.: Steinitz classes of cyclic extensions of prime degree. J. Reine Angew. Math. **250**, 87–98 (1971)
2624. Long, R.L.: The module structure of some tamely ramified extensions of algebraic number fields. In: Proceedings of the Number Theory Conference, Boulder, 139–141. Boulder (1972)
2625. Long, R.L.: Steinitz classes of cyclic extensions of degree l^ν. Proc. Amer. Math. Soc. **49**, 297–304 (1975)
2626. Long, R.L.: Algebraic Number Theory. Marcel Dekker (1977)
2627. Lorenz, F.: Über eine Verallgemeinerung des Hasseschen Normensatzes. Math. Z. **173**, 203–210 (1980)
2628. Lorenz, F.: Zur Theorie der Normenreste. J. Reine Angew. Math. **334**, 157–170 (1982)
2629. Lorenz, F.: Normal bases of units. In: Number Theory (Budapest, 1987), **2**, 851–869, Colloq. Math. Soc. János Bolyai **51**. North-Holland (1990)
2630. Lorenz, F.: Ein Scholion zum Satz 90 von Hilbert. Abh. Math. Semin. Univ. Hambg **68**, 347–362 (1998)
2631. Lorenz, F., Roquette, P.: Cahit Arf and his invariant. Mitt. Math. Ges. Hamburg **30**, 87–126 (2011) [[3509] 189–222.]
2632. Louboutin, R.: Sur la mesure de Mahler d'un nombre algébrique. C. R. Acad. Sci. Paris **296**, 707–708 (1983)
2633. Louboutin, S.: Minorations (sous l'hypothèse de Riemann généralisée) des nombres de classes des corps quadratiques imaginaires. Application. C. R. Acad. Sci. Paris **310**, 795–800 (1990)
2634. Louboutin, S.: On the class number one problem for non-normal quartic CM-fields. Tôhoku Math. J. (2) **46**, 1–12 (1994)
2635. Louboutin, S.: Determination of all quaternion octic CM-fields with class number 2. J. Lond. Math. Soc. (2) **54**, 227–238 (1996)
2636. Louboutin, S.: A finiteness theorem for imaginary abelian number fields. Manuscripta Math. **91**, 343–352 (1996)
2637. Louboutin, S.: Corps quadratiques à corps de classes de Hilbert principaux et à multiplication complexe. Acta Arith. **74**, 121–140 (1996)
2638. Louboutin, S.: The class number one problem for the non-abelian normal CM-fields of degree 16. Acta Arith. **82**, 173–196 (1997)
2639. Louboutin, S.: The imaginary cyclic sextic fields with class numbers equal to their genus class numbers. Colloq. Math. **75**, 205–212 (1998)
2640. Louboutin, S.: The class number one problem for the dihedral and dicyclic CM-fields. Colloq. Math. **80**, 259–265 (1999)
2641. Louboutin, S.: The nonquadratic imaginary cyclic fields of 2-power degrees with class numbers equal to their genus class numbers. Proc. Amer. Math. Soc. **127**, 355–361 (1999)
2642. Louboutin, S.: Explicit bounds for residues of Dedekind zeta functions, values of L-functions at $s = 1$, and relative class numbers. J. Number Theory **85**, 263–282 (2000)
2643. Louboutin, S.: Explicit upper bounds for residues of Dedekind zeta functions and values of L-functions at $s = 1$, and explicit lower bounds for relative class numbers of CM-fields. Canad. J. Math. **53**, 1194–1222 (2001)
2644. Louboutin, S.: Explicit lower bounds for residues at $s = 1$ of Dedekind zeta functions and relative class numbers of CM-fields. Trans. Amer. Math. Soc. **355**, 3079–3098 (2003)
2645. Louboutin, S.R.: Class numbers of real cyclotomic fields. Publ. Math. Debrecen **64**, 451–461 (2004)
2646. Louboutin, S.R.: On the use of explicit bounds on residues of Dedekind zeta functions taking into account the behavior of small primes. J. Théor. Nombres Bordeaux **17**, 559–573 (2005)
2647. Louboutin, S.R.: The Brauer-Siegel theorem. J. Lond. Math. Soc. (2) **72**, 40–52 (2005)
2648. Louboutin, S.R.: Efficient computation of root numbers and class numbers of parametrized families of real abelian number fields. Math. Comp. **76**, 455–473 (2007)

2649. Louboutin, S.R.: Simple proofs of the Siegel-Tatuzawa and Brauer-Siegel theorems. Colloq. Math. **108**, 277–283
2650. Louboutin, S.R.: On the Ono invariants of imaginary quadratic number fields. J. Number Theory **129**, 2289–2294 (2009)
2651. Louboutin, S. Okazaki, R., Determination of all non-normal quartic CM-fields and of all non-abelian octic CM-fields with class number one. Acta Arith. **67**, 47–62 (1994)
2652. Louboutin, S., Okazaki, R.: The class number one problem for some non-Abelian normal CM-fields of 2-power degrees. Proc. London Math. Soc. (3) **76**, 523–548 (1998)
2653. Louboutin, S., Okazaki, R.: Determination of all quaternion CM-fields with ideal class group of exponent 2. Osaka J. Math. **36**, 229–257 (1999)
2654. Louboutin, S., Okazaki, R.: Exponents of the ideal class groups of the CM number fields. Math. Z. **243**, 155–159 (2003)
2655. Louboutin, S., Okazaki, R., Olivier, M.: The class number one problem for some non-Abelian normal CM-fields. Trans. Amer. Math. Soc. **349**, 3657–3678 (1997)
2656. Low, M.E.: Real zeros of the Dedekind zeta function of an imaginary quadratic field. Acta Arith. **14**, 117–140 (1967/68)
2657. Lu, Y., Zhang, W.: On the Kummer conjecture. Acta Arith. **131**, 87–102 (2008)
2658. Lubelski, S.: Zur Theorie der höheren Kongruenzen. J. Reine Angew. Math. **162**, 63–68 (1930)
2659. Lubelski, S.: Über Kongruenzen mit Idealmoduln. In: C.R. Premier Congrés Math. Pays Slaves, 246–253 (1930)
2660. Lubelski, S.: Über Klassenzahlrelationen quadratischer Formen in quadratischen Zahlkörpern. J. Reine Angew. Math. **174**, 160–184 (1936)
2661. Lubelski, S.: Zur Gaussschen Kompositionstheorie der binären quadratischen Formen. J. Reine Angew. Math. **176**, 56–60 (1936)
2662. Lubelski, S.: Unpublished results on number theory, II. Composition theory of binary quadratic forms. Acta Arith. **7**, 9–17 (1961/62)
2663. Lubin, J.: The local Kronecker-Weber theorem. Trans. Amer. Math. Soc. **267**, 133–138 (1981)
2664. Lubin, J., Tate, J.: Formal complex multiplication in local fields. Ann. Math. (2) **81**, 380–387 (1965)
2665. Lubotzky, A.: Group presentation, p-adic analytic groups and lattices in $SL_2(C)$. Ann. Math. (2) **118**, 115–130 (1983)
2666. Luca, F.: On a question of G. Kuba. Arch. Math. (Basel) **74**, 269–275 (2000)
2667. Luca, F.: A note on the divisibility of class numbers of real quadratic fields. C.R. Math. Acad. Sci. Soc. R. Canada **25**, 71–75 (2003)
2668. Luca, F., Pizarro-Madariaga, A., Pomerance, C.: On the counting function of irregular primes. Indag. Math. (N.S.) **26**, 147–161 (2015)
2669. Lüneburg, H.: Resultanten von Kreisteilungspolynomen. Arch. Math. (Basel) **42**, 139–144 (1984)
2670. Lützen,vJ.: Joseph Liouville 1809–1882: Master of Pure and Applied Mathematics. Springer (1990)
2671. Lützen, J.: Why was Wantzel overlooked for a century? The changing importance of an impossibility result. Historia Math. **36**, 374–394 (2009)
2672. Lützen, J., Sabidusssi, G., Toft, B.: Julius Petersen (1839–1910). A biography. Discrete Math. **100**, 9–82 (1992)
2673. Maass, H.: Über die Darstellung total positiver Zahlen des Körpers $R(\sqrt{5})$ als Summe von drei Quadraten. Abh. Math. Sem. Hansischen Univ. **14**, 185–191 (1941)
2674. Macaulay, F.S.: On the resolution of a given modular system into primary systems including some properties of Hilbert numbers. Math. Ann. **74**, 66–121 (1913)
2675. [Macaulay, F.S.] F.S.M.: Max Noether. Proc. London Math. Soc. **21**, xxxvii–xlii (1923)
2676. MacDuffee, C.C.: A correspondence between matrices and quadratic ideals. Ann. Math. (2) **28**, 199–214 (1927/28)

2677. MacDuffee, C.C.: An introduction to the theory of ideals in linear associative algebras. Trans. Amer. Math. Soc. **31**, 71–90 (1929)
2678. MacDuffee, C.C.: A method for determining the canonical basis of an ideal in an algebraic field. Math. Ann. **105**, 663–665 (1931)
2679. MacDuffee, C.C.: Products and norms of ideals. Amer. J. Math. **64**, 646–652 (1942)
2680. MacDuffee, C.C., Jenkins, E.D.: A substitute for Euclid algorithm in algebraic fields. Ann. Math. (2) **36**, 40–45 (1935)
2681. Macintyre, A.J.: Abraham Robinson, 1918–1974. Bull. Amer. Math. Soc. **83**, 646–666 (1977)
2682. MacKenzie, R.E.: Class group relations in cyclotomic fields. Amer. J. Math. **74**, 759–763 (1952)
2683. MacKenzie, R.[E.], Scheuneman, J.: A number field without a relative integral basis. Amer. Math. Monthly **78**, 882–883 (1971)
2684. Mac Lane, S.: Note on some equations without affect. Bull. Amer. Math. Soc. **42**, 731–736 (1936)
2685. Mac Lane, S.: A construction for prime ideals as absolute values of an algebraic field. Duke Math. J. **2**, 492–510 (1936)
2686. Mac Lane, S.: The Schönemann-Eisenstein irreducibility criteria in terms of prime ideals. Trans. Amer. Math. Soc. **43**, 226–239 (1938)
2687. Mac Lane, S.: Steinitz field towers for modular fields. Trans. Amer. Math. Soc. **46**, 23–45 (1939)
2688. Mac Lane, S.: Subfields and automorphism groups of p-adic fields. Ann. Math. (2) **40**, 423–442 (1939)
2689. Mac Lane, S.: Note on the relative structure of p-adic fields. Ann. Math. (2) **41**, 751–753 (1940)
2690. Mac Lane, S.: Saunders Mac Lane — a mathematical autobiography. A.K. Peters (2005)
2691. Mac Lane, S., Schilling, O.F.G.: Infinite number fields with Noether ideal theories. Amer. J. Math. **61**, 771–782 (1939)
2692. Madan, M.L.: On class numbers of algebraic number fields. J. Number Theory **2**, 116–119 (1970)
2693. Magnus, W.: Über den Beweis des Hauptidealsatzes. J. Reine Angew. Math. **170**, 235–240 (1934)
2694. Mahler, K.: Ein Beweis der Transzendenz der P-adischen Exponentialfunktion. J. Reine Angew. Math. **169**, 61–66 (1932)
2695. Mahler, K.: Zur Approximation algebraischer Zahlen, I. Über den grössten Primteiler binärer Formen. Math. Ann. **107**, 691–730 (1933)
2696. Mahler, K.: Zur Approximation p-adischer Irrationalzahlen. Nieuw Arch. Wiskd. (2) **18**, 22–34 (1934)
2697. Mahler, K.: Über diophantische Approximation im Gebiet der p-adischen Zahlen. Jahresber. Deutsch. Math.-Verein. **44**, 250–255 (1934)
2698. Mahler, K.: On Hecke's theorem on the real zeros of the L-functions and the class number of quadratic fields. J. Lond. Math. Soc. **9**, 298–302 (1934)
2699. Mahler, K.: Über transzendente P-adische Zahlen. Compositio Math. **2**, 259–275 (1935); Corr.: Compositio Math. **9**, p. 112 (1949)
2700. Mahler, K.: Über Pseudobewertungen, I. Acta Math. **66**, 79–119 (1936)
2701. Mahler, K.: Über Pseudobewertungen, Ia. Zerlegungssätze. Proc. Nederl. Akad. Wet. Amsterdam **39**, 57–65 (1936)
2702. Mahler, K.: Über Pseudobewertungen, II. Über Pseudobewertungen eines endlichen algebraischen Zahlkörpers. Acta Math. **67**, 51–80 (1936)
2703. Mahler, K.: Über Pseudobewertungen, III. Die Pseudobewertungen der Hauptordnung eines endlichen algebraischen Zahlkörpers. Acta Math. **67**, 283–328 (1936)
2704. Mahler, K.: On a geometrical representation of p-adic numbers. Ann. Math. (2) **41**, 8–56 (1940)

2705. Mahler, K.: An application of Jensen's formula to polynomials. Mathematika **7**, 98–100 (1960)
2706. Mahler, K.: An inequality for the discriminant of a polynomial. Michigan Math. J. **11**, 257–262 (1964)
2707. Mai, L.: Values of L-functions at the critical point. Proc. Amer. Math. Soc. **122**, 415–428 (1994)
2708. Maier, H.: The coefficients of cyclotomic polynomials. Progr. Math. **85**, 349–366 (1990)
2709. Maier, H.: Cyclotomic polynomials with large coefficients. Acta Arith. **64**, 227–235 (1993)
2710. Maier, H.: Cyclotomic polynomials whose orders contain many prime factors. Period. Math. Hung. **43**, 155–164 (2001)
2711. Maillet, E.: Question 4269. L'Intermédiaire des Math. **20**, p. 218 (1913)
2712. Maire, C.: Un raffinement du théorème de Golod-Safarevic. Nagoya Math. J. **150**, 1–11 (1998)
2713. Maire, C.: On infinite unramified extensions. Pacific J. Math. **192**, 135–142 (2000)
2714. Mäki, S.: On the density of abelian number fields. Ann. Acad. Sci. Fenn. Math., Ser. A I **54**, 1–104 (1985)
2715. Maknis, M.: Zeros of Hecke Z-functions and the distribution of primes of an imaginary quadratic field. Lit. Mat. Sb. **15(1)**, 173–184 (1975) (Russian)
2716. Maknis, M.: Density theorems for Hecke Z-functions, and the distribution of the prime numbers of an imaginary quadratic field. Lit. Mat. Sb. **16(1)**, 173–180 (1976) (Russian)
2717. Maknis, M.: Refinement of the remainder term in the distribution law of prime numbers of an imaginary quadratic field in sectors. Lit. Mat. Sb. **17(1)**, 133–137 (1977) (Russian)
2718. Malinin, D.A.: On the existence of finite Galois stable groups over integers in unramified extensions of number fields. Publ. Math. Debrecen **60**, 179–191 (2002)
2719. Malle, G.: Polynomials with Galois groups Aut(M_{22}), M_{22}, and $PSL_3(F_4) \cdot 2_2$ over Q. Math. Comp. **51**, 761–768 (1988)
2720. Malle, G.: Exceptional groups of Lie type as Galois groups. J. Reine Angew. Math. **392**, 70–109 (1988)
2721. Malle, G.: Some unitary groups as Galois groups over Q. J. Algebra **131**, 476–482 (1990)
2722. Malle, G.: Darstellungstheoretische Methoden bei der Realisierung einfacher Gruppen vom Lie Typ als Galoisgruppen. Progr. Math. **95**, 443–459 (1991)
2723. Malle, G.: Disconnected groups of Lie type as Galois groups. J. Reine Angew. Math. **429**, 161–182 (1992)
2724. Malle, G.: Polynome mit Galoisgruppen $PGL_2(p)$ und $PSL_2(p)$ über $Q(t)$. Comm. Algebra **21**, 511–526 (1993)
2725. Malle, G.: On the distribution of Galois groups. J. Number Theory **92**, 315–329 (2002)
2726. Malle, G.: On the distribution of Galois groups. II. Experiment. Math. **13**, 129–135 (2004)
2727. Malle, G., Matzat, B.H.: Realisierung von Gruppen $PSL_2(F_p)$ als Galoisgruppen über Q. Math. Ann. **272**, 549–565 (1985)
2728. Malle, G., Matzat, B.H.: Inverse Galois Theory. Springer (1999)
2729. Malo, E.: L'Intermédiaire des Math. **21**, 173–176 (1914)
2730. Malyshev, A.V.: Yu.V. Linnik's works in number theory. Acta Arith. **27**, 3–10 (1975)
2731. Man, S.H.: A note on Dedekind domains. Rocky Mountain J. Math. **28**, 655–656 (1998)
2732. Mandl, M.: Ueber die Zerlegung ganzer, ganzzahliger Functionen in irreductible Factoren. J. Reine Angew. Math. **113**, 252–261 (1894)
2733. Mandl, M.: Ueber die Zerlegung von Funktionen mehrerer Variabeln in irreduktible Faktoren. J. Reine Angew. Math. **131**, 40–48 (1906)
2734. Mann, H.B.: Introduction to Algebraic Number Theory. Ohio State University Press (1955)
2735. Mann, H.B.: On integral bases. Proc. Amer. Math. Soc. **9**, 167–172 (1958)
2736. Mann, H.B., Vélez, W.Y.: Prime ideal decomposition in $F(\sqrt[m]{\mu})$. Monatsh. Math. **81**, 131–139 (1976)
2737. Manstavičius, E.: Jonas Kubilius (1921–2011). Acta Arith. **157**, 11–36 (2013)
2738. Márki, L., Steinfeld, O., Szép, J.: Short review of the work of László Rédei. Studia Sci. Math. Hungar. **16**, 3–14 (1981)

2739. Marko, F.: On the existence of p-units and Minkowski units in totally real cyclic fields. Abh. Math. Semin. Univ. Hambg **66**, 89–111 (1996)
2740. Marko, F.: On the existence of Minkowski units in totally real cyclic fields. J. Théor. Nombres Bordeaux **17**, 195–206 (2005)
2741. Markoff, A.A.: Sur une classe de nombres complexes. C. R. Acad. Sci. Paris **112**, 780–782, 1049–1050, 1123–1124 (1891)
2742. Markoff, A.A.: Sur les nombres entiers dépendant d'une racine cubique d'un nombre entier ordinaire, Mém. Acad. St. Petersb., (7) **38**, 1–37 (1892) [Russian translation: in: Markov, A.A.: Izbrannye trudy, 85–133, Izdat. AN SSSR (1951)]
2743. Marshall, M.A.: Ramification groups of abelian local field extensions. Canad. J. Math. **23**, 271–281 (1971)
2744. Marszałek, R.: Units in real abelian fields. Acta Arith. **146**, 115–151 (2011)
2745. Marszałek, R.: Units in real cyclic fields. Funct. Approx. Comment. Math. **45**, 139–153 (2011)
2746. Martin, K.: A symplectic case of Artin's conjecture. Math. Res. Lett. **10**, 483–492 (2003)
2747. Martin, K.: Modularity of hypertetrahedral representations. C. R. Acad. Sci. Paris **339**, 99–102 (2004)
2748. Martinet, J.: Sur l'arithmétique des extensions galoisiennes à groupe Galois diédral d'ordre $2p$. Ann. Inst. Fourier (Grenoble) **19(1)**, 1–80 (1969)
2749. Martinet, J.: Un contre-exemple à une conjecture d'E. Noether. Lecture Notes Math. **180**, 145–154 (1971)
2750. Martinet, J.: Modules sur l'algébre du groupe quaternionien. Ann. Sci. Éc. Norm. Supér. (4) **4**, 399–408 (1971)
2751. Martinet, J.: Character theory and Artin L-functions. In: [1281], 1–87
2752. Martinet, J.: Tours de corps de classes et estimations de discriminants. Invent. Math. **44**, 65–73 (1978)
2753. Martinet, J.: Petits discriminants. Ann. Inst. Fourier (Grenoble) **29(1)**, 159–179 (1979)
2754. Martinet, J.: Petits discriminants des corps de nombres. In: Journées Arithmétiques (Exeter 1980), 151–193. Cambridge University Press (1982)
2755. Martinet, J.: Methodes géometriques dans la recherche des petits discriminants. Progr. Math. **59**, 147–179 (1985)
2756. Martinet, J.: Anne-Marie Bergé — In memoriam. J. Théor. Nombres Bordeaux **20**, i–xi (2008)
2757. Marty, F.: Sur une généralisation de la notion de groupe. In: Proc. 8 Skand. Mat. Kongr., Stockholm, 45–49 (1934)
2758. Marty, F.: Rôle de la notion d'hypergroupe dans l'étude des groupes non abéliens. C. R. Acad. Sci. Paris **201**, 636–638 (1935)
2759. Masley, J.M.: On Euclidean rings of integers in cyclotomic fields. J. Reine Angew. Math. **272**, 45–48 (1974)
2760. Masley, J.M.: Solution of the class number two problem for cyclotomic fields. Invent. Math. **28**, 243–244 (1975)
2761. Masley, J.M.: Solution of small class numbers problems for cyclotomic fields. Compositio Math. **33**, 179–186 (1976)
2762. Masley, J.M.: Class numbers of real cyclic number fields with small conductor. Compositio Math. **37**, 297–319 (1978)
2763. Masley, J.M., Montgomery, H.L.: Cyclotomic fields with unique factorization. J. Reine Angew. Math. **286/287**, 248–256 (1976)
2764. Masuda, K.: On a problem of Chevalley. Nagoya Math. J. **8**, 59–63 (1955)
2765. Masuda, K.: Application of the theory of the group of classes of projective modules to the existence problem of independent parameters of invariant. J. Math. Soc. Japan **20**, 223–232 (1968)
2766. Mathews, G.B.: Irregular determinants and subtriplicate forms. Messenger Math. (2) **20**, 70–74 (1890)

2767. Mathews, G.B.: On binary quadratic forms with complex coefficients. Quart. J. Pure Appl. Math. **25**, 289–300 (1891)
2768. Mathews, G.B.: Note on Dirichlet's formula for the number of classes of binary quadratic forms for a complex determinant. Proc. London Math. Soc. **23**, 159–162 (1891/92)
2769. Mathews, G.B.: On the complex integers connected with the equation $\theta^3 - 2 = 0$. Proc. London Math. Soc. **24**, 319–327 (1892/93)
2770. Mathews, G.B.: The division of the lemniscate. Proc. London Math. Soc. **27**, 367–383 (1895/96)
2771. Mathews, G.B.: A theory of binary quadratic arithmetical forms, with complex integral coefficients. Proc. London Math. Soc. (2) **11**, 329–350 (1913)
2772. Mathews, G.B.: Division of the lemniscate into seven equal parts. Proc. London Math. Soc. (2) **14**, 464–466 (1915)
2773. Mathewson, L.C.: A simple proof of a theorem of Kronecker. J. Reine Angew. Math. **161**, p. 255 (1929)
2774. Mathieu, É.: Mémoire sur la théorie des résidus biquadratiques. J. Math. Pures Appl. (2) **12**, 377–438 (1867)
2775. Matsuda, R.: On purely-transcendency of certain fields. Tôhoku Math. J. (2) **16**, 189–202 (1964)
2776. Matsumoto, T.: On the sum of two quadratic rests with respect to a prime ideal. Tôhoku Math. J. **16**, 53–61 (1919)
2777. Matter, K.: Die den Bernoulli'schen Zahlen analogen Zahlen im Körper der dritten Einheitswurzeln. Vierteljahr. Naturforsch. Ges. Zürich **45**, 238–269 (1900)
2778. Matusita, K.: Ueber die Idealtheorie im Integritätsbereich mit dem eingeschränkten Vielfachenkettensatz. Proc. Phys.-Math. Soc. Japan (3) **23**, 8–12, 271–272 (1941)
2779. Matusita, K.: Über ein bewertungstheoretisches Axiomensystem für die Dedekind-Noethersche Idealtheorie. Jpn. J. Math. **19**, 97–110 (1944)
2780. Matveev, E.M.: On the size of integral algebraic numbers. Mat. Zametki **49**, 152–154 (1991) (Russian)
2781. Matveev, E.M.: On fundamental units of some fields. Mat. Sb. **183**, 35–48 (1992) (Russian)
2782. Matzat, B.H.: Konstruktion von Zahlkörpern mit der Galoisgruppe M_{11} über $Q(\sqrt{-11})$. Manuscripta Math. **27**, 103–111 (1979)
2783. Matzat, B.H.: Konstruktion von Zahl- und Funktionenkörpern mit vorgegebener Galoisgruppe. J. Reine Angew. Math. **349**, 179–220 (1984)
2784. Matzat, B.H.: Realisierung endlicher Gruppen als Galoisgruppen. Manuscripta Math. **51**, 253–265 (1985)
2785. Matzat, B.H.: Über das Umkehrproblem der Galoisschen Theorie. Jahresber. Deutsch. Math.-Verein. **90**, 155–183 (1988)
2786. Matzat, B.H.: Konstruktive Galoistheorie. Lecture Notes Math. **1284** (1987)
2787. Matzat, B.H.: Zöpfe und Galoissche Gruppen. J. Reine Angew. Math. **420**, 99–159 (1991)
2788. Matzat, B.H. Zeh-Marschke, A.: Realisierung der Mathieugruppen M_{11} und M_{12} als Galoisgruppen über Q. J. Number Theory **23**, 195–202 (1986)
2789. Matzat, B.H., Zeh-Marschke, A.: Polynome mit der Galoisgruppe M_{11} über Q. J. Symbolic Comput. **4**, 93–97 (1987)
2790. Maxwell, J.W.: William J. LeVeque. Notices Amer. Math. Soc. **55**, 1261–1262 (2008)
2791. Mayer, D.C.: Discriminants of metacyclic fields. Canad. Math. Bull. **36**, 103–107 (1993)
2792. Mayer, J.: Die absolut-kleinsten Diskriminanten der biquadratischen Zahlkörper. SBer. Kais. Akad. Wissensch. Wien **138**, 733–742 (1929)
2793. Mayr, K.: Zur Theorie des Reziprozitätsgesetzes für quadratische Reste in algebraischen Zahlkörpern. Monatsh. Math. Phys. **26**, 144–152 (1915)
2794. Maz'ya, V., Shaposhnikova, T.: Jacques Hadamard, a Universal Mathematician. Amer. Math. Soc. (1998)
2795. McCrea, W.H.: Edmund Taylor Whittaker. J. Lond. Math. Soc. **32**, 234–256 (1957)
2796. McCulloh, L.R.: Integral bases in Kummer extensions of Dedekind fields. Canad. J. Math. **15**, 755–765 (1963)

2797. McCulloh, L.R.: Cyclic extensions without relative integral bases. Proc. Amer. Math. Soc. **17**, 1191–1194 (1966)
2798. McCulloh, L.R.: Frobenius groups and integral bases. J. Reine Angew. Math. **248**, 123–126 (1971)
2799. McCulloh, L.R.: Galois module structure of abelian extensions. J. Reine Angew. Math. **375/376**, 259–306 (1987)
2800. McGown, K.J.: Norm-Euclidean cyclic fields of prime degree. Internat. J. Number Th. **8**, 227–254 (2012)
2801. McGown, K.J.: Norm-Euclidean Galois fields and the generalized Riemann hypothesis. J. Théor. Nombres Bordeaux **24**, 425–445 (2012)
2802. McKee, J.: Conjugate algebraic numbers on conics: a survey. In: Number Theory and Polynomials, 211–240. Cambridge University Press (2008)
2803. McKee, J.: Computing totally positive algebraic integers of small trace. Math. Comp. **80**, 1041–1052 (2011)
2804. McKee, J., Smyth, C.: Salem numbers of trace -2 and traces of totally positive algebraic integers. Lecture Notes in Computer Science **3096**, 327–337 (2004)
2805. McKee, J., Smyth, C.: There are Salem numbers of every trace. Bull. Lond. Math. Soc. **37**, 25–36 (2005)
2806. McKee, J., Smyth, C.: Integer symmetric matrices of small spectral radius and small Mahler measure. Int. Math. Res. Not. **2012**, 102–136
2807. McKenzie, R.G.: The ring of cyclotomic integers of modulus thirteen is norm-Euclidean. Dissertation, Michigan State Univ. (1988)
2808. McMullen, C.T.: Coxeter groups, Salem numbers and the Hilbert metric. Publ. Math. Inst. Hautes Études Sci. **95**, 151–183 (2002)
2809. McMullen, C.T.: Minkowski's conjecture, well-rounded lattices and topological dimension. J. Amer. Math. Soc. **18**, 711–734 (2005)
2810. Meissner, O.: Über die Darstellung der Zahlen einiger algebraischen Körper als Summen von Quadraten aus Zahlen des Körpers. Arch. Math. Phys. (3) **5**, 175–176 (1903)
2811. Meissner, O.: Über die Darstellung der Zahlen einiger algebraischen Zahlkörper als Summen von Quadratzahlen des Körpers. Arch. Math. Phys. (3) **7**, 266–268 (1904)
2812. Meissner, O.: Über die Darstellbarkeit der Zahlen quadratischer und kubischer Zahlkörper als Quadratsummen. Arch. Math. Phys. (3) **9**, 202–203 (1905)
2813. Menger, K.: Otto Schreier. Monatsh. Math. Phys. **37**, 1–6 (1930)
2814. Mertens, F.: Ueber einige asymptotische Gesetze der Zahlentheorie. J. Reine Angew. Math. **77**, 289–338 (1874)
2815. Mertens, F.: Ein Beitrag zur analytischen Zahlentheorie. J. Reine Angew. Math. **78**, 46–62 (1874)
2816. Mertens, F.: Ueber die Irreductibilität der Function $x^p - A$. Monatsh. Math. Phys. **2**, 291–292 (1891)
2817. Mertens, F.: Ueber die Fundamentalgleichung eines Gattungsbereiches algebraischer Zahlen. SBer. Kais. Akad. Wissensch. Wien **103**, 5–40 (1894)
2818. Mertens, F.: Ueber die Composition der binären quadratischen Formen. SBer. Kais. Akad. Wissensch. Wien **104**, 103–143 (1895)
2819. Mertens, F.: Über Dirichletsche Reihen. SBer. Kais. Akad. Wissensch. Wien **104**, 1093–1153 (1895)
2820. Mertens, F.: Beweis, dass jede lineare Function mit ganzen complexen teilerfremden Coefficienten unendlich viele complexe Primzahlen darstellt. SBer. Kais. Akad. Wissensch. Wien **108**, 517–556 (1899)
2821. Mertens, F.: Über einen Satz von Dirichlet. SBer. Kais. Akad. Wissensch. Wien **109**, 415–480 (1900)
2822. Mertens, F.: Ueber zyklische Gleichungen. SBer. Kais. Akad. Wissensch. Wien **114**, 105–148 (1905)
2823. Mertens, F.: Über den Dedekindschen Beweis der Irreduktibilität der Gleichung für die primitiven n-ten Einheitswurzeln. SBer. Kais. Akad. Wissensch. Wien **114**, 1293–1296 (1905)

2824. Mertens, F.: Ein Beweis des Satzes, dass jede Klasse von ganzzahligen primitiven binären quadratischen Formen des Hauptgeschlechts durch Duplikation entsteht. J. Reine Angew. Math. **129**, 181–186 (1905)
2825. Mertens, F.: Über zyklische Gleichungen. J. Reine Angew. Math. **131**, 87–112 (1906)
2826. Mertens, F.: Über die einfachen Einheiten des Bereichs $(\alpha, \sqrt{D})$, wo α eine primitive Einheitswurzel vom Primzahlgrad und D eine negative Zahl bezeichnen. SBer. Kais. Akad. Wissensch. Wien **116**, 1337–1342 (1907)
2827. Mertens, F.: Über die Irreduktibilität der Kreisteilungsgleichungen. SBer. Kais. Akad. Wissensch. Wien **117**, 689–690 (1908)
2828. Mertens, F.: Über Abelsche Gleichungen und den Satz von Kronecker über die Teilungsgleichungen der Lemniskate. SBer. Kais. Akad. Wissensch. Wien **118**, 245–291 (1909)
2829. Mertens, F.: Über die Zerfällung einer ganzen Funktion einer Veränderlichen in zwei Faktoren. SBer. Kais. Akad. Wissensch. Wien **120**, 1485–1502 (1911)
2830. Mertens, F.: Gleichungen 8-ten Grades mit Quaternionengruppe. SBer. Kais. Akad. Wissensch. Wien **125**, 735–740 (1916)
2831. Mertens, F.: Über die Bildung zyklischer Gleichungen in einem gegebenen Rationalitätsbereich. SBer. Kais. Akad. Wissensch. Wien **125**, 741–831 (1916)
2832. Mertens, F.: Gleichungen, deren Gruppe eine Quaternionengruppe ist. SBer. Kais. Akad. Wissensch. Wien **130**, 69–90 (1921)
2833. Mestre, J.-F.: Corps euclidiens, unités exceptionnelles et courbes elliptiques. J. Number Theory **13**, 123–137 (1981)
2834. Mestre, J.-F.: Courbes elliptiques et groupes de classes d'idéaux de certains corps quadratiques. J. Reine Angew. Math. **343**, 23–35 (1983)
2835. Mestre, J.-F.: Courbes de Weil de conducteur 5077. C. R. Acad. Sci. Paris **300**, 509–512 (1985)
2836. Meštrović, R.: A search for primes p such that the Euler number E_{p-3} is divisible by p. Math. Comp. **83**, 2967–2976 (2014)
2837. Metsänkylä, T.: Bemerkungen über den ersten Faktor der Klassenzahl des Kreiskörpers. Ann. Univ. Turku **105**, 1–15 (1967)
2838. Metsänkylä, T.: Note on the distribution of irregular primes. Ann. Acad. Sci. Fenn. Math., Ser. A I **492**, 1–7 (1971)
2839. Metsänkylä, T.: On the growth of the first factor of the cyclotomic class number. Ann. Univ. Turku **155**, 1–12 (1972)
2840. Metsänkylä, T.: Distribution of irregular prime numbers. J. Reine Angew. Math. **282**, 126–130 (1976)
2841. Metsänkylä, T.: Maillet's matrix and irregular primes. Ann. Univ. Turku **186**, 72–79 (1984)
2842. Metsänkylä, T.: Letter to the editor. J. Number Theory **64**, 162–163 (1997)
2843. Metsänkylä, T.: On the history of the study of ideal class groups. Expo. Math. **25**, 325–340 (2007)
2844. Meyer, A.: Ueber einen Satz von Dirichlet. J. Reine Angew. Math. **103**, 98–117 (1988)
2845. Meyer, C.: Die Berechnung der Klassenzahl Abelscher Körper über quadratischen Zahlkörpern. Akademie-Verlag (1957)
2846. Meyer, C.: Über die Bildung von elementar-arithmetischen Klasseninvarianten in reellquadratischen Zahlkörpern. Ber. Tagung Math. Forschungsinst. Oberwolfach **964**, 165–215, Bibliographisches Institut (1967)
2847. Meyer, C.: Bemerkungen zum Satz von Heegner-Stark über die imaginär-quadratischen Zahlkörper mit der Klassenzahl Eins. J. Reine Angew. Math. **242**, 179–214 (1970)
2848. Meyer, P.: Eine Charakterisierung vollständig regulärer, abelscher Erweiterungen. Abh. Math. Semin. Univ. Hambg **68**, 199–223 (1998)
2849. Meyer, Y.: Nombres de Pisot, nombres de Salem et analyse harmonique. Lecture Notes Math. **117** (1970)
2850. Michailov, I.[M.]: Noether's problem for some groups of order $16n$. Acta Arith. **143**, 277–290 (2010)

2851. Mignotte, M.: Approximation des nombres algébriques par des nombres algébriques de grand degré. Ann. Fac. Sci. Toulouse, Math. (5) **1**, 165–170 (1979)
2852. Mignotte, M., Ştefănescu, D.: La premiére méthode générale de factorisation des polynômes. Autour d'un mémoire de F.T. Schubert. Rev. Hist. Math. **7**, 67–89 (2001)
2853. Mignotte, M., Waldschmidt, M.: Linear forms in two logarithms and Schneider's method. Math. Ann. **231**, 241–267 (1977/78)
2854. Mignotte, M., Waldschmidt, M.: Linear forms in two logarithms and Schneider's method, II. Acta Arith. **53**, 251–287 (1989)
2855. Mignotte, M., Waldschmidt, M.: On algebraic numbers of small height: linear forms in one logarithm. J. Number Theory **47**, 43–62 (1994)
2856. Mihăilescu, P.: Turning Washington's heuristics in favor of Vandiver's conjecture. In: Essays in Mathematics and its Applications, 287–294. Springer (2012)
2857. Mihăilescu, P.: On Vandiver's best result on FLT1. Springer Optimization and its Applications **68**, 417–430 (2012)
2858. Miki, H.: On Grunwald-Hasse-Wang's theorem. J. Math. Soc. Japan **30**, 313–325 (1978)
2859. Miller, G.A.: Examples of normal domains of rationality belonging to elementary groups. Ann. Math. (2) **15**, 118–124 (1913/14)
2860. Miller, J.C.: Class numbers of totally real fields and an application to the Weber class number problem. Acta Arith. **164**, 381–397 (2014)
2861. Miller, J.C.: Class numbers of real cyclotomic fields of composite conductor. LMS J. Comput. Math. 17, suppl. A, 404–417 (2014)
2862. Miller, J.C.: Real cyclotomic fields of prime conductor and their class numbers. Math. Comp. **84**, 2459–2465 (2015)
2863. Miller, J.C.P.: Alfred Edward Western. J. Lond. Math. Soc. **38**, 278–281 (1963)
2864. Miller-Sims, L., Robertson, L.: Power integral bases for real cyclotomic fields. Bull. Aust. Math. Soc. **71**, 167–173 (2005)
2865. Mills, W.H.: The m-th power residue symbol. Amer. J. Math. **73**, 59–64 (1951)
2866. Mills, W.H.: Reciprocity in algebraic number fields. Amer. J. Math. **73**, 65–77 (1951)
2867. Milnor, J., Husemoller, D.: Symmetric Bilinear Forms. Springer (1973)
2868. Min, S.-H.: On the Euclidean algorithm in real quadratic fields. J. Lond. Math. Soc. **22**, 88–90 (1947)
2869. Min, S.-H.: Euclidean algorithm in real quadratic fields. Sci. Rep. Nat. Tsing-Hua Univ. **5**, 190–225 (1948)
2870. Mináč, J., Ware, R.: Pro-2-Demuškin groups of rank $\aleph_0$ as Galois groups of maximal 2-extensions of fields. Math. Ann. **292**, 337–353 (1992)
2871. Minkowski, H.: Ueber die Bedingungen, unter welchen zwei quadratische Formen mit rationalen Coefficienten in einander rational transformirt werden können. J. Reine Angew. Math. **106**, 5–26 (1890) [[2881], **1**, 219–239.]
2872. Minkowski, H.: Über die positiven quadratischen Formen und über kettenbruchähnlichen Algorithmen. J. Reine Angew. Math. **107**, 278–297 (1891) [[2881], **1**, 244–260.]
2873. Minkowski, H.: Théorèmes arithmétiques. C. R. Acad. Sci. Paris **112**, 209–212 (1891) [[2881], **1**, 261–263.]
2874. Minkowski, H.: Geometrie der Zahlen. Teubner (1896–1910) [Reprints: Teubner (1925); Chelsea (1953); Johnson (1968)]
2875. Minkowski, H.: Ein Kriterium für die algebraischen Zahlen. Nachr. Ges. Wiss. Göttingen **1899**, 64–88 [[2881], **1**, 293–315; reprint: in [2882], 15–37.]
2876. Minkowski, H.: Zur Theorie der Einheiten in den algebraischen Zahlkörpern. Nachr. Ges. Wiss. Göttingen **1900**, 90–93 [[2881], **1**, 316–319.]
2877. Minkowski, H.: Quelques nouveaux théorèmes sur l'approximation des quantités à l'aide de nombres rationals. Bull. Sci. Math. **25**, 72–76 (1901) [[2881], **1**, 353–356.]
2878. Minkowski, H.: Über periodische Approximationen algebraischer Zahlen. Acta Math. **26**, 333–352 (1902) [[2881], **1**, 357–371.]
2879. Minkowski, H.: Peter Gustav Lejeune Dirichlet und seine Bedeutung für die heutige Mathematik. Jahresber. Deutsch. Math.-Verein. **14**, 149–163 (1905) [[2881], **2**, 447–461.]

2880. Minkowski, H.: Diophantische Approximationen. Eine Einführung in die Zahlentheorie. Teubner (1907); 2nd ed. (1927). [Reprints: Chelsea (1957); Physica-Verlag (1961)]
2881. Minkowski, H.: Gesammelte Abhandlungen. **1,2**. Teubner (1911)
2882. Minkowski, H.: Ausgewählte Arbeiten zur Zahlentheorie und zur Geometrie. Teubner (1989)
2883. Mirimanoff, D.: Sur une question de la théorie des nombres. J. Reine Angew. Math. **109**, 82–88 (1891)
2884. Mirimanoff, D.: Sur l'équation $x^{37} + y^{37} + z^{37} = 0$. J. Reine Angew. Math. **111**, 26–30 (1893)
2885. Mirimanoff, D.: Racines cubiques de nombres entiers et multiplication complexe dans les fonctions elliptiques. Math. Ann. **56**, 115–128 (1903)
2886. Mirimanoff, D.: L'équation indéterminée $x^l + y^l + z^l = 0$ et le critérium de Kummer. J. Reine Angew. Math. **128**, 45–68 (1904)
2887. Mirimanoff, D.: L'équation $\xi^3 + \eta^3 + \zeta^3 = 0$ et la courbe $x^3 + y^3 = 1$. Comment. Math. Helv. **6**, 192–198 (1934)
2888. Mirimanoff, D., Hensel, K.: Sur la relation $\left(\frac{D}{p}\right) = (-1)^{n-h}$ et la loi de réciprocité. J. Reine Angew. Math. **129**, 86–87 (1905)
2889. Mitchell, H.H.: Proof that certain ideals in a cyclotomic realm are principal ideals. Trans. Amer. Math. Soc. **19**, 119–126 (1918)
2890. Mitchell, H.H.: On classes of ideals in a quadratic field. Ann. Math. (2) **27**, 297–314 (1926)
2891. Mitra, S.Ch.: On the division of the lemniscate into nine equal parts. Bull. Calcutta Math. Soc. **18**, 159–164 (1927)
2892. Mitsui, T.: Generalized prime number theorem. Jpn. J. Math. **26**, 1–42 (1956)
2893. Mitsui, T.: On the Goldbach problem in an algebraic number field, I. J. Math. Soc. Japan **12**, 290–324 (1960)
2894. Mitsui, T.: On the Goldbach problem in an algebraic number field, II. J. Math. Soc. Japan **12**, 325–372 (1960)
2895. Mitsui, T.: On the prime ideal theorem. J. Math. Soc. Japan **20**, 233–247 (1968)
2896. Miyada, I.: On imaginary abelian number fields of type $(2, 2, \dots, 2)$ with one class in each genus. Manuscripta Math. **88**, 535–540 (1995)
2897. Miyake, K.: Teiji Takagi, founder of the Japanese School of Modern Mathematics. Jpn. J. Math. **2**, 151–164 (2007)
2898. Miyanishi, M.: Masayoshi Nagata (1927–2008) and his mathematics. J. Math. Kyoto Univ. **50**, 645–659 (2010)
2899. Miyata, T.: Invariants of certain groups, I. Nagoya Math. J. **41**, 69–73 (1971)
2900. Molk, J.: Sur les unités complexes. Bull. Sci. Math. (2) **7**, 133–137 (1883)
2901. Molk, J.: Sur une notion qui comprend celle de la divisibilité et sur la théorie générale de l'élimination. Acta Math. **6**, 1–166 (1885)
2902. Möller, H.: Verallgemeinerung eines Satzes von Rabinowitsch über imaginär-quadratische Zahlkörper. J. Reine Angew. Math. **285**, 100–113 (1976)
2903. Mollin, R.A.: Class number one criteria for real quadratic fields, I. Proc. Japan Acad. Sci. **63**, 121–125 (1987)
2904. Mollin, R.A.: Necessary and sufficient conditions for the class number of a real quadratic field to be one, and a conjecture of S. Chowla. Proc. Amer. Math. Soc. **102**, 17–21 (1988)
2905. Mollin, R.A.: Orders in quadratic fields, III. Proc. Japan Acad. Sci. **70**, 176–181 (1994)
2906. Mollin, R.A.: Quadratics. CRC Press (1996)
2907. Mollin, R.A.: An elementary proof of the Rabinowitsch-Mollin-Williams criterion for real quadratic fields. J. Math. Sci. (Calcutta) **7**, 17–27 (1996)
2908. Mollin, R.A.: Quadratic polynomials producing consecutive, distinct primes and class groups of complex quadratic fields. Acta Arith. **74**, 17–30 (1996)
2909. Mollin, R.A.: A completely general Rabinowitsch criterion for complex quadratic fields. Canad. Math. Bull. **39**, 106–110 (1996)
2910. Mollin, R.A.: Class number one and prime-producing quadratic polynomials revisited. Canad. Math. Bull. **41**, 328–334 (1998)

2911. Mollin, R.A.: New prime-producing quadratic polynomials associated with class number one or two. New York J. Math. **8**, 161–168 (2002)
2912. Mollin, R.A., Goddard, B.: Quadratic prime–producing polynomials. J. Math. Sci. (Calcutta) **9**, 1–8 (1998)
2913. Mollin, R.A., Williams, H.C.: On prime valued polynomials and class numbers of real quadratic fields. Nagoya Math. J. **112**, 143–151 (1988)
2914. Mollin, R.A., Williams, H.C.: Computation of the class number of a real quadratic field. Utilitas Math. **41**, 259–308 (1992)
2915. Monna, A.F.: Zur Theorie des Masses im Körper der P-adischen Zahlen. Proc. Akad. Wet. Amsterdam **45**, 978–980 (1942)
2916. Monna, A.F.: Zur Geometrie der P-adischen Zahlen. Proc. Akad. Wet. Amsterdam **45**, 981–986 (1942)
2917. Monna, A.F.: Sur une transformation simple des nombres P-adiques en nombres réels. Indag. Math. **14**, 1–9 (1952)
2918. Montgomery, H.L.: Distribution of irregular primes. Illinois J. Math. **9**, 553–558 (1965)
2919. Montgomery, H.L., Vaughan, R.C.: The order of magnitude of the mth coefficients of cyclotomic polynomials. Glasg. Math. J. **27**, 143–159 (1985)
2920. Montgomery, H.L., Weinberger, P.J.: Notes on small class numbers. Acta Arith. **24**, 529–542 (1974)
2921. Montgomery, H.L., Weinberger, P.J.: Real quadratic fields with large class number. Math. Ann. **225**, 173–176 (1977)
2922. Moore, C.N.: Harris Hancock — In memoriam. Bull. Amer. Math. Soc. **50**, 812–815 (1944)
2923. Moravec, P.: Unramified Brauer groups of finite and infinite groups. Amer. J. Math. **134**, 1679–1704 (2012)
2924. Mordell, L.J.: The class number for definite binary quadratics. Messenger of Math. **47**, 138–142 (1918)
2925. Mordell, L.J.: On the representation of algebraic numbers as a sum of four squares. Proc. Cambridge Philos. Soc. **20**, 250–256 (1921)
2926. Mordell, L.J.: On the reciprocity formula for the Gauss's sums in the quadratic field. Proc. London Math. Soc. (2) **20**, 289–296 (1922)
2927. Mordell, L.J.: On trigonometric series involving algebraic numbers. Proc. London Math. Soc. (2) **21**, 493–496 (1923)
2928. Mordell, L.J.: Some applications of Fourier series in the analytic theory of numbers. Proc. Cambridge Philos. Soc. **24**, 585–596 (1928)
2929. Mordell, L.J.: On Hecke's modular functions, zeta functions, and some other analytic functions in the theory of numbers. Proc. London Math. Soc. (2) **32**, 501–556 (1931); Corr.: (2) **33**, p. 562 (1932)
2930. Mordell, L.J.: On the representation of a binary quadratic form as a sum of squares of linear forms. Math. Z. **35**, 1–15 (1932)
2931. Mordell, L.J.: On binary quadratic forms expressible as a sum of three linear squares with integer coefficients. J. Reine Angew. Math. **167**, 12–19 (1932)
2932. Mordell, L.J.: On the Riemann hypothesis and imaginary quadratic fields with a given class number. J. Lond. Math. Soc. **9**, 289–298 (1934)
2933. Mordell, L.J.: An application of quaternions to the representation of a binary quadratic form as a sum of four linear squares. Quart. J. Math., Oxford ser. **8**, 58–61 (1937)
2934. Mordell, L.J.: On the linear independence of algebraic numbers. Pacific J. Math. **3**, 625–630 (1953)
2935. Mordell, L.J.: On a Pellian equation conjecture. Acta Arith. **6**, 137–144 (1960)
2936. Mordell, L.J.: On a Pellian equation conjecture, II. J. Lond. Math. Soc. **36**, 282–288 (1961)
2937. Mordell, L.J.: The representation of a Gaussian integer as a sum of two squares. Math. Nachr. **40**, p. 209 (1967)
2938. Mordell, L.J.: On the magnitude of the integer solutions of the equation $ax^2 + by^2 + cz^2 = 0$. J. Number Theory **1**, 1–3 (1969)
2939. Mordell, L.J.: Harold Davenport (1907–1969). Acta Arith. **18**, 1–4 (1971)

2940. Mordell, L.J.: Some aspects of Davenport's work. Acta Arith. **18**, 5–11 (1971)
2941. Mori, S.: Axiomatische Begründung des Multiplikationsringes. J. Sci. Hiroshima Univ. **A3**, 43–59 (1932)
2942. Mori, S.: Über allgemeine Multiplikationsringe, I. J. Sci. Hiroshima Univ. **A4**, 1–26 (1934)
2943. Mori, S.: Über allgemeine Multiplikationsringe, II. J. Sci. Hiroshima Univ. **A4**, 99–109 (1934)
2944. Mori, S.: Über die Produktzerlegung der Hauptideale. J. Sci. Hiroshima Univ. **A8**, 7–13 (1938)
2945. Mori, S.: Über die Produktzerlegung der Hauptideale, II. J. Sci. Hiroshima Univ. **A9**, 145–155 (1939)
2946. Mori, S.: Über die Produktzerlegung der Hauptideale, III. J. Sci. Hiroshima Univ. **A10**, 85–94 (1940)
2947. Mori, S.: Allgemeine Z.P.I.-Ringe. J. Sci. Hiroshima Univ. **A10**, 117–136 (1940)
2948. Mori, S.: Über die Produktzerlegung der Hauptideale, IV. J. Sci. Hiroshima Univ. **A11**, 7–14 (1941)
2949. Mori, S., Dodo, T.: Bedingungen für ganze Abgeschlossenheit in Integritätsbereichen. J. Sci. Hiroshima Univ. **A7**, 15–27 (1937)
2950. Morisawa, T.: A class number problem in the cyclotomic Z_3-extension of Q. Tokyo J. Math. **32**, 549–558 (2009)
2951. Morisawa, T.: Mahler measure of the Horie unit and Weber's class number problem in the cyclotomic Z_3-extension of Q. Acta Arith. **153**, 35–49 (2012)
2952. Morishima, T.: Über den Fermatschen Quotienten. Jpn. J. Math. **8**, 159–173 (1931)
2953. Morishima, T.: Über die Fermatsche Vermutung, XI. Jpn. J. Math. **11**, 241–252 (1935)
2954. Moriya, M.: Über die Klassenzahl eines relativ-zyklischen Zahlkörpers vom Primzahlgrad. Proc. Imp. Acad. Tokyo **6**, 245–247 (1930)
2955. Moriya, M.: Über die Klassenzahl eines relativ-zyklischen Zahlkörpers vom Primzahlgrad. Jpn. J. Math. **10**, 1–18 (1933)
2956. Moriya, M.: Theorie der algebraischen Zahlkörper unendlichen Grades. J. Fac. Sci. Hokkaido Univ. **3**, 107–190 (1935); corr.: **4**, 121–122 (1936)
2957. Moriya, M.: Galoissche Theorie der algebraischen Zahlkörper unendlichen Grades. J. Fac. Sci. Hokkaido Univ. **4**, 67–120 (1936)
2958. Moriya, M.: Über einen Satz von Herbrand. J. Fac. Sci. Hokkaido Univ. **4**, 181–194 (1936)
2959. Moriya, M.: Klassenkörpertheorie im Kleinen für die unendlichen algebraischen Zahlkörper. J. Fac. Sci. Hokkaido Univ. **5**, 9–66 (1936)
2960. Moriya, M.: Klassenkörpertheorie im Grossen für unendliche algebraische Zahlkörper. Proc. Imp. Acad. Tokyo **12**, 322–324 (1936)
2961. Moriya, M.: Klassenkörpertheorie im Grossen für unendliche algebraische Zahlkörper. J. Fac. Sci. Hokkaido Univ. **6**, 63–101 (1937)
2962. Moriya, M.: Einige Eigenschaften der endlichen separablen algebraischen Erweiterungen über perfekten Körpern. Proc. Imp. Acad. Tokyo **17**, 405–410 (1941)
2963. Moriya, M.: Struktur der Divisionsalgebren über diskret bewerteten perfekten Körpern. Proc. Imp. Acad. Tokyo **18**, 5–11 (1942)
2964. Moriya, M.: Algebrenklassengruppen über diskret perfekten Körpern. Proc. Imp. Acad. Tokyo **18**, 37–38 (1942)
2965. Moriya, M.: Die Theorie der Klassenkörper im Kleinen über diskret perfekten Körpern, I. Proc. Imp. Acad. Tokyo **18**, 39–44 (1942)
2966. Moriya, M.: Die Theorie der Klassenkörper im Kleinen über diskret perfekten Körpern, II. Proc. Imp. Acad. Tokyo **18**, 452–459 (1942)
2967. Moriya, M.: Rein arithmetischer Beweis über die Unendlichkeit der Primideale 1 Grades aus einem endlichen algebraischen Zahlkörper. J. Fac. Sci. Hokkaido Univ. **11**, 164–166 (1950)
2968. Moriya, M.: Zur Theorie der Klassenkörper im Kleinen. J. Math. Soc. Japan **3**, 195–203 (1951)

2969. Moriya, M.: Eine notwendige Bedingung für die Gültigkeit der Klassenkörpertheorie im Kleinen. Math. J. Okayama Univ. **2**, 13–20 (1952)
2970. Moriya, M.: Theorie der Derivationen und Körperdifferenten. Math. J. Okayama Univ. **2**, 111–148 (1953)
2971. Moriya, M., Schilling, O.F.G.: Zur Klassenkörpertheorie über unendlichen perfekten Körpern. J. Fac. Sci. Hokkaido Univ. **5**, 189–205 (1937)
2972. Morton, P.: On Rédei's theory of the Pell equation. J. Reine Angew. Math. **307/308**, 373–398 (1979)
2973. Morton, P.: Density results for the 2-classgroups and fundamental units of real quadratic fields. Studia Sci. Math. Hungar. **17**, 21–43 (1982)
2974. Morton, P.: Density results for the 2-classgroups of imaginary quadratic fields. J. Reine Angew. Math. **323**, 156–187 (1982)
2975. Morton, P.: The quadratic number fields with cyclic 2-classgroups. Pacific J. Math. **108**, 165–175 (1983)
2976. Morton, P.: A correction to Hasse's version of the Grunwald-Hasse-Wang theorem. J. Reine Angew. Math. **659**, 169–174 (2011)
2977. Morton, P., Silverman, J.H.: Periodic points, multiplicities, and dynamical units. J. Reine Angew. Math. **461**, 81–122 (1995)
2978. Moser, C.: Représentation de -1 par une somme de carrés dans certains corps locaux et globaux, et dans certains anneaux d'entiers algébriques. C. R. Acad. Sci. Paris **271**, A1200–A1203 (1970)
2979. Moser, C.: Représentation de -1 comme somme de carrés d'entiers dans un corps quadratique imaginaire. Enseign. Math. (2) **17**, 279–287 (1971)
2980. Moser, C.: Nombre de classes d'une extension cyclique réelle de Q, de degré 4 ou 6 et de conducteur premier. Math. Nachr. **102**, 45–52 (1981)
2981. Moser, C., Payan, J.-J.: Majoration du nombre de classes d'un corps cubique cyclique de conducteur premier. J. Math. Soc. Japan **33**, 701–706 (1981)
2982. Mossinghoff, M.J.: Polynomials with small Mahler measure. Math. Comp. **67**, 1697–1705, S11–S14 (1998)
2983. Mossinghoff, M.J., Rhin, G., Wu, Q.: Minimal Mahler measures. Experiment. Math. **17**, 451–458 (2008)
2984. Mosunov, A.S., Jacobson, M.J.Jr.: Unconditional class group tabulation of imaginary quadratic fields to $|\Delta|<2^{40}$. Math. Comp. **85**, 1983–2009 (2016)
2985. Motoda, Y., Nakahara, T., Shah, S.I.A.: On a problem of Hasse for certain imaginary abelian fields. J. Number Theory **96**, 326–334 (2002)
2986. Mott, J.L.: Equivalent conditions for a ring to be a multiplication ring. Canad. J. Math. **16**, 429–434 (1964)
2987. Motzkin, T.: Sur l'équation irréducible $z^n + a_1 z^{n-1} + \cdots + a_0 = 0$, $n>1$. à coefficients complexes entiers, dont toutes les racines sont sur une droite. Les 11 classes de droites admissibles. C. R. Acad. Sci. Paris **221**, 220–222 (1945)
2988. Motzkin, T.: From among n conjugate algebraic integers, $n-1$ can be approximately given. Bull. Amer. Math. Soc. **53**, 156–162 (1947)
2989. Motzkin, T.: The Euclidean algorithm. Bull. Amer. Math. Soc. **55**, 1142–1146 (1949)
2990. Mouhib, A.: Infinite Hilbert 2-class field tower of quadratic number fields. Acta Arith. **145**, 267–272 (2010)
2991. Movahhedi, A.: Galois group of $X^p + aX + a$. J. Algebra **180**, 966–975 (1996)
2992. Movahhedi, A., Salinier, A.: The primitivity of the Galois group of a trinomial. J. Lond. Math. Soc. (2) **53**, 433–440 (1996)
2993. Movahhedi, A., Zahidi, M.: Symboles des restes quadratiques des discriminants dans les extensions modérément ramifiées. Acta Arith. **92**, 239–250 (2000)
2994. Mulholland, H.P.: On the product of n complex homogeneous linear forms. J. Lond. Math. Soc. **35**, 241–250 (1960)
2995. Müller, W.: On the distribution of ideals in cubic number fields. Monatsh. Math. **106**, 211–219 (1988)

2996. Mumford, D.: Oscar Zariski: 1899–1986. Notices Amer. Math. Soc. **33**, 891–894 (1986)
2997. Müntz, C.H.: Beziehungen der Riemannschen ζ-Funktion zu willkürlichen reellen Funktionen. Mat. Tidsskr. **B1922**, 39–47
2998. Müntz, C.H.: Der Summensatz von Cauchy in beliebigen algebraischen Zahlkörpern und die Diskriminante derselben. Math. Ann. **90**, 279–291 (1923)
2999. Müntz, C.H.: Allgemeine Begründung der Theorie der höheren ζ-Funktionen. Abh. Math. Semin. Univ. Hambg **3**, 1–11 (1924)
3000. Murty, M.R.: Exponents of class groups of quadratic fields. In: Topics in Number Theory, University Park PA 1997, 229–239. Kluwer (1999)
3001. Murty, M.R.: The work of K. Ramachandra in algebraic number theory. Hardy-Ramanujan J. **34**, 11–19 (2013)
3002. Murty, M.R., Petersen, K.L.: The Euclidean algorithm for number fields and primitive roots. Proc. Amer. Math. Soc. **141**, 181–190 (2013)
3003. Murty, M.R., Petridis, Y.N.: On Kummer's conjecture. J. Number Theory **90**, 294–303 (2001)
3004. Murty, M.R., Van Order, J.: Counting integral ideals in a number field. Expo. Math. **25**, 53–66 (2007)
3005. Murty, V.K.: Class numbers of CM-fields with solvable normal closure. Compositio Math. **127**, 273–287 (2001)
3006. Murty, V.K., Scherk, J.: Effective versions of the Chebotarev density theorem for function fields. C. R. Acad. Sci. Paris **319**, 523–528 (1994)
3007. Myerson, G.: Period polynomials and Gauss sums for finite fields. Acta Arith. **39**, 251–264 (1981)
3008. Myller-Lebedeff, V.: Sur les racines primitives et les systémes de bases et indices dans le corps quadratique général. J. Math. Pures Appl. (8) **2**, 81–98 (1919)
3009. Nagata, K.: On bases of purely cubic fields over quadratic fields. Tokyo J. Math. **8**, 121–131 (1985)
3010. Nagata, M.: Some remarks on Euclid rings. J. Math. Kyoto Univ. **25**, 421–422 (1985)
3011. Nagata, M.: On the definition of a Euclid ring. In: Commutative Algebra and Combinatorics (Kyoto, 1985), 167–171. North-Holland (1987)
3012. Nagata, M.: A pairwise algorithm and its application to $Z[\sqrt{14}]$. In: Algebraic Geometry Seminar (Singapore, 1987), 69–74. World Scientific Publishing (1988)
3013. Nagata, M., Nakayama, T., Tuzuku, T.: On an existence lemma in valuation theory. Nagoya Math. J. **6**, 59–61 (1953)
3014. Nagell, T. [Nagel,T.]: Über die Klassenzahl imaginär-quadratischer Zahlkörper. Abh. Math. Semin. Univ. Hambg **1**, 140–150 (1922)
3015. Nagell, T.: Über die Einheiten in reinen kubischen Zahlkörpern. Skrifter Videnskaps. Kristiania **1923(11)**, 1–34
3016. Nagell, T.: Solution complète de quelques équations cubiques à deux indéterminées. J. Math. Pures Appl. (9) **4**, 209–270 (1925)
3017. Nagell, T.: Über einige kubische Gleichungen mit zwei Unbestimmten. Math. Z. **24**, 422–447 (1925)
3018. Nagell, T.: Darstellung ganzer Zahlen durch binäre kubische Formen mit negativer Diskriminante. Math. Z. **28**, 10–29 (1928)
3019. Nagell, T.: Zur Theorie der kubischen Irrationalitäten. Acta Math. **55**, 33–65 (1930)
3020. Nagell, T.: Über algebraische Zahlkörper mit gegebener Diskriminante. Comment. Math. Helv. **2**, 169–173 (1930)
3021. Nagell, T.: Sätze über algebraische Ringe. Math. Z. **34**, 179–181 (1931)
3022. Nagell, T.: Zur Theorie der algebraischen Ringe. J. Reine Angew. Math. **164**, 80–84 (1931)
3023. Nagell, T.: Zur algebraischen Zahlentheorie. Math. Z. **34**, 183–193 (1932)
3024. Nagell, T.: Über die Lösbarkeit der Gleichung $x^2 - Dy^2 = -1$. Ark. Mat. **23(6)**, 1–5 (1933)
3025. Nagell, T.: Bestimmung des Grades gewisser relativ-algebraischer Zahlen. Monatsh. Math. Phys. **48**, 61–74 (1939)

3026. Nagell, T.: On the representations of integers as the sum of two integral squares in algebraic, mainly quadratic fields. Nova Acta Soc. Sci. Upsal. (4) **15(11)** 1–73 (1953)
3027. Nagell, T.: Les points exceptionnels rationnels sur certaines cubiques du premier genre. Acta Arith. **5**, 333–357 (1959)
3028. Nagell, T.: Les points exceptionnels sur les cubiques $ax^3 + by^3 + cz^3 = 0$. Acta Sci. Math. (Szeged) **21**, 173–180 (1960)
3029. Nagell, T.: Anders Wiman in memoriam. Acta Math. **103**, I–VI (1960)
3030. Nagell, T.: On the sum of two integral squares in certain quadratic fields. Ark. Mat. **4**, 267–286 (1961)
3031. Nagell, T.: On the number of representations of an A-number in an algebraic field. Ark. Mat. **4**, 467–478 (1962)
3032. Nagell, T.: On the A-numbers in the quadratic fields $K(\sqrt{\pm 37})$. Ark. Mat. **4**, 511–521 (1963)
3033. Nagell, T.: Thoralf Skolem in memoriam. Acta Math. **110**, I–XI (1963)
3034. Nagell, T.: Contributions à la théorie des corps et des polynomes cyclotomiques. Ark. Mat. **5**, 153–192 (1964)
3035. Nagell, T.: Sur une propriété des unités d'un corps algébrique. Ark. Mat. **5**, 343–356 (1964)
3036. Nagell, T.: Contributions à la théorie des modules et des anneaux algébriques. Ark. Mat. **6**, 161–178 (1965)
3037. Nagell, T.: Quelques résultats sur les diviseurs fixes de l'index des nombres entiers d'un corps algébrique. Ark. Mat. **6**, 269–289 (1966)
3038. Nagell, T.: Sur les discriminants des nombres algébriques. Ark. Mat. **7**, 265–282 (1967)
3039. Nagell, T.: Sur les unités dans les corps biquadratiques primitifs du premier rang. Ark. Mat. **7**, 359–394 (1968)
3040. Nagell, T.: Quelques propriétés des nombres algébriques du quatrième degré. Ark. Mat. **7**, 517–525 (1968)
3041. Nagell, T.: Quelques problèmes relatifs aux unités algébriques. Ark. Mat. **8**, 115–127 (1969)
3042. Nagell, T.: Sur un type particulier d'unités algébriques. Ark. Mat. **8**, 163–184 (1969)
3043. Nagell, T.: Über die Darstellung der Zahlen ± 1 als die Summe von zwei Quadraten in algebraischen Zahlkörpern. Arch. Math. (Basel) **23**, 25–29 (1972)
3044. Nakada, H.: On ergodic theory of A.Schmidt's complex continued fractions over Gaussian field. Monatsh. Math. **105**, 131–150 (1988)
3045. Nakada, H.: The metrical theory of complex continued fractions. Acta Arith. **56**, 279–289 (1990)
3046. Nakagawa, J.: On the Galois group of a number field with square free discriminant. Comment. Math. Univ. St. Pauli **37**, 95–98 (1988)
3047. Nakagoshi, N.: The structure of the multiplicative group of residue classes modulo $\mathfrak{p}^{N+1}$. Nagoya Math. J. **73**, 41–60 (1979)
3048. Nakahara, T.: On cyclic biquadratic fields related to a problem of Hasse. Monatsh. Math. **94**, 125–132 (1982)
3049. Nakahara, T.: On the indices and integral bases of noncyclic but Abelian biquadratic fields. Arch. Math. (Basel) **41**, 504–508 (1983)
3050. Nakano, N.: Über den Fundamentalsatz der Idealtheorie in unendlichen algebraischen Zahlkörpern. J. Sci. Hiroshima Univ. **15**, 171–175 (1952)
3051. Nakano, N.: Idealtheorie in einem speziellen unendlichen algebraischen Zahlkörper. J. Sci. Hiroshima Univ. **16**, 425–439 (1953)
3052. Nakano, N.: Über idempotente Ideale in unendlichen algebraischen Zahlkörpern. J. Sci. Hiroshima Univ. **17**, 11–20 (1953)
3053. Nakano, S.: Class numbers of pure cubic fields. Proc. Japan Acad. Sci. **59**, 263–265 (1983)
3054. Nakano, S.: On ideal class groups of algebraic number fields. J. Reine Angew. Math. **358**, 61–75 (1985)
3055. Nakatsuchi, S.: A note on Kronecker's "Randwertsatz". J. Math. Kyoto Univ. **13**, 129–137 (1973)

3056. Nakayama, T.: Über die Beziehungen zwischen den Faktorensystemen und der Normklassengruppe eines galoisschen Erweiterungskörpers. Math. Ann. **112**, 85–91 (1935)
3057. Nakayama, T.: On Frobeniusean algebras, II. Ann. Math. (2) **42**, 1–21 (1941)
3058. Nakayama, T.: A theorem on the norm group of a finite extension field. Jpn. J. Math. **18**, 877–885 (1943)
3059. Nakayama, T., Moriya, M.: Zur Theorie der Normenrestsymbole über diskret perfekten Körpern. Proc. Imp. Acad. Tokyo **19**, 129–131 (1943)
3060. Nakayama, T., Moriya, M.: Die Theorie der Klassenkörper im Kleinen über diskret perfekten Körpern, III. Proc. Imp. Acad. Tokyo **19**, 132–137 (1943)
3061. Narkiewicz, W.: On algebraic number fields with non-unique factorization. Colloq. Math. **12**, 59–67 (1964)
3062. Narkiewicz, W.: On algebraic number fields with non-unique factorization, II. Colloq. Math. **15**, 49–58 (1966)
3063. Narkiewicz, W.: On a theorem of A. Weil on derivations in number fields. Colloq. Math. **20**, 57–58 (1969)
3064. Narkiewicz, W.: Numbers with unique factorization in an algebraic number field. Acta Arith. **21**, 313–322 (1972)
3065. Narkiewicz, W.: Elementary and Analytic Theory of Algebraic Numbers. PWN, 1974. [2nd ed. PWN & Springer (1990); 3rd ed. Springer (2004)]
3066. Narkiewicz, W.: Classical Problems in Number Theory. PWN (1986)
3067. Narkiewicz, W.: Finite abelian groups and factorization problems. Colloq. Math. **42**, 319–330 (1979)
3068. Narkiewicz, W.: Hermann Weyl and the theory of numbers. In: Exact Sciences and their Philosophical Foundations, 51–60, P. Lang (1988)
3069. Narkiewicz, W.: Polynomial Mappings. Lecture Notes Math. **1600** (1995)
3070. Narkiewicz, W.: Euclidean algorithm in small abelian fields. Funct. Approx. Comment. Math. **37**, 337–340 (2007)
3071. Narkiewicz, W., Schinzel, A.: Ein einfacher Beweis des Dedekindschen Satzes über die Differente. Colloq. Math. **20**, 65–66 (1969)
3072. Narkiewicz, W., Śliwa, J.: Finite abelian groups and factorization problems, II. Colloq. Math. **46**, 115–122 (1982)
3073. Narkiewicz, W., Więsław, W.: Zenon Borewicz (1922–1995). Wiad. Mat. **36**, 65–72 (2000)
3074. Nart, E.: On the index of a number field. Trans. Amer. Math. Soc. **289**, 171–183 (1985)
3075. Naryškina, E.A.: On numbers analogous to Bernoulli numbers in the one-class quadratic rings of negative discriminant. Izv. Akad. Nauk SSSR, Ser. Mat. **19**, 145–176 (1925)
3076. Naryškina, E.A.: On numbers analogous to Bernoulli numbers in the one-class quadratic rings of negative discriminant, II. Izv. Akad. Nauk SSSR, Ser. Mat. **19**, 297–314 (1925)
3077. Nassirou, L.: Étude du niveau de certains corps. Bull. Belg. Math. Soc. Simon Stevin **6**, 131–146 (1999)
3078. Nastold, H.-J.: Wolfgang Krulls Arbeiten zur kommutativen Algebra und ihre Bedeutung für die algebraische Geometrie. Jahresber. Deutsch. Math.-Verein. **82**, 63–76 (1980)
3079. Nehrkorn, H.: Über absolute Idealklassengruppen und Einheiten in algebraischen Zahlkörpern. Abh. Math. Semin. Univ. Hambg **9**, 318–334 (1933)
3080. Neild, C., Shanks, D.: On the 3-rank of quadratic fields and the Euler product. Math. Comp. **28**, 279–29 (1974)
3081. Neiss, F.: Darstellung relativ Abelscher Zahlkörper durch Primkörper und Einheitskörper. J. Reine Angew. Math. **166**, 30–53 (1931)
3082. Nemenzo, F.R.: On a theorem of Scholz on the class number of quadratic fields. Proc. Japan Acad. Sci. **80**, 9–11 (2004)
3083. Nemenzo, F.[R.], Wada, H.: An elementary proof of Gauss' genus theorem. Proc. Japan Acad. Sci. **68**, 94–95 (1992)
3084. Netto, E.: Notiz über Gleichungen, deren Discriminante ein Quadrat ist. J. Reine Angew. Math. **95**, 237–239 (1883)

3085. Netto, E.: Über die Factorenzerlegung der Discriminanten algebraischer Gleichungen. Math. Ann. **24**, 579–587 (1884)
3086. Netto, E.: Ueber die Irreductibilität ganzzahliger ganzer Functionen. Math. Ann. **48**, 81–88 (1897)
3087. Netto, E.: Vorlesungen über Algebra. II. Teubner (1900)
3088. Neuhaus, F.W.: Affektlosigkeit der Gleichungen für fast alle Werte des linearen Koeffizienten. Deutsche Math. **7**, 87–116 (1942)
3089. Neuhaus, F.W.: Über die Verteilung aller ganzzahligen Gleichungen von mehr als zwei Unbestimmten auf ihre Galois'schen Gruppen. Math. Ann. **121**, 379–404 (1950)
3090. Neukirch, J.: Zur Differententheorie. Arch. Math. (Basel) **18**, 241–249 (1967)
3091. Neukirch, J.: Klassenkörpertheorie. Bibliographisches Institut (1969) [New ed.: Springer (2011); English translation: Class Field Theory. Springer (2013)]
3092. Neukirch, J.: Über das Einbettungsproblem der algebraischen Zahlentheorie. Invent. Math. **21**, 59–116 (1973)
3093. Neukirch, J.: Eine Bemerkung zum Existenzsatz von Grunwald-Hasse-Wang. J. Reine Angew. Math. **268/269**, 315–317 (1974)
3094. Neukirch, J.: On an existence theorem of Grunwald's type, Bol. Soc. Brasil. Mat. **5**, 79–83 (1974)
3095. Neukirch, J.: Zur Theorie der nilpotenten algebraischen Zahlkörper. Math. Ann. **219**, 65–83 (1976)
3096. Neukirch, J.: On solvable number fields. Invent. Math. **53**, 135–164 (1979)
3097. Neukirch, J.: Neubegründung der Klassenkörpertheorie. Math. Z. **186**, 557–574 (1984)
3098. Neukirch, J.: Class Field Theory. Springer (1986)
3099. Neukirch, J.: Algebraische Zahlentheorie. Springer (1992) [English translation: Algebraic Number Theory. Springer (1999)]
3100. Neukirch, J., Schmidt, A., Wingberg, K.: Cohomology of Number Fields. Springer (2000); 2nd ed. (2008)
3101. Neumann, O.: Zur Genesis der algebraischen Zahlentheorie, NTM Schriftenr. Gesch. Naturwiss., Technik, Med., Leipzig **16(2)**, 22–39 (1979)
3102. Neumann, O.: Zur Genesis der algebraischen Zahlentheorie, II, NTM Schriftenr. Gesch. Naturwiss., Technik, Med., Leipzig **17(1)** 32–48 (1980)
3103. Neumann, O.: Zur Genesis der algebraischen Zahlentheorie, III, NTM Schriftenr. Gesch. Naturwiss., Technik, Med., Leipzig, **17(2)** 38–58 (1980)
3104. Neumann, O.: Über die Anstöße zu Kummers Schöpfung der "idealen complexen Zahlen". In: Mathematical Perspectives, 179–199. Academic Press (1981)
3105. Neumann, O.: Two proofs of the Kronecker-Weber theorem "according to Kronecker, and Weber". J. Reine Angew. Math. **323**, 105–126 (1981)
3106. Neumann, O., Folkerts, M.: Carl Gustav Reuschle (1812–1875) — ein Stuttgarter Gymnasialprofessor für Mathematik, Physik und Geographie. In: Mathematik im Wandel, 220–227, Franzbecker (2001)
3107. Neumann, O., Purkert, W.: Richard Dedekind – zum 150. Geburtstag. Mitt. Math. Ges. DDR **1981(2)**, 2–4, 84–110
3108. Newman, M.: A table of the first factor for prime cyclotomic fields. Math. Comp. **24**, 215–219 (1970)
3109. Newman, M.H.A.: Hermann Weyl. J. Lond. Math. Soc. **33**, 500–511 (1958)
3110. Niklasch, G.: On the verification of Clark's example of a Euclidean but not norm-Euclidean number field. Manuscripta Math. **83**, 443–446 (1994)
3111. Niklasch, G.: Counting exceptional units. Collect. Math. **48**, 195–207 (1997)
3112. Niklasch, G., Smart, N.P.: Exceptional units in a family of quartic number fields. Math. Comp. **67**, 759–772 (1998)
3113. Niven, I.: Integers of quadratic fields as sums of squares. Trans. Amer. Math. Soc. **48**, 405–417 (1940)
3114. Niven, I.: Sums of n-th powers of quadratic integers. Duke Math. J. **8**, 441–451 (1941)

3115. Niven, I.: Sums of fourth powers of Gaussian integers. Bull. Amer. Math. Soc. **47**, 923–926 (1941)
3116. Niven, I.: Quadratic Diophantine equations in the rational and quadratic fields. Trans. Amer. Math. Soc. **52**, 1–11 (1942)
3117. Niven, I.: The Pell equation in quadratic fields. Bull. Amer. Math. Soc. **49**, 413–416 (1943)
3118. Noether, E.: Rationale Funktionenkörper. Jahresber. Deutsch. Math.-Verein. **22**, 316–319 (1913)
3119. Noether, E.: Gleichungen mit vorgeschriebener Gruppe. Math. Ann. **78**, 221–229 (1917)
3120. Noether, E.: Die arithmetische Theorie der algebraischen Funktionen einer Veränderlichen in ihrer Beziehung zu den übrigen Theorien und zu der Zahlkörpertheorie. Jahresber. Deutsch. Math.-Verein. **28**, 182–203 (1919)
3121. Noether, E.: Idealtheorie in Ringbereichen. Math. Ann. **83**, 24–66 (1921)
3122. Noether, E.: Der Diskriminantensatz für die Ordnungen eines algebraischen Zahl- oder Funktionenkörpers. J. Reine Angew. Math. **157**, 82–104 (1926)
3123. Noether, E.: Abstrakter Aufbau der Idealtheorie in algebraischen Zahl- und Funktionenkörpern. Math. Ann. **96**, 26–61 (1926)
3124. Noether, E.: Hyperkomplexe Grössen und Darstellungstheorie. Math. Z. **30**, 641–692 (1929)
3125. Noether, E.: Normalbasis bei Körpern ohne höhere Verzweigung. J. Reine Angew. Math. **167**, 147–152 (1932)
3126. Noether, E.: Der Hauptgeschlechtssatz für relativ-galoissche Zahlkörper. Math. Ann. **108**, 411–419 (1933)
3127. Noether, M.: Arthur Cayley. Math. Ann. **46**, 462–480 (1895)
3128. Noether, M.: Charles Hermite. Math. Ann. **55**, 337–385 (1902)
3129. Nörlund, N.E.: J.L.W.V. Jensen. Mat. Tidsskr. **B 1926**, 1–7
3130. Northcott, D.G.: An inequality in the theory of arithmetic on algebraic varieties. Proc. Cambridge Philos. Soc. **45**, 502–509 (1949)
3131. Northcott, D.G.: A further inequality in the theory of arithmetic on algebraic varieties. Proc. Cambridge Philos. Soc. **45**, 510–518 (1949)
3132. Northcott, D.G.: Ideal Theory. Cambridge University Press (1953)
3133. Northcott, D.G.: A note on classical ideal theory. Proc. Cambridge Philos. Soc. **51**, 766–767 (1955)
3134. Nowak, W.G.: On the distribution of integer ideals in algebraic number fields. Math. Nachr. **161**, 59–74 (1993)
3135. Nowlan, F.S.: Representation of integers by certain ternary cubic forms. Bull. Amer. Math. Soc. **32**, 374–380 (1926)
3136. Odai, Y.: On the index of the group generated by relative units. J. Number Theory **46**, 60–69
3137. Odlyzko, A.M.: Some analytic estimates of class numbers and discriminants. Invent. Math. **29**, 1975, 275–286 (1975)
3138. Odlyzko, A.M.: Lower bounds for discriminants of number fields. Acta Arith. **29**, 275–297 (1976)
3139. Odlyzko, A.M.: Lower bounds for discriminants of number fields, II. Tôhoku Math. J. **29**, 209–216 (1977)
3140. Odlyzko, A.M.: Bounds for discriminants and related estimates for class numbers, regulators and zeros of zeta functions; a survey of recent results. Sém. Théor. Nombres Bordeaux **2**, 119–141 (1990)
3141. Odoni, R.W.K.: On a problem of Narkiewicz. J. Reine Angew. Math. **288**, 160–167 (1976)
3142. Odoni, R.W.K.: Weil numbers and CM-fields, II. J. Number Theory **8**, 366–377 (1991)
3143. Oesterlé, J.: Travaux de Ferrero et Washington sur le nombre de classes d'idéaux des corps cyclotomiques. Lecture Notes Math. **770**, 170–182 (1980)
3144. Oesterlé, J.: Nombres de classes des corps quadratiques imaginaires. Astérisque **121/122**, 309–323 (1985)
3145. Oesterlé, J.: Le problème de Gauss sur le nombre de classes. Enseign. Math. (2) **34**, 43–67 (1988)

3146. Ojala, T.: Euclid's algorithm in the cyclotomic field $Q(\zeta_{16})$. Math. Comp. **31**, 268–273 (1977)
3147. Okada, S.: Generalized Maillet determinant. Nagoya Math. J. **94**, 165–170 (1984)
3148. Okutsu, K.: Integral basis of the field $Q(\sqrt[n]{a})$. Proc. Japan Acad. Sci. **58**, 219–222 (1982)
3149. Olivier, M.: Corps sextiques contenant un corps quadratique, I. Sém. Théor. Nombres Bordeaux (2) **1**, 205–250 (1989)
3150. Olivier, M.: Corps sextiques primitifs. Ann. Inst. Fourier (Grenoble) **40**, 757–767 (1990)
3151. Olivier, M.: The computation of sextic fields with a cubic subfield and no quadratic subfield. Math. Comp. **58**, 419–432 (1992)
3152. Olson, J.: Henry B. Mann. In: Number Theory and Algebra, xx–xxv. Academic Press (1977)
3153. O'Meara, O.T.: Basis structure of modules. Proc. Amer. Math. Soc. **7**, 965–974 (1956)
3154. O'Meara, O.T.: On the finite generation of linear groups over Hasse domains. J. Reine Angew. Math. **217**, 79–108 (1965)
3155. Ono, T.: Gauss transforms and zeta-functions. Ann. Math. (2) **91**, 332–361 (1970)
3156. Ono, T.: An Introduction to Algebraic Number Theory. Plenum Press (1990)
3157. Opolka, H.: Zur Auflösung zahlentheoretischer Knoten. Math. Z. **173**, 95–103 (1980)
3158. Opolka, H.: Auflösung zahlentheoretischer Knoten in Galoiserweiterungen von Q. Arch. Math. (Basel) **34**, 416–420 (1980)
3159. Opolka, H.: Some remarks on the Hasse norm theorem. Proc. Amer. Math. Soc. **84**, 464–466 (1982)
3160. Opolka, H.: Normenreste in relativ abelschen Zahlkörpererweiterungen und symplektische Paarungen. Abh. Math. Semin. Univ. Hambg **54**, 1–4 (1984)
3161. Opolka, H.: The norm exponent in Galois extensions of number fields. Proc. Amer. Math. Soc. **99**, 41–43 (1987)
3162. Opolka, H.: Norm exponents and representation groups. Proc. Amer. Math. Soc. **111**, 595–597 (1991)
3163. Oppenheim, A.: Quadratic fields with and without Euclid's algorithm. Math. Ann. **109**, 349–352 (1934)
3164. Orde, H.L.S.: On Dirichlet's class number formula. J. Lond. Math. Soc. **18**, 409–420 (1978)
3165. Ore, Ö.: Zur Theorie der algebraischen Körper. Acta Math. **44**, 219–314 (1923)
3166. Ore, Ö.: Die Dedekindschen Sätze über den Zusammenhang zwischen den Idealen und den höheren Kongruenzen. Skrifter Videns.-Akad. Oslo **1923(22)**, 1–11 (1923)
3167. Ore, O.: Zur Theorie der Irreduzibilitätskriterien. Math. Z. **18**, 278–288 (1923)
3168. Ore, Ö.: Ein Problem von Dedekind. Acta Litt. Sci. Reg. Univ. Hungar. Franc.-Jos., sect. Math. **2**, 15–17 (1924/26)
3169. Ore, Ö.: Algebraische Gleichungen mit primitiven Gruppen. Math. Z. **19**, 276–283 (1924)
3170. Ore, Ö.: Zur Theorie der Eisensteinschen Gleichungen. Math. Z. **20**, 267–279 (1924)
3171. Ore, Ö.: Note sur une identité dans la théorie des congruences supérieures. Rend. Circ. Mat. Palermo **48**, 37–40 (1924)
3172. Ore, Ö.: Weitere Untersuchungen zur Theorie der algebraischen Körper. Acta Math. **45**, 145–160 (1925)
3173. Ore, Ö.: Bestimmung der Diskriminanten algebraischer Körper. Acta Math. **45**, 303–344 (1925)
3174. Ore, Ö.: Verallgemeinerung des vorstehenden Satzes von Herrn Bauer. Acta Litt. Sci. Reg. Univ. Hungar. Franc.-Jos., sect. Math. **2**, 72–74 (1924/26)
3175. Ore, O.: Bestimmung der Differente eines algebraischen Zahlkörpers. Acta Math. **46**, 363–392 (1925)
3176. Ore, Ö.: Zur Theorie der Relativkörper. Pure Appl. Math. Q. **1925(10)**, 1–33
3177. Ore, O.: Bemerkungen zur Theorie der Differente. Math. Z. **25**, 1–8 (1926)
3178. Ore, Ö.: Existenzbeweise für algebraische Körper mit vorgeschriebenen Eigenschaften. Math. Z. **25**, 474–489 (1926)
3179. Ore, Ö.: Über den Zusammenhang zwischen den definierenden Gleichungen und der Idealtheorie in algebraischen Körpern. Math. Ann. **96**, 313–352 (1926)

3180. Ore, Ö.: Über den Zusammenhang zwischen den definierenden Gleichungen und der Idealtheorie in algebraischen Körpern, II. Math. Ann. **97**, 569–598 (1927)
3181. Ore, Ö.: Über zusammengesetzte algebraische Körper. Acta Math. **49**, 379–396 (1927)
3182. Ore, Ö.: Newtonsche Polygone in der Theorie der algebraischer Körper. Math. Ann. **99**, 84–117 (1928)
3183. Ore, O.: Some theorems on the connection between ideals and groups of a Galois field. Trans. Amer. Math. Soc. **30**, 610–620 (1928)
3184. Ore, Ö.: Abriß einer arithmetischen Theorie der Galoisschen Körper, I. Math. Ann. **100**, 650–673 (1928)
3185. Ore, Ö.: Abriß einer arithmetischen Theorie der Galoisschen Körper, II. Math. Ann. **102**, 283–304 (1930)
3186. Ore, O.: Hancock on algebraic numbers. Bull. Amer. Math. Soc. **39**, 645–647 (1935)
3187. Ore, O.: James Pierpont in memoriam. Bull. Amer. Math. Soc. **45**, 481–486 (1939)
3188. Ore, O.: Niels Henrik Abel: Mathematician extraordinary. University of Minnesota Press (1957)
3189. Oriat, B.: Étude arithmétique des corps cycliques de degré p^r sur le corps des nombres rationnels. Enseign. Math. **18**, 57–104 (1972)
3190. Oriat, B.: Sur la divisibilité par 8 et 16 des nombres de classes d'idéaux des corps quadratiques $Q(\sqrt{2p})$ et $Q(\sqrt{-2p})$. J. Math. Soc. Japan **30**, 279–285 (1978)
3191. Oriat, B.: Groupes des classes d'idéaux des corps quadratiques réels $Q(d^{1/2})$, $1<d<24572$. Publ. Math. Fac. Sci. Besançon **1986/87–1987/88(2)**, 65 pp. (1988)
3192. Oriat, B.: Groupes des classes d'idéaux des corps quadratiques imaginaires $Q(d^{1/2})$, $-24572<d<0$. Publ. Math. Fac. Sci. Besançon **1986/87–1987/88(2)**, 63 pp. (1988)
3193. Oriat, B., Satgé, P.: Un essai de généralisation du "Spiegelungssatz". J. Reine Angew. Math. **307/308**, 134–159 (1979)
3194. Ortiz, E.L., Pinkus, A.: Herman Müntz: a mathematicians Odyssey. Math. Intelligencer **27(1)**, 22–31 (2005)
3195. Osada, H.: The Galois groups of the polynomials $X^n + aX^l + b$. J. Number Theory **25**, 230–238 (1987)
3196. Osada, H.: The Galois groups of the polynomials $X^n + aX^l + b$. II. Tôhoku Math. J. (2) **39**, 437–445 (1987)
3197. Ostrowski, A.: Über einige Fragen der allgemeinen Körpertheorie. J. Reine Angew. Math. **143**, 255–284 (1913)
3198. Ostrowski, A.: Über sogenannte perfekte Körper. J. Reine Angew. Math. **147**, 191–204 (1917)
3199. Ostrowski, A.: Über einige Lösungen der Funktionalgleichung $\varphi(x)\varphi(y) = \varphi(xy)$. Acta Math. **41**, 271–284 (1917)
3200. Ostrowski, A.: Über ganzwertige Polynome in algebraischen Zahlkörpern. J. Reine Angew. Math. **149**, 117–124 (1919)
3201. Ostrowski, A.: Untersuchungen zur arithmetischer Theorie der Körper. Math. Z. **39**, 269–320 (1935)
3202. Ostrowski, A.: Untersuchungen zur arithmetischer Theorie der Körper, II, III. Math. Z. **39**, 321–404 (1935)
3203. Oswald, N.M.R., Steuding, J.: Complex continued fractions: early work of the brothers Adolf and Julius Hurwitz. Arch. Hist. Exact Sci. **68**, 499–528 (2014)
3204. Oswald, N.M.R., Steuding, J.: About the cover: zeta-functions associated with quadratic forms in Adolf Hurwitz's estate. Bull. Amer. Math. Soc. **53**, 477–481 (2016)
3205. Overholtzer, G.: Sum functions in elementary p-adic analysis. Amer. J. Math. **74**, 332–346 (1952)
3206. Ožigova, E.P.: Egor Ivanovič Zolotarev 1847–1878. Nauka (1966) (Russian)
3207. Pahlings, H.: Some sporadic groups as Galois groups. Rend. Semin. Mat. Univ. Padova **79**, 97–107 (1988)
3208. Pahlings, H.: Some sporadic groups as Galois groups, II. Rend. Semin. Mat. Univ. Padova **82**, 163–171 (1988); Corr.: **85**, 309–310 (1991)

3209. Pajunen, S.: Computations on the growth of the first factor for prime cyclotomic fields. Nordisk Tidskr. Inf. (BIT) **16**, 85–87 (1976)
3210. Pajunen, S.: Computations on the growth of the first factor for prime cyclotomic fields, II. Nordisk Tidskr. Inf. (BIT) **17**, 113–114 (1977)
3211. Pall, G.: The structure of the number of representations function in a positive binary quadratic form. Math. Z. **36**, 321–34 (1933)
3212. Pall, G.: The structure of the number of representations function in a binary quadratic form. Trans. Amer. Math. Soc. **35**, 491–509 (1933)
3213. Pall, G.: A new solution of the Gauss problem on $h(s^2d)/h(d)$. Bull. Amer. Math. Soc. **41**, 373–374 (1935)
3214. Pall, G.: Note on irregular determinants. J. Lond. Math. Soc. **11**, 34–35 (1936)
3215. Pall, G.: Note on factorization in a quadratic field. Bull. Amer. Math. Soc. **51**, 771–775 (1945)
3216. Pall, G.: Sums of two squares in a quadratic field. Duke Math. J. **18**, 399–409 (1951)
3217. Pappalardi, F.: On the exponent of the ideal class group of $Q(\sqrt{-d})$. Proc. Amer. Math. Soc. **123**, 663–671 (1995)
3218. Parikh, C.: The Unreal Life of Oscar Zariski. Academic Press (1991) [Reprint: Springer (2009)]
3219. Park, J.: Notes on quadratic integers and real quadratic number fields. Osaka J. Math. **53**, 983–1002 (2016)
3220. Park, P.-S.: Sums of distinct integral squares in $Q(\sqrt{5})$. C. R. Acad. Sci. Paris **346**, 723–725 (2008)
3221. Park, S.-M., Kwon, S.-H.: Class number one problem for normal CM-fields. J. Number Theory **125**, 59–84 (2007)
3222. Park, Y.-H.: The class number one problem for the non-abelian normal CM-fields of degree 24 and 40. Acta Arith. **101**, 63–80 (2002)
3223. Parnami, J.C., Agrawal, M.K., Rajwade, A.R.: On the 4-power Stufe of a field. Rend. Circ. Mat. Palermo (2) **30**, 245–254 (1981)
3224. Parnami, J.C., Agrawal, M.K., Rajwade, A.R.: On the stufe of quartic fields. J. Number Theory **38**, 106–109 (1991)
3225. Parry, C.J.: The $\mathfrak{p}$-adic generalisation of the Thue-Siegel theorem. Acta Math. **83**, 1–100 (1950)
3226. Parry, C.J.: Class number formulae for bicubic fields. Illinois J. Math. **21**, 148–163 (1977)
3227. Parry, W.: Growth series of Coxeter groups and Salem numbers. J. Algebra **154**, 406–415 (1993)
3228. Patterson, S.: Kurt Heegner — Biographical Notes. Report Math. Forschungsinstitut Oberwolfach **24/2008**
3229. Pauli, S.: Constructing class fields over local fields. J. Théor. Nombres Bordeaux **18**, 627–652 (2006)
3230. Pauli, S., Roblot, X.-F.: On the computation of all extensions of a p-adic field of a given degree. Math. Comp. **70**, 1641–1659 (2001)
3231. Payan, J.J.: Sur les classes ambiges et les ordres monogènes d'une extension cyclique de degré premier impair sur Q ou sur un corps quadratique imaginaire. Ark. Mat. **11**, 239–244 (1973)
3232. Peano, G.: Angelo Genocchi. Annuario Univ. degli Studi di Torino **1889/90**, 195–202
3233. Pellet, A.E.: Sur la décomposition d'une fonction entière en facteurs irréductibles suivant un module premier. C. R. Acad. Sci. Paris **86**, 1071–1072 (1878)
3234. Pellet, A.E.: Sur la réduction des fonctions entières algébriques. Bull. Soc. Math. France **19**, 48–52 (1891)
3235. Pépin, T.: Nombre des classes de formes quadratiques pour un déterminant donné. Ann. Sci. Éc. Norm. Supér. (2) **3**, 165–208 (1874)
3236. Pépin, T.: Sur certains nombres complexes compris dans la formule $a+b\sqrt{-c}$. J. Math. Pures Appl. (3) **1**, 317–372 (1875)

3237. Pépin, T.: Sur les lois de réciprocité dans la théorie des résidus de puissances. C. R. Acad. Sci. Paris **84**, 762–765 (1877)
3238. Pépin, T.: Mémoire sur les lois de réciprocité relatives aux résidus des puissances. Atti Accad. Pont. N. Lincei **31**, 40–149 (1880)
3239. Pépin, T.: Sur la classification des formes quadratiques binaires. Atti Accad. Pont. N. Lincei **33**, 354–391 (1882)
3240. Perlis, R.: On the equation $\zeta_K(s) = \zeta_{K'}(s)$. J. Number Theory **9**, 342–360 (1977)
3241. Perlis, R.: On the class numbers of arithmetically equivalent fields. J. Number Theory **10**, 489–509 (1978)
3242. Perlis, R.: On the analytic determination of the trace form. Canad. Math. Bull. **28**, 422–430 (1985)
3243. Perlis, S.: Normal bases of cyclic fields of prime-power degree. Duke Math. J. **9**, 507–517 (1942)
3244. Perott, J.: Sur la formation des déterminants irréguliers. J. Reine Angew. Math. **95**, 232–237 (1883)
3245. Perott, J.: Sur la formation des déterminants irréguliers. J. Reine Angew. Math. **96**, 327–348 (1884)
3246. Perret, M.: On the ideal class group problem for global fields. J. Number Theory **77**, 27–35 (1999)
3247. Perron, O.: Über eine Anwendung der Idealtheorie auf die Frage nach der Irreduzibilität algebraischer Gleichungen. Math. Ann. **60**, 448–458 (1905)
3248. Perron, O.: Grundlagen fuer eine Theorie des Jacobischen Kettenbruchalgorithmus. Math. Ann. **64**, 1–76 (1997)
3249. Perron, O.: Zur Theorie der Dirichletschen Reihen. J. Reine Angew. Math. **134**, 95–143 (1908)
3250. Perron, O.: Abschätzung der Lösung der Pellschen Gleichung. J. Reine Angew. Math. **144**, 71–73 (1914)
3251. Perron, O.: Über Gleichungen ohne Affekt. SBer. Heidelberg. Akad. Wiss. **1923(3)**, 1–13
3252. Perron, O.: Über die Approximation einer komplexen Zahl durch Zahlen des Körpers $\mathfrak{K}(i)$. Math. Ann. **103**, 533–544 (1930)
3253. Perron, O.: Über die Approximation einer komplexen Zahl durch Zahlen des Körpers $\mathfrak{K}(i)$, II. Math. Ann. **105**, 160–164 (1931)
3254. Perron, O.: Über einen Approximationssatz von Hurwitz und über die Approximation einer komplexen Zahl durch Zahlen des Körpers der dritten Einheitswurzeln. SBer. Bayer. Akad. Wiss. **1931**, 129–154
3255. Perron, O.: Quadratische Zahlkörper mit Euklidischem Algorithmus. Math. Ann. **107**, 489–495 (1932)
3256. Perron, O.: Diophantische Approximationen in imaginären quadratischen Zahlkörpern, insbesondere im Körper $\mathfrak{K}(i\sqrt{2})$. Math. Z. **37**, 749–767 (1988)
3257. Perron, O.: Alfred Pringsheim. Jahresber. Deutsch. Math.-Verein. **56**, 1–6 (1952)
3258. Perron, O.: Der Jacobi'sche Kettenalgorithmus in einem kubischen Zahlenkörper. SBer. Bayer. Akad. Wiss. **1971**, 13–49
3259. Perron, O.: Der Jacobi'sche Kettenalgorithmus in einem kubischen Zahlenkörper, II. SBer. Bayer. Akad. Wiss. **1973**, 9–22
3260. Perron, O.: Heinrich Tietze, 31.8.1880–17.2.1964. Jahresber. Deutsch. Math.-Verein. **83**, 182–191 (1981)
3261. Peters, M.: Die Stufe von Ordnungen ganzer Zahlen in algebraischen Zahlkörpern. Math. Ann. **195**, 309–314 (1971)
3262. Peters, M.: Quadratische Formen über Zahlringen. Acta Arith. **24**, 157–164 (1973)
3263. Peters, M.: Summen von Quadraten in Zahlringen. J. Reine Angew. Math. **268/269**, 318–323 (1974)
3264. Peters, M.: Einklassige Geschlechter von Einheitsformen in total reellen algebraischen Zahlkörpern. Math. Ann. **226**, 117–120 (1977)
3265. Petersen, J.: Theorie der algebraischen Gleichungen. Höst (1878)

3266. Petersson, H.: Das wissenschaftliche Werk von E. Hecke. Abh. Math. Semin. Univ. Hambg **16**, 7–31 (1949)
3267. Petersson, H.: Über eine Zerlegung des Kreisteilungspolynoms von Primzahlordnung. Math. Nachr. **14**, 361–375 (1955)
3268. Petersson, H.: Über Darstellungsanzahlen von Primzahlen durch Quadratsummen. Math. Z. **71**, 289–307 (1959)
3269. Pethő, A., Pohst, M.E.: On the indices of multiquadratic fields. Acta Arith. **153**, 393–414 (2012)
3270. Pethő, A., Ziegler, V.: On biquadratic fields that admit unit power integral basis. Acta Math. Hungar. **133**, 221–241 (2011)
3271. Petr, K.: O užití nauky o funkcích elliptických na theorii forem kvadratických záporného diskriminantu. Rozpravy České Akad. věd a uměni, Tř. II **9(38)** (1900)
3272. Petr, K.: O počtu tříd forem kvadratických záporného diskriminantu. Rozpravy České Akad. věd a uměni, Tř. II **10(40)** (1902)
3273. Petr, K.: O basi celých čisel v obecných telesach. Časopis mat. fys. **64(5)**, 62–72 (1935)
3274. Petri, B., Schappacher, N.: From Abel to Kronecker: episodes from 19th century algebra. In: The Legacy of Niels Henrik Abel, 227–266. Springer (2004)
3275. Pfeiffer, E.: Über die Periodicität in der Teilbarkeit der Zahlen und über die Verteilung der Klassen positiver quadratischer Formen auf ihre Determinanten. Jahresbericht der Pfeifferschen Lehr- und Erziehungs-Anstalt, Jena **1886**, 1–21
3276. Pfeuffer, H.: Quadratsummen in totalreellen algebraischen Zahlkörpern. J. Reine Angew. Math. **249**, 208–216 (1971)
3277. Pfeuffer, H.: On a conjecture about class numbers of totally positive quadratic forms in totally real algebraic number fields. J. Number Theory **11**, 188–196 (1979)
3278. Pfister, A.: Zur Darstellung von -1 als Summe von Quadraten in einem Körper. J. Lond. Math. Soc. **40**, 159–165 (1965)
3279. Pfister, A.: Zur Darstellung definiter Funktionen als Summe von Quadraten. Invent. Math. **4**, 229–237 (1967)
3280. Phragmén, E.: Sur un théorème de Dirichlet. Öfver. Kongl. Vet,-Akad. Förh. Stockholm **49**, 199–206 (1892)
3281. Piazza, P.: Egor Ivanovitch Zolotarev and the theory of ideal numbers for algebraic number fields. Rend. Circ. Mat. Palermo (2) **61**, 123–150 (1999)
3282. Piazza, P.: Zolotarev's theory of algebraic numbers. In: [1467], 453–462
3283. Picard, É.: L'oeuvre scientifique de Charles Hermite. Ann. Sci. Éc. Norm. Supér. (2) **18**, 9–34 (1901) [Rend. Circ. Mat. Palermo **15**, 132–155 (1901); Acta Math. **25**, 87–111 (1902)]
3284. Picard, É.: L'oeuvre de Henri Poincaré. Ann. Sci. Éc. Norm. Supér. (3) **10**, 463–482 (1913)
3285. Pick, H.: Über die complexe Multiplication der elliptischen Functionen. Math. Ann. **25**, 433–447 (1885)
3286. Pieper, H.: Die Einheitengruppe eines zahm-verzweigten galoisschen lokalen Körpers als Galois-Modul. Math. Nachr. **54**, 173–210 (1972)
3287. Pieper, H. (ed.): Korrespondenz A.-M. Legendre — C.G.J. Jacobi. Teubner (1998)
3288. Pierce, T.A.: The numerical factors of the arithmetic forms $\prod_{i=1}^{n}(1 \pm \alpha_i^m)$. Ann. Math. (2) **18**, 53–64 (1916)
3289. Pierce, T.A.: An approximation to the least root of a cubic equation with application to the determination of units in pure cubic fields. Bull. Amer. Math. Soc. **32**, 263–268 (1926)
3290. Pierpont, J.: On an undemonstrated theorem of the Disquisitiones Arithmeticae. Bull. Amer. Math. Soc. **2**, 77–83 (1895)
3291. Pierpont, J.: Weber's Algebra. Bull. Amer. Math. Soc. **4**, 200–234 (1898)
3292. Pierpont, J.: Galois' theory of algebraic equations, I, Rational resolvents. Ann. Math. (2) **1**, 113–143 (1899/1900)
3293. Pierpont, J.: Galois' theory of algebraic equations, II, Irrational resolvents. Ann. Math. (2) **2**, 22–56 (1900/01)
3294. Piltz, A.: Über das Gesetz, nach welchem die mittlere Darstellbarkeit der natürlichen Zahlen als Produkte einer gegebenen Anzahl Faktoren mit der Größe der Zahlen wächst. Dissertation, Univ. Berlin (1881)

3295. Pintz, J.: On Siegel's theorem. Acta Arith. **24**, 643–551 (1973/74)
3296. Pintz, J.: On the Brauer-Siegel theorem. In: Topics in Number Theory (Proc. Colloq., Debrecen, 1974), 259–265. North-Holland (1976)
3297. Pipping, N.: Un critérium pour les nombres algébriques réels, fondé sur une généralisation directe de l'algorithme d'Euclide. C. R. Acad. Sci. Paris **170**, 1155–1156 (1920)
3298. Pipping, N.: Ein Kriterium für die reellen algebraischen Zahlen. Auf eine direkte Verallgemeinerung des euklidischen Algorithmus gegründet. Acta Acad. Aboensis, M.Ph. **1(1)**, 1–16 (1921)
3299. Pipping, N.: Arithmetische Kriterien für reelle algebraische Zahlen. Acta Acad. Aboensis, M.Ph. **7(4)**, 1–32 (1933)
3300. Pipping, N.: Ein neues Kriterium für die reellen algebraischen Zahlen n-ten Grades. Acta Acad. Aboensis, M.Ph. **8(7)**, 1–14 (1935)
3301. Pisot, C.: Sur une propriété caractéristique de certaines entiers algébriques. C. R. Acad. Sci. Paris **202**, 892–894 (1936)
3302. Pisot, C.: Sur la répartition modulo 1 des puissances successives d'un même nombre. C. R. Acad. Sci. Paris **204**, 312–314 (1937)
3303. Pisot, C.: La répartition modulo 1 et les nombres algébriques. Ann. Scuola Norm. Sup. Pisa, Cl. Sci. (2) **7**, 205–248 (1938)
3304. Pisot, C.: Ein Kriterium für die algebraischen Zahlen. Math. Z. **48**, 293–323 (1942)
3305. Pisot, C.: Répartition (mod 1) des puissances successives des nombres réels. Comment. Math. Helv. **19**, 153–160 (1946)
3306. Plancherel, M.: Sur les congruences (mod. 2^m) relatives au nombre des classes des formes quadratiques binaires aux coefficients entiers. Riv. fis. mat. sc. nat. **17**, 265–280, 505–515, 585–596 (1908); **18**, 77–93, 179–196, 243–257 (1908)
3307. Plans, B.: Noether's problem for $GL(2, 3)$. Manuscripta Math. **124**, 481–487 (2007)
3308. Plans, B.: On Noether's problem for central extensions of symmetric and alternating groups. J. Algebra **321**, 3704–3713 (2009)
3309. Plans, B.: On Noether's rationality problem for cyclic groups over **Q**. Proc. Amer. Math. Soc. **145**, 2407–2409 (2017)
3310. Pleasants, P.A.B.: The number of generators of the integers of a number field. Mathematika **21**, 160–167 (1974)
3311. Plemelj, J.: Die Unlösbarkeit von $x^5 + y^5 + z^5 = 0$ im Körper $k\sqrt{5}$. Monatsh. Math. Phys. **23**, 305–308 (1912)
3312. Plemelj, J.: Die Irreduzibilität der Kreisteilungsgleichung. Publ. Math. Belgrade **2**, 163–165 (1934)
3313. Plouffe, S.: The first 498 Bernoulli numbers. www.gutenberg.org/ebooks/2586
3314. Pocklington, H.C.: The practical calculation of unit algebraic numbers. Proc. Cambridge Philos. Soc. **24**, 471–476 (1928)
3315. Pohst, M.: Invarianten des total reellen Körpers siebten Grades mit Minimaldiskriminante. Acta Arith. **30**, 199–207 (1976)
3316. Pohst, M.: On the computation of number fields with small discriminants including the minimum discriminants of sixth degree fields. J. Number Theory **14**, 99–117 (1982)
3317. Pohst, M.: Computational Algebraic Number Theory. Birkhäuser (1993)
3318. Pohst, M.: In memoriam: Hans Zassenhaus (1912–1991). J. Number Theory **47**, 1–19 (1994)
3319. Pohst, M., Martinet, J., Diaz y Diaz, F.: The minimum discriminant of totally real octic fields. J. Number Theory **36**, 149–159 (1990)
3320. Pohst, M., Zassenhaus, H.: Algorithmic Algebraic Number Theory. Cambridge University Press (1989)
3321. Poincaré, H.: Sur un mode nouveau de représentation géométrique des formes quadratiques définies ou indéfinies. J. École Polytech. **47**, 177–245 (1880)
3322. Poincaré, H.: Sur la représentation des nombres par les formes. C. R. Acad. Sci. Paris **92**, 777–779 (1881)
3323. Poincaré, H.: Sur la représentation des nombres par les formes. Bull. Soc. Math. France **13**, 162–194 (1885)

3324. Poincaré, H.: Extension aux nombres premiers complexes des théorèmes de M.Tchebicheff. J. Math. Pures Appl. (4) **8**, 25–68 (1892)
3325. Poincaré, H.: L'oeuvre mathématique de Weierstrass. Acta Math. **22**, 1–18 (1899)
3326. Poitou, G.: Sur l'approximation des nombres complexes par les nombres des corps imaginaires quadratiques dénués d'idéaux non principaux, particuliérement lorsque vaut l'algorithme d'Euclide. Ann. Sci. Éc. Norm. Supér. (3) **70**, 199–265 (1953)
3327. Poitou, G.: Sur les petits discriminants. Sém. Delange–Pisot–Poitou **18**, exp. 8 (1976/77)
3328. Poitou, G.: Minoration de discriminants (aprés Odlyzko). Lecture Notes Math. **567**, 136–153 (1977)
3329. Poli, A.: A deterministic construction for normal bases of abelian extensions. Comm. Algebra **22**, 4751–4757 (1994)
3330. Pollaczek, F.: Über die irregulären Kreiskörper der l-ten und l^2-ten Einheitswurzeln. Math. Z. **21**, 1–38 (1924)
3331. Pollaczek, F.: Über die Einheiten relativ-abelscher Zahlkörper. Math. Z. **30**, 520–551 (1929)
3332. Pollard, H.: The Theory of Algebraic Numbers. Math. Assoc. America (1950) [3rd ed. (with H.G. Diamond), Dover (1998)]
3333. Pólya, G.: Über die Verteilung der quadratischen Reste und Nichtreste. Nachr. Ges. Wiss. Göttingen **1918**, 21–29
3334. Pólya, G.: Über ganzwertige Polynome in algebraischen Zahlkörpern. J. Reine Angew. Math. **149**, 97–116 (1919)
3335. Popa, A.A., Zagier, D.: A combinatorial refinement of the Kronecker-Hurwitz class number relation. Proc. Amer. Math. Soc. **145**, 1003–1008 (2017)
3336. Porubský, Š.: Leben und Werk von Matyáš Lerch [1860–1922]. In: Mathematik im Wandel, 347–373. Franzbecker (2001)
3337. Pourchet, Y.: Sur la représentation en somme de carrés des polynômes à une indéterminée sur un corps de nombres algébriques. Acta Arith. **19**, 89–104 (1971)
3338. Porusch, J.: Die Arithmetik in Zahlkörpern, deren zugehörige Galoissche Körper spezielle metabelsche Gruppen besitzen, auf klassenkörpertheoretischer Grundlage. Math. Z. **37**, 134–160 (1933)
3339. Preuss, G., Schmidt, F.K.: Über den Hilbertschen Irreduzibilitätssatz. Math. Nachr. **4**, 348–365 (1951)
3340. Pringsheim, A.: Zur Theorie der Dirichlet'schen Reihen. Math. Ann. **37**, 38–60 (1890)
3341. Prudnikov, V.E.: Pafnutiĭ Lvovič Čebyšev (1821–1894). Nauka (1976) (Russian)
3342. Prüfer, H.: Neue Begründung der algebraischen Zahlentheorie. Math. Ann. **94**, 198–243 (1925)
3343. Prüfer, H.: Untersuchungen über Teilbarkeitseigenschaften in Körpern. J. Reine Angew. Math. **168**, 1–36 (1932)
3344. Puchta, J.-C.: On the class number of p-th cyclotomic field. Arch. Math. (Basel) **74**, 266–268 (2000)
3345. Pumplün, D.: Über Zerlegungen des Kreisteilungspolynoms. J. Reine Angew. Math. **213**, 200–220 (1963/64)
3346. Purkert, W., Ilgauds, H.-J.: Georg Cantor 1845–1918. Birkhäuser (1987)
3347. Putnam, T.M.: Derrick Norman Lehmer in memoriam. Bull. Amer. Math. Soc. **45**, 209–212 (1939)
3348. Quadri, M.A., Irfan, M.: A characterization of Dedekind domains. Tamkang J. Math. **10**, 165–167 (1979)
3349. Queen, C.S.: Euclidean subrings of global fields. Bull. Amer. Math. Soc. **79**, 437–439 (1973)
3350. Queen, C.S.: Euclidean-like characterizations of Dedekind, Krull, and factorial domains. J. Number Theory **47**, 359–370 (1994)
3351. Queen, C.S.: Factorial domains. Proc. Amer. Math. Soc. **124**, 11–16 (1996)
3352. Queen, C.S.: Arithmetic on groups of positive rationals. J. Number Theory **92**, 164–173 (2002)
3353. Quême, R.: A computer algorithm for finding new Euclidean number fields. J. Théor. Nombres Bordeaux **10**, 33–48 (1998)

3354. Quer, J.: Corps quadratiques de 3-rang 6 et courbes elliptiques de rang 12. C. R. Acad. Sci. Paris **305**, 215–218 (1987)
3355. Qin, H.: The 2-Sylow subgroups of the tame kernel of imaginary quadratic fields. Acta Arith. **69**, 153–169 (1995)
3356. Qin, H.: The 4-rank of $K_2 O_F$ for real quadratic fields F. Acta Arith. **72**, 323–333 (1995)
3357. Qin, H.: The sum of two squares in a quadratic field. Comm. Algebra **25**, 177–184 (1997)
3358. Rabinowitsch, G.: Eindeutigkeit der Zerlegung in Primzahlfaktoren in quadratischen Zahlkörpern. Proceedings of the Fifth International Congress of Mathematicians, **1**, 418–421. Cambridge University Press (1913)
3359. Rabinowitsch, G.: Eindeutigkeit der Zerlegung in Primfaktoren in quadratischen Zahlkörpern. J. Reine Angew. Math. **142**, 153–164 (1913)
3360. Rademacher, H.: Über die Anwendung der Viggo Brunschen Methode auf die Theorie der algebraischen Zahlkörper. SBer. Preuß. Akad. Wiss. Berlin **1923**, 211–218 [[3318], **1**, 259–279.]
3361. Rademacher, H.: Zur additiven Primzahltheorie algebraischer Zahlkörper, I. Über die Darstellung totalpositiver Zahlen als Summe von totalpositivon Primzahlen im reellquadratischen Zahlkörper. Abh. Math. Semin. Univ. Hambg **3**, 109–163 (1924) [[3318], **1**, 306–361.]
3362. Rademacher, H.: Zur additiven Primzahltheorie algebraischer Zahlkörper, II. Über die Darstellung von Körperzahlen als Summe von Primzahlen im imaginär-quadratischen Zahlkörper. Abh. Math. Semin. Univ. Hambg **3**, 331–378 (1924) [[3318], **1**, 362–412.]
3363. Rademacher, H.: Zur additiven Primzahltheorie algebraischer Zahlkörper, III. Über die Darstellung totalpositiver Zahlen als Summen von totalpositiven Primzahlen in einem beliebigen Zahlkörper. Math. Z. **27**, 321–426 (1927) [[3318], **1**, 445–553.]
3364. Rademacher, H.: Primzahlen reell-quadratischer Zahlkörper in Winkelräumen. Math. Ann. **111**, 209–228 (1935) [[3318], **2**, 38–58.]
3365. Rademacher, H.: Über die Anzahl der Primzahlen eines reellquadratischen Zahlkörpers, deren Konjugierte unterhalb gegebener Grenzen liegen. Acta Arith. **1**, 67–77 (1935) [[3318], **2**, 59–70.]
3366. Rademacher, H.: On Waring's problem in algebraic fields. Bull. Amer. Math. Soc. **42**, p. 634 (1936) [[3318], **2**, p. 616.]
3367. Rademacher, H.: On prime numbers of real quadratic fields in rectangles. Trans. Amer. Math. Soc. **39**, 380–398 (1936) [[3318], **2**, 80–99.]
3368. Rademacher, H.: Collected Papers. **1**,**2**. MIT Press (1974)
3369. Rado, R.: A proof of the basis theorem for finitely generated Abelian groups. J. Lond. Math. Soc. **26**, 74–75 (1951); Corr.: p. 160
3370. Rados, G.: Die Diskriminante der allgemeinen Kreisteilungsgleichung. J. Reine Angew. Math. **131**, 49–55 (1906)
3371. Rados, G.: Sur une identité remarquable de la théorie des congruences binomes. Rend. Circ. Mat. Palermo **46**, 308–314 (1922)
3372. Raĭkov, D.A.: On a property of polynomials of the division of the circle. Mat. Sb. (N.S.) **2**, 379–382 (1937) (Russian)
3373. Rajwade, A.R.: Note sur le théorème des trois carrés. Enseign. Math. (2) **22**, 171–173 (1976)
3374. Ramachandra, K.: One more proof of Siegel's theorem. Hardy-Ramanujan J. **3**, 25–40 (1980); Corr.: **4** Suppl., p. 12 (1981)
3375. Ramachandra, K.: A remark on Perron's formula. J. Indian Math. Soc. (N.S.) **65**, 145–151 (1998)
3376. Ramakrishnan, D.: On certain Artin L-series. In: L-functions and Arithmetic (Durham 1989), 339–352. Cambridge University Press (1991)
3377. Ramakrishnan, D.: Modularity of solvable Artin representations of $GO(4)$-type. Int. Math. Res. Not. **2002**, 1–54
3378. Ramakrishnan, D., Valenza, R.J.: Fourier Analysis on Number Fields. Springer (1999)
3379. Ramanathan, K.G. (ed.): Ramanujam — a tribute. Springer (1978)
3380. Ramanujam, C.P.: Sums of m-th powers in p-adic rings. Mathematika **10**, 137–146 (1963)

3381. Ranieri, G.: Générateurs de l'anneau des entiers d'une extension cyclotomique. J. Number Theory **128**, 1576–1586 (2008)
3382. Ranieri, G.: Power bases for rings of integers of abelian imaginary fields. J. Lond. Math. Soc. (2) **82**, 144–166 (2010)
3383. Rankin, R.A.: George Neville Watson. J. Lond. Math. Soc. **41**, 551–565 (1966)
3384. Ranum, A.: The group of classes of congruent quadratic integers with respect to a composite ideal modulus. Trans. Amer. Math. Soc. **11**, 172–198 (1910)
3385. Raulf, N.: Asymptotics of class numbers for progressions and for fundamental discriminants. Forum Math. **21**, 221–257 (2009)
3386. Rausch, U.: On a theorem of Dobrowolski about the product of conjugate numbers. Colloq. Math. **50**, 137–142 (1985)
3387. Rausch, U.: Eine Formel für die Koeffizientensumme von Potenzreihen mit Anwendung auf das Kreis- und Kugelproblem in total reellen algebraischen Zahlkörpern. Acta Arith. **50**, 381–404 (1988)
3388. Rausch, U.: A summation formula in algebraic number fields and applications, I. J. Number Theory **36**, 46–79 (1990)
3389. Rausch, U.: Zum Ellipsoidproblem in algebraischen Zahlkörpern. Acta Arith. **58**, 309–333 (1991)
3390. Rausch, U.: A further omega result for the ellipsoid problem in algebraic number fields. Acta Arith. **62**, 103–108 (1992)
3391. Rausch, U.: On the Piltz divisor problem in algebraic number fields. Acta Arith. **68**, 41–69 (1994)
3392. Rausch, U.: Character sums in algebraic number fields. J. Number Theory **46**, 179–195 (1994)
3393. Rauter, H.: Studien zur Theorie des Galoisschen Körpers über dem Körper der rationalen Funktionen einer Unbestimmten t mit Koeffizienten aus einem beliebigen endlichen Körper von p^{m_0} Elementen. J. Reine Angew. Math. **159**, 117–132 (1928); Add.: p. 228
3394. Rauter, H.: Höhere Kreiskörper. J. Reine Angew. Math. **159**, 220–227 (1928)
3395. Razar, M.J.: Central and genus class fields and Hasse norm theorem. Compositio Math. **35**, 281–298 (1977)
3396. Rédei, L.: Über die Klassenzahl des imaginären quadratischen Zahlkörpers. J. Reine Angew. Math. **159**, 210–219 (1928)
3397. Rédei, L.: Arithmetischer Beweis des Satzes über die Anzahl der durch vier teilbaren Invarianten der absoluten Klassengruppe im quadratischen Zahlkörper. J. Reine Angew. Math. **171**, 55–60 (1934)
3398. Rédei, L.: Eine obere Schranke der Anzahl der durch vier teilbaren Invarianten der absoluten Klassengruppe im quadratischen Zahlkörper. J. Reine Angew. Math. **171**, 61–64 (1934)
3399. Rédei, L.: Über die Grundeinheit und die durch 8 teilbaren Invarianten der absoluten Klassengruppe im quadratischen Zahlkörper. J. Reine Angew. Math. **171**, 131–148 (1934)
3400. Rédei, L.: Über die Pellsche Gleichung $t^2 - du^2 = -1$. J. Reine Angew. Math. **173**, 193–211 (1935)
3401. Rédei, L.: Über einige Mittelwertfragen im quadratischen Zahlkörper. J. Reine Angew. Math. **174**, 15–55 (1936)
3402. Rédei, L.: Ein neues zahlentheoretisches Symbol mit Anwendungen auf die Theorie der quadratischen Zahlkörper. J. Reine Angew. Math. **180**, 1–43 (1939)
3403. Rédei, L.: Über den Euklidischen Algorithmus in reellquadratischen Zahlkörpern. J. Reine Angew. Math. **183**, 183–192 (1941)
3404. Rédei, L.: Zur Frage des Euklidischen Algorithmus in quadratischen Zahlkörpern. Math. Ann. **118**, 588–608 (1942)
3405. Rédei, L.: Über die Klassengruppen und Klassenkörper algebraischer Zahlkörper. J. Reine Angew. Math. **186**, 80–90 (1944)
3406. Rédei, L.: Die 2-Ringklassengruppe des quadratischen Zahlkörpers und die Theorie der Pellschen Gleichung. Acta Math. Hungar. **4**, 31–87 (1953)

3407. Rédei, L., Reichardt, H.: Die Anzahl der durch 4 teilbaren Invarianten der Klassengruppe eines beliebigen quadratischen Zahlkörpers. J. Reine Angew. Math. **170**, 69–74 (1933)
3408. Redmond, D.: Omega theorems for a class of Dirichlet series. Rocky Mountain J. Math. **9**, 733–748 (1979)
3409. Redmond, D.: An O theorem for a class of Dirichlet series. Math. J. Okayama Univ. **28**, 151–158 (1986)
3410. Reichardt, H.: Zur Struktur der absoluten Idealklassengruppe im quadratischen Zahlkörper. J. Reine Angew. Math. **170**, 75–82 (1933)
3411. Reichardt, H.: Arithmetische Theorie der kubischen Körper als Radikalkörper. Monatsh. Math. Phys. **40**, 323–350 (1933)
3412. Reichardt, H.: Über Normalkörper mit Quaternionengruppe. Math. Z. **41**, 218–221 (1936)
3413. Reichardt, H.: Konstruktion von Zahlkörpern mit gegebener Galoisgruppe von Primzahlpotenzordnung. J. Reine Angew. Math. **177**, 1–5 (1937)
3414. Reichardt, H.: Über die 2-Klassengruppe gewisser quadratischer Zahlkörper. Math. Nachr. **46**, 71–80 (1970)
3415. Reichardt, H.: Über die Idealklassengruppe des Dirichletschen biquadratischen Zahlenkörpers. Acta Arith. **21**, 323–327 (1972)
3416. Reichardt, H., Wegner, U.: Arithmetische Charakterisierung von algebraisch auflösbaren Körpern und Gleichungen von Primzahlgrad. J. Reine Angew. Math. **178**, 1–10 (1937)
3417. Reichstein, Z.: On a theorem of Hermite and Joubert. Canad. J. Math. **51**, 69–95 (1999)
3418. Reichstein, Z.: Joubert's theorem fails in characteristic 2. C. R. Acad. Sci. Paris **352**, 773–777 (2014)
3419. Reichstein, Z., Youssin, B.: Conditions satisfied by characteristic polynomials in fields and division algebras. J. Pure Appl. Algebra **166**, 165–189 (2002)
3420. Reid, C.: Hilbert. Springer (1970) [Reprints: Hilbert–Courant. Springer (1986); Hilbert, Copernicus (1996)]
3421. Reid, C.: Julia: A Life in Mathematics. Math. Assoc. America (1996)
3422. Reid, L.W.: A table of class numbers for cubic number fields. Amer. J. Math. **23**, 68–84 (1901)
3423. Reid, L.W.: The Elements of the Theory of Algebraic Numbers. Macmillan (1910) [Reprints: Kessinger Publishing (2007); BiblioBazaar (2010)]
3424. Reidemeister, K.: Über die Relativklassenzahl gewisser relativquadratischer Zahlkörper. Abh. Math. Semin. Univ. Hambg **1**, 27–48 (1922)
3425. Reidemeister, K. (ed.): Hilbert Gedenkband. Springer (1971)
3426. Reiner, I.: On genera of binary quadratic forms. Bull. Amer. Math. Soc. **51**, 909–912 (1945)
3427. Reiner, I., Ullom, S.: Remarks on class groups of integral group rings. Symposia Math. **13**, 501–516 (1974)
3428. Reinhart, A.: A note on conductor ideals. Comm. Algebra **44**, 4243–4251 (2016)
3429. Reiter, C., Effective lower bounds on large fundamental units of real quadratic fields. Osaka J. Math. **22**, 755–765 (1985)
3430. Reiter, H.: Über den Satz von Weil-Cartier. Monatsh. Math. **86**, 13–62 (1978/79)
3431. Reiter, S.: Galoisrealisierungen klassischer Gruppen. J. Reine Angew. Math. **511**, 1999, 193–236 (1999)
3432. Rella, T.: Die Zerlegungsgesetze für die Primideale eines beliebigen algebraischen Zahlkörpers im Körper der l-ten Einheitswurzeln. Math. Z. **5**, 11–16 (1919)
3433. Rella, T.: Über die multiplikative Darstellung von algebraischen Zahlen eines Galoisschen Zahlkörpers für den Bereich eines beliebigen Primteilers. J. Reine Angew. Math. **150**, 157–174 (1920)
3434. Rella, T.: Bemerkungen zu Herrn Hensels Arbeit "Die Zerlegung der Primteiler eines beliebigen Zahlkörpers in einem auflösbaren Oberkörper". J. Reine Angew. Math. **153**, 108–110 (1923)
3435. Rella, T.: Zur Newtonschen Approximationsmethode in der Theorie der p-adischen Gleichungswurzeln. J. Reine Angew. Math. **153**, 111–112 (1923)

3436. Rella, T.: Ordnungsbestimmungen in Integritätsbereichen und Newtonsche Polygone. J. Reine Angew. Math. **158**, 33–48 (1937)
3437. Remak, R.: Abschätzung der Lösung der Pellschen Gleichung im Anschluss an den Dirichletschen Existenzsatz. J. Reine Angew. Math. **143**, 250–254 (1913)
3438. Remak, R.: Über die Zerlegung der kommutativen Gruppen in direkte unzerlegbare Faktoren. Math. Z. **10**, 12–16 (1921)
3439. Remak, R.: Verallgemeinerung eines Minkowskischen Satzes. Math. Z. **17**, 1–34 (1923)
3440. Remak, R.: Verallgemeinerung eines Minkowskischen Satzes, II. Math. Z. **18**, 173–200 (1923)
3441. Remak, R.: Elementare Abschätzungen von Fundamentaleinheiten und des Regulators eines algebraischen Zahlkörpers. J. Reine Angew. Math. **165**, 159–179 (1931)
3442. Remak, R.: Über den Euklidischen Algorithmus in reell-quadratischen Zahlkörpern. Jahresber. Deutsch. Math.-Verein. **44**, 238–250 (1934)
3443. Remak, R.: Über Grössenbeziehungen zwischen Diskriminante und Regulator eines algebraischen Zahlkörpers. Compositio Math. **10**, 245–285 (1952)
3444. Remak, R.: Über algebraische Zahlkörper mit schwachem Einheitsdefekt. Compositio Math. **12**, 35–80 (1954)
3445. Rémond, P.: Étude asymptotique de certaines partitions dans certaines semi-groupes. Ann. Sci. Éc. Norm. Supér. (3) **83**, 343–410 (1966)
3446. Reuschle, C.G.: Mathematische Abhandlung, enthaltend neue zahlentheoretische Tabellen. Programm zum Schlusse des Schuljahres 1855/56 am Kgl. Gymnasium zu Stuttgart (1856)
3447. Reuschle, C.G.: Tafeln complexer Primzahlen, welche aus Wurzeln der Einheit gebildet sind. Dümmler (1875)
3448. Revoy, Ph.: Sommes de bicarrés dans $Z(\sqrt{-1})$ et $Z(\sqrt[3]{1})$. Enseign. Math. (2) **25**, 257–260 (1979)
3449. Rhin, G.: A generalization of a theorem of Schinzel. Colloq. Math. **101**, 155–159 (2004)
3450. Rhin, G., Sac-Épée, J.-M.: New methods providing high degree polynomials with small Mahler measure. Experiment. Math. **12**, 457–461 (2003)
3451. Rhin, G., Wu, Q.: On the smallest value of the maximal modulus of an algebraic integer. Math. Comp. **76**, 1025–1038 (2007)
3452. Rhoades, S.L., Sandra L.: A generalization of the Aramata-Brauer theorem. Proc. Amer. Math. Soc. **119**, 357–364 (1993)
3453. Ribenboim, P.: Algebraic Numbers. Wiley-Interscience (1972)
3454. Ribenboim, P.: 13 Lectures on Fermat's Last Theorem. Springer (1979)
3455. Ribenboim, P.: Gauss and the class number problem. In: Proceedings of the International Symposium on Mathematics and Theoretical Physics (Guarujá, 1989), 3–62, Inst. Gaussianum (1990)
3456. Ribenboim, P.: The Theory of Classical Valuations. Springer (1999)
3457. Ribenboim, P.: Classical Theory of Algebraic Numbers. Springer (2001)
3458. Ribet, K.A.: A modular construction of unramified p-extensions of $Q(\mu_p)$. Invent. Math. **34**, 151–162 (1976)
3459. Richert, H.-E.: Zur Abschätzung der Riemannschen Zetafunktion in der Nähe der Vertikalen $\sigma = 1$. Math. Ann. **169**, 97–101 (1967)
3460. Richter, H.: Über die Lösbarkeit einiger nicht-Abelscher Einbettungsprobleme. Math. Ann. **112**, 69–84 (1936)
3461. Richter, H.: Über die Lösbarkeit des Einbettungsproblems für Abelsche Zahlkörper. Math. Ann. **112**, 700–726 (1936); Corr.: **113**, p. 628 (1937)
3462. Ridout, D.: The p-adic generalization of the Thue-Siegel-Roth theorem. Mathematika **5**, 40–48 (1958)
3463. Riebesehl, P.: E.Busche. Jahresber. Deutsch. Math.-Verein. **25**, p. 283 (1916)
3464. Riečan, B.: Štefan Schwarz (1914–1996). Czechoslovak Math. J. **47**, 375–382 (1997)
3465. Rieger, G.J.: Zum Waringschen Problem für algebraische Zahlen and Polynome. J. Reine Angew. Math. **195**, 108–120 (1956)

3466. Rieger, G.J.: Über die Anzahl der Teiler der Ideale in einem algebraischen Zahlkörper. Arch. Math. (Basel) **8**, 162–165 (1957)
3467. Rieger, G.J.: Verallgemeinerung der Selbergschen Formel auf Idealklassen mod f in algebraischen Zahlkörpern. Math. Z. **69**, 183–194 (1958)
3468. Rieger, G.J.: Ein weiterer Beweis der Selbergschen Formel für Idealklassen mod f in algebraischen Zahlkörpern. Math. Ann. **134**, 403–407 (1958)
3469. Rieger, G.J.: Über die Anzahl der Ideale in einer Idealklasse mod $\mathfrak{f}$ eines algebraischen Zahlkörpers. Math. Ann. **135**, 444–466 (1958)
3470. Rieger, G.J.: Zur Wienerschen Methode in der Zahlentheorie. Arch. Math. (Basel) **10**, 257–260 (1959)
3471. Rieger, G.J.: Eine Selbergsche Identität für algebraische Zahlen. Math. Ann. **145**, 77–80 (1961/62)
3472. Rieger, G.J.: Elementare Lösung des Waringschen Problems für algebraische Zahlkörper mit der verallgemeinerten Linnikschen Methode. Math. Ann. **148**, 83–88 (1962)
3473. Rieger, G.J.: Über die Darstellung ganzer algebraischer Zahlen durch Quadrate. Arch. Math. (Basel) **14**, 22–28 (1963)
3474. Riemann, B.: Ueber die Anzahl der Primzahlen unter einer gegebener Größe. Monatsber. Preuss. Akad. Wiss. **1860**, 671–680
3475. Riese, U.: Kronecker-Weber via Kummer. Expo. Math. **16**, 271–276 (1998)
3476. Risman, L.J.: A new proof of the three squares theorem. J. Number Theory **6**, 282–283 (1974)
3477. Roberts, D.P.: Density of cubic field discriminants. Math. Comp. **70**, 1699–1705 (2001)
3478. Robertson, L.: Power bases for cyclotomic integer rings. J. Number Theory **69**, 98–118 (1998)
3479. Robertson, L.: Power bases for 2-power cyclotomic fields. J. Number Theory **88**, 196–209 (2001)
3480. Robertson, L.: Monogeneity in cyclotomic fields. Internat. J. Number Th. **6**, 1589–1607 (2010)
3481. Robinson, R.M.: Intervals containing infinitely many sets of conjugate algebraic integers. In: Studies in Mathematical Analysis, 305–315. Stanford University Press (1962)
3482. Robinson, R.M.: Conjugate algebraic integers in real point sets. Math. Z. **84**, 415–427 (1964)
3483. Robinson, R.M.: Algebraic equations with span less than 4. Math. Comp. **18**, 547–559 (1964)
3484. Robinson, R.M.: Intervals containing infinitely many sets of conjugate algebraic units. Ann. Math. (2) **80**, 411–428 (1964)
3485. Robinson, R.M.: Some conjectures about cyclotomic integers. Math. Comp. **19**, 210–217 (1965)
3486. Robinson, R.M.: Conjugate algebraic integers on a circle. Math. Z. **110**, 41–51 (1969)
3487. Rogawski, J.D., Tunnell, J.B.: On Artin L-functions associated to Hilbert modular forms of weight one. Invent. Math. **74**, 1–42 (1983)
3488. Rogers, C.A.: The product of n real homogeneous linear forms. Acta Math. **82**, 185–208 (1950)
3489. Rogers, C.A.: A brief survey of the work of Harold Davenport. Acta Arith. **18**, 13–17 (1971)
3490. Rogers, C.A.: Obituary : Richard Rado. Bull. Lond. Math. Soc. **30**, 185–195 (1998)
3491. Rogers, C.A. et al.: Harold Davenport. Bull. Lond. Math. Soc. **4**, 66–99 (1972)
3492. Rogosinski, W.W.: Obituary: Michael Fekete. J. Lond. Math. Soc. **33**, 496–500 (1958)
3493. Rohrbach, H.: Bemerkungen zu einem Determinantensatz von Minkowski. Jahresber. Deutsch. Math.-Verein. **40**, 49–53 (1931)
3494. Rohrbach, H.: Erhard Schmidt. Ein Lebensbild. Jahresber. Deutsch. Math.-Verein. **69**, 209–224 (1967/68)
3495. Rohrbach, H.: Richard Brauer zum Gedächtnis. Jahresber. Deutsch. Math.-Verein. **83**, 125–134 (1981)
3496. Rohrbach, H.: Alfred Brauer zum Gedächtnis. Jahresber. Deutsch. Math.-Verein. **90**, 145–154 (1988)

3497. Roquette, P.: Riemannsche Vermutung in Funktionenkörpern. Arch. Math. (Basel) **4**, 6–16 (1953)
3498. Roquette, P.: Arithmetischer Beweis der Riemannschen Vermutung in Kongruenzfunktionenkörpern beliebigen Geschlechts. J. Reine Angew. Math. **191**, 199–252 (1953)
3499. Roquette, P.: On class field towers. In: [599], 231–249
3500. Roquette, P.: Nonstandard aspects of Hilbert's irreducibility theorem. Lecture Notes Math. **498**, 231–275 (1975)
3501. Roquette, P.: Über die algebraisch-zahlentheoretischen Arbeiten von Max Deuring. Jahresber. Deutsch. Math.-Verein. **91**, 109–125 (1989)
3502. Roquette, P.: History of Valuation Theory. In: Valuation Theory and its Applications (Saskatoon, SK, 1999), I. Fields Inst. Comm. Ser. **32**, 291–355 (2002)
3503. Roquette, P.: The Riemann hypothesis in characteristic p, its origin and development. Part 1. The formation of the zeta-function of Artin and F.K. Schmidt. Mitt. Math. Ges. Hamburg **21**, 79–157 (2002)
3504. Roquette, P.: The Riemann hypothesis in characteristic p, its origin and development. Part 2. The first steps by Davenport and Hasse. Mitt. Math. Ges. Hamburg **22**, 1–69 (2004)
3505. Roquette, P.: The Brauer-Hasse-Noether theorem in historical perspective. Schr. Math.-Naturwiss. Kl. Heidelberger AW **15**, 1–92 (2005) [[3509], 1–76.]
3506. Roquette, P.: The Riemann hypothesis in characteristic p, its origin and development. Part 3. The elliptic case. Mitt. Math. Ges. Hamburg **25**, 103–176 (2006)
3507. Roquette, P.: Ernst Steinitz and abstract field theory. J. Reine Angew. Math. **658**, 1–11 (2010) [[3509] 227–237.]
3508. Roquette, P.: The Riemann hypothesis in characteristic p, its origin and development. Part 4. Davenport-Hasse fields. Mitt. Math. Ges. Hamburg **32**, 145–210 (2012)
3509. Roquette, P.: Contributions to the History of Number Theory in the 20th Century. European Mathematical Society (2013)
3510. Roquette, P., Goss, D.: Heinrich-Wolfgang Leopoldt (22.8.1927–28.7.2011). J. Number Theory **132**, 1641–1644 (2012). [[3509] 239–243.]
3511. Roquette, P., Zassenhaus, H.: A class rank estimate for algebraic number fields. J. Lond. Math. Soc. **44**, 31–38 (1969)
3512. Rosen, M.: S-units and S-class group in algebraic function fields. J. Algebra **26**, 98–108 (1973)
3513. Rosen, M.: Elliptic curves and Dedekind domains. Proc. Amer. Math. Soc. **57**, 197–201 (1976)
3514. Rosen, M.: Abel's theorem on the lemniscate. Amer. Math. Monthly **88**, 387–395 (1981)
3515. Rosen, M.: An elementary proof of the local Kronecker-Weber theorem. Trans. Amer. Math. Soc. **265**, 599–605 (1981)
3516. Rosenbaum, K.: On the multiplicative group of cyclic extensions of a local field. Vestnik Leningrad. Univ. Ser. Mat. Mekh. Astronom. **21(1)**, 90–92 (1966) (Russian)
3517. Rosenbaum, K.: Über irreguläre zyklische Erweiterungen lokaler Körper. Math. Nachr. **43**, 143–159 (1970)
3518. Rosenbaum, K.: On the structure of p-adic closure of cyclic extensions of local fields. Indian J. Math. **20**, 255–264 (1978)
3519. Rosenblüth, E.: Die arithmetische Theorie und die Konstruktion der Quaternionenkörper auf klassenkörpertheoretischer Grundlage. Monatsh. Math. Phys. **41**, 85–125 (1934)
3520. Rosenhead, L.: Henry Cabourn Pocklington (1870–1952). Obituary Notices of Fellows of the Royal Society **8**, 555–565 (1953)
3521. Ross, P.M.: On Chen's theorem that each large even number has the form $p_1 + p_2$ or $p_1 + p_2 p_3$. J. Lond. Math. Soc. (2) **10**, 500–506 (1975)
3522. Rosser, B.: On the first case of Fermat's last theorem. Bull. Amer. Math. Soc. **45**, 636–640 (1939)
3523. Rosser, B.: Real roots of Dirichlet L-series. Bull. Amer. Math. Soc. **55**, 906–913 (1949)
3524. Rosser, B.: Real roots of real Dirichlet L-series. J. Res. Nat. Bur. Standards **45**, 505–514 (1950)

3525. Roth, K.F.: Rational approximations to algebraic numbers. Mathematika **2**, 1–20 (1955); corr.: p. 168
3526. Rothgiesser, H.: Zum Reziprozitätsgesetz für l^n. Abh. Math. Semin. Univ. Hambg **11**, 1–16 (1935)
3527. Ruban, A.A.: Certain metric properties of the p-adic numbers. Sibir. Mat. Zh. **11**, 222–227 (1970) (Russian)
3528. Rückle, G.: Quadratische Reziprozitätsgesetze in algebraischen Zahlkörpern. Dissertation, Univ. Göttingen (1901)
3529. Rudio, F.: Erinnerung an Moritz Abraham Stern. Vierteljahrsschr. Naturforsch. Ges. Zürich **39**, 133–143 (1894)
3530. Rudman, R.J., Steiner, R.P.: A generalization of Berwick's unit algorithm. J. Number Theory **10**, 16–34 (1978)
3531. Rumely, R.: Capacity theory on algebraic curves. Lecture Notes Math. **1378** (1989)
3532. Rumely, R.: The Fekete-Szegő theorem with splitting conditions, I. Acta Arith. **93**, 99–116 (2000)
3533. Rumely, R.: The Fekete-Szegő theorem with splitting conditions, II. Acta Arith. **103**, 347–410 (2002)
3534. Runge, C.: Ueber die Zerlegung ganzer ganzzahliger Functionen in irreductible Factoren. J. Reine Angew. Math. **99**, 89–97 (1886)
3535. Runge, C.: Ueber ganzzahlige Gleichungen ohne Affect. Nachr. Ges. Wiss. Göttingen **1899**, 89–93.
3536. Rush, D.E.: An arithmetic characterization of algebraic number fields with a given class group. Math. Proc. Cambridge Philos. Soc. **94**, 23–28 (1983)
3537. Rychlik, K.: O Henselových čislech. Rozpravy České Akad. věd a uměni, Tř. II **25(55)**, 1–16 (1916)
3538. Rychlik, K.: Dělitelnost v algebraických tělesech číselných vzhledem k racionálnému prvočíslu. Rozpravy České Akad. věd a uměni, Tř. II **28(14)**, 1–5 (1919)
3539. Rychlik, K.: Theorie dělitelnosti čísel algebraických. Rozpravy České Akad. věd a uměni, Tř. II **29(2)**, 1–6 (1920)
3540. Rychlik, K.: Eine stetige nicht differenzierbare Funktion im Gebiete der Henselschen Zahlen. J. Reine Angew. Math. **152**, 178–179 (1923)
3541. Rychlik, K.: Zur Theorie der Teilbarkeit in algebraischen Zahlkörpern. Vestnik ČSAV **9**, 1–36 (1923)
3542. Rychlik, K.: Zur Bewertungstheorie der algebraischen Körper. J. Reine Angew. Math. **153**, 94–107 (1924)
3543. Šafarevič, I.R.: On p-extensions. Mat. Sb. (N.S.) **20**, 351–363 (1947) (Russian)
3544. Šafarevič, I.R.: A general reciprocity law. Dokl. Akad. Nauk SSSR **64**, 25–28 (1949) (Russian)
3545. Šafarevič, I.R.: A general reciprocity law. Mat. Sb. **26**, 113–146 (1950) (Russian)
3546. Šafarevič, I.R.: A new proof of the Kronecker-Weber theorem. Trudy Mat. Inst. im. Steklova **38**, 382–387 (1951) (Russian)
3547. Šafarevič, I.R.: Construction of fields of algebraic numbers with given solvable Galois group. Izv. Akad. Nauk SSSR, Ser. Mat. **18**, 525–578 (1954) (Russian)
3548. Šafarevič, I.R.: On the problem of imbedding fields. Dokl. Akad. Nauk SSSR **95**, 459–461 (1954) (Russian)
3549. Šafarevič, I.R.: On the problem of imbedding of fields. Izv. Akad. Nauk SSSR, Ser. Mat. **18**, 389–418 (1954) (Russian)
3550. Šafarevič, I.R.: The imbedding problem for splitting extensions. Dokl. Akad. Nauk SSSR **120**, 1217–1219 (1958) (Russian)
3551. Šafarevič, I.R.: Algebraic number fields. Proc. ICM 1963, 163–176, Inst. Mittag-Leffler (1963)
3552. Šafarevič, I.R.: Factors of decreasing central series,Mat. Zametki **45**, 114–117 (1989) (Russian)
3553. Šafarevič, I.R., DmitriĭKonstantinovič Faddeev. Algebra i Analiz **2(5)**, 3–9 (1990) (Russian)

3554. Saint-Venant, A.J.-C.: Biographie: Wantzel. Nouv. Ann. Math. **7**, 321–331 (1848)
3555. Sairaiji, F., Shimizu, K.: A note on Ono's numbers associated to imaginary quadratic fields. Proc. Japan Acad. Sci. **77**, 29–31 (2001)
3556. Sairaiji, F., Shimizu, K.: An inequality between class numbers and Ono's numbers associated to imaginary quadratic fields. Proc. Japan Acad. Sci. **78**, 105–108 (2002)
3557. Salce, L., Zanardo, P.: Arithmetical characterization of rings of algebraic integers with cyclic ideal class group. Boll. Unione Mat. Ital. (6) **1**, 117–122 (1982)
3558. Salem, R.: Sets of uniqueness and sets of multiplicity. Trans. Amer. Math. Soc. **54**, 218–228 (1943)
3559. Salem, R.: A remarkable class of algebraic integers. Proof of a conjecture of Vijayaraghavan. Duke Math. J. **11**, 103–108 (1944)
3560. Salem, R.: Sets of uniqueness and sets of multiplicity, II. Trans. Amer. Math. Soc. **56**, 32–49 (1944)
3561. Salem, R.: Power series with integral coefficients. Duke Math. J. **12**, 153–172 (1945)
3562. Salem, R.: Rectifications to the papers: Sets of uniqueness and sets of multiplicity, I and II. Trans. Amer. Math. Soc. **63**, 595–598 (1948)
3563. Salem, R.: Algebraic Numbers and Fourier Analysis. D.C. Heath and Co., (1963) [Reprint: Wadsworth (1983)]
3564. Salem, R., Zygmund, A.: Sur les ensembles parfaits dissymétriques à rapport constant. C. R. Acad. Sci. Paris **240**, 2281–2283 (1955)
3565. Sall, O.: Points algébriques de degré au plus 10 sur la septique de Fermat. Afrika Mat. (3) **15**, 49–55 (2003)
3566. Saltman, D.J.: Generic Galois extensions and problems in field theory. Adv. Math. **43**, 250–283 (1982)
3567. Saltman, D.J.: Noether's problem over an algebraically closed field. Invent. Math. **77**, 71–84 (1984)
3568. Samet, P.A.: Algebraic integers with two conjugates outside the unit circle. Proc. Cambridge Philos. Soc. **49**, 421–436 (1953)
3569. Samet, P.A.: Algebraic integers with two conjugates outside the unit circle, II. Proc. Cambridge Philos. Soc. **50**, p. 346 (1954)
3570. Samuel, P.: About Euclidean rings. J. Algebra **19**, 282–301 (1971)
3571. Samuels, C.L.: The Weil height in terms of an auxiliary polynomial. Acta Arith. **128**, 209–221 (2007)
3572. Sands, J.W., Schwarz, W.: A Demjanenko matrix for abelian fields of prime power conductor. J. Number Theory **52**, 85–97 (1995)
3573. Santos, W.F., Moskowitz, M. (ed.): Gerhard Hochschild (1915–2010). Notices Amer. Math. Soc. **58**, 1089–1099 (2011)
3574. Saparnijazov, O.: Asymptotic equalities for the number of classes of ideals of an imaginary quadratic field. Lit. Mat. Sb. **5**, 303–305 (1965) (Russian)
3575. Saparnijazov, O., Faĭnleĭb, A.S.: Dispersion of real character sums, and the moments of $L(1,\chi)$. Izv. Akad. Nauk Uzb. SSR., ser. Fiz.-Mat. Nauk **1975(6)**, 24–29 (1975) (Russian)
3576. Sarbasov, G.: Sharpening of the remainder term in the asymptotical formula for the distribution of cyclic fields of prime degree. Izv. Akad. Nauk Kazah. SSR. **1**, 61–62 (1967) (Russian)
3577. Sarnak, P.: Class numbers of indefinite binary quadratic forms. J. Number Theory **15**, 229–247 (1982); Corr.: **16**, p. 284 (1982
3578. Sasaki, H.: Quaternary universal forms over $\mathbf{Q}(\sqrt{13})$. The Ramanujan J. **18**, 73–80 (2009)
3579. Sasaki, R.: On a lower bound for the class number of an imaginary quadratic field. Proc. Japan Acad. Sci. **62**, 37–39 (1986)
3580. Sasaki, R.: Criteria for the class number of real quadratic fields to be one. In: Number Theory (Banff, 1988), 501–508. de Gruyter (1990)
3581. Sasaki, S.: On Artin representations and nearly ordinary Hecke algebras over totally real fields. Doc. Math. **18**, 997–1038 (2013)

3582. Satake, I.: On a generalization of Hilbert's theory of ramification. Sci. Papers College Gen. Ed. Univ. Tokyo **2**, 25–39 (1952)
3583. Sato, K.: On the divisibility of $\zeta_{K_1K_2}(s)\zeta_{K_1\cap K_2}(s)/\zeta_{K_1}(s)\zeta_{K_2}(s)$. J. Coll. Eng. Nihon Univ. **23**, 41–46 (1982)
3584. Sato, K.: On a problem of R. Brauer on zeta functions. J. Coll. Eng. Nihon Univ. **24**, 1–5 (1983)
3585. Sato, K.: On a problem of R. Brauer on zeta functions, II. J. Coll. Eng. Nihon Univ. **25**, 33–39 (1984)
3586. Sato, K.: On a problem of R. Brauer on zeta-functions of algebraic number fields. Proc. Japan Acad. Sci. **61**, 305–307 (1985)
3587. Sato, K.: On Brauer's problem on zeta-functions of algebraic number fields. J. Coll. Eng. Nihon Univ. **27**, 1–4 (1986)
3588. Schaal, W.: Übertragung des Kreisproblems auf reell–quadratische Zahlkörper. Math. Ann. **145**, 273–284 (1961/62)
3589. Schaal, W.: On the expression of a number as the sum of two squares in totally real algebraic number fields. Proc. Amer. Math. Soc. **16**, 529–537 (1965)
3590. Schacher, M., Sonn, J.: Double covers of the symmetric groups as Galois groups over number fields. J. Algebra **116**, 243–250 (1988)
3591. Schäfer, W.: Beweis des Hauptidealsatzes der Klassenkörpertheorie für den Fall der komplexen Multiplikation. Dissertation, Univ. Halle (1929)
3592. Schaffstein, K.: Tafel der Klassenzahlen der reellen quadratischen Zahlkörper mit Primzahldiskriminante unter 12 000 und zwischen 100 000–101 000 und 1 000 000–1, 001, 000. Math. Ann. **98**, 745–748 (1928)
3593. Schappacher, N.: On the history of Hilbert's twelfth problem. A comedy of errors. In: Matériaux pour l'histoire des mathématiques au XXe siécle (Nice 1996), 243–273, Sémin. Congr. **3**, Soc. Math. France (1998)
3594. Scharaschkin, V.: The Hasse principle modulo nth powers. Acta Arith. **87**, 269–285 (1999)
3595. Scharlau, R.: On the Pythagoras number of orders in totally real number fields. J. Reine Angew. Math. **316**, 208–210 (1980)
3596. Scharlau, W.: Unveröffentlichte algebraische Arbeiten Richard Dedekinds aus seiner Göttinger Zeit 1855–1858. Arch. Hist. Exact Sci. **27**, 335–367 (1982)
3597. Scheidler, R., Stein, A.: Voronoi's algorithm in purely cubic congruence function fields of unit rank 1. Math. Comp. **69**, 1245–1266 (2000)
3598. Schenkman, E.: The basis theorem for finitely generated abelian groups. Amer. Math. Monthly **67**, 770–771 (1960)
3599. Schenkman, E.: On the multiplicative group of a field. Arch. Math. (Basel) **15**, 282–285 (1964)
3600. Schering, E.: Die Fundamentalklassen der zusammensetzbaren arithmetischen Formen. Abhandl. Kgl. Ges. Wiss. Göttingen **14**, 3–13 (1869) [[3602], **1**, 135–148.]
3601. Schering, E.: Beweis des Dirichletschen Satzes, dass durch jede eigentlich primitive quadratische Form unendlich viele Primzahlen dargestellt werden. In: [3602], **2**, 357–365
3602. Schering, E.: Gesammelte mathematische Werke. **1,2**. Mayer & Müller (1902–1909)
3603. Schertz, R.: Die singulären Werte der Weberschen Funktionen $\mathfrak{f}$, $\mathfrak{f}_1$, $\mathfrak{f}_2$, γ_2, γ_3. J. Reine Angew. Math. **286/287**, 46–74 (1976)
3604. Schertz, R.: Weber's class invariants revisited. J. Théor. Nombres Bordeaux **14**, 325–343 (2002)
3605. Schertz, R.: Complex Multiplication. Cambridge University Press (2010)
3606. Schilling, O.F.G.: Some remarks on class field theory over infinite fields of algebraic numbers. Amer. J. Math. **59**, 414–422 (1937)
3607. Schilling, O.F.G.: Class fields of infinite degree over p-adic number fields. Ann. Math. (2) **38**, 469–476 (1937)
3608. Schilling, O.F.G.: The structure of local class field theory. Amer. J. Math. **60**, 75–100 (1938)
3609. Schilling, O.F.G.: A generalization of local class field theory. Amer. J. Math. **60**, 667–704 (1938)

3610. Schilling, O.F.G.: Regular normal extensions over complete fields. Trans. Amer. Math. Soc. **47**, 440–454 (1940)
3611. Schilling, O.F.G.: The Theory of Valuations. Amer. Math. Soc. (1950)
3612. Schilling, O.F.G.: Necessary conditions for local class field theory. (Remarks to a paper of M. Moriya.) Math. J. Okayama Univ. **3**, 5–10 (1953)
3613. Schilling, O.F.G.: On local class field theory. J. Math. Soc. Japan **13**, 234–245 (1961)
3614. Schinzel, A.: On Hilbert's irreducibility theorem. Ann. Polon. Math. **16**, 333–340 (1965) [[3626], **1**, 838–858.]
3615. Schinzel, A.: On a theorem of Bauer and some of its applications. Acta Arith. **11**, 333–344 (1966); corr. **12**, p. 425 (1966/67) [[3626], **1**, 179–189.]
3616. Schinzel, A.: Reducibility of lacunary polynomials, I. Acta Arith. **16** 123–159 (1969/70) [[3626], **1**, 344–380.]
3617. Schinzel, A.: Wacław Sierpiński's papers on the theory of numbers. Acta Arith. **21** 7–13 (1972)
3618. Schinzel, A.: On a theorem of Bauer and some of its applications, II. Acta Arith. **23** 221–231 (1973) [[3626], **1**, 210–220.]
3619. Schinzel, A.: On the product of the conjugates outside the unit circle of an algebraic number. Acta Arith. **24** 385–399 (1947) Add.: Acta Arith. **26** 329–331 (1973) [[3626], **1**, 221–237.]
3620. Schinzel, A.: On linear dependence of roots. Acta Arith. **28** 161–175 (1975) [[3626], **1**, 238–252.]
3621. Schinzel, A.: Polynomials with Special Regards to Reducibility. Cambridge University Press (2000)
3622. Schinzel, A.: On values of the Mahler measure in a quadratic field (solution of a problem of Dixon and Dubickas). Acta Arith. **113** 401–408 (2004) [[3626], **1**, 272–279.]
3623. Schinzel, A.: Jerzy Urbanowicz's work in pure mathematics. In: Number Theory–Arithmetic in Shangri-La, 237–243. World Scientific Publications (2013)
3624. Schinzel, A.: Jerzy Browkin (1934–2015). Acta Arith. **172** 199–206 (2016)
3625. Schinzel, A., Zassenhaus, H.: A refinement of two theorems of Kronecker. Michigan Math. J. **12** 81–85 (1965) [[3626], **1**, 175–178.]
3626. Schinzel, A.: Selecta, **1,2**. European Mathematical Society (2007)
3627. Schleser, M.: Asymptotische Gesetze im kubischen Kreisteilungskörper. Monatsh. Math. Phys. **21** 61–102 (1910)
3628. Schlickewei, H.P.: The $\mathfrak{p}$-adic Thue-Siegel-Schmidt theorem. Arch. Math. (Basel) **29** 267–270 (1977)
3629. Schlickewei, H.P.: An explicit upper bound for the number of solutions of the S-unit equation. J. Reine Angew. Math. **406** 109–120 (1990)
3630. Schlickewei, H.P.: S-unit equations over number fields. Invent. Math. **102** (1990)
3631. Schlickewei, H.P.: The quantitative subspace theorem for number fields. Compositio Math. **82** 245–273 (1992)
3632. Schlickewei, H.P., Stepanov, S.A.: Algorithms to construct normal bases of cyclic number fields. J. Number Theory **44** 30–40 (1993)
3633. Schmeidler, W.: Zur Theorie der primären Punktmodulen. Math. Ann. **79** 56–75 (1918)
3634. Schmeidler, W.: Über Moduln und Gruppen hyperkomplexer Grössen. Math. Z. **3** 29–42 (1919)
3635. Schmidt, A.L.: Farey triangles and Farey quadrangles in the complex plane. Math. Scand. **21** 241–295 (1967)
3636. Schmidt, A.L.: Diophantine approximation of complex numbers. Acta Math. **134** 1–85 (1975)
3637. Schmidt, A.L.: Diophantine approximation in the field $\mathbf{Q}(i\sqrt{11})$. J. Number Theory **10** 151–176 (1978)
3638. Schmidt, A.L.: Ergodic theory for complex continued fractions. Monatsh. Math. **93** 39–62 (1982)
3639. Schmidt, A.L.: Diophantine approximation in the Eisensteinian field. J. Number Theory **16** 169–204 (1983)

3640. Schmidt, A.L.: Diophantine approximation in the field $Q(i\sqrt{2})$. J. Number Theory **131** 1983–2012 (2011)
3641. Schmidt, C.-G.: Größencharaktere und Relativklassenzahl abelscher Zahlkörper. J. Number Theory **11** 1979, 128–159 (1979)
3642. Schmidt, E.: Über die Anzahl der Primzahlen unter gegebener Grenze. Math. Ann. **57** 195–204 (1903)
3643. Schmidt, F.K.: Zur Theorie der algebraisch auflösbaren Polynome und Zahlkörper von Primzahlgrad. SBer. Heidelberg. Akad. Wiss. **1929(2)**, 1–10
3644. Schmidt, F.K.: Zur Klassenkörpertheorie im Kleinen. J. Reine Angew. Math. **162** 155–168 (1930)
3645. Schmidt, F.K.: Analytische Zahlentheorie in Körpern der Charakteristik p. Math. Z. **33** 1–32 (1931)
3646. Schmidt, F.K.: Die Theorie der Klassenkörper über einem Körper algebraischer Funktionen in einer Unbestimmten und mit endlichem Koeffizientenbereich. SBer. Erlangen **62** 267–284 (1931)
3647. Schmidt, W.M.: On simultaneous approximations of two algebraic numbers by rationals. Acta Math. **119**, 27–50 (1967)
3648. Schmidt, W.M.: Some diophantine equations in three variables with only finitely many solutions. Mathematika **14**, 113–120 (1967)
3649. Schmidt, W.M.: Linear forms with algebraic coefficients, I. J. Number Theory **3**, 253–277 (1971)
3650. Schmidt, W.M.: Linearformen mit algebraischen Koeffizienten, II. Math. Ann. **191**, 1–20 (1971)
3651. Schmidt, W.M.: Norm form equations. Ann. Math. (2) **96**, 526–551 (1972)
3652. Schmidt, W.M.: Zur Methode von Stepanov. Acta Arith. **24**, 347–367 (1972)
3653. Schmidt, W.M.: A lower bound for the number of solutions of equations over finite fields. J. Number Theory **6**, 1974, 446–480 (1974)
3654. Schmidt, W.[M.]: Equations over finite fields: an elementary approach. Lecture Notes Math. **536** (1976) [2nd ed.: Kendrick Press (2004)]
3655. Schmidt, W.M.: Diophantine Approximation. Lecture Notes Math. **785** (1980)
3656. Schmidt, W.M.: Number fields of given degree and bounded discriminant. Astérisque **228**, 189–195 (1995)
3657. Schmithals, B.: Konstruktion imaginärquadratischer Körper mit unendlichen Klassenkörperturm. Arch. Math. (Basel) **34**, 307–312 (1980)
3658. Schmithals, B.: Eine Verallgemeinerung der Klassenrangabschätzung für Zahlkörper von Roquette und Zassenhaus. Arch. Math. (Basel) **34**, 412–415 (1980)
3659. Schmitz, T.: Abschätzung der Lösung der Pellschen Gleichung. Arch. Math. Phys. **24**, 87–88 (1916)
3660. Schneider, T.: Über p-adische Kettenbrüche. Symposia Math. **4**, 181–189 (1970)
3661. Schneider, T.: Das Werk C.L. Siegels in der Zahlentheorie. Jahresber. Deutsch. Math.-Verein. **85**, 147–157 (1983)
3662. Schoeneberg, B.: Erich Hecke 1887–1947. Jahresber. Deutsch. Math.-Verein. **91**, 168–190 (1989)
3663. Schoeneborn, H.: In memoriam Wolfgang Krull. Jahresber. Deutsch. Math.-Verein. **82**, 51–62, 77–80 (1980)
3664. Scholz, A.: Zwei Bemerkungen zum Klassenkörperturm. J. Reine Angew. Math. **161**, 201–207 (1929)
3665. Scholz, A.: Reduktion der Konstruktion von Körpern mit zweistufiger (metabelscher) Gruppe. SBer. Heidelberg. Akad. Wiss. **1929(14)**, 3–15
3666. Scholz, A.: Über die Bildung algebraischer Zahlkörper mit auflösbarer Galoisscher Gruppe. Math. Z. **30**, 332–356 (1929)
3667. Scholz, A.: Über das Verhältnis von Idealklassen- und Einheitengruppe in Abelschen Körpern von Primzahlpotenzgrad. SBer. Heidelberg. Akad. Wiss. **1930(3)**, 31–55

3668. Scholz, A.: Die Abgrenzungssätze für Kreiskörper und Klassenkörper. SBer. Preuß. Akad. Wiss. Berlin **1931**, 417–426
3669. Scholz, A.: Über die Beziehung der Klassenzahlen quadratischer Körper zueinander. J. Reine Angew. Math. **166**, 201–203 (1932)
3670. Scholz, A.: Idealklassen und Einheiten in kubischen Körpern. Monatsh. Math. Phys. **40**, 211–222 (1933)
3671. Scholz, A.: Die Kreisklassenkörper von Primzahlpotenzgrad und die Konstruktion von Körpern mit vorgegebener zweistufiger Gruppe, I. Math. Ann. **109**, 161–190 (1934); corr. p. 764.
3672. Scholz, A.: Über die Lösbarkeit der Gleichung $t^2 - Du^2 = -4$. Math. Z. **39**, 95–111 (1934)
3673. Scholz, A.: Die Kreisklassenkörper von Primzahlpotenzgrad und die Konstruktion von Körpern mit vorgegebener zweistufiger Gruppe, II. Math. Ann. **110**, 633–649 (1935)
3674. Scholz, A.: Totale Reste, die keine Normen sind, als Erzeuger nichtabelscher Körpererweiterungen, I. J. Reine Angew. Math. **175**, 100–107 (1936)
3675. Scholz, A.: Konstruktion algebraischer Zahlkörper mit beliebiger Gruppe von Primzahlpotenzordnung, II. Math. Z. **42**, 161–188 (1937)
3676. Scholz, A.: Minimaldiskriminanten algebraischer Zahlkörper. J. Reine Angew. Math. **179**, 16–21 (1938)
3677. Scholz, A.: Abelsche Durchkreuzung. Monatsh. Math. Phys. **48**, 340–352 (1939)
3678. Scholz, A.: Zur Abelschen Durchkreuzung. J. Reine Angew. Math. **182**, p. 216 (1940)
3679. Scholz, A.: Totale Reste, die keine Normen sind, als Erzeuger nichtabelscher Körpererweiterungen, II. J. Reine Angew. Math. **182**, 217–234 (1940)
3680. Scholz, A.: Zur Idealtheorie in unendlichen algebraischen Zahlkörpern. J. Reine Angew. Math. **185**, 113–126 (1943)
3681. Scholz, A., Taussky, O.: Die Hauptideale der kubischen Klassenkörper imaginärquadratischer Zahlkörper: ihre rechnerische Bestimmung und ihr Einfluss auf den Klassenkörperturm. J. Reine Angew. Math. **171**, 19–41 (1934)
3682. Schönemann, T.: Von denjenigen Moduln, welche Potenzen von Primzahlen sind. J. Reine Angew. Math. **32**, 93–105 (1846)
3683. Schönemann, T.: Über einige vo Herrn Dr. Eisenstein aufgestellte Lehrsätze. J. Reine Angew. Math. 50, 185–187 (1850)
3684. Schönhage, A.: Polynomial root separation examples. J. Symbolic Comput. **41**, 1080–1090 (2006)
3685. Schoof, R.J.: Quadratic fields and factorization. In: Computational Methods in Number Theory, II, 235–286, Math. Centre Tracts **155**, Math. Centrum Amsterdam (1982)
3686. Schoof, R.J.: Class groups of complex quadratic fields. Math. Comp. **41**, 295–302 (1983)
3687. Schoof, R.[J.]: Infinite class field towers of quadratic fields. J. Reine Angew. Math. **372**, 209–220 (1986)
3688. Schoof, R.[J.]: Minus class groups of the fields of the lth roots of unity. Math. Comp. **67**, 1225–1245 (1998)
3689. Schoof, R.[J.], Washington, L.C.: Quintic polynomials and real cyclotomic fields with large class numbers. Math. Comp. **50**, 543–556 (1988)
3690. Schreier, O.: Über eine Arbeit von Herrn Tschebotareff. Abh. Math. Semin. Univ. Hambg **5**, 1–6 (1926)
3691. Schrutka v.Rechtenstamm, G.: Tabelle der (relativ)-Klassenzahlen der Kreiskörper, deren φ-Funktion des Wurzelexponenten (Grad) nicht grösser als 256 ist. Abh. Deutsch. Akad. Wiss., Berlin, Kl. Math. Nat. **2**, 1–64 (1964)
3692. Schubert, F.T.: De inventione divisorum. Nova Acta Acad. Sci. Petropol. **11**, 172–182 (1793)
3693. Schulz, W.: Über die Galoissche Gruppe der Hermiteschen Polynome. J. Reine Angew. Math. **177**, 248–252 (1937)
3694. Schulz-Arenstorff, R.: Über die zweidimensionale Verteilung der Primzahlen reellquadratischer Körper in Restklassen. J. Reine Angew. Math. **198**, 204–220 (1957)
3695. Schumann, H.-G.: Über Moduln und Gruppenbilder. Math. Ann. **114**, 385–413 (1937)

3696. Schumann, H.-G.: Zum Beweis des Hauptidealsatzes. Abh. Math. Sem. Hansischen Univ. **12**, 42–47 (1937)
3697. Schur, I.: Einige Bemerkungen zu der vorstehender Arbeit des Herrn Pólya: Über die Verteilung der quadratischen Reste und Nichtreste. Nachr. Ges. Wiss. Göttingen **1918**, 30–36 [[3707], **2**, 239–245]
3698. Schur, I.: Über die Verteilung der Wurzeln bei gewissen algebraischen Gleichungen mit ganzzahligen Koeffizienten. Math. Z. **1**, 377–402 (1918) [[3707], **2**, 213–238.]
3699. Schur, I.: Einige Bemerkungen zu der Arbeit "A.Speiser, Zahlentheoretische Sätze aus der Gruppentheorie". Math. Z. **5**, 7–10 (1919) [[3707], **2**, 276–279]
3700. Schur, I.: Beispiele für Gleichungen ohne Affekt. Jahresber. Deutsch. Math.-Verein. **29**, 145–150 (1920) [[3707], **2**, 280–285.]
3701. Schur, I.: Zur Irreduzibilität der Kreisteilungsgleichung. Math. Z. **29**, p. 463 (1929) [[3707], **3**, 234–239.]
3702. Schur, I.: Elementarer Beweis eines Satzes von L.Stickelberger. Math. Z. **29**, 464–465 (1929) [[3707], **3**, 87–88.]
3703. Schur, I.: Gleichungen ohne Affekt. SBer. Preuß. Akad. Wiss. Berlin **1930**, 443–449. [[3707], **3**, 191–197.]
3704. Schur, I.: Affektlose Gleichungen in der Theorie der Laguerreschen und Hermiteschen Polynome. J. Reine Angew. Math. **165**, 52–58 (1931) [[3707], **3**, 227–233.]
3705. Schur, I.: Einige Bemerkungen über die Diskriminante eines algebraischen Zahlkörpers. J. Reine Angew. Math. **167**, 264–269 (1932) [[3707], **3**, 234–239.]
3706. Schur, I.: Ein Beitrag zur elementaren Zahlentheorie. SBer. Preuß. Akad. Wiss. Berlin **1933**, 145-151 [[3707], **3**, 292–298.]
3707. Schur, I.: Gesammelte Abhandlungen. **1–3**. Springer (1973)
3708. Schuster, L.: Reellquadratische Zahlkörper ohne Euklidischen Algorithmus. Monatsh. Math. Phys. **47**, 117–127 (1938)
3709. Schwarz, Š.: Construction of normal bases in cyclic extensions of a field. Czechoslovak Math. J. **38**, 291–312 (1988)
3710. Schwarz, W.: John Knopfmacher, [abstract] analytic number theory, and the theory of arithmetical functions, Quaest. Math. **24**, 273–290 (2001)
3711. Schweiger, F.: The metrical theory of Jacobi-Perron algorithm. Lecture Notes Math. **334** (1973)
3712. Schwering, K.: Untersuchung über die fünften Potenzreste und die aus fünften Einheitswurzeln gebildeten ganzen Zahlen. Zeitschr. Math. Phys. **27**, 102–119 (1882)
3713. Schwering, K.: Multiplication der lemniskatischen Function sin am u. J. Reine Angew. Math. **107**, 196–240 (1891)
3714. Schwering, K.: Zerfällung der lemniskatischen Teilungsgleichung in vier Factoren. J. Reine Angew. Math. **110**, 42–72 (1892)
3715. Schwering, K.: Zur Auflösung der lemniskatischen Teilungsgleichungen. J. Reine Angew. Math. **111**, 170–204 (1893)
3716. Schwering, K.: Zusatz zur Abhandlung "Zerfällung der lemniskatischen Teilungsgleichung in vier Factoren". J. Reine Angew. Math. **112**, 37–38 (1893)
3717. Schwering, K.: Die Siebenteilung der lemniskatischen Funktion sin am (u). J. Reine Angew. Math. **152**, 40–48 (1922)
3718. Scott, Ch.A.: The binomial equation $x^p - 1 = 0$. Amer. J. Math. **8**, 261–264 (1886)
3719. Scott, L. et al.: Walter Feit (1930–2004). Notices Amer. Math. Soc. **52**, 728–735 (2005)
3720. Scriba, C.J.: Zur Erinnerung an Viggo Brun. Mitt. Math. Ges. Hamburg **11**, 271–290 (1985)
3721. Scriba, C.J.: Bartel Leendert van der Waerden (1903–1996). Mitt. Math. Ges. Hamburg **15**, 13–18 (1996)
3722. Seah, E., Washington, L.C.,Williams, H.C.: The calculation of a large cubic class number with an application to real cyclotomic fields. Math. Comp. **41**, 303–305 (1985)
3723. Segal, D.: Obituary: Fritz Grunewald, 1949–2010. Bull. Lond. Math. Soc. **44**, 183–197 (2012) [German translation: Jahresber. Deutsch. Math.-Verein. **113**, 3–20 (2011)]

3724. Seidelmann, F.: Die Gesamtheit der kubischen und biquadratischen Gleichungen mit Affekt bei beliebigen Rationalitätsbereich. Math. Ann. **78**, 230–233 (1917)
3725. Selberg, A.: An elementary proof of the prime-number theorem. Ann. Math. (2) **50**, 305–313 (1949)
3726. Selberg, S.: Ernst Jacobsthal. Norsk Mat. Tidsskr. **38**, 70–73 (1965)
3727. Selfridge, J.L., Nicol, C.A., Vandiver, H.S.: Proof of Fermat's last theorem for all prime exponents less than 4002. Proc. Nat. Acad. Sci. U.S.A. **41**, 970–973 (1955)
3728. Selling, E.: Ueber die idealen Primfactoren der complexen Zahlen, welche aus den Wurzeln einer beliebigen irreductiblen Gleichung rational gebildet sind. Zeitschr. Math. Phys. **10**, 17–47 (1865)
3729. Sen, S.: On explicit reciprocity laws, I. J. Reine Angew. Math. **313**, 1–26 (1980)
3730. Sen, S.: On explicit reciprocity laws, II. J. Reine Angew. Math. **323**, 68–87 (1981)
3731. Sengenhorst, P.: Über Körper der Charakteristik p. Math. Z. **24**, 1–39, 501 (1925)
3732. Serre, J.-P.: Classes des corps cyclotomiques (d'aprés K.Iwasawa). In: Sém. Bourbaki, 11ième année: 1958/59, exp. 174. [Sém. Bourbaki: Volume 1958/1959, Benjamin (1966); Séminaire Bourbaki **3**, 83–93, Soc. Math. Fr. (1995)]
3733. Serre, J.-P.: Corps locaux. Hermann (1962) [2nd ed. (1968); English translation: Local Fields. Springer (1979)]
3734. Serre, J.-P.: Zeta and L functions. In: Arithmetical Algebraic Geometry (Proc. Conf. Purdue Univ., 1963), 82–92. Harper & Row (1965)
3735. Serre, J.-P.: Cohomologie galoisienne. Lecture Notes Math. **5**, (1962/63); 3rd ed. (1965), 4th ed. (1973), 5th ed. (1994, 1997) [English translation: Galois cohomology. Springer (1997). reprint (2002)]
3736. Serre, J.-P.: Local class field theory. In: [599], 128–161
3737. Serre, J.-P.: Une "formule de masse" pour les extensions totalement ramifiées de degré donné d'un corps local. C. R. Acad. Sci. Paris **286**, A1031–A1036 (1978)
3738. Serre, J.-P.: Groupes de Galois sur Q. Astérisque **161/162**, 73–85 (1989)
3739. Serre, J.-P.: Topics in Galois theory. Jones & Bartlett (1992)
3740. Serre, J.-P.: Henri Cartan 1904–2008. Bull. Lond. Math. Soc. **46**, 211–216 (2014)
3741. Serret, J.-A.: Sur une question de théorie des nombres. J. Math. Pures Appl. (1) **15**, p. 296 (1850)
3742. Setzer, B.: The determination of all imaginary, quartic, abelian fields with class number 1. Math. Comp. **35**, 1383–1386 (1980)
3743. Shankar, A., Tsimerman, J.: Counting S_5-fields with a power saving error term. Forum Math. Sigma **2**, e13, 1–8 (2014)
3744. Shanks, D.: On Gauss's class number problems. Math. Comp. **23**, 151–163 (1969)
3745. Shanks, D.: Class number, a theory of factorization, and genera. Proc. Symposia Pure Math. **20**, 415–440 (1971)
3746. Shanks, D.: Gauss's ternary form reduction and the 2-Sylow subgroup. Math. Comp. **25**, 837–853 (1971); Corr.: **32**, 1328–1329 (1978)
3747. Shanks, D.: New types of quadratic fields having three invariants divisible by 3. J. Number Theory **4**, 537–556 (1972)
3748. Shanks, D., Serafin, R.: Quadratic fields with four invariants divisible by 3. Math. Comp. **27**, 183–187, 1012 (1973); Corr. p. 1012
3749. Shanks, D., Weinberger, P.[J.]: A quadratic field of prime discriminant requiring three generators for its class group, and related theory. Acta Arith. **21**, 71–87 (1972)
3750. Shapiro, H.N.: An elementary proof of the prime ideal theorem. Comm. Pure Appl. Math. **2**, 309–323 (1949)
3751. Sharp, R.Y.: Douglas Geoffrey Northcott, FRS, 1916–2005. Bull. Lond. Math. Soc. **41**, 164–178 (2009)
3752. Shephard, G.C., Todd, J.A.: Finite unitary reflection groups. Canad. J. Math. **6**, 274–304 (1954)
3753. Shih, K.-Y.: On the construction of Galois extensions of function fields and number fields. Math. Ann. **207**, 99–120 (1974)

3754. Shiina, T.: Rigid braid orbits related to $PSL_2(p^2)$ and some simple groups. Tôhoku Math. J. (2) **55**, 271–282 (2003)
3755. Shiina, T.: Regular Galois realizations of $PSL_2(p^2)$ over $Q(T)$. Develop. Math. **11**, 125–142 (2004)
3756. Shimura, G.: On canonical models of arithmetic quotients of bounded symmetric domains. Ann. Math. (2) **91**, 144–222 (1970)
3757. Shimura, G.: On canonical models of arithmetic quotients of bounded symmetric domains, II. Ann. Math. (2) **92**, 528–549 (1970)
3758. Shintani, T.: On zeta-functions associated with the vector space of quadratic forms. J. Fac. Sci. Univ. Tokyo Sect. 1 A Math. **22**, 25–65 (1975)
3759. Shintani, T.: On evaluation of zeta functions of totally real algebraic number fields at non-positive integers. J. Fac. Sci. Univ. Tokyo **23**, 393–417 (1976)
3760. Shiratani, K.: On the Gauss-Hecke sums. J. Math. Soc. Japan **16**, 32–38 (1964)
3761. Shiratani, K.: On some relations between Bernoulli numbers and class numbers of cyclotomic fields. Mem. Fac. Sci. Kyushu Univ. **A18**, 127–135 (1964)
3762. Shiratani, K.: Über den l^n-ten Potenzrestcharakter von l im Körper der l^n-ten Einheitswurzeln. J. Reine Angew. Math. **223**, 183–190 (1966)
3763. Shiratani, K.: Die Diskriminante der Weierstraßschen elliptischen Funktionen und das Reziprozitätsgesetz in besonderen imaginärquadratischen Zahlkörpern. Abh. Math. Semin. Univ. Hambg **31**, 51–61 (1967)
3764. Shiratani, K.: Über eine Anwendung elliptischer Funktionen auf das biquadratische Reziprozitätsgesetz. J. Reine Angew. Math. **268/269**, 203–208 (1974)
3765. Shokrollahi, M.A.: Relative class number of imaginary abelian fields of prime conductor below 10 000. Math. Comp. **68**, 1717–1728 (1999)
3766. Shparlinski, I.E.: Infinite Hilbert class field towers over cyclotomic fields. Glasg. Math. J. **50**, 27–32 (2008)
3767. Shyr, J.M.: Class numbers of binary quadratic forms over algebraic number fields. J. Reine Angew. Math. **307/308**, 353–364 (1979)
3768. Siegel, C.L.: Approximation algebraischer Zahlen. Math. Z. **10**, 173–213 (1921) [[3858], **1**, 6–46.]
3769. Siegel, C.L.: Darstellung total positiver Zahlen durch Quadrate. Math. Z. **11**, 246–275 (1921) [[3798], **1**, 47–76.]
3770. Siegel, C.L.: Über Näherungswerte algebraischer Zahlen. Math. Ann. **84**, 80–99 (1921) [[3798], **1**, 77–96.]
3771. Siegel, C.L.: Neuer Beweis für die Funktionalgleichung der Dedekindschen Zetafunktion. Math. Ann. **85**, 123–128 (1922) [[3798], **1**, 113–118.]
3772. Siegel, C.L.: Über die Diskriminanten total reeller Körper. Nachr. Ges. Wiss. Göttingen **1922**, 17–24 [[3798], **1**, 157–164.]
3773. Siegel, C.L.: Neuer Beweis für die Funktionalgleichung der Dedekindschen Zetafunktion, II. Nachr. Ges. Wiss. Göttingen **1922**, 25–31 [[3798], **1**, 173–179.]
3774. Siegel, C.L.: Additive Theorie der Zahlkörper, I. Math. Ann. **87**, 1–35 (1922) [[3798], **1**, 119–153.]
3775. Siegel, C.L.: Additive Zahlentheorie in Zahlkörpern. Jahresber. Deutsch. Math.-Verein. **31**, 22–26 (1922) [[3798], **1**, 168–172.]
3776. Siegel, C.L.: Additive Theorie der Zahlkörper, II. Math. Ann. **88**, 184–210 (1923) [[3798], **1**, 180–206.]
3777. Siegel, C.L.: Über einige Anwendungen diophantischer Approximationen. Abhandl. Preuss. Akad. Wiss. Berlin **1929(1)**, 1–70. [[3798], **1**, 209–266.]
3778. Siegel, C.L.: Über die Classenzahl quadratischer Zahlkörper. Acta Arith. **1**, 83–86 (1935) [[3798], **1**, 406–409.]
3779. Siegel, C.L.: Über die analytische Theorie der quadratischen Formen. Ann. Math. (2) **36**, 527–606 (1935) [[3798], **1**, 326–405.]
3780. Siegel, C.L.: Mittelwerte arithmetischer Funktionen in Zahlkörpern. Trans. Amer. Math. Soc. **39**, 219–224 (1936) [[3798], **1**, 453–458.]

3781. Siegel, C.L.: Über die analytische Theorie der quadratischen Formen, II. Ann. Math. (2) **37**, 230–263 (1936) [[3798], **1**, 410–443.]
3782. Siegel, C.L.: Über die analytische Theorie der quadratischen Formen, III. Ann. Math. (2) **38**, 212–291 (1937) [[3798], **1**, 469–548.]
3783. Siegel, C.L.: Equivalence of quadratic forms. Amer. J. Math. **63**, 658–680 (1941) [[3798], **2**, 217–239.]
3784. Siegel, C.L.: Generalization of Waring's problem to algebraic number fields. Amer. J. Math. **66**, 122–136 (1944) [[3798], **2**, 406–420.]
3785. Siegel, C.L.: On the theory of indefinite quadratic forms. Ann. Math. (2) **45**, 577–622 (1944) [[3798], **2**, 421–466.]
3786. Siegel, C.L.: The average measure of quadratic forms with given discriminant and signature. Ann. Math. (2) **45**, 667–685 (1944) [[3798], **2**, 473–491.]
3787. Siegel, C.L.: Algebraic integers whose conjugates lie in the unit circle. Duke Math. J. **11**, 597–602 (1944) [[3798], **2**, 467–472]
3788. Siegel, C.L.: The trace of totally positive and real algebraic integers. Ann. Math. (2) **46**, 302–312 (1945) [[3798], **3**, 1–11.]
3789. Siegel, C.L.: Sums of mth powers of algebraic integers. Ann. Math. (2) **46**, 313–339 (1945) [[3798], **3**, 12–38.]
3790. Siegel, C,L.: Über das quadratische Reziprozitätsgesetz in algebraischen Zahlkörpern. Nachr. Akad. Wiss. Göttingen **1960**, 1–13. [[3798], **3**, 334–349.]
3791. Siegel, C.L.: Advanced Analytic Number Theory. Tata Institute of Fundamental Research (1960) [Reprint (1980)]
3792. Siegel, C.L.: Zu zwei Bemerkungen Kummers. Nachr. Akad. Wiss. Göttingen **1964**, 51–57 [[3798], **3**, 436–442.]
3793. Siegel, C.L.: Zum Beweise des Starkschen Satzes. Invent. Math. **5**, 180–191 (1968) [[3798], **4**, 41–52.]
3794. Siegel, C.L.: Berechnung von Zetafunktionen an ganzzahligen Stellen. Nachr. Akad. Wiss. Göttingen **1969**, 87–102 [[3798], **4**, 82–97; English translation: in [3797], 249–268.]
3795. Siegel, C.L.: Über die Fourierschen Koeffizienten von Modulformen. Nachr. Akad. Wiss. Göttingen **1970**, 15–56 [[3798], **4**, 98–139.]
3796. Siegel, C.L.: Algebraische Abhängigkeit von Wurzeln. Acta Arith. **21**, 59–64 (1972) [[3798], **4**, 167–172.]
3797. Siegel, C.L.: Advanced Analytic Number Theory. Tata Institute of Fundamental Research (1980)
3798. Siegel, C.L.: Gesammelte Abhandlungen. **1–4**. Springer (1966–1979)
3799. Siegmund-Schutze, R.: Theodor Vahlen — zum Schuldanteil eines deutschen Mathematikers am faschistischen Mißbrauch der Wissenschaft, NTM Schriftenreihe zur Geschichte der Naturwissenschaften, Technik und Medizin **21**, 17–41 (1984)
3800. Sierpiński, W.: O pewnem zagadnieniu z rachunku funkcyj asymptotycznych. Prace Mat.-Fiz. **17**, 77–118 (1906) [French translation: Sierpiński, W., Oeuvres choisies, **1**, 73–108, PWN (1974)]
3801. Silverman, J.H.: Quantitative results in diophantine geometry. Preprint MIT (1983)
3802. Silverman, J.H.: Advanced Topics in the Arithmetic of Elliptic Curves. Springer (1994)
3803. Sinnott, W.: On the Stickelberger ideal and the circular units of a cyclotomic field. Ann. Math. (2) **108**, 107–134 (1988)
3804. Sinnott, W.: On the Stickelberger ideal and the circular units of an abelian field. Invent. Math. **62**, 181–234 (1980/81)
3805. Sinnott, W.: On the μ-invariant of the Γ-transform of a rational function. Invent. Math. **75**, 273–282 (1984)
3806. Sinnott, W.: Γ-transforms of rational function measures on Z_S. Invent. Math. **87**, 139–157 (1987)
3807. Sitaraman, S.: Vandiver revisited. J. Number Theory **57**, 122–129 (1966)
3808. Skolem, T.: Untersuchungen über die möglichen Verteilungen ganzzahliger Lösungen gewisser Gleichungen, Videnskapsselskapets Skr. **1(17)**, 1–57 (1921)

3809. Skolem, T.: Ein allgemeines quadratisches Reziprozitätsgesetz in denjenigen algebraischen Zahlkörpern, worin 2 voll zerfällt. Comment. Math. Helv. **5**, 305–318 (1933)
3810. Skolem, T.: Einige Sätze über p-adische Potenzreihen mit Anwendung auf gewisse exponentielle Gleichungen. Math. Ann. **111**, 399–424 (1935)
3811. Skolem, T.: Extension of two theorems of C.Størmer. Norsk Mat. Tidsskr. **26**, 85–95 (1944) (Norwegian)
3812. Skolem, T.: A theorem on the equation $\zeta^2 - \delta\eta^2 = 1$ where δ, ζ, η are integers in an imaginary quadratic field. Pure Appl. Math. Q. **1945(1)**, 1–13.
3813. Skolem, T.: A remark on the equation $\zeta^2 - \delta\eta^2 = 1$, δ>0.δ'.δ''<0 where δ, ζ, η belong to a total real number field. Pure Appl. Math. Q. **1945(12)**, 1–15.
3814. Skolem, T.: On certain exponential equations. Norske Vid. Selsk. Forh., Trondheim **18(18)**, 71–74 (1945)
3815. Skolem, T.: Solutions of the equation $axy + bx + cy + d = 0$ in algebraic integers. Pure Appl. Math. Q. **1946(3)**, 1–8
3816. Skolem, T. On the existence of a multiplicative basis for an algebraic number field. Norske Vid. Selsk. Forh., Trondheim **20**, 4–7 (1947)
3817. Skolem, T.: A proof of the irreducibility of the cyclotomic equation. Norsk Mat. Tidsskr. **31**, 116–120 (1949)
3818. Skolem, T.: On a certain connection between the discriminant of a polynomial and the number of its irreducible factors mod p. Norsk Mat. Tidsskr. **34**, 81–85 (1952)
3819. Skubenko, B.F.: A proof of Minkowski's conjecture on the product of n linear inhomogeneous forms in n variables for $n \leq 5$. Trudy Mat. Inst. im. Steklova **133**, 4–36 (1973) (Russian)
3820. Skula, L.: Another proof of Iwasawa's class number formula. Acta Arith. **39**, 1–6 (1981)
3821. Slavutskiĭ, I.Š.: On the class-number of ideals of a real quadratic field. Izv. Vysš. Učeb. Zaved. Mat. **1960(4)**, 173–177. (Russian)
3822. Slavutskiĭ, I.Š.: On Mordell's theorem. Acta Arith. **11**, 57–66 (1965)
3823. Slavutskiĭ, I.Š.: Upper bound and arithmetical determination of the class number of ideals in real quadratic fields. Izv. Vysš. Učeb. Zaved. Mat. **1965(2)**, 161–165. (Russian)
3824. Slavutskiĭ, I.Š.: Generalized Voronoi's congruence and the class-number of ideals of an imaginary quadratic field, II. Izv. Vysš. Učeb. Zaved. Mat. **1966(4)**, 118–126. (Russian)
3825. Slavutskiĭ, I.Š. [Slavutsky, I.Sh.]: The simplest proof of Vandiver's theorem. Acta Arith. **15**, 117–118 (1968/69)
3826. Slavutskiĭ, I.Š.: On the generalized Bernoulli numbers that belong to unequal characters. Rev. Mat. Iberoamericana **16**, 459–475 (2000)
3827. Śliwa, J.: Sums of distinct units. Bull. Acad. Pol. Sci., sér. sci. math. astr. phys. **22**, 11–13 (1972)
3828. Śliwa, J.: Factorizations of distinct lengths in algebraic number fields. Acta Arith. **31**, 399–417 (1976)
3829. Śliwa, J.: On the nonessential discriminant divisor of an algebraic number field. Acta Arith. **42**, 57–72 (1982)
3830. Small, C.: Sums of three squares and levels of quadratic number fields. Amer. Math. Monthly **93**, 276–279 (1986)
3831. Smith, H.J.S.: On the orders and genera of ternary quadratic forms. **157**, 255–298 (1867) [[3834], **1**, 455–509.]
3832. Smith, H.J.S.: Report on the theory of numbers. In: [3834], 38–364
3833. Smith, H.J.S.: On the present state and prospects of some branches of pure mathematics. Proc. London Math. Soc. **8**, 6–29 (1876) [[3834], **2**, 166–190.]
3834. Smith, H.J.S.: Collected Mathematical Papers. **1**,**2**. Oxford (1894)
3835. Smith, J.R.: On Euclid's algorithm in some cyclic cubic fields. J. Lond. Math. Soc. **44**, 577–582 (1969)
3836. Smyth, C.J.: On the product of the conjugates outside the unit circle of an algebraic integer. Bull. Lond. Math. Soc. **3**, 169–175 (1971)

3837. Smyth, C.J.: A Kronecker-type theorem for complex polynomials in several variables. Canad. Math. Bull. **24**, 447–452 (1981) Add. 25, 1982, p. 504 (1982)
3838. Smyth, C.J.: Conjugate algebraic numbers on conics. Acta Arith. **40**, 333–346 (1981/82)
3839. Smyth, C.J.: Totally positive algebraic integers of small trace. Ann. Inst. Fourier (Grenoble) **34(3)**, 1–28 (1984)
3840. Smyth, C.[J.]: The Mahler measure of algebraic numbers: a survey. In: Number Theory and Polynomials, 322–349. Cambridge University Press (2008)
3841. Snaith, V.P.: Galois Module Structure. Amer. Math. Soc. (1994)
3842. Sochocki, J.: The principle of greatest common divisor in its applicability to the divisibility of algebraic numbers. St. Petersburg (1893) (Russian) [Polish translation: Prace Mat.-Fiz. **4**, 95–153 (1893)]
3843. Sodaïgui, B.: Classes réalisables par des extensions métacycliques non abéliennes et éléments de Stickelberger. J. Number Theory **65**, 87–95 (1997)
3844. Sodaïgui, B.: Classes de Steinitz d'extensions galoisiennes relatives de degré une puissance de 2 et problème de plongement. Illinois J. Math. **43**, 47–60 (1999)
3845. Sodaïgui, B.: Relative Galois module structure and Steinitz classes of dihedral extensions of degree 8. J. Algebra **223**, 367–378 (2000)
3846. Sodaïgui, B.: Classes de Steinitz d'extensions galoisiennes à groupe de Galois un 2-groupe. Funct. Approx. Comment. Math. **48**, 183–196 (2013)
3847. Sodaïgui, B.: Classes de Steinitz d'extensions galoisiennes à groupe de Galois de centre non trivial. J. Number Theory **133**, 611–619 (2013)
3848. Söhne, P.: The Pólya-Vinogradov inequality for totally real algebraic number fields. Acta Arith. **65**, 197–212 (1993)
3849. Söhngen, H.: Zur komplexen Multiplikation. Math. Ann. **111**, 302–328 (1935)
3850. Sokolovskiĭ, A.V.: A theorem on the zeros of Dedekind's zeta-function and the distance between "neighboring" prime ideals. Acta Arith. **13**, 321–334 (1967/68) (Russian)
3851. Solderitsch, J.J.: Quadratic fields with special class groups. Dissertation, Lehigh University (1977)
3852. Solderitsch, J.J.: Quadratic fields with special class groups. Math. Comp. 59, 1992, 633–638 (1992)
3853. Sommer, J.: Vorlesungen über Zahlentheorie. Einführung in die Theorie der algebraischen Zahlkörper. Teubner (1907) [French translation: Introduction à la théorie des nombres algébriques, Hermann et Fils (1911)]
3854. Sonn, J.: On the embedding problem for nonsolvable Galois groups of algebraic number fields: Reduction theorems. J. Number Theory **4**, 411–436 (1972)
3855. Sonn, J.: Central extensions of S_n as Galois groups via trinomials. J. Algebra **125**, 320–330 (1989)
3856. Sono, M.: On congruences. II. Mem. Coll. Sci. Kyoto **3**, 113–149 (1919)
3857. Sono, M.: On congruences. III. Mem. Coll. Sci. Kyoto **3**, 189–197 (1919)
3858. Sono, M.: On the reduction of ideals. Mem. Coll. Sci. Kyoto **7**, 191–204 (1924)
3859. Soublin, J.-P.: Préhistoire des idéaux. Proc. Sem. Hist. Math. Inst. H. Poincaré, Paris **5**, 13–20 (1984)
3860. Soulé, C.: Perfect forms and the Vandiver conjecture. J. Reine Angew. Math. **517**, 209–221 (1999)
3861. Soundararajan, K.: Divisibility of class numbers of imaginary quadratic fields. J. Lond. Math. Soc. (2) **61**, 681–690 (2000)
3862. Soundararajan, K.: The number of imaginary quadratic fields with a given class number. Hardy-Ramanujan J. **30**, 13–18 (2007)
3863. Soverchia, E.: Steinitz classes of metacyclic extensions. J. Lond. Math. Soc. (2) **66**, 61–72 (2002)
3864. Späth, H.: Über die Irreduzibilität der Kreisteilungsgleichung. Math. Z. **26**, 442–444 (1927)
3865. Spearman, B.K.: Monogenic A_4 quartic fields. Int. Math. Forum **1**, 1969–1974 (2006)
3866. Spearman, B.K., Williams, K.S.: The simplest arithmetic proof of Jacobi's four squares theorem. Far East J. Math. Sci. **2**, 433–439 (2000)

3867. Spearman, B.K., Williams, K.S.: Cubic fields with a power basis. Rocky Mountain J. Math. **31**, 1103–1109 (2001)
3868. Speiser, A.: Die Theorie der binären quadratischen Formen mit Koeffizienten und Unbestimmten in einem beliebigen Zahlkörper. Dissertation, Univ. Göttingen (1909)
3869. Speiser, A.: Über die Komposition der binären quadratischen Formen. In: [3], 375–395
3870. Speiser, A.: Gruppendeterminante und Körperdiskriminante. Math. Ann. **77**, 546–562 (1916)
3871. Speiser, A.: Die Zerlegungsgruppe. J. Reine Angew. Math. **149**, 174–188 (1919)
3872. Speiser, A.: Zahlentheoretische Sätze aus der Gruppentheorie. Math. Z. **5**, 1–6 (1919)
3873. Speiser, A.: Die Zerlegung von Primzahlen in algebraischen Zahlkörpern. Trans. Amer. Math. Soc. **23**, 173–178 (1922)
3874. Speiser, A.: Rudolf Fueter, 1880–1950. Verhandl. Schweiz. Naturf. Ges. **130**, 399–404 (1950)
3875. Spencer, J.: An elementary proof of Kronecker's theorem. Fibonacci Quart. **15**, 9–10 (1977)
3876. Sprindžuk, V.G.: Hilbert's irreducibility theorem and rational points on algebraic curves. Dokl. Akad. Nauk SSSR **247**, 285–289 (1979) (Russian)
3877. Sprindžuk, V.G.: Reducibility of polynomials and rational points on algebraic curves. Dokl. Akad. Nauk SSSR **250**, 1327–1330 (1980) (Russian)
3878. Sprindžuk, V.G.: Reducibility of polynomials and rational points on algebraic curves. Progr. Math. **12**, 287–309 (1981)
3879. Sprindžuk, V.G.: Classical diophantine equations in two variables. Nauka (1982) (Russian) [English translation: Classical Diophantine equations. Lecture Notes Math. **1559** (1993)]
3880. Springer, T.A.: Note on quadratic forms over algebraic number fields. Indag. Math. **19**, 39–43 (1957)
3881. Springer, T.A.: H.D.Kloosterman and his work. Notices Amer. Math. Soc. **47**, 862–867 (2000)
3882. Srinivasan, A.: Computations of class numbers of real quadratic fields. Math. Comp. **67**, 1285–1308 (1998)
3883. Srinivasan, A.: Prime producing polynomials: proof of a conjecture by Mollin and Williams. Acta Arith. **89**, 1–7 (1999)
3884. Srinivasan, A.: Prime producing quadratic polynomials and class number one or two. The Ramanujan J. **10**, 5–22 (2005)
3885. Stafford, E.T., Vandiver, H.S.: Determination of the properly irregular cyclotomic fields. Proc. Nat. Acad. Sci. U.S.A. **16**, 139–150 (1930)
3886. Stankus, E.: The moments of $L(1, \chi_m)$ in arithmetic progressions. Lit. Mat. Sb. **16**, 207–227 (1976) (Russian)
3887. Stankus, E.: The mean value of $L(1/2, \chi_d)$ for $d \equiv 1 \pmod D$. Lit. Mat. Sb. **23(2)**, 169–177 (1983) (Russian)
3888. Stankus, E.: On the mean value of Dirichlet L-functions in the critical strip. Lit. Mat. Sb. **31**, 678–686 (1991) (Russian)
3889. Stark, H.[M.]: On complex quadratic fields with class number equal to one. Trans. Amer. Math. Soc. **122**, 112–119 (1966)
3890. Stark, H.M.: There is no tenth complex quadratic field with class-number one. Proc. Nat. Acad. Sci. U.S.A. **57**, 216–221 (1967)
3891. Stark, H.M.: A complete determination of the complex quadratic fields of class-number one. Michigan Math. J. **14**, 1–27 (1967)
3892. Stark, H.M.: On the "gap" in a theorem of Heegner. J. Number Theory **1**, 16–27 (1969)
3893. Stark, H.M.: The role of modular functions in a class-number problem. J. Number Theory **1**, 252–260 (1969)
3894. Stark, H.M.: A transcendence theorem for class-number problems. Ann. Math. (2) **94**, 153–173 (1971)
3895. Stark, H.M.: A transcendence theorem for class-number problems, II. Ann. Math. (2) **96**, 174–209 (1972)

3896. Stark, H.M.: Some effective cases of the Brauer–Siegel theorem. Invent. Math. **23**, 135–152 (1974)
3897. Stark, H.M.: On complex quadratic fields with class-number two. Math. Comp. **29**, 289–302 (1975)
3898. Stark, H.M.: L-functions at $s = 1$, II. Artin L-functions with rational characters. Adv. Math. **17**, 1975, 60–92 (1975)
3899. Stark, H.M.: L-functions at $s = 1$, III. Totally real fields and Hilbert twelfth problem. Adv. Math. **22**, 64–84 (1976)
3900. Stark, H.M.: Dirichlet's class-number formula revisited. Contemp. Math. **143**, 571–577 (1993)
3901. Staś, W.: Über eine Anwendung der Methode von Turán, auf die Theorie des Restgliedes im Primidealsatz. Acta Arith. **5**, 179–195 (1959)
3902. Staś, W.: On the order of Dedekind zeta–functions in the critical strip. Funct. Approx. Comment. Math. **4**, 19–26 (1976)
3903. Staś, W.: On the order of Dedekind zeta–functions near the line $\sigma = 1$. Acta Arith. 35, 1979, 195–202 (1979)
3904. Staś, W., Wiertelak, K.: Some estimates in the theory of Dedekind zeta-functions. Acta Arith. **23**, 127–135 (1973)
3905. Staś, W., Wiertelak, K.: On some estimates in the theory of $\zeta(s, \chi)$-functions. Acta Arith. **26**, 193–301 (1974/75)
3906. Staś, W., Wiertelak, K.: Further applications of Turán's methods to the distribution of prime ideals in ideal classes mod $\mathfrak{f}$. Acta Arith. **31**, 153–165 (1976)
3907. Stauffer, R.: The construction of a normal basis in a separable extension field. Amer. J. Math. **58**, 585–597 (1936)
3908. Steckel, H.-D.: Abelsche Erweiterungen mit vorgegebenem Zahlknoten. J. Reine Angew. Math. **330**, 93–99 (1982)
3909. Stečkin, S.B.: Mean values of the modulus of a trigonometric sum. Trudy Mat. Inst. im. Steklova **134**, 283–309 (1975) (Russian)
3910. Stein, A.: Die Gewinnung der Einheiten in gewissen relativquadratischen Zahlkörpern durch das J.Hurwitzsche Kettenbruchverfahren. J. Reine Angew. Math. **156**, 69–92 (1927)
3911. Steinbacher, F.: Abelsche Körper als Kreisteilungskörper. J. Reine Angew. Math. **139**, 85–100 (1911)
3912. Steinig, J.: On Euler's idoneal numbers. Elem. Math. **21**, 73–88 (1966)
3913. Steinitz, E.: Zur Theorie der Moduln. Math. Ann. **52**, 1–57 (1899)
3914. Steinitz, E.: Algebraische Theorie der Körper. J. Reine Angew. Math. **137**, 167–309 (1910)
3915. Steinitz, E.: Rechteckige Systeme und Moduln in algebraischen Zahlkörpern, I. Math. Ann. **71**, 1912, 328–254 (1912)
3916. Steinitz, E.: Rechteckige Systeme und Moduln in algebraischen Zahlkörpern, II. Math. Ann. **72**, 297–345 (1912)
3917. Steklov, V.A.: Andreĭ Andreevič Markov. Izv. Ross. Akad. Nauk, Ser. Mat. (6) **16**, 169–184 (1922) (Russian)
3918. Stepanov, S.A.: The number of points of a hyperelliptic curve over a finite prime field. Izv. Akad. Nauk SSSR, Ser. Mat. **33**, 1171–1181 (1969) (Russian)
3919. Stepanov, S.A.: Elementary method in the theory of congruences for a prime modulus. Acta Arith. **17**, 231–247 (1970)
3920. Stepanov, S.A.: An elementary proof of the Hasse-Weil theorem for hyperelliptic curves. J. Number Theory **4**, 1972, 118–143 (1972)
3921. Stepanov, S.A.: Congruences with two unknowns. Izv. Akad. Nauk SSSR, Ser. Mat. **36**, 683–711 (1972) (Russian)
3922. Stepanov, S.A.: The constructive method in the theory of equations over finite fields. Trudy Mat. Inst. im. Steklova **132**, 237–246 (1973) (Russian)
3923. Stender, H.J.: Einheiten für eine allgemeine Klasse total reeller algebraischer Zahlkörper. J. Reine Angew. Math. **257**, 151–178 (1972)

3924. Stephens, A.J., Williams, H.C.: Some computational results on a problem concerning powerful numbers. Math. Comp. **50**, 619–632 (1988)
3925. Steuding, J.: Voronoï's contribution to modern number theory. Šiauliai Math. Semin. **2**, 67–106 (2007)
3926. Steurer, A.: On the Galois groups of the 2-class towers of some imaginary quadratic fields. J. Number Theory **125**, 235–246 (2007)
3927. Stevenhagen, P.: Ray class groups and governing fields. Publ. Math. Fac. Sci. Besançon **1988/89**, 1–93.
3928. Stevenhagen, P.: The number of real quadratic fields having units of negative norm. Experiment. Math. **2**, 121–136 (1993)
3929. Stevenhagen, P.: Class number parity for the pth cyclotomic field. Math. Comp. **63**, 773–784 (1994)
3930. Stevenhagen, P.: A density conjecture for the negative Pell equation. In: Computational Algebra and Number Theory (Sydney, 1992), 187–200. Kluwer (1995)
3931. Stewart, C.L.: Algebraic integers whose conjugates lie near the unit circle. Bull. Soc. Math. France **106**, 169–176 (1978)
3932. Stewart, I., Tall, D.: Algebraic Number Theory Chapman and Hall, J.Wiley (1979). [2nd ed. Chapman and Hall (1987); 3rd ed. A.K. Peters (2002); 4th ed. CRC Press (2016)]
3933. Stickelberger, L.: Ueber eine Verallgemeinerung der Kreistheilung. Math. Ann. **37**, 321–367 (1890)
3934. Stickelberger, L.: Ueber eine neue Eigenschaft der Diskriminanten algebraischer Zahlkörper. In: Verhandlungen des ersten Internationalen Mathematiker-Kongresses in Zürich vom 9. bis 11. August 1897, 182–193. Teubner (1898)
3935. Stiemke, E.: Sur les modules dénombrables. C. R. Acad. Sci. Paris **157**, 273–274 (1913)
3936. Stiemke, E.: Über unendliche algebraische Zahlkörper. Math. Z. **25**, 9–39 (1926)
3937. Störmer, C.: Sur une équation indéterminée. C. R. Acad. Sci. Paris **127**, 752–754 (1898)
3938. Strade, H.: Hel Braun: 1914–1986. Mitt. Math. Ges. Hamburg **11**, 373–376 (1987)
3939. Strassmann, R.: Zur Theorie der π-adischen Zahlen. Dissertation, Univ. Göttingen (1926)
3940. Strassmann, R.: Über den Wertevorrat von Potenzreihen im Gebiet der p-adischen Zahlen. J. Reine Angew. Math. **159**, 13–28, 65–66 (1928)
3941. Stuart, D., Perlis, R.: A new characterization of arithmetic equivalence. J. Number Theory **53**, 300–308 (1995)
3942. Stubhaug, A.: Niels Henrik Abel and his Times. Springer (2000)
3943. Stuhler ,U.: Martin Kneser (21.1.1928–16.2.2004). Jahresber. Deutsch. Math.-Verein. **108**, 45–61 (2006)
3944. Styer, R.: Hecke theory over arbitrary number fields. J. Number Theory **33**, 107–131 (1989) Errata: J. Number Theory **62**, p. 220 (1997)
3945. Suetuna, Z.: Über die Maximalordnung einiger Funktionen in der Idealtheorie. J. Fac. Sci. Univ. Tokyo **1**, 105–153 (1925)
3946. Suetuna, Z.: On the product of L-functions. Jpn. J. Math. **2**, 19–37 (1925)
3947. Suetuna, Z.: Über die Maximalordnung einiger Funktionen in der Idealtheorie, II. J. Fac. Sci. Univ. Tokyo **1**, 249–283 (1926)
3948. Suetuna, Z.: Bemerkung über das Produkt von L-Funktionen. Tôhoku Math. J. **27**, 248–257 (1926)
3949. Suetuna, Z.: Über die Maximalordnung einiger Funktionen in der Idealtheorie, III. J. Fac. Sci. Univ. Tokyo **1**, 349–371 (1926)
3950. Suetuna, Z.: Die Idealnormen eines algebraischen Körpers. J. Fac. Sci. Univ. Tokyo **1**, 417–434 (1928)
3951. Suetuna, Z.: Bemerkung zu meiner Arbeit: "Uber die Maximalordnung einiger Funktionen in der Idealtheorie". J. Fac. Sci. Univ. Tokyo **1**, 435–437 (1928)
3952. Suetuna, Z.: Über die Nullstellen der Dedekindschen Zetafunktionen. Math. Z. **32**, 190–191 (1930)
3953. Suetuna, Z.: Über die Anzahl der Idealteiler. J. Fac. Sci. Univ. Tokyo **2**, 155–177 (1931)

3954. Suetuna, Z.: Über die L-Funktionen in gewissen algebraischen Zahlkörpern. Jpn. J. Math. **13**, 27–38 (1936)
3955. Suetuna, Z.: Abhängigkeit der L-Funktionen in gewissen algebraischen Zahlkörpern, I. J. Reine Angew. Math. **177**, 6–12 (1937)
3956. Suetuna, Z.: Abhängigkeit der L-Funktionen in gewissen algebraischen Zahlkörpern, II. J. Fac. Sci. Univ. Tokyo **3**, 223–252 (1937)
3957. Sueyoshi, Y.: A note on Miki's generalization of the Grunwald-Hasse-Wang theorem. Mem. Fac. Sci. Kyushu Univ. **A35**, 229–234 (1981)
3958. Sueyoshi, Y.: Infinite 2-class field towers of some imaginary quadratic number fields. Acta Arith. **113**, 251–257 (2004)
3959. Sugawara, M.: Über den Führer eines Relativ-Abelschen Zahlkörpers. Proc. Imp. Acad. Tokyo **2**, 366–367 (1926)
3960. Sugawara, M.: On the so-called Kronecker's dream in youngdays. Proc. Phys.-Math. Soc. Japan (3) **15**, 99–107 (1933)
3961. Sugawara, M.: Konstruktion gewisser algebraischer Zahlkörper durch die Modulfunktionen zweier Variablen, I. Jpn. J. Math. **11**, 131–184 (1935)
3962. Sugawara, M.: Zur Theorie der komplexen Multiplikation, I. J. Reine Angew. Math. **174**, 189–191 (1936)
3963. Sugawara, M.: Zur Theorie der komplexen Multiplikation, II. J. Reine Angew. Math. **175**, 65–68 (1936)
3964. Sury, B.: Arithmetic groups and Salem numbers. Manuscripta Math. **75**, 97–102 (1992)
3965. Swan, R.G.: Factorization of polynomials over finite fields. Pacific J. Math. **12**, 1099–1106 (1962)
3966. Swan, R.G.: Invariant rational functions and a problem of Steenrod. Invent. Math. **7**, 148–158 (1969)
3967. Swift, J.D.: Note on discriminants of binary quadratic forms with a single class in each genus. Bull. Amer. Math. Soc. **54**, 560–561 (1948)
3968. Szechtman, F.: Quadratic Gauss sums over finite commutative rings. J. Number Theory **95**, 1–13 (2002)
3969. Szegö, G., Walfisz, A.: Über das Piltzsche Teilerproblem in algebraischen Zahlkörpern, I. Math. Z. **26**, 138–156 (1927)
3970. Szegö, G., Walfisz, A.: Über das Piltzsche Teilerproblem in algebraischen Zahlkörpern, II. Math. Z. **26**, 467–486 (1927)
3971. Szekeres, G.: On the number of divisors of $x^2 + x + A$. J. Number Theory **6**, 434–442 (1974)
3972. Taelman, L.: A Herbrand-Ribet theorem for function fields. Invent. Math. **188**, 253–275 (2012)
3973. Takagi, T.: Über die im Bereiche der rationalen complexen Zahlen Abelschen Zahlkörper. J. Coll. Sci. Imp. Univ. Tokyo **19(5)**, 1–42 (1903) [[3979], 13–39.]
3974. Takagi, T.: Zur Theorie der relativ-Abel'schen Zahlkörper, I, II. Proc. Phys.-Math. Soc. Japan (2) **8**, 154–162, 243–254 (1915) [[3979], 43–60.]
3975. Takagi, T.: Zur Theorie der komplexen Multiplikation der elliptischen Funktionen, Proc. Phys.-Math. Soc. Japan Proc. Phys.-Math. Soc. Japan (2) **8**, 386–393 (1915) [[3979], 61–67.]
3976. Takagi, T.: Über eine Theorie des relativ-Abelschen Zahlkörpers. J. Coll. Sci. Imp. Univ. Tokyo **41(9)**, 1–133 (1920) [[3979], 73–167.]
3977. Takagi, T.: Über das Reciprocitätsgesetz in einem beliebigen algebraischen Zahlkörper. J. Coll. Sci. Imp. Univ. Tokyo **44(5)**, 1–50 (1922) [[3979], 179–216.]
3978. Takagi, T.: Zur Theorie des Kreiskörpers. J. Reine Angew. Math. **157**, 230–238 (1927) [[3979], 246–255.]
3979. Takagi, T.: Collected Papers. Springer (1990), 2nd ed. (2014)
3980. Takahashi, S.: An explicit representation of the generalized principal ideal theorem for the rational ground field. Tôhoku Math. J. (2) **16**, 176–182 (1964)

3981. Takenouchi, T.: On the classes of congruent integers in an algebraic Körper. J. Coll. Sci. Tokyo **36**, 1–13 (1913)
3982. Takenouchi, T.: On the classes of congruent integers in an algebraic Körper. Proc. Imp. Acad. Tokyo **1**, 122–128 (1914)
3983. Takenouchi, T.: On the relatively abelian corpora with respect to the corpus defined by a primitive cube root of unity. J. Coll. Sci. Tokyo **37(5)**, 1–70 (1916)
3984. Takeuchi, K.: Totally real algebraic number fields of degree 9 with small discriminant. Saitama Math. J. **17**, 63–85 (1999)
3985. Takeuchi, T.: Notes on the class field towers of cyclic fields of degree l. Tôhoku Math. J. (2) **31**, 301–307 (1979)
3986. Takeuchi, T.: On the ℓ-class field towers of cyclic fields of degree ℓ. Sci. Rep. Niigata Univ. **17**, 23–25 (1980)
3987. Takiff, S.J.: Rings of invariant polynomials for a class of Lie algebras. Trans. Amer. Math. Soc. **160**, 249–262 (1971)
3988. Tambs Lyche, R.: Un théorème sur les déterminants. Norske Vid. Selsk. Forh., Trondheim **1(41)**, 1–2 (1929)
3989. Taniguchi, T.: On proportional constants of the mean value of class numbers of quadratic extensions. Trans. Amer. Math. Soc. **359**, 5517–5524 (2007)
3990. Taniguchi, T., Thorne,F.: Secondary terms in counting functions for cubic fields. Duke Math. J. **162**, 2451–2508 (2013)
3991. Taniguchi, T., Thorne, F.: An error estimate for counting S_3-sextic number fields. Internat. J. Number Th. **10**, 935–948 (2014)
3992. Tannaka, T.: Über einen Satz von Herrn Artin. Proc. Imp. Acad. Tokyo **9**, 197–198 (1933)
3993. Tannaka, T.: Ein Hauptidealsatz relativ-Galoisscher Zahlkörper und ein Satz über den Normenrest. Proc. Imp. Acad. Tokyo **9**, 355–356 (1933)
3994. Tannaka, T.: Einige Bemerkungen zu den Arbeiten über den allgemeinen Hauptidealsatz. Jpn. J. Math. **10**, 163–167 (1934)
3995. Tannaka, T.: Ein Hauptidealsatz relativ-Galoisscher Zahlkörper und ein Satz über den Normenrest, II. Proc. Imp. Acad. Tokyo **10**, 183–189 (1934)
3996. Tannaka, T.: An alternative proof of a generalized principal ideal theorem. Proc. Japan Acad. Sci. **25**, 26–31 (1949)
3997. Tannaka, T.: On a generalization of the principal ideal theorem. Tôhoku Math. J. (2) **1**, 229–269 (1950)
3998. Tannaka, T.: Some remarks concerning principal ideal theorem. Tôhoku Math. J. (2) **1**, 270–278 (1950)
3999. Tannaka, T.: On the generalized principal ideal theorem. In: Proceedings of the International Symposium on Algebraic Number Theory, 65–77. Sci. Council of Japan (1956)
4000. Tannaka, T.: A generalized principal ideal theorem and a proof of a conjecture of Deuring. Ann. Math. (2) **67**, 574–589 (1958)
4001. Tannaka, T., Terada, F.: A generalization of the principal ideal theorem. Proc. Japan Acad. Sci. **25**, 7–8 (1949)
4002. Tanner, H.W.L.: On the binomial equation $x^p - 1 = 0$. Quinquisection. Proc. London Math. Soc. **18**, 214–234 (1886/87)
4003. Tanner, H.W.L.: On cyclotomic functions. Proc. London Math. Soc. **20**, 63–87, 258–296 (1888/89)
4004. Tanner, H.W.L.: On complex primes formed with the fifth roots of unity. Proc. London Math. Soc. **24**, 223–262, 263–272 (1892/93)
4005. Tanner, J.W., Wagstaff, S.S.Jr.: New congruences for the Bernoulli numbers. Math. Comp. **48**, 341–350 (1987)
4006. Tasaka, T.: Remarks on the validity of Hasse's norm theorem. J. Math. Soc. Japan **22**, 330–341 (1970)
4007. Tate, J.T.: Fourier analysis in number fields, and Hecke's zeta-functions. Dissertation, Princeton Univ. (1950) [[599], 305–347.]
4008. Tate, J.: Global class field theory. In: [599], 162–203

4009. Tate, J.: Problem 9: The general reciprocity law. In: [460], 311–322
4010. Tate, J.T.: Local constants. In: [1281], 89–131
4011. Tate, J.: Les conjectures de Stark sur les fonctions L d'Artin en $s = 0$. Birkhäuser (1984)
4012. Tateyama, K., Maillet's determinant. Sci. Papers College Gen. Ed. Univ. Tokyo **32**, 97–100 (1982)
4013. Tatuzawa, T.: On a theorem of Siegel. Jpn. J. Math. **21**, 163–178 (1951)
4014. Tatuzawa, T.: On the product of $L(1, \chi)$. Nagoya Math. J. **5**, 105–111 (1953)
4015. Tatuzawa, T.: Additive prime number theory in an algebraic number field. J. Math. Soc. Japan **7**, 409–423 (1955)
4016. Tatuzawa, T.: On the Waring problem in an algebraic number field. J. Math. Soc. Japan **10**, 322–341 (1958)
4017. Tatuzawa, T.: On the Hecke-Landau L-series. Nagoya Math. J. **16**, 11–20 (1960)
4018. Tatuzawa, T.: On the Waring problem in an algebraic number field. In: Seminar on Modern Methods in Number Theory, Inst. Statist. Math. Tokyo **35**, 1–8 (1971)
4019. Tatuzawa, T.: On Waring's problem in algebraic number fields. Acta Arith. **24**, 37–60 (1973)
4020. Taussky, O.: Über eine Verschärfung des Hauptidealsatzes für algebraische Zahlkörper. J. Reine Angew. Math. **168**, 193–210 (1932)
4021. Taussky, O.: A remark on the class field tower. J. Lond. Math. Soc. **12**, 82–85 (1937)
4022. Taussky, O.: On a theorem of Latimer and MacDuffee. Canad. J. Math. **1**, 300–302 (1949)
4023. Taussky, O.: Classes of matrices and quadratic fields. **1**, 127–132 (1951)
4024. Taussky-Todd, O. [Taussky, O.]: Arnold Scholz zum Gedächtniss. **7**, 379–386 (1952)
4025. Taussky, O.: Classes of matrices and quadratic fields, II. J. Lond. Math. Soc. **27**, 237–239 (1962)
4026. Taussky, O.: On matrix classes corresponding to an ideal and its inverse. Illinois J. Math. **1**, 108–113 (1957)
4027. Taussky, O.: Ideal matrices, I. Arch. Math. (Basel) **13**, 275–282 (1962)
4028. Taussky, O.: Ideal matrices, II. Math. Ann. 1 **150**, 218–225 (1963)
4029. Taussky, O.: A result concerning classes of matrices. J. Number Theory **6**, 64–71 (1974)
4030. Taussky, O.: Ideal matrices, III. Pacific J. Math. **118**, 599–601 (1985)
4031. Taussky, O., Todd, J.: A characterisation of algebraic numbers. Proc. Roy. Irish Acad. **A 46**, 1–8 (1940)
4032. Taylor, G.: Lehmer's conjecture for matrices over the ring of integers of some imaginary quadratic fields. J. Number Theory **132**, 590–607 (2012)
4033. Taylor, H., Taylor, L.: George Pólya, Master of discovery 1887–1985. Dale Seymour Publ. (1993)
4034. Taylor, H.S.: Obituary: Joseph Henry Maclagen Wedderburn (1882–1948). Obituary Notices of Fellows of the Royal Society **6**, 619–625 (1949)
4035. Taylor, M.J.: On Fröhlich's conjecture for rings of integers of tame extensions. Invent. Math. **63**, 41–79 (1981)
4036. Taylor, M.J.: Obituary: Albrecht Fröhlich, 1916–2001. Bull. Lond. Math. Soc. **38**, 329–350 (2006)
4037. Taylor, R.: On icosahedral Artin representations, II. Amer. J. Math. **125**, 549–566 (2003)
4038. Taylor, R., Wiles, A.: Ring-theoretic properties of certain Hecke algebras, Ann. Math. (2) **141**, 553–572 (1995)
4039. Taylor, S.J.: Paul Lévy. Bull. Lond. Math. Soc. **7**, 300–320 (1975)
4040. Teichmüller, O.: Diskret bewertete perfekte Körper mit unvollkommenem Restklassenkörper. J. Reine Angew. Math. **176**, 141–152 (1936)
4041. Teichmüller, O.: Über die Struktur diskret bewerteter perfekter Körper. Nachr. Ges. Wiss. Göttingen **1936**, 151–161
4042. Terada, F.: On a generalization of the principal ideal theorem. Tôhoku Math. J. (2) **1**, 229–269 (1950)
4043. Terada, F.: On the generalized principal ideal theorem. Tôhoku Math. J. (2) **6**, 95–100 (1954)
4044. Terada, F.: A generalization of the principal ideal theorem. J. Math. Soc. Japan **7**, 530–536 (1955)

4045. Terada, F.: A principal ideal theorem in the genus field. Tôhoku Math. J. (2) **23**, 697–718 (1971)
4046. Teranishi, Y.: A theorem on invariants of semi-simple Lie algebras. In: Perspectives in ring theory (Antwerp, 1987), 37–40. Kluwer (1988)
4047. Terjanian, G.: Sur la loi de réciprocité des puissances l-émes. Acta Arith. **54**, 87–125 (1989)
4048. Thaine, F.: On the ideal class groups of real abelian number fields. Ann. Math. (2) **128**, 1–18 (1988)
4049. Thévenaz, J.: Some remarks on G-functors and the Brauer morphism. J. Reine Angew. Math. **384**, 24–56 (1988)
4050. Thiebaud, C.: Sur la capitulation dans les corps de genres d'une extension abélienne d'un corps quadratique imaginaire. J. Number Theory **85**, 92–107 (2000)
4051. Thiele, R.: Leonhard Euler. Teubner (1982)
4052. von Thielmann, M.: Zur Pellschen Gleichung. Math. Ann. **95**, 635–640 (1926)
4053. Thomas, E.: Fundamental units for orders in certain cubic number fields. J. Reine Angew. Math. **310**, 35–55 (1979)
4054. Thome, A.: Existenz von Ganzheitsbasen bei Kummererweiterungen und Komposita. Arch. Math. (Basel) **51**, 523–531 (1988)
4055. Thompson, J.G.: Some finite groups which appear as $Gal(L/K)$, where $K \subseteq Q(\mu_n)$. J. Algebra **89**, 437–499 (1984)
4056. Thompson, W.R.: On the possible forms of discriminants of algebraic fields, I. Amer. J. Math. **53**, 81–90 (1931)
4057. Thompson, W.R.: On the possible forms of discriminants of algebraic fields, II. Amer. J. Math. **55**, 111–118 (1933)
4058. Thue, A.: Über Annäherungswerte algebraischer Zahlen. J. Reine Angew. Math. **135**, 284–305 (1909) [[4060], 232–253.]
4059. Thue, A.: Über eine Eigenschaft, die keine transzendente Grösse haben kann. Skrifter Videnskaps. Kristiania **191(20)** (1912) [[4060], 479–491.]
4060. Thue, A.: Selected Mathematical Papers, Universitetsforlaget Oslo (1977)
4061. Thurston, H.S.: The solution of p-adic equations. Amer. Math. Monthly **50**, 142–148 (1943)
4062. Tichy, R.F., Ziegler, V.: Units generating the ring of integers of complex cubic fields. Colloq. Math. **109**, 71–83 (2007)
4063. Tietze, H.: Über die Herstellung einer Basis für die ganzen Zahlen eines algebraischen Zahlkörpers. SBer. Bayer. Akad. Wiss. **1944**, 147–162
4064. Tietze, H.: Nachruf: Gustav Herglotz. Jahrbuch Bayer. Akad. Wiss. **1953**, 188–194
4065. Titchmarsh, E.C.: A divisor problem. Rend. Circ. Mat. Palermo **54**, 414–429 (1930)
4066. Titchmarsh, E.C.: Godfrey Harold Hardy. J. Lond. Math. Soc. **25**, 81–138 (1950)
4067. Toepken, H.: Zur Irreduzibilität der Kreisteilungsgleichung. Deutsche Math. **2**, 631–633 (1937)
4068. Tomanov, G.: On Grunwald-Wang's theorem. J. Reine Angew. Math. **389**, 209–220 (1988)
4069. Torelli, G.: Alfredo Capelli (1855–1910). Rend. Accad. Sci. fis. mat. Napoli (3) **16**, 20–30 (1910)
4070. Tornheim, L.: Minimal basis and inessential discriminant divisors for a cubic field. Pacific J. Math. **5**, 623–631 (1955)
4071. Touibi, C.: Une démonstration élémentaire du théorème des idéaux premiers "via une inégalité du type grand crible". Sém. Théor. Nombres Bordeaux (2) **2**, 333–348 (1990)
4072. Touibi, C., Zargouni, H.S.: Sur le théorème des idéaux premiers. Colloq. Math. **57**, 157–172 (1989)
4073. Towber, J.: Composition of oriented binary quadratic form-classes over commutative rings. Adv. Math. **36**, 1–107 (1980)
4074. Travesa, A.: Generating functions for the numbers of abelian extensions of a local field. Proc. Amer. Math. Soc. **108**, 331–339 (1990)
4075. Trojovský, P.: On divisibility of the class number h^+ of the real cyclotomic fields $Q(\zeta_p + \zeta_p^{-1})$ by primes $q<10000$. Math. Slovaca **50**, 541–555 (2000)

4076. Tschakaloff, L.: Unmöglichkeitsbeweis der Gleichung $\alpha^5 + \beta^5 = \eta\gamma^5$ im quadratischen Körper $K(\sqrt{5})$. Tôhoku Math. J. **27**, 189–194 (1926) Tschebotaröw, N., see Čebotarev, N.G.
4077. Tsumura, H.: On Demjanenko's matrix and Maillet's determinant for imaginary abelian number fields. J. Number Theory **60**, 70–79 (1996)
4078. Tunnell, J.[B.]: Artin's conjecture for representations of octahedral type. Bull. Amer. Math. Soc. (N.S.) **5**, 173–175 (1981)
4079. Tunnell, J.B.: A classical Diophantine problem and modular forms of weight 3/2. Invent. Math. **72**, 323–334 (1983)
4080. Turán, P.: On the remainder-term of the prime-number formula, II. Acta Math. Hungar. **1**, 155–166 (1950)
4081. Turán, P.: Eine neue Methode in der Analysis und deren Anwendungen. Akadémiai Kiadó (1953)
4082. Turán, P.: On a new method of analysis and its applications. J. Wiley (1984)
4083. Turán, P.: Commemoration on Stanisław Knapowski. Colloq. Math. **23**, 310–318 (1971)
4084. Turkstra, H.: Metrische bijdragen tot de theorie der diophantische approximaties in het lichaam der P-adische getallen, Dissertation, Amsterdam Univ. (1938)
4085. Tzermias, P.: Algebraic points of low degree on the Fermat curve of degree seven. Manuscripta Math. **97**, 483–488 (1998)
4086. Tzermias, P.: Low-degree points on Hurwitz-Klein curves. Trans. Amer. Math. Soc. **356**, 939–951 (2004)
4087. Tzermias, P.: Improved bounds on the number of low-degree points on certain curves. Acta Arith. **117**, 277–282 (2005)
4088. Uchida, K.: Unramified extensions of quadratic number fields, I. Tôhoku Math. J. (2) **22**, 138–141 (1970)
4089. Uchida, K.: Unramified extensions of quadratic number fields, II. Tôhoku Math. J. (2) **22**, 220–224 (1970)
4090. Uchida, K.: Class numbers of imaginary abelian number fields, I. Tôhoku Math. J. (2) **23**, 97–104 (1971)
4091. Uchida, K.: Class numbers of imaginary abelian number fields, II. Tôhoku Math. J. (2) **23**, 335–348 (1971)
4092. Uchida, K.: Class numbers of imaginary abelian number fields, III. Tôhoku Math. J. (2) **23**, 573–580 (1971)
4093. Uchida, K.: Imaginary abelian number fields with class number one. Tôhoku Math. J. (2) **24**, 487–499 (1972)
4094. Uchida, K.: Relative class numbers of normal CM-fields. Tôhoku Math. J. **25**, 347–353 (1973)
4095. Uchida, K.: Class numbers of cubic cyclic fields. J. Math. Soc. Japan **26**, 447–453 (1974)
4096. Uchida, K.: On Artin L-functions. Tôhoku Math. J. (2) **27**, 75–81 (1975)
4097. Uchida, K.: When is $Z[a]$ the ring of integers? Osaka J. Math. **14**, 155–157 (1977)
4098. Uchida, K.: Imaginary abelian number fields of degrees 2^m with class number one. In: Proc. International Conference on Class Numbers and Fundamental Units of Algebraic Number Fields (Katata 1986), 151–170, Nagoya Univ. (1986)
4099. Uehara, T.: On class numbers of imaginary quadratic and quartic fields. Arch. Math. (Basel) **41**, 256–260 (1983)
4100. Uehara, T.: Construction of certain real quadratic fields. Proc. Japan Acad. Sci. **59**, 390–392 (1983)
4101. Ulam, S.: John von Neumann, 1903–1957. Bull. Amer. Math. Soc. **64**, 1–49 (1958)
4102. Upadhyaya, P.O.: Cyclotomic quinque-section for every prime of the form $10n + 1$ between 100 and 500. Proc. London Math. Soc. (2) **20**, 494–496 (1922)
4103. Urazbaev, B.M.: On an asymptotical formula in algebra. Dokl. Akad. Nauk SSSR **95**, 935–938 (1954) (Russian)
4104. Urazbaev, B.M.: Asymptotical formula for the growth of the number of abelian fields of degree l^2. Dokl. Akad. Nauk SSSR **95**, 1145–1147 (1954) (Russian)

4105. Urazbaev, B.M.: Asymptotical formula for the growth of the number of abelian fields of type $(l, l, \ldots, l)$, Dokl. Akad. Nauk SSSR **105**, 659–661 (1955) (Russian)
4106. Urazbaev, B.M.: On the growth of the number of completely critical cyclic fields of degree l^h. Dokl. Akad. Nauk SSSR **113**, 1222–1223 (1957) (Russian)
4107. Urazbaev, B.M.: Asymptotical formulas in algebra. Kazah. Ped. Inst. Alma-Ata (1972) (Russian)
4108. Urazbaev, B.M.: The distribution of cyclic fields. Izv. Akad. Nauk Kazah. SSR. **1977(5)**, 66–70 (1977) (Russian)
4109. Urbanowicz, J., Williams, K.S.: Congruences for L-functions. Kluwer (2000)
4110. Urbjalis, I.: Distribution of the primes of the real quadratic field $K(\sqrt{2})$. Lit. Mat. Sb. **4**, 409–427 (1964) (Russian)
4111. Urbjalis, I.: Distribution of algebraic primes. Lit. Mat. Sb. **5**, 504–516 (1965) (Russian)
4112. Urbjalis, I.: Distribution of primes for a totally real field of algebraic numbers. Lit. Mat. Sb. **5**, 307–324 (1965) (Russian)
4113. [Uspensky, J.V.] Uspenskiĭ, J.: A remark on integral numbers depending on roots of unity of degree 5. Mat. Sb. **26**, 1–17 (1906) (Russian)
4114. [Uspensky, J.V.] Ouspensky, J.: Note sur les nombres entiers dépendant d'une racine cinquiéme de l'unité. Math. Ann. **66**, 109–112 (1909)
4115. [Uspensky, J.V.] Uspenskiĭ, J.: Arithmetical proof of Kronecker relations between the class-numbers of binary quadratic forms. Mat. Sb. **29**, 26–52 (1913) (Russian)
4116. [Uspensky, J.V.] Uspenskiĭ, J.: Sur les relations entre les nombres des classes des formes quadratiques binaires positives, 1_1, $1-2$. Izv. Akad. Nauk SSSR, Ser. Mat. (6) **19**, 599–620, 763–784 (1925)
4117. [Uspensky, J.V.] Uspenskiĭ, J.: Sur les relations entre les nombres des classes des formes quadratiques binaires positives, 2_1, 2_2, 3,4,5. Izv. Akad. Nauk SSSR, Ser. Mat. (6) **20**, 25–38, 175–196, 327–348, 547–564, 619–642 (1926)
4118. Uspensky, J.V.: A method of finding units in cubic orders of a negative discriminant. Trans. Amer. Math. Soc. **33**, 1–22 (1931)
4119. Vahlen, T.: Ueber reductible Binome. Acta Math. **19**, 195–198 (1895)
4120. de la Vallée-Poussin, C.J.: Recherches analytiques sur la théorie des nombres premiers, II, Ann. Soc. Sci. Bruxelles **20**, 281–362 (1896)
4121. de la Vallée-Poussin, C.J.: Recherches analytiques sur la théorie des nombres premiers, III, Ann. Soc. Sci. Bruxelles **20**, 363–397 (1896)
4122. de la Vallée-Poussin, C.J.: Recherches analytiques sur la théorie des nombres premiers, IV, Ann. Soc. Sci. Bruxelles **21**, 251–342 (1897)
4123. Valson, C.A.: La vie et les travaux du baron Cauchy. **1–2**. Gauthiers-Villars (1868)
4124. van der Kallen, W., Cohen, A., Strooker, J.R., Looijenga, E., Murre, J.: In memoriam Tonny Albert Springer: Dat is goede wiskunde!. Nieuw Arch. Wiskd. (5) **12**, 235–242 (2012)
4125. van der Linden, F.J.: Class number computations of real abelian number fields. Math. Comp. **39**, 693–707 (1982)
4126. van der Linden, F.J.: Euclidean rings with two infinite primes. Dissertation, Univ. Amsterdam (1984)
4127. van der Linden, F.J.: Euclidean rings of integers of fourth degree fields. Lecture Notes Math. **1068**, 139–148 (1984)
4128. van der Poorten, A.J.: Obituary of Kurt Mahler. J. Aus. Math. Soc. **A 51**, 343–365 (1991)
4129. van der Poorten, A.J., Schlickewei, H.P.: The growth conditions for recurrence sequences. Macquarie Math. Reports 82–0041 (1982)
4130. van der Poorten, A.J., Schlickewei, H.P.: Zeros of recurrence sequences. Bull. Aust. Math. Soc. **44**, 215–223 (1991)
4131. van der Poorten, A.J., te Riele, H.J.J., Williams, H.C.: Computer verification of the Ankeny-Artin-Chowla conjecture for all primes less than 100 000 000 000. Math. Comp. **70**, 1311–1328 (2001); Corr.: **72**, 521–523 (2003)
4132. van der Waall, R.W., Sato, K.: On a problem of R. Brauer for quotients of Dedekind zeta-functions. Indag. Math. (N.S.) **4**, 99–109 (1993)

4133. van der Waerden, B.L.: Ein logarithmenfreier Beweis des Dirichletschen Einheitensatzes. Abh. Math. Semin. Univ. Hambg **6**, 259–262 (1928)
4134. van der Waerden, B.L.: Moderne Algebra. **1–2**, Springer (1930–1931) [8th ed. Algebra, Springer (1967); English translations: Frederick Ungar (1970), Springer (1991)]
4135. van der Waerden, B.L.: Aufgabe 144. Jahresber. Deutsch. Math.-Verein. **42**, p. 71 (1932)
4136. van der Waerden, B.L.: Die Seltenheit der Gleichungen mit Affekt. Math. Ann. **109**, 13–16 (1934)
4137. van der Waerden, B.L.: Nachruf auf Emmy Noether. Math. Ann. **111**, 469–476 (1935)
4138. van der Waerden, B.L.: Die Zerlegungs- und Trägheitsgruppe als Permutationsgruppen. Math. Ann. **111**, 731–733 (1935)
4139. van der Waerden, B.L.: Die Seltenheit der reduziblen Gleichungen und der Gleichungen mit Affekt. Monatsh. Math. Phys. **43**, 133–147 (1936)
4140. van der Waall, R.W.: On a conjecture of Dedekind on zeta-functions. Indag. Math. **37**, 83–86 (1975)
4141. Vandiver, H.S.: The generalized Lagrange indeterminate congruence for a composite ideal modulus. Ann. Math. (2) **18**, 115–119 (1917)
4142. Vandiver, H.S.: Proof of a property of the norm of a cyclotomic integer. Bull. Amer. Math. Soc. **25**, 221–223 (1919)
4143. Vandiver, H.S.: On the first factor of the class number of a cyclotomic field. Bull. Amer. Math. Soc. **25**, 458–461 (1919)
4144. Vandiver, H.S.: A property of cyclotomic integers and its relation to Fermat's last theorem. Ann. Math. (2) **21**, 73–80 (1920)
4145. Vandiver, H.S.: On Kummer's memoir of 1857 concerning Fermat's last theorem. Proc. Nat. Acad. Sci. U.S.A. **6**, 266–269 (1920)
4146. Vandiver, H.S.: On the class number of the field $\Omega(e^{2i\pi/p^n})$ and the second case of Fermat's last theorem. Proc. Nat. Acad. Sci. U.S.A. **6**, 416–421 (1920)
4147. Vandiver, H.S.: Note on some results concerning Fermat's last theorem. Bull. Amer. Math. Soc. **28**, 258–260 (1922)
4148. Vandiver, H.S.: On Kummer's memoir of 1857 concerning Fermat's last theorem. Bull. Amer. Math. Soc. **28**, 400–407 (1922)
4149. Vandiver, H.S.: A new type of criteria for the first case of Fermat's last theorem. Ann. Math. (2) **26**, 88–94 (1924)
4150. Vandiver, H.S.: On the power characters of units in a cyclotomic field. Amer. J. Math. **47**, 140–147 (1925)
4151. Vandiver, H.S.: Laws of reciprocity and the first case of Fermat's last theorem. Proc. Nat. Acad. Sci. U.S.A. **11**, 292–298 (1925)
4152. Vandiver, H.S.: A property of cyclotomic integers and its relation to Fermat's last theorem. II. Ann. Math. (2) **26**, 217–232 (1925)
4153. Vandiver, H.S.: Summary of results and proofs concerning Fermat's last theorem. Proc. Nat. Acad. Sci. U.S.A. **12**, 106–109 (1926)
4154. Vandiver, H.S.: Transformation of the Kummer criteria in connection with Fermat's last theorem. Ann. Math. (2) **27**, 171–176 (1926)
4155. Vandiver, H.S.: Application of the theory of relative cyclic fields to both cases of Fermat's last theorem. Trans. Amer. Math. Soc. **28**, 554–560 (1926)
4156. Vandiver, H.S.: Summary of results and proofs concerning Fermat's last theorem. II. Proc. Nat. Acad. Sci. U.S.A. **12**, 767–772 (1926)
4157. Vandiver, H.S.: Transformation of the Kummer criteria in connection with Fermat's last theorem. II. Ann. Math. (2) **28**, 451–458 (1927)
4158. Vandiver, H.S.: Application of the theory of relative cyclic fields to both cases of Fermat's last theorem, II. Trans. Amer. Math. Soc. **29**, 154–162 (1927)
4159. Vandiver,H.: On Fermat's last theorem. Trans. Amer. Math. Soc. **31**, 613–642 (1929)
4160. Vandiver,H.: On the first case of Fermat's last theorem. Ann. Math. (2) **30**, 552–558 (1929)
4161. Vandiver, H.: An algorithm for transforming Kummer criteria in connection with Fermat's last theorem. Ann. Math. (2) **30**, 559–577 (1929)

4162. Vandiver, H.S.: Summary of results and proofs concerning Fermat's last theorem. III. Proc. Nat. Acad. Sci. U.S.A. **15**, 43–48 (1929)
4163. Vandiver, H.S.: Summary of results and proofs concerning Fermat's last theorem, IV. Proc. Nat. Acad. Sci. U.S.A. **15**, 108–109 (1929)
4164. Vandiver, H.S.: Some theorems concerning properly irregular cyclotomic fields. Proc. Nat. Acad. Sci. U.S.A. **15**, 202–207 (1929)
4165. Vandiver, H.S.: Summary of results and proofs concerning Fermat's last theorem, V. Proc. Nat. Acad. Sci. U.S.A. **16**, 298–304 (1930)
4166. Vandiver, H.S.: On the second factor of the class number of a cyclotomic field. Proc. Nat. Acad. Sci. U.S.A. **16**, 743–749 (1930)
4167. Vandiver, H.S.: On power characters of singular integers in a properly irregular cyclotomic field. Trans. Amer. Math. Soc. **32**, 391–408 (1930)
4168. Vandiver, H.S.: Summary of results and proofs concerning Fermat's last theorem. VI. Proc. Nat. Acad. Sci. U.S.A. **17**, 661–673 (1931)
4169. Vandiver, H.S.: On the method of infinite descent in connection with Fermat's last theorem for regular exponents. Comment. Math. Helv. **4**, 1–8 (1932)
4170. Vandiver, H.S.: On power characters in cyclotomic fields. Bull. Amer. Math. Soc. **40**, 111–117 (1934)
4171. Vandiver, H.S.: Fermat's last theorem and the second factor in the cyclotomic class number. Bull. Amer. Math. Soc. **40**, 118–126 (1934)
4172. Vandiver, H.S.: A note on units in super-cyclic fields. Bull. Amer. Math. Soc. **40**, 855–858 (1934)
4173. Vandiver, H.S.: Constructive derivation of the decomposition-field of a polynomial. Ann. Math. (2) **37**, 1–6 (1936)
4174. Vandiver, H.S.: On the ordering of real algebraic numbers by constructive methods. Ann. Math. (2) **38**, 7–16 (1936)
4175. Vandiver, H.S.: On Bernoulli's numbers and Fermat's last theorem. Duke Math. J. **3**, 569–584 (1937)
4176. Vandiver, H.S.: On Bernoulli's numbers and Fermat's last theorem, II. Duke Math. J. **5**, 418–427 (1939)
4177. Vandiver, H.S.: On the composition of the group of ideal classes in a properly irregular cyclotomic field. Monatsh. Math. Phys. **48**, 369–380 (1939)
4178. Vandiver, H.S.: On basis systems for groups of ideal classes in a properly irregular cyclotomic field. Proc. Nat. Acad. Sci. U.S.A. **25**, 586–591 (1939)
4179. Vandiver, H.S.: Note on Euler number criteria for the first case of Fermat's last theorem. Amer. J. Math. **62**, 79–82 (1940)
4180. Vandiver, H.S.: On improperly irregular cyclotomic fields. Proc. Nat. Acad. Sci. U.S.A. **27**, 77–83 (1941)
4181. Vandiver, H.S.: Fermat's last theorem. Its history and the nature of the known results concerning it. Amer. Math. Monthly **53**, 555–578 (1946)
4182. Vandiver, H.S.: Les travaux mathématiques de Dmitry Mirimanoff. Enseign. Math. **39**, 169–179 (1942/50)
4183. Vandiver, H.S.: Examination of methods of attack on the second case of Fermat's last theorem. Proc. Nat. Acad. Sci. U.S.A. **40**, 732–735 (1954)
4184. Vandiver, H.S.: Is there an infinity of regular primes?. Scripta Math. **21**, 306–309 (1955)
4185. Vandiver, H.S.: On the desirability of publishing classified bibliographies of the mathematics literature. Amer. Math. Monthly **67**, 47–50 (1960)
4186. Vandiver, H.S.: Some aspects of the Fermat problem, I. Proc. Nat. Acad. Sci. U.S.A. **47**, 202–209 (1961)
4187. Vandiver, H.S.: Some aspects of the Fermat problem, II. Proc. Nat. Acad. Sci. U.S.A. **47**, 585–590 (1961)
4188. Vandiver, H.S.: Some aspects of the Fermat problem, III. Proc. Nat. Acad. Sci. U.S.A. **47**, 1831–1838 (1961)

4189. Vandiver, H.S., Wahlin, G.E.: Algebraic numbers. Report of the Committee on algebraic numbers. **2**. Bull. National Research Council **62**, 1–111 (1928) [Reprint: Chelsea, 1967.]
4190. Varadarajan, V.S.: Euler through time: a new look at old themes. Amer. Math. Soc. (2006)
4191. Värmon, J.: Über Abelsche Körper, deren alle Gruppeninvarianten aus einer Primzahl l bestehen, und über Abelsche Körper als Kreiskörper, Dissertation, Univ. Lund (1925)
4192. Värmon, J.: Über die Klassenzahl Abelscher Körper. Ark. Mat. **22(13)**, 1–47 (1930)
4193. Varnavides, P.: The Euclidean real quadratic fields. Indag. Math. **14**, 111–122 (1952)
4194. Vaserstein, L.N.: Waring's problem for commutative rings. J. Number Theory **26**, 299–307 (1987)
4195. Vaserstein, L.N.: Waring's problem for algebras over fields. J. Number Theory **26**, 286–298 (1987)
4196. Vassiliou, P., Bestimmung der Führer der Verzweigungskörper relativabelscher Zahlkörper. Beweis der Produktformel für den Führer-Diskriminanten-Satz. J. Reine Angew. Math. **169**, 131–139 (1933)
4197. Vassiliou, P.: Über affektlose Gleichungen. J. Reine Angew. Math. **176**, 45–48 (1936)
4198. Vassiliou, P.: Über die Galoissche Gruppe einer Klasse von trinomischen Gleichungen. Math. Ann. **117**, 448–452 (1940)
4199. Vaughan, R.C.: Coefficients of cyclotomic polynomials and related topics. In: Proceedings of the Congress on Number Theory (Zarauz, 1984), 43–68, Univ. País Vasco (1989)
4200. Vaughan, R.C., Wooley, T.D.: Waring's problem: a survey. In: Number Theory for the Millennium (Urbana, 2000), 3, 301–340. A.K. Peters (2000)
4201. Vélez, W.Y.: Prime ideal decomposition in $F(\mu^{1/m})$, II. In: Number Theory and Algebra, 331–338. Academic Press (1977)
4202. Vélez, W.Y.: Prime ideal decomposition in $F(\mu^{1/p})$. Pacific J. Math. **75**, 589–600 (1978)
4203. Vélez, W.Y.: The factorization of p in $Q(a^{1/p^k})$ and the genus field of $Q(a^{1/n})$. Tokyo J. Math. **11**, 1–19 (1988)
4204. Velmin, V.P.: On the quadratic reciprocity law in an arbitrary quadratic field, Izv. Varš. Univ. **5(1)**, 1–139 (1914) (Russian)
4205. [Velmin, V.P.] Welmin, W.: Das quadratische Reziprozitätsgesetz im beliebigen quadratischen Zahlkörper. J. Reine Angew. Math. **149**, 147–173 (1919)
4206. Venkatachaliengar, K.: The method of finding the class-number and the structure of the class group of any algebraic field. Proc. Indian Acad. Sci., Math. Sci. **1**, 446–450 (1935)
4207. Venkov, B.A., On the number of classes of binary quadratic forms with negative determinants. Izv. Akad. Nauk SSSR, Ser. Mat. (7) **1928** 375–392 (Russian)
4208. Venkov, B.A., On the number of classes of binary quadratic forms with negative determinants. Izv. Akad. Nauk SSSR, Ser. Mat. (7) **1928** 455–480 (Russian)
4209. Venkov, B.A. [Wenkov, B.]: Über die Klassenanzahl positiver binärer quadratischer Formen. Math. Z. **33**, 350–374 (1931)
4210. Venkov, B.B., Koch, H.: The p-tower of class fields for an imaginary quadratic field. Zap. Naučn. Sem. LOMI **46**, 5–13 (1974) (Russian)
4211. Vennekohl, H.: Neuer Beweis für die explizite Reziprozitätsformel der l-ten Potenzreste im l-ten Kreiskörper. Math. Ann. **107**, 233–251 (1932)
4212. Vijayaraghavan, T.: On the fractional parts of powers of numbers. II. Proc. Cambridge Philos. Soc. **37**, 349–357 (1941)
4213. Vijayaraghavan, T.: On the fractional parts of powers of numbers. III. J. Lond. Math. Soc. (N.S.) **17**, 137–138 (1942)
4214. Vila, N.: On central extensions of A_n as Galois group over Q. Arch. Math. (Basel) **44**, 424–437 (1985)
4215. Vila, N.: On stem extensions of S_n as Galois group over number fields. J. Algebra **116**, 251–260 (1988)
4216. Vinberg, E.B.: On the theorem concerning the infinite-dimensionality of an associative algebra. Izv. Akad. Nauk SSSR, Ser. Mat. **29**, 209–214 (1932) (Russian)
4217. Vinogradov, A.I.: On the class number. Dokl. Akad. Nauk SSSR **146**, 274–276 (1962) (Russian)

4218. Vinogradov, A.I.: On the number of ideal classes and the group of divisor classes. Izv. Akad. Nauk SSSR, Ser. Mat. **27**, 1963, 561–576 (1963) (Russian)
4219. Vinogradov, A.I.: On Siegel's zeros. Dokl. Akad. Nauk SSSR **151**, 479–481 (1963) (Russian)
4220. Vinogradov, A.I.: The sieve method in algebraic fields. Lower bounds. Mat. Sb. **64**, 52–78 (1964) (Russian)
4221. Vinogradov, A.I. et al.: Mark Borisovič Barban: Obituary. Uspekhi Mat. Nauk **24(20)**, 213–216 (1969) (Russian)
4222. Vinogradov, A.I., Tahtadžyan, L.A.: On analogues of the Gauss-Vinogradov formula. Dokl. Akad. Nauk SSSR **254**, 1298–1301 (1980) (Russian)
4223. Vinogradov, A.I., Tahtadžyan, L.A.: Analogues of the Vinogradov-Gauss formula in the critical strip. Trudy Mat. Inst. im. Steklova **156**, 45–68 (1981) (Russian)
4224. Vinogradov, I.M.: A new method to find asymptotical expression for arithmetical functions. Izv. Ross. Akad. Nauk, Ser. Mat. (6) **11**, 1347–1378 (1917) (Russian)
4225. Vinogradov, I.M.: On the mean value of the class-number of primitive forms of negative discriminant, Comm. Soc. Math. Charkov (2) **16**, 10–38 (1918) (Russian)
4226. Vinogradov, I.M.: On an asymptotical equality of the theory of quadratic forms. Žurnal Fiz.-Mat. Obšč. Univ. Perm **1**, 18–28 (1918) (Russian)
4227. Vinogradov, I.M.: Sur la distribution des résidus et des nonrésidus des puissances. Žurnal Fiz.-Mat. Obšč. Univ. Perm **1**, 94–98 (1918)
4228. Vinogradov, I.M.: Nouvelles évaluations des sommes de Weyl. Dokl. Akad. Nauk SSSR **8**, 195–198 (1935)
4229. Vinogradov, I.M.: Improvement of the remainder term in an asymptotic formula. Izv. Akad. Nauk SSSR, Ser. Mat. **13**, 97–110 (1949) (Russian)
4230. Vinogradov, I.M.: Improvement of asymptotic formulas for the number of lattice points in a three-dimensional region. Izv. Akad. Nauk SSSR, Ser. Mat. **19**, 3–10 (1955) (Russian)
4231. Vinogradov, I.M.: A new estimate of the function $\zeta(1+it)$. Izv. Akad. Nauk SSSR, Ser. Mat. **22**, 161–164 (1958) (Russian)
4232. Vinogradov, I.M.: On the number of integral points in a given domain. Izv. Akad. Nauk SSSR, Ser. Mat. **24**, 777–786 (1960) (Russian)
4233. Vinogradov, I.M.: On the number of integral points in a three-dimensional domain. Izv. Akad. Nauk SSSR, Ser. Mat. **27**, 3–8 (1963) (Russian)
4234. Vinogradov, I.M.: On the number of integral points in a sphere. Izv. Akad. Nauk SSSR, Ser. Mat. **27**, 957–968 (1963) (Russian)
4235. Vlăduţ, S.G.: Kronecker's Jugendtraum and Modular Functions. Gordon and Breach (1991)
4236. Vogt, H.: Jules Molk, 8 décembre 1857–7 mai 1914. Enseign. Math. **16**, 380–383 (1914)
4237. Vojta, P.: Siegel's theorem in the compact case. Ann. Math. (2) **133**, 509–548 (1991)
4238. Völklein, H.: Groups as Galois groups. Cambridge University Press (1996)
4239. Völklein, H.: Rigid generators of classical groups. Math. Ann. **311**, 421–438 (1998)
4240. Voloch, J.F.: Chebyshev's method for number fields. J. Théor. Nombres Bordeaux **12**, 81–85 (2000)
4241. von Neumann, J.: Zur Prüferschen Theorie der idealen Zahlen. Acta Litt. Sci. Reg. Univ. Hungar. Franc.-Jos., sect. Math. **2**, 193–227 (1924/26)
4242. Voronina, M.M.: Gabriel Lamé, 1795–1870. Nauka (1987) (Russian)
4243. Voronoĭ, G.F.: On algebraic integers which depend on a root of an equation of third degree. St. Petersburg (1894) (Russian)
4244. Voronoĭ, G.F.: On a generalization of the continued fraction algorithm. Dissertation, Varšavskiĭ Univ. (1896) (Russian)
4245. Voronoĭ, G.[F.]: Sur une propriété du discriminant des fonctions entiéres. In: Verhandlungen des dritten Internationalen Mathematiker-Kongresses in Heidelberg vom 8. bis 13. August 1904. 186–189, Teubner (1905)
4246. Voskresenskiĭ, V.E.: On the question of the structure of the subfield of invariants of a cyclic group of automorphisms of the field $Q(x_1, \ldots, x_n)$. Izv. Akad. Nauk SSSR, Ser. Mat. **34**, 366–375 (1970) (Russian)

4247. Voskresenskiĭ, V.E.: Rationality of certain algebraic tori. Izv. Akad. Nauk SSSR, Ser. Mat. **35**, 1037–1046 (1971) (Russian)
4248. Voskresenskiĭ, V.E.: Fields of invariants of abelian groups. Uspekhi Mat. Nauk **28(4)**, 77–102 (1973) (Russian)
4249. Voss, A.: Heinrich Weber. Jahresber. Deutsch. Math.-Verein. **23**, 431–444 (1914)
4250. Vostokov, S.V.: An explicit form of the reciprocity law. Izv. Akad. Nauk SSSR, Ser. Mat. **42**, 1288–1321 (1978) (Russian)
4251. Voutier, P.: An effective lower bound for the height of algebraic numbers. Acta Arith. **74**, 81–95 (1996)
4252. Vulakh, L.Ya.: Diophantine approximation on Bianchi groups. J. Number Theory **54**, 73–80 (1995)
4253. Vulakh, L.Ya.: Farey polytopes and continued fractions associated with discrete hyperbolic groups. Trans. Amer. Math. Soc. **351**, 2295–2323 (1999)
4254. Vulakh, L.Ya.: Diophantine approximation in $Q(\sqrt{-5})$ and $Q(\sqrt{-6})$. Internat. J. Number Th. **2**, 25–48 (2006)
4255. Vulakh, L.Ya.: The Markov spectra for cocompact Fuchsian groups. Internat. J. Number Th. **5**. 679–718 (2009)
4256. Vulakh, L.Ya.: Diophantine approximation in imaginary quadratic fields. Internat. J. Number Th. **6**, 731–766 (2010)
4257. Vulakh, L.Ya.: Diophantine approximation in $\mathbf{Q}(\sqrt{-30})$, $\mathbf{Q}(\sqrt{-33})$ and $\mathbf{Q}(\sqrt{-57})$. Funct. Approx. Comment. Math. **47**, 183–205 (2012)
4258. Wada, H.: On cubic Galois extensions of $Q(\sqrt{-3})$. Proc. Japan Acad. Sci. **46**, 397–400 (1970)
4259. Wada, H.: A table of ideal class groups of imaginary quadratic fields. Proc. Japan Acad. Sci. **46**, 401–403 (1970)
4260. Wagner, C.: Class numbers 5, 6 and 7. Math. Comp. **65**, 785–800 (1996)
4261. Wagner, G.B.: Ideal matrices and ideal vectors. Math. Ann. **183**, 241–249 (1969)
4262. Wagstaff, S.S.Jr.: The irregular primes to 125 000. Math. Comp. **32**, 583–591 (1978)
4263. Wahlin, G.E.: On the base of a relative number-field, with an application to the composition of fields. Trans. Amer. Math. Soc. **11**, 487–493 (1910)
4264. Wahlin, G.E.: The equation $x^l - A = 0\,(p)$. J. Reine Angew. Math. **145**, 114–136 (1914)
4265. Wahlin, G.E.: A new development of the theory of algebraic numbers. Trans. Amer. Math. Soc. **16**, 502–508 (1915)
4266. Wahlin, G.E.: On the principal units of an algebraic domain $k(\mathfrak{p}, \alpha)$. Amer. Math. Monthly **23**, 450–455 (1917)
4267. Wahlin, G.E.: The factorization of the rational primes in a cubic domain. Amer. J. Math. **44**, 191–203 (1922)
4268. Wahlin, G.E.: On the application of the theory of ideals to diophantine analysis. Bull. Amer. Math. Soc. **30**, 140–154 (1924)
4269. Wahlin, G.E.: The number e in $k(p)$. J. Reine Angew. Math. **154**, 110–113 (1925)
4270. Wahlin, G.E.: On the solution of diophantine equations by means of ideals. Bull. Amer. Math. Soc. **31**, 430–444 (1925)
4271. Wahlin, G.E.: The multiplicative representation of the principal units of a relative cyclic field. J. Reine Angew. Math. **167**, 122–128 (1932)
4272. Walfisz, A.: Über die summatorischen Funktionen einiger Dirichletscher Reihen. Dissertation, Univ. Göttingen (1922)
4273. Walfisz, A.: Über das Piltzsche Teilerproblem in algebraischen Zahlkörpern. Math. Z. **22**, 153–188 (1925)
4274. Walfisz, A.: Über das Piltzsche Teilerproblem in algebraischen Zahlkörpern, II. Math. Z. **26**, 487–494 (1927)
4275. Walfisz, A.: Beiträge zur Theorie der Dedekindschen Zetafunktion, I: Abschätzung von $\zeta_{\mathfrak{K}}(1+it)$. Math. Ann. **97**, 624–634 (1927)
4276. Walfisz, A.: Wertevorrat der Dedekindschen Zetafunktion um $\sigma = 1$. Jahresber. Deutsch. Math.-Verein. **42**, 62–68 (1932)

4277. Walfisz, A.: Zur additiven Zahlentheorie, II. Math. Z. **40**, 592–607 (1936)
4278. Walfisz, A.: On the class-number of binary quadratic forms, Tr. Mat. Inst. Gruz. SSR **11**, 57–72, 173–186 (1942)
4279. Waldschmidt, M.: Les huit premiers travaux de Pierre Liardet. Unif. Distrib. Th. **11**, 169–177 (2016)
4280. Wallace, D.I.: Conjugacy classes of hyperbolic matrices in $Sl(n, Z)$ and ideal classes in an order. Trans. Amer. Math. Soc. **283**, 177–184 (1984)
4281. Walter, C.D.: A class number relation in Frobenius extensions of number fields. Mathematika **24**, 216–225 (1977)
4282. Walter, C.D.: Brauer's class number relation. Acta Arith. **35**, 33–40 (1979)
4283. Walter, C.D.: Kuroda's class number relation. Acta Arith. **35**, 41–51 (1979)
4284. Walter, W.: Das wissenschaftliche Werk von Erich Kamke. Jahresber. Deutsch. Math.-Verein. **69**, 193–205 (1967/68)
4285. Wang, K.: On Maillet determinant. J. Number Theory **18**, 306–312 (1984)
4286. Wang, L.X.: p-adic continued fractions, I. Sci. Sinica **A 28**, 1009–1017 (1985)
4287. Wang, L.X.: p-adic continued fractions, II. Sci. Sinica **A 28**, 1018–1023 (1985)
4288. Wang, S.: A counter-example to Grunwald's theorem. Ann. Math. (2) **49**, 1008–1009 (1948)
4289. Wang, S.: On Grunwald's theorem. Ann. Math. (2) **51**, 471–484 (1950)
4290. Wang, S.: An existence theorem for abelian extension over algebraic number fields, Sci. Record **3**, 25–27 (1950)
4291. Wang, Song: An effective version of the Grunwald-Wang theorem. Dissertation, Caltech (2002)
4292. Wang, T.-Z., Gong, K.: On the least primitive root in number fields. Sci. China Math. **53**, 2489–2500 (2010)
4293. Wang, Y.: Estimation and application of character sums, Shuxue Jinzhan **7**, 78–83 (1964) (Chinese) [English translation: Twelve papers in algebra. Amer. Math. Soc. Translations (2) **119**, 45–50 (1983)]
4294. Wang, Y.: Diophantine Equations and Inequalities in Algebraic Number Fields. Springer (1991)
4295. Wang, Y.: Hua Loo-Keng. Springer (1999)
4296. Wang, Y., Bauer, C.: The least primitive root in number fields. Acta Arith. **115**, 269–285 (2004)
4297. Wantzel, P.-L.: Recherches sur les moyens de reconnaître si un probléme de géométrie peut se résoudre avec la régle et le compas. J. Math. Pures Appl. **2**, 366–372 (1837)
4298. Wantzel, P.-L.: Note sur la théorie des nombres complexes à l'occasion de M.Lamé sur le théorème de Fermat. C. R. Acad. Sci. Paris **24**, 430–434 (1847)
4299. Ward, M.: On the factorization of polynomials to a prime modulus. Ann. Math. (2) **36**, 870–874 (1935)
4300. Warlimont, R.: Über die k-ten Mittelwerte der Klassenzahlen primitiven binären quadratischer Formen negativer Diskriminante. Monatsh. Math. **75**, 173–179 (1971)
4301. Washington, L.C.: The class number of the field of 5^nth roots of unity. Proc. Amer. Math. Soc. **61**, 205–208 (1976)
4302. Washington, L.C.: The non-p-part of the class number in a cyclotomic Z_p-extension. Invent. Math. **49**, 87–97 (1978)
4303. Washington, L.C.: Units of irregular cyclotomic fields. Illinois J. Math. **23**, 635–647 (1979)
4304. Washington, L.C.: Introduction to Cyclotomic Fields. Springer (1982), 2nd ed. (1997)
4305. Watabe, M.: An arithmetical application of elliptic functions to the theory of biquadratic residues. Abh. Math. Semin. Univ. Hambg **49**, 118–125 (1979)
4306. Waterhouse, W.C.: Pieces of eight in class groups of quadratic fields. J. Number Theory **5**, 95–97 (1973)
4307. Waterhouse, W.C.: The normal basis theorem. Amer. Math. Monthly **86**, p.212 (1979)
4308. Watkins, M.: Real zeros of real odd Dirichlet L-functions. Math. Comp. **73**, 415–423 (2004)
4309. Watkins, M.: Class numbers of imaginary quadratic fields. Math. Comp. **73**, 907–938 (2004)

4310. Weber, H.: Beweis des Satzes, dass jede eigentlich primitive quadratische Form unendlich viele Primzahlen darzustellen fähig ist. Math. Ann. **20**, 301–330 (1982)
4311. Weber, H.: Zur Theorie der elliptischen Functionen. Acta Math. **6**, 329–416 (1985)
4312. Weber, H.: Theorie der Abel'schen Zahlkörper. Acta Math. **8**, 193–263 (1986)
4313. Weber, H.: Theorie der Abel'schen Zahlkörper. Acta Math. **9**, 105–130 (1986/87)
4314. Weber, H.: Zur Theorie der elliptischen Functionen, II. Acta Math. **11**, 333–390 (1888)
4315. Weber, H.: Zur complexen Multiplication elliptischer Functionen. Math. Ann. **33**, 390–410 (1889)
4316. Weber, H.: Elliptische Functionen und algebraische Zahlen. Vieweg (1891), 2nd ed. (1908) (as the third volume of [4387])
4317. Weber, H.: Leopold Kronecker. Math. Ann. **43**, 1–25 (1893)
4318. Weber, H.: Die allgemeinen Grundlagen der Galois'schen Gleichungstheorie. Math. Ann. **43**, 521–549 (1893)
4319. Weber, H.: Zahlentheoretische Untersuchungen aus dem Gebiet der elliptischen Functionen, II. Nachr. Ges. Wiss. Göttingen **1893**, 138–153
4320. Weber, H.: Zahlentheoretische Untersuchungen aus dem Gebiet der elliptischen Functionen, II. Nachr. Ges. Wiss. Göttingen **1893**, 245–264
4321. Weber, H.: Lehrbuch der Algebra. **1–3**. Vieweg 1895–1908, 2nd ed. of volumes 1–2 (1898/99)
4322. Weber, H.: Über einen in der Zahlentheorie angewandten Satz der Integralrechnung. Nachr. Ges. Wiss. Göttingen **1896**, 275–281
4323. Weber, H.: Ueber Zahlengruppen in algebraischen Körpern. Math. Ann. **48**, 433–473 (1897)
4324. Weber, H.: Ueber Zahlengruppen in algebraischen Körpern, II. Math. Ann. **49**, 83–100 (1897)
4325. Weber, H.: Ueber Zahlengruppen in algebraischen Körpern, III, Anwendung an die complexen Multiplicationen und Theilung der elliptischen Functionen. Math. Ann. **50**, 1–26 (1898)
4326. Weber, H.: Complexe Multiplication. In: Encyklopädie der mathematischen Wissenschaften mit Einschluss ihrer Anwendungen, 1_2, 716–732. Teubner (1900)
4327. Weber, H.: Über komplexe Primzahlen in Linearformen. J. Reine Angew. Math. **129**, 35–62 (1905)
4328. Weber, H.: Über zyklische Zahlkörper. J. Reine Angew. Math. **132**, 167–188 (1907)
4329. Weber, H.: Zur Theorie der zyklischen Zahlkörper. Math. Ann. **67**, 32–60 (1909)
4330. Weber, H.: Zur Theorie der zyklischen Zahlkörper. II. Math. Ann. **70**, 459–470 (1911)
4331. Weber, H., Wellstein, J.: Der Minkowskische Satz über die Körperdiskriminante. Math. Ann. **73**, 275–285 (1913)
4332. Weber, H.: Über die Verteilung ganzer Zahlen mit ausgezeichneten Eigenschaften der Faktorzerlegung in algebraischen Zahlkörpern. Acta Arith. **44**, 215–239 (1984)
4333. Weber, W.: Bemerkungen zur arithmetischen Theorie der binären quadratischen Formen. Nachr. Ges. Wiss. Göttingen **1929**, 116–130
4334. Weber, W.: Idealtheoretische Deutung der Darstellbarkeit beliebiger natürlicher Zahlen durch quadratische Formen. Math. Ann. **102**, 740–767 (1930)
4335. Weber, W.: Umkehrbare Ideale. Math. Z. **34**, 131–157 (1931)
4336. Wedderburn, J.H.M.: On hypercomplex numbers. Proc. London Math. Soc. (2) **6**, 77–118 (1908)
4337. Wedderburn, J.H.M.: On division algebras. Trans. Amer. Math. Soc. **22**, 129–135 (1921)
4338. Wegner, U.: Ein Satz über auflösbare Polynome vom Primzahlgrad. Math. Ann. **105**, 256–261 (1931)
4339. Wegner, U.: Charakterisierung der binomischen Körper vom Primzahlgrad. Math. Ann. **105**, 262–266 (1931)
4340. Wegner, U.: Zur Theorie der auflösbaren Gleichungen von Primzahlgrad, I. J. Reine Angew. Math. **168**, 176–192 (1932)
4341. Wegner, U.: Ein Satz über die Zerlegung von Primzahlen bestimmter arithmetischer Progressionen in algebraischen Zahlkörpern. J. Reine Angew. Math. **168**, 231–232 (1932)

4342. Wegner, U.: Über trinomische Gleichungen von Primzahlgrad. Math. Ann. **111**, 734–737 (1935)
4343. Wegner, U.: Zur Theorie der affektlosen Gleichungen. Math. Ann. **111**, 738–742 (1935)
4344. Wegner, U.: Bestimmung eines auflösbaren Körpers von Primzahlgrad aus der Form seiner Diskriminante. J. Reine Angew. Math. **176**, 1–11 (1936)
4345. Wei, D.: On the sum of two integral squares in quadratic fields $\mathbf{Q}(\sqrt{\pm p})$. Acta Arith. **147**, 253–260 (2011)
4346. Wei, D.: On the sum of two integral squares in the imaginary quadratic field $Q(\sqrt{-2p})$. Sci. China Math. **57**, 49–60 (2014)
4347. Weierstrass, K.: Abhandlungen aus der Functionenlehre. Springer (1886)
4348. Weil, A.: L'arithmétique sur les courbes algébriques. Acta Math. **52**, 281–315 (1928) [[4425], **1**, 11–45.]
4349. Weil, A.: Zur algebraischen Theorie der algebraischen Funktionen. J. Reine Angew. Math. **179**, 129–133 (1938) [[4359], **1**, 227–231.]
4350. Weil, A.: Differentiation in algebraic number-fields. Bull. Amer. Math. Soc. **49**, p. 41 (1943) [[4359], **1**, p. 329.]
4351. Weil, A.: Sur les courbes algébriques et les variétés qui s'en déduisent, Publ. Inst. Math. Univ. Strasbourg **7**, 1–85 (1948) [[4359], **1**, 11–45.]
4352. Weil, A.: Arithmetic on algebraic varieties. Ann. Math. (2) **53**, 412–444 (1951) [[4359], **1**, 450–482.]
4353. Weil, A.: Sur certains groupes d'opérateurs unitaires. Acta Math. **111**, 143–211 (1964) [[4359], **3**, 1–69.]
4354. Weil, A.: Über die Bestimmung Dirichletscher Reihen durch Funktionalgleichungen. Math. Ann. **168**, 149–156 (1967) [[4359], 165–172.]
4355. Weil, A.: Basic Number Theory. Springer (1967)
4356. Weil, A.: Introduction. In: [2377], **1**, 1–14. [[4359], 3, 379–389.]
4357. Weil, A.: Review of [1088]. Bull. Amer. Math. Soc. **82**, 658–663 (1976)
4358. Weil, A.: Gauss et la composition des formes quadratiques binaires. In: Aspects of mathematics and its applications, 895–912. North-Holland (1986)
4359. Weil, A.: Oeuvres scientifiques **1–3**. Springer (1979) [Reprints: Springer (2009, 2014)]
4360. Weinberger, P.J.: On Euclidean rings of algebraic integers. Proc. Symposia Pure Math. **24**, 321–332 (1973)
4361. Weinberger, P.J.: Exponents of the class groups of complex quadratic fields. Acta Arith. **22**, 117–124 (1973)
4362. Weinberger, P.J.: Real quadratic fields with class numbers divisible by n. J. Number Theory **5**, 237–241 (1973)
4363. Weintraub, S.H.: Several proofs of the irreducibility of the cyclotomic polynomials. Amer. Math. Monthly **120**, 537–545 (2013)
4364. Weisner, L.: Quadratic fields in which cyclotomic polynomials are reducible. Ann. Math. (2) **29**, 377–381 (1927/28)
4365. Weiss, E.: Algebraic Number Theory. McGraw-Hill (1963) [Reprint: Chelsea (1976)]
4366. Weiss, M.J.: Fundamental systems of units in normal fields. Amer. J. Math. **58**, 249–254 (1936)
4367. Wendt, E.: Arithmetische Studien über den "letzten" Fermat'schen Satz, welcher aussagt, dass die Gleichung $a^n = b^n + c^n$ für $n>2$ in ganzen Zahlen nicht auflösbar ist. J. Reine Angew. Math. **113**, 335–347 (1894)
4368. Wendt, E.: Ueber die Zerlegbarkeit der Function $x^n - 1$ in einem beliebigen Körper. Math. Ann. **53**, 450–456 (1900)
4369. Western, A.E.: An extension of Eisenstein's law of reciprocity. Proc. London Math. Soc. (2) **6**, 2–6, 265–297 (1908)
4370. Western, A.E.: Some criteria for the residues of eighth and other powers. Proc. London Math. Soc. (2) **9**, 244–272 (1911)
4371. Western, A.E.: Allan Joseph Cunningham. J. Lond. Math. Soc. **3**, 317–318 (1928)

4372. Westlund, J.: On the congruence $x^{\Phi(P)} \equiv 1 \bmod P^n$. Bull. Amer. Math. Soc. **10**, 78–80 (1903)
4373. Westlund, J.: On the class number of the cyclotomic number field $k(e^{2\pi i/p^n})$. Trans. Amer. Math. Soc. **4**, 201–212 (1903)
4374. Westlund, J.: On the fundamental number of the algebraic number field $k(\sqrt[p]{m})$. Trans. Amer. Math. Soc. **11**, 388–392 (1910)
4375. Westlund, J.: Primitive roots of ideals in algebraic number fields. Math. Ann. **71**, 246–250 (1912)
4376. Westlund, J.: On the factorization of rational primes in cubic cyclotomic number fields. Jahresber. Deutsch. Math.-Verein. **22**, 135–140 (1913)
4377. Weyl, H.: Zur Abschätzung von $\zeta(1+it)$. Math. Z. **10**, 88–100 (1921)
4378. Weyl, H.: Algebraic Theory of Numbers. Princeton University Press (1940) [Reprint: Princeton University Press (1998); German translation: Bibliographisches Institut (1966)]
4379. Weyl, H.: David Hilbert and his mathematical work. Bull. Amer. Math. Soc. **50**, 612–654 (1944)
4380. Whaples, G.: Non-analytic class field theory and Grunwald's theorem. Duke Math. J. **9**, 455–473 (1942)
4381. Whaples, G.: On a conjecture about infinite class fields. Bull. Amer. Math. Soc. **53**, 377–380 (1947)
4382. Whaples, G.: Generalized local class field theory, I. Reciprocity law. Duke Math. J. **19**, 505–517 (1952)
4383. Whaples, G.: Existence of generalized local class fields. Proc. Nat. Acad. Sci. U.S.A. **39**, 1100–1103 (1953)
4384. Whaples, G.: Generalized local class field theory, II. Existence theorem. Duke Math. J. **21**, 247–255 (1954)
4385. Whaples, G.: Generalized local class field theory, III. Second form of existence theorem. Structure of analytic group. Duke Math. J. **21**, 757–581 (1954)
4386. Whaples, G.: Generalized local class field theory, IV. Cardinalities. Duke Math. J. **21**, 583–586 (1954)
4387. Whaples, G.: The generality of local class field theory (Generalized local class field theory. V.) Proc. Amer. Math. Soc. **8**, 137–140 (1957)
4388. Whaples, G.: Galois cohomology of additive polynomial and n-th power mappings of fields. Duke Math. J. **24**, 143–150 (1957)
4389. Whiteman, A.L.: Additive prime number theory in real quadratic fields. Duke Math. J. **7**, 208–232 (1940)
4390. Whiteman, A.L.: The sixteenth power residue character of 2. Canad. J. Math. **6**, 364–373 (1954)
4391. Whiteman, A.L.: The cyclotomic numbers of order sixteen. Trans. Amer. Math. Soc. **86**, 401–413 (1957)
4392. Whitford, E.E.: The Pell equation Dissertation. Columbia University (1912)
4393. Whitford, E.E.: Some solutions of the Pellian equations $x^2 - Ay^2 = \pm 4$. Ann. Math. (2) **15**, 157–160 (1915)
4394. Whittaker, E.T.: Obituary: George David Birkhoff. J. Lond. Math. Soc. **20**, 121–128 (1945)
4395. Whittaker, E.T., Watson, G.N.: A Course of Modern Analysis. 3rd ed. Cambridge University Press (1920) [4th ed. (1927); reprints: (1962, 1996)]
4396. Whyburn, W.M.: Raymond Joseph Garver — in memoriam. Bull. Amer. Math. Soc. **42**, p. 163 (1936)
4397. Wieferich, A.: Zum letzten Fermatschen Theorem. J. Reine Angew. Math. **136**, 293–302 (1909)
4398. Wielandt, H.: Hellmuth Kneser in memoriam. Aequationes Math. **11**, 120a–120c (1974)
4399. Wildanger, K.: Über das Lösen von Einheiten- und Indexformgleichungen in algebraischen Zahlkörpern. J. Number Theory **82**, 188–224 (2000)
4400. Wiles, A.: Modular elliptic curves and Fermat's last theorem, Ann. Math. (2) **141**, 443–551 (1995)

4401. Williams, H.C.: Improving the speed of calculating the regulator of certain pure cubic fields. Math. Comp. **35**, 1423–1434 (1980)
4402. Williams, H.C.: Some results concerning Voronoĭ's continued fraction over $\sqrt[3]{D}$. Math. Comp. **36**, 631–652 (1981)
4403. Williams, H.C.: Daniel Shanks (1917–1996). Math. Comp. **66**, 929–934 (1997)
4404. Williams, H.C., Broere, J.: A computational technique for evaluating $L(1, \chi)$ and the class number of a real quadratic field. Math. Comp. **30**, 887–893 (1976)
4405. Williams, H.C., Cormack, G., Seah, E.: Calculation of the regulator of a pure cubic field. Math. Comp. **34**, 567–611 (1980)
4406. Williams, H.C., Dueck, G.W., Schmid, B.K.: A rapid method of evaluating the regulator and class number of a pure cubic field. Math. Comp. **41**, 235–286 (1983)
4407. Williams, K.S.: On a theorem of Niven. Canad. Math. Bull. **10**, 573–578 (1967); Add.: **11**, p. 145 (1968)
4408. Williams, K.S.: On the size of a solution of Legendre's equation. Utilitas Math. **34**, 65–72 (1988)
4409. Williams, K.S.: Note on a theorem of Pall. Proc. Amer. Math. Soc. **28**, 315–316 (1971)
4410. Williams, K.S.: Representation of a binary quadratic form as a sum of two squares. Proc. Amer. Math. Soc. **32**, 368–370 (1972)
4411. Williams, K.S.: Another proof of a theorem of Niven. Math. Mag. **46**, p. 39 (1973)
4412. Williams, K.S.: Note on the supplement to the law of cubic reciprocity. Proc. Amer. Math. Soc. **47**, 333–334 (1975)
4413. Williams, K.S.: Note on non-Euclidean principal ideal domains. Math. Mag. **49**, 176–177 (1975)
4414. Williams, K.S.: Note on a result of Barrucand and Cohn. J. Reine Angew. Math. **285**, 218–220 (1976)
4415. Williams, K.S.: On the supplement to the law of biquadratic reciprocity. Proc. Amer. Math. Soc. **59**, 19–22 (1976)
4416. Williams, K.S.: On Eisenstein's supplement to the law of cubic reciprocity. Bull. Calcutta Math. Soc. **69**, 311–314 (1977)
4417. Williams, K.S.: On the class number of $Q(\sqrt{-p})$ modulo 16, for $p \equiv 1 \bmod 8$ a prime. Acta Arith. **39**, 381–398 (1981)
4418. Williams, K.S.: The class number of $Q(\sqrt{-2p})$ modulo 8, for $p \equiv 5 \bmod 8$ a prime. Rocky Mountain J. Math. **11**, 19–26 (1981)
4419. Williams, K.S.: The class number of $Q(\sqrt{p})$ modulo 4, for $p \equiv 5 \bmod 8$ a prime. Pacific J. Math. **92**, 241–248 (1981)
4420. Williams, K.S.: Congruences modulo 8 for the class numbers of $Q(\sqrt{\pm p})$, $p \equiv 3 \bmod 4$ a prime. J. Number Theory **15**, 182–198 (1982)
4421. Williams, K.S. Friesen, C.: Remark on the class number of $Q(\sqrt{2p})$ modulo 8 for $p \equiv 5 \bmod 8$ a prime. Proc. Amer. Math. Soc. **93**, 198–200 (1985)
4422. Wilson, J.S.: Finite presentations of pro-p groups and discrete groups. Invent. Math. **105**, 177–183 (1991)
4423. Wilson, N.R.: Integers and basis of a number field. Trans. Amer. Math. Soc. **29**, 111–126 (1927)
4424. Wilson, N.R.: On finding ideals. Ann. Math. (2) **30**, 411–428 (1928/29)
4425. Wiman, A.: Über die Ideale in einem algebraischen Zahlkörper, nach denen Primitivzahlen existieren. Öfversikt Svenska Vet. Akad. Förhandl. **56**, 879–885 (1899)
4426. Wingberg, K.: Der Eindeutigkeitssatz für Demuškinformationen. Invent. Math. **70**, 99–113 (1982/83)
4427. Winter, D.J.: A generalization of the normal basis theorem. Math. Nachr. **54**, 75–77 (1972)
4428. Wintner, A.: The densities of ideal classes and the existence of unities in algebraic number fields. Amer. J. Math. **67**, 235–238 (1948)
4429. Wintner, A.: A factorization of the densities of the ideals in algebraic number fields. Amer. J. Math. **68**, 273–284 (1946)

4430. Wisliceny, J.: Zur Darstellung von Pro-p-Gruppen und Lieschen Algebren durch Erzeugende und Relationen. Math. Nachr. **102**, 51–78 (1981)
4431. Witt, E.: Konstruktion von galoisschen Körpern der Charakteristik p zu vorgegebener Gruppe der Ordnung p^f. J. Reine Angew. Math. **174**, 237–245 (1936)
4432. Witt, E.: Zyklische Körper und Algebren der Charakteristik p vom Grad p^n. Struktur diskret bewerteter perfekter Körper mit vollkommenem Restklassenkörper der Charakteristik p. J. Reine Angew. Math. **176**, 126–140 (1936)
4433. Witt, E.: Bemerkungen zum Beweis des Hauptidealsatzes von S.Iyanaga. Abh. Math. Sem. Hansischen Univ. **11**, p. 221 (1911)
4434. Wohlfahrt, K.: Hans Petersson zum Gedächtnis. Jahresber. Deutsch. Math.-Verein. **96**, 117–129 (1994)
4435. Wolfe, C.: On the indeterminate cubic equation $x^3 + Dy^3 + D^2z^3 - 3Dxyz = 1$. California Univ. Publ. **16**, 359–369 (1923)
4436. Wolff, G.: Über Gruppen der Reste eines beliebigen Moduls m eines algebraischen Zahlkörpers. Dissertation, Univ. Giessen (1905)
4437. Wolfskehl, P.: Beweis, dass der zweite Factor der Klassenanzahl für die aus den elften und dreizehnten Einheitswurzeln gebildeten Zahlen gleich Eins ist. J. Reine Angew. Math. **99**, 173–179 (1985)
4438. Wolfskill, J.: On equivalence of binary forms. J. Number Theory **16**, 205–211 (1983)
4439. Wolfskill, J.: A note concerning reduced algebraic numbers and equivalence of binary forms. J. Number Theory **20**, 159–161 (1985)
4440. Wolfskill, J.: Reduced algebraic numbers in the complex plane. J. Number Theory **48**, 394–412 (1994)
4441. Wolke, D.: Moments of the number of classes of primitive quadratic forms with negative discriminant. J. Number Theory **1**, 502–511 (1969)
4442. Wolke, D.: Momente der Klassenzahlen. II. Arch. Math. (Basel) **22**, 65–69 (1971)
4443. Wolke, D.: Momente der Klassenzahlen. III. J. Number Theory **1**, 523–531 (1972)
4444. Wood, M.M.: Gauss composition over an arbitrary base. Adv. Math. **226**, 1756–1771 (2011)
4445. Wooley, T.D.: Large improvements in Waring's problem. Ann. of Math. (2) **135**, 131–164 (1992)
4446. Wooley, T.D.: Vinogradov's mean value theorem via efficient congruencing. Ann. Math. (2) **175**, 1575–1627 (2012)
4447. Wright, D.J.: Distribution of discriminants of abelian extensions. Proc. London Math. Soc. (3) **58**, 17–50 (1989)
4448. Yahagi, O.: Construction of number fields with prescribed l-class groups. Tokyo J. Math. **1**, 275–283 (1978)
4449. Yamagata, K., Yamagishi, M.: On the ring of integers of real cyclotomic fields. Proc. Japan Acad. Sci. **92**, 73–76 (2016)
4450. Yamamoto, Y.: On unramified Galois extensions of quadratic number fields. Osaka J. Math. **7**, 57–76 (1970)
4451. Yamamoto, Y.: Real quadratic number fields with large fundamental units. Osaka J. Math. **8**, 261–270 (1971)
4452. Yamamura, K.: The determination of the imaginary abelian number fields with class number one. Proc. Japan Acad. Sci. **68**, 21–24, 74 (1992)
4453. Yamamura, K.: The determination of the imaginary abelian number fields with class-number one. Math. Comp. **62**, 899–921 (1994)
4454. Yokoi, H.: On the distribution of irregular primes. J. Number Theory **7**, 71–76 (1975)
4455. Yokoyama, A.: On the Gaussian sum and the Jacobi sum with its application. Tôhoku Math. J. (2) **16**, 142–153 (1964)
4456. Yoshimura, Y.: Abelian number fields satisfying the Hilbert-Speiser condition at $p = 2$ or 3. Tokyo J. Math. **32**, 229–235 (2009)
4457. W.H.Y.[Young, W.H.]: Adolf Hurwitz. Proc. London Math. Soc. (2) **20**, xlviii–liv (1922)
4458. Yu, G.: A note on the divisibility of class numbers of real quadratic fields. J. Number Theory **97**, 35–44 (2002)

4459. Zagier, D.: A Kronecker limit formula for real quadratic fields. Math. Ann. **213**, 153–184 (1975)
4460. Zagier, D.: On the values at negative integers of the zeta-function of a real quadratic field. Enseign. Math. **22**, 55–95 (1976)
4461. Zagier, D.: Valeurs des fonctions zêta des corps quadratiques réels aux entiers négatifs. Astérisque **41/42**, 135–151 (1977)
4462. Zantema, H.: Integer-valued polynomials over a number field. Manuscripta Math. **40**, 155–203 (1982)
4463. Zantema, H.: Global restrictions on ramification in number fields. Manuscripta Math. **43**, 87–106 (1983)
4464. Zányi, L.: Zur Theorie der identischen Kongruenzen mit Idealmoduln, I. Acta Litt. Sci. Reg. Univ. Hungar. Franc.-Jos., sect. Math. **5**, 117–131 (1930/32)
4465. Zányi, L.: Zur Theorie der identischen Kongruenzen mit Idealmoduln, II. Acta Litt. Sci. Reg. Univ. Hungar. Franc.-Jos., sect. Math. **5**, 168–171 (1930/32)
4466. Zariski, O., Samuel, P.: Commutative Algebra. **1,2**. Van Nostrand (1958/60) [Reprint: Springer (1975)]
4467. Zhang, X.K.: Density of number fields of type $(2, 2, \dots, 2)$. Sci. Sinica **A 27**, 345–351 (1984)
4468. Zhang, X.K.: On number fields of type $(l, l, \dots, l)$. Sci. Sinica **A 27**, 1018–1026 (1984)
4469. Zhang, X.K.: Some composite relative extensions and relative integral bases. Algebra Colloq. **2**, 269–274 (1995)
4470. Zhuravskiĭ, A.M.: The cubic reciprocity law. J. Soc. Phys.-Math. Leningrad. **1**, 204–232 (1927) (Russian)
4471. Ziegler, V.: The additive unit structure of complex biquadratic fields. Glas. Mat. **43**, 293–307 (2008)
4472. Ziegler, V.: On unit power integral bases of $Z[\sqrt[4]{m}]$. Period. Math. Hung. **63**, 101–112 (2011)
4473. Zimmer, H.G.: Computational Problems, Methods, and Results in Algebraic Number Theory. Lecture Notes Math. **262** (1972)
4474. Zimmert, R.: Ideale kleiner Norm in Idealklassen und eine Regulatorabschätzung. Invent. Math. **62**, 367–380 (1981)
4475. Zink, E.-W.: Zum Hauptidealsatz von Tannaka-Terada. Math. Nachr. **67**, 317–325 (1975)
4476. Zlebov, E.D.: Pisot-Vijayaraghavan numbers and fundamental units of algebraic fields. Vesci AN BSSR, Ser. Fiz.–Mat. N. **1966(4)**, 110–112. (Russian)
4477. Zolotarev, E.I. [Zolotareff, E.I.]: On an indeterminate equation of the third degree. Dissertation, St. Petersburg Univ. (1869) (Russian)
4478. Zolotareff, E.I.: Théorie des nombres entiers complexes, avec une application au calcul intégral. St. Pétersbourg (1873)
4479. Zolotareff, G. (Although Zolotarev's first name was Egor Ivanovič, he used the initial "G." in his paper published in French.) [E.I.]: Sur la théorie des nombres complexes. J. Math. Pures Appl. (3) **6**, 51–84, 129–166 (1880)
4480. Zorn, M.: p-adic analysis and elementary number theory. Ann. Math. (2) **38**, 451–464 (1937)
4481. von Żyliński, E.: Zur Theorie der ausserwesentlichen Diskriminantenteiler algebraischer Körper. Math. Ann. **73**, 273–274 (1913)
4482. Zywina, D.: Hilbert's irreducibility theorem and the larger sieve. ArXiv:1011.6465, 1–28 (2010)

Author Index

W. Narkiewicz, *The Story of Algebraic Numbers in the First Half of the 20th Century*, Springer Monographs in Mathematics,
https://doi.org/10.1007/978-3-030-03754-3

G

H

L

M

Subject Index

W. Narkiewicz, *The Story of Algebraic Numbers in the First Half of the 20th Century*, Springer Monographs in Mathematics,
https://doi.org/10.1007/978-3-030-03754-3

D

E

Z

GPSR Compliance
The European Union's (EU) General Product Safety Regulation (GPSR) is a set of rules that requires consumer products to be safe and our obligations to ensure this.

If you have any concerns about our products, you can contact us on

ProductSafety@springernature.com

In case Publisher is established outside the EU, the EU authorized representative is:

Springer Nature Customer Service Center GmbH
Europaplatz 3
69115 Heidelberg, Germany

www.ingramcontent.com/pod-product-compliance
Ingram Content Group UK Ltd.
Pitfield, Milton Keynes, MK11 3LW, UK
UKHW021439280726
14060UKWH00001BA/152

* 9 7 8 3 0 3 0 0 3 7 5 3 6 *